CELL BIOLOGY IN MEDICINE

CELL BIOLOGY IN MEDICINE

Edited by

E. EDWARD BITTAR

Department of Physiology
University of Wisconsin
Madison, Wisconsin

A Wiley-Interscience Publication

JOHN WILEY & SONS
New York • London • Sydney • Toronto

Library of Congress Cataloging in Publication Data:

Bittar, E. Edward.
Cell biology in medicine.

Includes bibliographies.

1. Cell physiology. 2. Cytology. I. Title.
QH631.B53 574.8'76 72-3854

ISBN 0-471-07706-2

Printed in the United States of America

10 9 8 7 6 5 4 3 2 1

CONTRIBUTORS

C. C. Ashley, Department of Physiology, University of Bristol, Bristol, England

P. F. Baker, Physiological Laboratory, University of Cambridge, Cambridge, England

Lucille Bitensky, Division of Cellular Biology, The Mathilda and Terence Kennedy Institute of Rheumatology, London, England

R. H. Burdon, Department of Biochemistry, University of Glasgow, Glasgow, Scotland

J. Chayen, Division of Cellular Biology, The Mathilda and Terence Kennedy Institute of Rheumatology, London, England

A. S. G. Curtis, Department of Cell Biology, University of Glasgow, Glasgow, Scotland

David Epel, Marine Biology Research Division, Scripps Institution of Oceanography, University of California at San Diego, La Jolla, California

B. E. H. Maden, Department of Biochemistry, University of Glasgow, Glasgow, Scotland

Viviane Maggi, Faculty of Medicine, Memorial University of Newfoundland, St. John's Newfoundland, Canada

J. C. Marsden, Department of Biology, University of York, Heslington, York, England

A. A. Moscona, Department of Biology, The University of Chicago, Chicago, Illinois

D. H. Northcote, Department of Biochemistry, University of Cambridge, Cambridge, England

Margery G. Ord, Department of Biochemistry, University of Oxford, Oxford, England

E. C. Pollard, Department of Biophysics, The Pennsylvania State University, University Park, Pennsylvania

J. David Robertson, Department of Anatomy, Duke University School of Medicine, Durham, North Carolina

Michael A. Sleigh, Department of Zoology, University of Bristol, Bristol, England

J. Maynard Smith, School of Biological Sciences, University of Sussex, Brighton, England

L. A. Stocken, Department of Biochemistry, University of Oxford, Oxford, England

J. R. Tata, National Institute for Medical Research, London, England

D. D. Tyler, Department of Physiology, Royal Veterinary College, University of London, London, England

J. W. Watts, John Innes Institute, Norwich, England

PREFACE

The aim of this work is to set forth cell biology as a continuous and consistent whole and to do so with special reference to the medical sciences. It is meant for the educated student who aspires to become a medical scientist as well as for the student who is making his first acquaintance with modern cell biology and its broad implications. That medical education today lacks in several respects rational coherence is only too palpable. Thus the treatment as developed here is the result of a certain "point of view," namely that cell biology provides a sound basis for the study of medicine. The reader may wonder why a more complete account of the vast intellectual development which began with the cell doctrine of Schleiden and Schwann has not been written. The truth is that in a work of this scope the main intention has been, in Bacon's words, "to excite the judgment briefly, rather than to inform tediously."

E. EDWARD BITTAR

University of Wisconsin
July 1972

CONTENTS

PART 1

THE DYNAMICS OF CELL MEMBRANES

1

THE ORGANIZATION OF CELL MEMBRANES*

J. David Robertson

1. INTRODUCTION

This chapter deals with the molecular structure of cell membranes. It reviews some important aspects of the subject and supplements or extends reviews by the author (see for example Robertson, 1969) and by others (Sjöstrand, 1969; Stoeckenius and Engelman, 1969). Its emphasis is biophysical and it concentrates on contributions made by electron microscopy and x-ray diffraction analysis as well as other biophysical methods. It states the experimental basis of a theory of cell membrane structure which proposes that biological membranes have certain architectural

* An earlier version of this article was published in the *Archives of Internal Medicine*, Robertson, J. David, The structure of biological membranes, in press.

principles in common. A structure is defined which appears to constitute a fundamental repetitive unit of biological membranes and membranous organelles. This structure is called a "unit" membrane to distinguish it from all other uses of the word "membrane" in biology.

Two of the principal methods used for elucidating molecular arrangements in cells and tissues are electron microscopy and x-ray diffraction analysis. Light beams, electron beams, and x-ray beams have certain features in common. For example, they are all diffracted. Diffracted light can be gathered by glass lenses to produce images on photographic plates with a resolution of approximately 0.1–0.2 μ. Diffracted electrons can be gathered in a vacuum by magnetic lenses to produce similarly recordable images with a resolution of approximately 3–5 Å. Diffracted x-rays cannot be gathered in a practical manner by a lens, but they can be recorded directly and the image they would make can be synthesized mathematically to a resolution of less than 1 Å. An angstrom unit is about the size of an atom, and so x-ray diffraction and electron microscopy give information about molecular and atomic structure. X-ray beams, similar to light beams, travel in air or water a considerable distance without being absorbed, while electron beams must be handled in a vacuum.

Despite mathematical difficulties the methods of x-ray diffraction have been extensively applied over the last few decades to the molecular structure of biological entities. During recent years great strides have been made using this method in determining the relative positions of every atom in such large molecules as myoglobin and DNA, and in several others. Such analyses have depended on the availability of crystals of the compounds in question. In a crystal every atom has a position specified within certain limits, and within these limits it can be located to approximately 1 Å and identified.

It should be understood at the outset that no proposal of a detailed molecular model of a unit membrane in the above crystallographic sense has yet been made. The reason is simple. Unit membranes do not form solid crystals as do proteins, lipids, or nucleic acids. They are, however, to a large extent in the liquid crystalline state. Liquid crystals are aggregates of molecules in which classes of compounds associate in a regular and repetitive way. The individual molecules within each class, while they may be precisely arrayed, are usually in a fluid state and cannot be located exactly by x-ray diffraction methods; but the classes can be located usually to approximately 10 Å. The unit membrane model is to a large extent based on electron microscope and x-ray crystallographic studies of membranes and their constituents.

In dealing with membrane structure, it is useful to distinguish between general features common to all membranes or large classes of membranes as opposed to localized, special features of limited occurrence. Although whole membranes are not truly crystalline, some components of some membranes may be crystalline in the strictest sense, and it may well be that a detailed molecular model of such components is possible. For instance, there may exist a structural protein common to all membranes of an animal arranged in a crystalline array and having a relatively passive structural role. Indeed, there are crystalline arrays of molecules on or in membranes in special regions. However, solid crystallinity of whole membranes, if it occurs at all, seems to be localized and special.

2. EARLY CONCEPTS

In 1895, Overton published studies of the permeability of membranes to different substances and deduced that the plasma membrane must contain lipids. This was based on the fact that lipid-soluble substances seemed to penetrate much more effectively than lipid-insoluble substances. Ever since, it has been regarded as highly probable that cell membranes contain a large amount of lipid as an important constituent.

Later, Gorter and Grendel (1925), working with human red blood cell membranes, attempted to extract all the lipid from the membranes. They reasoned that they would then be able to determine the total area occupied by the lipid molecules by spreading them in a monomolecular film on a water surface and compressing the film to a minimal area. They extracted the lipid from a known number of red blood cells, measured the area the lipid occupied, and found it to be twice the area of the cells. This finding led them to propose that the red blood cell membrane contains a bilayer of lipid. They reasoned that since both sides of the red blood cell membrane presumably are aqueous, the polar hydrophilic heads of the lipid molecules must point outward and the nonpolar carbon chains must be directed toward one another or inside the bilayer. This was the first suggestion that any cell membrane might contain a lipid bilayer.

Cole (1932) measured the surface tension of starfish egg membranes. He measured the force required to compress an egg between two glass cover slips. This allowed him to calculate surface tension values of less than 1 dyne/cm. Even though we now know that some natural phospholipids may display such surface tension values, this low figure was considered at the time a surprising result assuming that the membranes were composed primarily of lipid because many lipids of natural origin were known to display surface tension values of 10–30 dynes/cm or more.

The next step came with the introduction of the centrifuge microscope. Harvey and Shapiro (1934) used this method to determine the surface tension of oil droplets within sea urchin eggs. It was possible to observe oil droplets elongating into dumbbell shapes within the cells under centrifugal force and breaking into droplets. From the force required to make the droplets break in two, surface tension values below 1 dyne/cm were obtained.

Danielli and Harvey (1935) and Danielli and Davson (1935) approached the problem by extracting oil from mackerel eggs and attempting to purify it and study its properties. They found that such oil had a very low surface tension. After purification the surface tension values rose to those usually found with pure oils. It was known at that time that proteins spread on a Langmuir trough and compressed to a monolayer tended to have fairly low surface tension values. Danielli and Davson concluded that the reason for the relatively low values displayed by the oil droplets was the presence on their surface of a protein absorbed onto the polar heads of the lipid molecules. They therefore put forward a proposal that the cell membrane consists of a lipid core, with the polar heads of the lipid molecules pointing outward toward the aqueous phase and covered by monolayers of protein as indicated in the diagram (Fig. 1) taken from their 1935 paper. This is the Danielli–Davson model which was to dominate thinking about the physical arrangement of the constituents of cell membranes until the mid 1950s.

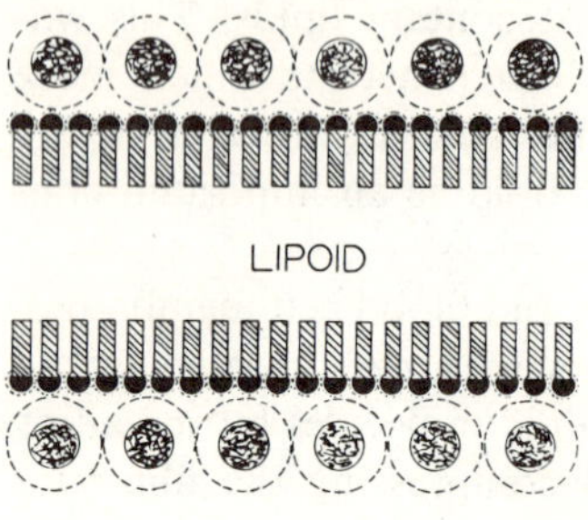

Figure 1. The Danielli-Davson membrane model. Note that the thickness of the lipid bilayer is not specified and that the model is symmetrical.

The Danielli–Davson model is a symmetrical one; the outer and inner surfaces are essentially the same. Furthermore, no specification at all was made about the thickness of the membrane. The reason for this is simply that the information available then was meager. In thus attempting to make their model as general as possible, Davson and Danielli wisely left open the question of thickness.

Waugh and Schmitt (1940) investigated the thickness of human red blood cell membranes by means of a special optical device called an analytical leptoscope. They obtained values of approximately 200 Å or less, but the problem of precisely determining membrane thickness remained without a definitive solution. Schmitt et al. (1936, 1938) reported studies of red blood cells by polarization optical analysis and concluded that the membranes exhibited a radially disposed optic axis and radially positive birefringence. This was consistent with their containing lipid molecules with their long axes radially oriented, but the thickness again was not determined. It was only by means of electron microscopy that the problem of membrane thickness was finally solved as is discussed in Section 5.

3. CHEMICAL COMPOSITION OF CELL MEMBRANES

A. Red Cell Ghosts

Red blood cells swell in water, lose their contents through several transient pores appearing in a patch in a small spot on the cell's surface (Seeman, 1967), and persist as simple membrane bags called ghosts. Ghosts are easily prepared by several methods, and it is possible to rid them of almost all nonmembranous constituents. A variety of chemical constituents have been defined within ghost preparations. It has been established that such membranes contain approximately 60% protein and 40% lipid by weight (Maddy and Malcolm, 1965). Although the composition of the lipid has been determined (Ways and Hanahan, 1964), there are considerable variations depending on technique. The lipid is primarily (~70%) phospholipid, and cholesterol makes up most of the remaining 30% by weight. This is a ratio of 1:1 on a molar basis, but the amount of cholesterol is quite variable. Other lipids, for example, glycolipids, are present in small amounts. Despite their paucity, these are thought to be very important in determining immunological properties. Table 1 summarizes the analyses of phospholipids of human erythrocytes reported by 10 different laboratories. The amount of cholesterol in different membranes varies widely. Thus in mitochondrial membranes the cholesterol concentration is very low (Green and Fleischer, 1964) and in frog retinal rod outer segments, at 1.7% of dry weight, it also is very low (Eichberg and Hess, 1967). In brain the cholesterol/phospholipid molar ratio as in erythrocytes may approach 1. Autilio et al. (1964), using purified myelin preparations, found cholesterol/galactolipids/phospholipids molar ratios of 4:2:3.

Table 1. Phospholipid Composition of Human Red Cells[a]

Component	Amount
Lecithin (phosphatidylcholine)	30–40%
Cephalin (phosphatidylethanolamine) including plasmalogens	23–28%
Phosphatidylserine	2–15%
Sphingomyelin	22–30%
Phosphatidic acid	Small amount
Inositol phosphoglycerides	Small amount
Lyso derivatives and unidentified compounds	Small amount

[a] Van Deenen (1965).

All the lipids of erythrocytes vary considerably according to species and are somewhat responsive to diet (Van Deenen, 1965). The neutral lipid chains also vary considerably in length and degree of unsaturation. The dominant lengths are approximately 16 to 18 carbons but may be as many as 26 carbons.

Many different proteins of varying molecular weight are found in red cell ghosts by the methods of gel filtration and electrophoresis, but none are lighter than hemoglobin monomer (molecular weight 16,000). There are two major peaks at approximately 200,000 and 100,000 which together contain over 70% of the proteins (Trayer et al., 1971). Recently, a so-called "miniprotein" of approximate molecular weight 5000–6000 has been reported in red cell ghosts (Laico et al., 1970), but it is believed by some that these proteins result from enzymic degradation of larger molecules during preparation. Reynolds (1971) found that prolonged washing of erythrocyte ghosts in water removes all the proteins of greater than 150,000 molecular weight, which constitute 40% of the total protein. She also found that about 85–95% of the total protein can be removed with 0.1 *M* tetramethylammonium chloride. These observations suggest that the protein and lipid may be electrostatically bonded. It is often said that if electrostatic bonding is important, the protein ought to separate in strong solutions of monovalent cations. If one assumes that divalent cations stabilize the bonding, this argument is not necessarily valid. In any case there is now good evidence that electrostatic lipid–protein bonding may be important in the erythrocyte membrane. Studies have also shown that cytochrome c is bonded electrostatically in mitochondrial membranes (Green and Fleischer, 1964) and in bilayers of phosphatidylinositide (Steinemann and Läuger, 1971). There is evidence that micellar interactions between cytochrome c and phosphatidylinositol are electrostatic (Gulik-Kryzwicki et al., 1969).

It is important to consider whether or not it is possible to construct a closely packed lipid bilayer model of the erythrocyte membrane which is covered on both sides completely by electrostatically bonded monomolecular layers of protein of the appropriate molecular weight and percentage by weight to satisfy the analytical chemical data cited above. A condensed film consisting of 70% phospholipid and 30% cholesterol has been found by De Bernard and Dervichian (cited in Van Deenen, 1965) to occupy on the average 65 Å^2 per molecule. We begin by assuming that most of the proteins in the red cell membrane are globular and that they are packed into the membrane in a manner analogous to the packing in hydrated globular protein crystals. Matthews (1968) found in a wide variety of globular protein crystals that the volume/molecular weight ratio is equal to approximately 2.4. Thus by multiplying the molecular weight in daltons by 2.4, one can obtain an approximate volume for any globular protein molecule. A low-molecular-weight protein of 17,000 daltons would thus occupy approximately 41,000 Å^3. The cube root of this figure, 34 Å, represents the edge of the area covered by the protein. The square of this figure, 1156 Å^2, is the area covered by the protein. This divided by 65 gives 18 as the maximum number of lipid molecules covered by this area in a monolayer. If we multiply this number by 70% times the average molecular weight of a phospholipid molecule, taking this as 900, we obtain 11,340 daltons of phospholipid. Using 387 as the molecular weight of cholesterol and repeating the operation with 30% gives 2087 daltons of cholesterol. The percentages by weight of associated lipid and protein are then 41 and 59%, respectively, very close to the analytical figures. However, if we perform the same operations with a protein molecule weighing 100,000 daltons, we obtain figures of approximately 70% protein and 30% lipid. At 200,000 daltons the protein reaches 74% and the lipid 26%. Clearly then, the model seems defective since high-molecular-weight proteins dominate in the erythrocyte membrane. One can overcome this difficulty by departing from the strict lipid bilayer configuration and adding lipid to the globular protein molecules hydrophobically in the manner described later for SDS–protein interaction. This allows one to obtain almost any percent of lipid desired while still maintaining the bilayer intact. It is interesting that having the protein intrude into the bilayer does not help. In fact this makes the calculations worse, because this effectively increases the area per molecule of lipid and decreases the total lipid percentage. Furthermore, adding some of the protein as an adsorbed component on top of the monolayer of protein next to the lipid, or elongating the protein molecules and standing them on end, also does not help. In either case one simply increases the percentage of protein. However, it is interesting that this problem is easily resolved while still maintaining the 1 : 1 protein monolayer to lipid polar surface relationship of the bilayer model by assuming that the large protein molecules are flattened out. If we assume that a 100,000-dalton protein is present as a square solid, approximately 78 Å on a side and 40 Å thick, we obtain from the same calculation figures of 41% phospholipid and 59% protein. Clearly, any flat shape would do as well. It is not unreasonable to assume that the protein has some such slightly flattened configuration in relation to the bilayer. For example, cytochrome c is egg-shaped, measuring 25 × 25 × 35 Å (Dickerson and Geis, 1969). This gives complete agreement with the analytical figures even when one takes into account the average molecular weights and proportions of the proteins present. This model has one important defect. The crystal molecular volumes used include solvent, and one would expect large cracks to be left between the protein molecules contrary to what might be expected in a membrane of selective ion permeability. This is, however, easily circumvented if one assumes asymmetric distribution of the protein on the two sides of the bilayer. Perhaps the protein is more closely packed on, for example, the inside, and quite loosely packed on the outside. The lipid bilayer model is thus quite adequate to explain the known facts about the erythrocyte membrane despite a different conclusion reached by Korn (1966).

B. Nerve Myelin Sheath

The nerve myelin sheath has been the subject of much chemical study. Myelin can be obtained in reasonably pure form from brain white matter. It is about 80% lipid and 20% protein by weight. The lipid composition resembles that of erythrocyte membranes, but there are also some prominent differences. Autilio et al. (1964), as well as O'Brien and Sampson (1965), have prepared highly purified myelin fractions; some of the results of their analyses are given in Tables 2 and 3. The latter is based on analysis of human brain myelin, whereas the former relates to young ox brain. Note that the two sets of figures are expressed differently. The numbers in parentheses after the percentage of total lipids in Table 2 are the values for percent total dry weight for comparison with Table 3.

Table 2. Average Composition of Purified Myelin[a,b]

Component	Light myelin			Heavy myelin		
	Total lipid (wt %)	Molar ratio	Molar percent	Total lipid (wt %)	Molar ratio	Molar percent
Cholesterol	26.7 (18.6)	3.90	43.2	25.9	3.84	42.7
Galactolipids	29.9 (20.8)	2.00	22.1	29.4	2.00	22.2
Phospholipids	42.9 (29.9)	3.14	34.7	42.6	3.16	25.3
Plasmalogens	13.8 (9.6)	1.00	11.1	13.7	1.02	11.3

	Light myelin (% dry wt)	Heavy myelin (% dry wt)
$CHCl_3$–CH_3-insoluble residue	1.0	4.9
PLP	21.3	22.8
Total lipid	77.7	72.3

[a] Autilio et al. (1964).
[b] Lipids are expressed as molar percentages.

Myelin is the source of two of the best defined membrane proteins. If myelin is extracted with chloroform–methanol (2:1 v/v) under certain conditions, a lipoprotein micellar constituent is obtained which is about 8% (w/w) protein and 92% (w/w) lipid. Folch-Pi and Lees (1951) called this special kind of lipoprotein a "proteolipid" because it is soluble in lipid solvents unlike most lipoproteins which are soluble in water (see also LeBaron and Folch, 1957; LeBaron, 1969). The purest preparations of myelin made by Autilio et al. (1964) from brain white matter were shown to be over 99% proteolipid.

Table 3. Concentrations of Myelin in Human Brains[a,b]

Component	10 months old	6 yr old	9 yr old	55 yr old
Water	—	—	—	—
Total lipid	78.0	80.9	78.0	78.0
Nonlipid residue	22.0	19.1	22.0	22.0
Total glycerophosphatides[c]	31.7	24.6	31.9	24.8
Total sphingolipids[d]	24.7	28.6	25.0	24.5
Unidentified	3.0	6.1	2.5	9.0
Cholesterol	18.6	21.5	18.6	19.7
Ethanolamine glycerophosphatides	14.2	11.3	14.2	11.2
Serine glycerophosphatides	5.5	4.2	5.5	5.3
Choline glycerophosphatides	12.1	9.1	12.2	8.3
Sphingomyelin	4.6	4.4	4.6	4.4
Cerebroside	13.7	19.2	14.0	16.0
Cerebroside sulfate	5.1	3.9	5.1	3.4
Ceramide	1.2	1.1	1.3	0.7

[a] O'Brien and Sampson (1965).
[b] All values except water are expressed as a percentage of the dry weight.
[c] Sum of EGP, SGP, and CGP.
[d] Excluding gangliosides.

Tenenbaum and Folch-Pi (1967) showed that the lipid can be stripped away from the protein of proteolipid without denaturing it. The resulting apoprotein is called proteolipid protein (PLP), and it has long been known that it is very easily denatured. It does not withstand drying and is disrupted by precipitation at a solvent–water interface. It is very heat-labile although very resistant to enzymic degradation. However, it is now possible to dialyze away the lipid with a special procedure that leaves the protein in solution in chloroform–methanol. By careful dialysis the solvent can be reversibly changed to water without precipitation of the protein. There is clearly a conformational change associated with this reversible transfer from an organic solvent to water that is manifested in part by circular dichroism changes (Sherman and Folch-Pi, 1970).

The amino acid composition of proteolipid protein is known (LeBaron and Folch, 1957), and its molecular weight is said to be approximately 36,000 (Folch-Pi, 1971). It comprises almost 50% of the total protein of the myelin sheath. In water solution the protein tends to form aggregates, and an electron micrograph of some of these aggregates varying in size from 80 to 350 Å in diameter is shown in Figure 2.

It is probable that studies of the combination of PLP with lipids known to be present in myelin will be revealing in terms of principles of organization of cell membranes. It appears that PLP molecules are able to display either a hydrophobic or a hydrophilic surface depending on conditions, and the mode of their association with particular lipids and combinations of lipids under varying conditions needs to be studied. Possibly, under some conditions they can associate with lipids hydrophobically, and under others electrostatically. It remains to be seen exactly how PLP is associated with the lipids in the myelin membrane, but there is much evidence supporting a hydrophilic micellar association, though some degree of hydrophobic action may occur. For instance, it would be quite reasonable to suppose that a

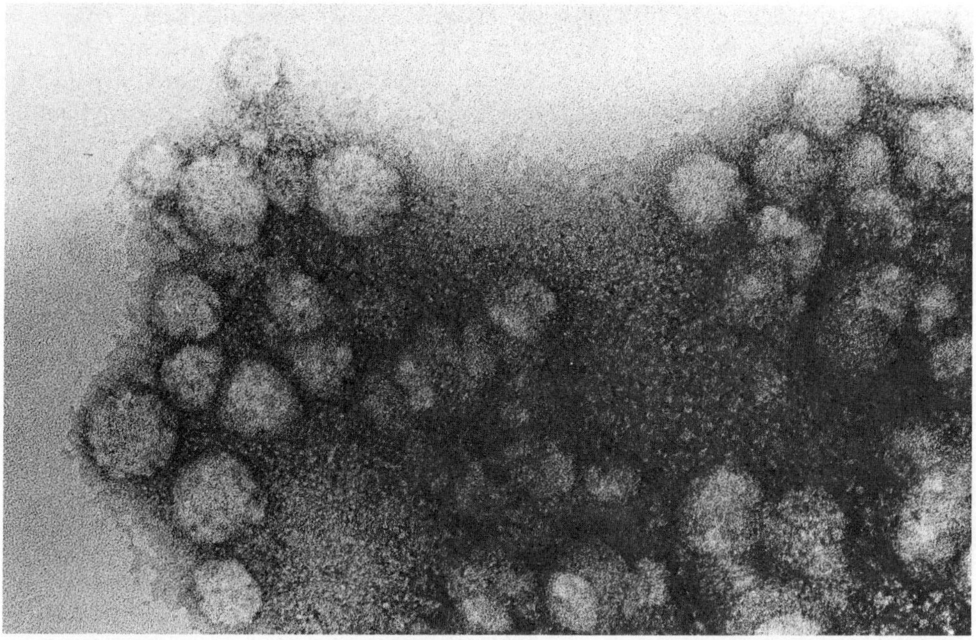

Figure 2. PTA-negative stain preparation of PLP. Each of the snowball-like structures is an aggregate of PLP molecules. The balls range in diameter from 80 to 350 Å. ×351,200.

hydrophobic part of this molecule might be associated with the first few CH_2 groups of lipid fatty acid chains in the bilayer. In fact, this could provide an alternative explanation for the shoulders in the electron density plate attributed by Caspar and Kirshner (1971) to an asymmetric distribution of cholesterol.

Another well-known protein of myelin is the "basic" protein. This comprises about 30% of its total protein. It may well be that it is the main or only protein located at the intraperiod line in myelin as claimed by Dickinson et al. (1970). These investigators observed that basic protein leaks out of myelin during preparation by the procedure of Autilio et al. (1964) and that this is accompanied by a "collapse" of the lamellae.

The basic protein has been studied extensively by Eylar (1970). It has a molecular weight of 18,000, and its amino acid composition and sequence are known. Its large fraction of basic as compared with acidic amino acids is of course the reason for its name. It is soluble in water without producing aggregates, but it is not soluble in organic solvents. It is very sensitive to enzymic degradation. When injected into the footpad of guinea pigs this protein is capable of producing experimental allergic encephalomyelitis, a demyelinating disease of the central nervous system. Eylar et al. (1970) found that a peptide consisting of a fragment of the polypeptide chain containing only nine amino acids is capable of producing this disease. An essential feature of these short peptides is a triple proline residue located near a threonine residue.

Baldwin and Carnegie (1971) produced evidence implicating a methylated arginine in position 107 on the chain near the encephalitogenic site in the function of the basic protein. A specific transferase is present in brain to transfer methyl groups from *S*-adenosylmethionine to this site. Vitamin B_{12} takes part in the metabolism of this compound, and deficiency of B_{12} results in the disease known as subacute combined degeneration in which an abnormality is found in myelin. It is inferred that this disease may be mediated by abnormal basic protein with unmethylated arginine in position 107. This raises interesting possibilities not only with regard to demyelinating diseases but to myelin formation and its stabilization.

The remaining 20% of the protein of the myelin sheath is grouped in a class called Wolfgram's protein (Wolfgram, 1966), and its properties are still largely unknown since it is insoluble.

C. Mitochondria

Mitochondria contain approximately 25% lipid by weight (Green and Fleischer, 1964). The remainder is mainly protein. About 95% of the lipid is phospholipid. The remaining 5% is neutral lipid including cholesterol, coenzyme Q (ubiquinone), tocopherol, carotenoids, and neutral fat. Table 4 shows a breakdown of the phospholipids. Cardiolipin occurs in very high concentration in mitochondria and appears only in small amounts in other membranes. Another striking feature of mitochondria is the relatively small amount of cholesterol as noted above.

Table 4[a]

Phospholipid	Total Phosphorus (%)
Phosphatidylcholine (lecithin)	42
Phosphatidylethanolamine (cephalin)	35
Cardiolipin	18
Others	5

[a] Green and Fleischer (1964).

The proteins of the mitochondrion include all those involved in oxidative phosphorylation including cytochromes b, c, a and a_3, ATPase, and succinic and NADH dehydrogenases. Some of these enzymes are attached to the membrane and some require lipid for function. The matrix contains fumarase, aconitase, and glutamate and malate dehydrogenases. DNA is also present. The inner and outer membranes can be separated, and the outer membrane has a higher content of lipids and is richer in cholesterol and phosphatidylinositol.

Green and his colleagues (1961, 1966) isolated an insoluble protein from mitochondrial preparations which they call "structural" protein. The name implies that this protein is involved in some way in maintaining the structure of the membrane. This sets it apart from other proteins which, although a part of the structure, are thought to be merely occupants rather than agents active in maintaining it. Although the composition of this protein is variable, and it is not certain that it is a fundamental constituent of the mitochondrial membrane, the general notion of a structural protein is an attractive one. However, there is as yet no clearly defined protein isolated in native form that has been proved unequivocally to perform such a function, and there is some reason to doubt that Green's structural protein fills this role. Indeed, a paper by Senior and MacLennan (1970) from Green's laboratory provides good evidence that this protein is a denatured heterogeneous mixture of several proteins. One of these is a denatured ATPase which is not normally associated with phospholipid but which binds phospholipid hydrophobically. It was shown that this is not a special property of this protein but one that other heat-denatured proteins may display. For example, heat-denatured conalbumen similarly binds phospholipid hydrophobically. Senior and MacLennan state, "We conclude therefore that these structure protein preparations are not relevant to a study of mitochondrial membrane proteins or organizer proteins because they contain denatured enzymic and non-membrane protein in high quantity." They stress, however, that their comments apply only to bovine heart mitochondrial structural proteins since they were the only ones studied and that their work does not invalidate the general concept of a "core" protein of nonenzymic nature in mitochondrial membranes. They leave the question open but note that in the future much better criteria must be applied in defining any such structural protein.

Conclusions similar to those of Senior and MacLennan were reached independently by Schatz and Saltzgaber (1969), who also found structural protein preparations to be heterogeneous. They used an entirely different approach, and this lends strength to the conclusions reached at both laboratories.

D. Interactions of Proteins and Lipids in Aqueous Media

Green makes much of the undisputed fact that some denatured proteins tend to interact with lipids by making hydrophobic bonds. One problem is an exact definition of the term hydrophobic bond. This term has a precise thermodynamic meaning (Kauzman, 1959), but it is sometimes used loosely to refer to the London–van der Waals dispersion forces that tend to bond any two molecules together. These forces are very weak and probably of relatively minor significance even though the work of Salem (1962) suggests that when operating between lipid carbon chains they may be important in stabilizing lipid bilayers. It is not correct, however, to refer to these forces as hydrophobic bonds.

Hydrophobic bonds are very important in the formation of lipid micelles in water and in determining the tertiary structure of proteins in an aqueous medium. Hydrophobic bonds are defined thermodynamically. Chemical reactions proceed when a net reduction in free energy results. Free energy change (ΔF) is defined as the sum ot change in heat content or enthalpy (ΔH) less the product of absolute temperature and change in entropy ($T\Delta S$). Thus $\Delta F = \Delta H - T\Delta S$. Entropy can be regarded as disorder. A chain of —CH_2 groups in water induces the formation of ordered cages of hydrogen-bonded water molecules around it analogous to the clathrate cages that form about noble gases in water. If the —CH_2 group is removed, these cages become disordered, that is, there is an increase in entropy. This is of greater magnitude than the decrease in entropy resulting from association of —CH_2 groups that occurs as they are removed from the water. At constant temperature the term $T\Delta S$ thus becomes larger. There is only a relatively small change in enthalpy in this reaction and so the free energy decreases. Hence the reaction proceeds. It has been estimated that there is a decrease in free energy of approximately 1000 cal/mole for each —CH_2 group transferred from water to a nonaqueous milieu (Tanford, 1961). Thus carbon chains tend to be driven out of the water. This is the main reason that lipid micelles form in water and the major factor in causing hydrophobic residues of polypeptide chains to be tucked inside most globular protein molecules. When a protein is denatured in any of a variety of ways, these hydrophobic residues may be returned to the surface of the molecule and render it insoluble in water. This is an abnormal relationship not found in most natural globular proteins which are for the most part water soluble. Since such proteins are often easily denatured, water insolubility is suggestive of denaturation unless there is good evidence to the contrary. Nevertheless, it is quite possible that there is a special class of proteins displaying amphipathic properties that may be natural membrane proteins that are normally components of membranes with their polar parts in the membrane surface and their nonpolar parts in the core of the bilayer. The proteolipid apoprotein would be one of the best characterized members of this class. If it is arranged in the myelin lamellae in this way its encroachment into the bilayer, however, must not be extensive. Rhodopsin might be another candidate, but again the x-ray data suggest no more than partial extension into the lipid bilayer core of the membrane. It is interesting in this connection to note that Folch-Pi (1971) recently stated he has evidence that there is a covalently bonded lipid component in the proteolipid apoprotein molecule. If this is correct it opens up the possibility that other proteins may be anchored in the bilayer core of unit membranes by a covalently bonded lipid component.

In considering how lipid and protein molecules interact with one another, one should consider the stoichiometry, that is, how much lipid is required to react completely with how much protein, and one should also note whether or not the lipid is charged. Green and Fleischer (1964) state that mitochondrial structural protein binds mitochondrial phospholipids to form a complex containing 25% phospholipid by weight. Reynolds and Tanford (1970a) report that proteins, after treatment with mercaptoethanol to break all disulfide bonds, generally bind 1.4 g of the lipid detergent sodium dodecyl sulfate (SDS) per gram of protein, and it appears probable that the proteins undergo a conformation change the nature of which is uncertain. The complex is 59% SDS by weight. Reynolds and Tanford (1970b) calculate that this corresponds to about one SDS molecule for every two amino acids. The charges of the attached SDS molecules swamp out any charges on the polypeptide chain. The number of SDS molecules attached, hence the net charge, is proportional to the length of the polypeptide chain, and this is the basis for the determination of molecular weight by gel electrophoresis (1970a).

Green and Fleischer (1964) remark that certain mitochondrial enzyme systems cease to work in a coordinated fashion when separated from lipid, but that function is restored if lipid is added. It should be noted, however, that in each case the lipid is added in excess as micelles. It may be that the lipid polar surfaces of the micelles simply provide a surface on which the enzymes can spread and orient so as to expose their reactive groups in a sterically favorable way, permitting function to be restored. It might also be that in some cases a few lipid molecules bound to the protein hydrophobically are required for activity. However, proof of any such binding in restoration of activity to enzymes is lacking. Furthermore, it is common knowledge that detergents are destructive to most enzymes and that glassware must always be free of them if the enzymes are to function properly. This may also be true of other amphiphiles. The separate facts that adding micelles of lecithin to some enzyme complexes restores activity and that some isolated denatured proteins may bind lecithin hydrophobically does not prove that an observed restoration of activity on addition of lecithin micelles occurs because of hydrophobic bonding. Green and Tzagoloff (1966) however, take these observations to be in favor of the existence of lipoprotein subunits in mitochondrial membranes with lipids bonded hydrophobically to the protein molecules. This explanation seems to have no compelling experimental basis.

Braun and Radin (1969) studied the interactions of PLP with certain phospholipids. They found that PLP is precipitated by relatively small quantities of phospholipids. Thus 80 μg of phosphatidylserine precipitated 290 μg of PLP completely. The complex contains approximately 22% phospholipid by weight. This is approximately one phospholipid molecule for every five or six amino acid residues. Usually, smaller amounts of the other lipids present in myelin precipitate this amount of PLP.

Limbrick and Robertson have conducted certain preliminary studies of PLP–phosphatidylserine complexes using the method of Braun and Radin. They examined the complexes by negative staining, and by fixation, embedding, and sectioning by the usual glutaraldehyde–osmium tetroxide–epoxy methods. There was some change in

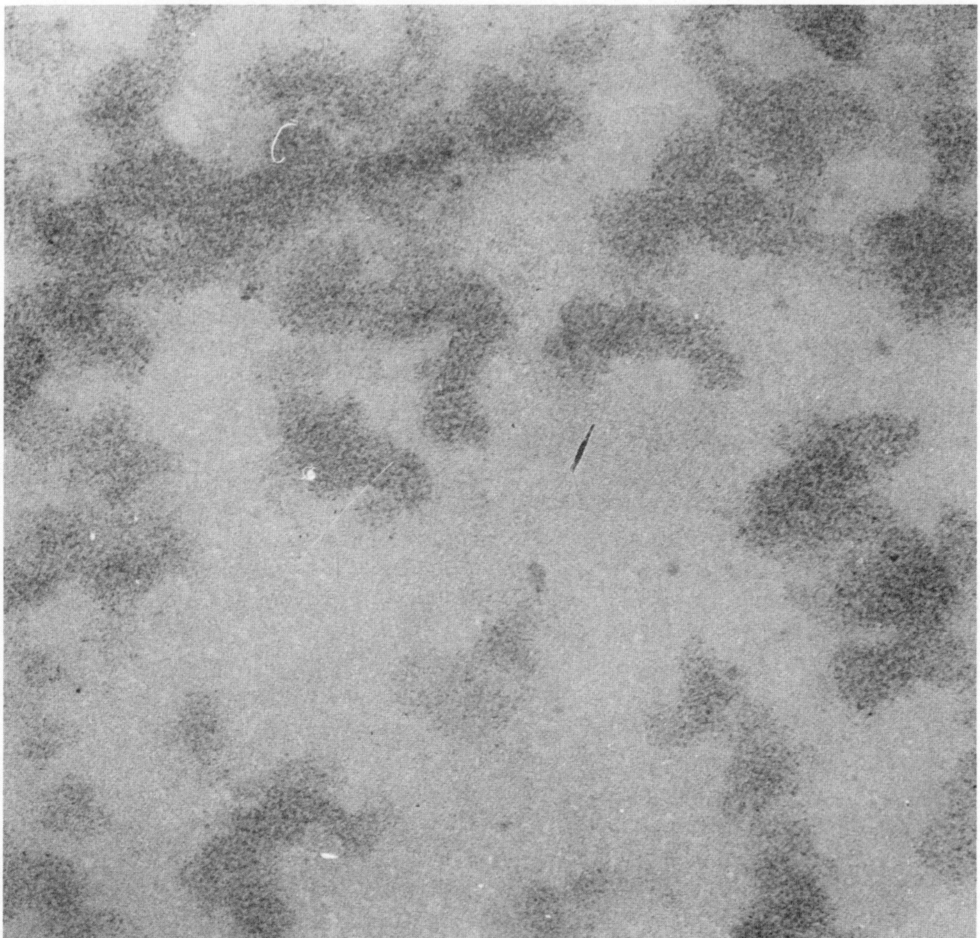

Figure 3. Preparation of PLP precipitated with phosphatidylserine, fixed with glutaraldehyde and osmium tetroxide, embedded in Araldite, and sectioned. ×221,000.

form of the protein in the negatively stained preparations, but there was no visible alteration in the sectioned material. Figure 3 shows a section of the fixed precipitate. The material appears as irregular clumps of amorphous patches which consist of dense spots 15–20 Å in diameter separated by irregular spaces of 10–50 Å. These amorphous aggregates are indistinguishable from pure PLP treated in this way as a control. No sign of lipid is present. Certainly lipid bilayers or other micellar forms if numerous would be identifiable. Indeed, if lipid is added in excess it is easily visible as lamellated bilayer micelles as in Figure 4. The repeat period of the lamellae is about 45 Å, about the same period found with pure lipid. Therefore, under the conditions of the reaction, the protein appears not to associate further with the lipid after it is precipitated by the initial, presumably mainly hydrophobic, reaction. Conditions under which it will associate further may exist. Gulik-Krzywicki et al. (1969), in studying interactions of ferrichrome c and lysozyme with phosphatidylinositol, cardiolipin, and mitochondria lipids, found the dominant association to be micellar, with one or two monolayers of protein becoming packed between lipid bilayers, and it may be that similar complexes between PLP and some phospholipids form under certain conditions. This matter is now under study.

None of the above nonmicellar hydrophobic interactions seem to be very relevant to how lipids and proteins are bonded together in native membranes. Indeed, there is considerable morphological evidence that in native membranes the association is micellar. Nonmicellar hydrophobic interactions of the kind described above are all qualitatively different. Their simple existence in model systems does not prove that they are important in native membranes any more than the existence of electrostatic interactions in model systems argues for their existence in native membranes. The central problem is to elucidate the nature of the interactions that occur in native membranes.

4. THE MEANING OF THE APPEARANCE OF BIOLOGICAL MEMBRANES IN ELECTRON MICROGRAPHS

Under certain conditions all biological membranes appear in electron micrographs of thin sections of fixed, heavy metal-treated material as a pair of dense lines approximately 20 Å thick separated by a light, central zone approximately 35 Å wide. This structure is called a unit membrane and the triple-layered appearance may be regarded as its fingerprint. Figure 5 shows a unit membrane in a section of a human red blood cell which was fixed with potassium permanganate, embedded in an epoxy resin, and sectioned with a diamond knife. Only a small segment of the surface of the cell is shown. The light, central zone is of very uniform thickness and the overall thickness of the structure is approximately 75 Å. Although the thickness tends to vary between 50 and 100 Å, the triple-layered appearance is characteristic of any unit membrane in a transverse section.

Since much of the evidence concerning the molecular structure of unit membranes depends on electron microscopy and the nature of this evidence is so often misunderstood, it seems worthwhile to digress briefly here to clarify a few technical points.

A specimen of biological material may be prepared for electron microscopy by fixation with formaldehyde or glutaraldehyde. The aldehydes cross-link proteins and make them insoluble. The tissue may then be dehydrated with acetone and embedded.

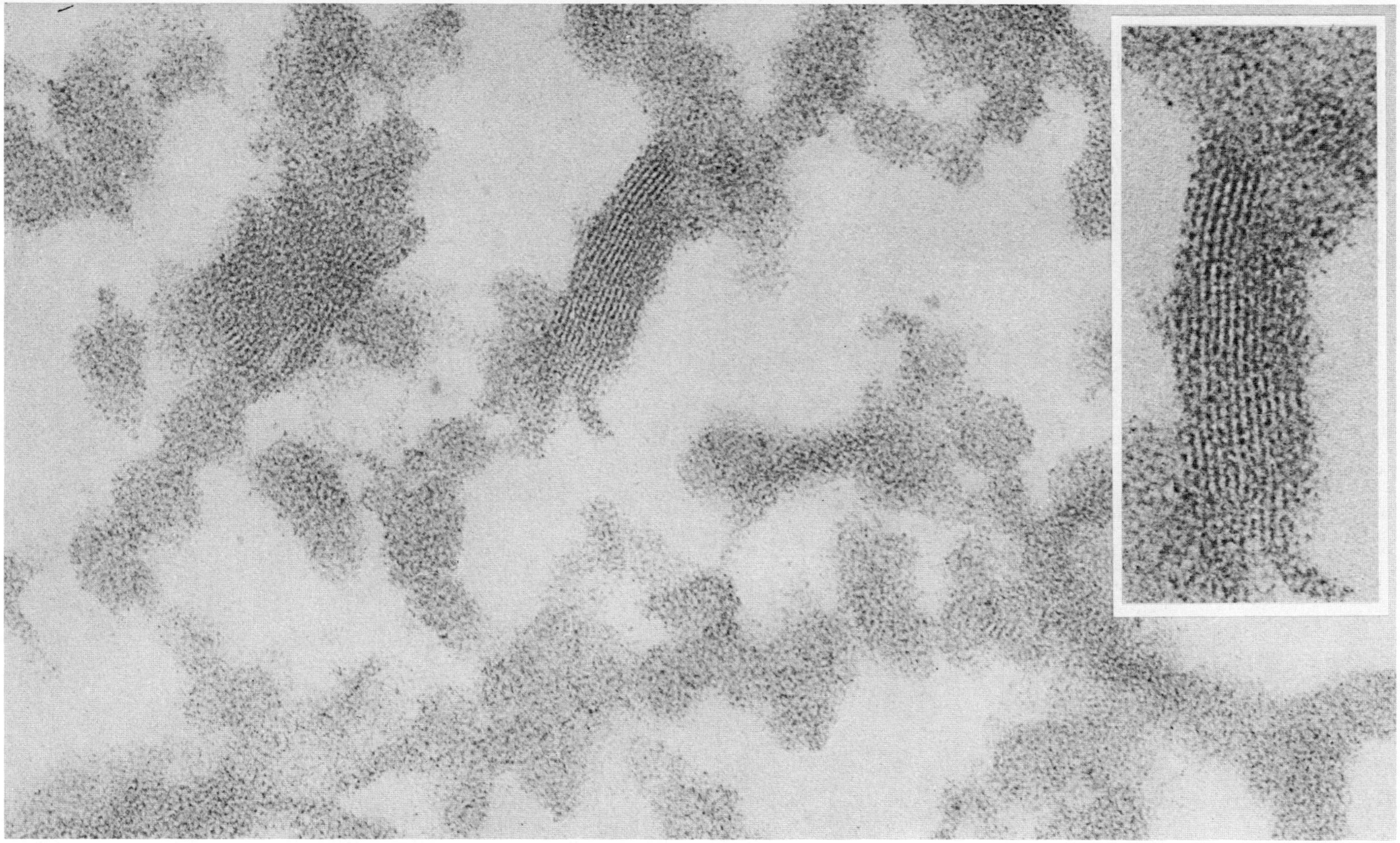

Figure 4. Preparation similar to that in Fig. 3 but with phosphatidylserine added in excess beyond the concentration needed for protein precipitation. ×221,000. The inset shows a typical lamellated lipid micelle. ×550,000.

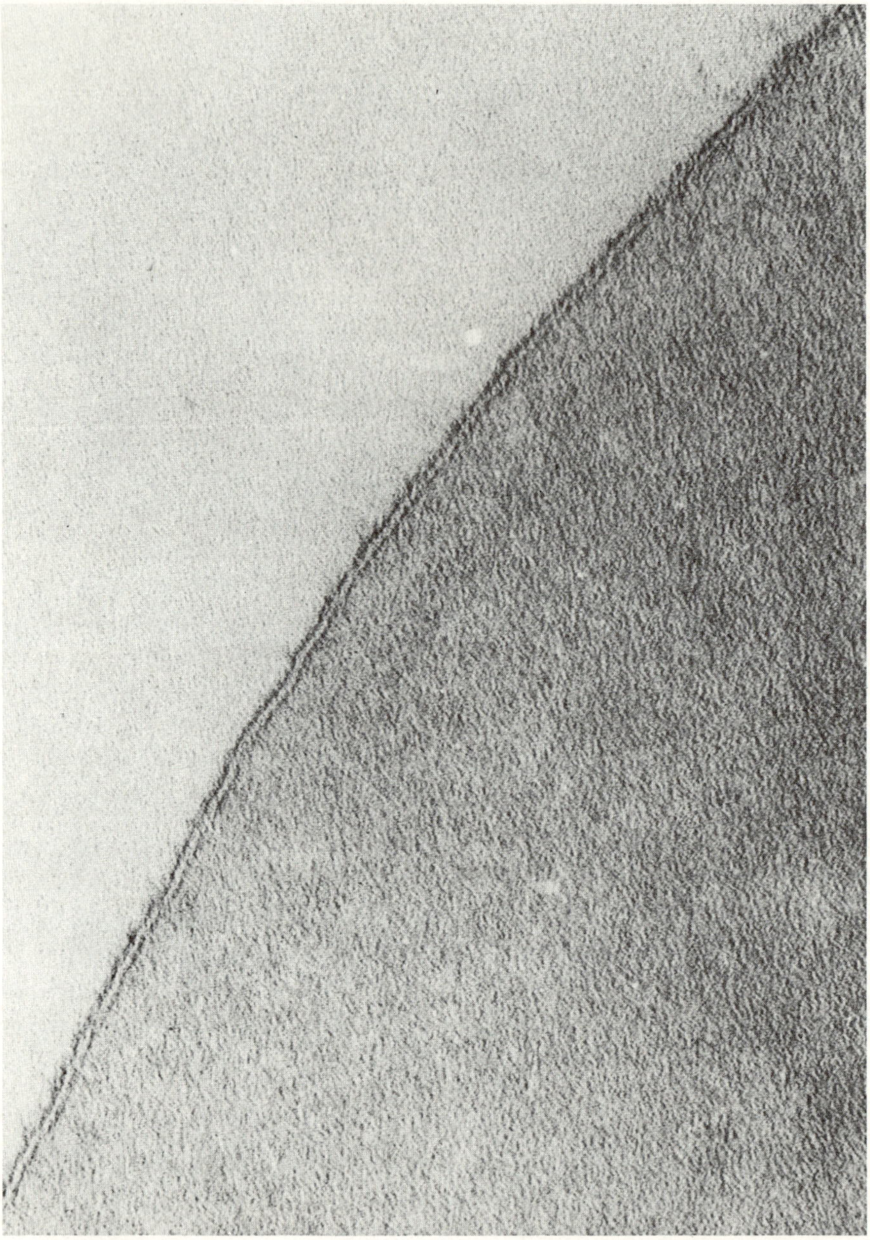

Figure 5. Electron micrograph of glutaraldehyde–osmium tetroxide-fixed human red blood cell membrane. ×280,000.

A thin (~300–1000 Å) section may then be cut and examined in the electron microscope. The image, however, will have almost no contrast. The scattering power of the biological material differs so slightly from that of the plastic embedding medium that essentially nothing will be seen. Shidlovsky (1965), for instance, in studying sections of multilayers of metallic salts of long-chain fatty acids embedded in methacrylate found that he could not visualize the fatty acid lamellae themselves but could see the superimposed metal atoms of the barium, lead, and uranyl salts. The atomic number of calcium, the lowest one visible, is 20, and atoms of this number and higher are visible in aggregates. Most tissue atoms have atomic numbers well under 20 and even in aggregates are invisible. However, if heavy metal atoms of atomic number greater than 20 are introduced into the specimen in aggregates a high-contrast image appears, especially if the lens is slightly underfocused. Obviously, the location of the aggregates of heavy metal atoms determines the final image.

There are several ways to insert the heavy metal atoms. The tissue may be fixed initially with osmium tetroxide. Here the immobilization and cross-linking of constituent molecules (fixing) is accomplished by the action of osmium tetroxide, and metallic osmium is deposited in the tissue as part of the reaction. The osmium does not distribute itself randomly but rather tends to aggregate in specific regions. The tissue may be fixed with potassium permanganate. Here manganese or manganese compounds aggregate selectively in certain regions and produce high contrast. Either of these methods may be preceded by initial aldehyde fixation, but this does not provide contrast. Contrast may, however, be provided by treating the tissue at some stage with a heavy metal compound such as uranyl acetate or lead hydroxide.

In all cases the problem is one of fixing the cellular components by some means that retains them in their native relationships and the introducing of heavy metal atoms either at the same time or later. It is crucial to understand that what is observed in the image of the specimen with the electron microscope is mainly the distribution of the heavy metal atoms; the tissue components themselves, dominantly comprised of atoms of atomic number below 20, contribute negligibly to the image. One must realize, however, that many of the original tissue molecules are still present, having been cross-linked into a stable structure, and that heavy metal atoms may be present in some locations without adding enough contrast to be seen. It is only *aggregates* of metal atoms that become visible.[1] Of course, most of the small ions and much lipid are washed out of the tissue during its dehydration with lipid solvents. However, significant amounts of lipid, protein, and carbohydrate remain behind, and even though they do not scatter electrons appreciably even when cross-linked by a few heavy metal atoms, they confer the order revealed by the aggregates of heavy metal atoms that tend to accumulate in some loci. The problem is to determine where the aggregated heavy metal atoms are located in terms of the original molecules comprising the membrane in its living state. It is essential that these points be understood in order to appreciate the following discussion.

How can we understand the meaning of the dense and light strata in the unit membrane image in terms of the underlying chemical components of the original membrane? In order to proceed further we digress and consider the structure of peripheral nerve fibers.

[1] Recently Crewe and coworkers (1970), using a special type of scanning-transmission electron microscope have succeeded in photographing spots reasonably interpreted as *individual* heavy metal atoms. However, we are dealing here only with standard transmission microscopes and standard techniques.

5. VERTEBRATE PERIPHERAL NERVE STRUCTURE

From studies of the structure of nerve fibers and of lipid and protein model systems, Robertson was able to advance the unit membrane theory in 1957 which with some minor modifications he still regards as essentially correct. The following discussion makes clear how this theory evolved. The illustrations are mainly diagrammatic since the relevant micrographs and more details are available in previous publications (see, for example, Robertson, 1969).

Peripheral nerve fibers are either myelinated or nonmyelinated. In both types the axons are associated with Schwann cells. Indeed, myelin is literally a part of the Schwann cell.

Figure 6 is a diagram of a nonmyelinated nerve fiber from mouse sciatic nerve. The Schwann cell is seen as a rather extended cytoplasmic mass with which several axons are associated. Some of these simply lie close to the Schwann cell and are connected to the outside world by a pathway between the two apposed Schwann cell membranes that are drawn down into a cellular trough as the axon is enveloped. Gasser (1955, 1956) called this pathway and the pair of membranes bordering it a mesaxon.

In 1954, Geren observed in young chick nerve fibers before myelin formation that there were a number of nonmyelinated fibers appearing very much like the ones studied by Gasser but differing in that there was only one axon per Schwann cell. She also noted that some of the mesaxons were wrapped around the axon in a spiral. This suggested to her that myelin might be simply a spirally wound-up mesaxon. About this time, Robertson (1955) found mature myelinated nerve fibers in lizard muscle that showed an outer and an inner mesaxon attached to the myelin sheath just as it would be if Geren's hypothesis were correct. However, the Schwann cell

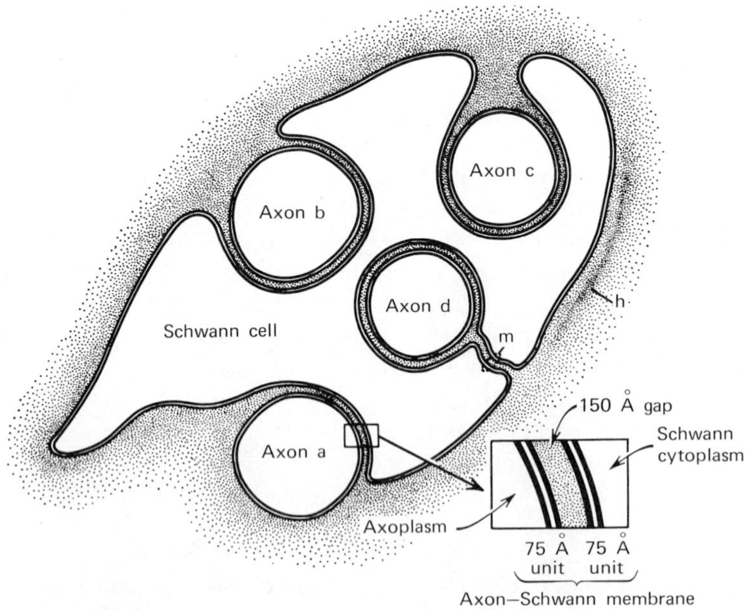

Figure 6. Diagram of nonmyelinated nerve fiber.

membrane was very poorly defined; it simply appeared as a vague, dense line, somewhat less than 60 Å in thickness. Furthermore, the junction of the mesaxon with the myelin sheath was not very clearly defined. Therefore, it was not possible to say that myelin was made *only* of Schwann cell membranes. It could merely be said that the sheath must form along the lines postulated by Geren by spiral evolution of the mesaxon. There remained a crucial question. Could some nonmembranous material be laid down, in particular between the outer surfaces of the Schwann cell membranes, and make a significant part of the myelin sheath?

About this time, potassium permanganate was introduced into electron microscopy as a fixing agent by John Luft (1956), and Araldite, an epoxy resin, was introduced as a new embedding material by Glauert et al. (1956). Robertson (1957) experimented with these two new methods and soon was able to produce unequivocal evidence that myelin was indeed nothing but Schwann cell membrane material down to the resolution of the micrographs.

Potassium permanganate-fixed myelin showed a much more regular and substantial intraperiod line structure than that fixed with osmium tetroxide. Another

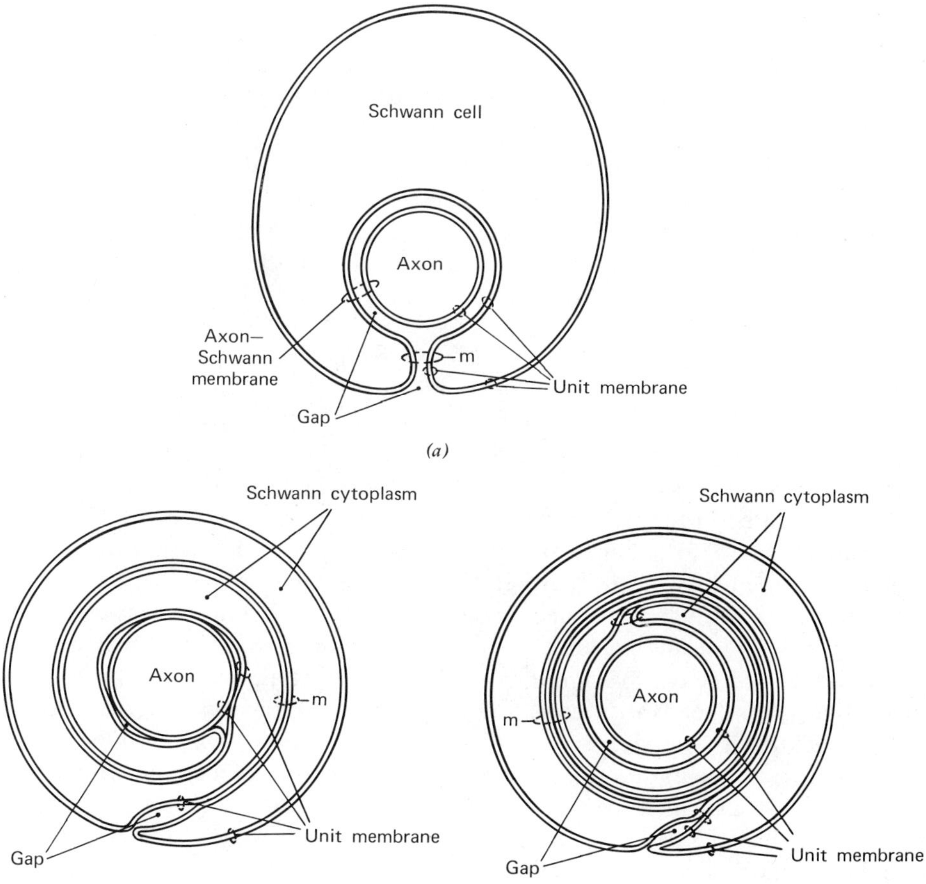

(a)

(b) *(c)*

Figure 7. Diagrams showing stages in the evolution of nerve myelin (m-mesaxon).

feature of potassium permanganate-fixed material was that the Schwann cell membrane appeared as a triple-layered structure, consisting of two parallel dense strata each about 20 Å thick, separated by a light interzone about 35 Å across. The mesaxon was found to consist of two of these layered structures in close apposition. It was apparent that the intraperiod line in myelin was formed by the intimate apposition of the two external surfaces of the membranes of the mesaxon and that the major dense line came about by the extrusion of cytoplasm between the cytoplasmic surfaces of the mesaxon loops. This arrangement was worked out in detail by studying mouse sciatic nerve fibers during the first week after birth. During this period the myelin sheath evolves, and many stages can be defined in any one section. These are shown diagrammatically in Figure 7. In the earliest stage a single Schwann cell with a single associated axon is seen as in Figure 7*a*. Later on, these protomyelinated fibers show elongated mesaxons extending in a spiral around the axon as in Figure 7*b*. Sometimes several turns in the spiral are visible in these intermediate fibers. The next stage comes with the apposition of the cytoplasmic surfaces of the spirally wound mesaxon loop as indicated in Figure 7*c*. This is arbitrarily defined as the earliest myelinated fiber. The evidence on which this diagram is based made it possible to state rigorously that myelin is made only of membranes that are closely packed. This allowed further deductions to be made as follows.

6. MOLECULAR STRUCTURE OF MYELIN

It was known in the last century that the myelin sheath was a fatty substance containing protein material called neurokeratin. It was also known from the work of Valentine (1861) that the intact fresh myelin sheath showed intrinsic radially positive birefringence. Schmidt (1936) restudied the myelin sheath by polarization optical techniques and showed that after lipid extraction the radially positive intrinsic birefringence reversed in sign and became of the "form" type. It is recalled that intrinsic or crystalline birefringence results from the oriented polarization tensors of molecules. It is not influenced by the ambient medium. Form birefringence is influenced by the medium. Schmidt's findings were taken to mean that there were alternating layers of material, thin with respect to the wavelength of light, alternately consisting of lipid and protein with the long axes of the lipid molecules radially oriented, and the protein molecules oriented in the tangential direction.

Schmitt et al. (1935) obtained the first small-angle x-ray diffraction patterns from myelinated nerve fibers. Their interpretation of these patterns was entirely in keeping with the findings of Schmidt. Their findings also indicated that there was a radially repeating structure in the myelin of 171 and 186 Å in fresh, unfixed frog and mammalian nerves, respectively (Schmitt et al., 1941). Bear et al. (1941) then proceeded to extract lipids from nerve, dry them, and obtain x-ray diffraction patterns from them. They found that the crude nerve lipids showed a pattern of layers repeating at a period of approximately 66 Å. They knew that protein molecules spread in a monolayer at an air–water interface on a Langmuir trough made a film approximately 15 Å thick. They thus postulated that the radially repeating unit in the myelin sheath probably contained two bimolecular leaflets of lipid with two to four interspersed monolayers of protein.

Finean (1955), taking advantage of further x-ray evidence and the electron microscope findings, suggested that the most probable repeating periodic structure was that shown in his diagram (Fig. 8). Here we see that the repeating unit is two bimolecular leaflets of lipid with the polar heads pointing outward and each polar surface covered by a monolayer of protein. It should be noted that since the repeating unit in the myelin sheath contains two of the apparently symmetrical membrane structures instead of one, there must be asymmetry either of structure or position in the membrane structure. Asymmetry of structure is suggested by the differing reactions of the two surfaces of the unit membrane to fixing and staining substances, but some asymmetry of position also exists. Finean used the term "difference factor" to refer to this asymmetry and noted that it probably was related to the fact that some difference would be expected in the intraperiod line where two membranes were back-to-back.

About this time, Robertson postulated that the difference factor might in fact be at least in part related to the presence of mucoproteins on the outside surface of the Schwann cell with dominance of pure protein on the inside surface. In any case it was then possible to take advantage of all of the information available. Assuming that the x-ray diffraction analysis of Finean was essentially correct, one could draw the diagram shown in Figure 9 and state that by extrapolation the Schwann cell membrane should have the general structural pattern shown in Figure 10. Here we see that since the repeating unit in the myelin sheath is the mesaxon, an identity can therefore be established between the molecular components in Finean's repeating unit and each of the dense and light strata seen with the electron microscope as indicated. Extrapola-

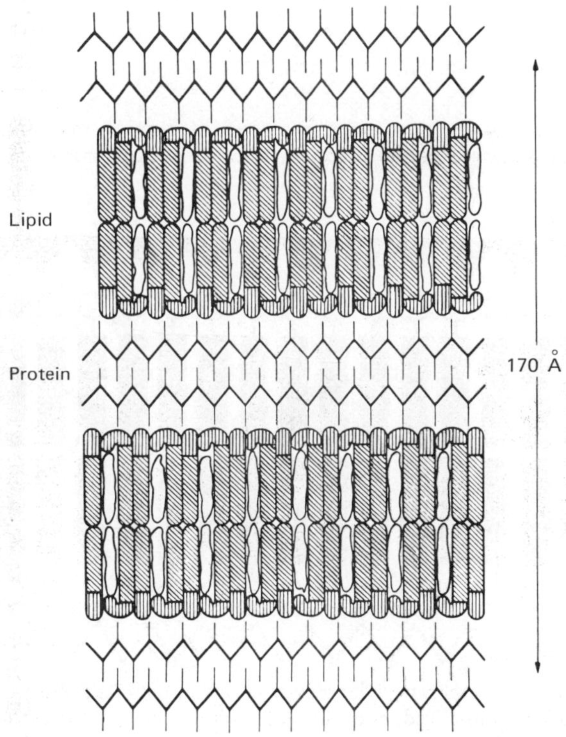

Figure 8. Diagram from Finean showing molecular model of radial repeating unit in myelin.

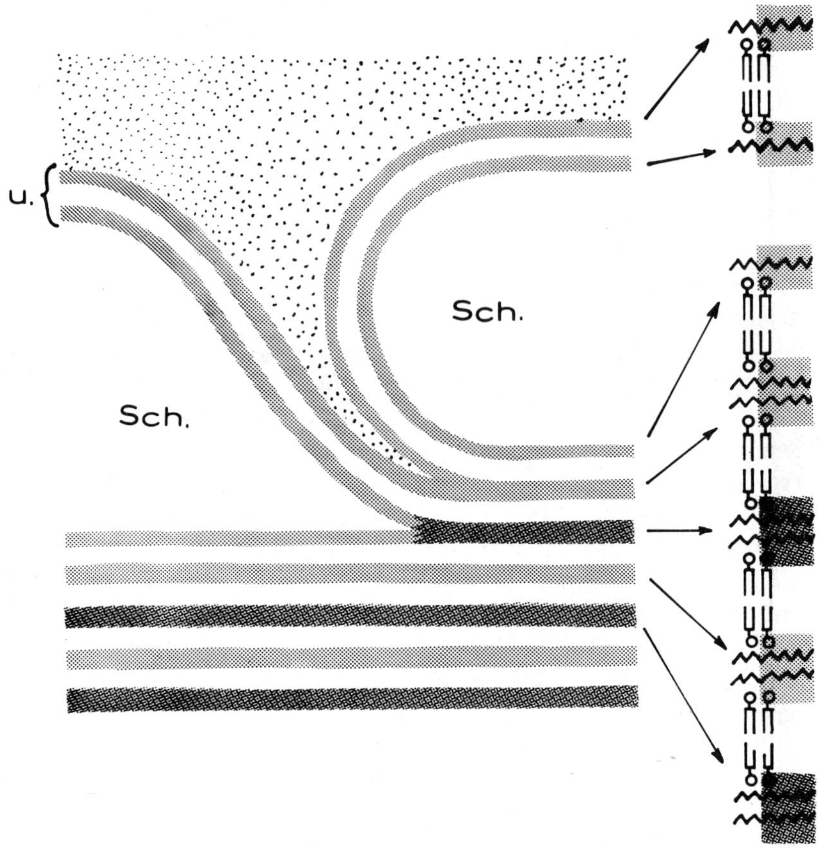

Figure 9. Diagram of myelin–mesaxon junction as seen by electron microscopy (left) with superimposition of molecular pattern deduced mainly by x-ray diffraction (right).

Figure 10. Diagram of unit membrane. The bilayer is shown with its polar heads pointing outward and covered by monolayers of nonlipid, one of which is chemically different from the other. The diagram is highly schematic and is not meant to imply anything about molecular conformation nor a particular molecular species.

tion was then made to the surface of the Schwann cell, and the molecular structure underlying the triple-layered pattern was specific. Assuming that Finean's analysis was correct, the Schwann cell membrane must consist of a bilayer of lipid with the polar heads of the lipids pointing outward and covered by monolayers of nonlipid as indicated in Figure 9. Furthermore, the structure is chemically asymmetric with some important differences existing between the components at the inside and outside surfaces.

Robertson (1957b, 1959) then went to some length to show that the triple-layered appearance of the Schwann cell membrane was not unique to the Schwann cell. It was shown that biological membranes in general presented this appearance if treated with potassium permanganate and embedded in epoxy resins. It was also shown that the membranes of all membranous organelles, including mitochondria, Golgi apparatus, nuclear membranes, and so on, displayed this typical pattern (Fig. 11). As mentioned earlier, because of its widespread occurrence and its involvement in the myelin sheath as a fundamental element, and also to distinguish it from other kinds of membranes such as a basement membrane, this structure was called a unit membrane. It was considered to be the fundamental unit of all biological membrane structures; thus mitochondria were shown to consist of two unit membranes with one next to the cytoplasm and an inner one thrown into folds to make the *cristae mitochondriales* with a matrix material between the folds. The nuclear membrane was shown to consist of a flattened unit membrane-bounded sac with pores connecting the nuclear matrix with the cytoplasmic matrix. The nuclear membrane may be regarded as a modified spherical endoplasmic reticulum (ER) cistern with pores. The ER was found to be a set of unit

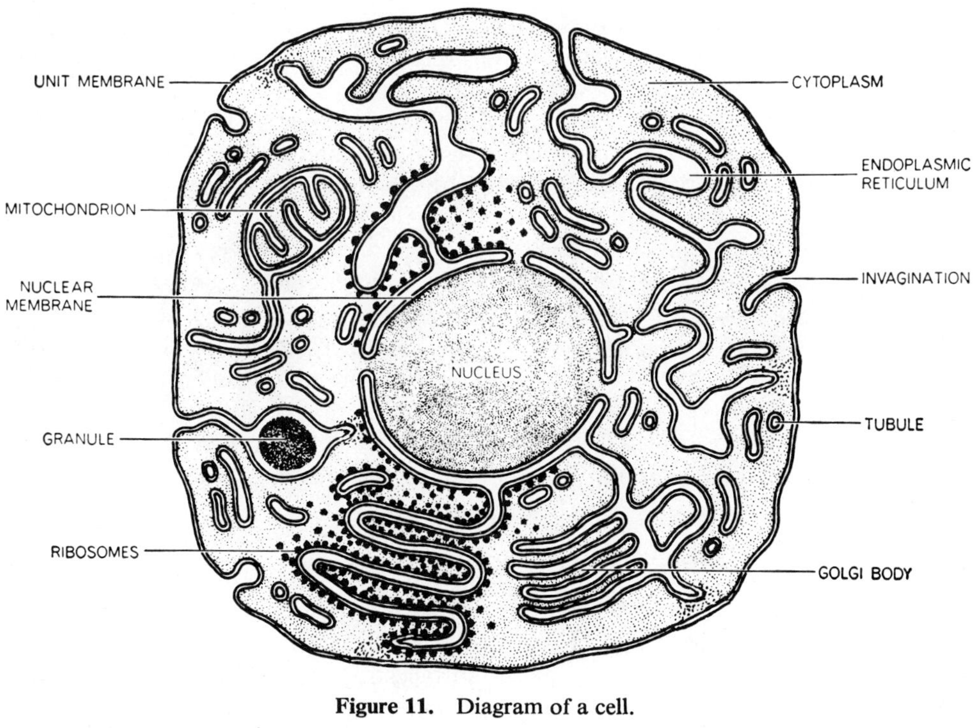

Figure 11. Diagram of a cell.

membrane-limited sacs or cisterns, to a large extent interconnected, and the Golgi apparatus was found to be a stack of unit membrane-bounded cisterns. A survey of many different kinds of cells in animals, plants, bacteria, and certain viruses showed that all displayed unit membranes. It thus appeared that a generalization of far-reaching nature could be made about cells; it seemed reasonable to state that biological membranes in general displayed a similar architectural pattern. This is not to say that all biological membranes are the same. Indeed, they differ widely chemically. The theory states that they are the same only in the sense that they contain a bimolecular leaflet of unspecified lipid at the center—with the polar heads pointing outward and covered by monolayers of nonlipids. The theory says nothing about the particular molecular species making up the lipid bilayer, nor the particular ones making up the nonlipid layers. These vary widely depending upon the membrane concerned. The theory states only a general pattern of molecular organization. It also says nothing about the possibility that there might be local deviations from this pattern such as the establishment of protein lined pores in some regions of membranes. However, since such pores were not seen, it seemed unlikely that they were present. Perhaps some will be found, but so far they have eluded definitive detection. The encroachment of portions of amphipathic protein molecules into the bilayer in some regions, and even the full incorporation in limited regions of hydrophobic protein molecules into the bilayer, as implied already, remains an open though as yet unproven possibility. The assumption in the original theory that the membrane proteins were opened out and fully spread in a monolayer was based on the dimensions that could be assigned to them in myelin. The possibility that they might be predominantly in a globular conformation, while left open, was not emphasized. It now appears that the latter is more prevalent but there still is no hard evidence known to the author that allows the exact conformation of an intrinsic membrane protein to be specified in its native state.

It was possible at this stage, with the facts available, to make another useful generalization about the overall topological organization of cells. Figure 11 illustrates this. A cell can be considered a three-phase system. First, there is a nucleocytoplasmic matrix phase (stippled) continuous through the nuclear pores. This might be called the "endophase." Second, there is a phase consisting of material that is in some places continuous with the external material around the cell which we might call the "exophase." Finally, there is a continuous membrane phase between these two continuous phases. The important point here is that the exo- and endophases are never continuous. They are always separated by an intervening unit membrane. A corollary of this is that unit membranes are never bounded on both sides by cytoplasm. They form closed loops with endophase on one side and exophase on the other. While this last feature is usually valid and a useful one in defining unit membranes, it is not inviolate. For instance, in capillaries adjacent pinocytotic vesicles can be seen with their membranes fused to make a triple-layered structure bounded on both sides by exophase. The inverse of this with endophase on both sides has not been observed but its existence can be inferred as a stage in cell fusion and division.

The above rules apply only to normal undisrupted cells. Cells may be disrupted in such a way that unit membrane fragments are released that do not obey the rules. Thus unit membrane-limited vesicles may occur that are inside-out and free of most of the material normally bordering either side. If such vesicles fuse, a unit membrane might occur bounded on both sides by the surfaces formerly bounded by cytoplasm. This kind of arrangement may also occur in some phases of formation of autophagic vacuoles, in which whole unit membrane organelles may become incorporated into a unit membrane-limited cavity for digestion.

The degree of continuity between the membranous elements of cells shown in Figure 11 is of course rather exaggerated. However, all of the continuities shown have been seen at some time or other in some cells. It is useful to refer to the external surface membrane as the "plasma" membrane. Material can enter the cell by engulfment in plasma membrane-limited pouches (micropinocytosis) and leave the cell by extrusion involving eversion of such pockets (exocytosis). The exophase material, in the sense that it may be entirely beneath the plasma membrane after pinocytosis, is inside the cell, but since it is membrane limited and never continuous with the endophase material it is in one sense still outside the cell. In order to enter the endophase, it must transverse the membrane. As material enters a cell by pinocytosis, the patch of plasma membrane incorporated with it is no longer referred to as plasma membrane. The membrane-limited vesicles containing the material detach from the plasma membrane and move into the cell, coalesce with other membranous organelles, and perhaps deposit their exophase contents in a cavity of the Golgi complex or the ER or into a lysosome where it may be digested. In a reverse manner secretory products are thought to be formed by synthesis on the membranes of the rough ER (RER) by bound ribosomes. The newly synthesized molecules are thought to pass through the membrane from the endophase to the exophase by some unknown mechanism. They accumulate in RER cisternal cavities and flow into the smooth ER (SER). Material accumulates here in pools until the membrane containing it pinches off to make an isolated vesicle. This moves over to the Golgi apparatus and coalesces with it, effecting the transfer of the newly formed exophase material from the RER to the Golgi apparatus. This is referred to as "bulk transport." It is believed by some investigators that the secretory material at this step flows into the Golgi cisterns and the patch of membrane that bounded the vesicle that brought it there from the RER pinches off and returns as a "shuttle vesicle" to pick up more material at the RER. The Golgi cisterns feed the secretory product into new SER vesicles that become very large and become secretion granules. These coalesce with the apical surface of the cell and the material is released. There are some obvious logical difficulties in this scheme because it implies that whole membranes are easily transferred from one membrane system in the cell to another. This could be taken to imply chemical identity of the different membranes, which certainly is not true. Nevertheless, this general scheme is widely believed to be correct.

7. MODEL SYSTEMS OF PROTEIN AND LIPID

It was stated that the unit membrane theory depended upon the x-ray diffraction analysis being correct. In 1955, Finean's x-ray model itself represented only a theory, since it was not possible at that time to determine the phases of the diffracted intensities with any degree of certainty. A theory based on a theory was clearly inadequate. Without the verified x-ray analysis, other models had to be considered. It is important to realize, however, that the number of possible models was restricted. The three most reasonable ones are shown in Figure 12. We chose the particular model we selected with a high degree of certainty because of some studies of model lipid systems. These involved studies of pure proteins and pure phospholipids by the same methods used to study cells.

It was already known that pure protein in osmium tetroxide-fixed material usually appeared more or less homogeneously dense in thin sections. This seemed quite reasonable since unpublished work in our laboratory as well as studies by Bahr (1954) had shown that most amino acids as well as proteins themselves react with osmium tetroxide. Hence protein molecules would be expected to become diffusely dense because of osmium atoms distributed rather diffusely among many of their constituent amino acids. However, Robertson (1960a, 1960b, 1963) studied some pure proteins to be certain about this point. For example, crystals of chymotrypsinogen were fixed, embedded, and sectioned in the same way as tissues. The protein crystals

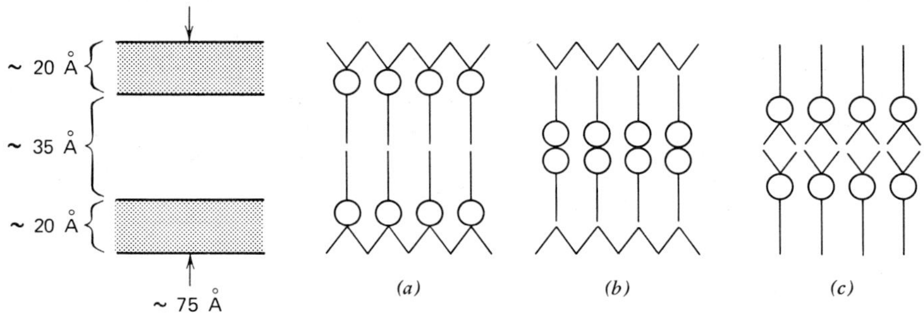

Figure 12. Diagram showing appearance of a unit membrane in electron micrograph (left) and three possible underlying molecular patterns in (*a*) through (*c*).

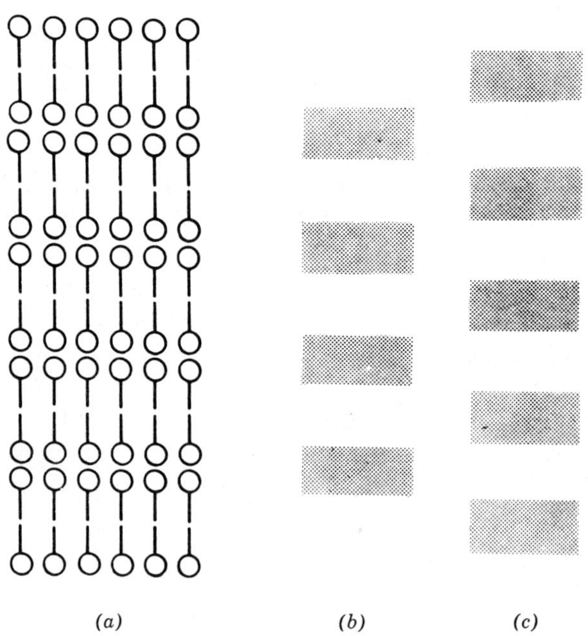

Figure 13. Diagram of molecular arrangement in a smectic lipid micelle (*a*); (*b*) and (*c*) are two possible interpretations of electron micrographs showing dense and light strata in terms of the underlying lipid molecules.

appeared as diffusely dense bodies. Later with improved methods overlapping groups of individual protein molecules in glutaraldehyde fixed catalase crystals were shown up in heavy metal treated sections as dense spots (Longley, 1967). The uniformly dense spots could clearly be identified as rows of overlapping protein molecules in crystal lattice planes and it was apparent that each protein molecule was entirely shown up as a uniformly dense spot of approximately the size of the molecule. Phospholipids studied by similar techniques, that is osmium tetroxide or potassium permanganate fixation followed by embedding and sectioning appeared different. Patterns of repeating, dense and light strata were observed similar to those in the lipid micelles in Figure 4. The dense strata repeated with a period of approximately 40–45 Å. This was known to be approximately the repeat period of such lipids in the dry state from x-ray diffraction studies. Furthermore, from polarization optical studies of such systems in addition to x-ray studies, it was known that the lipid was in the smectic fluid crystalline state. In other words, the lipid was arranged as in Figure 13*a*. From the observed dimensions it was apparent that each of the dense strata repressed only half the thickness of one monolayer of lipid in edge-on view added to the half thickness of another opposed monolayer. It was deduced that unlike protein molecules the lipid molecules were not uniformly dense. Now, there was a straightforward problem to solve. If it could be decided which part of the lipid molecule was producing the dense strata, that is, between (*b*) and (*c*) in Figure 13, a choice could be made among the three possible molecular models for the unit membrane shown in Figure 12*a–c*. On symmetry grounds and considering the dimensions, the dense strata in Figure 13 had to represent aggregated osmium atoms either along the polar ends of the molecules as in (*b*) or the nonpolar carbon chains as in (*c*). In order to decide between these two possibilities, advantage was taken of an experiment by Bear et al. (1941). These

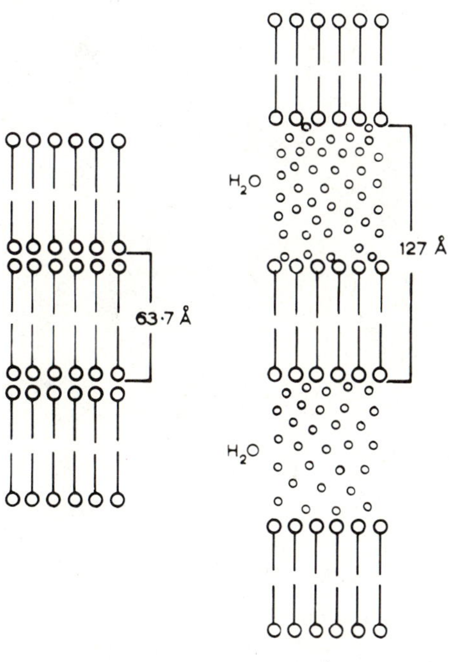

Figure 14. (*a*) A smectic array of lipid molecules. The 63.7Å spacing detected by x-ray diffraction is indicated. (*b*) The structure after hydration. The lipid bilayers are highly ordered and x-ray diffraction studies indicate a spacing of 127 Å. (From Bear et al., 1941.)

investigators showed that lipids in the smectic fluid crystalline state when wet with water are hydrated so that individual lipid bilayers split apart as in Figure 14. This was a reversible phenomenon, and the fact that diffraction patterns could be obtained showed that the bilayers were regularly spaced. This experiment was repeated using several phospholipids and it was found that hydrated individual lipid bilayers could be fixed and displayed in thin sections of electron micrographs as in Figure 15. Although the order between layers was poorly preserved, each individual bilayer appeared in the electron micrographs as a pair of dense strata separated by a light, central zone, making a layer approximately 60–70 Å thick. It was apparent then that the polar ends of the lipid molecules must be producing the densities in the electron micrographs. If the nonpolar carbon chains had been responsible, a single dense band would have been observed for each bilayer instead of a triple-layered structure.

Taken together, these results allow model (*a*) in Figure 12 to be chosen as the correct one. In effect, the x-ray diffraction problem was partly solved to a resolution of approx-

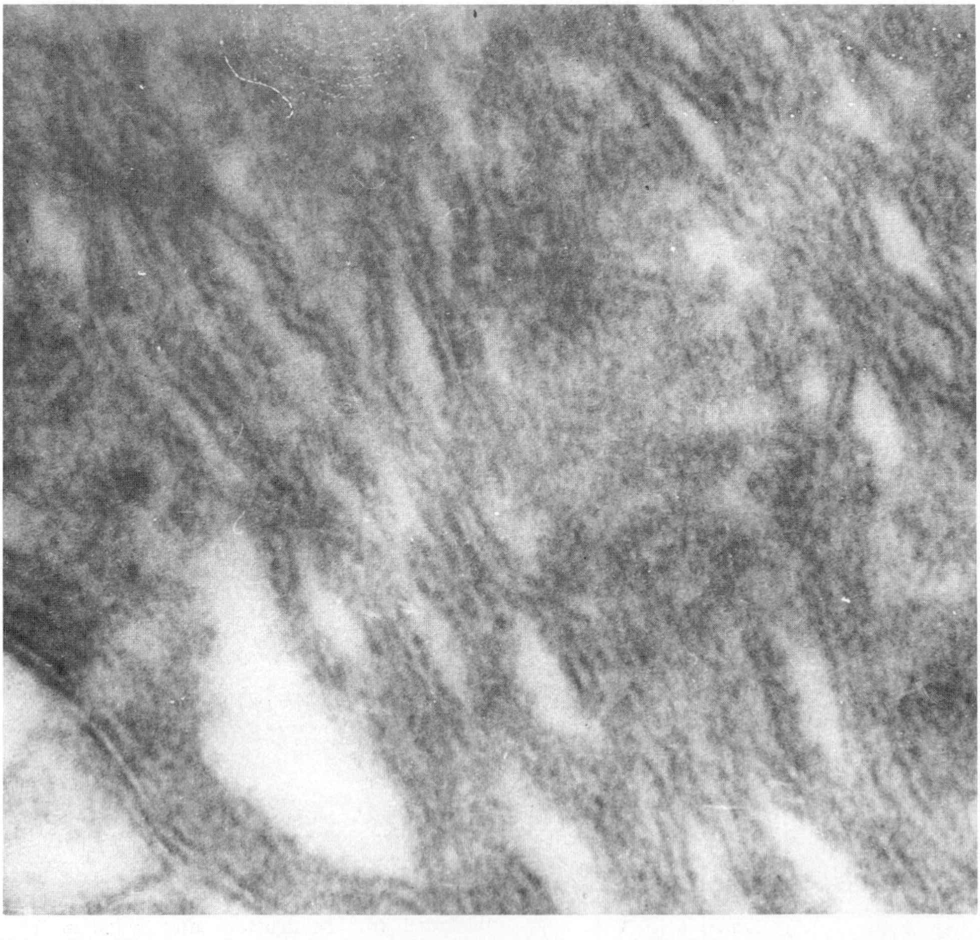

Figure 15. Electron micrograph of section of egg cephalin (phosphatidylethanolamine) after fixation and staining to show individual bilayers as pairs of closely spaced dense strata separated by light, central zones. Each bilayer measures approximately 60–70 Å in overall thickness. ×462,000.

imately 30 Å from the studies of lipid and protein model systems; a rough electron density plot was constructed to approximately 30 Å resolution. Clearly model (*a*) was the only one that could explain all the facts. Model (*b*) would have resulted in three dense strata and two light zones, and model (*c*) would have resulted in a single, broad, dense band for the bilayer. Therefore it could be stated with reasonable certainty that the electron density plot must have a low trough in the center of each unit membrane, just as x-ray diffractionists were able to prove in the next decade by assigning correct phases as discussed below.

Subsequently, Stoeckenius (1962), although at first differing (1959), confirmed Robertson's (1960a) results in studies on lipid model systems carried out in collaboration with Luzzati and Husson (1962). The latter had shown that the smectic lamellar phase was only one of the possible phases lipid and water systems could assume. They showed that under certain conditions of hydration and temperature a hexagonal phase formed in which lipid molecules were found with their polar heads enclosing columns of water, each column being a part of a hexagonal array. Stoeckenius succeeded in fixing such a system and in demonstrating dense spots for each of the columns of water containing the polar heads of the molecules. In the inverse structure in which the polar heads pointed outward, he demonstrated dense walled honeycomb-like patterns. These findings confirmed our results completely, and we thus felt additionally secure in our choice of the general molecular pattern of the unit membrane structure.

At this time, Moody (1963) proposed a method of determining the phases of the diffraction orders he had obtained from the myelin sheath. By swelling and shrinking experiments, he limited the phase assignments of the first five orders. Finean and Burge (1962), by similar swelling and shrinking experiments, made a single choice of phases and calculated an electron density plot of the myelin sheath mesaxon to a resolution of approximately 30 Å. This was confirmed and extended by Worthington and Blaurock (1969*a*, b). Akers and Parsons (1970) published an analysis leading to quite different conclusions, but Worthington (1970) indicated where their analysis was incorrect. Subsequently, Caspar and Kirschner (1971) assigned more phases, including higher orders, and calculated a more precise electron density plot to a resolution of approximately 10 Å. Except for that of Akers and Parsons, all of the electron density plots are consistent with the molecular arrangement postulated in Finean's 1955 model, although additional important details are added by the more recent analyses. Caspar and Kirschner's model, for instance, introduces the additional point of a shoulder on the internal (cytoplasmic) half of each unit membrane. They believe this shoulder is attributable to a selective concentration of cholesterol in this part of the membrane. This, as well as differences in the nonlipid components, may contribute to the asymmetry of the membrane. Thus we know now that the molecular model chosen for the unit membrane in 1957 is in general correct.

The generalization of the unit membrane theory was originally based on observations of many different kinds of membranes in different animals, plants, and even some viruses. Essentially similar triple-layered patterns were always found to be demonstrable wherever there was a lipid–protein biological membrane. However, for some time only the analysis of myelin and the model studies on lipids by electron microscopy formed the main basis for the molecular interpretation. In 1963, Robertson chose another regular membrane system on which a similar x-ray study could be carried out. The retinal rod outer segment is such a structure. It consists of regular stacks of flattened disc-shaped membrane sacs piled up like coins. Each disc is made of one folded unit membrane with its two inner sides apposed. The discs originate from inpocketing of the outer rod membrane, and so these inner sides correspond to external cell surfaces. The discs absorb light and this sets up electrical charges in the receptor cell. In frogs each rod is on the order of 40–50 μ long and about 5–7 μ in diameter. They are stacked side by side in palisade fashion in the retina. Finean et al. (1953) obtained x-ray diffraction patterns from osmium tetroxide-fixed retinal rods, which served to confirm the lamellar repeat period of approximately 300 Å seen by electron microscopy, but they were not detailed enough to be very helpful in a molecular analysis. Blaisie et al. (1965) reported x-ray studies of isolated rod outer segment membranes that were claimed to support the presence of a square array of particles in the rod outer segment membranes, but their evidence for this was not entirely convincing and their conclusions have not been confirmed by other investigators.

Robertson (1964, 1966) recorded two x-ray diffraction orders of the lamellar repeat in fresh unfixed rods *in situ* in the intact retina. Subsequent work by Corless (1970, 1971a, b) in our laboratory has now resulted in the recording of 11 discrete orders and additional continuous diffractions out to 17 Å from frog retina. Figure 16 is one of Corless' patterns. He succeeded in assigning phases to these orders and in constructing an electron density plot to a resolution of 26 Å as in Figure 17. While this work was in progress, Blaurock and Wilkins (1969) and Gras and Worthington (1969, 1970) also calculated electron density profiles for the frog rod outer segment disc membranes which were published independently. The conclusions were all in essential agreement with regard to the presence of density near the center of the membrane significantly lower than water and consistent with abutting $—CH_3$ groups. This is in complete accord with the unit membrane theory, for in each case the low-density regions occurred in the center of each unit membrane exactly as predicted by the theory. There are, to be sure, important differences between these three analyses in

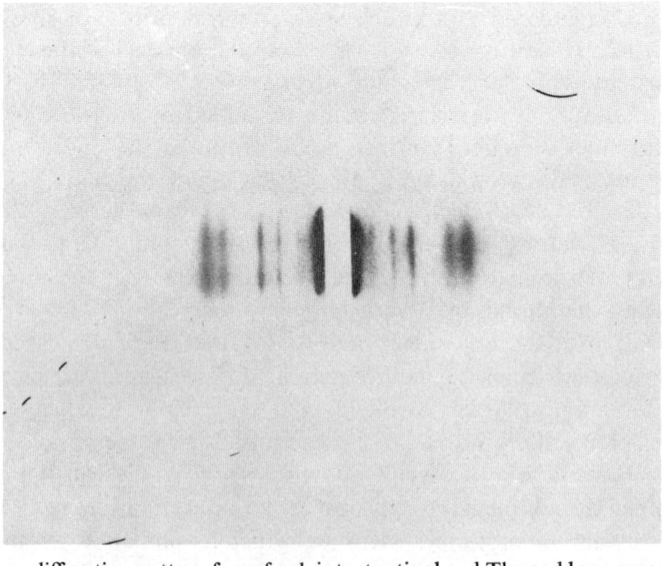

Figure 16. X-ray diffraction pattern from fresh intact retinal rod. The rod long axes run horizontally.

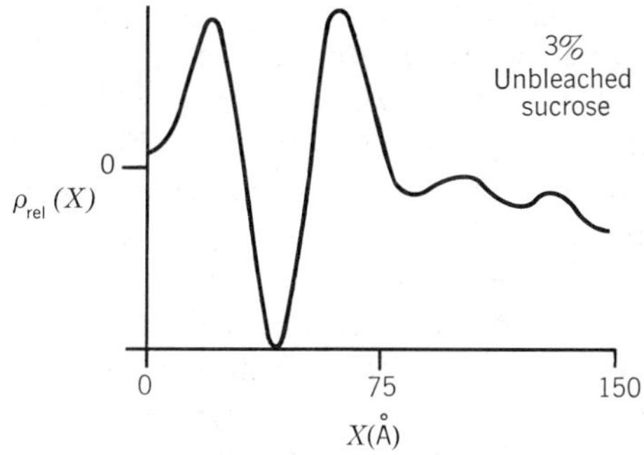

Figure 17. Electron density plot of one unit membrane in the rod outer segment. The center of a disc is the origin (0). $\rho_{rel}(X)$ is the electron density and the abscissa is parallel to the long axis oı the rod. Unbleached 3% sucrose.

other aspects that need not be discussed here. The point to be emphasized is that there are now two well-studied regular membrane systems in which the general molecular architecture has been determined by two independent methods and confirmed in several different laboratories.

Chloroplasts represent another system that can be studied by combined x-ray diffraction and electron microscopy in this way. Analyses of chloroplasts by Kreutz (1966) led to a somewhat contradictory proposal, but the analysis has not been confirmed by any other laboratory and there is reason to believe it may be incorrect in some important details. Studies are now in progress in our laboratory to perform an independent analysis of chloroplasts by combined electron microscopy and x-ray diffraction. Meanwhile we are withholding judgment on the chloroplast problem. There is also important evidence from freeze-fracture studies of chloroplasts (Branton and Park, 1967), but this is such a large subject that it cannot be discussed adequately here. Suffice it to say that in general the freeze-fracture interpretations agree with the unit membrane theory as regards the presence of the lipid bilayer as the core of the membrane and add the important fact thai the particles ~80 Å in diameter appear within the lipid bilayer in certain locations. Some interpret these particles being entirely protein molecules but they could be special aggregates of lipid and protein molecules with special bonding between the lipid and protein moieties like that postulated above for proteolipid protein and lipid in myelin. It would be beyond the scope of this article to go into this large subject further at present.

X-ray diffraction analyses have been carried out in recent years on several kinds of isolated membranes. The membranes are first isolated and packed by centrifugation. They are then partially dehydrated so that they pack more closely and become oriented well enough to give diffraction patterns showing enough orders at different degrees of hydration to allow electron density plots to be made in conjunction with electron microscope evidence. Such studies by Finean and his collaborators (1969, 1970) have led to the conclusion that the membrane architecture is essentially as specified in the unit membrane theory for rat erythrocyte, muscle microsomal, and brush border membranes of guinea pig epithelial cells. Similar studies by Engelman (1971) of membranes of the organism *Mycoplasma laidlawii* have led to similar conclusions. The unit membrane theory thus is withstanding increasing numbers of tests and is being shown to be applicable to many different membranes. The triple-layered fingerprint in electron micrographs has proved to be an excellent indicator of the underlying molecular architecture.

There is an interesting case in which a membrane appeared to represent an exception but was finally proved simply not to be a unit membrane, emphasizing again the usefulness of the term. This is the gas vacuole membrane of *Halobacterium halobium* studied by Stoeckenius and Rowen (1967) and Stoeckenius and Kunau (1968). In an elegant series of experiments, it was first shown that these bacteria contained membranes that appeared in thin sections to be unit membranes in that they showed the characteristic triple-layered pattern of the correct dimensions. However, there were two very important features of these membranes that suggested that they were not unit membranes. These membranes did not obey all the rules used in defining unit membranes outlined earlier. Thus they were bounded on both sides by cytoplasm and had free ends. They were fairly numerous and measured only a few tenths of a micron in length. When isolated, they were found to contain no lipid but consisted rather of a very regular lattice of protein molecules comprising sheets about 40 Å thick. The membranes were soon recognized as collapsed gas vacuoles. When gas is present in the vacuole, its wall appears as a 40-Å-thick, single, dense band between the lumen and the cytoplasmic matrix. The triple-layered appearance in this case proved to be produced by apposition of the two 40-Å-thick protein membranes making the wall. This provided one of the first instances in which an intracellular membrane less than 100 Å thick displaying the triple-layered pattern was proved unequivocally not to be a unit membrane.

It is possible that another kind of membrane present in *H. halobium* is also not a unit membrane. A peculiar purple membrane fraction isolated from this organism by the above investigators and studied further by Stoeckenius and Oesterhelt (1971) and Blaurock and Stoeckenius (1971) was found to contain a single protein of molecular weight 26,000. It also contains retinaldehyde in a 1:1 molar ratio to the protein, and 25% of the membrane dry weight is lipid. Blaurock obtained elegant x-ray diffraction patterns from aggregates of this membrane that show two or three orders of a bilayer lamellar repeat and several orders of a hexagonal lattice within the plane of the membrane. He proposed a model in which the protein is packed as four molecules in a hexagonal prism 50 Å thick with two molecules facing each membrane surface. Essentially, the protein lattice is visualized as being contained within a lipid bilayer. It is claimed that this membrane is in continuity with the plasma membrane and that the plasma membrane displays the unit membrane pattern in sections while this purple membrane does not. If this is correct and the model is correct, this membrane represents a modified unit membrane in which the central bilayer is penetrated by protein. However, the model is not yet proved and the continuity between this membrane and a true unit membrane claimed has not been unequivocally demonstrated. If this is done, we shall have an instance of localized departure in a plasma membrane from the unit membrane architectural plan in the sense that the continuity of the lipid bilayer is interrupted. Another instance of this kind may be present in synaptic discs (Robertson, 1963) and other localized external compound membranes or nexuses (McNutt and Weinstein, 1970; Sommer and Johnson, 1970). To discuss this topic adequately is beyond the scope of this chapter. Suffice it to say that these departures from the unit membrane architectural principles if proved will probably be relatively localized and restricted in extent. So far, whenever the simple triple-layered pattern is observed and the other essential criteria satisfied, it seems to represent the general molecular architecture specified by the unit membrane theory, and the uncomplicated unit membrane pattern remains by far the dominant one in biological membranes. We now consider the general topic of membrane models in more detail.

8. MEMBRANE MODELS

It is economical in protein synthesis for the cell to synthesize components of large protein molecules as relatively short-chain protamers of covalently bonded amino acids which can be put together to make larger structures by a subassembly process. This tends to minimize the effects of mistakes in genetic transcription. There have been many attempts to deal with membrane biogenesis in similar terms. However, important components of membranes are not precisely definable in the way a protein protamer can be defined; for example, the lipids are in a liquid state and the ratios of the different lipids are variable. The lipids are micellar and there is no precise way to relate them to the proteins except in paucimolecular complexes as described above. This relationship probably is not relevant to membrane structure. Lipoprotein molecules tend to have the lipid in micellar form, and no precise genetically determined molecular subunit of this kind can be reasonably defined. Hence considering membrane assembly in terms of lipoprotein subunit models is of questionable use.

Nevertheless, there is a way in which subunits might be involved in membrane assembly. It is conceivable that a kind of protein molecule synthesized by cells has the property of self-assembly on the polar surface of a lipid bilayer to make a closely knit monolayer bonded firmly in a regular crystallographic lattice. This monolayer could serve the function of a structural protein on one or both sides of the bilayer. It could also contain interstices into which specific protein molecules could be fitted to confer specific properties. The lipid bilayer itself would form spontaneously by hydrophobic bonding. Divalent cations might mediate the association of the lipid polar heads and the protein molecules.

A structure such as this is useful in considering problems of membrane permeability to small ions. It is believed that carrier molecules sometimes bind ions and literally transport them to the other side of a membrane. In fact, it is not necessary for this to be true. It may be possible to explain membrane permeability in the following way. Assume that a tight monolayer of protein molecules exists on one of the surfaces of a lipid bilayer. Suppose that some of the molecules of this protein monolayer are capable of binding an ion at a certain site on the membrane surface away from the bilayer core. Suppose that this binding of itself causes an allosteric change in the molecule which puts the ion into an internalized position next to the lipid polar groups. In this position the binding capacity for the ion may drop precipitously with release of the ion, reversal of the conformation change, and return of the bonding group to its former position and affinity. The ion will then be located in a position next to a lipid polar group where it cannot diffuse back through the protein monolayer. If this process is repeated, ions will build up locally around the polar group and cause a breakdown of the bilayer. In fact, only a very few such ions will raise the potential across a bilayer to about 400 mV. It is known that a lipid bilayer may be gradually brought to almost this potential in a model black membrane bilayer system but that it then breaks down (Tosteson, 1971). The ion may thus be driven through the bilayer as soon as two or three accumulate. If one includes enzyme reactions between the protein monolayer and the surrounding medium, including ATP to provide energy, one can easily see how active transport can occur by such a mechanism. Most if not all of the specificity of this system resides in one of the protein monolayers of the membrane. The lipid bilayer serves mainly as a barrier to free diffusion of ions. One of the features of this model of permeability is that only one monolayer of protein is needed with specific properties, the other could be nonspecific, perhaps serving some entirely unrelated function. This fits with the generalization in the unit membrane theory that the molecular architecture of membranes is chemically asymmetric.

9. GLOBULAR SUBSTRUCTURE

Following the demonstration by Luzzati and Husson (1962) of phase changes in lipids with departures from the lamellar state and the demonstration that the lipid polar heads could be shown in sections, Sjöstrand (1963) again proposed a new model of membrane structure. He had observed along with others that several membranes seemed under certain conditions to display regular transverse densities. These were seen to repeat at a period of approximately 80–90 Å, and the overall effect was to give an appearance of a globular nature to membranes. More recently, Sjöstrand (1969) showed that with very short permanganate fixation mitochondrial membranes sometimes give a particularly clear globular appearance in sections. This kind of observation was reported by Fernández-Morán (1969) in retinal rod outer segment membranes in thin sections several years earlier and also by Nilsson (1964) in retinal rod outer segment membranes in which the sections were somewhat thinner and the phenomenon could be seen more clearly. Sjöstrand referred to this appearance as globular substructure and postulated that what this meant in terms of molecular architecture of the membrane was that the lipid had been converted into some kind of non-lamellar phase such as that observed by Luzzati and Husson (1962). It was thus postulated that the lipid in the core of the membrane was converted into microspheres or microcolumns with the polar heads pointing outward. In either event, the polar heads, it was argued, would appear as densities running across the light, central zones of the membranes. Sjöstrand went further to postulate that not only was the lipid in this state but also the lipid polar surfaces were completely covered by protein. According to this model, the protein molecules project across the membrane in a repetitive fashion at a period of approximately 100 Å or less. This model is similar to ones proposed by Lenard and Singer (1966, 1968) and by Wallach and Zahler (1966) based on interpretations of molecular spectroscopic studies which can just as well be interpreted in other ways.

A somewhat similar proposal was advanced about the same time by Glauert and Lucy (1968), who studied the appearances of lipids in model systems by negative staining and by fixing and sectioning the material. They concluded that the lipid could exist in the various states found by Luzzati and proposed that these states might occur in membranes. However, these workers neglected the fact that membranes are lipoprotein structures, and Luzzati's group had shown the phase changes in question in pure lipids. They had shown also that whenever proteins were present the lamellar phase dominated. Only rarely do departures from the lamellar phase occur in lipid–protein micelles. Gulik-Kryzywicki et al. (1969), however, found two-dimensional square and hexagonal lattices in lyzozyme–phosphatidylinositol water systems at pH

4–7 and temperatures from 0 to 35°C, and so the possibility of the existence of such phase changes in membrane lipids must be considered. However, they have not been found in any native membrane or in a protein–lipid complex involving membrane proteins.

The retinal rod outer segment provided a test object for deciding whether or not the transverse densities observed by Nilsson were attributable to lipid phase changes or transverse protein molecules or some other feature of their organization. It was apparent that if the arguments were correct and the general kind of substructure involving departure from the lamellar state by the lipid were present generally in the rod outer segment membranes, it would be possible to detect this physical state by applying x-ray diffraction analysis. It seemed reasonable to assume that if x-ray patterns could be obtained from fresh or fixed rod outer segments in which the lamellar repeat, known to be about 300 Å, could be detected, then this would appear in the direction parallel to the long axes of the rods. Transverse regularity *within* the plane of the lamellae would give rise to reflections in the equatorial direction perpendicular to the rod axis. These latter were expected to be approximately 80–100 Å, and so should be much easier to detect than the axial ones because of the reciprocal nature of Bragg's law. It was this reasoning that led Robertson to the experiments described above in which x-ray diffraction patterns were obtained from fresh retinal rods. No such equatorial diffractions were found in preparations in which the lamellar repeat pattern was clear, and no such diffractions have been found in later studies in which point collimation was used and the diffractions should have appeared (1971). The experiments thus failed to support the globular substructure theory of Sjöstrand as applied to retinal rod membranes, for the lipid globules and protein lattice proposed as an explanation for the transverse densities in rod membranes and others, would most likely have resulted in detectable diffractions. Some kind of repetitive organization might still be present within the planes of the membranes of the rods, but it probably is more subtle than the one postulated in the globular substructure theory.

This analysis left one problem completely unanswered. How could one explain the apparent transverse densities in the electron micrographs? It was possible to provide a reasonable explanation. They may simply represent image overlap artifacts resulting from a slight tilt of the membranes in the sections. The evidence for this interpretation has been presented elsewhere (Robertson, 1969). This evidence confirmed the obvious fact that one cannot always believe too literally what one sees in electron micrographs. It is always of the utmost importance to be certain that the interpretation is supported by other evidence from as many other different techniques as possible. It is the failure to observe this rule that has resulted in the production of some membrane models that are not compatible with all the evidence. One of these was recently published by Sjöstrand (1969). It is similar to a model proposed by Benson (1964, 1966) based on indirect chemical evidence. Benson's model shows the membrane to be made of loosely packed globules of protein with lipid molecules attached hydrophobically in a rather random fashion. Sjöstrand's model is based mainly on observations of fixed material in thin sections and indeed the material was treated in a way that might well give questionable fixation. The main point is that apart from Benson's chemical arguments there is no acceptable structural evidence on which such a model can be based other than simple observation of electron micrographs in thin sections. This kind of electron microscopy is exceedingly dangerous and leads more often than not to completely incorrect conclusions. Thus the globular subunit models of membranes and especially those based solely on electron microscope observations are not well supported. There is, however, another kind of substructure that might be discussed. This has been called granulofibrillar substructure and has been dealt with in some detail elsewhere (Robertson, 1969).

10. SPECIAL MOLECULAR ARRAYS IN MEMBRANES

Porter et al. (1967) observed an unusual membrane in the apical cells of the epithelium lining the mouse urinary bladder. They noted that oblique sections of these membranes sometimes displayed a polygonal pattern. This membrane has been studied further in our laboratory (Vergara et al., 1969; Knutton et al., 1971) and by Hicks (1965, 1970), Staehelin et al. (1970), and Chlapowski et al. (1970). The hexagonal pattern appears clearly in glutaraldehyde–osmic-fixed material that has been uranyl acetate block-stained and sectioned. Where the membranes are sectioned obliquely, a hexagonal pattern can be seen. Its striations are approximately 140 Å apart. When observed in cross section, these membranes are thicker than the usual unit membrane. The thickness of the inner leaflet has the usual value of approximately 20–30 Å, but the outer one is about 50 Å thick. The overall thickness of the membrane is thus increased to approximately 120 Å. When these membranes are perpendicularly sectioned, one sometimes sees regularly spaced rows of dots approximately 150 Å apart. It is possible to free these membranes by simply scraping the mucosal surface or sonicating the bladder lightly. Fragments displaying a regular hexagonal array are then found readily in PTA negatively stained preparations. Such an array is shown in Figure 18. The purification may be enhanced by treating the membrane fraction with the detergent Triton X-100. This in low concentration (0.1%) destroys all membranes except ones with the hexagonal patterns. The distinctive hexagonal array is made by particles approximately 30 Å in diameter in a regular lattice. This feature is made particularly clear by obtaining optical diffraction patterns from the electron micrographs, filtering the diffraction image through a mask to reduce background "noise," and reconstituting the image. Such a reconstituted image obtained by Dr. W. Longley from a micrograph made by Dr. J. Vergara is shown in Figure 19. The structure of one hexagon is shown in the inset. It consists of 12 particles each approximately 30 Å in diameter. This agrees with the conclusions of Hicks (1970) based on image enhancement by the Markham integration method. On the basis of such reconstituted images, a detailed model of the structure has been constructed in Figure 20. The lattice has been found to belong to the crystallographic class p6.

Lanthanum hydroxide preparations of these bladder membranes have proven quite interesting. At high magnification one sees a light zone between a dense layer of lanthanum in which there are rows of dots. The spacing of these rows is approximately 140 Å, and it appears that the subunits observed in the negative stain preparations lie upon the external dense stratum of the unit membrane. Clearly, they project above the surface of the membrane, and there is no density in the light zone to suggest that they extend into the membrane bilayer core at all.

Recently, the technique of freeze-etch introduced several years ago by Moor and Muhlethaler (1963) based on earlier studies of Steere (1963) has come into prominence in membrane studies. In this technique a small piece of tissue is frozen in liquid nitrogen, fractured in a vacuum chamber, and etched by allowing some water to

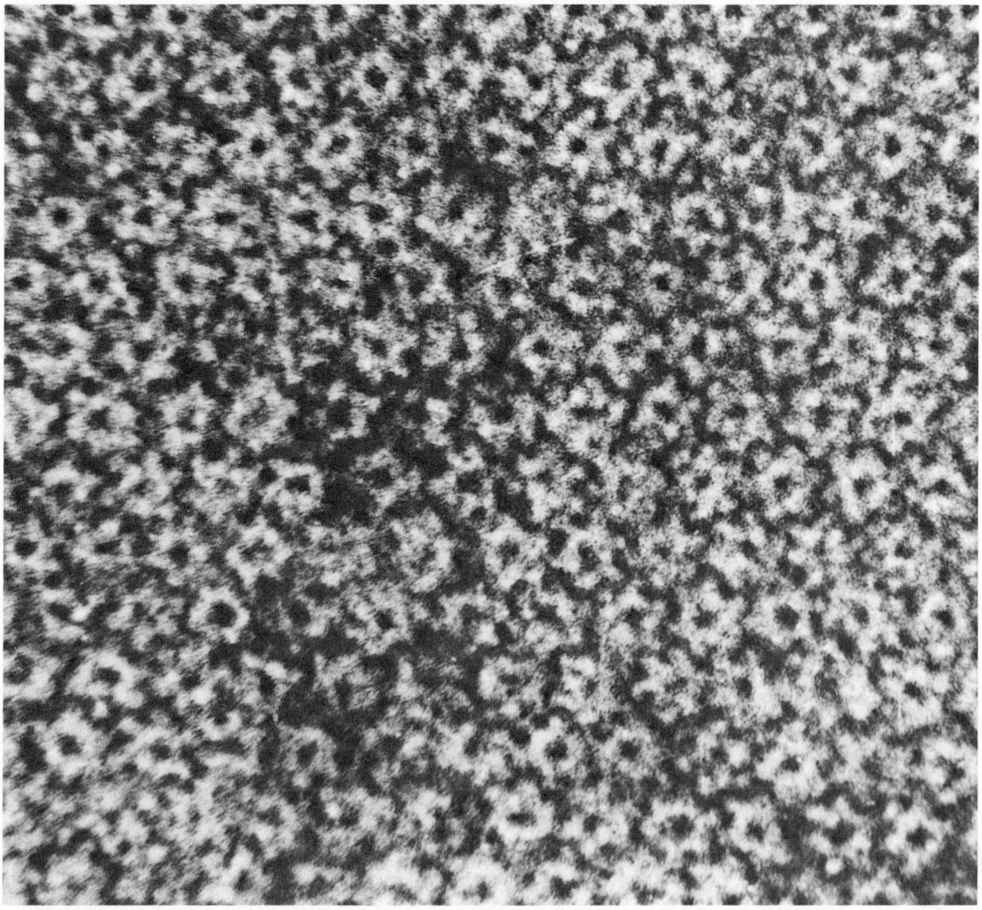

Figure 18. PTA-negatively stained fragment of mouse urinary bladder membrane. ×560,000.

sublime. The surface is sprayed with evaporated platinum and carbon from an angle of 45° and the replica is strengthened by further vertical carbon deposition. The tissue is dissolved away and the replica is then studied by electron microscopy.

Such freeze-etch studies have been conducted on mammalian urinary bladder in our laboratory. The hexagonal facets in the bladder membranes appear as in Figure 21. After deep etching it becomes apparent that the hexagonal arrays are present in roughly circular patches on the surface of the bladder membrane. These patches are several tenths of a micron in diameter and irregular in size. They do not overlap, and there is invariably a small amount of smooth membrane between each patch. Freeze-etch preparations, according to Staehelin et al. (1970), are said to show in some places penetration of the particles of the hexagonal array through the membrane bilayer. At first our results did not seem compatible with those of Staehelin et al. since the patterns were seen only in etched preparations. However, these were all glycerinated and more recent studies of nonglycerinated preparations in which definite etching is demonstrable revealed that the pattern seen in Figure 21 is in a fracture face. A similar

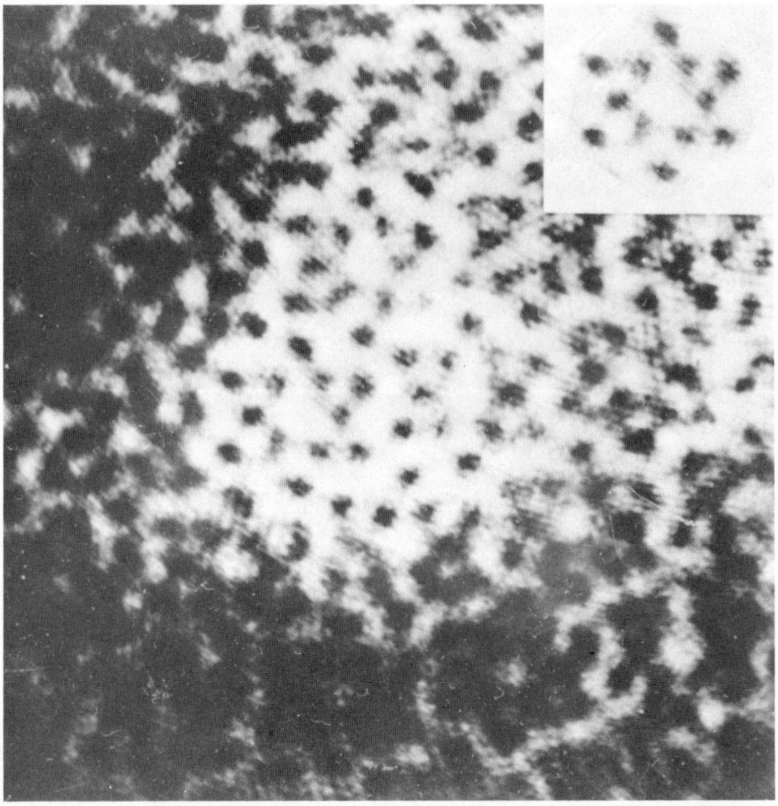

Figure 19. Optically reconstructed image made by mask filtration of diffraction pattern from a micrograph similar to that in Fig. 18. Note the substructure of each hexagon. One of the hexagons has been cut out and placed in the inset to the left to show that it is really a dodecamer. Each of the dense spots is believed to represent a protein molecule and to measure about 30Å.

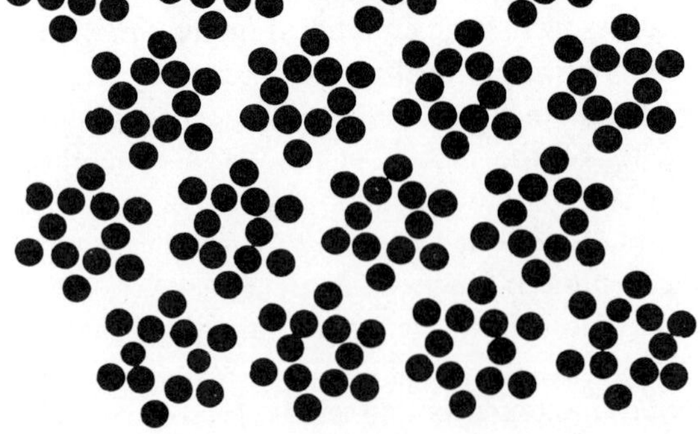

Figure 20. Model of the structure in Fig. 19 showing the detailed arrangements of the molecules. Each hexamer has been slightly separated for artistic accentuation.

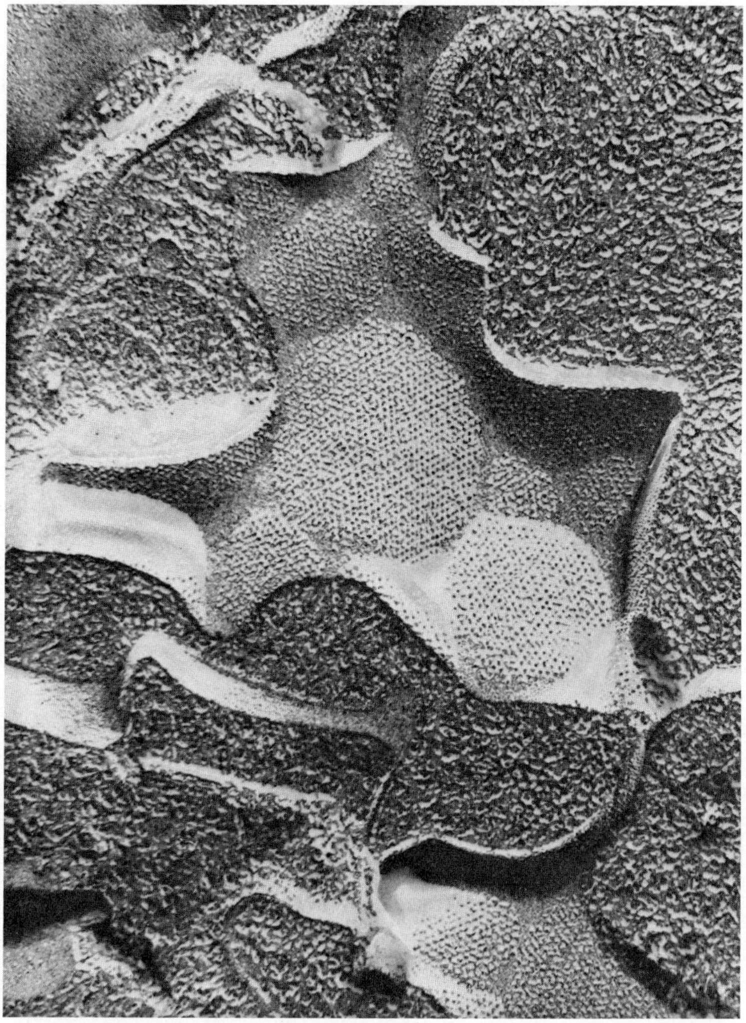

Figure 21. Electron micrograph of a freeze-fracture preparation of a portion of a mouse urinary bladder epithelial cell showing the patch of particulate arrays of a hexagonal lattice believed to represent the same structure shown in Fig. 18. ×73,000.

pattern also appears in etch faces but it is less distinctive. It thus now appears that the particles extend into the fracture face which, in keeping with the findings of Branton et al. (1967), probably follows the core of the bilayer. If this is the case the particles penetrate deeply into the bilayer. The sectioning evidence shows decisively that the portion of the particles appearing uniformly dense is outside the bilayer. Since there is only uniform low density within the core of the unit membrane under the particles it seems reasonable to suppose that this is primarily lipid. The possibility exists, of course, that the particle is an amphipathic protein or glycoprotein with a hydrophobic part embedded in the bilayer and reacting like lipid to the fixatives and stains. Until we have more definitive evidence all these possibilities remain open. We tend to favor the notion that the outer particles are nonlipid and the part in the bilayer is a special lipid very firmly bound to it, but present evidence does not allow us to make a decision at present. Similarly, the so-called nexus particles seen in freeze-etch preparations might also lie on the surface of adjacent membranes rather than penetrating through them, although this is not yet clear and they have been interpreted as at least partially penetrating the bilayer by McNutt and Weinstein (1970) and by Sommer et al. (1970). If the particles represent nonlipid running through the membranes, then the nexus represents a significant departure from the general unit membrane architectural pattern similar to the one postulated by Blaurock and Stoeckenius discussed above. However, the interpretation of the freeze-etch micrographs is by no means certain. Much further study is required.

11. OTHER BIOPHYSICAL STUDIES OF MEMBRANES

Recently, the unit membrane model has received support from molecular spectroscopy. Compounds have been synthesized to which have been attached radicals such as the nitroxide group containing an unpaired electron. This unpaired electron gives a very high magnetic moment to the radical and it precesses in an applied magnetic field. Electron spin resonance measurements can be made on material containing such compounds, which determine the orientation and mobility of the radical. Hubbell and McConnell (1968, 1969) and McConnell (1970) synthesized lipids in which a nitroxide radical is placed either at the polar end of the molecule, in the middle of the nonpolar carbon chain, or at the nonpolar end of the chain. They introduced such tagged lipid molecules into oriented membranes and tested the unit membrane theory. They found that molecules with the spin label on the polar end after introduction into membranes rotate but that the direction of orientation of the radicals remains fixed. If the spin label is located in the nonpolar carbon chain, about half way down, the radical still rotates but its orientation is not well maintained; the radical "tumbles" to some extent. When the spin label is placed at the extreme end of the nonpolar carbon chain, the radical still spins and on the average maintains some orientation similar to that of the ones placed on the polar end. However, it now has much greater mobility and tumbles about a great deal. The spin-labeling experiments have been interpreted as indicating that the lipid molecules are in a lipid bilayer configuration. In addition, this work provides good evidence that the core of the membrane is relatively liquid. Hubbell and McConnell (1968) introduced their spin-labeled compounds into nerve and muscle membranes and found them to be more fluid than others they have studied. This is in keeping with some recent work by Fischer et al. (1970) in Santiago. These investigators showed that nerve membranes contain a higher degree of unsaturation in their lipid carbon chains than do glia membranes. This study was made on molluscan optic nerve material in which it is possible to obtain an almost pure axon membrane fraction because the glia content is relatively low.

Further studies using another method support the bilayer model and suggest that the bilayer is quite liquid. Coleman et al. (1970) studied rat and human erythrocyte ghosts by treating them with phospholipase c. It was shown in agreement with Lenard and Singer (1968) that phospholipase c can act on the membrane lipids without destroying the membrane. Evidently, there are some places where the outer monolayer of nonlipid either is discontinuous or can be penetrated by this enzyme enough to react with the underlying lipid so as to release breakdown products. Coleman et al. found that as the reaction proceeded oil droplets 0.3–1.0 μ in diameter appeared on the external surface of the membrane and that the membrane area was gradually reduced as the phospholipids were broken down. These findings suggest that the bilayer is quite fluid since a large proportion of the membrane phospholipid can be removed in this way without disrupting the membrane. The reduction in membrane area that occurs is suggestive not only of fluidity in the lipid but of a certain degree of fluidity in all the membrane components. In any case these investigators interpreted their findings as being compatible with the bilayer model.

Finally, it should be noted that another piece of supporting biophysical evidence comes from the use of a completely different technique. Steim et al. (1969) used calorimetric techniques to study the heat absorption by lipids in the smectic or bilayer state. Certain characteristic curves were obtained which they related to melting of the carbon chains in the bilayers. They then subjected isolated preparations of *Mycoplasma* membranes to the same procedures. The characteristics of the curves were the same, and they concluded that the membranes contain lipid bilayers. Their findings seem to support the bilayer model, but certain assumptions implicit in the interpretations have since been challenged by Chapman (1971), and other interpretations may be reasonable.

12. CONCLUSION

In conclusion, it seems fair to say that the unit membrane theory provides a useful membrane structure theory because it is one that can be subjected to experimental tests and stimulates the development and application of new tests. It is open to disproof. It has been subjected to a number of different kinds of experimental tests by widely varying techniques and it still appears to be the most widely applicable theory. There probably are departures from this architectural pattern in certain specific membranes or regions of membranes and these departures may be of great interest and importance. Nevertheless, the bilayer pattern still seems most generally applicable to biological membranes. It should be said again that the theory deals only with the architectural principles whereby lipids and proteins are generally arranged in membranes. It does not deal with the details of the particular molecular species involved, nor does it specify the exact conformation of any proteins or other components although in some membranes certain well-defined patterns are seen. These patterns may represent special aggregates of protein molecules. Perhaps similar more subtle aggregates are present but not recognized.

The unit membrane model is not a truly molecular model in the crystallographic sense. Indeed, it is almost surely impossible ever to construct a general model of cell membranes of this type that is broadly applicable to many different membranes. Nevertheless, there may be some undisclosed skeletal element of membranes of general significance that has such precision but which so far has escaped detection. It is the quest for knowledge of this kind applicable generally to membranes that provides perhaps the greatest stimulus to research in this field at present. Perhaps of equal importance is the unequivocal determination of the chemical nature of intra-membrane particles seen in freeze-fracture preparations.

REFERENCES

Akers, C. C., and D. F. Parsons (1970). X-ray diffraction of myelin membrane. II. Determination of the phase angles of the frog sciatic nerve by heavy atom labeling and calculation of the electron density distribution of the membrane. *Biophys. J.*, **10,** 116.

Autilio, L. A., W. T. Norton, and R. D. Terry (1964). The preparation and some properties of purified myelin from the central nervous system. *J. Neurochem.*, **11,** 17.

Bahr, G. F. (1954). Osmium tetroxide and ruthenium tetroxide and their reactions with biologically important substances. Electron stains. III. *Exptl. Cell Res.*, **7,** 457.

Baldwin, G. S., and P. R. Carnegie (1971). Specific enzymic methylation of an arginine in the experimental allergic encephalomyelitis protein from human myelin. *Science*, **171,** 579.

Bear, R. S., K. J. Palmer, and F. O. Schmitt (1941). X-ray diffraction studies of nerve lipids. *J. Cellular Comp. Physiol.*, **9,** 289.

Benson, A. A. (1964). Plant membrane lipids. *Ann. Rev. Plant Physiol.*, **15,** 1.

Benson, A. A. (1966). On the orientation of lipids in chloroplast and cell membranes. *J. Am. Oil Chemists Soc.*, **43,** 265.

Blaisie, F. K., M. M. Dewey, A. E. Blaurock, and C. R. Worthington (1965). Electron microscope and low angle x-ray diffraction studies on outer segment membranes. *J. Mol. Biol.*, **14,** 143.

Blaurock, A. E., and W. Stoeckenius (1971). A quasi-crystalline membrane. Biophysical Society Abstracts, *Biophys. J.*, **11,** 115a.

Blaurock, A. E., and M. H. F. Wilkins (1969). Structure of frog photoreceptor membranes. *Nature*, **223,** 906.

Branton, D., and R. B. Park (1967). Subunits in chloroplast lamellae. *J. Ultrastruct. Res.*, **19,** 283.

Braun, P. E., and N. S. Radin (1969). Interactions of lipids with a membrane structural protein from myelin. *Biochemistry*, **8,** 4310.

Bruns, R. R., and G. E. Palade (1968). Studies on blood capillaries. I. General organization of mouse capillaries. *J. Cell Biol.*, **37,** 244.

Caspar, D. L. D., and D. A. Kirschner (1971). Myelin membrane structure at 10 Å resolution. *Nature*, in press.

Chapman, D., and H. Urbina (1971). Phase transitions and bilayer structure of *Mycoplasma laidlawii* B. *FEBS Letters*, **12,** 169.

Chlapowski, F. J., M. A. Bonneville, and L. A. Stoekelin (1970). Luminal plasma membrane of the urinary bladder. II. Isolation and structure of membrane components. *J. Cell Biol.*, **47,** 34a.

Cole, K. C. (1932). Surface forces of the *Arbacia* egg. *J. Cellular Comp. Physiol.*, **1,** 1.

Coleman, R., J. B. Finean, S. Knutton, and A. R. Limbrick (1970). A structural study of the modification of erythrocyte ghosts by phospholipase c. *Biochem. Biophys. Acta*, **219,** 81.

Corless, J. M. (1971a). X-ray diffraction analysis of the lamellar structure of bleached and unbleached photoreceptor membranes. *Biophys. J.*, **11,** 248a.

Corless, J. M. (1972). The lamellar structure of bleached and unbleached rod photoreceptor membranes. *Nature*, in press.

Corless, J. M., and W. Longley (1970). The structure of retinal photoreceptor membranes. *Anat. Record*, **166,** 293.

Crewe, A. V., J. Wall, and J. Langmore (1970). Visibility of single atoms. *Science*,**168,** 1338.

Danielli, J. F., and H. A. Davson (1935). A contribution to the theory of permeability of thin films. *J. Cell. Comp. Physiol.*, **5,** 495.

Danielli, J. F., and E. N. Harvey (1935). The tension at the surface of mackerel egg oil with remarks on the nature of the cell surface. *J. Cell. Comp. Physiol.*, **5,** 483.

Dickerson, R. E., and I. Geis (1969). *The Structure and Action of Proteins*, p. 63. Harper, New York.

Dickinson, J. P., K. M. Jones, S. R. Aparicio, and C. E. Lumsden (1970). Localization of encephalitogenic basic protein in the intraperiod line of lamellar myelin. *Nature*, **227,** 1133.

Eichberg, J., and H. H. Hess (1967). The lipid composition of frog retinal rod outer segments. *Experientia*, **23,** 993.

Engelman, D. M. (1971). Lipid bilayer structure in the membrane of *Mycoplasma laidlawii*. *J. Mol. Biol.*, **58,** 153–165.

Eylar, E. H. (1970). Amino acid sequence of the basic protein of the myelin membrane. *Proc. Natl. Acad. Sci., U.S.*, **67,** 1425.

Eylar, E. H., J. Salk, G. C. Beveridge, and L. V. Braun (1969). Experimental allergic encephalomyelitis and encephalitogenic basic protein from bovine myelin. *Arch. Biochem. Biophys.*, **132,** 34.

Fernandez-Moran, H. (1962). In *Ultrastructure and Metabolism of the Nervous Systems*, Research Publications, Association for Research in Nervous and Mental Disorders, **40,** 235.

Finean, J. B. (1955). *Proceedings of the 2nd International Conference on Biochemical Problems of Lipids, University of Ghent*, p. 129. Butterworths, London.

Finean, J. B. (1969). Biophysical contributions to membrane structure. *Quart. Rev. Biophys.*, **2,** 1

Finean, J. B., and R. E. Burge (1962). The determination of the Fourier transformation of the myelin layer from a study of swelling phenomena. *J. Mol. Biol.*, **7,** 672.

Finean, J. B., F. S. Sjöstrand, and E. Steinmann (1953). Submicroscopic organization of some layered lipoprotein structures. (Nerve myelin, retinal rods, and chloroplasts). *Exptl. Cell Res.*, **5,** 557.

Fischer, S., M. Cellino, F. Zambrano, G. Zampighi, M. T. Nagel, D. Marcus, and M. Canessa-Fischer (1970). The molecular organization of nerve membranes. I. Isolation and characterization of the plasma membranes from the retinal axons of the squid: An axolemma-rich preparation. *Arch. Biochem. Biophys.*, **138,** 1.

Folch, J., and M. Lees (1951). Proteolipids, a new type of tissue lipoproteins. *J. Biol. Chem.*, **191,** 807.

Folch-Pi, J. (1971). Personal communication.

Gasser, H. S. (1955). Properties of dorsal root unmedullated fibers on the two sides of the ganglion. *J. Gen. Physiol.*, **38,** 709.

Gasser, H. S. (1956). Olfactory nerve fibers. *J. Gen. Physiol.*, **39,** 473.

Geren, B. B. (1954). The formation from the Schwann cell surface of myelin in the peripheral nerves of chick embryos. *Exptl. Cell Res.*, **7,** 558.

Glauert, A. M., and J. S. Lucy (1968). In *Ultrastructure in Biological Systems* (A. J. Dalton and F. Hagenau, eds.), Vol. 1, pp. 1–32. Academic, New York.

Glauert, A. M., G. E. Rogers, and R. H. Glauert (1956). A new embedding medium for electron microscopy. *Nature*, **178,** 803.

Gorter, E., and R. Grendel (1925). On bimolecular layers of lipoid on the chromocytes of the blood. *J. Exptl. Med.*, **41,** 439.

Gras, W. J., and C. R. Worthington (1969). X-ray analysis of retinal photoreceptors. *Proc. Natl. Acad. Sci., U.S.A.*, **63,** 233.

Gras, W. J., and C. R. Worthington (1970). Membrane and lipid structure. *Biophys. J.*, **10,** 52a.

Green, D. E., and S. Fleischer (1964). In *Metabolism and Physiological Significance of Lipids* (R. M. C. Dawson, and D. M Rhodes, eds.), pp. 580–618. Wiley, New York.

Green, D. E., and T. Oda (1961). On the unit of mitochondrial structure and function. *J. Biochem.*, **49,** 742.

Green, D. E., and J. F. Perdue (1966). Membranes as expressions of repeating units. *Proc. Natl. Acad. Sci., U.S.*, **55,** 1295.

Green, D. E., and A. Tzagoloff (1966). Role of lipids in the structure and function of biological membranes. *Lipid Res.*, **7,** 587–602.

Gulik-Kryzywicki, T., E. Schacter, and V. Luzzati (1969). Interactions of proteins and lipids: Structure and polymorphism of protein-lipid-water phases. *Nature*, **223,** 1116.

Harvey, E. N., and H. Shapiro (1934). The interfacial tension between oil and protoplasm within living cells. *J. Cellular Comp. Physiol.*, **5,** 255.

Hicks, R. M. (1965). The fine structure of the transitional epithelium of rat ureter. *J. Cell Biol.*, **26,** 25.

Hicks, R. M. (1970). Structure of the subunits in the thick luminal membrane of rat urinary bladder. *Nature*, **227,** 280.

Hubbell, W. L., and H. M. McConnell (1968). Spin-label studies of the excitable membranes of nerve and muscle. *Proc. Natl. Acad. Sci., U.S.*, **61,** 12.

Hubbell, W. L., and H. M. McConnell (1969). Motion of steroid spin labels in membranes. *Proc. Natl. Acad. Sci., U.S.*, **63,** 16.

Kauzman, W. (1959). Some factors in the interpretation of protein denaturation. *Advan. Protein Chem.*, **14,** 1.

Knutton, S., A. R. Limbrick, W. Longley, J. D. Robertson, and J. Vergara (1971). Regular structures in membranes of urinary bladder and intestinal epithelial cells. Biophysical Society Abstract, *Biophys. J.*, **11,** 290a.

Korn, E. D. (1966). Structure of biological membranes. *Science*, **153,** 11491.

Kreutz, W. (1964). Über die Struktur der Lipoproteinlamellen in Chloroplasten lebender Zellen. *Ber. Bunsenges. Physik. Chem.*, **68,** 820.

Kreutz, W. (1966). Röntgenographische Strukturuntersuchungen in der Photosyntheseforschung (X-ray structural analyses in photosynthetic research). *Umschau Wiss. Technik*, **24,** 806.

Laico, M. T., E. I. Ruoslahti, D. S. Papermaster, and W. J. Dreyer (1970). Isolation of the fundamental polypeptide subunits of biological membranes. *Proc. Natl. Acad. Sci., U.S.*, **67,** 120.

LeBaron, F. N. (1969). In *Structural and Functional Aspects of Lipoproteins in Living Systems* (E. Tria and A. M. Scann, eds.), pp. 201–226. Academic, New York.

LeBaron, F. N., and J. Folch (1957). Structure of brain tissue lipids. *Physiol. Rev.*, **37,** 539.

Lenard, J., and S. J. Singer (1966). Protein conformation in cell membrane preparations as stuided by optical rotatory dispersion and circular dichroism. *Proc. Natl. Acad. Sci., U.S.*, **56,** 1828.

Lenard, J., and S. J. Singer (1968). Structure of membranes: Reaction of red blood cell membranes with phospholipase c. *Science*, **159,** 738.

Limbrick, A. R., and J. B. Finean (1970). X-ray diffraction and electron microscopic studies of the brush border membrane of tinea pig intestinal epithelial cells. *J. Cell Sci.*, **7,** 373.

Longley, W. (1967). The structure of bovine liver catalase: A combined study of x-ray diffraction and electron microscopy. *J. Molec. Biol.*, **30,** 323.

Luft, J. H. (1956). Permanganate—A new fixative for electron microscopy. *J. Biophys. Biochem. Cytol.*, **2,** 799.

Luzzati, V., and R. Husson (1962). The structure of the liquid-crystalline phases of lipid water systems. *J. Cell Biol.*, **12,** 207.

McConnell, H. M. (1970). In *The Neurosciences* (F. O. Schmitt, ed.), pp. 697–714. Rockefeller Univ. Press, New York.

McNutt, N. S., and R. S. Weinstein (1970). The ultrastructure of the nexus, a correlated thin section and freeze-cleave study. *J. Cell Biol.*, **47,** 666.

Maddy, A. H., and B. R. Malcolm (1965). Protein conformations in the plasma membrane. *Science*, **150,** 1616.

Matthews, B. W. (1968). Solvent content of protein crystals. *J. Mol. Biol.*, **33,** 491.

Menke, W. (1966). The molecular structure of photosynthetic lamellar systems. *Brookhaven Symp. Biol.*, **19,** 328.

Moody, M. (1963). X-ray diffraction pattern of nerve myelin: A method for determining the phases. *Science*, **142,** 1173.

Moor, H., and K. Mühlethaler (1963). Fine structure in frozen-etched yeast cells. *J. Cell Biol.*, **17,** 609.

Nilsson, S. E. G. (1964). A globular substructure of the retinal receptor outer segment membranes and some other cell membranes in the tadpole. *Nature*, **202**(4931), 509.

O'Brien, J. S., and E. L. Sampson (1965). Lipid composition of the normal human brain: Gray matter, white matter and myelin. *J. Lipid Res.*, **6,** 537.

Overton, E. (1895). Über die osmotischen Eigenshaften der lebenden Pflanzen und Tierzelle. *Vjschr. Naturforsch. Ges. Zurich*, **40,** 159.

Porter, K. R., K. Kenyon, and S. Badenhausen (1967). Specializations of the unit membrane. *Protoplasma*, **63,** 262.

Reynolds, J. Personal communication.

Reynolds, J. A., and C. Tanford (1970a). Binding of dodecyl sulfate to proteins at high binding ratios. Possible implications for the state of proteins in biological membranes. *Proc. Natl. Acad. Sci., U.S.*, **66,** 1002.

Reynolds, J. A., and C. Tanford (1970b). The gross conformation of protein–sodium dodecyl sulfate complexes. *J. Biol. Chem.*, **24,** 5161.

Richardson, D. C., and J. S. Richardson. Personal communication.

Robertson, J. D. (1955). The ultrastructure of adult vertebrate peripheral myelinated fibers in relation to myelinogenesis. *J. Biophys. Biochem. Cytol.*, **1**(4), 271.

Robertson, J. D. (1957a). New observations on the ultrastructure of the membranes of frog peripheral nerve fibers. *J. Biophys. Biochem. Cytol.*, **3,** 1043.

Robertson, J. D. (1957b). The cell membrane concept. *J. Physiol.*, **140,** 58.

Robertson, J. D. (1959). The ultrastructure of cell membranes and their derivatives. *Biochem. Soc. Symposia*, **16,** 3.

Robertson, J. D. (1960a). A molecular theory of cell membrane structure. *Proc. Fourth Intern. Conf. Electron Microscopy*, Berlin, Springer, 1968, pp. 159–171.

Robertson, J. D. (1960b). The molecular structure and contact relationships of cell membranes, in *Progress in Biophysics* (B. Katz and J. A. V. Butler, eds.). Pergamon Press, New York, pp. 343–418.

Robertson, J. D. (1963). Unit membranes: A review with recent new studies of experimental alterations and a new subunit structure in synaptic membranes, in Cellular Membranes in Development. *Proc. of the XXII Symposium of the Society for the Study of Development and Growht* (Michael Locke, ed.) Academic, New York, pp. 1–046.

Robertson, J. D. (1963). The occurrence of a subunit pattern in the unit membranes of club endings in Mauthner cell synapses in goldfish brains. *J. Cell Biol.*, **19,** 201

Robertson, J. D. (1964). Addendum to summary of NRP work session on cell membranes, including certain new x-ray diffraction findings on retinal rod outer segment membranes. *Neurosci. Res. Program Bull.*, **2,** 75.

Robertson, J. D. (1966). In *Nerve as a Tissue* (K. Rodahl, ed.). pp. 11–48. Harper, New York.

Salem, L. (1962). Attractive forces between long saturated chains at short distances. *J. Chem. Phys.*, **37,** 2100.

Schatz, G., and J. Saltzgaber (1969). Identification of denatured mitochondrial ATPase in "structural protein" from beef heart mitochondria. *Biochem. Biophys. Acta*, **180,** 186.

Schmidt, W. J. (1936). Doppelbrechung und Feinbau der Markscheide der Nervenfassern. *Z. Zellforsch. Mikroskop. Anat.*, **23,** 657.

Schmitt, F. O., R. S. Bear, and G. L. Clark.(1935). X-ray diffraction studies on nerve. *Radiology*, **25,** 131.

Schmitt, F. O., R. S. Bear, and E. Ponder (1936). Optical properties of the red cell membrane. *J. Cellular Comp. Physiol.*, **9,** 89.

Schmitt, F. O., R. S. Bear, and E. Ponder (1938). The red cell envelope considered as a Wiener mixed body. *J. Cellular Comp. Physiol.*, **11,** 309.

Schmitt, F. O., R. S. Bear, and K. J. Palmer (1941). X-ray diffraction studies on the structure of the nerve myelin sheath. *J. Cellular Comp. Physiol.*, **18,** 31.

Seeman, P. (1967). Transient holes in the erythroycte membrane during hypotonic hemolysis and stable holes in the membrane after lysis by saponin and lysolecithin. *J. Cell. Biol.*, **32,** 55.

Senior, A. E., and D. H. MacLennan (1970). Mitochondrial "structural protein," a reassessment. *J. Biol. Chem.*, **245,** 5086.

Sherman, G., and J. Folch-Pi (1970). Rotatory dispersion and circular dichrosim of brain "proteolipid" protein. *J. Neurochem.*, **17,** 597.

Shidlovsky, G. (1965). Contrast in multilayer systems after various fixations. *Lab. Invest.*, **14,** 1213.

Sjöstrand, F. S. (1963). A new ultrastructural element of the membranes in mitochondria and of some cytoplasmic membranes. *J. Ultrastruc. Res.*, **9,** 340.

Sjöstrand, F. S. (1969). In *Structural and Functional Aspects of Lipoprotein in Living Systems* (E. Tria and A. M. Scann, eds.), pp. 73–139. Academic, New York.

Sommer, J. R., and E. A. Johnson (1970). A comparative ultrastructure of cardiac cell membrane specializations. A review. *Am. J. Cardiol.*, **25,** 184.

Staehelin, L. A., and F. J. Chlapowski (1970). Luminal plasma membrane of the urinary bladder. I. Three-dimensional reconstruction from freeze-etch images. *J. Cell Biol.*, **47,** 200a.

Steere, R. L. (1957). Electron microscopy of structural detail in frozen biological specimens. *J. Biophys. Biochem. Cytol.*, **3**(1), 45.

Steim, J. M., M. E. Tourtellottee, J. C. Reinert, R. N. McElhaney, and R. L. Rader (1969). Calorimetric evidence for the liquid crystalline state of lipids in a biomembrane. *Proc. Natl. Acad. Sci., U.S.*, **63,** 104.

Steinemann, A., and P. Läuger (1971). Interaction of cytochrome c with phospholipid monolayers and bilayer membranes. *J. Mem. Biol.*, **4,** 74.

Stoeckenius, W. (1959). An electron microscope study of myelin figures. *J. Biophys. Biochem. Cytol.*, **5,** 491.

Stoeckenius, W. (1962). Some electron microscopical observations on lipid-crystalline phases in liquid-water systems. *J. Cell Biol.*, **12,** 221.

Stoeckenius, W., and D. M. Engelman (1969). Current models for the structure of biological membranes. *J. Cell Biol.*, **42,** 613.

Stoeckenius, W., and W. Kunau (1968). Further characterization of particulate fractions from lysed cell envelopes of *Halobacterium halobium* and isolation of gas vacuole membranes. *J. Cell Biol.*, **38,** 337.

Stoeckenius, W., and D. Oesterholt (1971). Isolation of a membrane containing a rhodopsin-like protein from *Halobacterium halobium*. Biophysical Society Abstracts, *Biophys., J.*, **11,** 114a.

Stoeckenius, W., and R. Rowen (1967). A morphological study of *Halobacterium halobium* and its lysis in media of low salt concentration. *J. Cell Biol.*, **34,** 365.

Tanford, C. (1961). *Physical Chemistry of Macromolecules*, p. 130. Wiley, New York.

Tenenbaum, D., and J. Folch-Pi (1966). The preparation and characterization of water-soluble proteolipid protein from bovine brain white matter. *Biochem. Biophys. Acta*, **115,** 141.

Tosteson, D. C. Personal communication.

Trayer, H., Y. Nozaki, J. Reynolds, and C. Tanford (1971). Polypeptide chains from human red blood cells. *Biochemistry*, in press.

Valentine, G. (1861). *Die Huteronchung der Pflanzen—und der Thiergemche in polarizienten Lichte.* Leipzig.

Van Deenen, L. L. M. (1965). Phospholipids and biomembranes. *Prog. Chem. Fats Other Lipids*, **8**(1), 1.

Vanderkooi, G., and M. Sundaralingum (1970). Biological membrane structure. II. A detailed model for the retinal rod outer segment membrane. *Proc. Natl. Acad. Sci., U.S.*, **67,** 233.

Vergara, J., W. Longley, and J. D. Robertson (1969). A hexagonal arrangement of subunits in membrane of mouse urinary bladder. *J. Mol. Biol.*, **46,** 593.

Wallach, D. F. H., and P. H. Zahler (1966). Protein conformations in cellular membranes. *Proc. Natl. Acad. Sci., U.S.*, **56,** 1552.

Waugh, D. F., and F. O. Schmitt (1940). Investigations of the thickness and ultrastructure of cellular membranes by the analytical leptoscope. *Cold Spring Harbor Symp. Quant. Biol.*, **8,** 233.

Ways, P., and D. J. Hanahan (1964). Characterization and quantification of red cell lipids in normal man. *J. Lipid Res.*, **5,** 318.

Wolfgram, F. (1966). A new proteolipid fraction of the nervous system. I. Isolation and amino acid analyses. *J. Neurochem.*, **13,** 461.

Worthington, C. R. (1970). On the interpretation of x-ray diffraction intensities from chemically treated frog sciatic nerve. *Biophys. J.*, **10,** 675.

Worthington, C. R., and A. E. Blaurock (1969a). A structural analysis of nerve myelin. *Biophys. J.*, **9**(2), 970.

Worthington, C. R., and A. E. Blaurock (1969b). A low- angle x-ray diffraction study of the swelling behavior of peripheral nerve myelin. *Biochem. Biophys. Acta*, **173,** 427.

2

THE THERMODYNAMICS OF METABOLISM

J. C. Marsden

Be not afraid of thermodynamics; some men understand thermodynamics, some achieve an understanding of thermodynamics, and most have thermodynamics thrust upon them.

—With apologies to Shakespeare (1623)

1. INTRODUCTION

Undoubtedly much of the spectacular progress achieved in biology over the past quarter-century, particularly at the molecular level, has come from the application of mathematical, physical, and chemical techniques to living processes. It is a silver lining with a cloud. The biologist must now familiarize himself with the outlines of such techniques fully to appreciate his subject. The application of thermodynamics to biology is just such an area, and one that causes both student and specialist considerable difficulty.

It must be admitted at the outset that the results of thermodynamic studies in biology have provided comfort rather than fundamental advance. Living organisms, despite their complexity, do not seem to violate, either as a whole or in their individual parts, the laws of thermodynamics that are among the most basic in Nature.

For the biologist a drawback to studying thermodynamics is that the subject is so often approached from the standpoint of physical rather than biological experience. The immediate relevance of, for example, heat engines and Carnot's theorem, to the living cell or to a particular biochemical process is not immediately apparent. However, to use even the simplest biochemical process to illustrate the uses of thermodynamics is to court disaster. Biochemical processes may be subtle; they are not simple.

This chapter attempts to select a few cases, some perhaps of only marginal biological interest, to illustrate the concepts and uses of thermodynamics in biology. Many texts have covered the subject, including those of Glasstone (1947), Krebs and Kornberg (1957), Lehninger (1965), and Klotz (1967).

2. ENERGY CHANGES IN CHEMICAL REACTIONS

A. Internal Energy

The chemical fixation of nitrogen proceeds according to reaction 1.

$$N_2 + 3H_2 \rightarrow 2NH_3 \qquad (1)$$

At 25°C, in a closed container in which the partial pressure of each substance involved in reaction 1 is atmospheric, the reaction is accompanied by the production of 41.4 kJ* of heat for every mole of ammonia formed under these conditions. This heat is *lost* to the system (Eq. 1) and so is reckoned a negative asset as in Eq. 2

$$\frac{1}{2}N_2 + \frac{3}{2}H_2 \rightarrow NH_3 \qquad \Delta E_1 = -41.4 \text{ kJ} \qquad (2)$$

where ΔE_1 is the internal energy change taking place during the reaction.

This internal energy change arises from the redistribution of chemical bonds during the reaction. It can be considered as arising from the dissociation of the reactants, nitrogen and hydrogen, into atoms which recombine to form ammonia. Reaction 2 can be broken down into three parts, reactions 3 through 5.

$$N_2 \rightarrow 2N \qquad \Delta E_2 = +940.6 \text{ kJ} \qquad (3)$$

$$H_2 \rightarrow 2H \qquad \Delta E_3 = +432.2 \text{ kJ} \qquad (4)$$

$$N + 3H \rightarrow NH_3 \qquad \Delta E_4 = -1160.0 \text{ kJ} \qquad (5)$$

Addition of $\frac{1}{2}\Delta E_2 + \frac{3}{2}\Delta E_3 + \Delta E_4$, provides ΔE_1 for the synthesis of ammonia. The

* 4.18 kJ are equivalent to 1 kcal.

energy changes involved in reactions 3 and 4 are *bond dissociation* energies since only bonds between two atoms are broken during these changes; bond dissociation energies are best obtained spectroscopically.

B. Activation Energy

Despite the favorable energy change involved in the synthesis of ammonia—energy is *released* in the reaction—it is common knowledge that this reaction does not proceed at all in the laboratory at 25°C and with the constituents at atmospheric pressure. The value of the internal energy change given is obtained by extrapolation from much more extreme conditions. The Haber process for the manufacture of ammonia according to reaction 1 requires elevated temperatures and pressures together with a catalyst. Energetically, the course of the reaction can be represented by Figure 1. This indicates that a certain *activation energy*, subsequently recovered, must be supplied to the system before reaction can take place. Increasing the temperature, which increases the kinetic energy of the reacting molecules, helps to provide the activation energy, although it has a similar effect on the reverse reaction also, the tendency of ammonia to decompose. A catalyst, by combining chemically with reactants, reduces the size of the activation energy, which increases the rate of reaction, since, at a given temperature, more molecules possess the lower energy necessary for reaction. As catalysts, the enzymes of living systems are truly remarkable; they reduce the activation energies of many reactions to values that enable the reactions to take place at ordinary temperatures. The decomposition of hydrogen peroxide is a reaction requiring an activation energy of 75 kJ; at normal temperature the molecules of peroxide have energies of the order of 60 kJ. Colloidal platinum reduces the activation energy to 42 kJ—a sufficient reduction to enable the reaction to proceed rapidly; the enzyme catalase reduces the activation energy to less than 8 kJ. Microorganisms and plants carry out a process very similar to the synthesis of ammonia—biological nitrogen fixation (Bayliss, 1956)—at temperatures less than 25°C without the difficulties encountered in the Haber process.

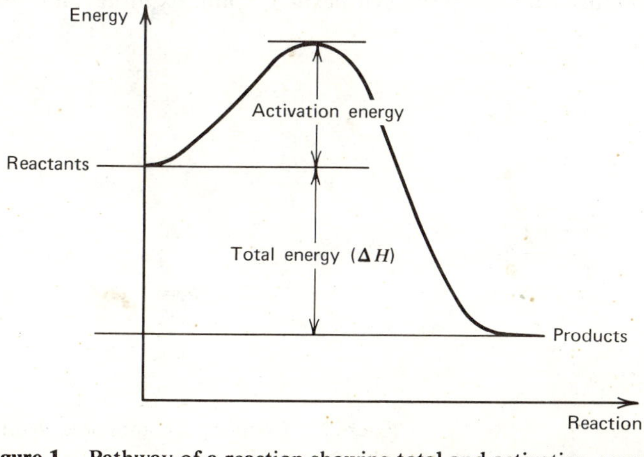

Figure 1. Pathway of a reaction showing total and activation energies.

C. Volume Changes in Chemical Reactions: Enthalpy

The synthesis of ammonia, or any other chemical reaction for that matter, is not generally carried out in an isolated container, whose volume is taken as constant, but under conditions of *constant pressure*, either of the prevailing atmosphere or externally applied. During the formation of ammonia under such conditions, a volume contraction of 22.4 liters occurs at atmospheric pressure; such a contraction *increases* the energy released by the system (a volume *increase* means work must be done by the system against the prevailing pressure). In general, at a pressure P and for a volume *change* ΔV (positive for an increase and negative for a decrease), the work done by the reaction is $P\Delta V$. At 25°C, $P\Delta V$ is -2.5 kJ for each gram-molecule of ammonia generated, so at constant pressure the energy change, now denoted by ΔH_1 is $-41.4 - 2.5$, or -43.9 kJ. This is known as the *enthalpy* change; in general, $\Delta H = \Delta E + P\Delta V$ and since most reactions are carried out under conditions of constant pressure, the enthalpy change is of more immediate usefulness than the internal energy change. Equation 2 may be rewritten as Eq. 6.

$$\frac{1}{2}N_2 + \frac{3}{2}H_2 \rightarrow NH_3 \qquad \Delta H_1 = \quad -43.9 \text{ kJ} \tag{6}$$

The negative value of ΔH_1 indicates heat evolution at constant pressure; the reaction is *exothermic.*

D. Free Energy

The enthalpy change occurring in reaction 1 is manifest as *heat* when the reaction takes place. In living systems heat production is at least partially wasteful since no physiological mechanism exists whereby heat can be converted directly back into the chemical energy needed to support life. It must be said that some heat production during biochemical reactions is necessary, first to maintain body temperatures and, second, because heat lost during a reaction cannot be recouped directly and so the heat loss confers some degree of irreversibility on the reaction. Metabolism, in general, consists of an interlocking array of chemical reactions, a few of which lead to the direct production of mechanical work (such as muscular contraction or flagellar movement) but mainly leading to the synthesis of complex molecules, large and small,

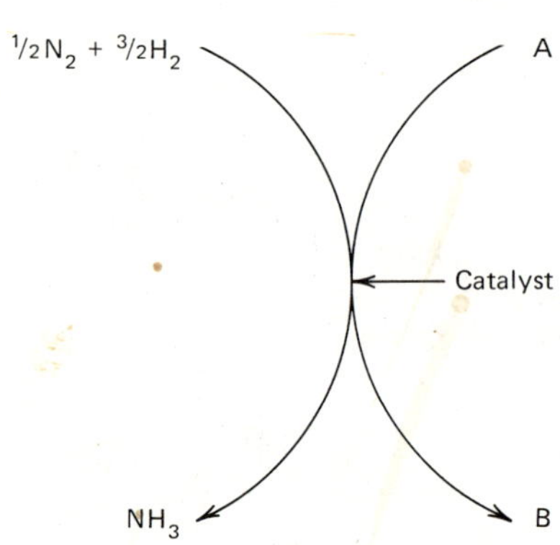

Figure 2. Coupling of ammonia synthesis to another reaction.

essential for survival and reproduction. The chemical energy available from one reaction is passed on to other reactions by coupling an energetically favorable reaction with an unfavorable one; such coupling is achieved by the enzyme catalysts of living tissues. Reaction 1 might so be coupled to a hypothetical reaction as in Figure 2. The catalyst is such that it permits only the formation of ammonia if, simultaneously, A is converted to B. Ammonia synthesis *releases* energy; how much of this energy can be passed on to the conversion of A to B? The *maximum* amount of energy that can be passed on can be precisely evaluated; it is termed the *free energy change* ΔF_1, and for the ammonia reaction is 17.7 kJ or just 39.4% of the enthalpy change ΔH_1. The remaining 60% of the energy is still lost as heat.

E. Entropy

The difference between the free energy change and the enthalpy change occurring in a chemical reaction is related to the *entropy change* ΔS_1. The precise relationship is Eq. 7

$$\Delta F_1 = \Delta H_1 - T\Delta S_1 \tag{7}$$

from which it can be seen that entropy itself is not an energy measurement; it is measured in joules per degree Kelvin, T being the temperature. The concept of entropy is troublesome in biology and is perhaps best seen in terms of molecular energy levels. In a chemical reaction not only do atoms freed from combination recombine again but, in the new molecules formed, there are differences in the rotations and vibrations of the atoms about each other, as well as their translational energies. There is a general redistribution of energy levels, which is expressed quantitatively by the entropy change. In the ammonia reaction the redistribution of energy levels leads to a *negative* entropy change; the ammonia molecule is in fact *more complex* than the two molecules from which it is derived. A *decrease* in entropy is seen as an *increase in the ordering* of the system. In this case the entropy contribution to the free energy ($T\Delta S_1$) is a measure of the *inefficiency* of utilization of the enthalpy change ΔH_1, since the $T\Delta S_1$ term represents chemical energy from one reaction which cannot be passed on to another.

Entropy changes may be positive, that is, the system is less ordered as a result of a chemical reaction. For example, the decomposition of liquid hydrogen peroxide at 25°C (Eq. 8)

$$H_2O_2 \rightarrow H_2O + \frac{1}{2}O_2 \tag{8}$$

involves an enthalpy change of −98.2 kJ but a free energy change of −119.1 kJ. If the reaction is carried out in an open vessel without coupling to another reaction, 98.2 kJ of heat are produced. If, however, the reaction is coupled to another reaction (Fig. 3), 119.1 kJ of energy are passed on to the conversion C → D. The additional 20.9 kJ is *absorbed from the surroundings* and corresponds to the $T\Delta S$ term of Eq. 7; if the coupling is carried out adiabatically (no heat supplied), the temperature must fall. An extreme situation of this kind occurs in endothermic reactions, in which the enthalpy change is actually positive but is offset by a positive entropy change, so that reaction can take place only with the absorption of heat. The denaturation of globular proteins (Tanford, 1968) brought about by heat or chemical agents is an example of

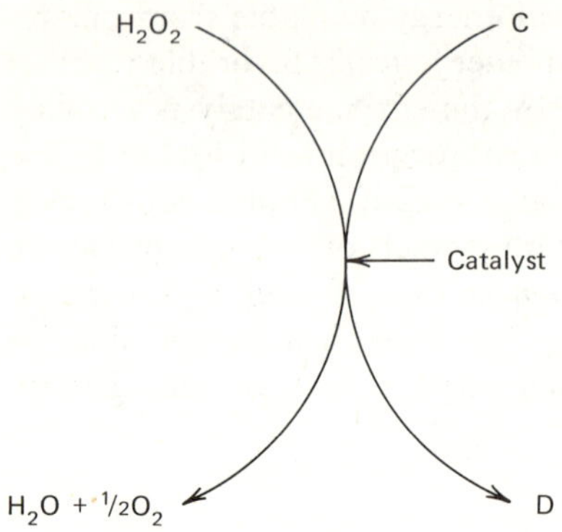

Figure 3. Coupling of hydrogen peroxide decomposition to another reaction.

such a change. The long amino acid chain of a globular protein, such as the enzyme RNase, is held in a highly specific configuration in the native molecules, which constitutes the active enzyme, as a result of a large number of weak interactions. Denaturation, represented by Eq. 9,

$$\text{RNase (native)} \leftrightarrows \text{RNase (denatured)} \tag{9}$$

involves the breakage of these weak interactions, which cumulatively contribute to a large unfavorable enthalpy change (for RNase denaturation, ΔH at 30°C is +23.8 kJ/mole). However, the resulting amino acid chain, which represents the denatured substance, lacks the highly ordered structure of the native enzyme; the entropy change [77.3 J/(°K) (mole)] is large and positive; its contribution to the free energy change (−23.2 kJ) ensures that even though the denaturation process is marginally unfavorable energetically at 30°C, a solution of RNase contains approximately 20% of the denatured form at equilibrium.

3. PRESSURES, CONCENTRATIONS, FUGACITIES, AND ACTIVITIES

A. The Effects of Pressure on Gas Reactions

The free energy change for reaction 1 is 17.7 kJ for each gram-molecule of ammonia formed. Nitrogen, hydrogen, and ammonia are all assumed to be at atmospheric pressure and at 25°C. The free energy change under these conditions is referred to as the *standard free energy change* and symbolized by ΔF_1°. Atmospheric pressure* and a temperature of 25°C represent the energy base line or *standard state* for gas reactions. Assuming that the pressures of the substances involved in reaction 1 are p_{N_3}, p_{H_2}, and p_{NH_2}, what is the value of ΔF_1? The effects of these pressures can be analyzed from Figure 4; nitrogen and hydrogen are allowed to come to atmospheric pressure before reacting, while ammonia, generated at atmospheric pressure, is allowed to come

*1.01.10^5 newtons/m^2 (S.I. units: *Nature* (1967), **216**, 1272).

to pressure p_{NH_3}. The work involved in changing the pressure of a gas by an amount P

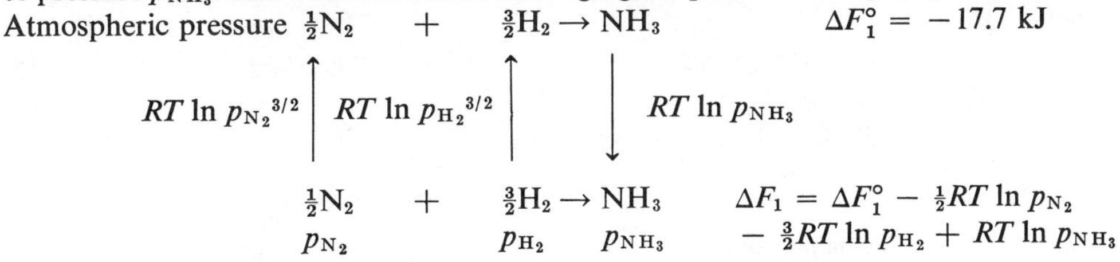

Figure 4. The effects of pressure on the ammonia reaction.

is $RT \ln P$ per mole of gas, where R is the gas contant and T the temperature; whence ΔF_1 is given by Eq. 10.

$$\Delta F_1 = \Delta F_1^\circ - \frac{1}{2}RT \ln p_{N_2} - \frac{3}{2}RT \ln p_{H_2} + RT \ln p_{NH_3} \tag{10}$$

$$= \Delta F_1^\circ - RT \ln \frac{p_{N_2}^{1/2} \times p_{H_2}^{3/2}}{p_{NH_3}} \tag{11}$$

From Eq. 11 an increase in the reactant pressures (p_{N_2}, p_{H_2}) leads to an increase in energy *available* from the reaction; a similar effect occurs on reducing the pressure of the ammonia formed.

If in Eq. 11, ΔF_1 is set at zero, no energy can be obtained from the reaction; this is the *equilibrium condition*, and the values of p_{N_2}, p_{H_2}, and p_{NH_3} correspond to those at equilibrium. The equilibrium constant K_p is given by Eq. 12,

$$K_p = \frac{p_{NH_3}}{p_{N_2}^{1/2} \times p_{H_2}^{3/2}} \tag{12}$$

so that an expression for the standard free energy change Eq. (13) relates ΔF_1° to K_p.

$$\Delta F_1^\circ = -RT \ln K_p \tag{13}$$

This relationship provides an important means for evaluation of ΔF° values if equilibrium constants can be determined.

B. Reactions in Solution

So far, consideration of a gas reaction has allowed definition of simple thermodynamic quantities. The behavior of dilute solutions bears a close resemblance to that of a gas as far as the solute is concerned. Gaseous pressure is equivalent to solute concentration in dilute solution and Eqs. 11 and 12 can be rewritten accordingly for reactions in solution, substituting concentrations for pressures. The energy base line for solutions is not, however, the same as for gases. A pressure of $1.01.10^5$ newtons/m^2 corresponds to a molar concentration of $4.4.10^{-2}$ M; *standard* free energy changes in solution are related to molar concentrations of the species involved in the reaction.* For solids such as glycogen, the saturated solution (at 25°C) defines the standard state. Thus for reaction 15, the conversion of fumaric acid and ammonia to aspartic acid catalyzed by the enzyme aspartase, the standard free energy change ΔF_3° is 38.9 kJ

* Or better, unit *activities*; see Section D.

at 25°C, all the species involved being considered to have molar concentrations during reaction 15.

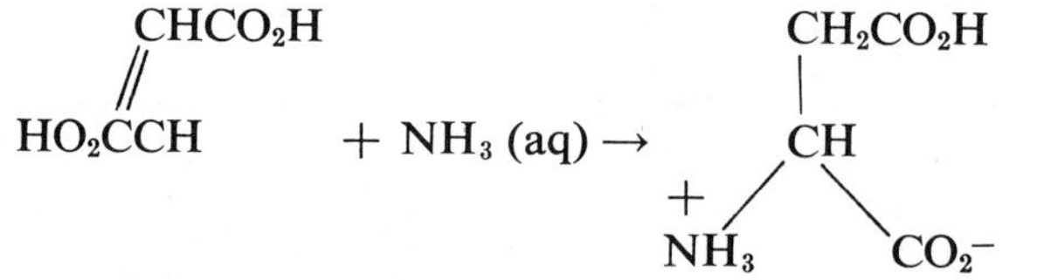

(15)

Fumarate (aq) L-Aspartic acid (aq)

At any other concentrations the free energy change ΔF_3 is given by Eq. 16.

$$\Delta F_3 = \Delta F^\circ - RT \ln (\text{fumaric acid})(\text{NH}_3) + RT \ln (\text{aspartic acid}) \qquad (16)$$

At equilibrium the free energy change ΔF_3 is zero, so that ΔF_3° is given by Eq. 17,

$$\Delta F_3 = -RT \ln \frac{(\text{aspartic acid})}{(\text{fumaric acid})(\text{NH}_3)} = -RT \ln K_c \qquad (17)$$

where K_c is the equilibrium constant in solution.

C. Free Energies of Formation and Standard States

Reaction 1 represents the formation of ammonia from its constituent elements in their standard states. The attendant standard free energy change is also known as the *standard free energy of formation of ammonia* from its constituent elements in their standard states; standard free energies of formation are designated ΔF_f°. ΔF_f° values are more useful than the ΔF° values of individual reactions, since the number of reactions available to a particular compound may be virtually limitless, but if the ΔF_f° values for all the substances involved in a reaction are known then the standard free energy change of the reaction is readily calculated. For hydrogen peroxide ΔF_f° is -117.9 kJ; for water the value is -237.0 kJ. For reaction 8, redrawn in Figure 5, ΔF° for the reaction is -119.1 kJ. Oxygen, being generated under standard conditions, does not contribute to the free energy change of the reaction.

For pure elements and compounds, the standard state is defined as pure at atmospheric pressure and 25°C. Many reactions, especially in living systems, take place

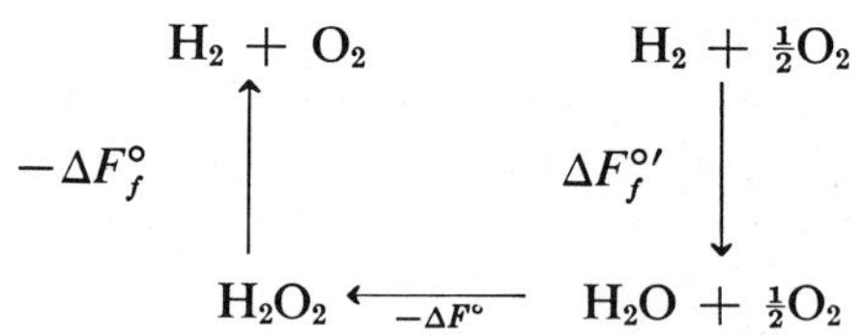

$$\Delta F_f^{\circ\prime} - \Delta F^\circ - \Delta F_f^\circ = 0$$

$$\text{that is, } \Delta F^\circ = \Delta F_f^{\circ\prime} - \Delta F_f^\circ$$

$$= -237.0 + 117.9 = -119.1 \text{ kJ}$$

Figure 5. Derivation of the standard free energy change of a reaction from the standard free energies of formation of reactant and products.

in solutions in which standard states are 1 M. The two sets of conditions are not the same. For ammonia, for example, ΔF_f° (gas) is -17.7 kJ; in aqueous solution the value is -26.6 kJ.* Additionally, the pH of the solution may affect the species involved. At low pH ammonia exists in solution as NH_4^+, for which ΔF_f° (aq) is -79.4 kJ. For reaction 15, the conversion of solid fumaric acid and gaseous ammonia to solid aspartic acid, ΔF° is -61.4 kJ. Under hypothetical conditions in which both the acids and ammonia are fully ionized, the value is -15.5 kJ.

D. Fugacities and Activities

In the discussion so far, it has been assumed that the behavior of gases and solutions is ideal, that is, that relationships such as Eqs. 11 or 16 are quantitatively obeyed no matter what the pressures of the gases or concentrations of the solutions involved. This is true only when the thermodynamically active quantity of material participating in a reaction is the same as the known concentration. It ignores the fact that gas molecules at high pressures may interact with each other; solute molecules may interact both with themselves and with the solvent. Such interactions generally reduce the *effective* pressures or concentrations. The effective pressures are known as *fugacities* and effective concentrations as *activities*. Activities in solution can be measured by studying so-called colligative phenomena—osmotic pressures, freezing point depressions, vapor pressures, and so on—which depend upon the effective concentrations. The activity a is related to the concentration c by Eq. 18,

$$a = \gamma c \tag{18}$$

where γ is termed the activity coefficient. In dilute solutions γ approaches unity; it is greater for charged than uncharged species (Tables 1 and 2). For charged species it is possible to calculate γ. In view of the sometimes large corrections necessary to measure concentrations, standard free energy changes in solution are best defined in terms of *solutions of unit activity* rather than molar concentrations.

Table 1. Activity Coefficients of Glucose Solutions[a]

Measured concentration (M)	Activity coefficient
0.55	1.02
1.10	1.11
1.76	1.16
2.48	1.30
3.03	1.36

[a] Burton and Krebs (1953).

* The difference between these two values (-8.9 kJ) is the *free energy of solution* ΔF_s. This is not the same as the *heat of solution* ΔH_s, which is relatively easily measured and in this case is -34.7 kJ. The equation $\Delta F_s = \Delta H_s - T\Delta S_s$ can be applied to solution processes; the entropy term in such cases is usually large and negative, suggesting considerable ordering is taking place, probably as a result of specific hydration effects.

Table 2. Activity Coefficients of Citrate ($C_6H_5O_7^{3-}$) Solutions[a]

Measured concentration (M)	Activity coefficient
10^{-3}	0.796
10^{-2}	0.51
10^{-1}	0.18

[a] Kielland (1937).

4. ATP

A. ATP in Metabolism

ATP occupies a key position in metabolism. Much of the chemical energy released during the breakdown of foodstuffs is conserved in this substance. Oxidation of amino acids, carbohydrates, and fats takes place with the simultaneous reduction of coenzymes, such as nicotinamide adenine dinucleotide (NAD^+) or flavin mononucleotide (FMN). During oxidative phosphorylation the reduced forms of these coenzymes are reoxidized by molecular oxygen in several stages—the electron transport chain—during some of which phosphorylation of ADP to ATP occurs (Griffiths, 1965). For NADH, the reaction can be summarized by Eq. 19,

$$NADH + H^+ + \frac{1}{2}O_2 + 3ADP + 3P_i \rightarrow NAD^+ + H_2O + 3ATP \qquad (19)$$

where P_i is inorganic phosphate. During photosynthesis ATP is generated in a rather similar way in a process known as photophosphorylation (Hill, 1965).

Once formed, ATP serves as a direct source of chemical energy in muscle contraction (Mommaerts, 1969; Young, 1969) and flagellar movement (Holwill, 1966); in moving ions such as sodium and potassium against concentration gradients to maintain the very different intracellular levels of these ions to those prevailing outside the cell (Garrahan and Glynn, 1967); and in numerous biosynthetic reactions, such as the formation of starch in plants (Nordin and Kirkwood, 1965). Alternatively, via enzymes that catalyze the exchange of the phosphate groups of ATP with other nucleoside mono- or diphosphates—those of cytosine, guanosine, or uridine—other triphosphates are formed which participate directly in protein synthesis, fatty acid biosynthesis, and in the building up of both simple and complex carbohydrates, such as glycogen or the mucopolysaccharides of connective tissue (Leloir et al., 1960).

Not surprisingly, then, the substance ATP has excited very considerable thermodynamic interest (George and Rutman, 1960). A knowledge of the energy that can be passed on via ATP to other substances and systems—the free energy—provides information on the energy requirements of the living cell and has suggested at which stages in the electron transport chain ATP synthesis can take place (Slater, 1958).

A typical involvement of ATP in metabolism is the phosphorylation of glucose, catalyzed by the enzyme hexokinase, which is the first stage of utilization of this sugar in tissues. The overall reaction (Eq. 22) can be subdivided into Eqs. 20 and 21.

$$\alpha\text{-D-Glucose} + P_i \rightarrow \alpha\text{-D-Glucose 6-phosphate} + H_2O \qquad \Delta F + ve \quad (20)$$

$$ATP + H_2O \rightarrow ADP + P_i \qquad \Delta F - ve \quad (21)$$

Overall reaction:

$$\alpha\text{-D-Glucose} + ATP \rightarrow \alpha\text{-D-Glucose 6-phosphate} + ADP \qquad \Delta F - ve \quad (22)$$

In the subdivision reaction 20 is energetically unfavorable but, coupled with the very exergonic reaction 21, leads to an overall reaction (Eq. 22) during which energy is also released. The hydrolysis of ATP can be seen as the driving reaction for the phosphorylation of glucose, which would otherwise be so energetically disfavored as not to occur at all. This is a typical example of the utilization of triphosphates in metabolism.

B. The Free Energy of ATP Hydrolysis

1. Experimental Approaches. Interest has centered on reaction 21, the hydrolysis of ATP. Early observations indicated an abnormally large *enthalpy* change during the hydrolysis of ATP and similar substances (Table 3), suggesting some special structural features or bonding in these compounds. Since the free energy change is related to the enthalpy change (see Eq. 7), high free energy changes might be expected. The term "energy-rich" was coined for these substances by Lipmann (1941), and is, with the invaluable benefit of hindsight, something of a misnomer. The high enthalpy changes recorded—and now known to be in error—persuaded some investigators that the binding of the phosphate units was somehow unstable, perhaps because of the mutual repulsions of adjacent phosphate groups or perhaps because of some kind of unfavorable folding of the phosphate side chain. In any event, observations that polyphosphate chains are highly flexible structures in solvents of high dielectric constant seems to vitiate the former idea of the inherent instability of ATP, while the three-dimensional structure of ATP (Fig. 6), obtained by x-ray crystallography (Kennard et al., 1970), dispels the latter. Moreover, the enthalpy of hydrolysis of ATP, although now thought to be a little larger than the values for other phosphates and contributing considerably to the ΔF value (Tables 4 and 5), is not thought to be decisive in explaining the high ΔF values obtained for reaction 21. Earlier estimates of ΔH had suffered from the crudeness of the calorimetric methods used at the time.

Table 3. Early Data on $-\Delta H_{obs}$ for the Hydrolysis of Organic Phosphate Esters[a]

Compound	$-\Delta H_{obs}$ (kJ)
Creatine phosphate	44.7
Phosphoenolpyruvate	35.5
ATP	50.2
Glucose 6-phosphate	5.4
Glycerol 1-phosphate	5.4

[a] George and Rutman (1960).

Table 4. Current Values of $-\Delta H_{obs}$ for the Hydrolysis of Organic Phosphate Esters[a]

Compound	$-\Delta H_{obs}$ (kJ)[b]
Creatine phosphate	20.1
ATP	19.9
Glucose 6-phosphate	5.4

[b] All values at pH 7.5. (From George and Rutman, 1960).

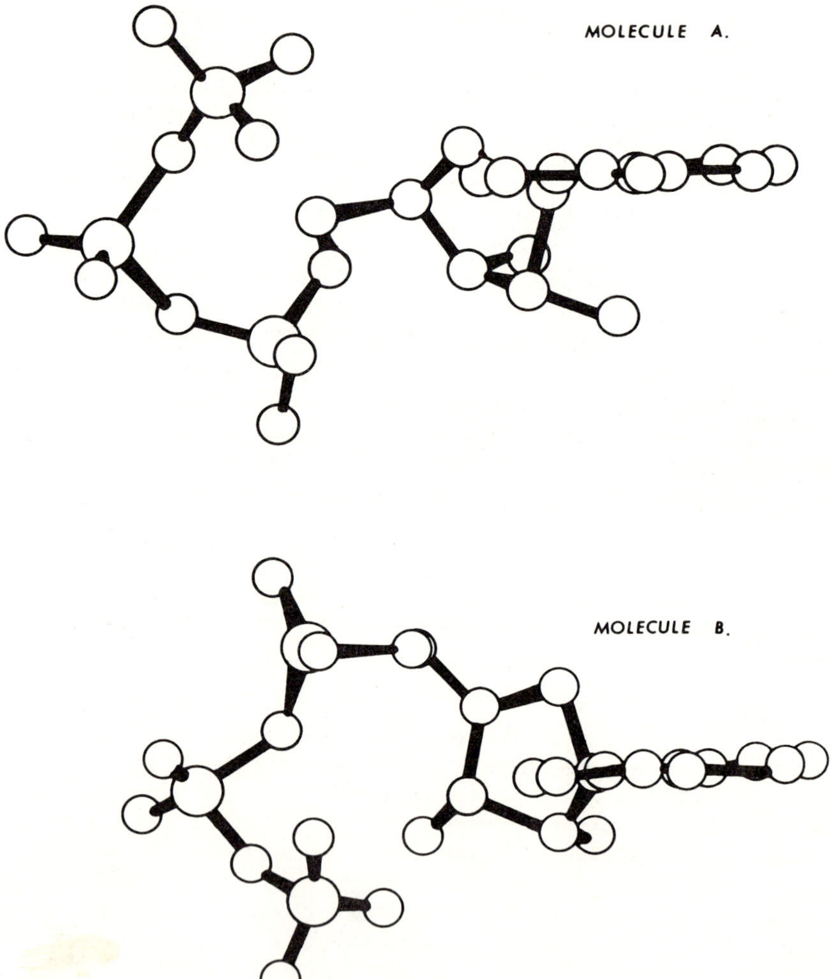

Figure 6. Molecule A and molecule B together comprise the two-molecule-unit crystal of ATP. The folded triphosphate chains are on the left; the plane of the adenine bases (right and into the page) is perpendicular to that of the five-membered ribose sugar rings. (Courtesy of Mrs. Olga Kennard, University of Cambridge; from Kennard, et al., 1971.)

Table 5. Values of $-\Delta F_{obs}$ for the Hydrolysis of Organic Phosphate Esters[a]

Compound	$-\Delta F_{obs}$ (kJ)
Creatine phosphate	44.0
Phosphoenolpyruvate	53.5
ATP (to ADP)	28.8
Glucose 6-phosphate	13.8
Glycerol 1-phosphate	9.6

[a] All values at 25°C, pH 7.0, and 25 m*M* Mg^{2+}. (From Atkinson and Morton, 1960).

The free energy of hydrolysis of ATP has been the subject of much ingenious experimentation and discussion. The basic reaction (Eq. 21) has an equilibrium constant K_{obs} given by Eq. 23.

$$K_{obs} = \frac{[ADP]\,[P_i]}{[ATP]} \tag{23}$$

This definition of K_{obs} is important in connection with its determination; H_2O is omitted from the definition. The ΔF_{obs} value obtained by substituting K_{obs} in Eq. 17 is not a *standard* free energy change, only the value prevailing under the conditions of measurement of K_{obs}.

The equilibrium suggested by Eq. 21 is not easy to examine. First, K_{obs} is large (ca. 10^6), so that at equilibrium [ATP] is very small compared to [ADP], and accurate evaluation of [ATP] is very difficult in the presence of much larger quantities of the chemically very similar ADP. Second, the equilibrium is difficult to establish; no enzyme can be expected to catalyze this process alone, since this would be a completely metabolically wasteful expenditure of ATP.

Determination of K_{obs} illustrates some of the problems of quantitative bioenergetics. One of the methods—that of Burton and Krebs (1953)—involves consideration of a sequence of reactions in which ATP is formed. The reaction sequence 24 through 31 is summarized by Eq. 32 in Table 6; glycerol is converted to pyruvate and ATP is formed. The free energy changes of reactions 24 through 31 are known; if the standard free energies of formation of glycerol and pyruvate are also available, ΔF_{obs} is accessible. The various free energies involved in this determination of ΔF_{obs} are listed in Table 5; ΔF_{obs} is -39.3 kJ. Later, Burton (1955) amended this on the basis of a better value of the free energy change of reaction 31 of -65.5 to -37.2 kJ. The value of ΔF_{obs} obtained has certain drawbacks: (1) the free energy changes of reactions 24 through 31 were obtained by different investigators sometimes under unspecified or disparate conditions, and (2) the values of ΔF_f° for glycerol and pyruvate in aqueous solution are not of the highest order of accuracy (Stull et al., 1969); small errors in these values, -488.0 and -473.7 kJ, respectively, variously estimated at ca. 0.5%, have a disproportionately large effect on ΔF_{obs}.

Table 6. Sequence of Reactions Involved in Determining ΔF_{obs} for ATP Hydrolysis[a]

Reaction		ΔF_{obs}	
Glycerol + P_i	→ DL-α-Glycerophosphate	+9.4 kJ	(24)
DL-α-Glycerophosphate	→ L-α-Glycerophosphate	+1.7 kJ	(25)
L-α-Glycerophosphate	→ Dihydroxyacetone phosphate + H_2	+42.8 kJ	(26)
Dihydroxyacetone phosphate	→ 3-Phosphoglyceric aldehyde	+7.6 kJ	(27)
3-Phosphoglyceric aldehyde + P_i	→ 1-3-Diphosphoglycerate + H_2	+24.4 kJ	(28)
1-3-Diphosphoglycerate + ADP	→ 3-Phosphoglycerate + ATP	−20.0 kJ	(29)
3-Phosphoglycerate	→ 2-Phosphopyruvate + H_2O	+1.8 kJ	(30)
2-Phosphopyruvate + ADP + H^+	→ Pyruvate + ATP	−61.9 kJ	(31)
Overall:			
Glycerol + $2P_i$ + 2ADP + H^+	→ Pyruvate + 2ATP + $2H_2O$ + $2H_2$	+5.8 kJ	(32)
Glycerol	→ Pyruvate (from ΔF_f° data)	−14.3	
whence P_i + ADP + H^+	→ ATP	+4.3	

[a] At pH 7.5, ΔF_{obs} becomes −39.3 kJ correcting for H^+. (From Burton and Krebs, 1953).

The technique of Levintow and Meister (1954) involves experimental study of the reaction catalyzed by the enzyme glutaminyl transferase from pea seedlings. In this reaction glutamate is converted to glutamine using ATP (Eq. 33).

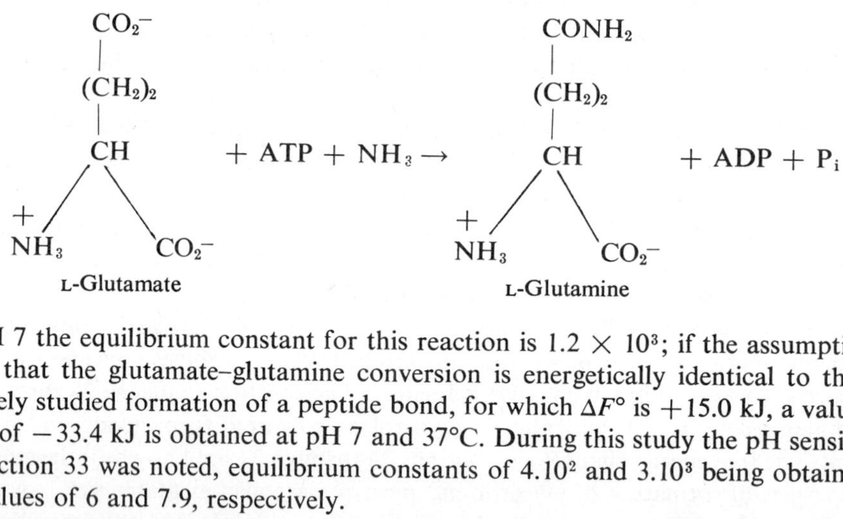

(33)

L-Glutamate L-Glutamine

At pH 7 the equilibrium constant for this reaction is 1.2×10^3; if the assumption is made that the glutamate–glutamine conversion is energetically identical to the extensively studied formation of a peptide bond, for which ΔF° is +15.0 kJ, a value for ΔF_{obs} of −33.4 kJ is obtained at pH 7 and 37°C. During this study the pH sensitivity of reaction 33 was noted, equilibrium constants of 4.10^2 and 3.10^3 being obtained at pH values of 6 and 7.9, respectively.

A more accurate reevaluation of these results was made possible by the use of the enzyme glutaminase from the bacterium *Clostridium welchii* by Benzinger et al. (1959). This enzyme catalyzes the glutamate–glutamine interconversion without the subvention of ATP. ΔF° for the glutaminase reaction was found to be 14.3 kJ and combined with the data on the glutaminyl transferase reaction has yielded the most reliable estimate of ΔF_{obs} under defined conditions (Table 7). During this study the marked effect of magnesium ions on the ATP–ADP reaction was noted; in the presence of an excess of Mg^{2+}, the value of ΔF_{obs} dropped to −29.3 kJ. Mg^{2+} forms a stronger complex with ATP than ADP, thus hindering its hydrolysis (Alberty, 1969a).

Table 7. Values of $-\Delta F_{obs}$ Obtained by Various Investigators for the Hydrolysis of ATP

Reference	pH	Temperature (°C)	$-\Delta F_{obs}$ (kJ)
Meyerhof (1944)	7.8	20	50.2
Burton and Krebs (1953)	7.5	25	39.3
Burton (1955)	7.5	25	37.2
Levintow and Meister (1954)	7.0	37	33.4[a]
Morales et al. (1955)	7.0	37	29.3
Robbins and Boyer (1957)	7.0	30	31.8[b]
Benzinger et al. (1959)	7.5	37	38.9[b]
Atkinson et al. (1959)	7.0	25	28.4[c]
Alberty (1968)	7.0	37	37.54[d]

[a] Mg^{2+} = 50 m*M*.
[b] Mg^{2+} = 0.
[c] Mg^{2+} = 25 m*M*.
[d] Mg^{2+} = 35 m*M*.

The use of radioisotopes permits the quantitative estimation of minute quantities of material. For the hexokinase reaction (Eq. 22), which has a sufficiently large equilibrium constant (ca. 10^5) to make direct chemical determination of glucose, ATP, ADP, and P_i impossible, the use of ^{14}C-labeled glucose permits its direct estimation in the equilibrium mixture. The equilibrium mixture is "frozen" by inactivating the hexokinase and the [^{14}C]-glucose isolated by removing the other constituents of the mixture by ion exchange. This ingenious method, attributable to Robbins and Boyer (1957), provided $\Delta F°$ values for the hexokinase reaction between -18.8 and -19.6 kJ, depending on the magnesium ion concentration; coupled with a $\Delta F°$ value for the phosphatase reaction (Eq. 37) of -13.0 kJ, this gave a ΔF_{obs} value of -31.8 kJ under conditions of zero Mg^{2+}, 30°C, and pH 7.

Galactokinase differs from hexokinase in causing the phosphorylation of galactose at the 1-position (Fig. 7). The free energy change for this reaction is less (7.9 kJ) than for the hexokinase reaction, enabling Atkinson et al. (1959) to study the equilibrium (Eq. 34) directly

$$\alpha\text{-D-Galactose} + \text{ATP} \rightarrow \alpha\text{-D-Galactose 1-phosphate} + \text{ADP} \quad (34)$$

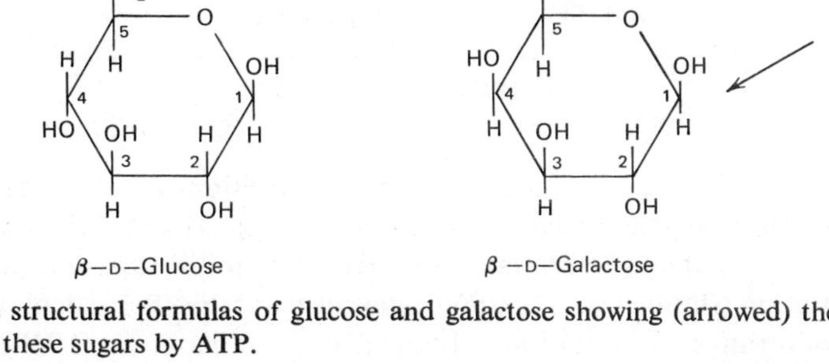

Figure 7. The structural formulas of glucose and galactose showing (arrowed) the sites of phosphorylation of these sugars by ATP.

coupled with three other reactions, for which free energy data are available, Eqs. 35 through 37,

$$\text{Glucose 1-phosphate} \longrightarrow \alpha\text{-D-Galactose 1-phosphate} \qquad \Delta F = +2.9 \text{ kJ} \quad (35)$$

$$\text{Glucose 1-phosphate} \longrightarrow \alpha\text{-D-Glucose 6-phosphate} \qquad \Delta F = -7.1 \text{ kJ} \quad (36)$$

$$\text{Glucose 6-phosphate} + H_2O \longrightarrow \alpha\text{-D-Glucose} + P_i \qquad \Delta F = -13.8 \text{ kJ} \quad (37)$$

ΔF_{obs} is found to be 28.4 kJ at pH 7, 25°C, and 25 m*M* Mg^{2+}.

The values of ΔF_{obs} by these various methods (Table 7) are in reasonable agreement, suggesting that little controversy should surround the results. Recently, efforts have been made to extrapolate values for ΔF_{obs} to a wide variety of conditions of temperature, pH, Mg^{2+}, and ionic strength (Alberty, 1968, 1969b; Phillips et al., 1969); since uncertainty still remains as to the precise values of the last three of these parameters in living tissues. In the course of these calculations, the value of ΔF_{obs} obtained by Benzinger et al. (1959) has been adjusted to take account of the effects of Mg^{2+} on the glutaminyl transferase reaction. In considering the effects of pH, the ionizations of all the species involved in the hydrolysis of ATP must be considered. At pH 6 the predominant reaction is Eq. 38, while at pH 9, the major reaction is Eq. 39.

$$HATP^{3-} + H_2O \longrightarrow HADP^{2-} + H_2PO_4^- \quad (38)$$

$$ATP^{4-} + H_2O \longrightarrow ADP^{3-} + H^+ + HPO_4^{2-} \quad (39)$$

The appearance of a hydrogen ion in Eq. 39 suggests an immediate reason for the pH dependence of K_{obs}, but the more subtle changes in ionization of ATP, ADP, and P_i clearly affect the overall equilibrium. For the two extreme reactions of Eqs. 38 and 39, Phillips et al. (1969) have calculated $\Delta F°$ (*standard* free energy changes—for Eq. 39 $[H^+] = 1\ M$) values of -41.4 and -3.3 kJ, respectively.

Table 8. Additional Reactions Involved in the Hydrolysis of ATP in the Presence of Mg^{2+} Ions[a]

$HATP^{3-}$	$\rightarrow H^+$	+	ATP^{4-}
H_2ATP^{2-}	$\rightarrow H^+$	+	$HATP^{3-}$
$MgATP^{2-}$	$\rightarrow Mg^{2+}$	+	ATP^{4-}
$MgHATP^-$	$\rightarrow Mg^{2+}$	+	$HATP^{3-}$
$HADP^{2-}$	$\rightarrow H^+$	+	ADP^{3-}
H_2ADP^-	$\rightarrow H^+$	+	$HADP^{2-}$
$MgADP^-$	$\rightarrow Mg^{2+}$	+	ADP^{3-}
$MgHADP$	$\rightarrow Mg^{2+}$	+	$HADP^{2-}$
$H_2PO_4^-$	$\rightarrow H^+$	+	HPO_4^{2-}
$MgHPO_4$	$\rightarrow Mg^{2+}$	+	HPO_4^{2-}

[a] Alberty (1968).

To provide values of ΔF_{obs} over a wide range of conditions involves consideration of several other equilibria. These have been summarized by Alberty (1968) and are listed in Table 8. Unfortunately, determinations of these equilibrium constants have not proved concordant (Phillips et al., 1966); because of this the calculated values of ΔF_{obs} for a wide range of conditions obtained by Phillips et al. (1969) and Alberty

(1968) are not quite in agreement (Table 9); it is clear, however, that discrepancies are, in general, small. Tables 10 and 11 show the effects of temperature, pH, and Mg^{2+} on ΔF_{obs} and on the enthalpy change.

Table 9. Computed Values of $-\Delta F_{obs}$ for the Hydrolysis of ATP at Various pH Values and Mg^{2+} Concentrations[a]

		Phillips et al. (1969)	Alberty (1968)
pH 7.0			
	$Mg^{2+} = 0$	42.6	39.8
	$Mg^{2+} = 10^{-3}$ *M*	37.6	36.8
	$Mg^{2+} = 10^{-2}$ *M*	36.8	35.9
pH 7.5			
	$Mg^{2+} = 0$	44.7	42.2[b]
	$Mg^{2+} = 10^{-3}$ *M*	39.3	39.3[b]
	$Mg^{2+} = 10^{-2}$ *M*	39.3	39.3[b]

[a] All values are at 25°C, ionic strength 0.2 and are given in kilojoules.
[b] Interpolated from graph.

Table 10. Computed Values of $-\Delta F_{obs}$ for the Hydrolysis of ATP at Two Temperatures[a]

pH	25°C	37° C
pH 7.0	42.6	42.2
pH 7.5	44.7	44.3

[a] All values are at $Mg^{2+} = 0$, ionic strength 0.2 and are given in kilojoules. (From Phillips et al., 1969.)

Table 11. Values of $-\Delta H_{obs}$ and $-\Delta F_{obs}$ for the Hydrolysis of ATP Showing the effects of pH and Mg^{2+}

	Mg^{2+}	pH 6	pH 7	pH 8
$-\Delta H_{obs}$	10^{-3} *M*	27.2	22.2	20.9
	0	21.7	20.1	19.6
$-\Delta F_{obs}$	10^{-3} *M*	37.6	39.7	47.7
	0	40.1	45.1	53.1

[a] ΔF_{obs} at 25°C; all values given in kilojoules. (From Phillips et al., 1969).

2. The Hydrolysis of ATP in vivo. The values of ΔF_{obs}, measured or computed, relate to the reaction carried out with a ratio of the activities of ATP (a_{ATP}) to ADP (a_{ADP}) being unity and the activity coefficient of inorganic phosphate (γ_{P_i}) also being unity. In living systems the latter condition relating to P_i is likely to be invalid. The metabolic value of ΔF_{obs}, ΔF_{met}, can be found from Eq. 40 (Burton and Krebs, 1953).

$$\Delta F_{met} = \Delta F_{obs} - RT \ln \frac{a_{ATP}}{a_{ADP}} + RT \ln a_{P_i} \tag{40}$$

Values of ATP/ADP do not seem to vary widely in several tissues studied. Values between 1 and 10 have generally been recorded (see, for example, Hess and Chance, 1961; Regen et al., 1964; Ballard, 1969; Veech et al., 1970) at levels (micromoles per gram of tissue) where a_{ATP} = [ATP] and a_{ADP} = [ADP] can be reasonably assumed. Clearly, since ATP is involved so ubiquitously in cellular metabolism, wide variations in its concentration might be considered very undesirable since very many different reactions and pathways would be affected by such variations. It has been suggested that the reaction catalyzed by adenylate kinase (Eq. 41)

$$\text{ATP} + \text{AMP} \rightleftharpoons 2\text{ADP} \tag{41}$$

may act to control ATP and ADP levels; the much smaller quantities of AMP present as a result of this equilibrium mean that relatively a small reduction in ATP produces a much larger increase in AMP; AMP is known to activate enzymes such as phosphofructokinase which induces more glucose breakdown, hence more ATP formation (Atkinson and Walton, 1967).

Inorganic phosphate levels from 10^{-2} to 10^{-4} *M* have been recorded *in vivo*, so that ΔF_{met} may vary from -50.7 to -67.7 kJ at pH 7.5 and 1 m*M* Mg^{2+}. These limits can be compared to those put forward by Burton and Krebs (1953). Ignoring the effects of Mg^{2+} and with far fewer, less accurate data, they suggested an *in vivo* value between -52.3 and -66.9 kJ at pH 7, perhaps an indication of just how far sideways our knowledge of ATP hydrolysis has progressed.

For the purpose of providing a value for ΔF_{met}, the living cell is assumed to be a homogeneous solution of the species involved. There are strong grounds for disbelieving this (Gumaa et al., 1971). Ultrastructurally, the cell cytoplasm is seen to be divided up by the endoplasmic reticulum (ER), through or around which diffusion is unlikely to be sufficiently rapid to make any assumptions of a homogeneous solution valid. Such *compartmentation* of the cell inevitably casts an air of unreality over ΔF_{met}, since local variations in a_{ATP}, a_{ADP}, and a_{P_i} may greatly alter the value of ΔF_{met} prevailing in a particular region of the cell. Some indication of the need for compartmentation within the cell is gained from studies of the breakdown of glucose, glycolysis, which yields some ATP and the reverse process, gluconeogenesis, which requires rather more. Enzymes necessary for both processes are found in the same tissue preparations, although in amounts that depend upon the nutritional or hormonal state of the animal (Weber et al., 1967). Without some separation of these two pathways within the cell, the result must be a cycle of synthesis and degradation with the continuous and wasteful hydrolysis of ATP, an energetically very favorable reaction. Conceding the compartmentation of these two pathways within the cell must limit the usefulness of attempts to provide a precise value for ΔF_{met}.

C. Why ATP?

Little has so far been suggested for the undoubtedly high value of ΔF_{obs} for ATP hydrolysis as compared to values for other phosphate esters (Table 5). In the absence of magnesium ions (Table 11), about half of ΔF_{obs} is accounted for by the enthalpy change, therefore about half by the entropy change. Some idea of the complexities involved can be gleaned from Table 11; the effects of Mg^{2+} must be in opposite directions on ΔH_{obs} and ΔS_{obs}, resulting in rather smaller values of ΔF_{obs}. The effect of hydrogen ions is also difficult to gauge; in Eq. 39 formation of H^+ at pH 7 must play an important part in ΔF_{obs} at this pH, yet in Eq. 38 in which no hydrogen ions are involved ΔF_{obs} remains high. Concluding their classic review of this subject, George and Rutman (1960) state ". . . the especially favorable free energy change, which characterizes the 'high energy' compounds can arise through the predominance of any one of three factors, ΔH, ΔS or the hydrogen ion term. Furthermore, ΔS and the hydrogen ion term are very dependent upon pH. However, the values of ΔH to which the bond energy terms contribute, bear no systematic relation to the values of ΔF, and likewise, neither ΔS nor the hydrogen ion term provide a fundamental criterion for distinguishing the 'high energy' from the 'low energy' compound." Whatever the reasons for its curious thermodynamic behavior, there is little doubt of its fundamental importance in living organisms.

5. OXIDATION–REDUCTION POTENTIALS

A. General Principles

Reactions involving ATP and similar substances belong to a class of reactions known as esterifications and hydrolyses. Energy changes in such reactions are small compared with, for example, the free energy change for the complete oxidation of glucose (Eq. 42) of 2867 kJ.

$$C_6H_{12}O_6 + 6O_2 \rightarrow 6CO_2 + 6H_2O \tag{42}$$

During glycolysis the metabolic breakdown of glucose to pyruvic acid, a minor part (ca. 7%) of this energy is converted directly to ATP via the reaction sequence 24 through 31 in Table 6. The continued oxidation of pyruvate in the tricarboxylic acid cycle (Fig. 8) leads to the formation of reduced coenzymes and guanosine triphosphate (GTP). The tricarboxylic acid cycle can be summarized by Eq. 43.

$$CH_3COCO_2^- + 3NAD^+ + FAD + GDP + P_i + 2H_2O \rightarrow 3CO_2 + 3NADH + FADH_2 + 2H^+ + GTP \tag{43}$$

During oxidative phosphorylation each gram-molecule of NADH is reoxidized with the release of 218 kJ; three molecules of ATP are formed during this process and conserve a substantial part of this energy. $FADH_2$ oxidation yields only two ATP molecules.

Although thermodynamically oxidation–reduction reactions (the two cannot be separated) can be treated exactly similarly to esterifications and hydrolyses—which are indeed intertwined during oxidative phosphorylation—yet because oxidation–reduction reactions can be made the basis of electrochemical cells, in which chemical and electrical energy are interchangeable, a rather different system of treatment has arisen.

Cytochrome b_5 is an iron-containing protein firmly bound to the ER of animal cells. It participates in a variety of oxidation–reduction reactions within the cell cytoplasm (Strittmayer, 1968); the basis of these reactions is the interconversion of

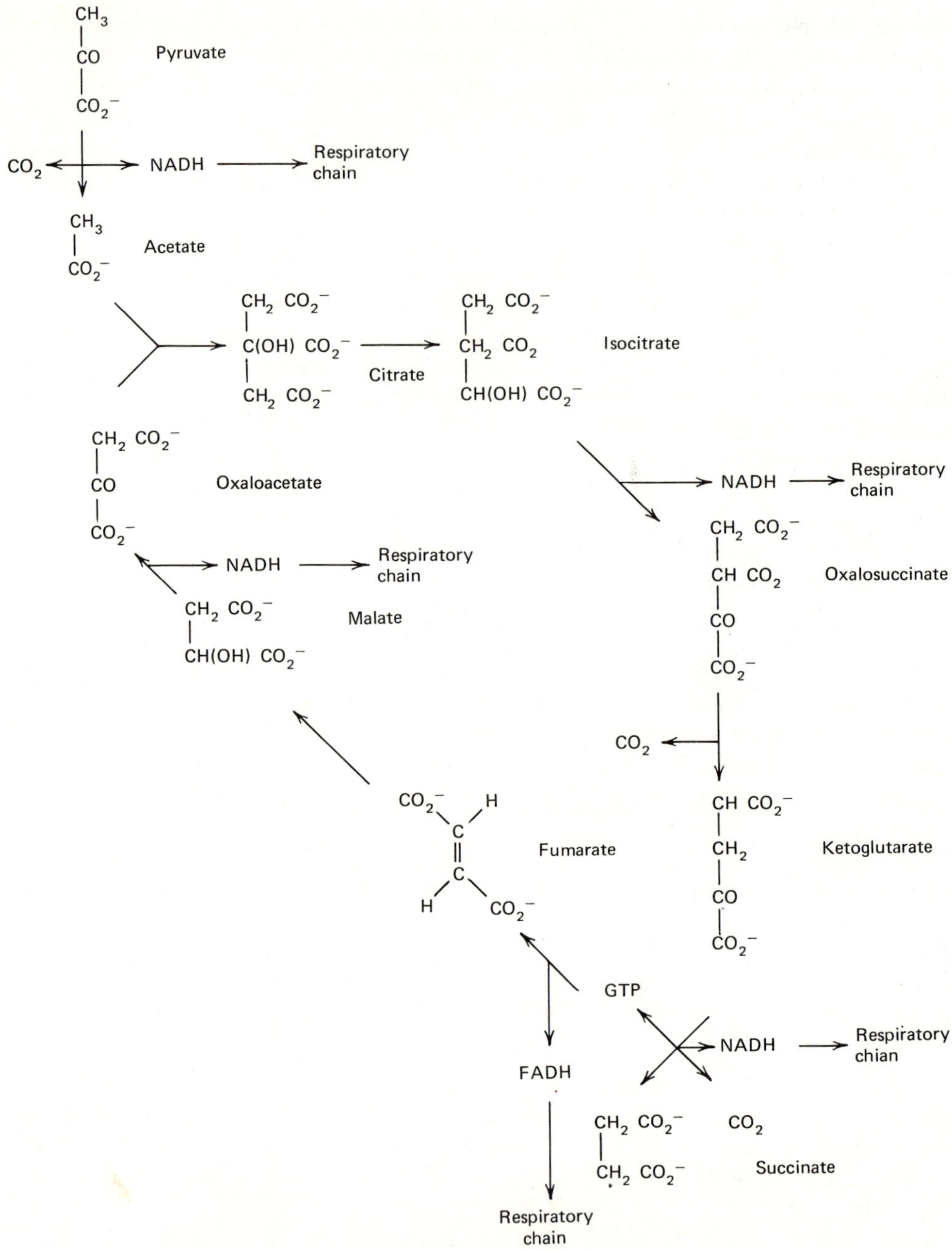

Figure 8. Simplified outline of the tricarboxylic acid cycle.

ferrous to ferric iron between, respectively, the reduced and oxidized forms (cyt b_5^{2+} and cyt b_5^{3+}). Isolated, reduced cytochrome b_5 reacts with ferricyanide ions according to Eq. 44.

$$\text{Cyt } b_5^{2+} + \underset{\text{Ferricyanide}}{Fe(CN)_6^{3-}} \rightarrow \text{Cyt } b_5^{3+} + \underset{\text{Ferrocyanide}}{Fe(CN)_6^{4-}} \tag{44}$$

The reaction can be split into two *half-reactions*, Eqs. 45 and 46.

$$\text{Cyt } b_5^{2+} \rightarrow \text{Cyt } b_5^{3+} + \epsilon \tag{45}$$

$$Fe(CN)_6^{3-} + \epsilon \rightarrow Fe(CN)_6^{4-} \tag{46}$$

Here ϵ represents an electron, and it can be seen that oxidation represents a *loss of electrons* while reduction represents a gain. If an inert electrode (usually of platinum) is placed in a solution containing both the reduced and oxidized forms of cytochrome b_5, an electrical potential develops upon it, representing the tendency of the half-reaction 45 to take place. A similar effect might be noted in a mixed solution of ferri- and ferrocyanides. The *absolute* size of such potentials—*electrode potentials* or *redox potentials*—is both valueless and impossible to measure, but if the two electrodes—A and B in Figure 9—are connected, then the measurable potential *difference* ΔE_1 represents the tendency of the overall reaction (Eq. 44) to take place. The *sign* of this potential difference—which electrode is the most positive—indicates the direction of electron flow; for the two electrodes A and B, the ferricyanide–ferrocyanide half-reaction provides the most positive potential, indicating that reduced cytochrome b_5, becoming oxidized, is providing electrons necessary to reduce ferricyanide. A potential difference remains between the two electrodes and allows electrons to flow until equilibrium is reached, at which stage no further energy can be tapped from the reaction.

The potential difference ΔE_1 is dependent upon a variety of factors; its disappearance at equilibrium suggests variation in the concentrations of reduced and oxidized forms. It is also affected by the temperature and, if gases are involved at the electrode, by

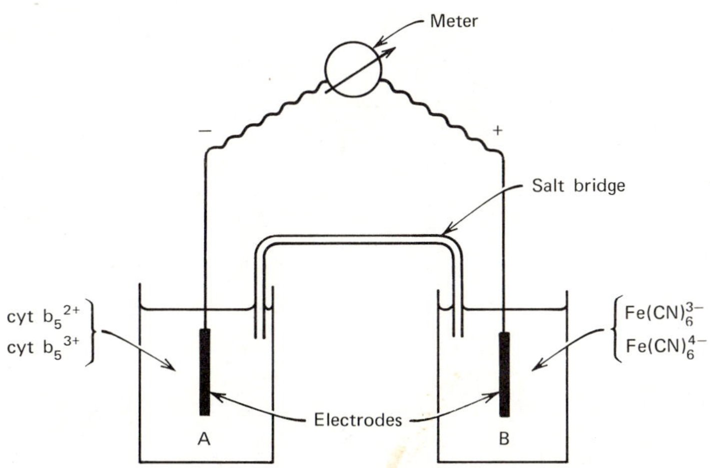

Figure 9. The principle of redox potential difference measurement.

their pressures. A *standard* potential *difference* ΔE° is defined when the activities of the substances involved in the two electrode reactions are unity in solution (for gases, atmospheric pressure) at 25°C. If, instead of connecting the two electrodes A and B to each other as in Figure 9, they had been connected to a common *standard* electrode, the standard hydrogen electrode (Eq. 47)

$$\tfrac{1}{2}H_2 \rightarrow H^+ + \epsilon \qquad (47)$$

consisting of an activated platinum electrode immersed in a solution of hydrogen ions of unity activity at 25°C and over which gaseous hydrogen at atmospheric pressure is bubbled, then the potential differences (in volts) are designated *standard electrode potentials* E_A° and E_B°; they represent the tendency of reactions 48 and 49 to occur.

$$\text{Cyt } b_5^{3+} + \tfrac{1}{2}H_2 \rightarrow \text{Cyt } b_5^{2+} + H^+ \qquad E_A^\circ = +0.02 \text{ V} \qquad (48)$$

$$Fe(CN)_6^{3-} + \tfrac{1}{2}H_2 \rightarrow Fe(CN)_6^{4-} + H^+ \qquad E_B^\circ = +0.36 \text{ V} \qquad (49)$$

Subtraction of Eq. 48 from Eq. 49 leads to Eq. 44; ΔE_1° becomes $E_B^\circ - E_A^\circ$, or 0.34 V. The use of a common standard electrode permits immediate evaluation of the combination of any two particular half-reactions and can be compared to the use of standard free energies of formation, although the two are not otherwise similar. A list of electrode potentials of some of the intermediates in oxidative phosphorylation, simplified in Figure 10, appears in Table 12; the greater the positive potential, the greater the oxidizing tendency; molecular oxygen (reduced to water) oxidizes reduced cytochrome a; oxidized cytochrome a oxidizes reduced cytochrome c; finally, the reduced coenzyme, NADH, is reoxidized. Thus the overall reaction (Eq. 50)

$$NADH + H^+ + \tfrac{1}{2}O_2 \rightarrow NAD^+ + H_2O \qquad (50)$$

takes place in several stages, during some of which (indicated in Fig. 10) energy con-

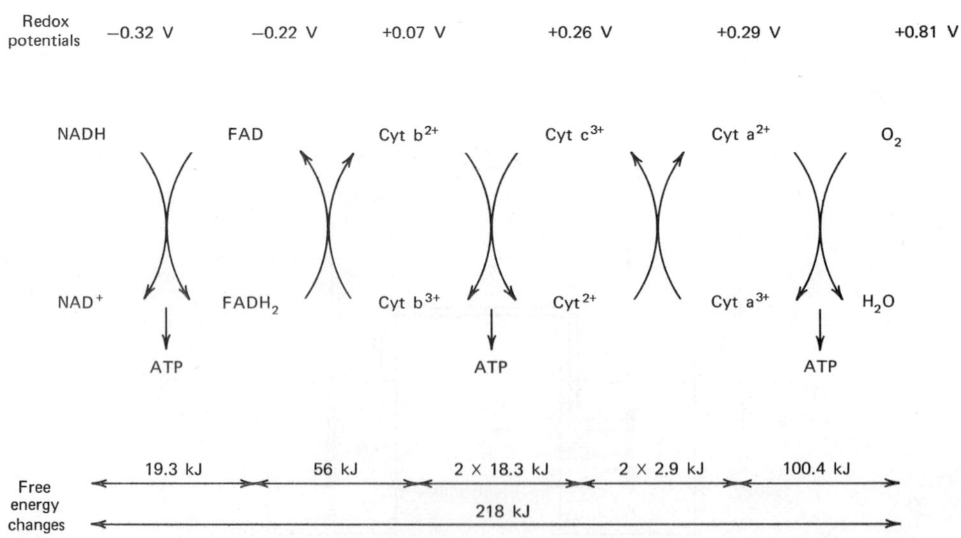

Figure 10. Outline of the respiratory chain showing probable sites of oxidative phosphorylation.

servation in the form of ATP synthesis may take place. The electrode potentials listed in Table 12 are not strictly standard potentials, since they are measured under nonstandard conditions, namely, at pH 7. Clearly, such potentials are used more in biological work; they are sometimes designated $E^{\circ\prime}$ values.

Table 12. Oxidation–Reduction Potentials of Various Half-Reactions of the Respiratory Chain[a]

Half-reaction		$E^{\circ\prime}$ at pH 7 (V)
$NAD^+ + H^+ + 2\epsilon$	$\rightarrow$ NADH	−0.32
$FAD + 2H^+ + 2\epsilon$	$\rightarrow FADH_2$	−0.22
Cyt $b^{3+} + \epsilon$	$\rightarrow$ cyt b^{2+}	+0.07
Cyt $c_1{}^{3+} + \epsilon$	$\rightarrow$ cyt $c_1{}^{2+}$	+0.22
Cyt $c^{3+} + \epsilon$	$\rightarrow$ cyt c^{2+}	+0.26
Cyt $a^{3+} + \epsilon$	$\rightarrow$ cyt a^{2+}	+0.29
$\frac{1}{2}O_2 + 2H^+ + 2\epsilon$	$\rightarrow H_2O$	+1.23

[a] Griffiths (1965).

If in Figure 9, 96,500 coulombs (the Faraday constant, F) of electricity are permitted to flow between the electrodes A and B, 1 mole of cytochrome b^5 will become oxidized as 1 mole of ferricyanide becomes reduced. The electrical energy *released* in this process, $\Delta E_1 F$ J, is the free energy change ΔF taking place in reaction 44. The *standard* conditions for electrode potential measurement are the same as those defining the standard free energy change of a reaction; for an oxidation–reduction reaction in which n electrons are involved per gram-molecular conversion, the standard free energy change is related to the standard electrode potential difference by the general equation 51,

$$\Delta F^\circ = -nF\Delta E^\circ \tag{51}$$

the negative sign indicating that energy released during reaction is conventionally a negative asset.

Combination of Eq. 41 with a relationship between ΔF° and ΔF, the free energy change under nonstandard conditions, exemplified by Eq. 11 provides values of ΔE_1 at any particular activities of the reduced and oxidized forms participating in reaction 44; ΔE_1 is given by Eq. 52.

$$\Delta E_1 = \Delta E_1^\circ + \frac{RT}{nF} \ln \frac{{}^{A}\mathrm{Fe(CN)}_6^{4-} \times {}^{A}\mathrm{cyt\ b}_5^{3+}}{{}^{A}\mathrm{Fe(CN)}_6^{3-} \times {}^{A}\mathrm{cyt\ b}_5^{2+}} \tag{52}$$

For the general half-reaction given by Eq. 53

$$\mathrm{ox} + n\epsilon \rightleftharpoons \mathrm{red} \tag{53}$$

in which ox and red represent, respectively, oxidized and reduced species, the electrode potential E at any value of a_{ox} and a_{red} is related to the standard electrode potential E° by Eq. 54 at 25°C.

$$E = E^\circ + \frac{RT}{nF} \ln \frac{a_{ox}}{a_{red}} \doteqdot E^\circ + \frac{0.059}{n} \log \frac{\mathrm{ox}}{\mathrm{red}} \tag{54}$$

For the hydrogen electrode at pH 7 the electrode potential becomes −0.42 V at 25°C (Eq. 55).

$$E = 0.059 \log 10^{-7} = -0.42 \text{ V} \tag{55}$$

since $E°$ for the standard hydrogen electrode, the basis for comparison of oxidation–reduction potentials, is conventionally zero.

B. Experimental Approaches

The graphic representation of E as a function of ox/red for cytochrome b_5 is shown in Figure 11. Such a curve is entirely hypothetical since oxidation of one substance cannot occur independently of the reduction of another. However, if the addition of ferricyanide to reduced cytochrome b_5 is taken as an example (Fig. 12), it is clear that the E values obtained (with respect to the standard hydrogen electrode) do not depart significantly from those in Figure 11, the ideal curve. This is known as a *redox titration* curve and can be used to determine E_A°. At the equivalence point, where the molar quantity of added ferricyanide equals that of the reduced cytochrome b_5 originally

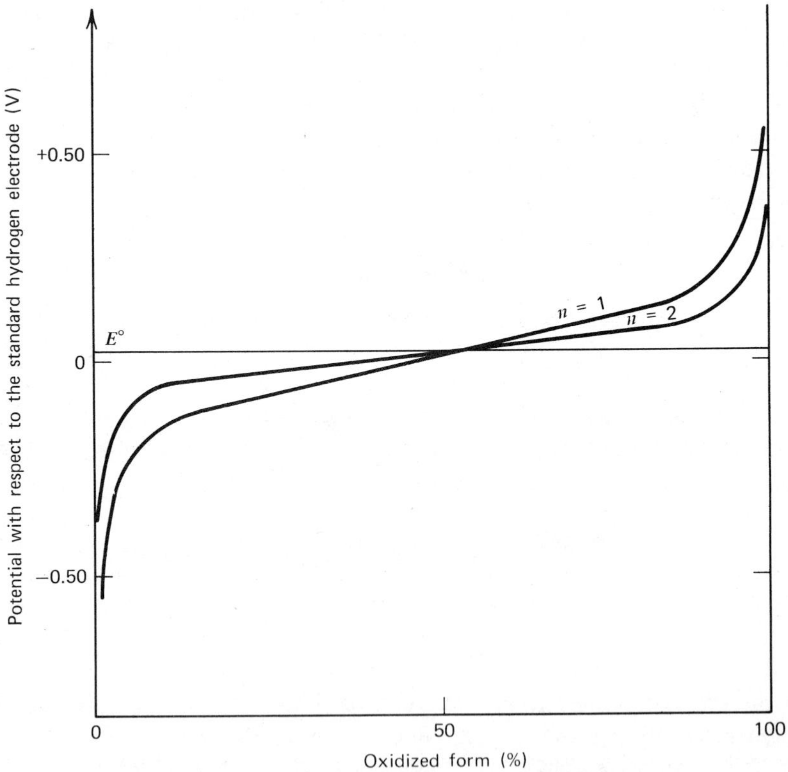

Figure 11. Ideal electrode potential curve ($n = 1$) for cytochrome b_5 according to Eq. 54 at 25°C. The effect of increasing n is also shown.

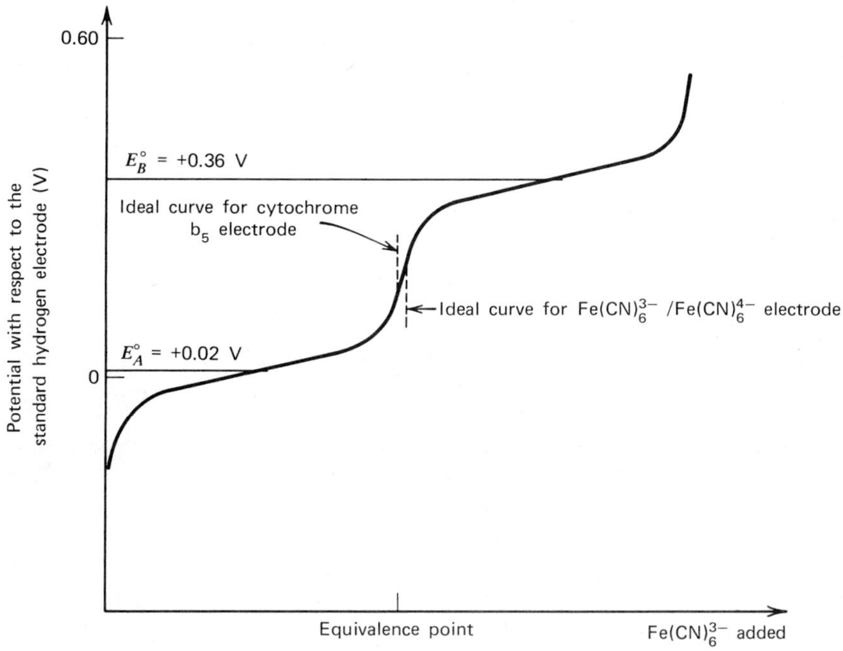

Figure 12. Redox titration of cyt b_5^{2+} with ferricyanide at 25°C. Deviations from ideal curves, represented by Eq. 54, are shown by dotted lines.

present, the concentrations (activities) of all the various species involved in Eq. 44 are given by Eqs. 56 and 57

$$[Fe(CN)_6^{4-}] = [\text{cyt } b_5^{3+}] \tag{56}$$

$$[Fe(CN)_6^{3-}] + [Fe(CN)_6^{4-}] = [\text{cyt } b_5^{2+}] + [\text{cyt } b_5^{3+}] \tag{57}$$

and the ratios of the concentrations of reduced and oxidized forms by Eq. 58.

$$\frac{[Fe(CN)_6^{3-}]}{[Fe(CN)_6^{4-}]} = \frac{[\text{cyt } b_5^{2+}]}{[\text{cyt } b_5^{3+}]} \tag{58}$$

Remembering that the system is at equilibrium, we can also add Eq. 59 from Eqs. 52 and 58,

$$-\Delta E_1^\circ = 0.059 \log \frac{[Fe(CN)_6^{4-}]^2}{[Fe(CN)_6^{3-}]^2} = -0.34 \tag{59}$$

whence the ratio of ferri- to ferrocyanide is 10^{-3}; in other words, at the equivalence point 99.9% of the ferricyanide has been reduced and 99.9% of the cytochrome b_5 has been oxidized. Reaction is virtually complete.

The completeness of reaction at the equivalence point is dependent upon ΔE°, the difference between the standard electrode potentials of the two half-reactions involved. A difference of approximately 0.1 V means that at the equivalence point the reaction is less than 90% complete. The form of a redox titration curve in such a case is exemplified by that for cytochrome c_1, for which E_c° is 0.22 V, with ferricyanide

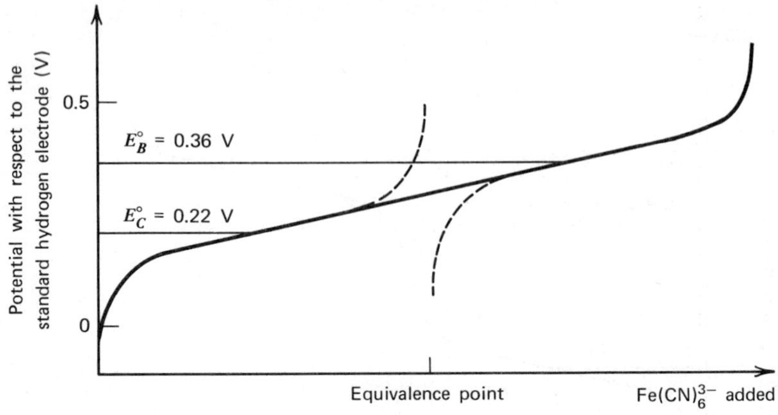

Figure 13. Redox titration of cyt c_1^{2+} with ferricyanide at 25°C showing the much greater departure from the ideal curve (dotted line) when the standard electrode potentials ($E_B°$ and $E_C°$) are close together.

(Fig. 13; Green et al. 1959). In order to obtain E_c° from such a curve, mathematical rectification is necessary (Clark, 1960).

If, instead of finding the $E°$ value by redox titration, some estimate is required of the amount of a particular species, for example, reduced cytochrome b_5, it becomes necessary to find the equivalence point. For biological systems this does not in general present too much difficulty since the reduced and oxidized forms of substances such as cytochromes show marked spectral differences; cytochrome b_5 , for example, has absorption maxima at 423, 526, and 556 nm in the reduced state, and the oxidized form has a single maximum at 413 nm. Alternatively, because a rapid increase in electrode potential takes place around the equivalence point, a *redox indicator* can be used. A small quantity of cresyl violet, with $E^{\circ\prime} = 0.173$ V, changes from its colorless reduced form to the intensely colored oxidized state over the end point region of the ferricyanide–cytochrome b_5 titration. A suitable apparatus for redox titrations is outlined in Figure 14; details of the precise experimental technique are given by Henderson and Rawlinson (1956) and Vogel (1961).

C. Activation of Oxidation–Reduction Reactions

That the redox potential for a particular half-reaction is different from that for another does not imply that interaction between these two half-reactions is automatic. As is the case in reactions for which a favorable free energy change is predicted, in the absence of a suitable catalyst to lower the activation energy, no chemical change need occur. During the isolation of cytochrome c_1 from beef heart mitochondria, the reduced form of this protein ($E^{\circ\prime} = 0.22$ V) persists despite the ubiquitous presence of molecular oxygen ($E^{\circ\prime} = 1.23$ V). Cytochrome c_1 cannot react directly with oxygen. Catalytic activation of electrodes used for redox measurement may also be necessary to set up a true potential reflecting the oxidation–reduction reaction; an account of the problems of this nature encountered in studying the succinate–fumarate reaction (Eq. 60) is given by Borsook and Schott (1932).

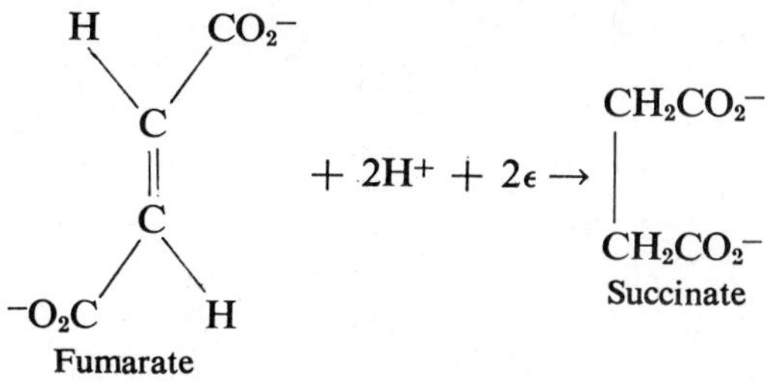

(60)

D. Applications of Oxidation–Reduction Potential Measurement

Undoubtedly, the greatest value of redox measurements lies in the extreme precision with which such electrical measurements can be made. They have provided some of the most accurate thermodynamic data available to the biochemist. For example, the free energy change in reaction 60, provided by redox potential measurements, is known with considerably greater accuracy than the free energies of formation of either fumarate or succinate (Burton cited in Krebs and Kornberg, 1957); it is unfortunate that the very precise value of $\Delta F°$ for reaction 60 cannot be used to improve the ΔF_f° values of the participants.

Redox potential measurements have been used in assessing the possible sites of ATP synthesis in the respiratory chain, from a knowledge of the $E°'$ values (Table 12) of the various participants (Slater, 1958). One drawback to these deductions is that a ΔF_{obs} value rather lower than that currently accepted and certainly much smaller than ΔF_{met} for ATP hydrolysis was used to "fit" the potential phosphorylation sites into the

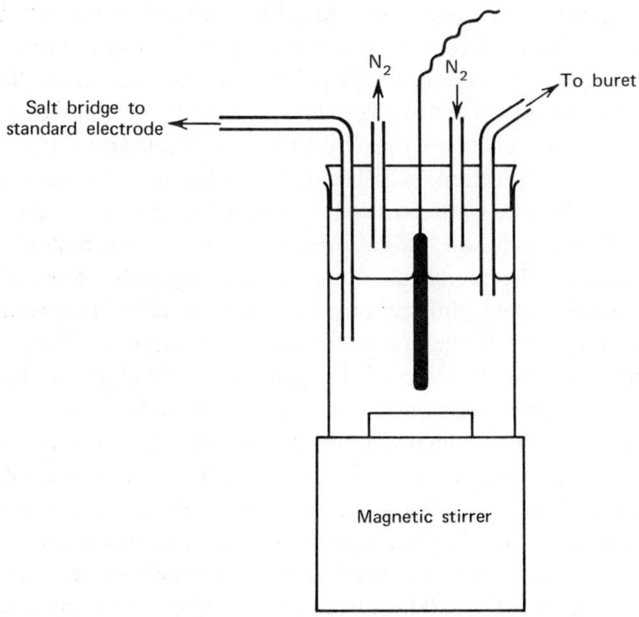

Figure 14. Apparatus suitable for redox titrations.

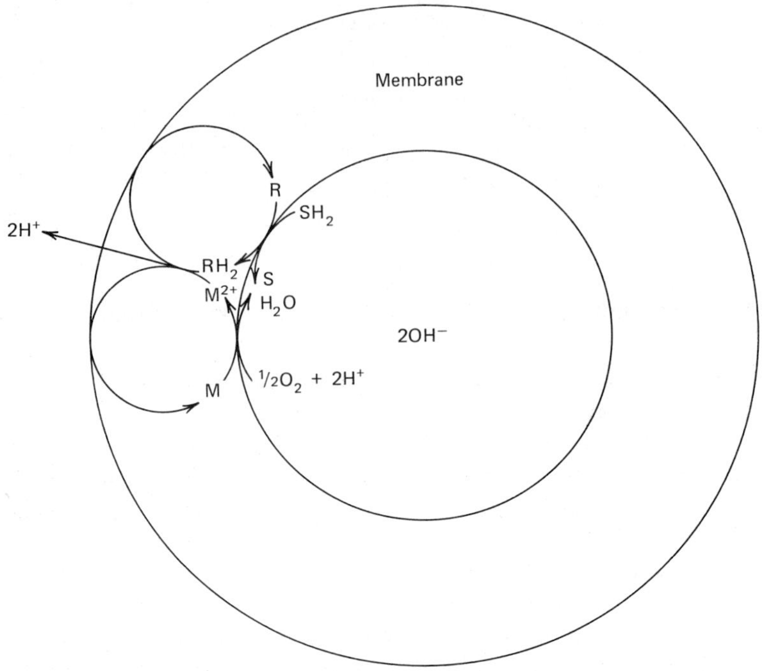

Figure 15. Schematic representation of a membrane-bounded organelle showing how two separated half-reactions generate a proton gradient from the outside to the inside of the organelle. SH_2 is a reduced substrate; RH_2 and M are the reduced parts of the two half-reactions. (After Mitchell, 1966.)

respiratory chain. However, values of the absolute concentrations of the various respiratory carriers and the ratios of reduced to oxidized forms are still the subject of dispute and may well have a considerable effect on the $E^{\circ\prime}$ values prevailing in the mitochondrion during phosphorylation. However, allowing too much leeway in the $E^{\circ\prime}$ values throws the whole object of the exercise, namely, to estimate the sites of phosphorylation from differences in the $E^{\circ\prime}$ values, into question; the reader may well prefer an alternative hypothesis which eliminates the need for such sites. The *chemiosmotic theory* of Mitchell (1966, 1968) is based on the idea that, if the two half-reactions participating in an oxidation–reduction are located at different points within, for example, the membrane of a cell or organelle, some degree of charge separation can be achieved. Subsequently, a *proton gradient* may be developed across the membrane (Fig. 15). If the enzyme for ATP synthesis is also located within the membrane, such a gradient can be used to generate ATP (Fig. 16). Such a mechanism does not require that phosphorylation occur at any specific sites, for which the experimental evidence remains inconclusive despite two decades of intensive investigation; there is, moreover, evidence of such proton movement during phosphoryaltion.

Many oxidation–reduction reactions in living cells involve the participation of metal ions, taken with the increased awareness of the importance of metal ions in biological systems this must inevitably lead to more study of the mechanisms of many metal ion-mediated reactions (Hamilton, 1969), which in many cases must involve considerations of their behavior during oxidation and reduction.

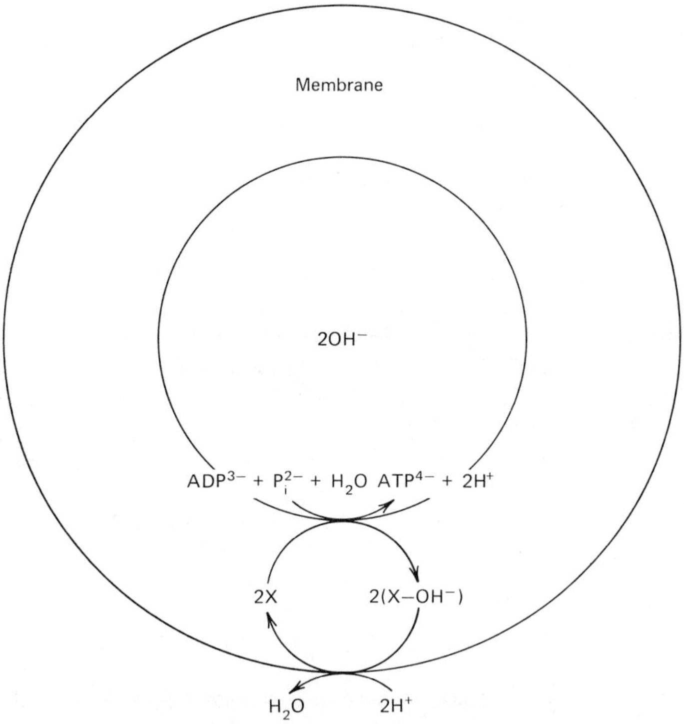

Figure 16. As Fig. 14 showing how the proton gradient generated by the two redox systems can be utilized for the generation of ATP within the membrane-bounded organelle. X represents the membrane-bound enzyme catalyzing ATP formation. (After Mitchell, 1966.)

REFERENCES

Alberty, R. A. (1968). Effect of pH and metal ion concentration on the equilibrium hydrolysis of adenosine triphosphate to adenosine diphosphate. *J. Biol. Chem.*, **243,** 1337.

Alberty, R. A. (1969a). Thermodynamics of the hydrolysis of adenosine triphosphate. *J. Chem. Educ.*, **46,** 713.

Alberty, R. A. (1969b). Standard Gibbs free energy, enthalpy and entropy changes as a function of pH and pMg for several reactions involving adenosine phosphates. *J. Biol. Chem.*, **244,** 3290.

Atkinson, D. E., and G. M. Walton (1967). Adenosine triphosphate conservation in metabolic regulation. Rat liver citrate cleavage enzyme. *J. Biol. Chem.*, **242,** 3239.

Atkinson, M. R., E. Johnson, and R. K. Morton (1959). Equilibrium constant of the galactokinase reaction and free energy of hydrolysis of adenosine triphosphate. *Nature*, **184,** 1925.

Ballard, F. J. (1970). Adenine nucleotides and the adenylate kinase equilibrium in livers of foetal and newborn rats. *Biochem. J.*, **117,** 231.

Bayliss, N. S. (1956). The thermochemistry of biological nitrogen fixation. *Australian J. Biol. Sci.*, **9,** 364.

Benzinger, T., C. Kitzinger, R. Hems, and K. Burton (1959). Free energy changes of the glutaminase reaction and the hydrolysis of the terminal pyrophosphate bond of adenosine triphosphate. *Biochem. J.*, **71,** 400.

Borsook, H., and H. F. Schott (1932). The role of the enzyme in the succinate-enzyme-fumarate equilibrium. *J. Biol. Chem.*, **92,** 535.

Burton, K. (1955). The free energy change associated with the hydrolysis of the thiol ester bond of acetyl coenzyme A. *Biochem. J.*, **59**, 44.

Burton, K., and H. A. Krebs (1953). The free energy changes associated with the individual steps of the tricarboxylic acid cycle, glycolysis and alcoholic fermentation and with the hydrolysis of the pyrophosphate groups of adenosine triphosphate. *Biochem. J.*, **54**, 94.

Clark, W. M. (1960). *Oxidation-Reduction Potentials of Organic Systems.* Williams and Wilkins, Baltimore.

Garrahan, P. J., and I. M. Glynn (1967). The stoicheiometry of the sodium pump. *J. Physiol.*, **192**, 217.

George, P., and R. J. Rutman. The "high energy phosphate bond" concept. *Prog. Biophys. Biophys. Chem.*, **10**, 1.

Glasstone, S. (1947). *Thermodynamics for Chemists.* Van Nostrand Reinhold, New York.

Green, D. E., J. Järnefelt, and H. D. Tisdale (1959). Studies on the electron transport system. 14. The isolation and properties of soluble cytochrome c_1. *Biochim. Biophys. Acta*, **31**, 34.

Griffiths, D. E. (1965). Oxidative phosphorylation. *Essays Biochem.*, **1**, 91

Gumaa, K. A., P. McLean, and A. L. Greenbaum (1971). Compartmentation in relation to metabolic control in liver. *Essays Biochem.*, **7**, 39.

Hamilton, G. H. (1969). Mechanisms of two- and four-electron oxidations catalyzed by some metallo-enzymes. *Advan. Enzymol.*, **32**, 55.

Henderson, R. W., and W. A. Rawlinson (1956). Potentiometric and other studies on preparations of cytochrome c from ox- and horse-heart preparations. *Biochem. J.*, **62**, 21.

Hess, B., and B. Chance (1961). Metabolic control mechanisms. VI. Chemical events after glucose addition to ascites tumor cells. *J. Biol. Chem.*, **236**, 239.

Hill, R. (1965). The biochemists' green mansions: the photosynthetic electron transport chain in plants. *Essays Biochem.*, **1**, 121.

Holwill, M. E. J. (1966). Physical aspects of flagellar movement. *Physiol. Rev.*, **46**, 696.

Kennard, O., et al. (1970). Three dimensional structure of adenosine triphosphate. *Nature*, **225**, 333.

Kennard, O. et al. (1971). The crystal and molecular structure of adenosine triphosphate. *Proc. R. Soc. Lond. A.*, **325**, 401.

Kielland, J. (1937). Individual activity coefficients in aqueous solution. *J. Am. Chem. Soc.*, **59**, 1675.

Klotz, I. M. (1967). *Energy Changes in Biochemical Reactions.* Academic, New York.

Krebs, H. A., and H. L. Kornberg (1957). *Energy transformations in living matter.* Springer, Berlin.

Lehninger, A. L. (1965). *Bioenergetics.* Benjamin, New York.

Levintow, L., and A. Meister (1954). Reversibility of the enzymatic synthesis of glutamine. *J. Biol. Chem.*, **209**, 265.

Leloir, L. F., C. E. Cardini, and E. Cabib. (1960). In *Comparative Biochemistry* (M. Florkin and H. Mason, eds.), Vol. II, pp. 97–138. Academic, New York.

Lipman, F. (1941). Metabolic generation and utilization of phosphate bond energy. *Advan. Enzymol.*, **1**, 99.

Mitchell, P. (1966). *Chemiosmotic Coupling in Oxidative and Photosynthetic Phosphorylation.* Glynn Research, Bodmin, England.

Mommaerts, W. F. H. M. (1969). Energetics of muscle contraction. *Physiol. Rev.*, **49**, 427.

Nordin, J. H., and S. Kirkwood (1965). Biochemical aspects of plant polysaccharides. *Ann. Rev. Plant Physiol.*, **16**, 393.

Phillips, R. C., P. George, and R. J. Rutman (1966). Thermodynamic studies of the formation and ionization of the magnesium (II) complexes of ADP and ATP over the pH range 5 to 9. *J. Am. Chem. Soc.*, **88**, 2631.

Phillips, R. C., P. George, and R. J. Rutman (1969). Thermodynamic data for the hydrolysis of adenosine triphosphate as a function of pH, Mg^{2+} ion concentration and ionic strength. *J. Biol. Chem.*, **244**, 3330.

Regen, D. M., W. W. Davis, H. E. Morgan, and C. R. Park (1964). The regulation of hexokinase and phosphofructokinase activity in heart muscle. *J. Biol. Chem.*, **239**, 43.

Robbins, E. A., and P. D. Boyer (1959). Determination of the equilibrium constant of the hexokinase reaction and the free energy of hydrolysis of adenosine triphosphate. *J. Biol. Chem.*, **224,** 121.

Shakespeare, W. (1623). *Mr. William Shakespeares Comedies, Histories and Tragedies*, p. 264. Jaggard and Blount, London.

Slater, E. C. (1958). Mechanism of oxidative phosphorylation. *Ann. Rev. Pure Appl. Chem.*, **8,** 221.

Strittmayer, P. (1968). In *Biological Oxidations* (T. P. Singer, ed.), pp. 171–191. Wiley, New York.

Stull, D. R., E. F. Westman, and G. C. Sinke (1969). *The Chemical Thermodynamics of Organic Compounds.* Wiley, New York.

Tanford, C. (1968). Protein denaturation. *Advan. Protein Chem.*, **23,** 122.

Veech, R. L., L. Raijman, and H. A. Krebs (1970). Equilibrium relations between the cytoplasmic adenine nucleotide system and nicotinamide-adenine nucleotide system in rat liver. *Biochem. J.*, **117,** 499.

Vogel, A. I. (1961). *A Textbook of Quantitative Inorganic Analysis.* Longmans, London.

Weber, G., M. A. Lea, H. J. H. Convery, and N. B. Staith (1969). *Advan. Enzyme Regulation*, **4,** 257.

Young, M. (1969). The molecular basis of muscle contraction. *Ann. Rev. Biochem.*, **38,** 913.

3

MEMBRANE TRANSPORT

P. F. Baker

1. INTRODUCTION: CRITERIA FOR DISTINGUISHING BETWEEN ACTIVE AND PASSIVE TRANSPORT

Membranes present a barrier which must be crossed by any substance entering or leaving cells and intracellular organelles. In consequence, the composition of the intracellular environment is very largely determined by the properties of the membranes that surround it. Substances are transported across membranes either in bulk or as individual molecules. Bulk transfer normally involves some form of membrane fusion and includes such apparently diverse phenomena as pinocytosis, phagocytosis, fertilization, and many forms of secretion. Although a detailed discussion of membrane fusion is beyond the scope of this chapter, some of the mechanisms involved in secretion are discussed in Section 4.

Transport of individual molecules and ions can be divided into two types: passive and active. For the transport of a substance to be described as passive, net movement must always be *down* the chemical gradient or, in the case of charged molecules, the electrochemical gradient. For transport to be described as active, it must be capable of effecting net transfer against an electrochemical gradient. Active transport need not always take place against an electrochemical gradient, although it must be capable of doing so, whereas passive transport can never effect net transfer against an electrochemical gradient.

Passive transport is primarily a diffusion phenomenon deriving its energy from the kinetic energy of the solute particles. For an uncharged particle, Fick's law of diffusion states that the net flux per unit area is

$$\text{flux}_{1-2}/\text{unit area} = \frac{-D(C_1 - C_2)}{s} \tag{1}$$

where D is the diffusion coefficient, s the thickness of the membrane, and $C_1 - C_2$ the difference in concentration or strictly activity on the two sides (1 and 2) of the membrane. The negative sign appears because the flux is down the concentration gradient ($C_1 > C_2$). The diffusion coefficient of a substance in the membrane depends not only on its mobility in aqueous solution but also on the ease with which it can penetrate the membrane. Thus D contains a partition coefficient term. This term is determined by the structure and composition of the membrane. For most cell membranes uncharged lipophilic materials penetrate more readily than charged or highly hydrated ones. As s is unknown, but is assumed to be constant, it is convenient to include it with D, and the ratio D/s is known as the permeability constant which is expressed in cm sec^{-1}.

This treatment of passive transport is strictly applicable only when the rate of transport is proportional to the concentration gradient. In several cases passive transport seems to involve interaction with some specific component of the membrane, and under these conditions the rate of transport frequently shows saturation kinetics. Passive transport of this type is called facilitated diffusion, and an example is the uptake of glucose by red blood cells and muscle.

For charged particles it is also necessary to take into consideration the membrane potential. For instance, if the interior of the cell is negative with respect to the outside, positively charged ions will tend to accumulate in the cell and negatively charged ions will be expelled. This occurs because the inward diffusion of positively charged ions is assisted by the electric field and the outward diffusion is hindered. Equilibrium is achieved when the tendency for the ion to move down its chemical gradient is just balanced by the electric field. In quantitative terms, if we neglect frictional forces in the membrane, equilibrium is achieved when the electrical work required to move a charged particle against the electric field just equals the osmotic work available from the movement of the particle down the chemical gradient. This can be described as:

$$\text{osmotic work} = RT \ln C_2/C_1$$

and

$$\text{electrical work} = zF(E_2 - E_1)$$

where $E_2 - E_1$ is the potential difference across the membrane. At equilibrium,

$$RT \ln \frac{C_2}{C_1} + zF(E_2 - E_1) = 0$$

and

$$(E_1 - E_2) = \frac{RT}{zF} \ln \frac{C_2}{C_1} \tag{2}$$

which is known as the Nernst equation.

If a substance is distributed passively across a membrane, its activity on the two sides of the membrane should approach that predicted by the Nernst equation. Unfortunately, distribution in accordance with the Nernst equation is not of itself adequate evidence for purely passive transport, although failure to distribute in accordance with the Nernst equation strongly suggests that some form of active transport must be operating.

A very useful criterion in deciding whether transport of a solute is active or passive was put forward by Ussing. Making two assumptions: (1) that no energy is fed into the system, that is, the energy is derived entirely from the kinetic energy of the particles themselves, and (2) that the chance of a solute particle passing through the membrane is unaffected by the presence of other particles of the same species in the membrane, that is, there is no interaction in the membrane, Ussing showed that the unidirectional fluxes $M_{1\to2}$ and $M_{2\to1}$ are related in a very simple way

$$\frac{M_{1\to2}}{M_{2\to1}} = \frac{C_1}{C_2} \exp^{zEF/RT} \tag{3}$$

where C_1 and C_2 are the activities of solute on sides 1 and 2, respectively.

If the Ussing criterion does not hold, either the movement of each particle is not unaffected by the presence of others, that is, movement is not independent, or the movement is not passive. The nature of the deviation of the observed behavior from that predicted by the Ussing equation may give some clue as to the mode of permeation. Two examples are:

1. In some instances the flux ratio $M_{1\to2}/M_{2\to1}$ is always unity. This suggests that some form of one-for-one exchange is occurring across the membrane, and Ussing has termed this exchange diffusion.

2. Interaction between the fluxes of K^+ ions in fully poisoned nerves is suggested by the following data of Hodgkin and Keynes. When the membrane potential was held at 30 mV (inside negative) and external potassium was raised by a factor of 10, the efflux was diminished by 2.6 times and the influx was increased 30 times. As the membrane potential was kept constant, the influx should have been proportional to the external concentration and the efflux should have remained constant. An explanation advanced by Hodgkin and Keynes is that the movement of K^+ ions across the membrane takes place in single file through narrow pores which contain more than one K^+ ion. If, for example, three ions occupy the channel, for a radioactive ion added at one end of the channel to emerge at the other end it must receive three successive collisions from the same side from which it was added without a single collision occurring at the other end. It follows that raising the potassium concentration on one side of the membrane markedly increases the flux measured from that side and decreases that measured from the other side.

2. ACTIVE TRANSPORT AND THE MAINTENANCE OF THE STEADY STATE

A. Measurement of Ion Activities inside Cells

Comparison of the ionic composition of most animal cells with that of the solution bathing them reveals that the interior of the cell has an excess of K^+ ions and a deficit of Na^+, Ca^{2+}, Mg^{2+}, and Cl^- ions. Two additional pieces of evidence are required to determine whether this distribution is likely to involve an expenditure of energy or is a purely passive phenomenon. Measurements must be made of the transmembrane potential and also of the ion activities on the two sides of the membrane. The internal potential can be measured by inserting a microelectrode into the cell. It is always found to be negative with respect to the outside. Activity coefficients for ions in the salt solution bathing the nerve can be measured by conventional methods, but the measurement of activity inside cells poses a difficult problem.

In the squid axon, which is up to 1 mm in diameter and many centimeters in length, it is possible to inject into the axoplasm a small radioactive sample of the ion under investigation and examine its mobility along the long axis of the fiber under the influence of an applied electric field. Using this technique, Hodgkin and Keynes showed that the mobilities of Na^+ and K^+ ions were only slightly less (one-half to two-thirds) than their values in free solution, whereas the mobility of Ca^{2+} was very considerably reduced and a patch of Ca^{2+} ions showed virtually no movement in an electric field.

These results have been confirmed and extended by other techniques including the use of both sodium- and potassium-selective glass electrodes and chloride electrodes. The ionized calcium inside large nerve and muscle cells has been measured using the calcium-sensitive protein aequorin. Aequorin, which is obtained from jellyfish of the genus *Aequorea*, reacts with Ca^{2+} ions to produce light. It can be concentrated and injected into a squid axon in which the rate of light emission gives a measure of the intracellular ionized calcium concentration. Injection of Ca–EGTA buffers can either enhance or depress the light emission. The buffer, which caused no change in light output, was assumed to have an ionized calcium concentration identical to that in axoplasm. The value obtained was 0.3 μM. Measurements of the rate at which calcium can be dialysed from axoplasm suggest that the concentration of diffusible calcium is approximately 10 μM, so it seems that of a total intracellular calcium concentration of 400 μM only 10 μM is diffusible and of this only 0.3 μM is ionized. The bulk of the calcium seems to be sequestered in intracellular organelles, probably mitochondria. This sequestration process requires energy because metabolic inhibitors such as cyanide or dinitrophenol (DNP) result in a rise in the ionized calcium concentration in the axoplasm, apparently through the release of calcium from intracellular stores.

In conclusion, although the available methods of measuring intracellular ion activities can all be criticized on the grounds that they involve some interference with the cell, there can be little doubt that in the few cells in which measurements are possible the activities of Na^+, K^+, and Cl^- are not less than half their values in free solution, whereas the activity of Ca^{2+} is very markedly less. As the interior of the cell is negative with respect to the outside, exclusion of the cations Na^+ and Ca^{2+} from the interior of the cell must involve the expenditure of energy. The same probably applies to Mg^{2+} and H^+ ions. The position for K^+ and Cl^- is less clear. In some cells separate inwardly directed K^+ and Cl^- pumps seem to be required, whereas in others the distribution of these ions could be achieved by purely passive means.

B. The Na^+–K^+ Exchange Pump

Radioactive tracers have been used to investigate the processes responsible for maintaining the differences in Na^+ ion activity across the membrane. In a squid axon at 20°C, the influx and efflux of Na^+ are both approximately 40 pmoles cm^2 sec. For an ion moving across the membrane in a purely passive manner, provided there is no interaction between the ions, the ratio of influx to efflux can be calculated from the Ussing equation (Eq. 3). For Na^+ ions the negative internal potential and the higher activity of Na^+ outside the cell should result in a flux ratio much greater than the observed value of unity. This provides further evidence that the efflux of Na^+ cannot be purely passive and must involve some form of active transport.

The properties of this system have been investigated intensively in red blood cells and nerve, especially squid axons. The results obtained from these two dissimilar tissues are surprisingly similar and suggest that the Na^+–K^+ exchange pump is a constant feature of cell membranes. In the description that follows, the data refer to the squid axon unless otherwise stated.

1. Dependence on ATP. The Na^+ efflux, but not the Na^+ influx, is reduced by poisoning with metabolic inhibitors such as cyanide or DNP. Injection of ATP or the phosphagen arginine phosphate into these cyanide-poisoned axons restores the Na^+ efflux (but see Section 2.B.7). It now seems that the primary source of energy for the Na^+ efflux, which is commonly referred to as the Na^+ pump, is ATP, and substances such as arginine phosphate merely serve to regenerate ATP inside the axon. The best evidence for this comes from the work of Brinley and Mullins, who inserted a porous capillary along the length of a squid axon and dialyzed most of the low-molecular-weight materials from the axoplasm. In this preparation the Na^+ efflux fell to a low level, and of the various energy-rich phosphate compounds tested only ATP and to a lesser extent dATP could restore a normal Na^+ efflux. The dependence of transport on ATP can be shown in another way. Most nervous tissue is aerobic, and the consumption of ATP by the pump leads to a rise in ADP and P_i, which in turn stimulates oxidative phosphorylation, leading to an increased uptake of oxygen. In some nerves more than 50% of the resting oxygen consumption is associated with pumping.

2. Inhibition by Cardiac Glycosides. The energy-dependent extrusion of Na^+ is very specifically inhibited by external application of such cardioactive glycosides as ouabain and strophanthidin. Injection of very high concentrations of these glycosides into the cell has no effect on the Na^+ pump. This inhibitory action of externally applied glycosides is antagonized by external K^+ ions apparently because K^+ ions impede the binding of glycoside to the pumping sites. The binding of labeled glycoside indicates that there may be as many as 1000 pumping sites per square micron of axon membrane.

3. Linkage of Na^+ efflux to K^+ Influx. The efflux of Na^+ is reduced in the absence of external K^+ ions. Hyperpolarization of the membrane does not produce a comparable effect, which rules out the possibility that the apparent coupling between Na^+ and K^+ is electrical, the hyperpolarization produced in K^+-free conditions leading to a reduction in Na^+ efflux. It thus seems that some form of chemical coupling is more likely, and tracer experiments have established that Na^+ efflux and K^+ influx both require ATP and are both inhibited by ouabain. The parallel behavior of these fluxes suggests that they are linked, Na^+ moving out of the cell in exchange for K^+ moving in. High-speed measurements show that removal of external potassium causes an almost immediate reduction in Na^+ efflux, which suggests that there is a rather tight linkage between Na^+ and K^+ movements.

The dependence of Na^+ efflux on external potassium is shown in Figure 1. In this figure the ouabain-sensitive Na^+ efflux is plotted for a range of external sodium concentrations as a function of the external potassium concentration. It should be noted that in the presence of external sodium the Na^+ efflux is activated along a sigmoid curve which saturates at high potassium concentrations. Progressive replacement of external sodium by choline or dextrose (*a*), but not lithium (*b*), shifts the curve to the left without appreciably changing the maximum ouabain-sensitive Na^+ efflux. In the absence of sodium the curves are not sigmoid. It has proved difficult experimentally to determine whether this change in shape is genuine or whether K^+ ions leaking from the axon and Schwann cells into the small space immediately external to the axolemma maintain the pumping rate even in nominally K^+-free conditions above the range at which sigmoid behavior would be expected.

4. Stoichiometry of the Pump: Evidence for Electrogenic Behavior. The ouabain-sensitive Na^+ efflux is always greater than the ouabain-sensitive K^+ influx. Na^+/K^+ ratios of approximately 2:1 or 3:2 seem most probable; 3:2 is preferred because the sigmoid shape of the K^+-activation curve suggests that at least two K^+ ions must cooperate in activating the Na^+ efflux. Measurements in both crab and squid axons suggest that three Na^+ ions are extruded per energy-rich phosphate bond split. The

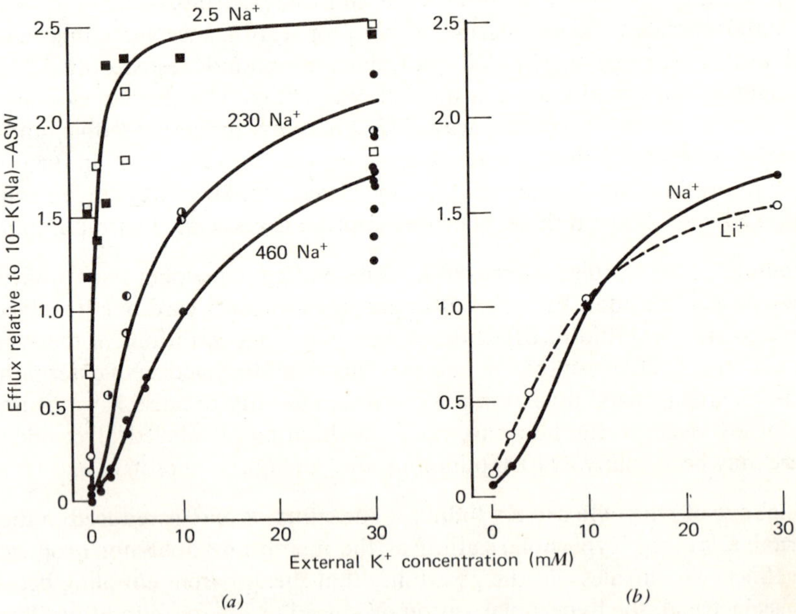

Figure 1. The potassium dependence of the ouabain-sensitive Na^+ efflux from squid axon showing the effects of replacing external sodium by choline or dextrose (*a*) or lithium (*b*). In both graphs the ordinate is the ouabain-sensitive Na^+ efflux relative to that in 10-K^+(Na^+)-ASW (artificial sea water) and the abscissa is the external potassium concentration (millimolar) ●, 460 Na-ASW; ○, 460 Li-ASW; □, choline-ASW; ■, dextrose-ASW; ◑, 230 Na^+–choline-ASW; ◐, 230 Na^+–dextrose-ASW. (From Baker et al., *J. Physiol.*, **200**, 431, 1969).

overall stoichiometry thus seems close to: three Na^+ pumped out/two K^+ pumped in/one energy-rich phosphate bond split.

If more Na^+ ions are pumped out than K^+ ions reabsorbed, unless there is some other ouabain-sensitive ion movement which has not been detected, the Na^+ pump should be electrogenic; that is, it should contribute directly to the membrane potential. The size of this contribution of course depends on the rate of pumping and on the membrane resistance; but under favorable conditions it should be detectable experimentally. The squid axon is an unfavorable preparation in this respect—it can be calculated that if one-third of the sodium emerged as a current of Na^+ ions, because of the relatively low membrane resistance in its tissue it would contribute only 1.8 mV to the resting potential. There is growing evidence for the electrogenic behavior of the Na^+ pump in nervous tissue. Perhaps the clearest demonstration is that of Thomas (1969), who worked with a snail neuron. He showed that injection of sodium (but not lithium or potassium) resulted in hyperpolarization of the cell, that is, the interior of the cell became more negative with respect to the outside. This effect required external K^+ ions and was inhibited by application of ouabain. By voltage-clamping the neuron and simultaneously monitoring the internal sodium concentration by a sodium-selective glass electrode, Thomas was able to show that the hyperpolarization that follows injection of sodium was produced by an outward current of Na^+ ions which amounted quantitatively to about one-third of the total sodium extruded from the cell. In other words, of the sodium being pumped out of the cell, one-third emerged as a stream of Na^+ ions. Coupled with the strong evidence for extrusion of three Na^+ ions per energy-rich phosphate bond split, this is consistent with an overall stoichiometry of three Na^+/two K^+/one $\sim$P.

The observation that the Na^+ pump is electrogenic has considerable physiological significance because it means that apart from the permeability mechanisms, which are discussed later in this chapter, in some cells—or parts of cells—changes in pumping rate may lead *directly* to alterations in membrane potential as well as indirectly through changes in the ion gradients.

5. Selectivity toward Ions. No ions have been found to replace completely internal sodium in activating the Na^+ pump. In most tissues lithium is a very poor substitute. At the external activation site, K^+ can be replaced by a range of other cations. Of the ions tested Tl^+, Rb^+, NH_4^+, Cs^+, and Li^+ all can replace K^+ to varying extents, and their effectiveness is in the order listed with Tl^+ slightly more effective than K^+, Rb^+ about the same as K^+ and Cs^+ much less effective. It is of interest that this sequence is the same as that for the effectiveness of different externally applied cations in depolarizing nerve. The relation between these observations is not clear; but they suggest some similarity between the mechanisms that effect K^+ uptake by the Na^+ pump and passive movement of K^+ down its electrochemical gradient.

6. Enzymic Basis of the Na^+ Pump. The biochemical study of the Na^+ pump was started in 1957 by Skou, who isolated from crab nerve an ATPase that required Na^+ and K^+ ions for maximum activity and was inhibited by ouabain. This enzyme is present in the membrane of intact crab nerve where it requires *internal* sodium and *external* potassium for activation. It has not yet proved possible to purify this enzyme free from other membrane materials, and it seems that some lipid may be required for enzyme activity. During each pumping cycle the enzyme is alternately phosphorylated and dephosphorylated. In the presence of sodium and ATP, the enzyme is phosphorylated, whereas potassium favors dephosphorylation. The phosphate seems to be attached to the membrane as an acyl phosphate.

7. Partial Reactions. The Na^+ pump does not always exchange Na^+ for K^+. Caldwell et al. first noted in 1960 that under certain conditions, for instance, after the injection of ATP into a cyanide-poisoned axon, the Na^+ efflux was unaffected by removal of external potassium. These conditions were ones in which the ATP/ADP ratio was lowered, and they suggested that under these conditions internal sodium may exchange with external sodium. This has been confirmed by Baker et al., who showed that in partially poisoned axons there is an approximately 1 : 1 exchange of internal sodium for external sodium by a ouabain-sensitive route. This mechanism serves no useful purpose but needs to be borne in mind when interpreting flux data. It seems that the mechanism that normally uses ATP to drive Na^+ ions outward across the membrane can be reversed if the ATP/ADP ratio falls to a low enough value; but so far no one has demonstrated that Na^+–Na^+ exchange involves the alternate breakdown and resynthesis of ATP.

In red cells a high inorganic phosphate level in the cell promotes exchange of internal potassium for external potassium by a ouabain-sensitive route. It is not known whether or not a similar process can occur in nerve.

8. Reversal of the Na Pump. In theory it should be possible to reverse the Na^+ pump, each reversed cycle resulting in the synthesis of ATP. This was first demonstrated in red cells by Garrahan and Glynn. They made the ion gradients steeper than normal; that is, high potassium and low sodium inside the cell and high sodium and nominally zero potassium in the external medium. Under these conditions there was incorporation of inorganic phosphate into ATP which could be blocked either by cardiac glycosides or by addition of potassium to the external medium. Reversal is associated with a ouabain-sensitive efflux of K^+ and a ouabain-sensitive influx of Na^+.

C. Calcium Pumps

If the movement of Ca^+ across the cell membrane were purely passive, the free intracellular concentration of calcium should be approximately 0.1–1.0 *M*, whereas in those cells in which measurements have proved possible the concentration inside the cell is approximately 10^{-7}–10^{-6} *M*, suggesting again, some form of pump.

In general, there seem to be two types of process maintaining the low intracellular ionized calcium concentration: (1) intracellular buffers, and (2) extrusion of calcium by pumps located in the surface membrane. Of the buffering mechanisms by far the most important quantitatively seems to be uptake of calcium by mitochondria. A second buffering system of importance in muscle is uptake of calcium by the sarcoplasmic reticulum.

Schatzmann and Vincenzi demonstrated in red cell membranes an ATP-dependent calcium-extrusion mechanism which has many features in common with the Ca^{2+} uptake system of the sarcoplasmic reticulum. Both processes require ATP and are associated with calcium-activated ATPases in the membrane, and both are insensitive to cardiac glycosides and to changes in the sodium and potassium content of the bathing solution. It is not clear whether or not a similar system exists in nerve; but if Ca^{2+} extrusion from squid axons were solely dependent on ATP, application of a metabolic inhibitor such as cyanide should result in a fall in the Ca^{2+} efflux. In practice cyanide has a dual action; producing first a small fall and later a marked rise in the

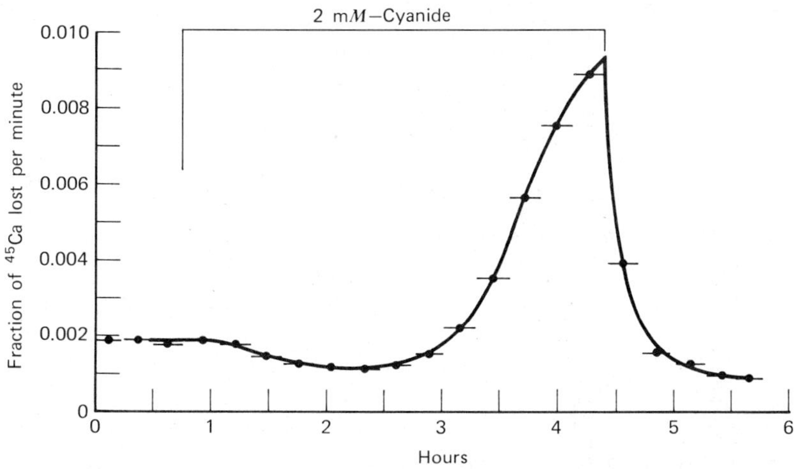

Figure 2. Effect of 2 m*M* cyanide on rate constant of Ca^{2+} efflux from squid axon. Abscissa: time. Ordinate: rate constant of Ca^{2+} efflux. Axon diameter, 865 μ. The delay in the onset of the cyanide effect was 50–100% greater than average in this experiment. (From Blaustein and Hodgkin, *J. Physiol.*, **200,** 497, 1969).

Ca^{2+} efflux (Fig. 2). The early fall may reflect cessation of an ATP-dependent extrusion mechanism; but the rise, which occurs when the ATP level is probably less than 100 μM, seems difficult to reconcile with a dependence of Ca^{2+} extrusion on ATP. The explanation for the rise in Ca^{2+} efflux seems to be the release of calcium from intracellular binding sites, probably mitochondria. Energy for the efflux seems to be provided at least in part by the downhill movement of Na^+ and possibly Mg^{2+} ions into the axon. Only part of the efflux is an exchange of internal calcium for external calcium.

The suggestion that sodium can exchange with calcium across the axon membrane is strengthened by the observation that reversing the Na^+ gradient by replacing external sodium by lithium, choline, or dextrose results in an efflux of Na^+ coupled to an influx of Ca^{2+}. Thus, depending on the magnitude and direction of the electrochemical gradient of Na^+, Ca^{2+} movements can either be into or out of the axon.

Under physiological conditions the electrochemical gradient for Na^+ is maintained fairly constant by operation of the Na^+ pump and it seems quite likely that the potential energy available in this gradient is used to maintain the low level of ionized calcium in excitable cells. This hypothesis can be examined quantitatively. If two Na^+ ions were to move into the cell in exchange for one Ca^{2+} moving out, the final intracellular calcium concentration would be given by

$$\frac{[\text{Ca}]_i}{[\text{Ca}]_o} = \frac{[\text{Na}]_i^2}{[\text{Na}]_o^2}$$

Thus if the ratio $Na_i/Na_o = 1/10$, the ratio Ca_i/Ca_o will be 1/100. This is not adequate to account for the observed values, but if three Na^+ ions moved in exchange for one Ca^{2+} with either one Na^+ ion moving down the electrochemical gradient or in exchange for K^+, the intracellular calcium would be given by either

$$\frac{[\text{Ca}]_i}{[\text{Ca}]_o} = \frac{[\text{Na}]_i^3}{[\text{Na}]_o^3}\, e^{EmF/RT}$$

or

$$\frac{[Ca]_i}{[Ca]_o} = \frac{[Na]_i^3[K]_o}{[Na]_o^3[K]_i}$$

both of which are capable of achieving the observed range of intracellular calcium concentrations.

A notable feature of sodium-dependent Ca^{2+} transport is that reducing the external sodium concentration not only reduces the Ca^{2+} efflux, but it also increases the Ca^{2+} influx. It must therefore lead to a rise in intracellular calcium. In squid axons these changes in intracellular calcium have been observed directly by introducing the calcium-sensitive protein aequorin into the cell. Ca^{2+} influx is also increased by raising the internal sodium concentration. Of particular interest is the observation that the Ca^{2+} influx increases at least as the square of the internal sodium concentration. Thus a small rise in internal sodium results in a large increase in Ca^{2+} influx.

These observations have important physiological and pharmacological implications. For instance, cardiac glycosides increase the force of contraction of the heart and also the resting and stimulus-dependent secretions of various nervous transmitters and hormones. Both actions require sodium, and the glycosides are only effective at concentrations that effect partial or complete inhibition of the Na^+ pump. It seems likely that the glycoside first produces a rise in internal sodium, which leads in turn to a rise in internal calcium. Coupled with the markedly nonlinear dependence of secretion and contraction on calcium, this mechanism seems to give a fairly good account of the action of cardiac glycosides on these cells (Baker, 1971). It should be stressed that Na^+–Ca^{2+} counter-transport seems to be a quite distinct process from Na^+–K^+ exchange (Table 1).

Table 1. Comparison of the Calcium-Dependent and Potassium-Dependent Components of the Na^+ efflux from Squid Axons

Treatment	Ouabain-insensitive Na^+–Ca^{2+} exchange	Ouabain-sensitive Na^+–K^+ exchange
Ouabain	No effect at 10^{-3} *M*	Inhibited by 10^{-7} *M*
External monovalent cations	K^+: activates with low affinity	K^+: activates with high affinity
	Na^+: activates with low affinity up to 100 m*M* and then inhibits	Na^+: inhibits
	Li^+: activates with low affinity	Li^+: inhibits
	Choline: no effect	Choline: no effect
Ca^{2+}-free media	Markedly reduced	Slightly increased
Internal Na^+	Approximately proportional to $[Na]_i^2$	Approximately proportional to $[Na]_i$
Lanthanum (0.1 m*M*)	Inhibits	No effect
Tetrodotoxin (10^{-6} g/ml)	No effect	No effect
Cyanide (2 m*M*)	Inhibits	Inhibits
DNP (0.4 m*M*) pH 8.0	No effect	Increases Na^+–Na^+ exchange in K^+-free seawater
Internal EGTA (5 m*M*)	Inhibits	No effect

D. Other Transport Mechanisms

Little is known of the mechanisms that maintain the gradients of inorganic ions other than Na^+, K^+, and Ca^{2+}. The mechanism of ion gradient coupling just discussed for Ca^{2+} seems to be quite common and there is evidence that many organic substances, for instance, amino acids, are actively transported into cells using the energy conserved in the gradients of Na^+ and K^+ ions. Thus uptake of many amino acids is dependent on the external sodium concentration and is reduced if external sodium is replaced by choline, lithium, or tris. The uptake of amino acids is a saturable process with groups of related amino acids competing for the same carrier, and kinetic analysis shows that Na^+ ions increase the affinity of the transport process for the amino acid. Externally applied K^+ ions have the opposite effect, reducing the affinity for amino acid to a value lower than that in the absence of sodium. Putting the data together, it seems that in the external medium where the sodium concentration is high the binding of amino acid is facilitated, whereas once inside the cell the high concentration of K^+ ions should effect release of the amino acid into the cytosol.

3. PASSIVE TRANSPORT: GENERATION OF MEMBRANE POTENTIALS

A. The Resting Potential

The energy conserved in the ion gradients can also be used to generate a potential difference (PD) across the cell membrane. Each ion gradient represents a battery capable of generating a PD; but whether or not such a PD is established depends on the other ion gradients present and the relative permeability of the membrane to these ions. Transport across the membrane is purely passive. Thus if we consider only the relative permeability to Na^+ and K^+ ions, and the permeability to other ions is assumed to be very low, the potential difference is given by Eq. 3

$$E = 58 \log \frac{[\mathrm{K}]_o + b[\mathrm{Na}]_o}{[\mathrm{K}]_i + b[\mathrm{Na}]_i} \tag{3}$$

where b is the selectivity ratio $P_{\mathrm{Na}}/P_{\mathrm{K}}$ and [ ] are activities.

If the membrane is completely impermeable to Na^+ ions, the potential will be determined solely by potassium, and as there is more potassium inside the cell than outside the internal potential will be negative with respect to the outside. If, on the other hand, the membrane is impermeable to K^+ ions, the potential will be determined solely by sodium, and because there is more sodium outside the cell than inside the internal potential will be positive. These two examples represent the extremes of potential that can be obtained using the Na^+ and K^+ gradients alone. By varying the permeability ratio b, all intermediate potentials can be obtained. Under physiological conditions the permeability to anions (usually chloride) must also be considered. In general, most animal cells are more permeable to K^+ and Cl^- than to Na, and the internal potential of these cells is negative with respect to the external medium.

The dependence of the resting potential of nerve on K^+ ions can be most readily demonstrated in the perfused squid axon. In this preparation the axoplasm is squeezed out and replaced by a salt solution of known composition. Figure 3 shows the transmembrane potential—determined with an intracellular electrode—at a range of different internal and external potassium concentrations. When the potassium concen-

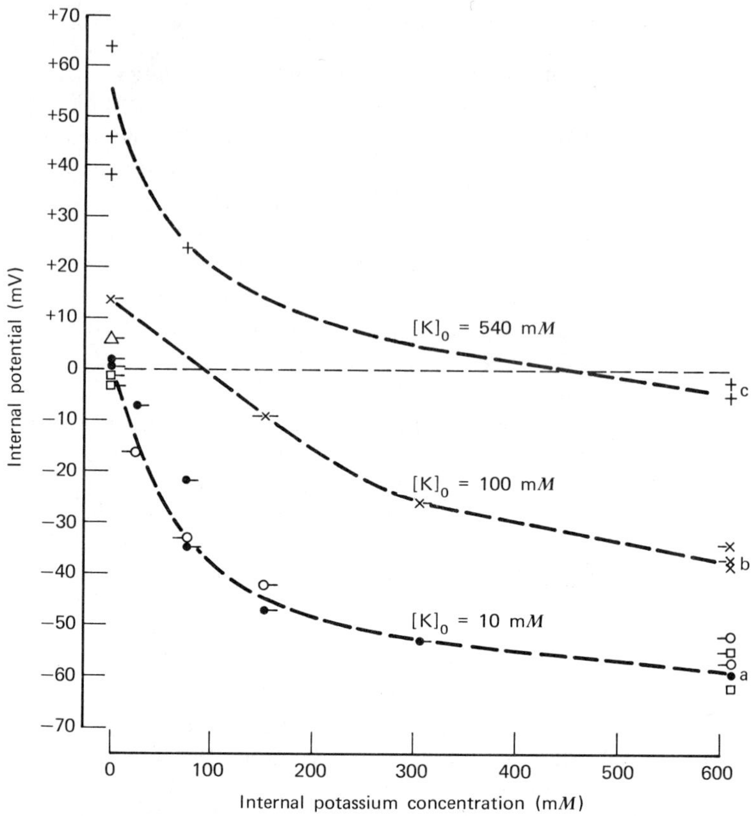

Figure 3. Effect of varying the internal potassium concentration by replacing isotonic potassium chloride with isotonic sodium chloride in perfused axons surrounded by solutions containing 10 (*a*), 100 (*b*), or 540 m*M* (*c*) potassium. Each symbol represents a different axon. (From Baker, Hodgkin, and Shaw, *J. Physiol.*, **164,** 355, 1962).

tration inside exceeds that outside, the interior of the axon is negative; when the concentrations on both sides are the same, the PD is close to zero and, when there is more potassium outside than inside, the interior of the axon is positive. These results show quite clearly that the magnitude and size of the resting potential depend solely on the steepness and direction of the K^+ ion gradient.

An interesting feature of Figure 3 is that as the potassium concentration gradient is increased the potential does not increase as much as would be predicted by the Nernst equation. The explanation seems to be that the potassium permeability falls as the resting potential becomes more negative, the potential ultimately reached being limited by the small inward leak of Na^+ ions.

B. The Action Potential: Voltage-Dependent Changes in Permeability

Although under identical conditions a certain selectivity ratio $P_{Na}/P_K/P_{Cl}$ may be characteristic of a particular cell, it is not necessarily a fixed property of the cell membrane. Thus the relative permeabilities may change as a function of membrane potential or following application of various chemicals such as transmitter substances or drugs. A particularly striking example of such changes in permeability is the action potential of excitable tissues. Conduction along nerve and muscle is essentially electrical in nature; but without some boosting mechanism a current fed in at one end of a nerve would be attenuated long before it reached its destination. The mechanism by which the flow of current is boosted is called the action potential. It results from three temporally distinct changes in permeability which occur in response to a sudden reduction in membrane potential (depolarization). These are: (1) a rapid increase in the permeability to Na^+ ions; (2) a less rapid blocking or inactivation of the increase in permeability to Na^+ ions; and (3) a slow, maintained increase in the permeability to K^+ ions. The result of these permeability changes is that the internal potential swings from negative to positive and back again to negative, the period of positivity serving to boost the flow of electric current along the fiber (Hodgkin, 1964; Katz, 1966).

The dependence of excitability on Na^+ or Li^+ ions was first described by Overton; but it was Hodgkin and Katz who showed that during the action potential the membrane becomes *selectively* permeable to Na^+ ions. The potential at the crest of the spike can be calculated from the Nernst equation with a selectivity ratio P_{Na}/P_K of approximately 10:1. Reducing the external sodium concentration or raising the internal sodium concentration both result in action potentials of smaller size (Fig. 4) and ultimately block nervous conduction.

The nature of the permeability changes occurring during the action potential has been investigated using a technique known as the voltage clamp. In this method, current-passing and potential-measuring electrodes are inserted into the squid axon thus allowing the membrane potential to be displaced to a new value and held there by electronic feedback. The current that flows through a definite area of membrane under the influence of the applied voltage is measured with a separate amplifier.

In the absence of feedback the immediate response of the membrane to a short shock is the attainment of a new and transient potential difference which corresponds to the charging of a condenser by the stimulus current. From the ratio (charge passed)/(change in PD), the capacity of the nerve membrane is found to be 0.9 $\mu F\ cm^{-2}$.

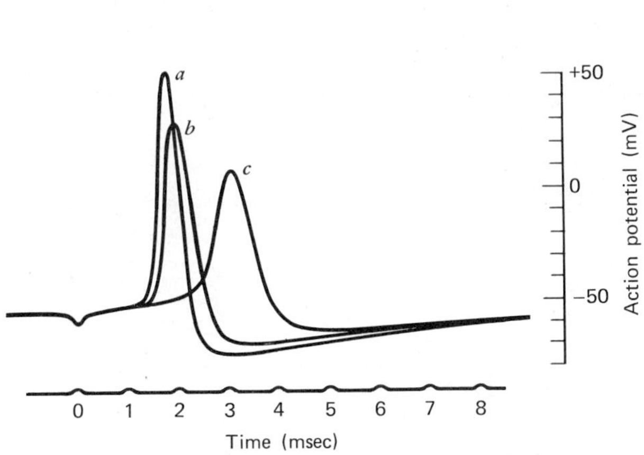

Figure 4. Effect on action potential of perfused squid axon of replacing internal potassium with Na^+ ions. (*a*) Perfusion fluid isotonic potassium sulfate; (*b*) one-quarter of the K^+ ions replaced by Na^+ ions; (*c*) half the K^+ ions replaced by Na^+ ions. The records were obtained in the order *B*, *A*, *C*. (From Baker, Hodgkin, and Shaw, *Nature*, **190,** 885, 1961).

Although the initial charging process is linear, the subsequent behavior of the membrane potential varies with the strength of shock. If the potential is increased (hyperpolarized) or decreased (depolarized) by not more than 15 mV, the PD decays back to the original value in a few milliseconds. The time course of the decay is much slower than for the initial capacity surge because now the current is not flowing through the wires of the stimulating circuit but is carried by ions through the membrane.

For depolarizations much greater than 15 mV, an action potential is produced. The displacement of the potential becomes regenerative and the membrane continues to depolarize until the internal potential is approximately +40 mV. The potential does not stay at this value but slowly declines to its resting level.

The voltage-clamp technique provides information on the currents flowing through the membrane in response to different impressed voltages. Figure 5 shows the membrane currents for depolarizations of 91–143 mV. For depolarizations of 20–110 mV, the ionic current consists of two phases: an early inward current and a later outward current. The early inward current must represent either an inward flow of positive ions or an outward flow of negative ions. In practice it depends on Na^+ ions and is abolished or reversed in solutions in which the bulk of the sodium is replaced by choline or dextrose. Li^+ ions can substitute for Na^+. Further evidence that the early inward current is carried by Na^+ ions is that when the potential is made equal to that of the sodium battery (as predicted by the Nernst equation), there is no early current and only the delayed (potassium) current is seen. For voltage displacements beyond 110 mV, both components of the membrane current are outward. The internal potential at which there is no early current is known as the "sodium equilibrium potential" and is normally approximately 40–50 mV inside positive (i.e., a total displacement of membrane potential of 100–110 mV from a resting value of approximately 60 mV inside negative). Changing $[Na]_o$ or $[Na]_i$ alters the equilibrium potential in accord with the Nernst equation. Apart from the electrical evidence, there is also tracer data showing a net entry of sodium during the action potential. The main conclusion from these observations is that the early current is carried by Na^+ ions and that the magnitude and direction of the current depends on the concentration gradient for this ion and the potential difference across the membrane.

Hodgkin and Huxley made use of the observation that the early current depends on external sodium to separate the membrane current into its two components. Measurements were made of membrane current for the same displacement of membrane

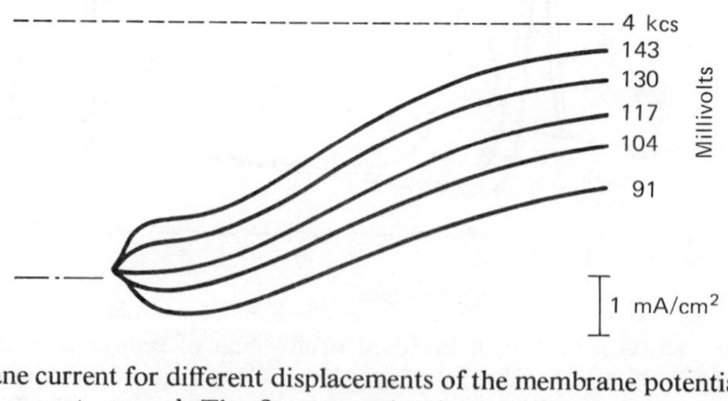

Figure 5. Membrane current for different displacements of the membrane potential at a temperature of 3.5°C; outward current upward. The figures at the right give the change in internal potential. (Hodgkin, Huxley, and Katz, *J. Physiol.*, **116**, 424, 1952).

potential, first in the presence of Na^+ ions, then with most of the sodium replaced by choline, and finally in sodium again. On the assumption that choline does not affect the time courses of either current, they obtained separate curves for the Na^+ and K^+ currents. Nowadays, tetrodotoxin (TTX) is preferred to choline as a means of separating sodium and potassium currents (see Section 3.C).

Currents obtained in these ways do not immediately provide information about the permeability of the membrane, for instance, at the sodium equilibrium potential there is no sodium current; but the membrane is permeable to Na^+ ions. The current carried by a particular ion depends on two factors: (1) the driving force on the ion which is itself the resultant of the concentration difference of the ion on the two sides of the membrane and the PD across the membrane, and (2) the permeability of the membrane to the ion in question. A measure of the driving force on an ion is the difference between the membrane potential (V) and the equilibrium potential (V_E) for the ion, $V - V_E$. The permeability of the membrane to the ion can be expressed as a conductance (current divided by PD). The Na^+ conductance g_{Na} is measured by $I_{Na}/(V - V_{Na})$, and the K^+ conductance g_K by $I_K/(V - V_K)$. Figure 6 shows how g_{Na} and g_K vary with time following a step depolarization of 56 mV. It illustrates several important features:

1. Both g_{Na} and g_K rise along sigmoid curves, g_K rising 10–30 times more slowly than g_{Na}. The maximum conductance for both ions is approximately 20–30 mmho/cm².
2. The increase in g_{Na} is transient. In less than 1 msec it begins to decline exponentially. This decline is called inactivation.
3. The increase in g_K does not show any appreciable inactivation.
4. Returning the membrane potential to its resting level cuts off both currents exponentially. The resting value of g_K is greater than that of g_{Na}.

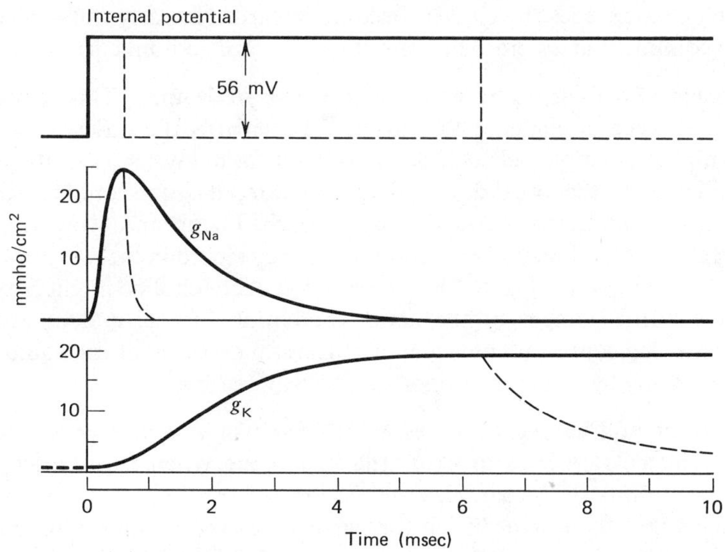

Figure 6. Time course of sodium conductance (g_{Na}) and potassium conductance (g_K) associated with depolarization of 56 mV. The continuous curves are for a maintained depolarization; broken curves give the effect of repolarizing the membrane after 0.6 or 6.3 msec. (From Hodgkin, *Proc. Roy. Soc. Ser. B.*, **148,** 1, 1958).

To this list should be added three further observations not illustrated in Figure 6:

5. The rate of change of both g_{Na} and g_K increases with increasing depolarization, but the maximum conductance attained remains fairly constant for depolarization greater than 50 mV.

6. Temperature markedly alters the rates of change of conductance but has rather little effect on their final values. Cooling a nerve greatly prolongs the action potential.

7. The size of g_{Na} in response to a large depolarization is very dependent on the potential from which the depolarization is made. A small, maintained depolarization markedly reduces (inactivates) the maximum sodium current that can be elicited, whereas a small hyperpolarization allows a larger sodium current up to some maximum value.

At this stage it is perhaps worth considering some physical models that have been put forward to explain the above data. The steep dependence of permeability on potential suggests that the permeability changes must arise from some effect of the electric field on the orientation of charged particles within the membrane. The striking influence of potential on g_{Na} and g_K is shown by the observation that g_K increases *e*-fold when the PD is reduced by 5–6 mV and g_{Na} is increased *e*-fold by a reduction of 4 mV. The actual processes involved are not known, but various ideas have been put forward.

1. Movement of a Charged Carrier to Which the Ions Are Attached. If the ion–carrier complex bears several negative charges, in the resting state it will be attracted to the outside of the membrane. Depolarization will allow the carrier to move and the extent of the redistribution will depend on the total negative charge of the complex. Attractive as this hypothesis is, it is very unlikely, as it predicts that there should be an initial inward movement of negative charge, that is, an outward current, carried by the carrier molecules. Only when Na^+ ions are released into the axon and the carriers return to the outside does the current become inward. No early outward current has been observed immediately preceding the rise in inward sodium current.

2. Movement of a Charged Substrate or Enzyme Activator. This remains a possibility; but one piece of evidence that argues against it is the observation that axons perfused with only simple salt solutions conduct as many as 5×10^5 impulses. If conduction involves the breakdown of a substrate, it seems unlikely that enough substrate for 5×10^5 impulses could be stored in the membrane. However, some form of cyclic breakdown and resynthesis using the energy stored in the ion gradients cannot be ruled out. It has been suggested in the past that acetylcholine might play some part in the permeability changes of the action potential, just as it does at cholinergic synapses and at the neuromuscular junction. Careful experiments on squid axons have failed to reveal any evidence in support of this suggestion.

3. Movement of Charged Particles Which Do Not Act as Carriers in the Usual Sense But Which Allow Ions to Cross the Membrane When They Occupy Particular Sites. One can think of the particles as creating a channel across the membrane. This model predicts that the rate at which the particles move determines the rate at which the ionic current approaches its maximum; but it should have little or no effect on the maximum current. This is in agreement with experiment. Temperature has a large effect on the rate at which the sodium current increases but has little effect on its maximum value. Further, many particles may be required to form a channel; but the removal of only one blocks it. This is consistent with the sigmoid rise in g_{Na} and g_K after depolarization, and their exponential decline after repolarization. The transient nature of the sodium current can be explained by assuming either that after forming a channel the activating particles undergo some change that leads to blockage of the channel, or that there is a slowly moving inhibitory particle which blocks the sodium channel when it reaches a certain position. This model has many points in its favor, and it is relevant to note that the empirical equations used by Hodgkin and Huxley to reconstruct the conducted action potential can be given a physical basis in terms of a model of this kind. An essentially similar analysis has been applied successfully to action potentials of amphibian myelinated nerve, amphibian skeletal muscle, and cardiac muscle.

C. Agents That Block or Modify the Permeability Changes

There is now much evidence in support of the analysis of Hodgkin and Huxley. Perhaps the clearest evidence for the existence of separate sodium and potassium channels is the discovery of agents that block g_{Na} without affecting g_K, and vice versa (see reviews by Baker, 1968; Hille, 1970). The puffer fish poison TTX is the best example of a substance that blocks g_{Na} without affecting g_K. TTX, which is effective only when applied to the outer face of the membrane, does not interfere with the kinetics of either opening or closing the sodium channels; it seems simply to block the sodium channels as soon as they are formed. The interaction between TTX and the membrane is so powerful and specific that it has proved possible to use the binding of TTX to estimate the number of Na^+ ion channels in the axon membrane. Between 10 and 50 molecules of TTX are bound per square micron nerve membrane. If it is assumed that one TTX binds per sodium channel, this gives an upper limit to the number of channels per square micron of membrane. Although there are a number of errors inherent in an estimate of this kind, the result emphasizes how thinly scattered the sodium channels are throughout the membrane material. It underlines the problems that must be faced in attempting to isolate the molecules concerned. Measurement of the number of Na^+ pumping sites gives about $1000/\mu^2$, so the pumping sites are 20–100 times as numerous as the sodium channels.

While TTX is the best example of a chemical that blocks specifically the increase in Na^+ ion permeability, it is not the only one. The paralytic shellfish poison saxitoxin, the anesthetics procaine, xylocaine, and urethan and the tranquillizer Compazine (prochlorperazine) all act in a similar way although, in many tissues, they also exert some effect on the K^+ ion permeability mechanism.

Any process tending to block specifically the K^+ ion channels would be expected to slow down the repolarization phase of the action potential. Long-lasting action potentials are seen to some extent when the K^+ ions inside a perfused axon are replaced by Rb^+ ions and very dramatically when they are partially replaced by Cs^+ ions. Voltage-clamp experiments show that internal, but not external, Cs^+ ions block the delayed currents normally carried by K^+ ions without much affecting the transient currents carried by Na^+ ions.

In several excitable tissues, including the node of Ranvier, the action potential can be prolonged by inclusion of tetraethylammonium ions (TEA) in the medium. In the squid a similar prolongation is seen only when TEA is applied internally. In both cases the results are consistent with the idea that the TEA blocks the K^+ ion channels without affecting the transient rise in Na^+ ion permeability. Inclusion of TEA and TTX in the medium blocks both the sodium and potassium currents. In squid, but not at the node, TEA blocks only outward currents flowing through the K^+ ion channel. These results suggest that internal TEA might be swept into the membrane by outward potassium currents producing a block and be flushed out again by a current of K^+ ions flowing in the opposite direction. At the node of Ranvier, the results suggest that the binding of one TEA molecule blocks the flow of ions in either direction through one K^+ ion channel. The binding of TEA is not affected by the presence of TTX; this suggests that these drugs act at different sites on the membrane.

Once the Na^+ ion channels are open, they normally become closed by inactivation. Anything that keeps the Na^+ ion channels open, either by slowing the process of inactivation or by acting in some other way, tends to prolong the action potential and depolarize the membrane.

A toxin that seems to act by blocking inactivation specifically has been isolated from the sea anemone *Condylactis gigantea*. In the presence of condylactis toxin (CTX), the action potential is greatly prolonged, and under voltage-clamp conditions depolarization of the membrane produces a maintained sodium current which can be turned off only by repolarizing the membrane to the resting potential. This toxin has a molecular weight of 10,000–15,000, and it is interesting that it affects inactivation specifically with little or no effect on the processes turning on g_{Na} and g_K. A component of scorpion venom also has an action very similar to CTX.

The sodium channel can also be held open in other ways. The alkaloid veratridine and the insecticide DDT both interfere with the closing of the sodium channels without much affecting the kinetics of activation of either g_{Na} or g_K. These substances seem to work not solely by slowing the process of inactivation. They also block the closing of the sodium channels that normally follows repolarization of the membrane. Their mode of action is therefore to stabilize the sodium channel once it is formed.

D. Potential-Dependent Changes in Calcium Permeability

Although the majority of the inward current in squid axons is carried by Na^+ ions, the use of aequorin has shown that there is an increase in P_{Ca} in response to depolarization. Although this does not contribute significantly to the inward current in squid axons, in some tissues, for example, certain smooth muscles, the action potential seems to be dependent on Ca^{2+} ions. Rather little is known about these calcium channels except that in squid axons calcium seems able to pass both through the TTX-sensitive sodium channel and also to enter through a TTX-insensitive channel which has much the same opening characteristics as the potassium channel, but seems pharmacologically distinct from that system. This calcium channel may be involved in the release of neurotransmitter substances.

E. Nature of the Permeability Changes

The biochemical basis of permeability changes is still a mystery. Perhaps the most striking recent discovery is that various cyclic peptides and related substances can act as carriers for ions both in artificial membranes and natural ones. Examples of these ionophores are valinomycin, alamethicin and nystatin. It is not difficult to imagine that such carriers might represent the particles postulated by Hodgkin and Huxley.

For instance, if the ionophores are negatively charged, depolarization might allow numerous particles to form into a column stretching across the membrane, thus providing a channel for the passage of ions. As yet, substances of this kind have not been isolated from excitable tissue.

4. BULK TRANSPORT: THE RELEASE OF NERVOUS TRANSMITTERS

Although it is still too early to be certain, the release of nervous transmitters seems to involve a basically similar mechanism in all nerves, the only variable being the transmitter released. Transmission has been most studied at the neuromuscular junction, and the squid giant synapse and has recently been reviewed by Katz. The following are a few important features:

1. The transmitter substance seems to be packaged in small vesicles approximately 500 Å in diameter. In the resting nerve there is a slow random release of quantal units of transmitter molecules, which is presumed to reflect the loss of the contents of individual vesicles. The rate of release is greatly increased for a brief period following an action potential or electrical depolarization of the terminal.
2. Ca^{2+} ions are essential for transmitter release. Mg^{2+} ions can replace Ca^{2+} only to a very limited extent and they antagonize the action of calcium. At low calcium concentrations there is a sigmoid relation between the end plate potential which is a measure of transmitter released and the external calcium concentration. At higher calcium concentrations the relation approaches a limiting value. When plotted on double logarithmic coordinates, the initial part of this relation gives a straight line with a slope of nearly 4. Addition of a constant amount of magnesium reduces the end plate potential but does not alter the slope of the relation between log Ca and log end plate potential. These results can be explained quantitatively if it is assumed that a cooperative action of about four Ca^{2+} ions is necessary for the release of each packet of transmitter by the nerve impulse. At the squid giant synapse transmitter release involves cooperation between two to four Ca^{2+} ions. It is not known where these Ca^{2+} ions are required, but it is an attractive idea that they may be required to facilitate fusion of a vesicle with the surface membrane.
3. By analogy with experiments on squid axons, it seems most likely that Ca^{2+} ions enter the nerve terminal; but this is still not proved. Transmitter release in response to depolarizing pulses applied to the presynaptic terminal is unaffected by TTX and by internally applied TEA, but is blocked by external Mn^{2+} ions. These properties are surprisingly similar to those of the second phase of calcium entry during the action potential of squid axons (Section 3.D), and it seems likely that calcium entering the terminal by this route leads to the release of transmitter. Two further pieces of evidence favor calcium entry triggering transmitter release: (1) when the currents of sodium and potassium have been reduced to low levels by application of TTX and TEA, depolarization of a presynaptic terminal may lead to a calcium-based action potential—implying an inward current carried by Ca^{2+} ions; and (2) if Ca^{2+} ions are entering the terminal down their electrochemical gradient, by making the internal potential very positive it should be possible to stop calcium entry and so block transmitter release. This has been achieved at the squid giant synapse. When the internal potential is pulsed to about $+130$ mV, no transmitter is released during the pulse.

An equilibrium potential of +130 mV is consistent with an internal calcium concentration of 4×10^{-7} *M* which is very close to that measured in squid axons.

4. The mechanism of fusion between the vesicle and the surface membrane is unknown.

5. TRANSPORT ACROSS EPITHELIA

So far we have considered some mechanisms for the transport of substances across the membranes of individual cells; but in the body absorption of materials from the gut or reabsorption of substances filtered by the kidney involves transport across an epithelial cell layer. Transport across epithelia seems to take place through rather than between the cells, which means that substances must first enter the cell and then be expelled on the other side; the complete traverse involves crossing two cell membranes.

A. Transport of Salt and Water

Many epithelia are capable of transporting salt and water against an adverse gradient. The gut and kidney can absorb Na^+ and Cl^- ions from a solution more dilute than blood and can absorb water against an adverse osmotic gradient; frogs are able to take up salt through their skins from very dilute solutions. When transport is taking place, the inside face of the epithelium, that is, the side to which salt and water are being carried, usually becomes positive with respect to the outside. This means that Na^+ ions can be moved against both an adverse chemical and electrical gradient and some form of active transport of Na^+ ions is indicated. The movement of Cl^- ions is assisted by the potential and may be passive. The transport of salt has been investigated most extensively in the frog skin.

It is possible to short-circuit an isolated piece of skin by flowing current from an external battery in the direction opposite that being generated by the skin so that the potential difference across the skin is maintained at zero. The current required to keep the skin short-circuited is a measure of the current being generated by the skin. The experimental observations are: (1) that the current generated by the skin is dependent on the presence of Na^+ ions at the outer face of the skin, but is insensitive to Na^+ ions at the inner face of the skin, and (2) radioactive tracer measurements show that the transport of Na^+ ions across the skin accounts for almost all of the short-circuit current. Provided short-circuiting does not affect the properties of the skin, these experiments show that the skin can generate an inwardly directed current of Na^+ ions.

How is this vectorial movement of Na^+ achieved? The following observations are relevant:

1. The short-circuit current is reduced or abolished by metabolic inhibitors applied to either side of the skin.

2. The short-circuit current is markedly reduced by specific inhibitors of the Na^+ pump, for instance, cardiac glycosides, only when applied to the inner face of the skin. High concentrations applied to the outer face have no effect.

3. During transport the skin consumes oxygen and the relation between extra oxygen consumed and Na transported gives $Na/O_2 = 18$. Assuming a *P*/O ratio of 3, three Na^+ ions are being transported per energy-rich phosphate bond split, which is consistent with the operation of the Na^+ pump (see Section 2.B.4).

4. In the absence of a short-circuiting current, the skin potential is sensitive to changes in the ionic medium, but the two faces differ in their sensitivity. The outer face is permeable to Na^+ ions and replacement of external sodium by choline reduces the potential. The inner face is virtually insensitive to sodium but is sensitive to K^+ ions and raising the potassium concentration bathing the inner face of the skin reduces the potential. The outer face is insensitive to potassium. These experiments show that the two faces of the epithelium differ in their permeability to ions. The outer face is permeable to sodium and the inner face to potassium.

These observations can be combined to provide a mechanism for the transport of Na^+ across epithelia. Na^+ ions enter the epithelial cells at their outer side moving into the cell down the electrochemical gradient maintained by the Na^+ pump. At the inner face the energy-dependent Na^+ pump expels the Na^+ ions in exchange for K^+, the K^+ ions tending to leak out again and be recycled by the pump. The net effect is that a stream of Na^+ ions is moved inward across the epithelium.

The potential developed by this process tends to draw negatively charged ions after the Na^+, resulting in a net inward movement of sodium chloride. Reduction of the chloride permeability or use of a large anion such as sulfate increases the skin potential because the shunting effect of the anion on the potential is reduced. Although chloride can be transported passively, active transport systems for Cl^- probably exist in many epithelia.

We now come to the more difficult problem of water transport. In some systems the inward transport of sodium chloride leads to a difference in osmotic pressure (activity of water) on the two sides of the membrane, and water tends to follow the sodium chloride moving passively down its activity gradient; but in some epithelia salt and water transport can occur in the absence of any difference in the activity of the bulk solutions bathing the two faces of the epithelium. Under these conditions transport is still dependent on Na^+ ions and closely resembles that in frog skin except that the potential generated may be quite small. This problem has been resolved by showing that the Na^+ pumping sites accessible to substances supplied at the inner face of the epithelium are located between the epithelial cells. Where two epithelial cells meet, their inner faces are in contact for a considerable distance, forming a long intercellular space (Fig. 7). The dimensions of this space depend on the pumping activity of the epithelium. In an actively pumping epithelium, the space is dilated, whereas in one that is not pumping, for instance, because of the absence of Na^+ ions or the presence of a metabolic inhibitor, the space collapses down to a few hundred angstroms.

The importance of these observations is that the pumping of Na^+ into this restricted space followed presumably by Cl^- can lead to the production of a region of osmotic pressure locally higher than in either of the bulk solutions. Water will tend to move into this region and there will be bulk flow of fluid out of the intercellular spaces into the inside solution. The osmotic pressure of the emerging fluid will depend on the dimensions of the intercellular space. If it is long and thin, the emerging fluid may be isotonic, thus accounting for the ability of certain epithelia to absorb isotonic fluid.

If the pumps are located deep within the epithelium at the end of long, blind channels their influence on the skin potential will depend on the cable properties of the system, and the measured skin potential may be only a small fraction of that generated at the pumping sites, which may help explain the small potentials observed in some epithelia.

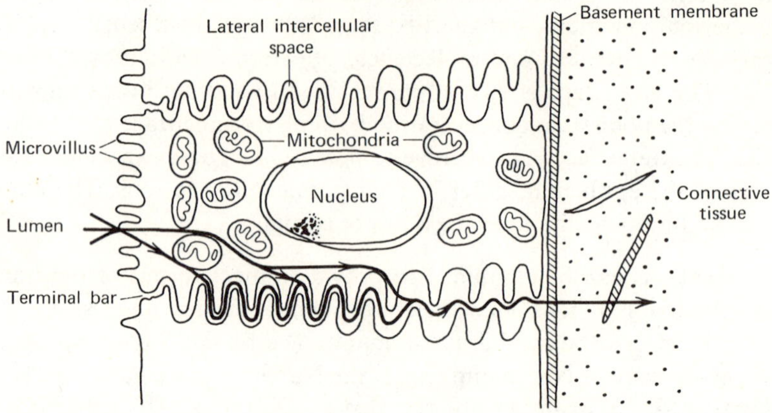

Figure 7. Route of water transport across rabbit gallbladder, as deduced from combined physiological and anatomical experiments. The movement of water from left to right. Salt is pumped from the cells into the lateral intercellular spaces and makes them hypertonic. Water enters osmotically along the length of the spaces and distends them; an isotonic sodium chloride solution emerges from the ends of the spaces facing the connective tissue. (From Diamond, and Tormey, *Nature*, **210,** 817, 1966).

B. Transport of Sugars and Amino Acids

Although sugars enter most cells by facilitated diffusion, the epithelia of the gut and kidney are capable of actively transporting certain sugars. The mechanism involved is similar for both sugars and amino acids.

Transport involves two steps: (1) concentrative uptake into the epithelial cell, and (2) passive exit, possibly carrier mediated, out of the epithelial cell into the extracellular fluid. Again vectorial transport depends on the two faces of the epithelium having different properties. The concentrative uptake is dependent on Na^+ ions at the outer face of the epithelium and has the properties of an ion gradient-coupled process (see Section 2.D). Thus sugars and amino acids enter at the outer face of the epithelium in association with Na^+ ions, the Na^+ ions being pumped out at the inner face. The extent to which sugars and amino acids are accumulated by epithelial cells depends on the electrochemical gradient for Na^+ ions. Accumulation is markedly reduced in sodium-free media or in the presence of cardiac glycosides which lead to accumulation of sodium inside the cells. Na^+ ions increase the affinity of the transport process for organic solutes, whereas K^+ ions reduce the affinity below that in the absence of sodium. As described in Section 2.D, this favors the accumulation of solute inside the cell. Two other observations are consistent with the above scheme: (1) uptake of solute is associated with a stoichiometric uptake of sodium, and (2) the short-circuit current is increased during transport of sugars and amino acids. This is not secondary to making more energy available as the same effect is seen with nonmetabolizable sugars. The simplest explanation is that the organic solutes are facilitating the transport of sodium into the cells so making more sodium available to the pumping sites. The mechanism of passive exit from the cells is not known.

REFERENCES

General

Davson, H. (1970). *A Textbook of General Physiology*. Churchill, London.

Stein, W. D. (1967). *The Movement of Molecules across Cell Membranes*. Academic, New York.

Sodium Pump

Baker, P. F. (1966). *Endeavour,* **25,** 166.

Glynn, I. M. (1968). *Brit. Med. Bull.,* **24,** 165.

Calcium Pumps

Ebashi, S., and M. Endo (1968). *Prog. Biophys.,* **18,** 123.

Baker, P. F. (1972). *Prog. Biophys.,* (in press).

Transport of Organic Solutes

Schultz, S., and P. F. Curran (1970). *Physiol. Rev.,* **50,** 637.

Membrane Potentials

Hodgkin, A. L. (1964). *The Conduction of the Nervous Impulse*. Liverpool Univ. Press, Liverpool.

Katz, B. (1966). *Nerve, Muscle and Synapse*. McGraw-Hill, New York.

Modification of Permeability

Baker, P. F. (1968). *Brit. Med. Bull.,* **24,** 179.

Hille, B. (1970). *Prog. Biophys.,* **21,** 1.

Transmitter Release

Katz, B. (1969). *The Release of Neural Transmitter Substances*. Liverpool Univ. Press, Liverpool.

Baker, P. F. (1972). In *Recent Advances in Physiology*, Churchill, London.

Transport across Epithelia

Crane, R. K. (1968). In *Handbook of Physiology*, Section 6, Vol. III, p. 1323. American Physiological Society, Washington, D.C.

Diamond, J. M. (1968). In *Handbook of Physiology*, Section 6, Vol. V, p. 2451. American Physiological Society, Washington, D.C.

Keynes, R. D. (1969). *Quart. Rev. Biophys.,* **2,** 177.

Ussing, H. H. (1965). *Harvey Lectures,* **59,** 1.

PART 2

THE ORGANIZATION OF CELLULAR FUNCTION

4

THE MITOCHONDRION

D. D. Tyler

1. INTRODUCTION

Mitochondria are lipoprotein particles of characteristic shape, size, structure, and staining properties. They are present in the cytoplasm of all aerobic cells of higher animals and plants and also in certain microorganisms including algae, protozoa, and fungi. Extensive studies by cytologists in the period 1870–1915 established the ubiquitous presence and invariant chemical nature of mitochondria and their distinctive arrangement within particular types of cells (Cowdry, 1918). At first, the use of fixatives and stains suitable for revealing nuclear detail did not permit a precise distinction to be made between mitochondria, other cytoplasmic granules, and artifacts of fixation. Benda (1898) developed more suitable staining methods for mitochondria and introduced the word "mitochondrion" from the Greek *mitos*, a thread, and *chondros*, a granule. Early studies of mitochondria in living cells were made by Michaelis (1899), who found that mitochondria promote oxidoreduction changes in the dye Janus green, and by Lewis and Lewis (1915), who described continual changes in the shape, size, and location of mitochondria in tissue culture cells. Kolliker (1888) was probably the first person to isolate mitochondria when he teased granules from insect muscle, in which mitochondria are abundant, and showed that they swell in water and possess a membrane.

By the time Cowdry (1918) wrote his detailed review, it had become apparent that mitochondria must perform a fundamental function in cellular activity. Kingsbury (1912) observed that the chief ingredients of mitochondrial fixatives then in use were oxidizing agents. He assumed that their efficiency as fixatives depended on their reaction with reducing substances in the cytoplasm, and hence became the first to suggest that mitochondria were the sites of cellular oxidations. Early observations consistent with this conclusion include those of Battelli and Stern (1912) and Warburg (1913), who discovered that the respiratory activity of disintegrated tissues was firmly bound to granular material. Among the many speculations on the origin and physiological functions of mitochondria arising from these early studies, it is possible to discern important ideas that have been amply confirmed by more modern techniques.

Further progress has depended on the isolation of mitochondria from cells in amounts sufficient for biochemical analysis. It was also essential to relate the enzymic activities of isolated mitochondria to those of the whole cell or the unfractionated cytoplasm on a strictly quantitative basis. Experiments that satisfied these requirements were first made by Claude (1946, 1948), who successfully developed a combination of biochemical, microscopic, and differential centrifugation methods for the isolation and study of subcellular fractions from animal tissues. By applying these methods to rat liver, Hogeboom et al. (1946) established that some important components of the respiratory chain were situated in the mitochondria. Studies by Belitzer (1937) and Kalckar (1937) had indicated that energy released during the oxidation of cell metabolites by the respiratory chain was linked or "coupled" to the synthesis of relatively large amounts of the high-energy phosphate bonds of adenosine triphosphate (ATP). Claude was therefore able to identify mitochondria as the "power plants" of the cell. The conversion of respiratory energy into phosphate bond energy is usually the major source of cellular ATP, the form in which chemical energy is supplied to energize many kinds of physiological activity (Krebs and Kornberg, 1957). It is probable that mitochondria also play an important role in the movement of ions and water both within and across cells and in the regulation of the ionic composition of the cytoplasm, and that the use of respiratory energy for these functions occurs concurrently with ATP synthesis. Mitochondria are the exclusive site of citric acid cycle activity, which may be regarded as the hub of several metabolic pathways in the cell. Many of the carbon atoms of cellular constituents pass through the citric acid cycle at some stage during cell metabolism. Mitochondria are therefore not only the site of catabolic reactions coupled to energy conservation but also play a central role in the supply of intermediates for anabolic processes and in the regulation of cell metabolism.

A new and very important aspect of mitochondrial research began in the early 1950s when the fine structure of mitochondria was revealed by electron microscopy. This development has led to many discoveries concerning the relationship between mitochondrial structure and function and to the observation in mitochondria of DNA, which had usually been regarded as an exclusive constituent of the cell nucleus.

These and other studies have allowed some progress to be made toward understanding the origin and biogenesis of mitochondria in the living cell.

Until recently, development of ideas on the mechanism of respiratory energy coupling had been slow and inconclusive. This situation has been changed profoundly by the appearance of the chemiosmotic hypothesis of Mitchell (1961, 1966). Although the validity of the hypothesis is still a matter of dispute, it has already had a remarkable influence on ideas and investigations into the mechanism of action of uncoupling agents, the transport of anions and cations across the mitochondrial membrane, and the spatial arrangement of membrane-bound components of mitochondria.

The purpose of this chapter is to survey the properties of mitochondria that are most significant in understanding the functions of mitochondria in the physiology of the cell, and briefly to summarize current ideas on the mechanisms by which these functions are carried out. A comprehensive review of mitochondrial research can be found in Lehninger (1964). Recent developments are discussed in a monograph edited by Racker (1970a).

2. MITOCHONDRIAL STRUCTURE

A. Electron Microscope Observations

Under microscopic examination, mitochondria often appear in fixed tissues as rods approximately 1–4 μ long and 0.3–1 μ wide, but many variations are found, mainly in length which may be up to 10 μ in muscle cells. The rather constant width suggests that mitochondrial growth may occur by elongation resulting from the addition of material at their ends. Electron micrographs tend to give a misleading impression of their shape because the mitochondria are usually cut transversely or obliquely in the thin sections used and are only rarely cut along their entire length. Knowledge of the ultrastructure of mitochondria was dependent on the development of the electron microscope and originated from the work of Palade (1956) and Sjöstrand (1956). The main features of mitochondrial structure revealed by electron microscopy and the terms commonly used to identify them are summarized in Figure 1. Positive staining with osmium tetroxide shows that the mitochondrion is bounded by a double membrane and that the interior or matrix of the mitochondrion is penetrated by *cristae mitochondriales* which are formed by invaginations of the inner membrane (Fig. 2). Both the outer and inner membranes appear to have a fairly constant thickness of 60–70 Å, separated by an intermembrane space 60–80 Å in width. The cristae are often arranged in parallel and at right angles to the long axis of the mitochondrion but occur in various other conformations in different cell types and even within a single mitochondrion. In many tissues the cristae are usually lamellar or platelike, but in many protozoa and in steroid-synthesizing tissues including the adrenal cortex and corpus luteum, they occur as regularly packed tubules (Fig. 3). The tubular form of cristae may represent a phylogenetically basic type of intramitochondrial structure (Powers et al., 1956).

Fernández-Morán (1962) applied the negative staining technique to disrupted mitochondria and discovered spherical structures approximately 85Å in diameter (named "elementary particles" or "inner membrane subunits") attached to the matrix face of the cristae in regular array (Fig. 4). In some electron micrographs the

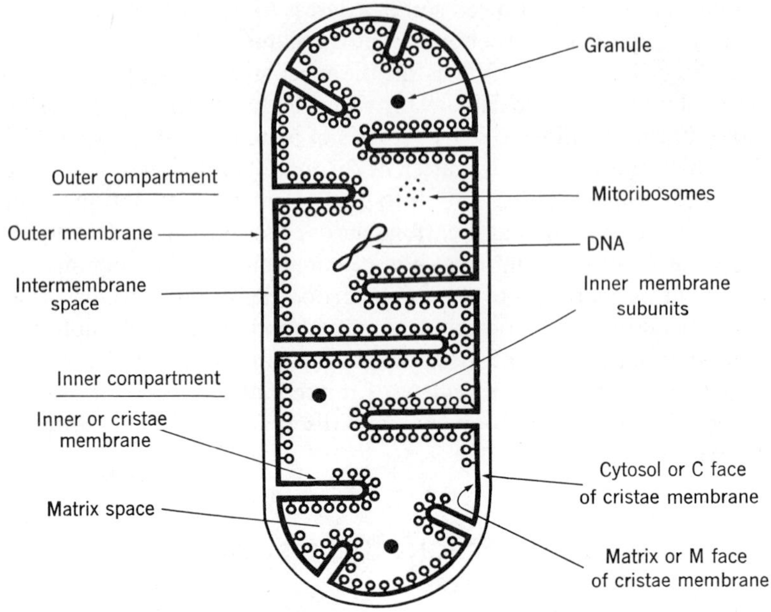

Figure 1. Composite diagram of mitochondrial fine structure observed by electron microscopy.

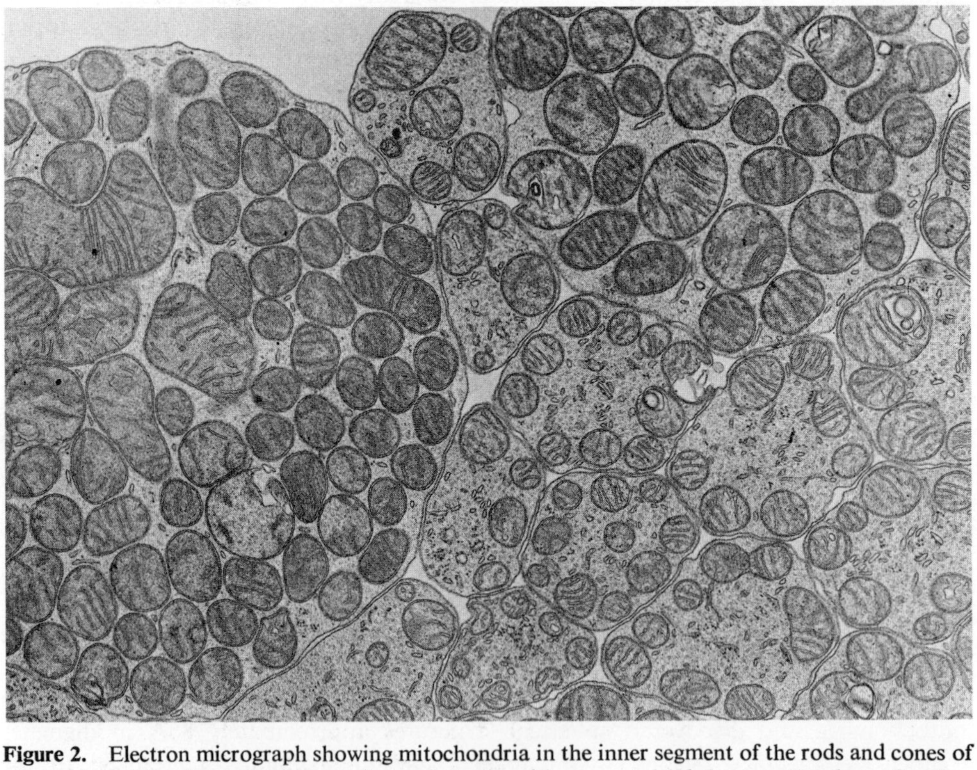

Figure 2. Electron micrograph showing mitochondria in the inner segment of the rods and cones of sheep retina. The cristae membranes of some mitochondria are not visible because the plane of section is parallel to the orientation of the cristae. ×10,900. (Courtesy of Dr. Alphonse Leure Dupree.)

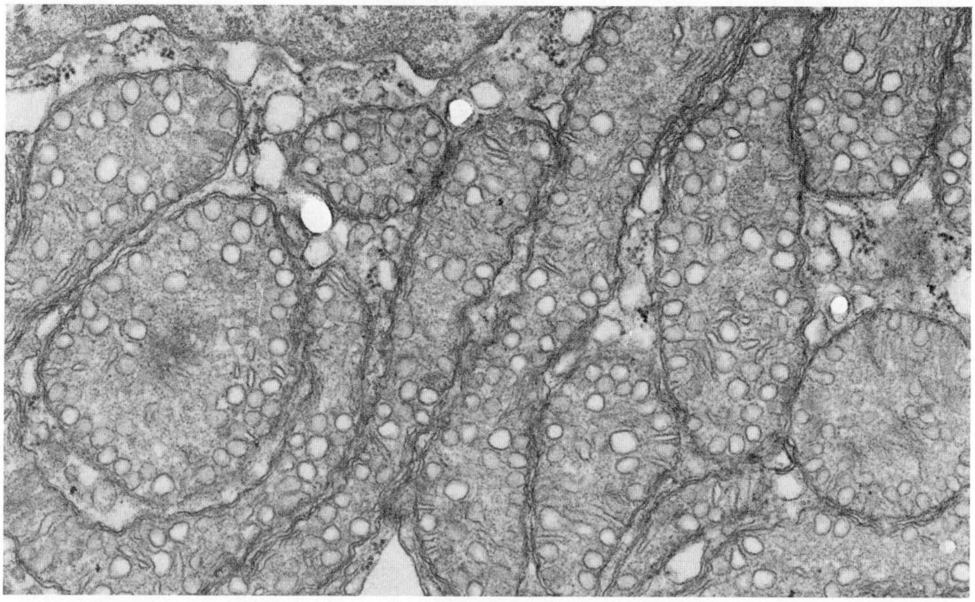

Figure 3. Electron micrograph of beef adrenal cortex mitochondria from a cell of the fasciculata layer. The micrograph shows the tubular cristae characteristic of mitochondria engaged in hydroxylation reactions (steroid synthesis). ×30,600. (Courtesy of Prof. A. Claude.)

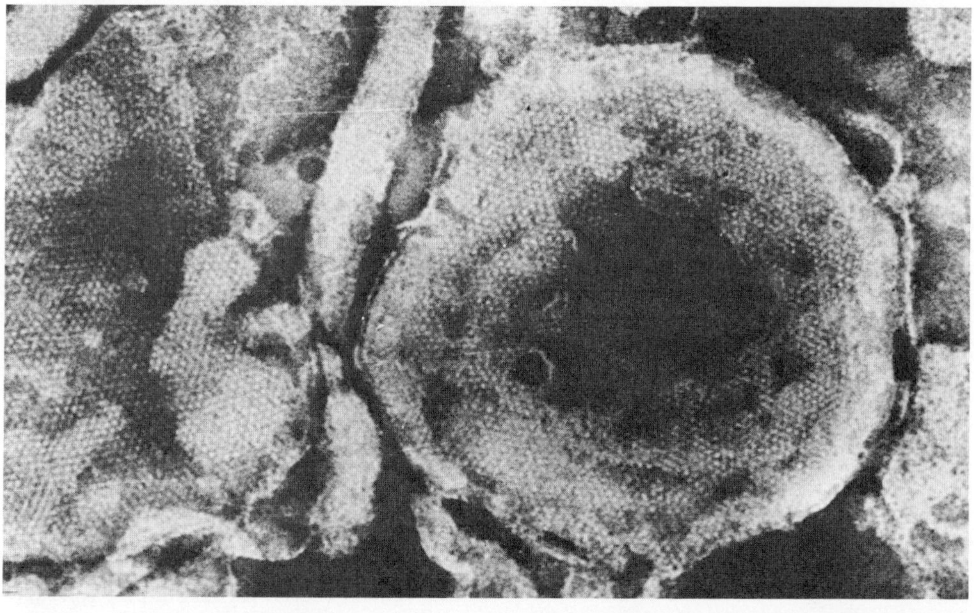

Figure 4. Electron micrograph of isolated rat liver mitochondria negatively stained with phosphotungstic acid. The micrograph shows a front view of the M face of the cristae membrane covered with inner membrane subunits and the thin outer membrane devoid of subunits. ×142,000. (Courtesy of Prof. A. Claude.)

subunits appear to be connected to the cristae membrane by a stalk. Whether this appearance is an artifact induced by the negative stain, or is attributable to the presence of a connecting structure between the subunits and the membrane under physiological conditions, remains to be established. It is uncertain whether or not the inner membrane subunits are attached to those portions of the inner membrane that run parallel to the outer membrane, but studies of mitochondria subjected to freeze-etching suggest that the units are present over the entire inner cristae surface of the mitochondrion (Wrigglesworth et al., 1970). Negative staining of submitochondrial particles ("sonic particles") isolated from mitochondria disrupted by sonic vibrations shows that inner membrane subunits are attached to the outer surface of the particles. This observation has led to the concept (Lee and Ernster, 1966) that the membrane of sonic particles is inside-out with respect to the inner membrane, so that the M face of these particles is presented to the suspending medium (Fig. 5).

The matrix often contains granules that are possibly related to the ability of mitochondria to accumulate ions. Granules are prominent in the mitochondria of cells concerned with the transport of ions and water, including kidney tubule cells, epithelial cells of the small intestine, and the osteoclasts of bone-forming cells. Mitochondrial granules occur more frequently in mouse duodenal cells when the salt

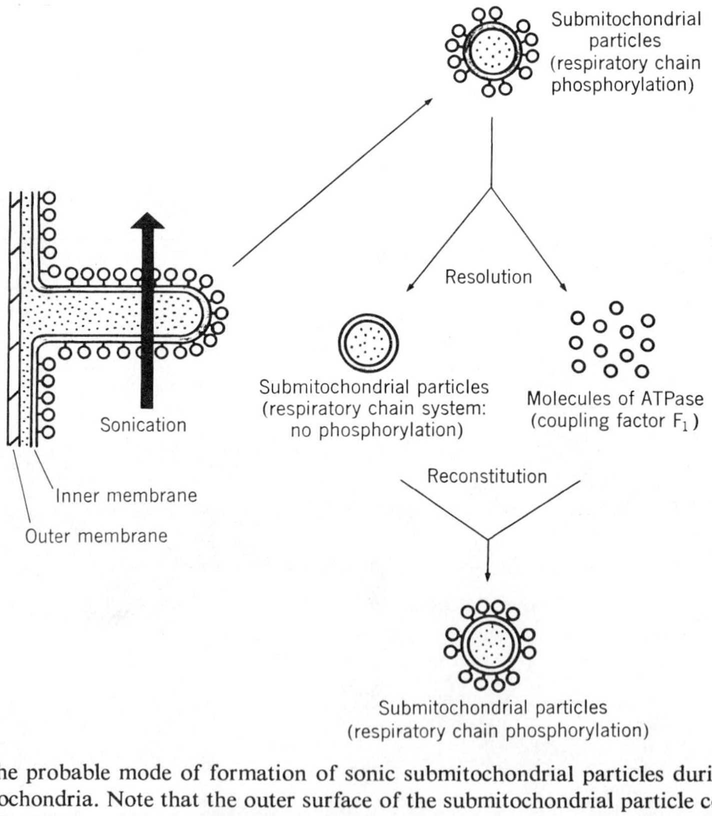

Figure 5. The probable mode of formation of sonic submitochondrial particles during sonication of whole mitochondria. Note that the outer surface of the submitochondrial particle corresponds to the inner surface of the cristae membrane of mitochondria, that is, the membrane is morphologically inverted by sonication. The figure also shows the resolution and reconstitution of phosphorylating respiration catalyzed by submitochondrial particles.

content of the diet is increased (Weiss, 1955). Electron microscopy of a wide variety of cell types has shown the presence of DNA (Nass et al., 1965) and ribosomes (see Rabinowitz and Swift, 1970) in the matrix space of mitochondria.

B. Fundamental Structure–Function Relationships

Palade (1956) noted that the surface area of cristae in various types of mitochondria is roughly proportional to respiratory activity. For example, heart muscle mitochondria have a much higher content of cristae material and of respiratory enzymes compared to liver mitochondria. He therefore concluded that the cristae were the sites of the respiratory enzymes and represented a method for increasing the capacity of mitochondria for the respiratory chain system. Lehninger (1964) estimated that the cristae surface area in the liver cell is about 10 times that of the cell membrane. The conclusion of Palade was supported by the observations of Watson and Siekevitz (1956). They found that the respiratory activity of suspensions of mitochondria disrupted completely by deoxycholate was recovered in small membrane fragments. By layering deoxycholate over a mitochondrial pellet they observed that the cristae membranes of intact mitochondria seen at the bottom of the pellet, where deoxycholate failed to penetrate, became progressively more disrupted by deoxycholate at higher levels in the pellet. Only small membrane fragments were present at the surface of the pellet where the mitochondria had been exposed completely to deoxycholate.

Sonic submitochondrial particles retain the ability to catalyze the coupled oxidation of reduced nicotinamide adenine dinucleotide ($NADH_2$) and succinate. Racker and his colleagues (Racker et al., 1965; Kagawa and Racker, 1966) achieved the resolution and recombination of the inner membrane subunits of sonic particles and thus identified the subunits as molecules of the reversible ATPase enzyme of mitochondria (coupling factor F_1) responsible for catalyzing the terminal reaction step of ATP synthesis (Fig. 5). Since the subunits are attached specifically to the cristae membrane, the presence of these subunits on the surface of submitochondrial particles, such as sonic particles, that contain all the enzymes of the respiratory chain system originally present in the parent mitochondria, established conclusively that the system is located in the cristae membrane. The suggestion of Slater and Cleland (1953) that the Keilin–Hartree heart muscle preparation (Keilin, 1929; Keilin and Hartree, 1940) consists of membrane fragments derived from heart mitochondria has been confirmed by electron microscope observation, which shows such preparations to consist of particles bearing typical inner membrane subunits (Slater, 1966a). The respiratory enzymes probably form a considerable part of the total cristae membrane protein. Estabrook and Holowinsky (1961) estimated that about 8.5% of the total protein of rat liver mitochondria consists of flavoproteins and cytochromes, which corresponds to about 17% of the total cristae protein. This percentage may be even higher in mitochondria particularly rich in respiratory enzymes.

3. MITOCHONDRIA IN THE CELL

A. Number and Distribution

Wide variations are found in the number of mitochondria per cell (Lehninger, 1964). In rat liver, the tissue used most frequently as a source of isolated mitochondria, the number ranges from 500 to 2500, with an average value of approximately 800. Sperm cells contain relatively few mitochondria, usually less than 100, and at the other extreme there may be as many as 500,000 mitochondria in the giant ameba cell *Chaos chaos*. Mitochondria often comprise a substantial proportion of the cellular material and volume. For example, rat liver mitochondria contain about one-fifth of the total cell protein and occupy about one-fifth of the cell volume. These proportions are even higher in cells containing large numbers of mitochondria. As much as one-third of the volume of insect flight muscle consists of mitochondria (Bücher, 1965).

Many features of the general distribution and arrangement of mitochondria in cells were observed with the light microscope (Cowdry, 1918). The electron microscope has permitted a detailed study to be made of the physical relationships between mitochondria and other subcellular structures and inclusions. Mitochondria are often found close to energy-requiring sites in cells, so that ATP synthesized by the mitochondria is made available through a very short diffusion path. This generalization is well illustrated by the close proximity of mitochondria to structures involved in the motor activities of cells. For example, mitochondria are found between and around the contractile fibers of skeletal muscle and packed profusely in close contact with the contractile fibers of continuously working heart muscle and of flight muscles that impose sudden and severe energy requirements on the cell. Mitochondria are situated close to the ciliated border of epithelial cells and certain protozoa. The arrangement of mitochondria is highly developed in mature spermatozoa, which contain mitochondria in the form of an end-to-end spiral around the midpiece, adjacent to the upper region of the sperm tail. In all these examples the mitochondria lie close to ATP-requiring structures.

The mitochondria of epithelial cells are often oriented with their long axes lying in the direction of the secretion or transport process (Cowdry, 1918). Electron microscope studies have revealed that the basal membrane area of certain epithelial cells active in ion and water transport is increased tremendously by folds that penetrate deep into the cytoplasm. Numerous mitochondria are lodged in narrow compartments formed by the folds, so that the mitochondria are brought into contact with a large area of the cell surface (Fawcett, 1966). This arrangement is seen particularly clearly in kidney tubule cells, the avian salt gland, and the secretory ducts of the salivary gland.

Mitochondria are numerous in the parietal or oxyntic cells of the stomach where they lie close to the extensive smooth membrane system involved in hydrochloric acid secretion, and near the synaptic junctions of axons where ion exchanges occur during transmission of the nerve impulse. In the exocrine cells of the pancreas, which synthesize digestive enzymes, and in liver cells that synthesize plasma proteins, mitochondria are found close to the sites of ATP-dependent protein synthesis in the endoplasmic reticulum (ER). Mitochondria are also abundant in the inner segment of the retina (Fig. 2) and in the oxyphil cells of the parathyroid gland, but the exact significance of these observations is unknown. Brown adipose tissue is well suited as a thermogenic source by virtue of its high concentration of mitochondria, in which the cristae are numerous and densely packed. The abundance of respiratory chain components in these mitochondria presumably contributes to the brown color of the tissue. The high metabolic potential of brown adipose tissue is enhanced by chronic cold exposure through induction of hyperplasia and an associated differential increase in mitochondrial material. Although initial studies suggested that brown fat mitochondria might be incapable of respiratory chain phosphorylation, it is now clear that they do have a functional system at least under conditions in which brown fat metabolism is not stimulated. The norepinephrine-induced stimulation of brown fat heat production may be attributable either to an uncoupling of respiratory chain phosphorylation by increased intracellular levels of free fatty acids, or to increased substrate availability and ATP usage (Smith and Horwitz, 1969).

B. Synthesis, Life, and Death of Mitochondria

Prior to 1965, cytological studies had indicated that mitochondria might originate either by the growth and division of preexisting mitochondria, by *de novo* synthesis, or by differentiation from other membrane structures of the cell. Conclusive proof for the first mechanism was obtained by Luck (1963) during experiments with exponentially growing *Neurospora crassa* cells. Choline-deficient cells were supplied with labeled choline, and the label was incorporated into the phosphatidylcholine of the mitochondrial membrane. When the cells were subsequently incubated in a nonlabeled growth medium, a random distribution of label was found over all the mitochondria and the average grain count per mitochondrion, determined by autoradiography, diminished as a function of growth. Luck therefore concluded that most or all of the mitochondria formed during exponential growth in the nonlabeled medium had arisen by growth and division of preexisting mitochondria. A similar conclusion was reached by Frédéric (1958), who observed that mitochondria from a parent fibroblast tissue culture cell were distributed equally to the daughter cells during cell division and later underwent lengthening and division. Unfortunately, there is little evidence to show whether or not a similar mechanism is operating in mammalian cells *in vivo*. However, the Poisson distribution of liver mitochondrial grain counts observed after injecting rats with tritiated thymidine suggests that this may indeed be the case (Droz and Bergeron, 1965).

The term "*de novo* formation" implies the synthesis of mitochondria exclusively from individual molecules. This type of formation was proposed, on the basis of electron microscope studies, to account for the appearance of typical mitochondria during respiratory adaptation in yeast (Wallace and Linnane, 1964). However, it was later shown that anaerobically grown yeast cells contain undifferentiated "promitochondria" which differentiate into fully developed mitochondria in the presence of oxygen (see Schatz, 1970). Promitochondria are readily observed in electron micrographs of freeze-etched anaerobic yeast cells (Plattner and Schatz, 1969). The cause of previous failures to detect promitochondria with osmic acid staining may be related to the profound alteration in the lipids of yeast cells induced by anaerobic growth.

Mitochondria contain several components required for protein synthesis, including DNA, DNA-dependent RNA polymerase, RNA, ribosomes ("mitoribosomes"), and amino acid-activating enzymes (Ashwell and Work, 1970). The presence of DNA in mitochondria has been shown by electron microscopy, by autoradiography of cells labeled with tritiated thymidine, and by the isolation from mitochondria of DNA molecules differing in buoyant density from nuclear DNA of the same species. The mitochondrial DNA of vertebrate and some nonvertebrate species exists in various simple and linked circular forms (Rabinowitz and Swift, 1970). Chick liver mitochondria contain about 1% of the total cell DNA. Estimates of the maximum amount of coding information that can be carried by the quantity of mitochondrial DNA present in cells indicate that a large proportion of the mitochondrial proteins must be coded by the nuclear genes and are probably synthesized on the ribosomes ("cytoribosomes") of the ER. This conclusion is supported by studies indicating that the synthesis of several specific mitochondrial proteins, including cytochrome c, malate dehydrogenase, and ATPase, occurs on the cytoribosomes.

Isolated mitochondria appear to be capable of protein synthesis because they incorporate added radioactive amino acids into mitochondrial protein. The labeled protein is similar to the hydrophobic material ("structural protein") isolated by Criddle et al. (1962). Part of this material has since been shown to consist of denatured ATPase, and it may also contain protein derived from the transporting systems (Section 4.E). Nevertheless, it is possible that part of the structural protein fraction consists of protein ("organizer protein") containing binding sites which determine the spatial relationships between enzymes in the mitochondrial membrane (Kaplan and Criddle, 1971). The cristae membrane can be resolved into several protein and phospholipid fractions, and many respiratory chain and phosphorylation activities can be reconstituted by suitable recombination of these fractions (Section 5). It has therefore been suggested that the functional cristae membrane is formed *in vivo* by the spontaneous association of catalytic components synthesized on cytoribosomes and that the arrangement of components to form a functional system may be dependent on the availability of organizer protein coded by mitochondrial DNA (see Ashwell and Work, 1970). However, this interpretation assumes that isolated mitochondria are capable of synthesizing at least one discrete inner membrane protein, and this assumption has yet to be proved. Since several mitochondrial enzymes have a structural role that can be clearly distinguished from their catalytic role (Racker, 1970b), it remains possible that the various enzymes and phospholipids of the membrane possess the property of self-assembly without the aid of an organizer protein.

Very little is known about the regulation of mitochondrial growth and division. Mitoribosome RNA may be of crucial importance in the reading of messages from the nucleus and thus in the synthesis of new mitochondrial material (Ashwell and Work, 1970). The synthesis of mitochondria in rat kidney cells is greatly stimulated by long-term feeding of the herbicide morfamquat (Ferguson et al., 1969). The study of this type of phenomenon may be of great value in future studies of the regulation of organelle biogenesis.

In the early isotopic experiments of Fletcher and Sanadi (1961) on the turnover and fate of mitochondria, the half-lives of three protein fractions and mixed lipid of liver mitochondria each had a value of approximately 10 days, suggesting that the various mitochondrial components form a unit with a fixed rate of turnover. The half-life of liver mitochondrial DNA was subsequently found to be very similar. Since the half-life of liver nuclear DNA may be 350 days or longer, it is probable that at least the major components of mitochondria are replaced many times during the lifetime of the liver cell. Similar conclusions apply in the case of kidney, heart, and brain cells (Gross et al., 1969). More recent values for the half-life of various mitochondria constituents, including individual enzymes and phospholipids, and inner and outer membrane proteins, have been tabulated by Gross et al. (1969) and by Ashwell and Work (1970). Although the absolute half-life values depend upon the method of measurement and factors such as the extent of reuse of isotope, it is clear that considerable differences exist between the rates of turnover of individual constituents, and that the proteins of the outer membrane and the ER turn over substantially faster than those of the inner membrane. It seems probable that although a major part of the inner compartment of the liver mitochondrion does turn over as a unit, some matrix components including alanine aminotransferase and δ-aminolevulinic acid synthetase, which have estimated half-lives of about 1 day and 70 min, respectively, can be inserted and removed independently of the synthesis of the major structural unit. Individual mitochondrial phospholipids also have widely varying half-lives and therefore do not turn over as a unit. The outer membrane may well be synthesized independently from the inner compartment and in a manner similar to ER synthesis. If this were true, it would help to explain the similarities between the outer mitochondrial membrane and the ER noted elsewhere (Section 5).

The ultimate fate of mitochondria is uncertain. Lysosomes in tissue sections prepared from several cell types, including liver, kidney, and pancreas, are sometimes found to contain mitochondria. It is possible that these lysosomes are in the process of digesting old or damaged mitochondria. This activity could account for the turnover of at least a major part of the mitochondrion as a unit, but the quantitative importance of lysosomal activity in mitochondrial "death" remains unknown.

C. Evolutionary Origin of Mitochondria

A comparison between the properties of mitochondria and bacteria has revealed many similarities, indicating that an evolutionary relationship may exist. Some of the most interesting comparisons are listed in Table 1. On this basis it has been suggested that mitochondria may have evolved from a symbiotic association between aerobic microorganisms and large anaerobic host cells. This might have conferred evolutionary advantage by enabling these cells to make use of the more efficient aerobic metabolism. Such cells would have contained larger amounts of respiratory enzymes and shorter ATP diffusion paths than aerobic cells in which these enzymes were confined to the limited surface area of the cell membrane. It is possible that primitive mitochondria possessed a wider range of biochemical activities than contemporary mitochondria, and that the genetic apparatus of mitochondria has become diminished progressively during evolution. Viewed in this way, the DNA of contemporary mitochondria might represent a residue of genetic information essential for the survival of the mitochondrion (Ashwell and Work, 1970).

4. METABOLIC ACTIVITY OF MITOCHONDRIA

A. Isolation of Mitochondria

The most fruitful approach to the study of mitochondria has been made possible by their isolation from tissues and subsequent investigation in the separated state. Attempts to isolate mitochondria in large numbers began in the 1930s. Progress in the isolation technique has been dependent on recognition that the sedimentation behavior and metabolic properties of isolated mitochondria are strongly influenced by the pH, tonicity, and chemical nature of the fluid in which they are suspended.

The mitochondria and other cell particles are first liberated by disrupting the cell membrane. This is usually achieved by exposing a suspension of finely chopped or minced tissue to the liquid shearing forces that occur during homogenization. The disruption of tough tissues and cells (e.g., muscle and yeast cells) is sometimes aided

Table 1. Some Comparisons Between Mammalian Mitochondria and Bacteria[a]

Mitochondria	Bacteria
Size of rod-shaped mitochondria: width 0.3–1.0 μ; length 1.0–4.0 μ	Size of typical rod-shaped bacteria: width 0.2–2.0 μ; length 0.3–10.0 μ
Respiratory chain phosphorylation system embedded in lipoprotein membrane 60–70 Å wide; total activity of system dependent on complexity of cristae; 90-Å subunits (ATPase) on inner side of cristae membrane	Respiratory chain phosphorylation system embedded in lipoprotein membrane 70–80 Å wide; total activity of system dependent on complexity of invaginations (mesosomes) in plasma membrane; 90-Å subunits on inner face of plasma membrane
ATP synthesis separated from cytosol by permeability barrier (cristae membrane)	ATP synthesis directly available to cytosol
Transporting systems in cristae membrane; mercurial-sensitive phosphate transporter	Transporting systems in plasma membrane; mercurial-sensitive phosphate transport
Marked respiratory control and high efficiency of ATP synthesis	Respiratory control low or absent; efficiency of ATP synthesis uncertain
Fairly constant composition of respiratory chain components; single terminal oxidase	Very variable composition of respiratory enzymes; often more than one terminal oxidase
Respiratory chain phosphorylation inhibited strongly by classic respiratory inhibitors (CN^-, CO, azide, antimycin) and by uncouplers (DNP) and oligomycin	Inhibitors and uncouplers of mitochondrial respiratory chain phosphorylation often less effective or ineffective
Closed circular configuration of mitochondrial DNA	Closed circular configuration of DNA often found in bacteria
Mitoribosomal protein synthesis inhibited by chloramphenicol but resistant to cycloheximide which blocks cytoribosomal protein synthesis in mammalian tissues	Protein synthesis inhibited by chloramphenicol but resistant to cycloheximide

[a] For further details, see Hughes et al. (1970); Ashwell and Work (1970); and Rabinowitz and Swift (1970).

by adding a suitable digestive enzyme to the homogenizing fluid. With some tissues, particularly plants, precautions must be taken to protect mitochondria against the harmful effects of certain cell constituents liberated during cell breakage. The resultant homogenate is mainly a suspension of cell particles together with cell debris and unbroken cells. The homogenate is centrifuged in tubes at a speed that throws down the heavier constituents such as nuclei, cells, and cell debris. The supernatant fluid is removed and spun at a speed sufficient to sediment mitochondria while leaving most smaller particles in suspension. By resuspending and recentrifuging the sediment several times, a preparation of washed mitochondria is obtained relatively free from other organelles. The chief contaminants of mitochondrial preparations are usually lysosomes, peroxisomes, nuclear fragments, microsomes, and bacteria. Their presence often complicates attempts to establish whether or not a particular enzyme activity is truly localized in the mitochondria.

The size and staining properties of isolated mitochondria are often similar to those of mitochondria in the cell, but they may possess an altered morphology as a result of isolation. For example, the common use of 0.25 *M* sucrose as homogenizing and washing medium yields isolated liver mitochondria that are spherical and not rod-shaped. However, these preparations still possess an intact membrane system. Many of the coordinated metabolic activities of mitochondria are situated in the inner compartment (Section 5) and are dependent on the integrity of the inner membrane. Thus the shape of isolated mitochondria appears not to affect their enzymic activities provided that the membranes remain intact. It is probable that the most significant alteration of mitochondria occurring during standard isolation procedures is the loss of materials from the intermembrane space (see Section 5).

B. Metabolism of Substrates

Studies since 1948 by Green, Lardy, Lehninger, Slater, and others (see Lehninger, 1964) have shown that isolated mitochondria invariably contain the complete multi-enzyme system and cofactors required to catalyze coupled oxidation by molecular oxygen of acetyl-CoA or intermediates of the citric acid cycle. These relatively simple compounds arise from preliminary steps in the metabolism of carbohydrate, proteins, and fats of the diet (Krebs and Kornberg, 1957; Fig. 6). Several citric acid cycle enzymes have a dual localization, appearing in both the mitochondria and the soluble fraction of the cell. However, α-oxoglutarate dehydrogenase and succinic dehydrogenase appear to be exclusively mitochondrial enzymes, indicating that a functional citric acid cycle is active only in mitochondria. Work in the laboratories of Lynen, Green, Lehninger, and others has shown that mitochondria from many tissues also contain enzyme systems that catalyze the coupled oxidation of fatty acids and ketone bodies (see Lynen, 1955). The oxidation of fatty acids has a "sparing" action on the oxidation of carbohydrates and amino acids, especially when carbohydrate is in short supply. It is estimated that long-chain fatty acid oxidation satisfies about half the energy requirement of man in the resting state (Fritz, 1961). Short-chain fatty acids and ketone bodies provide an important part of the fuel used by the tissues of ruminants, which receive a continual influx of these materials as products of rumen fermentation.

Some citric acid cycle intermediates lie on pathways for the interconversion of carbohydrates, fats, and proteins, and therefore the activity of mitochondrial enzymes can contribute to important biosynthetic processes. Some reactions involved in these interconversions are shown in Figure 6 (see also Fig. 14). Many of the carbon skeletons of the nonessential amino acids can be produced from carbohydrate via the citric acid cycle and transamination reactions. In liver and kidney mitochondria, amino acids are converted into cycle intermediates which are then readily converted to glucose and glycogen via enzymic pathways outside the mitochondria. The key enzymes involved in amino acid metabolism are glutamate dehydrogenase, which is localized exclusively in mitochondria, and glutamate-linked aminotransferases. Together they catalyze the deamination of amino acids:

$$\alpha\text{-Amino acids} + \alpha\text{-Oxoglutarate} \rightleftharpoons \alpha\text{-Oxo acid} + \text{Glutamate} \quad (1)$$

$$\text{Glutamate} + \text{NADP} \rightleftharpoons \alpha\text{-Oxoglutarate} + \text{NADPH}_2 + \text{NH}_3 \quad (2)$$

Isolated glutamate dehydrogenase is active with either NAD or NADP as hydrogen acceptor, but it is probable that in mitochondria the reaction is NADP-specific

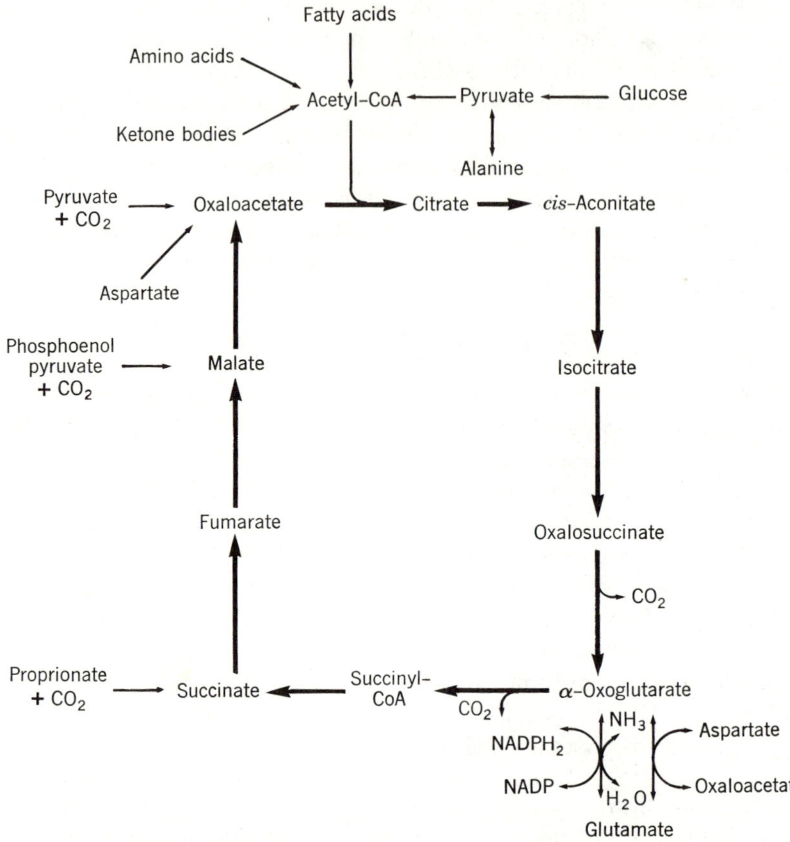

Figure 6. Simplified scheme of the citric acid cycle and associated reactions in liver motochondria. There appear to be species differences in the subcellular distribution of phosphopyruvate carboxykinase, which is predominantly mitochondrial in rabbit and guinea pig liver and almost entirely cytoplasmic in rat, mouse, and hamster. (Nordlie and Lardy, 1963.)

(Tager, 1966). Since reactions 1 and 2 are both reversible, they can also contribute to biosynthetic activities. A second pathway for the conversion of glutamate to α-oxoglutarate is provided by a transamination reaction:

$$\text{Glutamate} + \text{Oxaloacetate} \rightleftharpoons \alpha\text{-Oxoglutarate} + \text{Aspartate} \qquad (3)$$

The aspartate formed may be used for protein or urea synthesis. Chappell (1968) suggested that reaction 3 is part of the mechanism for cytoplasmic $NADH_2$ oxidation by mitochondria (Section 4.E). Acetyl-CoA, arising from the oxidative decarboxylation of pyruvate, provides two-carbon fragments for the mitochondrial enzyme system responsible for the elongation of existing fatty acids and for the extramitochondrial system that catalyzes the synthesis of fatty acids (Hülsmann et al., 1966). Mitochondrial reactions responsible for the fixation of CO_2 to form cycle intermediates are important in gluconeogenesis. For example, the pyruvate carboxylase reaction is a key regulatory step in carbohydrate formation from pyruvate (Greville, 1966). Proprionate enters the gluconeogenic pathway via the cycle after conversion to succinyl-CoA. This reaction sequence provides a major source of glucose in ruminants, using proprionate obtained as a principal end product of rumen fermentation.

The regulation of mitochondrial metabolism is a very complex subject. In addition to regulation imposed by the permeability properties of the mitochondrial membrane (Section 4.E) and the regulation of respiration by the availability of ADP and phosphate (Section 4.C), there are many other factors to be considered (Lehninger, 1964). For example, the citric acid cycle activity of isolated mitochondria can be influenced by feedback inhibition, enzyme activation by precursors and other metabolites, mass action effects at reversible steps, removal of intermediates, and competition for cofactors and hydrogen acceptors (Greville, 1966). Superimposed on these regulatory mechanisms are the effects of hormones, notably the iodinated thyroid hormones which regulate basal oxygen consumption of the tissues. Physiological amounts of these hormones influence the rate of tissue respiration by promoting the synthesis of respiratory enzymes. Respiratory chain phosphorylation in mitochondria isolated from thyrotoxic animals is partially or totally uncoupled, and a similar effect may occur in the tissues when excessive amounts of hormone are secreted by the thyroid (Tata, 1964).

The metabolism of mitochondria may also be influenced by intracellular oxygen concentration, since the oxygen tension gradient between the blood and tissues of animals can be very high, and therefore some regions of cells may exist close to the anaerobic state. According to Warburg (1956), the development of cancer cells in tissues is caused by an impairment of respiration, as a result of oxygen starvation or for other reasons, and a consequent replacement of the aerobic energy production typical of normal body cells by the anaerobic energy production on which cancer cells depend. The ability of cytochrome oxidase to maintain catalytic activity even at very low oxygen tensions (half-maximal activity is observed at an oxygen tension of about 0.1 mm Hg) is evidently of crucial importance to the many functions supported by cell respiration and possibly also to the protection of the cell against the onset of pathological change.

C. Respiratory Chain Phosphorylation

Studies of oxidizing mechanisms in animal tissues during the period 1885–1920 led to two rival theories of tissue respiration. The theory of Wieland and Thunberg emphasized the role of dehydrogenases in the activation of hydrogen atoms in substrate molecules. However, according to Warburg, the essential feature of tissue respiration was the activation of oxygen by an iron-containing catalyst, the "respiratory ferment" (see Keilin, 1966). These ideas were unified in the concept of Keilin (1925, 1929) which stated that the majority of cellular respiration was attributable to the activity of a respiratory chain system of dehydrogenases and cytochromes, and by evidence (Keilin and Hartree, 1939) that the Warburg "respiratory ferment" was identical with cytochrome oxidase (cytochrome a_3), the terminal component of the cytochrome system. Between 1934 and 1937 it became clear, chiefly from the work of Warburg, that many substrates require in addition the nicotinamide adenine dinucleotide coenzymes (NAD, NADP) and flavoprotein for their oxidation. Ball (1938) measured the oxidation–reduction potentials of cytochromes a, b, and c and obtained values that placed respiratory chain components in the sequence:

$$\text{Substrate} \longrightarrow \text{b} \longrightarrow \text{c} \longrightarrow \text{a} \longrightarrow \text{Oxygen}$$

A more detailed sequence was proposed by Keilin and Hartree (1940) following their extensive studies on the effects of urethan and of cytochrome c depletion and re-addition on the oxidase activities of a heart muscle particle preparation, and their identification of cytochrome oxidase (cytochrome a_3). Their results indicated the scheme:

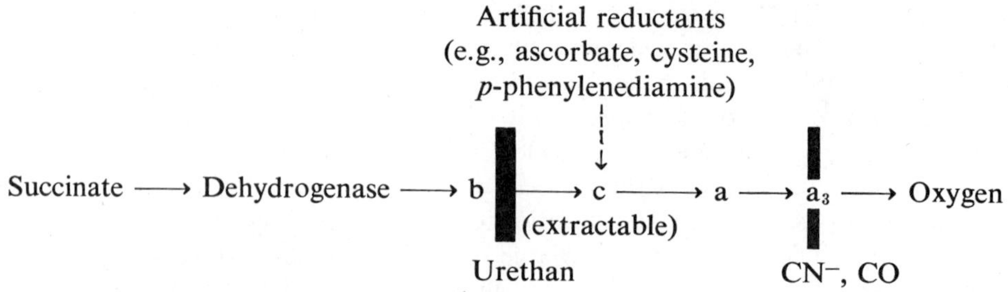

The presence of cytochrome c_1 between cytochromes b and c was established by Yakashiji and Okunuki (1940). Chance and his colleagues later made fundamental contributions to the development of rapid and sensitive spectrophotometric measurements of the kinetics of oxidoreduction reactions in the respiratory chain. With the aid of these techniques, Chance examined the time sequence of reactions between cytochrome components (Chance and Williams, 1955a; Chance et al., 1965). These studies confirmed the cytochrome reaction sequence proposed by Keilin and Hartree and showed that the reactivity of each cytochrome component of the phosphorylating respiratory chain of rat liver mitochondria was sufficiently rapid to function in the overall oxidase activity of the preparation.

Chance and Williams (1956) found that the molecular amounts of each type of cytochrome present in mitochondria are approximately equal. This feature, and the ordered distribution of ATPase molecules (Fig. 4), suggests that the respiratory enzymes are organized into respiratory assemblies that contain stoichiometric amounts of the molecules necessary for the coupled oxidation of cell metabolites. This arrangement would greatly increase the efficiency of respiratory metabolism by minimizing the distance that each respiratory chain component would have to move or rotate before reacting with the next member of the chain. Estabrook and Holowinsky (1961) estimated that about 17,000 respiratory assemblies are present in an "average" liver mitochondrion. Many of the soluble enzymes of the citric acid cycle, and at least one of the transporter systems, the adenine nucleotide transporter (Section 4.E), also appear to be present in constant proportions (Klingenberg and Pette, 1962; Klingenberg, 1970a), suggesting that all these components are synthesized in specific molecular ratios. Certain mitochondria also contain significant amounts of other respiratory chain-linked flavoproteins, including α-glycerophosphate dehydrogenase (liver, kidney, brain, sperm, skeletal muscle, and insect flight muscle), and choline dehydrogenase (liver and kidney). Adrenal cortex mitochondria are unusual in containing relatively large amounts of cytochrome P_{450} in addition to the usual respiratory chain cytochrome components. Cytochrome P_{450}, which is functional in many microsomal mixed function oxidations, acts as a catalyst in certain steroid hydroxylation reactions occurring in adrenal cortex mitochondria (Cammer and Estabrook, 1967).

A reaction scheme of the respiratory chain, presented in a generally accepted form, is shown in Figure 7. Although there appears to be good evidence that coenzyme Q (ubiquinone) is a member of the respiratory chain (Szarkowska, 1966), there is still some disagreement concerning the role of coenzyme Q in mitochondria because under certain conditions the major part of the quinone pool does not seem to undergo rapid enough oxidoreduction to function in the main oxidation pathway (F. L. Crane and Löw, 1966; Chance et al., 1967b). Hence the exact role of this component remains the subject of much discussion (Klingenberg and Kröger, 1967; Ernster et al., 1969; Racker, 1970b). The question is of particular interest in relation to the chemiosmotic hypothesis which postulates the presence of quinone or other hydrogen carriers in the NADH-cytochrome b and cytochrome b–cytochrome c regions of the respiratory chain (Fig. 8). Sweetman and Griffiths (1970) postulated a specific role for coenzyme Q in the energy-linked transhydrogenase reaction.

Recent work has shown that certain respiratory chain components exhibit a distinct "sidedness" in their catalytic activity (Racker, 1970c; Klingenberg and Von Jagow, 1970). The respiratory chain system therefore appears to be organized asymmetrically in the cristae membrane, as originally proposed by Mitchell (1961). A tentative scheme for the respiratory chain, showing the spatial arrangement of the components, is presented in Figure 8.

Since the early studies of Green, Lardy, Ochoa, and others (see Slater, 1966b), nearly all investigators have found that a maximum of three molecules of ATP are synthesized per atom of oxygen consumed during the oxidation of NAD-linked substrates by the respiratory chain, that is, the P/O ratio

$$\frac{\text{Molecules of phosphate esterified}}{\text{Atoms of oxygen consumed}}$$

(Belitzer and Tzibakova, 1939) is 3.0. The oxidation of α-oxoglutarate, which includes a substrate phosphorylation reaction (Hunter, 1949), gives a value approaching 4.0, while the oxidation of succinate yields a value up to 2.0. However, some significantly higher P/O ratios have been reported (Lynn and Brown, 1965; Sarkissian and Srivastava, 1970).

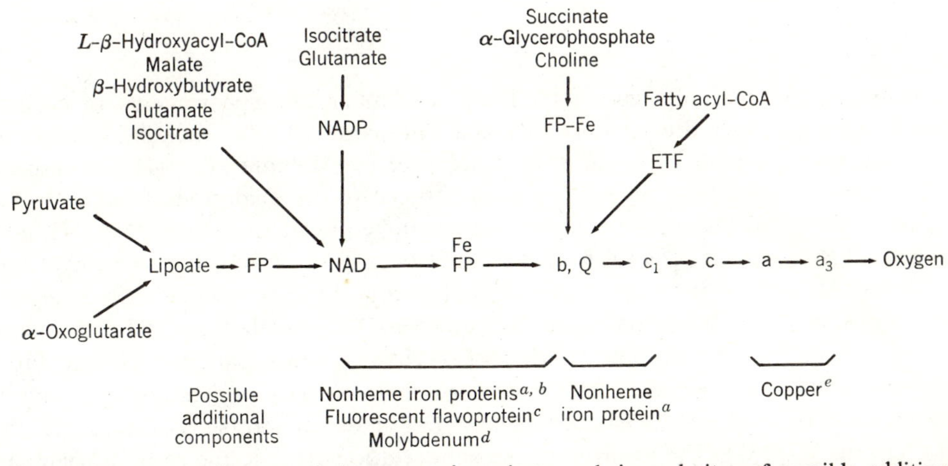

Figure 7. Schematic diagram of the accepted respiratory chain and sites of possible additional components. ETF, Electron transferring flavoprotein. References: (*a*) Beinert and Lee (1961); (*b*) Hatefi (1968); (*c*) Chance et al. (1967a); (*d*) Albracht and Slater (1970); (*e*) Beinert and Palmer (1965).

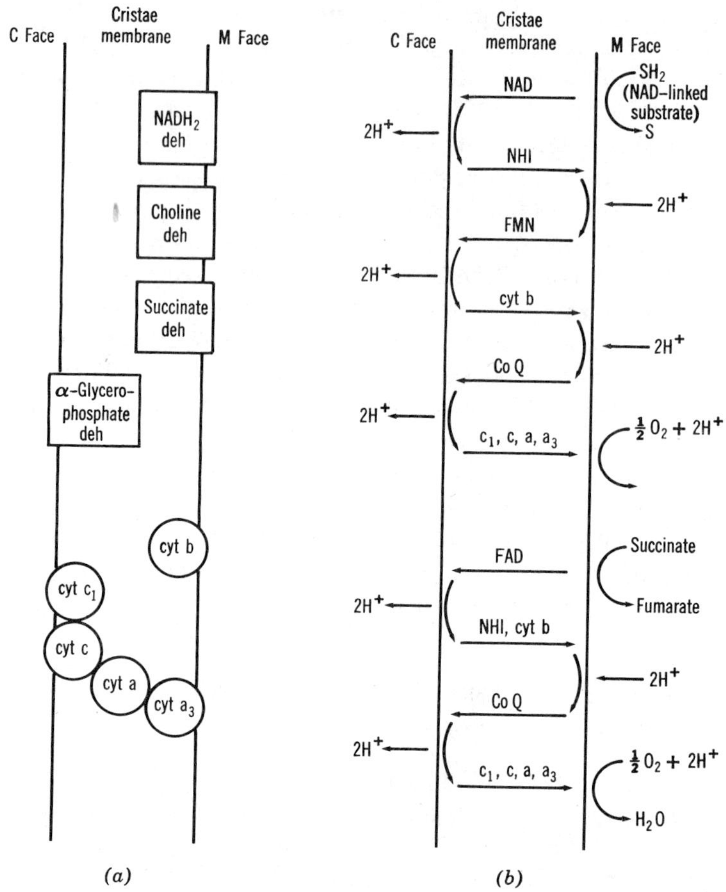

Figure 8. Spatial arrangement of respiratory chain components in mammalian mitochondria. (*a*) Probable location of dehydrogenases (deh) and cytochromes (cyt). (*b*) Scheme for a proton-translocating respiratory chain system similar to that postulated by the chemiosmotic hypothesis. NHI, Nonhaeme iron protein.

Lehninger and his colleagues introduced a number of improvements in experimental technique, including the use of washed preparations of mitochondria and the use of adenine nucleotides as phosphate acceptor (see Lehninger, 1964). Lehninger (1951) showed that all the phosphorylation coupled to the oxidation of NAD-linked substrates occurred during the transfer of reducing equivalents between $NADH_2$ and oxygen. This process is referred to as oxidative or respiratory chain phosphorylation to distinguish it from the ATP synthesis that occurs during the anaerobic or glycolytic conversion of glucose to lactate. As much as 95% of the total ATP synthesis during glucose metabolism, and all the ATP synthesis during amino acid and fatty acid metabolism, is dependent on respiratory chain activity, indicating that energy conservation is far more efficient aerobically than under anaerobic conditions (Krebs and Kornberg, 1957). The location of phosphorylation steps in the respiratory chain (Fig. 9) was established mainly from thermodynamic calculations and by measurements of phosphorylation coupled to isolated portions of the chain (Slater, 1966b).

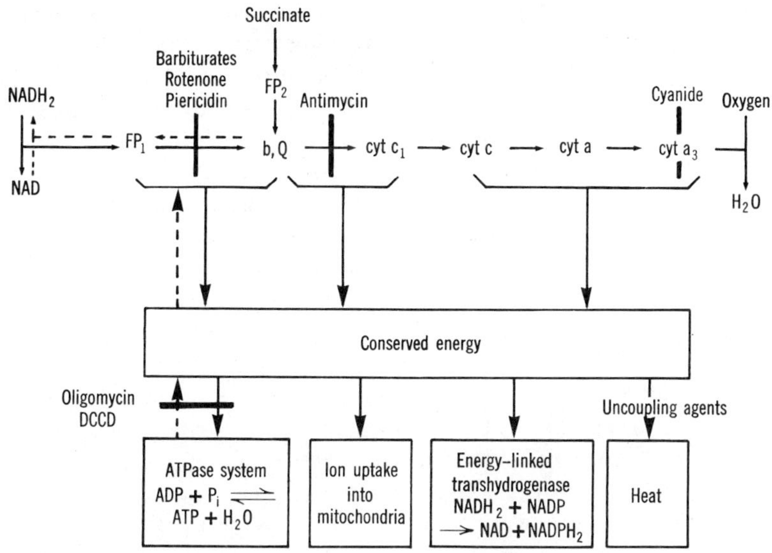

Figure 9. Schematic diagram of the phosphorylating respiratory chain and the sites of action of some ATP synthesis inhibitors. FP_1, $NADH_2$ dehydrogenase; FP_2, succinic dehydrogenase; cyt, cytochrome; P_i, inorganic phosphate; DCCD, dicyclohexylcarbodiimide.

The energy conserved during respiratory chain activity can be used not only for ATP synthesis but also in other energy-linked reactions including ion uptake (Section 4.D), reduction of NAD by succinate (Chance and Hollunger, 1961; Löw and Vallin, 1963), and energy-dependent transhydrogenase activity (Klingenberg and Slenczka, 1959; Danielson and Ernster, 1963). It appears that the use of the energy-linked reduction of NAD for reductive biosyntheses can play only a limited role in animal tissues, although it may be very important in autotrophic microorganisms (Krebs, 1967). The energy-linked transhydrogenase activity in mitochondria ensures the provision of $NADPH_2$ for reductive biosyntheses even under conditions favoring substrate oxidation by the respiratory chain.

The rate of oxidation of added substrates by suspensions of isolated mitochondria is stimulated strongly by the addition of adenosine diphosphate (ADP) and inorganic phosphate, and continues rapidly until either ADP or phosphate has been removed by conversion to ATP through phosphorylation reactions (Lardy and Wellman, 1952; Chance and Williams, 1956). These observations suggest that in the cell the rate of oxidation of foodstuffs is critically dependent on the supply of ADP and phosphate which are released only when the energy of ATP is expended. Hence the consumption of ATP to energize cell metabolism promotes coupled substrate oxidation until ATP is resynthesized. This phenomenon, termed "respiratory control," is believed to be a fundamental mechanism for the continuous and automatic regulation of oxidative metabolism to the energy requirement of the cell, and to underlie the marked increase in the respiration rate of intact tissues that occurs during the transition from the resting state to the active state of metabolism (Lardy, 1956). The resting and active states of mitochondrial respiration have been termed state 4 and state 3, respectively (Chance and Williams, 1956).

Reagents such as 2,4-dinitrophenol (DNP) uncouple the linkage between oxidation and energy-linked reactions, including phosphorylation, and abolish respiratory control in isolated mitochondria. Under these conditions rapid uncontrolled oxidation of substrates occurs, and the energy of oxidation is dissipated as heat instead of being conserved as useful chemical energy (Poe and Estabrook, 1968). A similar uncoupling action is believed to take place in the tissues of animals treated with DNP and to cause the increase in oxygen consumption and body temperature found in DNP poisoning (Slater, 1966b). Uncoupling agents have often been used to test whether or not individual physiological functions are dependent on energy metabolism. The nature of the "conserved energy" formed during respiratory chain activity (Fig. 9) is discussed in Section 6.

One important general factor that influences the respiration rate of cells through the respiratory control mechanism is the activity of a ouabain-sensitive Na^+- and K^+-dependent ATPase ("transport ATPase") present in the plasma membrane of many different tissues (Fig. 10). This enzyme acts as an ion pump responsible for maintaining the asymmetric distribution of Na^+ and K^+ between the intracellular and extracellular spaces of cells (Skou, 1965). When the enzyme present in slice preparations from various tissues, including brain, liver, kidney, and thyroid, is blocked by ouabain, the respiration rate of the preparations is inhibited by about 40%. These results suggest that a considerable part of the ATP synthesized by mitochondria in cells is consumed by the transport ATPase (Whittam and Blond, 1964). By energizing Na^+ and K^+ movements across the plasma membrane, mitochondria may also energize other transport processes, as the activity of the transport ATPase appears to be specifically

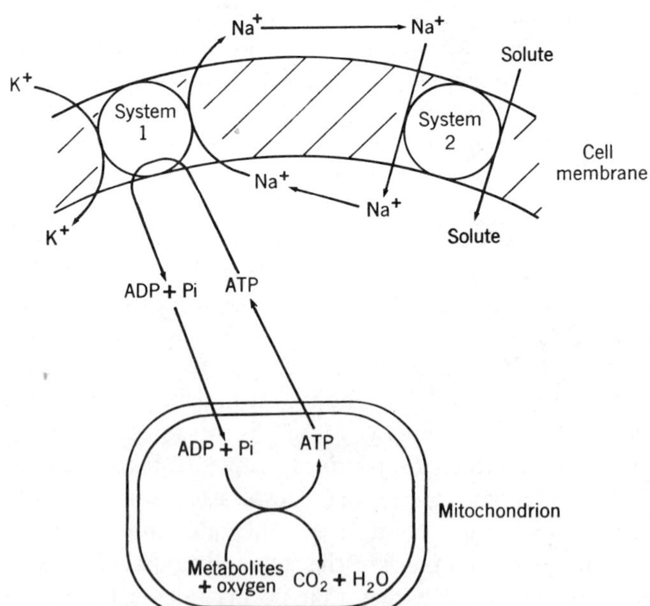

Figure 10. Simplified scheme showing the linkage between respiratory chain phosphorylation, transport ATPase activity, and Na^+-dependent solute uptake. System 1, Na^+/K^+ transport ATPase; system 2, Na^+-dependent solute transporter.

linked to the uptake of a variety of solutes by tissues (R. K. Crane, 1965). A simplified scheme for this process (Na^+–solute symport) is included in Figure 10. The Crane hypothesis was originally proposed to account for the Na^+-linked energy-dependent uptake of glucose by intestinal cells. Other solutes that may be transported in a similar manner into a variety of tissues include amino acids, sulfate, phosphate, iodide, ascorbic acid, bile salts, thiamine, and uracil (Schultz and Curran, 1970). The energy inherent in the sodium concentration gradient can thus be used via the coupled transport systems to bring about active transport of solutes without the requirement for separate and direct coupling of each of these systems to ATP.

A large variety of chemical compounds, when added to suspensions of isolated mitochondria, have been found to inhibit the reactions of oxidative phosphorylation. Particular attention is given to the effects of compounds that appear to have a specific site of action. Citric acid cycle activity is suppressed by fluoroacetate as a result of the lethal synthesis of fluorocitrate which blocks citrate metabolism; by arsenicals which combine with the disulfide group of lipoic acid, a cofactor of pyruvate and α-oxoglutarate metabolism; and by malonate, a competitive inhibitor of succinic dehydrogenase. Reagents that abolish the coupled oxidation of reduced flavoproteins by the respiratory chain include uncoupling agents, inhibitors of the respiratory chain, and inhibitors of the ATPase system. The scheme in Figure 9 shows the sites of action of some of these reagents. The similarities between the action of cyanide on cell respiration and on the respiratory activity of the cytochrome system in isolated cell particles provided early important evidence in favor of a major role of the cytochromes in cell respiration (Keilin, 1925).

The antibiotic oligomycin, organotin compounds, and dicyclohexylcarbodiimide (DCCD) all inactivate ATP synthesis without influencing the ability of mitochondria to catalyze energy-linked ion uptake reactions. Their effects indicate that respiratory energy can be used directly for ion uptake by mitochondria without having first to be converted into ATP.

In addition to the inhibitors mentioned above, a new class of inhibitors has been discovered recently that block oxidative phosphorylation by inactivating transporter systems that catalyze the transport of metabolites across the inner mitochondrial membrane (Section 4.E).

D. Ion and Water Uptake

It was long suspected that mitochondria within the cell possessed the ability to concentrate a variety of substances, a view expressed in the term "eclectosome" applied to mitochondria by Regaud (1911). Recent observations on isolated mitochondria indicate that they are capable of accumulating ions from the suspending medium, including several major ionic constituents of cells (e.g., K^+, Na^+, Ca^{2+}, Mg^{2+}, and phosphate) in a process that is abolished by inhibitors of respiratory chain phosphorylation and which therefore depends on respiratory energy. The ability of mitochondria to segregate ions and to energize the transport of ions and other solutes across the cell membrane, and the distinctive arrangement of mitochondria in cells engaged in the transport or secretion of fluids, all support the idea that mitochondria participate in the regulation of the ionic composition of cells and in the transport of solutes and water across them.

Mitochondria isolated from a wide variety of vertebrate tissues are capable of accumulating Ca^{2+} very efficiently from a suspending medium that also contains phosphate ions (Carafoli and Lehninger, 1971). The Ca^{2+}/O ratio is approximately 6.0 during NAD-linked oxidation and approximately 4.0 when succinate is the substrate, indicating that two Ca^{2+} are taken up at each coupling site spanned by an oxidoreduction reaction in the respiratory chain. When a mixture of Ca^{2+} and ADP at physiological concentrations is added to mitochondria in the presence of phosphate, the uptake of Ca^{2+} takes priority over ATP synthesis, suggesting that mitochondrial calcium transport may serve an extremely important function in the cell (Rossi and Lehninger, 1964). Calcium uptake is accompanied by a strong stimulation of state-4 respiration to a rate greater than the state-3 rate (Chance, 1963), and both calcium uptake and the stimulated phase of respiration continue until the concentration of Ca^{2+} in the suspending medium is less than 10^{-6} *M*. This value lies within the range of free Ca^{2+} concentration found in the cytoplasm of animal cells (10^{-5}–10^{-8} *M*). Calcium ions probably enter cells by passive diffusion down a concentration gradient and are pumped out of cells by an ATP-dependent calcium pump, thus establishing a continuous Ca^{2+} exchange across cell membranes (Rasmussen, 1970). Calcium efflux from heart muscle and nerve appears to depend specifically on external Na^+ (Reuter and Sietz, 1968), suggesting that calcium movements across cell membranes in these tissues may be energized by a sodium gradient (Na^+–Ca^{2+} antiport).

These observations indicate that mitochondria are probably involved in maintaining a low cytoplasmic Ca^{2+} concentration, both by segregating Ca^{2+} in the matrix space and by supplying ATP to energize a calcium pump. The existence of mechanisms such as these is essential for the effective response of cell structures and ion-sensitive regulatory enzymes to small changes in cytoplasmic Ca^{2+} concentration. The most widely accepted roles of such changes are in the excitation–relaxation cycle of muscle, and in excitation–secretion processes at nerve endings and in exocrine and endocrine glands (Lehninger, 1970a; Rasmussen, 1970). For example, muscle contraction is initiated when the intracellular free Ca^{2+} concentration rises above 10^{-7} *M* (Weber et al., 1964). In addition, low concentrations of Ca^{2+} also have marked effects on the activity of key enzymes in glycogenolysis [phosphorylase b kinase (Ozawa and Ebashi, 1967)], glycolysis [pyruvate kinase (Gevers and Krebs, 1966)], and gluconeogenesis [pyruvate carboxylase (Kimmich and Rasmussen, 1969)]. A change in the intracellular concentration or distribution of calcium ions may therefore play a role in regulating the activities of metabolic pathways in the cell. The activation of Ca^{2+}-sensitive processes in cells probably occurs in several ways, for example: by the release of Ca^{2+} from the sarcoplasmic reticulum, to initiate contraction in white muscle; by an increased entry of calcium into the cell, as in situations in which both Ca^{2+} and 3′,5′-cyclic AMP appear to act as second messengers in response to the hormonal excitation of cells (Rasmussen, 1970); and possibly also by the release of stored calcium from mitochondria. Little is known about the last-mentioned process, but a large fraction of the calcium taken up by mitochondria can be released rapidly by uncoupling agents, indicating that a pool of intramitochondrial calcium remains very mobile (Lehninger, 1970a). Adrenal corticoid hormone treatment decreases the ability of liver mitochondria to accumulate and retain Ca^{2+}, and the secretion of these hormones might therefore cause a redistribution of intracellular calcium (Rasmussen and Tenenhouse, 1968). The release of calcium from muscle mitochondria might be caused by a collapse of the electrochemical gradient across the mitochondrial membrane in response to a depolarization of the sarcoplasmic reticulum, with which mitochondria appear to make contact in some red muscles. These tissues contain abundant mitochondria, but the sarcoplasmic reticulum is poorly developed compared to that in white muscles. It is therefore possible that in red muscle the mitochondria perform a supplementary or even dominant role in calcium segregation during excitation–relaxation (Lehninger, 1970a).

Electron micrographs of isolated mitochondria loaded with Ca^{2+} and phosphate ions show them to contain a large number of electron-dense granules in the matrix space. Weinbach and Von Brand (1967) isolated these granules and found them to consist mainly of Ca^{2+}, Mg^{2+}, and phosphate. Similar calcium phosphate deposits probably occur in the mitochondrial matrix in a variety of tissues. These deposits may serve as essential precursors for calcification in cells, both under normal conditions, as in the formation of bone and egg shells and the hardening of stalks and stems of higher plants, and also under pathological conditions, as in the calcification of brain, liver, and kidney that can occur in response to injury (Lehninger, 1970a). Contrary to earlier suggestions, it is unlikely that experimentally induced calcium phosphate deposits arise from the endogenous granules present in liver mitochondria. The real nature and function of the last-mentioned granules remains unknown (Pasquali-Ronchetti et al., 1969).

Monovalent cation uptake by isolated mitochondria is usually a slow process compared to calcium uptake, but is stimulated strongly by cation-transporting antibiotics consisting of both cyclic peptides (e.g., valinomycin, enniatins) and open-chain peptides (e.g., gramicidins) (see Pressman, 1970). The first class, including valinomycin (K^+-, Rb^+-, and Cs^+-binding) and gramicidin (also H^+-, Li^+-, and Na^+-binding), are lipid-soluble compounds lacking ionizable basic or acidic groups. They increase the permeability of natural membranes (mitochondria, erythrocytes) or artificial lipid membranes to alkali metal ions by carrying these cations across as positively charged complexes. Provided that an uncoupling agent is present to conduct protons in the reverse direction, valinomycin catalyzes a rapid K^+/H^+ exchange across lipid membranes. The second class of antibiotics, typified by nigericin, possess an ionizable carboxyl group and probably cross lipid membranes in one direction as a neutral cation complex and return in the —COOH form. Nigericin can therefore catalyze K^+/H^+ or Na^+/H^+ exchange in the absence of an uncoupler. Blondin et al. (1971) have concluded that a very small quantity of a cyclic peptide, which acts as a monovalent cation transporter, is present in beef heart mitochondria.

The role of mitochondria in the energy-dependent uptake of ions and other solutes across the cell membrane is discussed in Section 4.C (Fig. 10). If the solute transporter and the energy-linked ion pump present in an epithelial cell are situated in the mucosal and serosal surfaces, respectively, then events similar to those depicted in Figure 10 could account for the transport of solutes and water across the cell. This arrangement appears to mediate the transport of alanine across the epithelium of isolated rabbit ileum (Chez et al., 1967; Fig. 11).

Diamond (1971) noted that virtually all epithelia engaged in fluid transport possess long and narrow dead-end channels, with various regions of the channels often in close contact with mitochondria. Examples of this arrangement include: the lateral intercellular spaces in gallbladder, intestine, and kidney proximal tubule; the intracellular infolding of the basal membrane in the kidney distal tubule, salivary gland ducts, and nasal salt gland of birds; and intracellular canaliculi involved in bile secretion from liver and hydrochloric acid secretion from parietal cells. Diamond and Bossert (1967) suggested that the function of these channels is to allow the operation

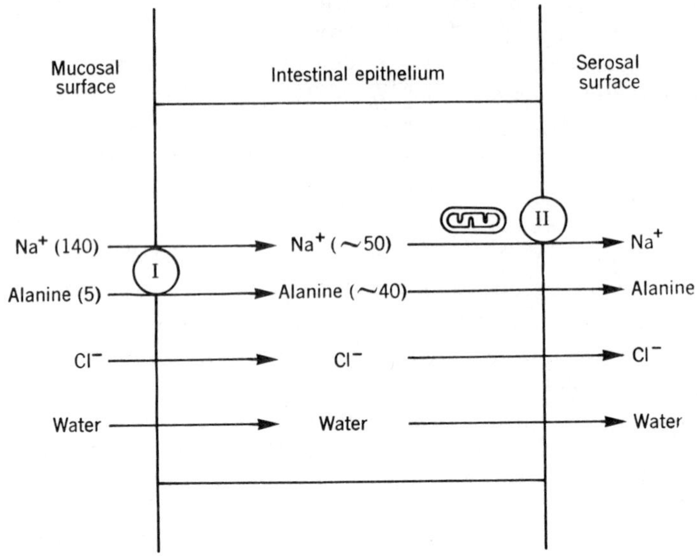

Figure 11. Alanine, salt, and water fluxes across the intestinal mucosa. The numbers indicate the millimolar concentration of alanine and sodium. I, Na^+-dependent alanine transporter; II, Na^+/K^+ transport ATPase. For further details, see Chez et al., (1967).

of a standing gradient osmotic flow induced by the presence of solute pumps in the channels (Fig. 12). The system consists of a long, narrow channel closed at one end. Solute is actively transported into the channel, making the channel fluid hypertonic. It is assumed that the channel contents remain effectively unstirred. As solute diffuses toward the open mouth, more water enters the channel passively as a result of the osmotic gradient. Fluid emerging from the open end may be either hypertonic or isotonic, depending upon a number of factors including the length, radius, and water permeability of the walls of the channel and the location of the solute pumps. The transported fluid is most nearly isotonic when solute transport is confined to the closed end of the channel, and most hypertonic when solute transport occurs close to the open end. The hypothesis is consistent with the observation that in biological systems the primary transported fluid may be either hypertonic or isotonic to plasma. Hypotonic secretions in vertebrates appear to result from isotonic secretion followed by hypertonic reabsorption. If the location of mitochondria indicate the location of solute pumps, then the presence of gallbladder mitochondria near the closed end of the channel would indicate that the solute pumps are sited most effectively to ensure the isosmotic transport of fluid observed to occur across the gallbladder epithelium. A more uniform distribution of mitochondria and a relatively short closed channel as in the kidney distal tubule and the salivary gland duct would favor the transport of hypertonic fluid occurring in these tissues. In epithelia such as gallbladder, intestine, and kidney, the channels open in the forward direction of fluid transport. In some epithelia, however, including those of the choroid plexus and the salt gland of birds, the channels open in the direction from which fluid is being taken up and transported. The operation of a standing-gradient flow system can also account for "backward" fluid transport, provided that the solute pumps operate in the reverse direction, making

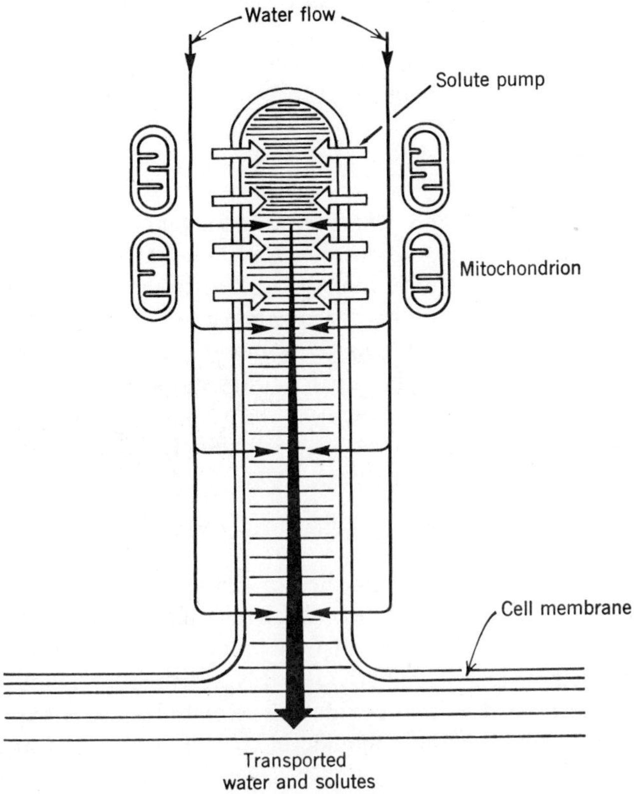

Figure 12. Diagram of a standing gradient osmotic flow system. The density of shaded lines indicates the solute concentration.

the channel fluid hypotonic, and provided also that major solutes of the bathing solution or plasma can be transported, thus causing water uptake (Diamond and Bossert, 1968).

The swelling of mitochondria, both in cells and in the isolated state, has been studied very extensively. Lehninger (1962) noted that the contractile systems of isolated mitochondria and myofibrils share some common properties and suggested that changes in mitochondrial volume might be a result of changes in the physical conformation of a mechanoprotein situated in the mitochondrial membrane and driven by respiratory energy. This view is consistent with evidence suggesting that an actomyosinlike protein can be extracted from liver mitochondria (Ohnishi and Ohnishi, 1962), but these observations appear to be of questionable significance (Vignais et al., 1964; Conover and Bárány, 1966). Hackenbrock (1966) and Green et al. (1968) observed cyclic changes in the morphology of mitochondrial membranes, dependent upon the metabolic state of the mitochondria. These investigators postulated that the conserved energy of oxidoreductions (Fig. 9) was a form of "conformational" energy, but it is doubtful that the changes occur sufficiently rapidly to reflect an intermediate state in oxidative phosphorylation (Chance et al., 1969).

Chappell and Crofts (1965) showed that an osmotic type of mitochondrial swelling occurs during the uptake of ions by isolated liver mitochondria. The energy-dependent uptake of cations was accompanied by anion entry and a consequent increase in osmotic pressure inside the mitochondria, causing water entry and mitochondrial swelling. Similar events within the cell may enable the mitochondria to act as temporary stores of ions and water during the transport of these materials across the cell.

E. Transporting Systems

Studies by Bartley, Tedeschi, and others (see Lehninger, 1962) of the space occupied by sucrose and other substances present in suspensions of isolated mitochondria, in relation to the total water space of mitochondria, led to the conclusion that the inner membrane possesses only a very limited permeability to most substances of molecular weight higher than 100 to 150, whereas the outer membrane is freely permeable to many substances of molecular weights up to approximately 10,000. This conclusion was supported by observations of "latency" in the activity of certain enzymes in intact mitochondria toward added substrates (Chappell and Greville, 1963) even when only small molecules were involved, as in the reaction

$$S_2O_3^{2-} + HCN \rightleftharpoons SO_3^{2-} + CNS^- + H^+$$

catalyzed by rhodanese. The latent enzymes of mitochondria are now known to be situated in the matrix space, and their activity is restrained by the low permeability of the inner membrane to their substrates. Virtually all mammalian mitochondria behave as if they are impermeable to external $NADH_2$ (Lehninger, 1964), indicating that other mechanisms must exist to allow the reoxidation of cytoplasmic $NADH_2$. Remarkable differences have been noted in the behavior of liver and flight muscle mitochondria toward added citric acid cycle intermediates. Several of these compounds are readily oxidized by liver mitochondria but are oxidized very slowly by flight muscle mitochondria until the inner membranes of the latter preparation are disrupted by sonication (Van Den Bergh and Slater, 1962). An extreme example of the low permeability of the inner membrane was shown by the behavior of mitochondria to protons. The backflow of protons into mitochondria after they have been translocated outward across the membrane has a half-life of approximately 90 sec (Mitchell and Moyle, 1967). The proton conductance of the mitochondrial membrane and of artificial membranes is greatly increased by uncoupling agents (Mitchell and Moyle, 1967; Skulachev et al., 1967), as predicted by the chemiosmotic hypothesis.

The development of methods for the study of metabolite transport in mitochondria (Chappell, 1968; Klingenberg, 1970b) has led to the discovery of several transporting systems that catalyze the movement of ions in either direction across the inner mitochondrial membrane (Table 2). The most detailed kinetic and inhibitor studies have been made on the adenine nucleotide transporter system (Klingenberg, 1970a). Lehninger (1970b) isolated a calcium-binding protein from rat liver mitochondria that appears to be closely related to, or identical with, the calcium transporter. The use of thiol-binding reagents, which inhibit the phosphate transporter (Fonyo, 1968; Tyler, 1968, 1969) has demonstrated the key importance of phosphate entry into mitochondria, not only for ATP synthesis and during cation transport but also for the entry of dicarboxylic and tricarboxylic acids (Chappell, 1969).

The movement of fatty acids across the inner membrane is also catalyzed by a transport system, involving carnitine (Greville and Tubbs, 1968; Klingenberg, 1970a). A possible reaction pathway for fatty acid oxidation by mitochondria, based on the

Table 2. Transporting Systems of Mitochondria[a]

Substance transported	Occurrence of transporter	Exchange reactions catalyzed	Inhibitors	Comments
Phosphate (arsenate)	Probably all mitochondria	Phosphate–hydroxyl (?)	Thiol-binding reagents, organic mercurials, *N*-ethylmaleimide (NEM)	—
ADP and ATP	Probably all mitochondria	ADP–ATP	Atractyloside, bongkrekic acid	Inactive with AMP
Tricarboxylates: citrate, isocitrate, *cis*-aconitate	Liver, kidney, white fat; low in heart, brain; absent from blowfly flight muscle	Tricarboxylate exchange; tricarboxylate–malate	2-Ethylcitrate[b]	—
Dicarboxylates: succinate, malate	Mammalian tissues; absent from blowfly flight muscle	Dicarboxylate exchange; dicarboxylate–phosphate; phosphate exchange	Butyl-malonate	Inactive with fumarate; organic mercurials inhibit phosphate movements; NEM has no effect[c]
Oxoglutarate	Many vertebrate tissues	Oxoglutarate–malate; oxoglutarate exchange (?)	Aspartate	Aspartate inhibits α-oxoglutarate entry but not efflux
Glutamate	Many vertebrate tissues	Glutamate–hydroxyl (?)	Avenaciolide	—
Aspartate	Liver, kidney, heart	—	—	Activated by glutamate
Calcium	Many vertebrate tissues; absent from some yeasts, plants, and insect flight muscles[d]	Calcium ion–proton	La^{3+}, Pr^{3+}	Also active with Mn^{2+}, Sr^{2+}; not active with Mg^{2+}

[a] For other details, see Chappell (1968, 1969).
[b] Robinson et al. (1970).
[c] Meijer et al. (1970).
[d] Carafoli and Lehninger (1971).

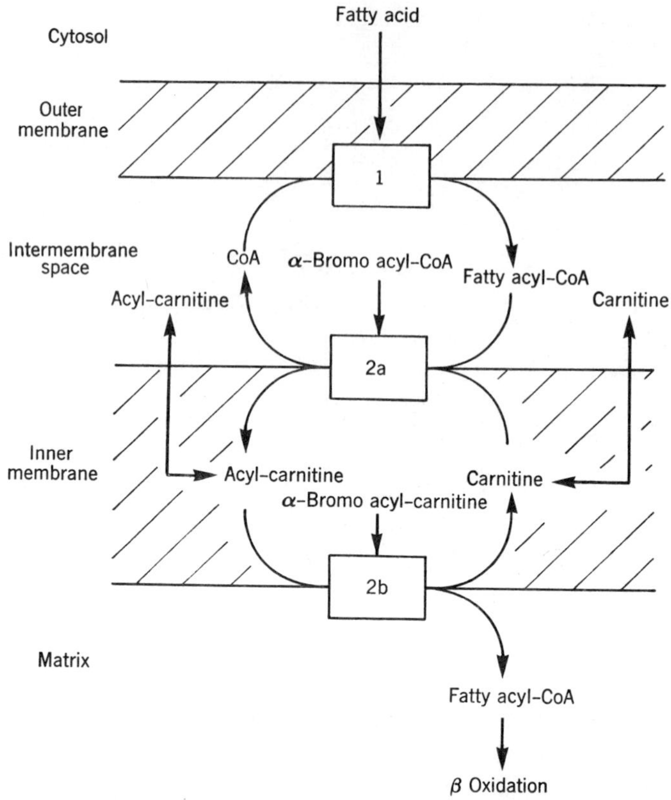

Figure 13. Fatty acid oxidation in mitochondria and the role of carnitine. 1, ATP-dependent fatty acyl-CoA synthetase; 2a, 2b, outer and inner pools of carnitine acryl transferase. The scheme accounts for the following features of fatty acid oxidation: (1) the intramitochondrial location of the enzymes concerned; (2) the effects of bromo analogs that inhibit fatty acid oxidation; (3) the failure of carnitine and acyl carnitine to penetrate the inner membrane; (4) the existence of two transferase pools as indicated by differential inhibition and differential extraction studies. For further details, see Tubbs and Garland (1968) and Klingenberg (1970a).

intramitochondrial localization of the enzymes involved, is shown in Figure 13. There is some evidence for the existence of a monovalent cation transporter or transporters (Mitchell and Moyle, 1969) and a pyruvate transporter (Papa et al., 1971) in liver mitochondria.

The physiological significance of the anion-transporting systems of mitochondria has been discussed by Chappell (1968, 1969). With flight muscle mitochondria, which appear to function solely as a site of ATP synthesis, the absence of transporting systems for citric acid cycle intermediates ensures the retention of intermediates in the matrix space, hence their rapid and efficient use in oxidative phosphorylation reactions. In contrast, liver mitochondria have a more complex metabolic function, since in addition to ATP synthesis they supply the cytoplasm with intermediates for a variety of biosynthetic processes, including lipogenesis and gluconeogenesis (Lardy et al., 1965). The role of transporting systems in these processes is shown in Figure 14.

Chappell (1968) has suggested that in many mammalian tissues the oxidation of cytoplasmic $NADH_2$ is catalyzed by the combined operation of the malate-, aspartate-,

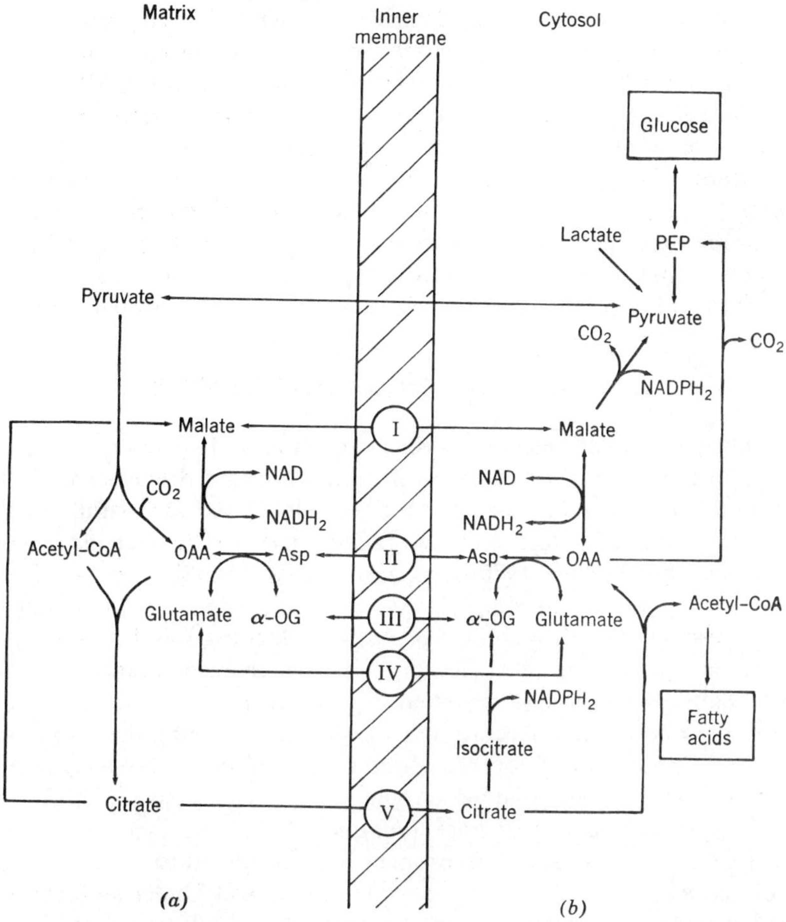

Figure 14. Role of the transporting systems of liver mitochondria in (*a*) the supply of metabolites for lipogenesis and gluconeogenesis, and (*b*) the oxidation of cytoplasmic $NADH_2$ by mitochondria. The exit of citrate from mitochondria permits the transfer of a two-carbon unit (acetyl-CoA) across the inner membrane for use in fatty acid synthesis. Isocitrate and malate from mitochondria act in the cytosol as reductants for NADP which can then be used in biosynthetic reactions. The exit of malate, aspartate, α-oxoglutarate and glutamate from mitochondria provides precursors for gluconeogenesis via their conversion to oxaloacetate and phosphoenolpyruvate. The entry of malate together with glutamate into mitochondria together with the exit of α-oxoglutarate and aspartate provides a catalytic mechanism for the transfer of reducing equivalents from cytoplasmic $NADH_2$ into mitochondria. I, Dicarboxylate transporter; II, aspartate transporter; III, α-oxoglutarate transporter; IV, glutamate transporter; V, tricarboxylate transporter.

glutamate-, and α-oxoglutarate-transporting systems, in which reducing equivalents enter mitochondria in the form of malate. In tissues such as flight muscle, in which the potential activity of this mechanism is low or absent, cytoplasmic $NADH_2$ is probably oxidized via the glycerol–phosphate shuttle in which reducing equivalents enter mitochondria in the form of α-glycerophosphate (Estabrook and Sacktor, 1958).

The presence of transporting systems in the mitochondrial inner membrane is an essential postulate of the chemiosmotic hypothesis. If the membrane were unspecifically permeable to anions and cations, their movement across the membrane would collapse the membrane potential used to drive ATP synthesis. The proposal (Mitchell, 1961) that transporting systems are present in mitochondria to regulate the electroneutral entry and exit of metabolites at first received little attention. Although Mitchell (1954) demonstrated the presence of a mercurial-sensitive phosphate transporter in bacteria, it was another 14 yr before the inhibition of respiratory chain phosphorylation in mitochondria by mercurials (Scott and Gamble, 1961) was shown by Fonyo (1968) and by Tyler (1968) to be attributable to a similar inhibition of phosphate transport. The inhibition of phosphate movements by protein-binding agents (Tyler, 1968) strongly suggests that the phosphate transporter of mitochondria is a protein, and the properties of the other transporters indicate that they are also probably proteins.

5. FRACTIONATION OF MITOCHONDRIA

Some fundamental relationships between mitochondrial structure and the respiratory chain phosphorylation reactions of mitochondria are discussed in Section 2.B. Although soluble multienzyme systems have been elucidated successfully by fractionation methods, comparable work on membrane-bound enzyme systems has been only moderately successful. Racker (1970b) has summarized the difficulties as being attributable to the strong interactions holding components together, the changes in the properties of resolved proteins, and the problem of devising suitable assay systems for individual components. The most notable attempts to resolve and reconstitute the respiratory chain phosphorylation system have been made by Green, Hatefi, King, Racker, and their colleagues. According to Green et al. (1965), the respiratory chain can be resolved into four lipoprotein complexes, namely, $NADH_2$–coenzyme Q reductase (complex I); succinate–coenzyme Q reductase (complex II); reduced coenzyme Q–cytochrome c reductase (complex III); and cytochrome c oxidase (complex IV). Coenzyme Q and cytochrome c are considered to act as mobile carriers that shuttle reducing equivalents between the complexes. Under suitable conditions the complexes can recombine to form a membrane that catalyzes a noncoupled oxidation of $NADH_2$ and succinate by oxygen, apparently through the same sequence of reactions found in the native inner membrane. Phospholipids are found to be essential for the activity of each of the complexes. Studies on the reconstitution of the succinate oxidation system have been described by Takemori and King (1964) and Yamashita and Racker (1969). Hatefi introduced the use of chaotropic agents, such as thiocyanate, perchlorate, and urea, to resolve the complexes of the respiratory chain system. Chaotropic agents change the structure and lipophilicity of water such that membrane hydrophobic bonds are weakened (Hatefi and Hanstein, 1969). These studies indicate that the respiratory chain system contains iron sulfur proteins distinct from succinic and NADH dehydrogenase flavoproteins which themselves contain iron (Hatefi and Stempel, 1969). Despite the great amount of effort and ingenuity represented by all these investigations, a genuine reconstruction of the respiratory chain employing only highly purified native components entirely free from membrane structure has not yet been achieved. It is therefore still not known whether or not a functional membrane can be formed from molecularly dispersed components by a self-assembly process in the absence of an existing membrane pattern or organizer protein (see also Section 3.B).

Various procedures have been used to separate the coupling system from the respiratory chain. Usually, coupling proteins have been solubilized, while most respiratory chain components have remained attached to the inner membrane. The solubilized proteins have been termed coupling factors to indicate that at least some of them may be structural components of the membrane, regulating the organization of catalytic components rather than being catalysts themselves. The numerous coupling factor preparations described in the literature can be divided into two groups, those with ATPase activity, and those with no demonstrable enzyme activity. The latter group stimulates one or more of the various energy-dependent reactions, including succinate-linked reduction of NAD, and transhydrogenase activity, that are catalyzed slowly by deficient submitochondrial particles. Among the ATPase coupling factors, the most highly purified and characterized are factor F_1 (Pullman et al., 1958) and factor A (Warshaw et al., 1968). Penefsky et al. (1960) proposed that F_1 acts as a phosphotransferase catalyzing the terminal reaction in the synthesis of ATP from ADP and phosphate. More recent studies suggest that it may also have a structural role in energy-linked reactions (Racker, 1970b). The ATPase activity catalyzed by the soluble F_1 protein is insensitive to oligomycin, DCCD, and other compounds that inhibit ATP hydrolysis in mitochondria and submitochondrial particles. When F_1 is added to deficient particles, the sensitivity of the ATPase activity to these inhibitors can be restored by various protein factors, including factor F_c (Bulos and Racker, 1968) and an oligomycin sensitivity-conferring protein (OSCP) (MacLennan and Tzagoloff, 1968).

The most highly purified proteins of the second (nonenzymic) group of coupling factors include factor B (Sanadi et al., 1968) and OSCP. It is possible that the coupling factor activity of many of the less purified proteins in this group is attributable to the presence of the OSCP protein (Beechey, 1970).

Few attempts have been made to identify the site of action of specific inhibitors by radioactive labeling techniques. The use of labeled rotenone appears to show that respiratory chain-linked $NADH_2$ dehydrogenase is the site of rotenone binding (Gutman et al., 1970). Cattell et al. (1970) isolated a DCCD-binding protein from mitochondria that may be involved in the coupling mechanism.

Many attempts to determine the site of mitochondrial enzymes in the membranes and spaces of the mitochondrion by fractionation procedures were complicated by the partial release of enzymes from the matrix space. However, the isolation of an intact inner membrane–matrix fraction after controlled lysis of the outer membrane by digitonin (Levy et al., 1967; Schnaitman and Greenawalt, 1968) has initiated a very successful period of research into structure–function relationships (Ernster and Kuylenstierna, 1970; Smoly et al., 1970). Information is now available on the intramitochondrial distribution of almost all the enzymes and other chemical constituents known to be present in liver mitochondria. Some of the results obtained are presented in Table 3. The inner compartment is found to contain most of the integrated metabolic activities characteristic of whole mitochondria, including the citric acid cycle, oxidative phosphorylation and other energy-linked functions, anion and cation translocation, and mitochondrial nucleic acid and protein synthesis. In contrast, the group of enzymes present in the outer compartment has very diverse functions (Ernster and Kuylenstierna, 1970). It includes a rotenone- and antimycin-insensitive $NADH_2$–cytochrome c reductase, distinct from the respiratory chain, phosphotransferases, and several enzymes concerned in phospholipid metabolism, including acyl transferases for glycerol–phosphate, lysophosphatidate, and lysolecithin. The outer membrane is the principal site of incorporation of labeled choline and serine into the phospholipids of rat liver mitochondria (Kaiser and Bygrave, 1968). Mitochondrial fatty acid-elongating activity appears to be present in both membranes (Colli et al., 1968). Monoamine oxidase activity is probably localized exclusively in the outer membrane (Schnaitman et al., 1967) and has frequently been used as a marker enzyme for this membrane. It is notable that a number of enzymes associated with the outer membrane of liver mitochondria, including rotenone- and antimycin-insensitive NADH-cytochrome c reductase, monoamine oxidase, and ATP-dependent fatty acyl-CoA-synthetase are also present in liver microsomes. Other liver microsomal enzymes such as NADPH–cytochrome c reductase, cytochrome P_{450} and glucose-6-phosphatase seem to be absent from liver mitochondria. As mentioned in Section 4.E, the permeability properties of the inner and outer membranes of mitochondria differ remarkably. These differences in permeability are paralleled by differences in the lipid composition

Table 3. Localization of Enzymes in Mitochondria[a]

Outer compartment	
Outer membrane	**Intermembrane space**
NADH–cytochrome c reductase (rotenone-insensitive), NADH–cytochrome b_5 reductase, cytochrome b_5 Monoamine oxidase ATP-dependent fatty acyl-CoA synthetase Elongation of C_{14} and C_{16} fatty acids Enzymes of phospholipid metabolism: Acyl transferases for glycerolphosphate, lysophosphatidate, and lysolecithin; cholinephosphotransferase; phosphatidate phosphatase; phospholipase A_{II}	Phosphotransferases; adenylate kinase; nucleoside mono- and diphosphokinases; creatine kinase (heart, skeletal muscle, brain)
Inner compartment	
Inner (cristae) membrane	**Matrix**
Respiratory chain system: cytochromes b, c_1, c, a, a_3; ubiquinone; dehydrogenases for NADH, succinate, β-hydroxybutyrate, α-glycerophosphate, and choline Respiratory-chain phosphorylation system (reversible ATPase) Anion and cation transporting systems Elongation of C_{10} fatty acids Carnitine palmityl transferase Cytochrome P_{450} (adrenal cortex)	Citric acid cycle enzymes (except succinic dehydrogenase)[b] Enzymes of nucleic acid and protein synthesis Glutamate dehydrogenase[b] Aspartate aminotransferase[b] Fatty acyl-CoA synthetase Fatty acid β-oxidation system

[a] The table refers to the enzymes of liver mitochondria except where indicated. [b] Examples of "latent" enzymes in intact mitochondria. Latent citric- acid cycle enzymes include isocitrate dehydrogenase, malate dehydrogenase, aconitate hydratase, and fumarate hydratase.

of the two membranes. For example, the outer membrane has a low cardiolipin content and a high phosphatidylinositol and cholesterol content compared to the inner membrane. In lipid composition, enzyme content, and turnover rate, the outer membrane resembles the ER membrane more closely than it does the mitochondrial inner membrane.

The permeability properties of the outer membrane suggest that under physiological conditions the intermembrane space contains low-molecular-weight cytoplasmic components which are probably lost during the washing procedures used to prepare isolated mitochondria. Thus it may prove difficult to study the physiological functions of the outer compartment by using preparations of isolated mitochondria (Smoly et al., 1970).

6. MECHANISMS OF ENERGY COUPLING

The two most popular hypotheses proposed to explain the mechanism of respiratory chain phosphorylation have been the chemical coupling hypothesis of Slater (1953, 1971) and the chemiosmotic hypothesis of Mitchell (1961, 1966). In both mechanisms the energy liberated during oxidoreduction reactions between adjacent carriers A_{red} and B_{ox} in the respiratory chain is at some stage conserved in the form of a high-energy intermediate (X$\sim$I) which can then be used for ATP synthesis. These reactions can be represented (Greville, 1969) as:

$$A_{red} + B_{ox} + X + I \rightleftharpoons A_{ox} + B_{red} + X{\sim}I \qquad (4)$$

$$X{\sim}I + ADP + P_i \rightleftharpoons X + I + ATP + H_2O \qquad (5)$$

$$\text{Sum:} \quad A_{red} + B_{ox} + ADP + P_i \rightleftharpoons A_{ox} + B_{red} + ATP + H_2O \qquad (6)$$

In the Slater hypothesis oxidoreduction is linked to X$\sim$I formation through high-energy intermediates of the respiratory chain ($A_{ox}{\sim}I$ or $B_{red}{\sim}I$) according to the type of reaction:

$$A_{red} + B_{ox} + I \rightleftharpoons A_{ox}{\sim}I + B_{red} \qquad (7)$$

$$A_{ox}{\sim}I + X \rightleftharpoons A_{ox} + X{\sim}I \qquad (8)$$

$$\text{Sum:} \quad A_{red} + B_{ox} + X + I \rightleftharpoons A_{ox} + B_{red} + X{\sim}I \qquad (4)$$

Uncoupling agents are presumed to promote the hydrolysis of X$\sim$I to X and I, thus stimulating respiration through reaction 4 and abolishing ATP synthesis by making X$\sim$I unavailable for reaction 5. The 1953 Slater hypothesis can be modified to accommodate the view that uncoupling agents act by virtue of their proton-conducting activity (Section 4.E) by postulating that X$\sim$I is able to energize an outward-facing proton pump. If this were true, then uncoupling agents would indirectly promote the hydrolysis of X$\sim$I, by continually conducting back into mitochondria the protons pumped out at the expense of X$\sim$I bonds. Energy-dependent cation uptake is explained by further postulating that the proton pump is linked to a cation transporter (Slater, 1967). Alternatively, cation uptake might be explained by postulating the formation of a mobile carrier intermediate containing the compound cation (X$\sim$I).

The energy of the X$\sim$I bond could then be used for energy-dependent cation binding, followed by the release of cations into the matrix space, provided that a suitable anion (e.g., phosphate or acetate) is available (Chance, 1965).

Attempts to isolate and identify compounds with the properties expected of a respiratory chain high-energy intermediate have met, in the words of Slater (1966c), with a "conspicuous non-success." The possible existence of these intermediates was first examined by Chance and Williams (1955b), whose spectrophotometric studies led them to conclude that "our data afford no support for the idea that a special respiratory chain acts in oxidative phosphorylation and confirm the idea that the known elements of the chain are the active participants." Recent studies have revealed some unexplained anomalies in the spectrophotometric behavior of cytochrome b that occur in response to changes in the metabolic state of mitochondria (Chance and Schoener, 1966; Chance et al., 1970). On the basis of these phenomena, Slater (1971) has proposed a more detailed chemical coupling mechanism for energy conservation in the cytochrome b region of the respiratory chain.

The chemiosmotic hypothesis postulates the presence of two separate proton-translocating systems, the respiratory chain system and the reversible ATPase system, which are inserted in an ion-impermeable coupling membrane (the inner mitochondrial membrane) and coupled together by a proton current. An electrochemical potential difference (proton motive force or PMF) is developed across the membrane during respiratory chain activity, and this force drives a proton current through the ATPase system, causing the ATPase reaction to operate in the direction of ATP synthesis (Fig. 15). The difference in the free energy of hydrolysis of compound XI during translocation across the membrane is presumed to result from the difference in H^+ ion activity, on either side of the membrane, that is generated by proton translocation during respiratory chain activity. Mitchell (1961, 1966) calculated that a PMF of about 250 mV across the cristae membrane is required to drive ATP synthesis. The total PMF (Δp), expressed in millivolts, is made up of electrical and chemical components ($\Delta\psi$ and $-Z\Delta$pH, respectively) corresponding to a membrane potential and a pH difference across the membrane according to the equation:

$$\Delta p = \Delta\psi - Z\Delta\text{pH}$$

where Z is approximately 59 at 25°C, and Δ pH is the pH difference across the membrane. Mitchell and Moyle (1969) obtained experimental values of Δp by separate measurements of the two components, using mitochondria in various metabolic states. Some of these values, which showed fairly good agreement with the theoretical value required, are shown in Table 4. The proton-conducting activity of uncoupling agents was found to collapse both the pH differential and the membrane potential, indicating that the latter component is largely caused by a disequilibrium of protons across the membrane during respiratory activity.

Cation uptake by mitochondria is simply explained as being attributable to cation movement down the electrical potential gradient created by respiration or by the ATPase, provided that a natural cation carrier (e.g., the calcium transporter) or an artificial cation carrier (e.g., valinomycin) is present in the cristae membrane. Mitchell proposed that the apparent failure of sonic submitochondrial particles to accumulate cations is attributable to the reversed polarity of the particle membrane. However, Loyter et al. (1969) have since observed that suitably prepared submitochondrial particles from beef heart, although possessing a membrane that is morphologically

Table 4. Some Estimates of the Proton Motive Force
($\Delta p = \Delta\psi - Z\Delta pH$)[a,b]

Metabolic state	Membrane potential, $\Delta\psi$	pH difference, $-Z\Delta pH$	PMF, Δp
State 4 (ADP-limited)	169	56	225
State 3 (phosphorylating)	~145	~56	~201
Uncoupling agent present	67	5	72
State 5 (anaerobic)	33	−33	0

[a] Values are given in millivolts.
[b] For other details, see Mitchell and Moyle (1969). The values quoted in the table are estimates based on a particular experimental rationale and should not be regarded as definitive.

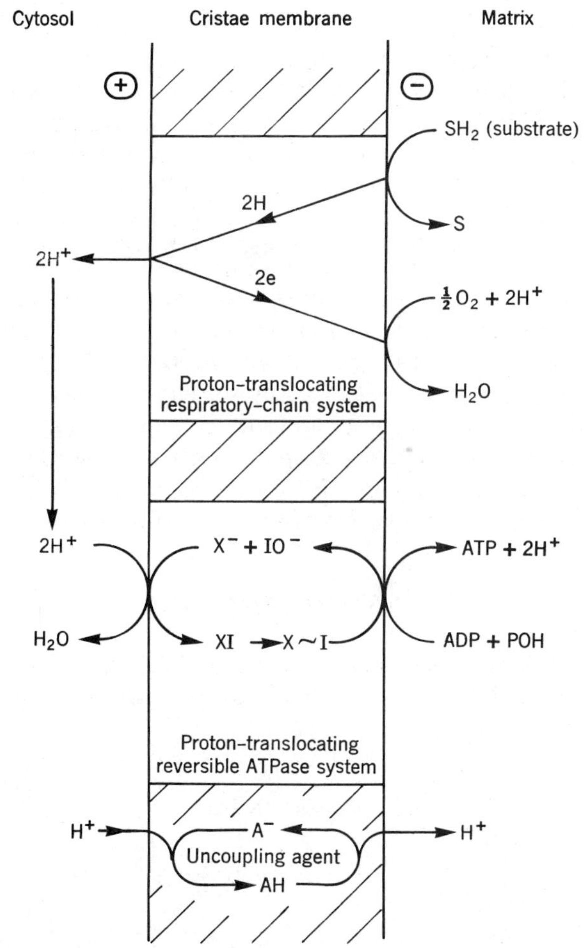

Figure 15. The coupling of respiratory chain oxidoreductions and ATP synthesis according to the chemiosmotic hypothesis. The upper part of the figure shows one oxidoreduction loop, corresponding to oxidoreduction across one coupling site. Similar loops are postulated to occur at each of the three coupling sites in the respiratory chain (see Fig. 8*b*). The figure also shows the mechanism of action of uncoupling agents attributable to the proton-conducting activity of their weak acid (AH) and anionic base (A^-) forms.

inverted compared to the parent mitochondrial membrane, nevertheless are capable of catalyzing an energy-dependent uptake of calcium ions. These observations are difficult to explain by current formulations of either the chemical coupling or chemiosmotic hypotheses. Evidence indicating the separateness of the respiratory chain and ATPase systems has been obtained from studies of [^{32}P] P_i–ATP and [^{18}O] H_2O–P_i exchange reactions in mitochondria (Boyer et al., 1966) and in promitochondria (Groot et al., 1971). The latter study strongly suggests that a mitochondrial energy transfer system can exist even when virtually all respiratory chain components are absent.

The present situation with regard to energy-coupling mechanisms has been aptly summarized by Greville (1969) in his very helpful scrutiny of the chemiosmotic hypothesis. He wrote:

"The Mitchell hypothesis assigns functions to certain definite entities such as PMF, membrane potential, and pH gradient, and postulates that certain known carriers and enzymes act vectorially. It leads to various predictions which are susceptible to verification or disproof. It is another matter, however, to establish the function of chemical intermediates which have not been identified and, indeed, have not even been shown to exist. It is, likewise, difficult to study a putative proton pump driven by a hypothetical $X \sim I$. . . . The versatility of the chemical hypothesis permits in most instances an alternative explanation to observations consistent with the Mitchell hypothesis. Such observations are less precise and more hypothetical, since they involve postulated but undiscovered entities; nevertheless they can usually be advanced."

This situation is likely to prevail until more precise information is available, especially data on the exact composition and spatial arrangement of the respiratory chain system, the function of lipids in the cristae membrane, and the identities of X and I.

Dr. Samuel Johnson declared that "It is by studying little things that we attain to the great art of having as little misery and as much happiness as possible." Although he was not referring to mitochondria, his sentiments are shared by physiologists and biochemists inspired by the many notable advances made during the past century in understanding the place of the mitochondrion in the fundamentals of cell biology.

ACKNOWLEDGMENTS

The author is grateful for the support of a Senior Wellcome Trust Research Fellowship and to Mr. John Newton for his helpful criticism and assistance during the preparation of the manuscript.

REFERENCES

Albracht, S. P. J., and E. C. Slater (1970). Molybdenum associated with NADH dehydrogenase in complex I. *Biochim. Biophys. Acta*, **223**, 457.

Ashwell, M., and T. S. Work (1970). The biogenesis of mitochondria. *Ann. Rev. Biochem.*, **39**, 251.

Ball, E. G. (1938). Über die Oxydation und Reduktion der drei Cytochrom-Komponenten. *Biochem. Z.*, **295**, 262.

Battelli, F., and L. Stern (1912). Die Oxydationsfermente. *Ergeb. Physiol.*, **12**, 96.

Beechey, R. B. (1970). Mitochondrial coupling factors. *Biochem. J.*, **116**, 6P.

Beinert, H., and W. Lee (1961). Evidence for a new type of iron containing electron carrier in mitochondria. *Biochem. Biophys. Res. Commun.*, **5**, 40.

Beinert, H., and G. Palmer (1965). In *Oxidases and Related Redox Systems* (T. E. King, H. S. Mason, and M. Morrison, eds.), Vol. 2, p. 567. Wiley, New York.

Belitzer, V. A. (1937). Effect of creatine on the respiration of muscle tissue. *Biokimiya*, **2,** 332.

Belitzer, V. A., and E. T. Tzibakova (1939). On the mechanism of phosphorylation coupled with respiration. *Biokimiya*, **4,** 516.

Benda, C. (1898). Über die Spermatogenese der Vertebraten und hoherer Evertebraten. *Verhandl. Physiol. Ges. (Berlin)*, p. 14.

Blondin, G. A., A. F. DeCastro, and A. E. Senior (1971). The isolation and properties of a peptide ionophore from beef heart mitochondria. *Biochem. Biophys. Res. Commun.*, **43,** 28.

Boyer, P. D., L. L. Bieber, R. A. Mitchell, and G. Szabolcsi (1966). The apparent independence of the phosphorylation and water formation reactions from the oxidation reactions of oxidative phosphorylation. *J. Biol. Chem.*, **241,** 5384.

Bucher, T. (1965). Formation of the specific structural and enzymic pattern of the insect flight muscle. *Biochem. Soc. Symp.*, **25,** 15.

Bulos, B., and E. Racker (1968). Partial resolution of the enzymes catalyzing oxidative phosphorylation. XVII. Further resolution of the rutamycin-sensitive adenosine triphosphatase. *J. Biol. Chem.*, **243,** 3891.

Cammer, W., and R. W. Estabrook (1967). Spectrophotometric studies of the pigments of adrenal cortex mitochondria. *Arch. Biochem. Biophys.*, **122,** 735.

Carafoli, E., and A. L. Lehninger (1971). A survey of the interaction of calcium ions with mitochondria from different tissues and species. *Biochem. J.*, **122,** 681.

Cattell, K. J., I. G. Knight, C. R. Lindop, and R. B. Beechey (1970). The isolation of dicyclohexylcarbodi-imide binding proteins from mitochondrial membranes. *Biochem. J.*, **117,** 1011.

Chance, B. (1963). In *Energy-Linked Functions of Mitochondria* (B. Chance, ed.), p. 253. Academic, New York.

Chance, B. (1965). The energy-linked reaction of calcium with mitochondria. *J. Biol. Chem.*, **240,** 2729.

Chance, B., and G. Hollunger (1961). The interaction of energy and electron transfer reactions in mitochondria. I. General properties and nature of the products of succinate-linked reduction of pyridine nucleotide. *J. Biol. Chem.*, **236,** 1534.

Chance, B., and B. Schoener (1966). High and low energy states of cytochromes. I. In mitochondria. *J. Biol. Chem.*, **241,** 4567.

Chance, B., and G. R. Williams (1955a). Respiratory enzymes in oxidative phosphorylation. IV. The respiratory chain. *J. Biol. Chem.*, **217,** 429.

Chance, B., and G. R. Williams (1955b). Respiratory enzymes in oxidative phosphorylation. III. The steady state. *J. Biol. Chem.*, **217,** 409.

Chance, B., and G. R. Williams (1956). The respiratory chain and oxidative phosphorylation. *Advan. Enzymol.*, **17,** 65.

Chance, B., B. Schoener, and D. De Vault (1965). In *Oxidases and Related Redox Systems* (T. E. King, H. S. Mason, and M. Morrison, eds.), Vol. 2, p. 907. Wiley, New York.

Chance, B., L. Ernster, P. B. Garland, C. P. Lee, P. A. Light, T. Ohnishi, C. I. Ragan, and D. Wong (1967a). Flavoproteins of the mitochondrial respiratory chain. *Proc. Natl. Acad. Sci., U.S.*, **57,** 1498.

Chance, B., C. P. Lee, and L. Mela (1967b). Control and conservation of energy in the cytochrome chain. *Federation Proc.*, **26,** 1341.

Chance, B., A. Azzi, I. Y. Lee, C. P. Lee, and L. Mela (1969). In *Mitochondria—Structure and Function* (L. Ernster and Z. Drahota, eds.), p. 233. Academic, New York.

Chance, B., D. F. Wilson, P. L. Dutton, and M. Erecinska (1970). Energy coupling mechanisms in mitochondria: Kinetic, spectroscopic and thermodynamic properties of an energy-transducing form of cytochrome *b*. *Proc. Natl. Acad. Sci., U.S.*, **66,** 1175.

Chappell, J. B. (1968). Systems used for the transport of substances into mitochondria. *Brit. Med. Bull.*, **24,** 150.

Chappell, J. B. (1969). In *Inhibitors—Tools in Cell Research* (T. Bucher and H. Sies, eds.), p. 335. Springer, Berlin.

Chappell, J. B., and A. R. Crofts (1965). Gramicidin and ion transport in isolated liver mitochondria. *Biochem. J.*, **95,** 393.

Chappell, J. B., and G. D. Greville (1963). The influence of the composition of the suspending medium on the properties of mitochondria. *Biochem. Soc. Symp.*, **23,** 39.

Chez, R. A., R. R. Palmer, S. G. Schultz, and P. F. Curran (1967). Effect of inhibitors on alanine transport in isolated rabbit ileum. *J. Gen. Physiol.*, **50,** 2357.

Claude, A. (1946). Fractionation of mammalian liver cells by differential centrifugation. *J. Exptl. Med.*, **84,** 51.

Claude, A. (1948). Studies on cells: Morphology, chemical constitution and distribution of biochemical functions. *Harvey Lectures*, **43,** 121.

Colli, W., P. Hinkle, and M. E. Pullman (1968). Elongation of fatty acids in rat liver mitochondria. *Federation Proc.*, **27,** 648.

Conover, T. E., and M. Bárány (1966). The absence of a myosin-like protein in liver mitochondria. *Biochim. Biophys. Acta*, **127,** 235.

Cowdry, E. V. (1918). The mitochondrial constituents of protoplasm. *Contrib. Embryol.*, (*Washington*), **8,** 39.

Crane, F. L., and H. Low (1966). Quinones in energy-coupling systems. *Physiol. Rev.*, **46,** 662.

Crane, R. K. (1965). Na^+-dependent transport in the intestine and other animal tissues. *Federation Proc.*, **24,** 1000.

Criddle, R. S., R. M. Bock, D. E. Green, and H. Tisdale (1962). Physical characteristics of proteins of the electron transfer system and interpretation of the structure of the mitochondria. *Biochemistry*, **1,** 827.

Danielson, L., and L. Ernster (1963). In *Energy-Linked Functions of Mitochondria* (B. Chance, ed.), p. 157, Academic, New York.

Diamond, J. M. (1971). Standing-gradient model of fluid transport in epithelia. *Federation Proc.*, **30,** 6.

Diamond, J. M., and W. H. Bossert (1967). Standing-gradient osmotic flow: A mechanism for coupling of water and solute transport in epithelia. *J. Gen. Physiol.*, **50,** 2061.

Diamond, J. M., and W. H. Bossert (1968). Functional consequences of ultrastructural geometry in "backwards" fluid-transporting epithelia. *J. Cell Biol.*, **37,** 694.

Droz, B., and M. Bergeron (1965). Autoradiographic detection of newly synthesized proteins from leucine-^3H in rat liver and kidney mitochondria. *Compt. Rend.*, **261,** 2757.

Ernster, L., and B. Kuylenstierna (1970). In *Membranes of Mitochondria and Chloroplasts* (E. Racker, ed.), A. C. S. Monograph 165, p. 172. Van Nostrand Reinhold, New York.

Ernster, L., I. Y. Lee, B. Norling, and B. Persson (1969). Studies with ubiquinone-depleted submitochondrial particles. *FEBS Letters*, **3,** 21.

Estabrook, R. W., and A. Holowinsky (1961). Studies on the content and organization of the respiratory enzymes of mitochondria. *J. Biophys. Biochem. Cytol.*, **9,** 19.

Estabrook, R. W., and B. Sacktor (1958). α-Glycerophosphate oxidase of flight muscle mitochondria. *J. Biol. Chem.*, **233,** 1014.

Fawcett, D. W. (1966). *The cell; Its organelles and inclusions.* Saunders, Philadelphia.

Ferguson, D. M., J. E. Etherton, and M. J. Hayes (1969). Mitochondrial increase after long-term feeding of morfamquat. *Nature*, **224,** 83.

Fernández-Morán, H. (1962). Cell membrane ultrastructure. Low temperature electron microscopy and X-ray diffraction studies of lipoprotein components in lamellar systems. *Circulation*, **26,** 1039.

Fletcher, M. J., and D. R. Sanadi (1961). Turnover of rat liver mitochondria. *Biochim. Biophys. Acta*, **51,** 356.

Fonyo, A. (1968). Phosphate carrier of rat liver mitochondria. Its role in phosphate outflow. *Biochem. Biophys. Res. Commun.*, **32,** 624.

Frédéric, J. (1958). Recherches cytologiques sur le chondriome normal ou soumis á l'expérimentation dans les cellules vivantes cultivées *in vitro*. *Arch. Biol.* (*Liège*), **69,** 167.

Fritz, I. B. (1961). Factors influencing the rates of long-chain fatty acid oxidation and synthesis in mammalian systems. *Physiol. Rev.*, **41,** 52.

Gevers, W., and H. A. Krebs (1966). The effects of adenine nucleotides on carbohydrate metabolism in pigeon-liver homogenates. *Biochem. J.*, **98**, 720.

Green, D. E., J. Asai, R. A. Harris, and J. T. Penniston (1968). Conformational basis of energy transformations in membrane systems. III. Configurational changes in the mitochondrial inner membrane induced by changes in functional states. *Arch. Biochem. Biophys.*, **125**, 684.

Green, D. E., D. C. Wharton, A. Tzagoloff, J. S. Rieske, and G. P. Brierley (1965). In *Oxidases and Related Redox Systems* (T. E. King, H. S. Mason, and M. Morrison, eds.), Vol. 2, p. 1032. Wiley, New York.

Greville, G. D. (1966). In *Regulation of Metabolic Processes in Mitochondria* (J. M. Tager, S. Papa, E. Quagliariello, and E. C. Slater, eds.), p. 86. Elsevier, Amsterdam.

Greville, G. D. (1969). A scrutiny of Mitchell's chemiosmotic hypothesis of respiratory chain and photosynthetic phosphorylation. *Curr. Top. Bioenerg.*, **3**, 1.

Greville, G. D., and P. K. Tubbs (1968). The catabolism of long chain fatty acids in mammalian tissues. *Essays Biochem.*, **4**, 155.

Groot, G. S. P., L. Kovac, and G. Schatz (1971). Promitochondria of anaerobically grown yeast. V. Energy transfer in the absence of an electron transfer chain. *Proc. Natl. Acad. Sci., U.S.*, **68**, 308.

Gross, N. J., G. S. Getz, and M. Rabinowitz (1969). Apparent turnover of mitochondrial deoxyribonucleic acid and mitochondrial phospholipids in the tissues of the rat. *J. Biol. Chem.*, **244**, 1552.

Gutman, M., T. P. Singer, H. Beinert, and J. E. Casida (1970). Reaction sites of rotenone, piericidin A and amytal in relation to the non-heme iron components of NADH dehydrogenase. *Proc. Natl. Acad. Sci., U.S.*, **65**, 763.

Hackenbrock, C. R. (1966). Ultrastructural bases for metabolically linked mechanical activity in mitochondria. I. Reversible ultrastructural changes with change in metabolic steady state in isolated liver mitochondria. *J. Cell Biol.*, **30**, 269.

Hatefi, Y. (1968). Flavoproteins of the electron transport system and the site of action of amytal, rotenone and piericidin A. *Proc. Natl. Acad. Sci., U.S.*, **60**, 733.

Hatefi, Y., and W. G. Hanstein (1969). Solubilization of particulate proteins and non-electrolytes by chaotropic agents. *Proc. Natl. Acad. Sci., U.S.*, **62**, 1129.

Hatefi, Y., and K. E. Stempel (1969). Isolation and enzymatic properties of the mitochondrial reduced diphosphopyridine nucleotide dehydrogenase. *J. Biol. Chem.*, **244**, 2350.

Hogeboom, G. H., A. Claude, and R. D. Hotchkiss (1946). The distribution of cytochrome oxidase and succinoxidase in the cytoplasm of the mammalian liver cell. *J. Biol. Chem.*, **165**, 615.

Hughes, D. E., D. Lloyd, and R. Brightwell (1970). Structure, function and distribution of organelles in prokaryotic and eukaryotic microbes. *Symp. Soc. Gen. Microbiol.*, **20**, 295.

Hulsmann, W. C., E. M. Wit-Peeters and C. Benckhuysen (1966). In *Regulation of Metabolic Processes in Mitochondria* (J. M. Tager, S. Papa, E. Quagliariello, and E. C. Slater, eds.), p. 460. Elsevier, Amsterdam.

Hunter, F. E. (1949). Anaerobic phosphorylation due to a coupled oxidation-reduction between α-ketoglutaric acid and oxaloacetic acid. *J. Biol. Chem.*, **177**, 361.

Kagawa, Y., and E. Racker (1966). Partial resolution of the enzymes catalysing oxidative phosphorylation. X. Correlation of morphology and function in submitochondrial particles. *J. Biol. Chem.*, **241**, 2475.

Kaiser, W., and F. L. Bygrave (1968). Incorporation of choline into the outer and inner membranes of isolated rat liver mitochondria. *European J. Biochem.*, **4**, 582.

Kalckar, H. (1937). Phosphorylation in kidney tissue. *Enzymologia*, **2**, 47.

Kaplan, D. M., and R. S. Criddle (1971). Membrane structural proteins. *Physiol. Rev.*, **51**, 249.

Keilin, D. (1925). On cytochrome, a respiratory pigment, common to animals, yeasts, and higher plants. *Proc. Roy. Soc. (London)*, **B98**, 312.

Keilin, D. (1929). Cytochrome and respiratory enzymes. *Proc. Roy. Soc. (London)*, **B104**, 206.

Keilin, D. (1966). *The History of Cell Respiration and Cytochrome*. Prepared for publication by J. Keilin. Cambridge Univ. Press, Cambridge.

Keilin, D., and E. F. Hartree (1939). Cytochrome and cytochrome oxidase. *Proc. Roy. Soc.* (*London*), **B127,** 167.

Keilin, D., and E. F. Hartree (1940). Succinic dehydrogenase-cytochrome system of cells. Intracellular respiratory enzyme system catalysing aerobic oxidation of succinic acid. *Proc. Roy. Soc.* (*London*), **B129,** 277.

Kimmich, G. A., and H. Rasmussen (1969). Regulation of pyruvate carboxylase activity by calcium in intact rat liver mitochondria. *J. Biol. Chem.*, **244,** 190.

Kingsbury, B. F. (1912). Cytoplasmic fixation. *Anat. Record*, **6,** 39.

Klingenberg, M. (1970a). Metabolite transport in mitochondria: An example for intracellular membrane function. *Essays Biochem.*, **6,** 119.

Klingenberg, M. (1970b). Mitochondrial metabolite transport. *FEBS Letters*, **6,** 145.

Klingenberg, M., and A. Kröger (1967). In *Biochemistry of Mitochondria* (E. C. Slater, Z. Kaniuga, and L. Wojtczak, eds.), p. 11. Academic, New York.

Klingenberg, M., and D. Pette (1962). Proportions of mitochondrial enzymes and pyridine nucleotides. *Biochem. Biophys. Res. Commun.*, **7,** 430.

Klingenberg, M., and W. Slenczka (1959). Pyridinnucleotide in Leber-Mitochondrien. Eine Analyse ihrer Redox-Beziehungen. *Biochem. Z.*, **331,** 486.

Klingenberg, M., and G. Von Jagow (1970). In *Electron Transport and Energy Conservation* (J. M. Tager, S. Papa, E. Quagliariello, and E. C. Slater, eds.), p. 281. Adriatica Editrice, Bari.

Kölliker, A. (1888). Zur Kenntnis der quergestreiften Muskelfasern. *Z. Wiss. Zool.*, **47,** 689.

Krebs, H. A. (1967). In *Biochemistry of Mitochondria* (E. C. Slater, Z. Kaniuga, and L. Wojtczak, eds.), p. 105. Academic, New York.

Krebs, H. A., and H. L. Kornberg (1957). Energy transformations in living matter. *Ergeb. Physiol.*, **49,** 212. Also published as a monograph by Springer, Berlin.

Lardy, H. A. (1956). Energetic coupling and the regulation of metabolic rates. In *Proceedings of the 3rd International Congress of Biochemistry, Brussels*, (C. Liebecq, ed.), p. 287. Vaillant-Carmanne, Liège.

Lardy, H. A., and H. Wellman (1952). Oxidative phosphorylations: Role of inorganic phosphate and acceptor systems in control of metabolic rates. *J. Biol. Chem.*, **195,** 215.

Lardy, H. A., V. Paetkau, and P. Walter (1965). Paths of carbon in gluconeogenesis and lipogenesis. The role of mitochondria in supplying precursors of phosphoenol pyruvate. *Proc. Natl. Acad. Sci., U.S.*, **53,** 1410.

Lee, C. P., and L. Ernster (1966). In *Regulation of Metabolic Processes in Mitochondria* (J. M. Tager, S. Papa, E. Quagliariello, and E. C. Slater, eds.), p. 218. Elsevier, Amsterdam.

Lehninger, A. L. (1951). Phosphorylation coupled to oxidation of dihydrodiphosphopyridine nucleotide. *J. Biol. Chem.*, **190,** 345.

Lehninger, A. L. (1962). Water uptake and extrusion by mitochondria in relation to oxidative phosphorylation. *Physiol. Rev.*, **42,** 467.

Lehninger, A. L. (1964). *The Mitochondrion.* Benjamin, New York.

Lehninger, A. L. (1970a). Mitochondria and calcium ion transport. *Biochem. J.*, **119,** 129.

Lehninger, A. L. (1970b). A soluble, heat-labile, high-affinity Ca^{2+}-binding factor extracted from rat liver mitochondria. *Biochem. Biophys. Res. Commun.*, **42,** 312.

Lévy, M., R. Toury, and J. André (1967). Séparation des membranes mitochondriales. Purification et caractérisation enzymatique de la membrane externe. *Biochim. Biophys. Acta*, **135,** 599.

Lewis, M. R., and W. H. Lewis (1915). Mitochondria (and other cytoplasmic structures) in tissue cultures. *Am. J. Anat.*, **17,** 339.

Löw, H., and I. Vallin (1963). Succinate-linked diphosphopyridine nucleotide reduction in submitochondrial particles. *Biochim. Biophys. Acta*, **69,** 361.

Loyter, A., R. O. Christiansen, H. Steensland, J. Saltzgaber, and E. Racker (1969). Energy-linked ion translocation in submitochondrial particles. *J. Biol. Chem.*, **244,** 4422, 4428.

Luck, D. J. L. (1963). Genesis of mitochondria in *Neurospora crassa*. *Proc. Natl. Acad. Sci., U.S.*, **49,** 233.

Lynen, F. (1955). Lipid metabolism. *Ann. Rev. Biochem.*, **24,** 653.

Lynn, W. S., and R. H. Brown (1965). Efficiency of energy conservation in oxidative phosphorylation. *Biochim. Biophys. Acta,* **105,** 15.

MacLennan, D. H., and A. Tzagoloff (1968). Studies on the mitochondrial adenosine triphosphatase system. IV. Purification and characterization of the oligomycin sensitivity conferring protein. *Biochemistry,* **7,** 1603.

Meijer, A. J., G. S. P. Groot, and J. M. Tager (1970). Effect of sulphydryl-blocking reagents on mitochondrial anion-exchange reactions involving phosphate. *FEBS Letters,* **8,** 41.

Michaelis, L. (1899). Die vitale Färbung, eine Darstellungs-Methode der Zellgranula. *Arch. Mikroskop. Anat.,* **55,** 558.

Mitchell, P. (1954). Transport of phosphate across the osmotic barrier of *Micrococcus pyogenes:* Specificity and kinetics. *J. Gen. Microbiol.,* **11,** 73.

Mitchell, P. (1961). A chemiosmotic hypothesis for the mechanism of oxidative and photosynthetic phosphorylation. *Nature,* **191,** 144.

Mitchell, P. (1966). Chemiosmotic coupling in oxidative and photosynthetic phosphorylation. *Biol. Rev.,* **41,** 445.

Mitchell, P., and J. Moyle (1967). Acid-base titration across the membrane system of rat-liver mitochondria. Catalysis by uncouplers. *Biochem. J.,* **104,** 588.

Mitchell, P., and J. Moyle (1969). Estimation of membrane potential and pH difference across the cristae membrane of rat liver mitochondria. *European J. Biochem.,* **7,** 471.

Nass, M. M. K., S. Nass, and B. A. Afzelius (1965). The general occurrence of mitochondrial DNA. *Exptl. Cell Research,* **37,** 516.

Nordlie, R. C., and H. A. Lardy (1963). Mammalian liver phosphoenolpyruvate carboxykinase activities. *J. Biol. Chem.,* **238,** 2259.

Ohnishi, T., and T. Ohnishi (1962). Extraction of contractile protein from liver mitochondria. *J. Biochem. (Tokyo),* **51,** 380.

Ozawa, E., and S. Ebashi (1967). Requirement of calcium ion for the stimulating effect of cyclic 3′,5′-AMP on muscle phosphorylase b kinase. *J. Biochem. (Tokyo),* **62,** 285.

Palade, G. E. (1956). Electron microscopy of cytoplasmic structures. In *Enzymes: Units of Biological Structure and Function* (O. H. Gaebler, ed.), p. 185. Academic, New York.

Papa, S., A. Francavilla, G. Paradies, and B. Meduri (1971). The transport of pyruvate in rat liver. *FEBS Letters,* **12,** 285.

Pasquali-Ronchetti, I., J. W. Greenawalt, and E. Carafoli (1969). On the nature of the dense matrix granules of normal mitochondria. *J. Cell Biol.,* **40,** 565.

Penefsky, H. S., M. E. Pullman, A. Datta, and E. Racker (1960). Partial resolution of the enzymes catalyzing oxidative phosphorylation. II. Participation of a soluble adenosine triphosphatase in oxidative phosphorylation. *J. Biol. Chem.,* **235,** 3330.

Plattner, H., and G. Schatz (1969). Promitochondria of anaerobically grown yeast. III. Morphology. *Biochemistry,* **8,** 339.

Poe, M., and R. W. Estabrook (1968). Kinetic studies of temperature changes and oxygen uptake concomitant with substrate oxidation by mitochondria: The enthalpy of succinate oxidation during ATP formation by mitochondria. *Arch. Biochem. Biophys.,* **126,** 320.

Powers, E. L., C. F. Ehret, L. E. Roth, and O. T. Minick (1956). The internal organization of mitochondria. *J. Biophys. Biochem. Cytol., Suppl.,* **2,** 341.

Pressman, B. C. (1970). In *Membranes of Mitochondria and Chloroplasts* (E. Racker, ed.), A. C. S. Monograph 165, p. 213. Van Nostrand Reinhold, New York.

Pullman, M. E., H. Penefsky, and E. Racker (1958). A soluble protein fraction required for coupling phosphorylation to oxidation in submitochondrial fragments of beef heart mitochondria. *Arch. Biochem. Biophys.,* **76,** 227.

Rabinowitz, M., and H. Swift (1970). Mitochondrial nucleic acids and their relation to the biogenesis of mitochondria. *Physiol. Rev.,* **50,** 376.

Racker, E. (1969). Resolution and reconstitution of a mammalian membrane. *J. Gen. Physiol.,* **54,** 38s.

Racker, E., ed. (1970a). *Membranes of Mitochondria and Chloroplasts.* A. C. S. Monograph 165, Van Nostrand Reinhold, New York.

Racker, E. (1970b). Function and structure of the inner membrane of mitochondria and chloroplasts. In *Membranes of Mitochondria and Chloroplasts* (E. Racker, ed.), A.C.S. Monograph 165, p. 127. Van Nostrand Reinhold, New York.

Racker, E. (1970c). The two faces of the inner mitochondrial membrane. *Essays Biochem.*, **6,** 1.

Racker, E., D. D. Tyler, R. W. Estabrook, T. E. Conover, D. F. Parsons, and B. Chance (1965). In *Oxidases and Related Redox Systems* (T. E. King, H. S. Mason, and M. Morrison, eds.), Vol. 2, p. 1077. Wiley, New York.

Rasmussen, H. (1970). Cell communication, calcium ion, and cyclic adenosine monophosphate. *Science*, **170,** 404.

Rasmussen, H., and A. Tenenhouse (1968). Cyclic adenosine monophosphate, Ca^{++}, and membranes. *Proc. Natl. Acad. Sci., U.S.*, **59,** 1364.

Regaud, C. (1911). Les mitochondries. *Rev. Méd.*, **31,** 681.

Reuter, H., and N. Seitz (1968). The dependence of calcium efflux from cardiac muscle on temperature and external ion composition. *J. Physiol.*, **195,** 451.

Robinson, B. H., G. R. Williams, M. L. Halperin, and C. C. Leznoff (1970). The effects of 2-ethylcitrate and tricarballylate on citrate transport in rat liver mitochondria and fatty acid synthesis in rat white adipose tissue. *European J. Biochem.*, **15,** 263.

Rossi, C. S., and A. L. Lehninger (1964). Stoichiometry of respiratory stimulation, accumulation of Ca^{++} and phosphate, and oxidative phosphorylation in rat liver mitochondria. *J. Biol. Chem.*, **239,** 3971.

Sanadi, D. R., K. W. Lam, and C. K. Ramakrishna Kurup (1968). The role of factor B in the energy transfer reactions of oxidative phosphorylation. *Proc. Natl. Acad. Sci., U.S.*, **61,** 277.

Sarkissian, I. V., and H. K. Srivastava (1970). High efficiency of oxidative phosphorylation in mitochondria of wheat. *Can. J. Biochem.*, **48,** 692.

Schatz, G. (1970). In *Membranes of Mitochondria and Chloroplasts* (E. Racker, ed.), A. C. S. Monograph 165, p. 251. Van Nostrand Reinhold, New York.

Schnaitman, C., and J. W. Greenawalt (1968). Enzymatic properties of the inner and outer membranes in rat liver mitochondria. *J. Cell Biol.*, **38,** 158.

Schnaitman, C., V. G. Erwin, and J. W. Greenawalt (1967). The submitochondrial localization of monoamine oxidase. An enzymatic marker for the outer membrane of rat liver mitochondria. *J. Cell Biol.*, **32,** 719.

Schultz, S. G., and P. F. Curran (1970). Coupled transport of sodium and organic solutes. *Physiol. Rev.*, **50,** 637.

Scott, R. L., and J. L. Gamble (1961). Effect of mercurial compounds on potassium binding by mitochondria. *J. Biol. Chem.*, **236,** 570.

Sjöstrand, F. S. (1956). The ultrastructure of cells as revealed by the electron microscope. *Intern. Rev. Cytol.*, **5,** 455.

Skou, J. C. (1965). Enzymatic basis for active transport of Na^+ and K^+ across cell membrane. *Physiol. Rev.*, **45,** 596.

Skulachev, V. P., A. A. Sharaf, and E. A. Liberman (1967). Proton conductors in the respiratory chain and artificial membranes. *Nature*, **216,** 718.

Slater, E. C. (1953). Mechanism of phosphorylation in the respiratory chain. *Nature*, **172,** 975.

Slater, E. C. (1966a). In *Regulation of Metabolic Processes in Mitochondria* (J. M. Tager, S. Papa, E. Quagliariello, and E. C. Slater, eds.), p. 528. Elsevier, Amsterdam.

Slater, E. C. (1966b). In *Comprehensive Biochemistry* (M. Florkin and E. H. Stotz, eds.), Vol. 14, p. 327. Elsevier, Amsterdam.

Slater, E. C. (1966c). In *Regulation of Metabolic Processes in Mitochondria* (J. M. Tager, S. Papa, E. Quagliariello, and E. C. Slater, eds.), p. 166. Elsevier, Amsterdam.

Slater, E. C. (1967). An evaluation of the Mitchell hypothesis of chemiosmotic coupling in oxidative and photosynthetic phosphorylation. *European J. Biochem.*, **1,** 317.

Slater, E. C. (1971). The coupling between energy-yielding and energy-utilizing reactions in mitochondria. *Quart. Rev. Biophys.*, **4,** 35.

Slater, E. C., and K. W. Cleland (1953). The effect of calcium on the respiratory and phosphorylative activities of heart-muscle sarcosomes. *Biochem. J.*, **55,** 566.

Smith, R. E., and B. A. Horwitz (1969). Brown fat and thermogenesis. *Physiol. Rev.*, **49,** 330.

Smoly, J. M., B. Kuylenstierna, and L. Ernster (1970). Topological and functional organization of the mitochondrion. *Proc. Natl. Acad. Sci., U.S.*, **66,** 125.

Sweetman, A. J., and D. E. Griffiths (1970). Energy-linked reactions in mitochondria: A requirement for ubiquinone after pentane extraction. *FEBS Letters*, **10,** 92.

Szarkowska, L. (1966). The restoration of DPNH oxidase activity by coenzyme Q (ubiquinone). *Arch. Biochem. Biophys.*, **113,** 519.

Tager, J. M. (1966). In *Regulation of Metabolic Processes in Mitochondria* (J. M. Tager, S. Papa, E. Quagliariello, and E. C. Slater, eds.), p. 202. Elsevier, Amsterdam.

Takemori, S., and T. E. King (1964). Reconstitution of respiratory-chain enzyme systems. XIII. Sequential fragmentation of succinic oxidase. *J. Biol. Chem.*, **239,** 3546.

Tata, J. R. (1964). In *Actions of Hormones on Molecular Processes* (G. Litwack and D. Kritchevsky, eds.), p. 58. Wiley, New York.

Tubbs, P. K., and P. B. Garland (1968). Membranes and fatty acid metabolism. *Brit. Med. Bull.*, **24,** 158.

Tyler, D. D. (1968). The inhibition of phosphate entry into rat liver mitochondria by organic mercurials and by formaldehyde. *Biochem. J.*, **107,** 121.

Tyler, D. D. (1969). Evidence of a phosphate transporter system in the inner membrane of isolated mitochondria. *Biochem. J.*, **111,** 665.

Van Den Burgh, S. G., and E. C. Slater (1962). The respiratory activity and permeability of housefly sarcosomes. *Biochem. J.*, **82,** 362.

Vignais, P. M., P. V. Vignais, and A. L. Lehninger (1964). Identification of phosphatidylinositol as a factor required in mitochondrial contraction. *J. Biol. Chem.*, **239,** 2011.

Wallace, P. G., and A. W. Linnane (1964). Oxygen-induced synthesis of yeast mitochondria. *Nature*, **201,** 1191.

Warburg, O. (1913). Über sauerstoffatmende Körnchen aus Leberzellen und über sauerstoffatmung in Berkfeld-Filtraten wässriger Leber-extrackte. *Arch. Ges. Physiol.*, **154,** 599.

Warburg, O. (1956). On the origin of cancer cells. *Science*, **123,** 309.

Warshaw, J. B., K. W. Lam, B. Nagy, and D. R. Sanadi (1968). Studies on oxidative phosphorylation. XV. Latent adenosine 5′-triphosphatase activity of factor A. *Arch. Biochem. Biophys.*, **123,** 385.

Watson, M. L., and P. Siekevitz (1956). Cytochemical studies of mitochondria. I. The separation and identification of a membrane fraction from isolated mitochondria. *J. Biophys. Biochem. Cytol.*, **2,** 639.

Weber, A., R. Herz, and I. Reiss (1964). The regulation of myofibrillar activity by calcium. *Proc. Roy. Soc.* (*London*), **B160,** 489.

Weinbach, E. C., and T. Von Brand (1967). Formation, isolation and composition of dense granules from mitochondria. *Biochim. Biophys. Acta*, **148,** 256.

Weiss, J. M. (1955). Mitochondrial changes induced by potassium and sodium in the duodenal absorptive cell as studied with the electron microscope. *J. Exptl. Med.*, **102,** 783.

Whittam, R., and D. M. Blond (1964). Respiratory control by an adenosine triphosphatase involved in active transport in brain cortex. *Biochem. J.*, **92,** 147.

Wrigglesworth, J. M., L. Packer, and D. Branton (1970). Organization of mitochondrial structure as revealed by freeze etching. *Biochim. Biophys. Acta*, **205,** 125.

Yakushiji, E., and K. Okunuki (1940). Über eine neue Cytochromkomponente und ihre Funktion. *Proc. Imp. Acad. Japan*, **16,** 299.

Yamashita, S., and E. Racker (1969). Resolution and reconstitution of the mitochondrial electron transport system. II. Reconstitution of succinoxidase from individual components. *J. Biol. Chem.*, **244,** 1220.

5

THE NUCLEUS

Margery G. Ord and L. A. Stocken

1. INTRODUCTION

The nucleus was first described by Leeuwenhoek at the end of the seventeenth century (see Hughes, 1959), and early studies on its structure and significance were made primarily by microscopists. In 1869, Miescher isolated nuclei from pus cells and provided the first description of nucleoprotein. For many years following, attention centered on nuclear isolation and analysis, but an understanding both of the constituents and especially of protein–nucleic acid interrelationships is still far from complete. Recently, techniques for such investigations have been greatly extended by nuclear transplantation (Gurdon, 1964) and cell hybridization (Harris, 1970), and attention is now focused on the dynamic state of the nucleus and its interactions with the cytoplasm.

Our considerations here begin with an account of the techniques for examining nuclei, and turn next to the properties of the principal nuclear constituents and present knowledge of the organization of chromatin at a molecular level. After a brief discussion of nuclear enzymes, nuclear organelles are reviewed, and an introduction is given to the modifications of nuclear structure and function that may be brought about by cytoplasmic factors.

When possible, references to review articles are given rather than to the contributory papers; we are indebted to many previous reviewers and hope that reference has been made to their contributions in the sections in which their articles are particularly helpful.

2. TECHNIQUES FOR THE EXAMINATION OF NUCLEI

Isolation techniques in general are devised to produce high yields of cell organelles, free from contaminating cell fragments, under conditions in which the morphological appearance and function of the organelle is retained and redistribution of cell contents between organelles has not occurred. These objectives are rarely satisfied in full. The principal difficulties in the case of nuclei are as follows. First, the nuclear membrane is more permeable to macromolecules than, for example, the membranes surrounding the mitochondrion. Second, conditions such as the use of hyperosmolar or nonaqueous media, which are required to produce nuclei that under electron microscopy appear free from contamination, inevitably produce unpredictable changes in the function or structure of the nucleus. Third, the common use of divalent ions (1–5 mM Ca^{2+} or Mg^{2+}) to maintain the morphological integrity of the nucleus may alter the properties of other organelles necessarily present during the initial stages of isolation, hence alter the properties of the nuclei.

Much of the early work on the isolation of nuclei was carried out with thymus gland with its relatively low cytoplasmic volume. Most isolation procedures now in use provide fairly clean preparations of thymus nuclei, and by comparison of the properties shown by nuclei prepared in different media, some idea of the limitations of the various procedures can be obtained. It is, however, now becoming clearer that the thymocyte is a highly specialized cell whose nuclei, especially in relation to permeability to proteins and also in relation to ATP generation, may be quite atypical.

Two general procedures can be distinguished: first, the use of nonaqueous solvents (Behrens, 1932); and, second, various methods involving aqueous media at neutral or acid pH. In the nonaqueous techniques, the tissue is first frozen, lyophilized, and the powder ground mechanically until it is found empirically that the majority of nuclei are free, albeit still associated with tags of cytoplasm. The material is then dispersed in organic solvent, such as a carbon tetrachloride–cyclohexane–petroleum ether mixture (Siebert, 1961), so that the nuclei can be sedimented free from cytoplasm and ultimately often floated off above aggregated or contaminating particles.

Nuclei obtained in this way appear in the electron microscope to be relatively free of cytoplasmic contamination (Siebert and Humphrey, 1965). Part or all of the nuclear membranes may, however, be lost, and the appearance of the nucleoplasm is more homogeneous than usual. Not all enzymic activity detectable after isolating nuclei in aqueous media is still retained, but control experiments with liver and kidney nuclei (Siebert and Humphrey, 1965) have indicated enzymes that are unaffected by the organic solvents and which may therefore be used as biochemical indices of contamination. As might have been predicted, lysosomal and microsomal contamination is the most serious. There is also a possibility of adsorption of other cytoplasmic material on the nuclei.

Analysis by sucrose gradients indicates that nuclei isolated by nonaqueous procedures contain very little ribosomal RNA (rRNA) (Siebert and Humphrey, 1965). Most significantly, they show a higher cation concentration per gram dry weight than with nuclei isolated through aqueous media (Itoh and Schwartz, 1957; Langendorf et al., 1961). With nuclei from calf thymus, for example, only 37% of the intranuclear Na^+ and K^+ was retained after isolation through a sucrose medium in contrast to those prepared by nonaqueous techniques. Similarly, nonaqueous nuclei retain DNA polymerase otherwise lost into aqueous media (Bollum and Potter, 1958; Davidson et al., 1958; Mantsavinos and Cannelakis, 1959; Keir et al., 1962).

This relatively tedious procedure is therefore valuable for analytic purposes, but enzyme activity may not be preserved and the organization of the nucleoprotein is certainly altered by the organic solvents.

The oldest (Miescher, 1897) and simplest of the aqueous techniques is the use of dilute acid as isolation medium. Cytoplasm is stripped off, and clean nuclei can be obtained rapidly in good yield. Use of acid pH has been particularly developed by Dounce and his colleagues (Dounce and Umaña, 1962). More commonly, neutral sucrose solutions of molarities ranging from 0.25 (Hogeboom et al., 1948) to 2.2 *M* (Chauveau et al., 1956) are used, often with the addition of Ca^{2+} (Hogeboom et al., 1952) or Mg^{2+} to improve the yield and morphology of the nuclei. With tissues such as thymus (Allfrey et al., 1957), and even pancreas (Klyszejko-Stefanowicz and Khouvine, 1968; Fitzgerald et al., 1970), isosmolar solutions yield nuclei free from cytoplasmic contamination, but purification through sucrose gradients of greater molarity is commonly employed to remove nuclei with cytoplasmic tags and to allow sedimentation of the dense nuclei. Nuclei prepared in this way usually have good morphology and the nuclear membrane is well preserved (Chauveau et al., 1956), but nuclei prepared in hyperosmolar solutions are much less active metabolically than nuclei prepared in 0.25 *M* sucrose medium (Allfrey, 1959). The implications of these alterations of metabolic properties are discussed in Section 4.

Dounce and Ickowicz (1969) showed that the ratio of protein to DNA is profoundly affected by the presence of divalent ions, 3–5 m*M* Ca^{2+} or Mg^{2+}, producing histone/DNA ratios of 1:1, whereas in the absence of the metal ions higher proportions of protein to DNA are present. Earlier work from this group (Dounce and Umaña, 1962) had shown that proteolysis could markedly reduce the protein content of liver nuclei and that this reduction is minimal at pH 5.8. Redistribution of proteins between cytoplasm and nucleus, however, can still occur and has been demonstrated particularly at a pH below 3.5 (Hancock and Amos, 1968).

The presence of surface-active substances in the medium has been introduced to simplify isolation procedures. Anionic detergents have been employed (Hodes et al., 1958), but nonionic polyoxyethylene derivatives are more usual (Gilbert and Radley, 1964; O'Brien, 1964). If hyposmolar solutions are used (Fisher and Harris, 1962), the nuclei become swollen and some loss of protein occurs (Crawley and Harris, 1963). Changes in nuclear membranes are very probable even in isosmolar solutions, but the methods provide clean nuclei very rapidly, so that they can be used, for example, in obtaining undegraded RNA (Dabeva and Tsanev, 1966). Zonal centrifugation

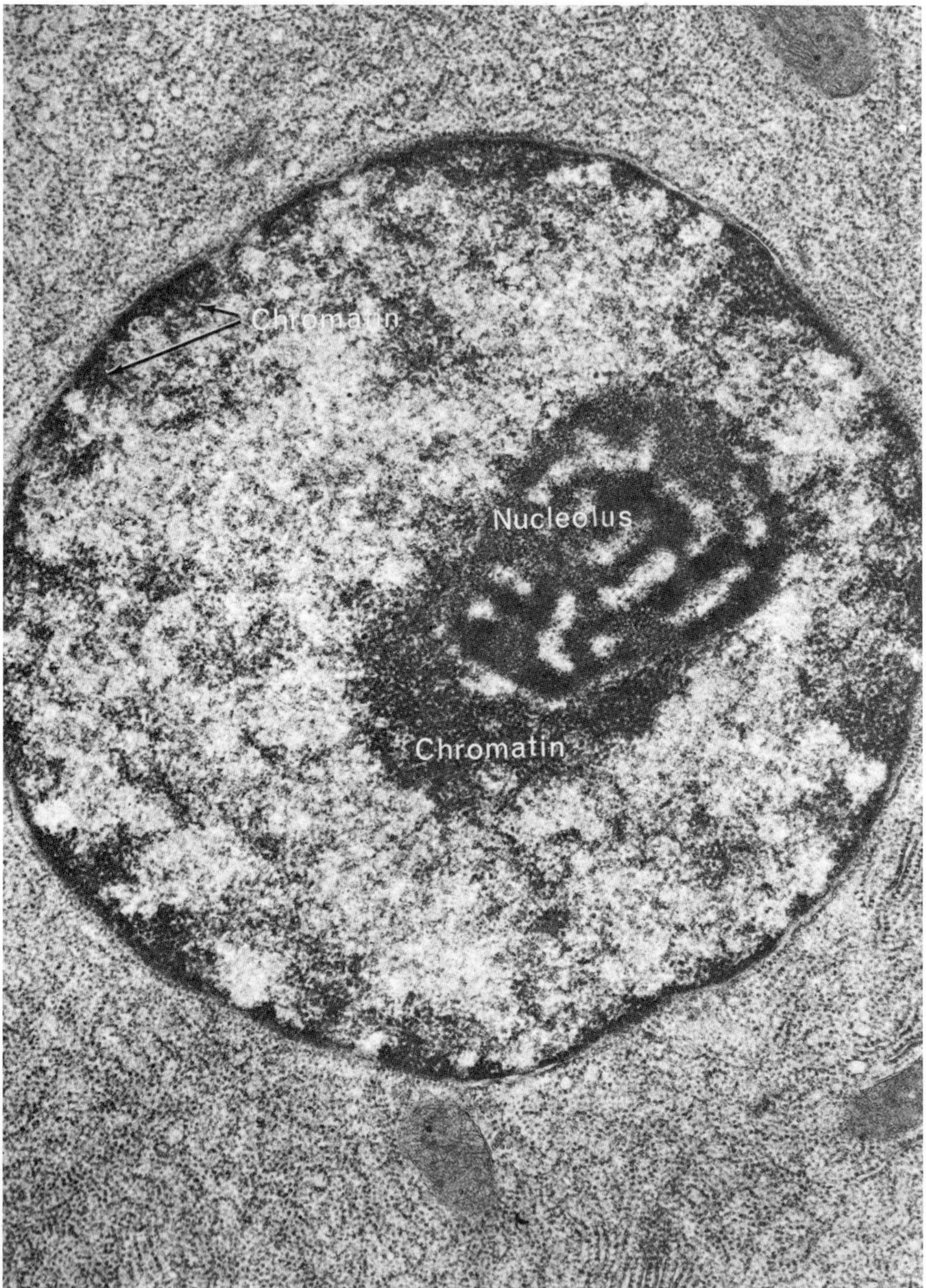

Figure 1. Nucleus of acinar cell from bat pancreas. (From Fawcett, 1966b.)

(Anderson, 1968) is a further important modification being developed to provide large numbers of clean nuclei, and the technique allows good separation of diploid and tetraploid classes (Johnston et al., 1968).

The commonest assessment of the isolation procedures is by electron microscopy (see Fig. 1). The nuclei should have intact membranes with minimal adherent microsomal material. The nucleoli are usually recognizable with their associated heterochromatin. Excessive condensation of chromatin around the periphery of the nucleus or homogeneity of nuclear contents is undesirable. Particularly critical evaluations have been made of calf thymus nuclei isolated in isosmolar sucrose by Kodama and Tedeschi (1963); these investigators concluded that intact or disrupted cells might contribute as much as 30% of the population and that intact nuclei free from tags of cytoplasm only account for a further 30%. When nuclei were passaged through 1.9 *M* sucrose, contamination with cells was greatly reduced, but rupturing of the nuclei was increased. Liver nuclei may be more robust since Maggio et al. (1963a) showed good morphology in their preparations obtained by the Chauveau method (Fig. 2). Nuclear membranes were intact; there was relatively little vesicular membrane adherent from the endoplasmic reticulum (ER) and the granulation within the nucleus did not indicate abnormal chromatin condensation around the periphery.

Assessment of the composition or function of the isolated nucleus is even more difficult because of the absence of independent means by which nuclear parameters may be measured in the intact cell. Spectrophotometric or cytochemical measurement can sometimes be used, for example, for DNA, and levels of constituents may be compared in nuclei obtained by different methods. Interpretation of the data is, however, often controversial, and proper functioning of the nuclei may require the presence of diffusable cytoplasmic factors.

3. CHARACTERIZATION OF NUCLEAR CONSTITUENTS

A. Nucleic Acids

1. DNA. Following the discovery of DNA in the nucleus by Miescher (see Hughes, 1959; Davidson, 1969), Flemming (1882) suggested that it might be a constituent of chromatin, which he defined as the substance making up the "framework" of the nucleus. Loops of chromatin that became visible during cell division were called chromosomes by Waldeyer (1888) and their importance in heredity was discussed by Weissman (1892). The chromosomal organization of genetic factors became recognized mainly as a result of Morgan's work on *Drosophila* (1919). Understanding of the function of DNA in inheritance was initiated by the demonstration by Avery et al. (1944) that transformation in *Pneumococcus* was brought about by DNA; the requisite constancy of the amount of DNA per somatic nucleus of cells of the same species was shown by Boivin et al. (1948) and Mirsky and Ris (1949).

Very considerable variations occur between organisms in amounts of DNA per nucleus (Table 1), and the average base composition of the DNA is also variable (Table 1). A rough correlation exists between DNA content and the complexity of the organism, but there are striking exceptions among the Amphibia, especially the newts. Certain functions, such as the synthesis of DNA itself, must be performed by all cell types. Closely related organisms may also be expected to have closely related DNAs.

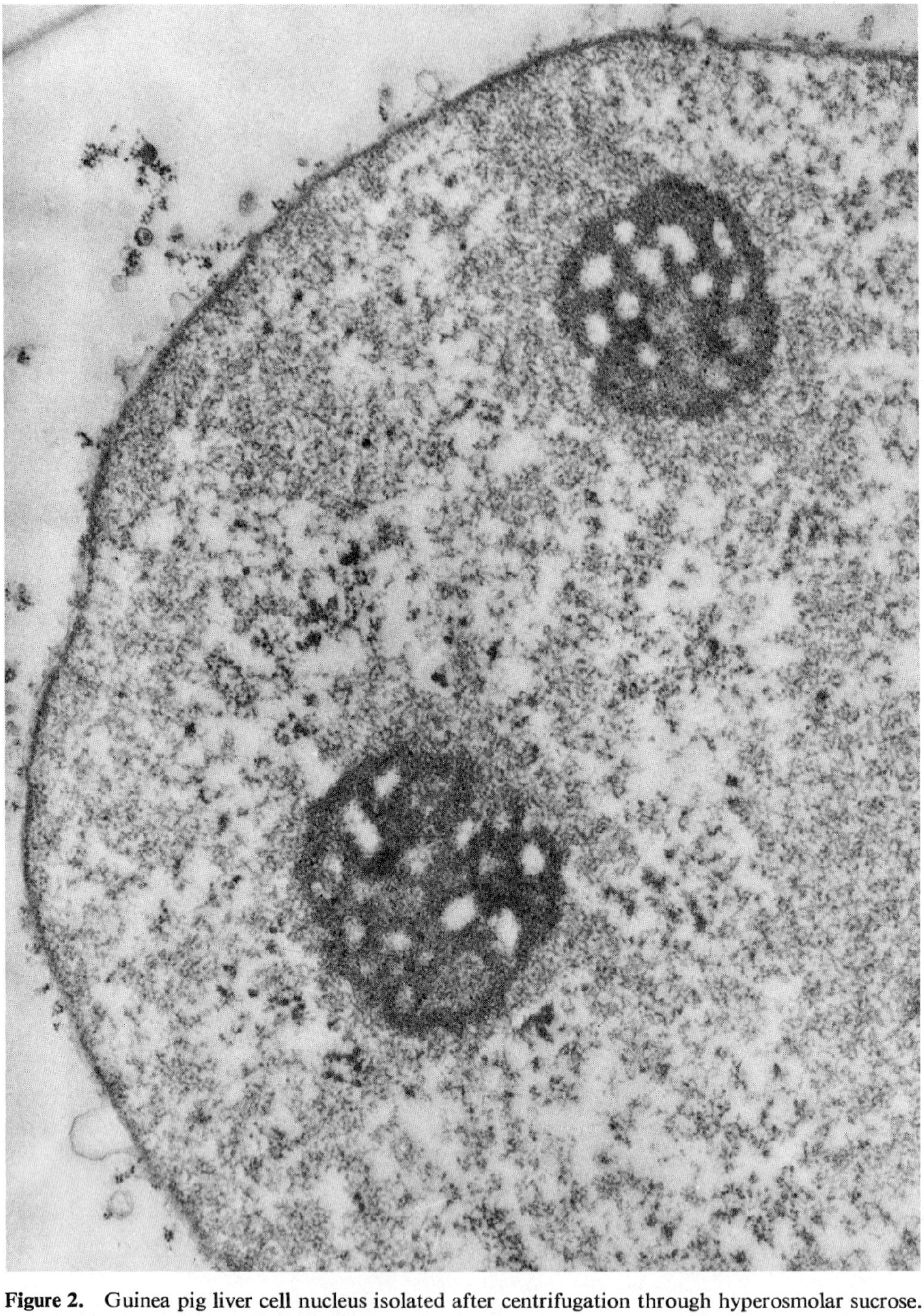

Figure 2. Guinea pig liver cell nucleus isolated after centrifugation through hyperosmolar sucrose. (From Maggio et al., 1963a.) ×30,000.

Homologies in DNA are beginning to be explored by determining the stability of duplexes formed between single-stranded DNA and DNA or oligodeoxyribonucleotides from various organisms (McCarthy, 1969). A recent finding is that the DNA of higher organisms contains nucleotide sequences that repeat much more frequently than can be expected by chance (Britten and Kohne, 1969a, b). Thus far this is a feature of the DNA of eukaryotic cells and has not been found in prokaryotes. In certain cases (Walker, 1971) repeated sequences are found sequestered into satellite DNA which is separable by density gradient centrifugation from the rest of the DNA because of differences in base composition; this satellite DNA may account for 10–14% of the genome. The high DNA content of some amphibian cells may be attributable to a high frequency of repetition (Britten and Kohne, 1969b).

Table 1. DNA Content and Composition of Various Organisms[a]

	DNA content (pg)		Composition
	Sperm	Diploid cell	(A + T/G + C)
T2 bacteriophage	—	(0.0002)	2[b]
Escherichia coli	—	(0.009)	1
Micrococcus lysodeikticus	—	—	0.32
Micrococcus pyogenes	—	—	2.1–2.2
Euglena gracilis	—	2.9	0.75
Arbacia punctulata	0.67–0.79	—	1.5–1.6
Herring	6.85	—	1.2–1.4
Rainbow trout	2.4–2.5	5.0–5.6	1.3
Rana pipiens	6.5	13.0–16.0	1.1
Triturus cristatus	—	45.0	—
Chicken	0.95–1.5	1.7–3.3	1.3–1.4
Rat	3.1–3.9	6–7	1.2–1.5
Man	1.15–3.5	5–7	1.2–1.7
Vicia faba	—	60–70	—
Lupinus albus	—	5.5	—

[a] From data compiled by H. S. Shapiro in Sober (1968).
[b] 5-Hydroxymethylcytosine included.

Various theories are being considered at present to explain the repetitions. Britten and Kohne (1969b) suggest that the repeated sequences would provide through "mutation, translocation and recombination . . . a whole range of potential genetic activity which would perhaps come into view at times long after the appearance [of the copies]." Walker and his colleagues, however (Walker et al., 1969; Hennig and Walker, 1970), believe that the sequences are preserved because they are in use during the life cycle of the organisms. It is important that reiterated sequences appear to be different in different organisms (Walker et al., 1969). Instances in which sequences are expected to be repeated are known; a well-established case of replication (Birnstiel et al., 1969) is the ribosomal cistron of *Xenopus* which is 800-fold reiterated; during oogenesis, rRNA genes are selectively replicated several hundred times in association with intense rRNA synthesis in the developing oocyte (Brown, 1967).

Although it is assumed that DNA in all cells of the organism is identical in composition and amount, the data available is not yet sufficiently rigorous to establish this. Pelc (1968) observed greater [^{3}H]thymidine uptake in some cells than can be accounted for by their known mitotic index. He suggests that this represents "metabolic DNA" which consists of extra copies of the genes active in the cell; this DNA is broken down and replaced outside the replication of the normal genome. Uncertainty still exists concerning the presence of excisases and ligases in eukaryotic cells; if they are present, their action could largely explain the phenomenon.

A role of reiteration other than in purely transcriptional events has been advanced by Callan (1967) and elaborated by Edström (1968). It is suggested that master genes may exist in germ lines and that they do not engage directly in RNA synthesis which is the function of the reiterated (slave) sequences. The master genes are thus sequestered from the remaining genetic material and are less exposed to degradation. The slaves may be represented in the chromosome as the chromomeres. Certain details of meiosis can be conveniently expressed by this concept (Callan, 1967), but the molecular mechanism by which master sequences are recognized is not at present evident.

A further suggestion is that reiterated sequences are concerned with control processes for transcription. Associated with this Crick (1971) has suggested a new model for eukaryotic chromosomes in which transcribed DNA is represented by the interband regions seen in polytene chromosomes and reiterated DNA is located in the chromomeres.

Organization of the DNA into chromosomes is considered after discussion of the protein constituents of the nucleus.

2. RNA. RNA was cytochemically detected in nucleoli by Caspersson and Schultz (1939) and by Brachet (1940). It was soon recognized that the number of nucleoli and their RNA content could be correlated with the extent of protein synthesis in the cell, as could the level of precursor uptake into nuclear RNA (see Brachet, 1957). RNA is also found in chromatin (Brachet, 1942; Monneron and Bernard, 1969), and although the function of the nucleolus in the synthesis of rRNA (Vincent and Baltus, 1960; Chipchase and Birnstiel, 1963) is becoming clarified (vide infra) (Section 5.B), identification of the molecular species of chromatin-associated RNAs and their significance is still incompletely understood.

Apparently undegraded RNA can be obtained from nuclei by extraction with phenol (Dabeva and Tsanev, 1966) and can be analyzed by sucrose gradients or electrophoresis (Dingman and Peacock, 1968). Different species of nuclear RNA can also be obtained by extraction with carefully defined phenolic solutions at different temperatures (Sibatani et al., 1959; and see Georgiev, 1967). In this way rRNA associated with the nucleolus can be separated from RNA that has a base composition similar to DNA (dRNA; Georgiev, 1967) and hybridizes effectively with DNA. This dRNA rapidly incorporates RNA precursors and is heterogeneous. It therefore has some properties consistent with messenger RNA (mRNA); DNA-like RNA has been isolated from cytoplasm (Samarina, 1964). One of the major problems in interpreting studies of rapidly labeled DNA-like RNA in the nucleus is that the kinetics of precursor incorporation indicate that a large proportion of nuclear RNA never enters the cytoplasm but is rapidly broken down again within the nucleus (see Harris, 1968). Various needs for species of nuclear RNA other than mRNA, tRNA, or rRNA have been proposed from time to time, such as a role in specific site recognition by

RNA covalently bound to protein (see Section 3.C). (Huang and Bonner, 1965). Alternatively, since under some conditions transfer of information from the nucleus to the cytoplasm is associated with availability of ribosomes and thus the synthetic activity of the nucleolus, the excessive turnover of nuclear RNA may represent RNA that has been synthesized but has failed to find transporters to the cytoplasm (Harris, 1970).

A very different and powerful technique for examining nuclear RNA molecules is that developed by Edström and his colleagues (see Beerman, 1963; Edström et al., 1969), who have combined microdissection and microanalytical procedures to investigate the biosynthesis of RNA formed in association with specific regions of DNA, even at the level of individual puffs. Direct investigation of the formation, processing, and transport of different types of RNA and their control is thus possible.

B. Nuclear Proteins

The isolation and characterization of the proteins bound in the nucleus have proved extremely troublesome. This results mainly from the overall similarity in physicochemical properties among the histones on the one hand and between the acidic or residual proteins on the other. There is also a tendency for the proteins to aggregate and to form rather intractable complexes with the nucleic acids.

1. The Protamines. Protamines are found in the sperm of many fish and invertebrates and in some birds. They are isolated by acid extraction of sperm heads or of nucleoprotamine and fractionated by the methods used for the histones (vide infra) (Section 2.B.2). In general, they are characterized by their low molecular weight (ca. 5000) and high arginine content. The three components of clupeine are defined in terms of their amino acid sequence (Table 2) and are notable for the similar clustering of the arginine residues (Ando et al., 1967).

Table 2. Amino Acid Sequences in the Clupeines[a]

Z	H_2N-	*Ala-*	Arg-	Arg-	Arg-	*Arg-*	*Ser-*	*Arg-*	*Arg-*	*Ala-*	
YII	H_2N-	*Pro-*	Arg-	Arg-	Arg-		*Thr-*	*Arg-*	*Arg-*	*Ala-*	
YI	H_2N-	*Ala-*	Arg-	Arg-	Arg-	*Arg-*	*Ser-*			*Ser-*	
	10										
Z	Ser-	Arg-	Pro-	*Val-*	Arg-	Arg-	Arg-	Arg-	Pro-	Arg-	
YII	Ser-	Arg-	Pro-	*Val-*	Arg-	Arg-	Arg-	Arg-	Pro-	Arg-	
YI	Ser-	Arg-	Pro-	*Ile-*	Arg-	Arg-	Arg-	Arg-	Pro-	Arg-	
	20										
Z	Arg-		*Val-*	*Ser-*	Arg-	Arg-	Arg-	Arg-	Ala-		Arg-
YII	Arg-		*Val-*	*Ser-*	Arg-	Arg-	Arg-	Arg-	Ala-		Arg-
YI	Arg-	*Arg-*	*Thr-*	*Thr-*	Arg-	Arg-	Arg-	Arg-	Ala-	*Gly-*	Arg-
		30									
Z	Arg-	Arg-	Arg-	CO_2H							
YII	Arg-	Arg-	Arg-	CO_2H							
YI	Arg-	Arg-	Arg-	CO_2H							

[a] Ando et al. (1967).

2. Histones. Histones have been isolated from whole tissues, isolated nuclei, and from nucleoprotein or chromatin. Care is required to avoid proteolytic degradation which can occur when prolonged purification of nuclei or chromatin is used, but considerable advantage accrues from such preliminary treatment. (For a full discussion of histones see Phillips, 1971.)

Daly and Mirsky (1955) were the first to realize that histones could be separated into a lysine-rich and an arginine-rich fraction by showing that the former could readily be removed from thymus nuclei by extraction with 0.5 *M* sodium chloride containing citric acid. They observed that the extent of dissociation of the histone from the nucleoprotein increased with the citric acid concentration and was independent of the sodium chloride. They also noted that although the nuclei were isolated in very dilute citric acid some histone could have been lost.

The use of mineral acid of increasing strength for the selective extraction of histones was introduced by Ui (1957), who treated calf thymus nuclei with 0.1, 0.2, and 0.5 *N* sulfuric acid and then precipitated the histones with ethanol.

These early techniques have been refined by Murray (1966), Dounce et al. (1966), and Murray et al. (1968), who have used acid at carefully adjusted pH values and obtained a considerable degree of selective extraction of the various histones. The lysine-rich histone is extracted with the weakest and the arginine-rich with the strongest acids.

More selective extractants than either hydrochloric or sulfuric acid are available for the isolation of the lysine-rich histone F1. Based on the observation by Bijvoet (1957) that the histone F1 was not precipitated by 5% trichloroacetic acid, Nooij and Westenbrink (1962) used it to obtain the lysine-rich fraction from calf thymus, and Johns and Butler (1962) had similar success with 5% perchloric acid.

The use of ethanolic hydrochloric acid was introduced by Johns et al. (1960), who showed that 0.25 *M* hydrochloric acid in high ethanol concentrations preferentially extracted the arginine-rich histones from calf thymus nuclei and the histones could then be precipitated by dialysis against ethanol. This technique was elaborated (Johns, 1964) to yield histones F2b, F2a, and F3. The 2a fraction was further subdivided into 2a1 and 2a2 by acetone fractionation (Phillips and Johns, 1965), or by the selective extraction of calf thymus deoxyribonucleoprotein with ethanolic guanidinium chloride and subsequent fractional precipitation with acetone (Johns, 1967).

These methods depending on selective extraction or fractional precipitation provide material which, although not without some cross-contamination, can be readily separated into homogeneous fractions by column chromatography on Amberlite, Sephadex, or carboxymethyl cellulose or by electrophoretic procedures (see Table 3).

Some of the earliest work using ion exchange resins was performed by Crampton et al. (1955), who attempted to resolve calf thymus histones by elution with calcium or barium acetate from Amberlite IRC-50. Although this method did not yield high recoveries, these investigators concluded in a later paper (1957) that there was little difference in the composition of histones from calf and pig kidney, thymus, and testis. They also described the difficulties caused by proteolysis.

A significant improvement in the resolution of histones by ion exchange chromatography was obtained by Rasmussen et al. (1962), who introduced gradient elution from Amberlite IRC-50 or CG-50 with guanidinium chloride. More recently, Kinkade and Cole (1966a), by using a lower starting concentration of guanidinium chloride in 0.1 *M* phosphate buffer at pH 6.8, resolved calf thymus histone F1 into four sub-

Table 3. Amino Acid Composition of Thymus and Erythrocyte (F2c or V) Histones[a]

	Thymus						RBC
	F1[b]	F2a1[c]	F2a2[d]	F2b1[e]	F2b2[d]	F3[f]	F2c[g]
Lys	*25.7*	9.8	*12.6*	*16.0*	*14.3*	8.9	*22.6*
His	0.1	2.0	2.8	2.4	2.4	2.4	1.6
Arg	2.5	*13.7*	9.5	6.4	8.7	*12.6*	11.5
Asp	3.1	4.9	5.6	4.8	5.7	4.0	2.3
Thr	5.6	6.9	4.9	6.4	4.8	6.6	3.4
Ser	6.4	2.0	5.0	*12.0*	3.7	3.8	10.3
Glu	5.3	5.9	8.9	8.0	8.8	12.2	4.6
Pro	*9.7*	1.0	4.2	4.8	4.4	4.7	7.0
Gly	7.4	*16.7*	9.1	5.6	9.9	5.4	5.4
Ala	*23.3*	6.9	*13.1*	10.4	*14.3*	12.9	*17.0*
Cys	0	0	0	0	0	*1.1*	(+)
Val	4.5	8.8	6.0	7.2	5.7	4.6	3.8
Met	0	1.0	0.3	1.6	0.3	1.1	0.6
Ile	0.9	5.9	4.3	4.8	3.9	5.1	3.2
Leu	4.0	7.8	*10.3*	4.8	*10.1*	9.2	4.3
Tyr	0.6	3.9	2.3	4.0	2.0	2.4	1.6
Phe	0.7	2.0	1.0	1.6	1.0	3.1	0.8

[a] Values are given in moles per 100 moles.
[b] Kinkade and Cole (1966a).
[c] Smith et al. (1970).
[d] Hnilica (1967).
[e] Iwai et al. (1970).
[f] J. Hilton, unpublished.
[g] Vidali and Neelin (1968).

fractions. The four subfractions were distinguished by their different arginine and valine contents (Kinkade and Cole, 1966b).

A very sensitive method for the qualitative resolution of histone mixtures is based on the work of Reisfeld et al. (1962), who separated basic proteins by disc electrophoresis on polyacrylamide gel (see also Shepherd and Gurley, 1966). This method has been adapted to the separation of rat thymus histone F1 into four subfractions which differ in their content of threonine, valine, and isoleucine (Buckingham and Stocken, 1970).

One of the disappointing features in any attempt to determine what differences there are among histones in different tissues of the same species or among the same tissues in different species is that no one method is certain to give the same separation when applied under different circumstances. Failure to recognize this has led to a certain amount of conflict in the literature, and the only completely satisfactory identification of a histone seems to lie in the determination of its amino acid sequence. With the aid of chromatographic procedures, two histones have been isolated in a form sufficiently pure for their amino acid sequences to be established. These are histone F2a1, described by Busch and colleagues and Bonner and colleagues (cited in Smith et al., 1970) and histone F2b (Iwai et al., 1970). Considerable progress has also been made with both F1 and F3 (DeLange and Smith, 1971).

Table 4. Amino Acid Sequences of Calf Thymus Histone F2a1 and F2b1

F2a1[a]

Acetyl- Ser- Gly- Arg- Gly- Lys- Gly- Gly- Lys-
10
Gly- Leu- Gly- Lys- Gly- Gly- Ala- Lys-
20
Arg- His- Arg- Lys- (Me)- Val- Leu- Arg- Asp-
30
Asn- Ile- Gln- Gly- Ile- Thr- Lys- Pro- Ala-
40
Ile- Arg- Arg- Leu- Ala- Arg- Arg- Gly- Gly-
50
Val- Lys- Arg- Ile- Ser- Gly- Leu- Ile- Tyr-
60
Glu- Glu- Thr- Arg- Gly- Val- Leu- Lys- Val-
Phe- Leu- Glu- Asn- Val- Ile- Arg- Asp- Ala-
70
Val- Thr- Tyr- Thr- Glu- His- Ala- Lys- Arg-
80
Lys- Thr- Val- Thr- Ala- Met- Asp- Val- Val-
90
Tyr- Ala- Leu- Lys- Arg- Gln- Gly- Arg- Thr-
100
Leu- Tyr- Gly- Phe- Gly- Gly- CO_2H

F2b1[b]

H_2N- Pro- Gln- Pro- Ala- Lys- Ser- Ala- Pro- Ala-
10
Pro- Lys- Lys- Gly- Ser- Lys- Ala- Val- Thr- Lys-
20
Lys- Ala- Gln- Lys- Lys- Asp- Gly- Lys- Lys- Arg-
30
Lys- Arg- Ser- Arg- Lys- Glu- Ser- Tyr- Ser- Val-
40
Tyr- Val- Tyr- Lys- Val- Leu- Lys- Gln- Val- His-
50
Pro- Asp- Thr- Gly- Ile- Ser- Ser- Lys- Ala- Met-
60
Gly- Ile- Met- Asn- Ser- Phe- Val- Asn- Asp- Ile-
70
Phe- Glu- Arg- Ile- Ala- Gly- Glu- Ala- Ser- Arg-
80
Leu- Ala- His- Tyr- Asn- Lys- Arg- Ser- Thr- Ile-
90
Thr- Ser- Arg- Glu- Ile- Gln- Thr- Ala- Val- Arg-
100
Leu- Leu- Leu- Pro- Gly- Glu- Leu- Ala- Lys- His-
100
Ala- Val- Ser- Glu- Gly- Thr- Lys- Ala- Val- Thr-
120
Lys- Tyr- Thr- Ser- Ser- Lys- CO_2H

Lysine residues 5, 8, 12, and 16 may be acetylated in F2a1.

[a] Smith et al. (1970).

[b] Iwai et al. (1970).

As can be seen from the data in Tables 4 and 5, the interesting feature is the uneven distribution of basic, acidic, and hydrophobic amino acids. The inference is that the neutralization of the phosphate group of DNA must be satisfied by cations other than the basic groups of the histone and that interactions of loops of hydrophobic amino acids make significant contributions to the stability of the nucleoprotein complex.

Table 5. Distributions of Residues in Histones F1, F2a1 and F2b1[a]

Histone	N-terminal region	Middle region	C-terminal region
Histone F1			
Number of residues	51	49	116
Basic	10	9	45
Acidic	6	7	3
Hydroxy amino	6	10	7
Hydrophobic	5	11	6
Proline	6	0	16
Histone F2a1			
Number of residues	45	—	57
Basic	17	—	10
Acidic	1	—	6
Hydroxy amino	2	—	7
Hydrophobic	8	—	22
Proline	1	—	0
Histone F2b1			
Number of residues	34	45	46
Basic	15	6	10
Acidic	1	5	3
Hydroxy amino	4	7	11
Hydrophobic	1	17	12
Proline	4	1	1

[a] From Iwai et al. (1970), incorporating data from Smith et al. (1970) and Bustin et al. (1969).

The elucidation of the amino acid sequence of F2a1 in different species lends support to the idea that their primary structure is conserved to a very considerable degree during the evolution of higher organisms (see Smith et al., 1970). There are, however, differences in the amino acids in corresponding histones in different species which may or may not have biological significance; for example, pea histone F3 has one cysteine residue as compared with two in calf thymus. It is well known (Rapkine, 1931) that the oxidoreduction state of the thiol groups changes during the cell cycle, and clearly an increasing complexity of polymers between the histone–DNA moiety could be formed in the thymus.

Other kinds of differences exist as shown by the different chromatographic profiles of the subfractions of histone F1 from different tissues, and the varying rates of biosynthesis of the subfractions in mammary gland at different stages of development and lactation (Bustin and Cole, 1968; Stellwagen and Cole, 1969).

3. Micromodifications in Histone Structure. Micromodifications in histone structure have been known for some time. The first to be described was acetylation of the N-terminal groups of serine (Phillips, 1963) and of the ϵ-amino group of lysine (Gershey et al., 1968). Acetylation of the arginine-rich histones has been shown to relieve the inhibition of the DNA-dependent RNA polymerase (Allfrey et al., 1964a) and, since a diminished removal of acetate groups from liver histones occurred during the first hour after partial hepatectomy, it was (Pogo et al., 1968) suggested that acetylation altered the state and function of the chromatin.

These ideas have been extended by further work from the Rockefeller group (Allfrey, 1970). Marzluff and McCarty (1970) distinguished between the stable N-terminal acetylation of histones F1 and F2a and the variable acetylation of the ϵ-amino group of lysine in histones F2a1 and F3 in mouse mammary gland explants which they too associate with the control of transcription.

The presence of ϵ-*N*-methyllysine in the arginine-rich histones of calf thymus was first noted by Murray (1964), who also found it in liver, kidney, and spleen of calf and rabbit and in lamb and rat liver. The dimethyl derivative has been found in calf thymus and chicken erythrocytes, and two monomethyl derivatives of arginine are present in calf thymus. The Rockefeller group (Tidwell et al., 1968) studied the timing of incorporation of methyl groups into histones other than the lysine-rich histone and, since the peak of incorporation was just subsequent to the peak of DNA synthesis, they again suggested that methylation was associated with the structural and functional changes known to take place just before mitosis.

Similar timing for the methylation of histones F2a and F3 in Chinese hamster ovary cell cultures was found by Shepherd et al., (1971a). Methylation of histone F2b was maximal in S phase.

Acetylation of the amino terminal serine residues on the completed protein has been shown to take place in the nucleus (Allfrey et al., 1964a), and a partial purification of two transacetylases from liver nuclei has been achieved by Gallwitz (1970). The methylation of histones has been studied extensively by Paik and Kim (1969), who obtained their enzyme from the cytosol. No fluctuations in the activities of these enzymes during the cell cycle have yet been reported.

It must be mentioned that the evidence explaining the role of acetylation and methylation in biological terms is correlative rather than causative. A similar situation exists for the phosphorylation of histones and protamines.

Histone F1 reaches a peak of phosphorylation approximately 1 hr before the peak of DNA synthesis in regenerating liver (Ord and Stocken, 1967; Stevely and Stocken, 1968). The histone kinase behaves (Pawse et al., 1969) similarly to the thymidylate kinase in its temporal relationship with DNA synthesis in the cell cycle. If the animals are exposed to ionizing radiation shortly after partial hepatectomy, there is an inhibition of F1 phosphorylation concomitantly with the inhibition of DNA synthesis. Comparable findings have been reported in CHO cell cultures (Gurley and Walters, 1971) in which thymidine uptake into DNA, but not ^{32}P incorporation or measured DNA increase, was also inhibited.

This association between histone phosphorylation and DNA synthesis is also present during the maturation of sperm. Ingles and Dixon (1967) showed that both histone and protamine are phosphorylated in trout testes during spermatogenesis: 74% of the serine in protamine is phosphorylated in the early stages, and this falls to 9% in the late stages. Two interpretations have been made of these data. First, phosphorylation may occur during or shortly after the synthesis of F1 and may be concerned with the transport of the histone into the nucleus (Dixon and Smith, 1968). Alternatively, or additionally, the phosphorylation may weaken the association of the histone with DNA and so facilitate replication (Sung and Dixon, 1970). A nuclear location (Siebert et al., 1971) for the kinase phosphorylating histone F1 (Langan and Smith, 1967), as well as the earlier established soluble form of the enzyme, has now been described. Circular dichroism and immunological studies have indicated that phosphorylated histone F1 is less effective than the nonphosphorylated histone in producing conformational changes in DNA (Adler et al., 1971).

Much of the work on micromodifications has relied on isotopic evidence. Some caution is required in its interpretation and this is evident from the careful study by Shepherd et al. (1970), who found small quantities of nucleic acids in all histone fractions. Claims for synthesis based on isotopic incorporation must be validated either by showing a net phosphate increase or by a demonstration that no phosphate exchange takes place. Extensive turnover in phosphoprotein in intact nuclei has been found by Kleinsmith et al. (1966) and in histone F1 phosphate (Pawse et al., 1971).

Some differences are found between the phosphorylation of F1 in nuclei and in a soluble system. The four subfractions of F1 are each phosphorylated when the fractions are treated separately, but when phosphorylation takes place in nuclei only three of the four accept phosphate. This indicates that *in vivo* steric factors are important in any role played by histone phosphorylation. Variations in the phosphorylated peptides from histone F1 from different sources have been reported by Jergil et al. (1970), and from rabbit thymus F1 a form has been isolated in which alanine replaces serine, thus preventing phosphorylation (Langan et al., 1971). These subclasses of F1 and the further differences now emerging suggest that the highly conserved sequence found in histone F2a1 may not be a general feature of all species of histones.

Evidence for the phosphorylation of histones other than F1 (Dixon et al., 1969; Ord and Stocken, 1969; Turkington and Riddle, 1969; Fitzgerald et al., 1970) indicates that the phosphorylation of these histones takes place in liver, mammary gland, and pancreas when a surge of new RNA is being made. Phosphorylation of histone F2b (Shepherd et al., 1971b; Sung et al., 1971) or F2a2 (Gurley and Walters, 1971) just before S phase has also been described.

All the micromodifications so far described can be demonstrated in isolated nuclei. Marushige et al. (1969) found that histones are phosphorylated as nucleohistone in chromatin. All the major histone fractions were phosphorylated at the terminal stages of spermatogenesis and it was speculated that, since the phosphate is carried as the terminal *N*-acetyl-*O*-phosphoseryl-glycyl-arginine, a localized change in charge could be an unzipping point for the dissociation of histone–DNA prior to conversion to protamine–DNA.

At present, it is not clear whether the phosphorylation of the different histones is catalyzed by the same kinase (Langan and Smith, 1967) or what relation this kinase has to other protein kinases. The activation of histone kinase by cyclic AMP (Langan, 1969) offers a means whereby extracellular agents influencing adenyl cyclase can affect nuclear proteins and thus, potentially, the organization of the genome.

One of the situations most readily altered by normal metabolic processes is the redox state of a thiol–disulfide system. Glutathione reductase occurs in nuclei, and by altering the status of glutathione a simple thiol–disulfide exchange alters the equilibrium of the sulfur-containing proteins. Relevant to this is the observation (Ord and Stocken, 1966) that active chromatin has a higher thiol content than dense chromatin. Vidali and Neelin (1968) also showed that the histone of the mature avian erythrocyte is in the disulfide form.

Changes in the thiol content of histone F3 occur during the mitotic cycle in *Echinus* and lymphocyte cultures (Ord and Stocken, 1968; Cross and Ord, 1970) and are at a maximum when DNA synthesis and structural reorganization of the nucleus are taking place. Sakai (1967) has described in sea urchins a sulfur-containing contractile protein from the cell cortex and another from the mitotic apparatus whose oxido-reduction states change reciprocally. What connection there might be between these proteins in developing sea urchin eggs and the thiol histones is not clear.

From this evidence it appears that the disulfide-to-thiol shift in histones takes place at times of DNA synthesis and intense structural reorganization in the nucleus.

4. **Histone Deviations.** There are examples in the literature both of unusual histones and of unusual distribution of histones in certain plant and animal species. Reference has already been made to the pea bud histone F3 which contains only one cysteinyl thiol group as compared with two in F3 from thymus. The mature sea urchin egg lacks F1 (Backström, 1965; Ord and Stocken, 1968), and a histone containing a large amount of serine and basic amino acids has been found in nucleated erythrocytes of birds, fish, and frogs (Table 3) (Edwards and Hnilica, 1968). Sea urchin sperm chromatin has a protein containing high lysine, proline, and alanine (Paoletti and Huang, 1969), but whose other amino acids distinguish it from F1. Panyim and Chalkley (1969) found a very lysine-rich histone in tissues with low mitotic activity, such as lung, which was not present in thymus and tumor tissue; Sheridan and Stern (1967) discovered a "meiotic" histone in lily and tulip anthers, which was practically absent from somatic tissues of the same species. A similar notable difference is found in the newt, in which F1, F2, and F3 are present in adult liver but absent in the preearly gastrula stage (Asao, 1969).

These deviations seem to involve histones whose individual or combined composition results in the occurrence of strongly basic proteins in situations in which there is minimal mitotic and metabolic activity; possible micromodifications of their structure to release their repressor ability have yet to be discovered.

5. **Nonhistone Proteins.** Some difficulty has been experienced in correlating information in the literature about proteins other than histones because of the widely different experimental techniques used to obtain material that is heterogeneous and not well characterized (see Busch, 1965). We have therefore been compelled to use the terms "nonhistone protein" and "acidic protein" indiscriminately. It would be helpful if in the future the fractions could be related to "nonhistone protein," "alkali-soluble protein," and "residual protein," as described below.

To obtain these proteins nuclei are commonly extracted with isotonic saline or dilute tris buffer at pH 7.6 to remove nuclear ribonucleoproteins. Chromatin is then extracted by 1 *M* sodium chloride (Allfrey et al., 1955; Wang and Johns, 1968) (or 2 *M* sodium chloride; see Busch, 1965) and the extract dialyzed against 0.15 *M* sodium chloride to precipitate DNA–histone. Nonhistone proteins remain in solution. Further acidic protein (alkali-soluble protein) may be obtained by treating the nuclear residue with dilute sodium hydroxide. The final sediment (residual protein) contains hydroxyproline, probably derived from collagen contamination. Johns and Forrester

(1969) showed, by using calf thymus nuclei washed with isotonic saline, that in this tissue 0.35 *M* sodium chloride extracts nonhistone protein similar to the alkali-soluble protein described above. Detergents have also been used by some workers to solubilize nonhistone proteins (Patel et al., 1968). Nonhistone proteins contain tryptophan (Wang, 1966) and large amounts of cysteine.

The functional importance of acidic proteins resides in their heterogeneity and in their capacity to form insoluble complexes with histones, which may facilitate information transfer from DNA. Frenster (1965) reported that metabolically active, diffuse chromatin contained significantly more nonhistone protein than repressed chromatin. Clever (1967) demonstrated cytochemically that chromosome puffs, with their high rate of RNA synthesis, are enriched with respect to acidic proteins. Acidic protein added together with histone to a DNA-dependent RNA polymerase system significantly reduced the repression normally caused by the presence of histones (Wang, 1968).

It is noteworthy that chromatin-associated nonhistone proteins were found by Wang and Johns (1968) to be soluble in 0.25 *M* hydrochloric acid, as were the complexes of nonhistone protein and histone. These complexes remained near the anode on gel electrophoresis and probably contribute to the slow-moving components found on electrophoresis of 0.25 *M* hydrochloric acid extracts of total histones from various sources (for example Echinus, Ord, and Stocken, 1970). The use of 0.2 *N* sulfuric acid rather than 0.25 *M* hydrochloric acid may reduce the amount of contaminatory nonhistone–histone complex, especially with metaphase nuclei (Sadgopal and Bonner, 1970).

The preparations of nonhistone protein so far described are all heterogeneous; reports are now being made of variations in electrophoretic pattern of nuclear acidic proteins among different tissues of the same species or among the same tissues of different species (Loeb and Creuzet, 1969). Still more significant are the hybridization studies of Paul and Gilmour (1968), who showed that specificity in RNA synthesis could be demonstrated using chromatin as template, provided that the nonhistone proteins were present. If these were removed and only histone–DNA complexes reconstituted, nonspecific transcription occurred. Techniques for demonstrating the presence in nuclei of extremely small amounts of specific proteins are only starting to be used, but the results reported thus far suggest that such experiments are very valuable (Kleinsmith et al., 1970; Teng et al., 1971).

A special class of nonhistone proteins comprise the phosphoprotein studied by Langan (1967). Following the earlier investigations of Lipmann of the enzymology of phosvitin formation, Langan and Smith (1966) isolated a phosphoprotein from liver nuclei and noted that it formed an insoluble complex with histones. They also showed that the 80% inhibition produced by histones on DNA-dependent RNA synthesis was reduced to 20% when the phosphoprotein was added to the system.

A similar phosphoprotein was isolated from calf thymus nuclei by Kleinsmith et al. (1966), who showed that it had a high rate of phosphorus turnover. These investigators and Ord and Stocken (1966) observed that histones were also phosphorylated.

The significance of phosphoproteins as opposed to phosphohistones in nuclear metabolism is not yet completely understood, although many conjectures have been made. Diffuse or active chromatin, so defined because of its ready capacity for incorporating precursors into RNA, contains very much higher amounts of phosphoprotein as compared with dense or inactive chromatin (Frenster, 1965). Because of this, and since polyanions act as depressors of RNA synthesis, it has been suggested that their presence is associated with active protein and RNA synthesis (Langan, 1968). Evidence indicating the heterogeneity and tissue specificity of phosphoproteins is emerging (Platz, et al., 1970).

More tangible evidence has been provided by Kleinsmith et al. (1966), who stimulated lymphocytes with phytohemagglutinin and noted that the increased RNA synthesis was accompanied by increased uptake of $^{32}PO_4$ into phosphoprotein. A similar temporal relationship between histone and nonhistone protein phosphorylation and RNA synthesis was observed by Turkington and Riddle (1969) when mammary epithelial cells in organ culture were stimulated by insulin.

C. Chromatin

Chromatin is the complex association of DNA with proteins, RNA, and possibly lipid, whose composition is thought to be related to that of the chromosomes, which are normally isolated only for analysis at metaphase (see below). By weight the proportions of DNA/histone/nonhistone protein/RNA are about 1:1.5:0.3–0.6:0.03–0.05 (see Fambrough, 1969), higher levels of nonhistone protein and RNA being found in nuclei from cells active in protein synthesis (see Frenster, 1965). We are concerned with the molecular interaction among DNA, proteins, and RNA in this material under different conditions of genetic activity.

Our information comes principally from physicochemical studies on soluble nucleoprotein from calf thymus employing the method of Zubay and Doty (1959). They obtained 85% of the DNA in the sol form, nonsedimentable after 30 min at 70,000× *g*. Great difficulty, however, may be experienced in obtaining high yields of material as sol (for example, Champagne and Mazen, 1966) and with other tissues, for example, liver; degradation of high-molecular-weight material is easily produced by nuclease and/or protease attack, so devaluing subsequent physicochemical studies.

The nucleoprotein obtained from calf thymus has been shown by x-ray diffraction (Pardon et al., 1967), sedimentation (Ohba, 1966), optical rotatory dispersion, and other studies (Bradbury et al., 1965; Bradbury et al., 1967) to exist as a supercoiled helix. 0.5 *M* sodium chloride selectively removes histone F1 (Ohba, 1966; Ohlenbusch et al., 1967) and, by raising the sodium chloride concentration stepwise to 2 *M*, it is possible to remove increasing amounts of histone from the nucleoprotein complex, with histone F2a1 probably the last to be dissociated (Henson and Walker, 1970). Thermal denaturation, ORD, viscosity, and sedimentation measurements indicate (Bradbury et al., 1967; Bradbury, 1969; Henson and Walker, 1970) that the most easily removed histone, F1, contributes to the thermal stability but has little effect on the supercoiling. This is most profoundly affected by the removal of histones of the F2a–F3 group, and Henson and Walker (1970) suggest that histone F2a1 may be distributed at regular intervals along the DNA molecule, providing the necessary symmetry for supercoiling to occur. These findings are consistent with studies on individual histones in which present evidence suggests that F1 probably exists in the β-chain conformation (Boublik et al., 1970) and that the other histones may assume up to 30% helical content as in nucleoprotein (Bradbury, 1969).

Sequence studies referred to earlier suggest regions on the histones which may be expected to "stick" to the DNA: Cohen and Kidson (1968) showed that polylysine preferentially stabilizes A–T regions of DNA, but we have no indications that specific interactions occur between the basic residues and regions on the DNA helix although methyl groups from thymine project into the grooves and could provide means of recognition. It is also probable that the hydrophobic regions on the histones could allow specific interactions between nuclear proteins, their positively charged regions then binding less specifically but in an ordered manner to the DNA. The presence of thiol groups on histones and nonhistone proteins suggest means by which interhelical cross-bridges can be built up.

A different approach to the study of chromatin has come from attempts to produce chromatin preparations *in vitro* that mimic cytochemical or biochemical properties *in vivo*. Littau et al. (1964), using autoradiography, reported that RNA synthesis occurs in diffuse regions in the nucleus, as distinct from regions that appear dense in the electron microscope and which contain the bulk of the DNA. When washed nuclei from isotopically labeled animals were sonicated, it was possible to obtain two fractions of nucleoprotein, one sedimenting after $1000\times g$ for 10 min, which was inactive with respect to a precursor uptake, and a much smaller fraction of nucleoprotein which sedimented at 60 min at $78{,}000\times g$ and which showed high radioactivity (Frenster et al., 1963). Extracting nuclei with 0.1 *M* citric acid to remove F1 reversibly converted the dense clumps of chromatin in thymus nuclei to a diffuse network of fibers (Littau et al., 1965), indicating that the very lysine-rich histone may be involved in cross-linking nucleoprotein helices in "switched-off" chromatin.

Many workers have shown that metabolically active chromatin has histones that differ only quantitatively, if at all, from those in inactive chromatin. The microstructure of the histones is however, different; active chromatin contains histones that are more phosphorylated and more reduced than inactive chromatin (Ord and Stocken, 1966; Stevely and Stocken, 1968). As discussed in Section 3.B.5, active chromatin also contains a higher concentration of various polyanions (Frenster, 1965) to which is currently attributed its greater capacity for supporting RNA synthesis.

Duerkson and McCarthy (1971) showed, using controlled conditions for shearing chromatin, that inactive, easily sedimentable fractions from crab testicular cells are enriched with redundant satellite DNA.

A further approach to the study of protein–DNA interactions in chromatin has been the attempts to reconstitute chromatin from DNA and nuclear proteins with the same chemicophysical characteristics and the same properties with respect to RNA synthesis as shown by native chromatin. Success was first reported by Huang and Bonner (1962), but many workers encountered difficulties in making soluble preparations reconstituted from histone and DNA. The association appears random (Johns and Butler, 1964), the solubility of the complex being critically affected by the divalent ion concentration and the relative proportions of histone and DNA (Johns and Hoare, 1970). Specificity of RNA synthesis supported by reconstituted chromatin has been described by Paul and Gilmour (1968) when nonhistone protein was retained in the preparation (see Section 3.B.5), and by Bonner and his colleagues when dissociation and reconstitution were performed in the presence of 5 *M* urea which prevented hydrolysis of RNA covalently associated with acidic protein in the preparation (Bekhor et al., 1969; Huang and Huang, 1969; but see Heyden and Zachau, 1971). The virtues of RNA associated with protein as a means of providing specificity of association to DNA are obvious; extension of this work is urgently required.

4. NUCLEAR ENZYMES

Nuclear enzymes fall into three classes: (1) enzymes such as the nucleic acid polymerases which are believed to occur principally in the nucleus and are involved in the metabolism of nuclear constituents themselves; (2) enzymes such as NMN adenyl transferase, concentrated significantly in the nucleus, which are important in extranuclear metabolism, and (3) enzymes that are detectable in the nucleus but which probably arise from contamination and play no special part in nuclear function (see Roodyn, 1959; Siebert, 1968). We consider here only the first two groups, and the problems of ATP and protein synthesis by certain types of nuclei.

A. Nucleic Acid Polymerases

DNA polymerase (DNA nucleotidyl transferase, EC 2.7.7.7) is the enzyme most obviously expected to be concentrated within the nucleus. Conditions necessary to demonstrate this are discussed in Section 2. The properties of polymerase I in *Escherichia coli* are being clarified (see Richardson, 1969); for replication of biologically active DNA, the enzyme requires single-stranded DNA as a template and a 3′-hydroxy-terminated primer onto which the deoxynucleoside phosphates are added. DNA polymerase in calf thymus was first studied by Bollum (1960); the use of undenatured DNA as template has been reported for polymerases from various nucleated cells (see Keir, 1965; Adams and Lindsay, 1969; Howk and Wang, 1970) but, in general, the mechanism by which undirectional replication of chromosomes is achieved is still speculative.

The periodicity of DNA synthesis in the cell cycle has suggested that there may be fluctuations in the level of the synthesizing enzyme. DNA synthesis (the S period) commonly occurs about the middle of interphase, but in certain types of cells DNA is made immediately following mitosis, so that the G_1 period, if any, is very short. In regenerating liver and some synchronized cell cultures, increased levels of DNA polymerase have been reported in the nucleus during S phase, while the level of the enzyme found in the cytosol may be invariant (Friedman, 1970), or may increase (Adams and Lindsay, 1969). It must be mentioned, however, that interpretation of levels of soluble enzyme that may traverse the nuclear membrane presents difficulties when one is interested in their strict localization. Induction of DNA polymerase, and/or its nuclear location, may therefore be a required step for DNA synthesis although increased activity of the enzyme is not usually thought to be the critical rate-determining stage.

In addition to DNA polymerase, nuclei may also contain a terminal nucleotidyl transferase catalyzing the addition of single nucleoside triphosphates to 3′-hydroxy ends of DNA strands. Both deoxyribonucleoside and ribonucleoside triphosphates can be utilized (Krakow et al., 1961). The significance of this enzyme is uncertain.

A further enzyme that may be involved in DNA synthesis is polynucleotide ligase which covalently links single oligodeoxynucleotide strands. Again, the enzyme in *E. coli* and in organisms able to repair damage produced by ultraviolet radiations has been extensively studied. Repair of DNA following ultraviolet and x-irradiation in eukaryotic cells is also believed to occur (Lett et al., 1967) and might be expected to require the presence of a ligase and/or excisase. The former enzyme has now been described in spleen nuclei (Lindahl and Edelman, 1968); few details of its mechanism are yet available.

RNA polymerase was first studied in nuclei by Weiss (1960). The nuclear enzyme is firmly bound; two types of RNA synthesis occur. An ammonium sulfate-activated enzyme II is present, which is stimulated by Mn^{2+} (Chambon et al., 1965) and synthesizes RNA with a base composition similar to DNA (Widnell and Tata, 1966). In the absence of ammonium sulfate, Mg^{2+} is the optimal divalent cation and rRNA is formed (polymerase I). The high concentration of ammonium sulfate may be expected to cause dissociation of protein and thus to expose long regions of DNA as template. The difference in divalent cation requirement and the pH optimum of the two activities, however, may indicate that different enzymes are concerned in the syntheses (Widnell and Tata, 1966; Roeder and Rutter, 1969). Consistent with the existence of at least two forms of the enzyme, polymerase II is preferentially inhibited by α-amanitin (Novello and Stirpe, 1970). rRNA-synthesizing ability is found in the nucleolus (Hurlburt et al., 1964), and the Mn^{2+}, ammonium sulfate-activated synthesis of dRNA on chromosomes occurs in the extranucleolar region (Maul and Hamilton, 1967; Roeder and Rutter, 1970).

A highly purified soluble enzyme has also been isolated from nuclei. This enzyme requires divalent ions and shows a slight preference for Mn^{2+}; it is also stimulated by ammonium sulfate (Siefart and Sekeris, 1969). It may therefore represent the species concerned in the synthesis of dRNA or, alternatively, properties of the enzyme may be altered by its binding *in vivo*.

Chesterton and Butterworth (1971) made a detailed study of two forms of RNA polymerase I from rat liver nuclei, which differ in their extractability at different sodium chloride concentrations. Our lack of understanding of the different species of nuclear RNA and of the factors involved in their syntheses however, precludes conclusions about the number of different forms of RNA polymerase involved.

A further process associated with nucleic acid synthesis is methylation. Methylation of RNA was first reported by Borek and his associates (see Starr and Sells, 1969) and occurs both on the bases and on the 2′-hydroxy group of the ribose. Methylation of tRNA is predominantly cytoplasmic, while that of rRNA occurs in the nucleolus. The specificity of the methylases and the significance of the methyl groups on rRNA are under investigation.

B. NMN Adenyl Transferase (EC 2.7.7.1)

The nuclear site for NAD synthesis was established by Hogeboom and Schneider in 1952; later studies showed that the enzyme was more precisely nucleolar (Baltus, 1954). Such a location for an enzyme prominently involved in intermediary metabolism is somewhat surprising, but it has been suggested (Haines et al., 1969) that its presence might provide a supply of NAD which may, by analogy with the better understood enzyme in *E. coli*, be essential for the action of polynucleotide ligase, and so be required for DNA synthesis. NAD is also the source of poly ADPribose which has been suggested as a regulator of DNA synthesis (Burzio and Koide, 1972).

C. ATP- and Protein-Synthesizing Enzymes

In most cell types oxidative ATP generation is an exclusive function of the mitochondrion, but in 1957, Osawa et al. reported that calf thymus nuclei isolated in isosmolar sucrose retained nucleoside phosphates and could oxidatively convert AMP to ATP. This observation has been extensively confirmed both in thymus nuclei and in nuclei from, for example, lymph nodes and bone marrow (Creasey and Stocken, 1959). The reaction does not occur in liver, kidney, or brain nuclei isolated in aqueous media (Creasey and Stocken, 1959; Klouwen et al., 1965) and has been most intensively studied in thymus. It is now agreed (see Conover, 1970b) that it does not result from contamination with whole cells or mitochondria.

Three aspects of nuclear phosphorylation are considered: the mechanism of the reaction, its importance in thymocytes, and the possibility of more widespread distribution of the reaction. Although thymus nuclei isolated both by nonaqueous and by aqueous techniques contain enzymes and substrates associated with glycolysis, the pentose phosphate pathway, and the citric acid cycle (Ord and Stocken, 1962a; McEwen et al., 1963b); the substrates actually used in the reactions *in vitro* are uncertain. Levels of endogenous glycogen and glucose in isolated nuclei seemed too low (McEwen et al., 1963b). Similarly, although exogenous pyruvate, in contrast to the di- and tricarboxylic acids, can penetrate the nuclear membrane and become oxidized (McEwen et al., 1963c) and extramitochondrial glucose and pyruvate might be available to the thymus nuclei *in vivo*, they cannot account for the endogenous oxygen uptake of isolated nuclei. Recent work by Betel (1969) and Dancheva (1969) suggests that ribose phosphate may be a significant substrate *in vitro*.

The mechanism of oxygen utilization has now been clarified; Betel and Klouwen (1967) obtained clear evidence for the existence of a respiratory chain, and this has now been conclusively demonstrated (Betel, 1967a; Conover, 1970a). The components are very similar to those in thymus mitochondria and support the conclusion advanced by Betel and Klouwen in 1967 that the mechanisms of nuclear phosphorylation do not differ essentially from the mechanism of oxidative phosphorylation in mitochondria.

One further problem, however, remains—the origin of nucleotide acceptors. From the first observations by Osawa et al. (1957), it was clear that exogenous adenine nucleotides were very poorly utilized and, indeed, are normally degraded by isolated nuclei to adenosine and adenine (McEwen et al., 1963a). Although nucleoside phosphates are not found in liver, kidney, or brain nuclei when they are prepared by aqueous procedures, phosphates are commonly present when nuclei are isolated by nonaqueous techniques (Siebert, 1961) and must be available for nucleic acid synthesis within the nucleus. Atractyloside, which blocks adenine nucleotide transport in mitochondria, did not affect nuclear phosphorylation (Betel and Klouwen, 1967), but it is not known whether or not this inhibitor has any effect on the accumulation of endogenous adenine nucleotide which is then phosphorylated.

Most of the factors involving anion transport that have been studied in connection with nuclear phosphorylation suggest that the nuclear membrane behaves similarly to that of the mitochondrion in these respects (see also Section 5).

According to Conover (1970b), it is likely that the respiratory components are located in the nuclear membrane. This investigator has suggested that the nuclear respiratory and energy-generating systems may have arisen simply from the fact that 60% of the thymus cell volume is occupied by the nucleus. As far as is known at present, some components of the electron transport system are synthesized extramitochondrially and subsequently packaged into mitochondrial membranes. When the nucleus is so prominent, packaging into its membranes could also occur. Alternatively, when there are so few mitochondria, such an intranuclear location in the thymocyte and lymphocyte may be a means of increasing the capacity of the cell to generate ATP; careful studies with liver nuclei have failed to provide convincing evidence for the existence in their membranes of a similar electron transport system which is unlikely to be required in cells with higher mitochondrial contents. It is interesting that nuclear generation of ATP in contrast to that of mitochondria is extremely sensitive to x-irradiation (Creasey and Stocken, 1959; Klouwen and Betel, 1963). The phosphorylation mechanism is unaffected, but the amount of nucleotide bound by the nucleus is reduced (Ord and Stocken, 1962b; Betel, 1967b) so that phosphate acceptors are no longer available. The reason for this is still unknown.

Protein synthesis in nuclei presents a problem similar to that of ATP generation. Isolated thymus nuclei incorporate amino acids into nuclear protein (Allfrey et al., 1957, 1964b); in most respects the requirements are identical to those of the cytoplasmic system, but there is a marked Na^+ dependency which appears to be the result of a Na^+ requirement for amino acid uptake by the nuclei (Allfrey et al., 1961). The incorporation is associated with ribonucleoprotein particles which are particularly evident in and around the nucleolus (Frenster et al., 1960). The particles can be extracted with neutral salt solutions (Frenster et al., 1960; Wang, 1961) and have a high $G + C/A + T$ ratio similar to that of cytoplasmic ribosomes (Wang, 1962). An ability to incorporate amino acids into nuclear proteins has been found in cells from tissues other than the thymus, such as those from developing peas (Birnstiel et al., 1962).

The proteins that most obviously might be made within the nucleus are histones or other nuclear proteins; it is likely, however, that histones are synthesized on cytoplasmic disomes and subsequently migrate to the nucleus (Borun et al., 1967; Gallwitz and Mueller, 1969). The origins of the nonhistone proteins are unknown.

Ribosomes are also found on the cytoplasmic surface of the nuclear membrane. Again, there is no evidence for the synthesis of any particular type of protein at these sites and, indeed, the presence of ribosomes within the nucleus might be entirely coincidental or consequential upon the assembly of ribosomes in the nucleolus and have no special significance in protein synthesis. It is possible, however, that the intranuclear location is important in some cell types because of the paucity of cytoplasmic ribosomes and/or ER.

5. NUCLEAR ORGANELLES

The relation between the chemical constituents of the nucleus and their organization into functioning organelles is one of the largest gaps in current eukaryotic cell biology. This is further compounded by difficulties in relating genetic information of considerable sophistication, which is available for some organisms, to the molecular structure of the chromosome. In this section we attempt only to draw attention to the varying approaches being used to study three intranuclear organelles: the chromosomes, the nucleolus, and the nuclear membrane. Structural and functional changes in nuclei during mitosis and meiosis have been authoritatively reviewed by Mazia (1961) and Rhoades (1961) and, more recently, by Forer (1969), Stern and Hotta (1969), and Callan (1969) and so are not considered here.

A. Chromosomes

In Section 3.C evidence was presented that isolated deoxyribonucleoprotein exists in solution as a supercoil, with histones F2a and F3 being essential for the maintenance of the coil. At present we do not know whether or not different chromosomes differ in the families of histone and nonhistone protein with which they are associated, although the presence of increasing amounts of nonhistone protein and polyanions is thought to be a feature of euchromatin (Frenster, 1969). Metaphase chromosomes can now be isolated (Maio and Schildkraut, 1967) and separated into groups differing by size; preliminary experiments (Huberman and Attardi, 1967) have so far shown that rRNA hybridizes exclusively with the group of small chromosomes known to carry the nucleolar organizer. Annealing studies have also been carried out on mouse chromosomes using mouse satellite DNA (Jones, 1970); satellite DNA is bound preferentially around the centromere. Other experiments have indicated satellite binding in the nucleolus (Schildkraut and Maio, 1968), suggesting that satellite DNA is concentrated close to regions of heterochromatin (Jones, 1970).

Most of our information on the molecular structure of chromosomes has come from electron microscopy and cytochemistry of giant polytene chromosomes in dipteran larval and imaginal tissues (for references prior to 1957, see Ris, 1957). The interphase polytene chromosome has a banded structure with bands (chromomeres) staining strongly for DNA. There is very good evidence indicating that the chromomeres represent the genetic functioning unit (Pelling, 1966; Berendes and Beerman, 1969). Phenotypic alterations in morphology and function can be correlated with changes in individual bands. These bands may also be selectively activated at different periods of development as shown with lampbrush chromosomes in the diplotene stage in amphibian oocytes (Callan, 1963; 1969; Gall, 1963) and in the puffs associated with polytene chromosomes in *Chironomus* (Clever, 1961, 1968; Beerman, 1963).

Early electron microscopy indicated the presence of DNA–protein fibers about 200 Å in cross section (Ris, 1957). Experiments with Feulgen staining suggested that the interband regions in polytene chromosomes did not contain DNA. During the 1950s the molecular weight of DNA from higher organisms was thought to be only approximately $1–10 \times 10^6$ daltons, so that to obtain the known dimensions of the chromosome a multistranded model was proposed, the DNA-containing chromomeres being linked by exclusively protein regions.

During the past 10 yr evidence has accumulated, however, suggesting that a multistranded model is not essential and that DNA might run continuously through the length of the chromosome (for discussion see Swift, 1965). Lezzi (1965) found that DNAase treatment fragmented unfixed polytene chromosomes at the interbands, suggesting that the DNA molecule continued through those regions although in lower concentration than in the chromomeres and possibly without the protection from nuclease attack afforded by the presence of histones. Similar results were obtained for lampbrush chromosomes (Callan, 1963; Gall, 1963). Electron microscopy of chromatin spread on surface films (Cairns, 1963; Kleinschmidt, 1967) has been extended to eukaryotic cells by Huberman and Riggs (1966), and very long filamentous molecules have been seen with cross-sectional diameters of 100 Å; when protein is removed by treatment with pronase, DNAase-sensitive fibers of 20–30 Å remain (Ris, 1969). Isolation of DNA under conditions of minimal shear has led to higher and higher molecular weights ($>10^9$ daltons). Evidence from autoradiographic studies can be accommodated in either model. Discontinuous uptake of [^{3}H]thymidine is firmly established (Plaut, 1963), and several replicating points have now been shown for mammalian chromosomes (Huberman and Riggs, 1966, 1968).

Closely related subspecies may differ sharply in the details of their chromosome structure. In *Chironomus thummi thummi* and *C. thummi piger*, homologous chromomeres differ in their contents of DNA, which are, however, geometrically related (Keyl, 1965); differences in amount and type of reiterated DNA were discussed earlier. It is also possible to find closely related species differing in their number of chromosomes, for example the cotton rat, *Sigmodon hispidus* in the southeastern United States has $2n = 52$ and in Arizona $2n = 22$ (Zimmerman and Lee, 1968). Indeed, at present we are very ignorant about significant factors in packaging DNA into chromosomes and, although linkage maps are becoming available for increasing numbers of higher organisms, interpretation in terms of identifiable protein synthesis is far from complete [see, for example, Renwick (1969) for human chromosomes; Mitchell (1967) for *Drosophila*; and Gross (1969) for fungi).

Recently, two new techniques have been introduced which may offer additional means of relating genetic control of enzymes to different chromosomes. Freed and Mezger-Freed (1970) developed two haploid lines from cells of frog embryos. Once pure stable cultures of this type become available, analysis of recessive mutants will be greatly facilitated. Another approach is to fuse cell lines carrying different characters and to study the properties of the resulting heterokaryons. Chromosomal instability is found, and provided unequivocal means of distinguishing donor and host chromosomes are available, assignment of enzymes to particular loci becomes possible (for discussion of this topic, see Harris, 1970).

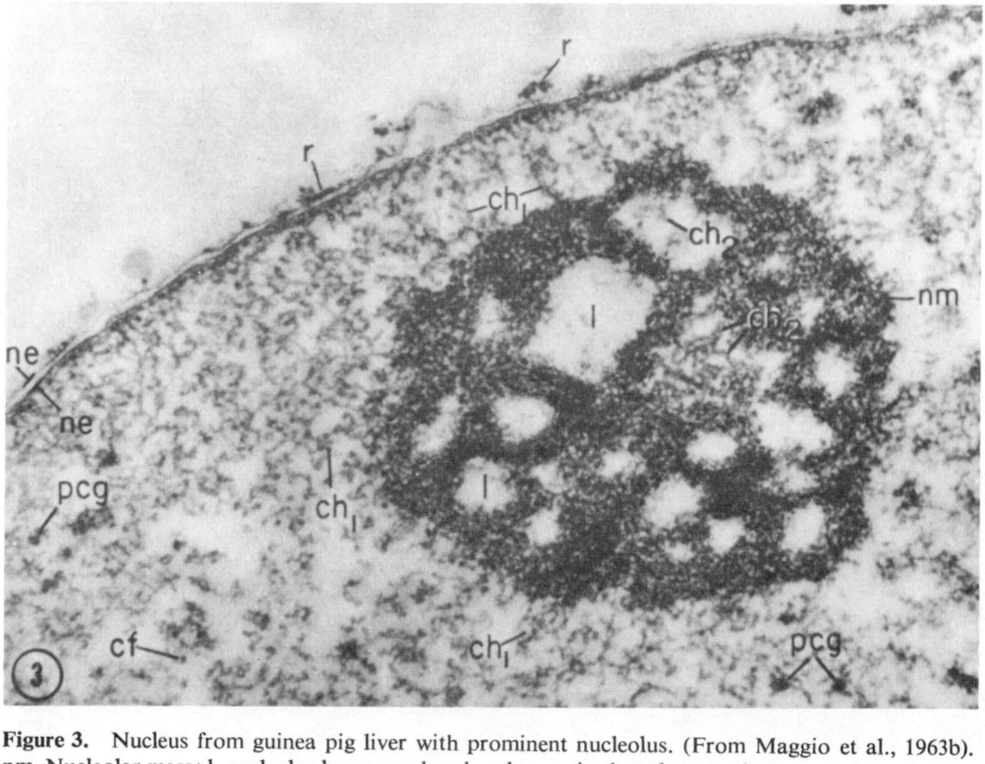

Figure 3. Nucleus from guinea pig liver with prominent nucleolus. (From Maggio et al., 1963b). nm, Nucleolar mass; l, nucleolar lacunae; ch_1, ch_2, chromatin threads; pcg, dense granules; cf, finer-textured material; ne, nuclear envelope with ribosomes (r) on outer membrane.

B. The Nucleolus

Prominent nucleoli are seen in nuclei of cells actively engaged in protein synthesis (Fig. 3). They are generally absent from nuclei of cells in early cleavage stages. Nucleoli can be isolated from many cell types by differential centrifugation, usually after disrupting the nuclear structure by sonication (Fig. 4) (Maggio et al., 1963b; Murumatsu et al., 1963; see also Busch, 1968), although early methods used hypotonic media (Vincent, 1952) and detergents (Penman et al., 1966) to lyse the nuclei. The greater resistance of nucleoli to disruption as compared to nuclei may be related to their higher protein content (Vincent, 1952; Birnstiel et al., 1964). Some histone is present, but the bulk of the nucleolar protein is more acidic than are histones (Birnstiel et al., 1964; see also Busch, 1965) and is thought to be related to ribosomal proteins associated with rRNA in the nucleolus. The little DNA that is present (Vincent, 1952) is derived from nucleolar-associated chromatin (Birnstiel et al., 1964) and represents the genetic material programming the synthesis of rRNA. The fine structure of the nucleolus has been considered by La Cour (1966) and Miller and Beatty (1969); the nucleolar location of NAD synthesis and rRNA methylases are discussed in Section 4.

There is unequivocal evidence that the nucleolus is the site of biosynthesis of rRNA. This was elegantly demonstrated by Perry and his colleagues (Perry et al., 1961) by using ultraviolet microbeam irradiation of nucleoli in HeLa cell cultures, by Brown and Gurdon (1964) in *Xenopus* mutants homozygous for the inability to make nucleoli, and by isotopic demonstration of rRNA precursor relationships in rat liver

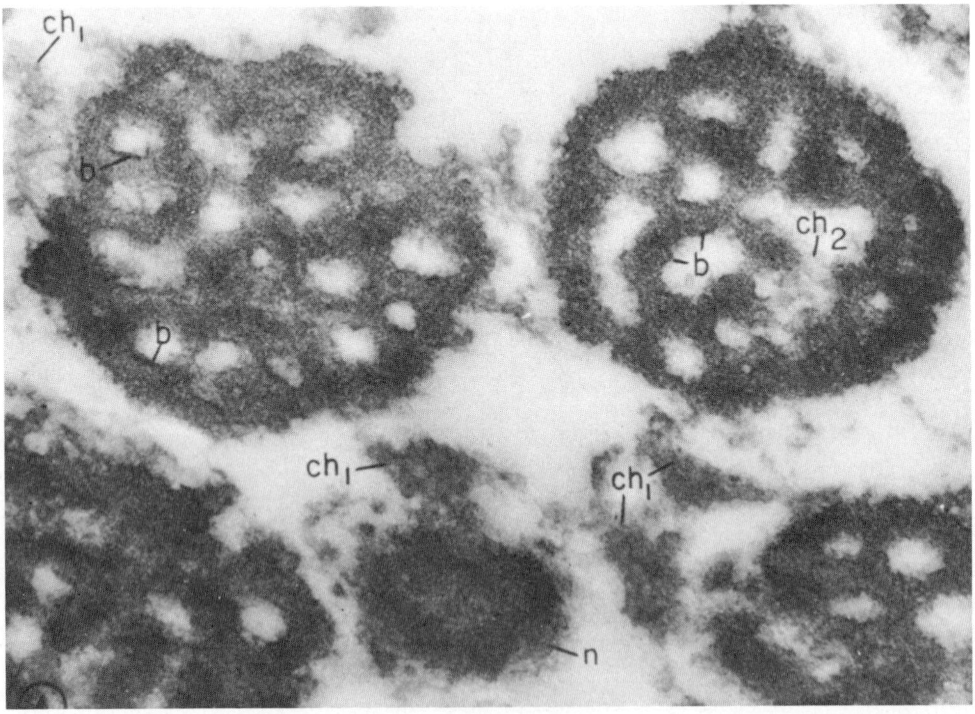

Figure 4. Isolated nucleoli from guinea pig liver. n, Nucleolus just grazed by section; ch_1, ch_2, chromatin threads; b, boundary of lacunae.

nucleoli (Murumatsu et al., 1966). It was mentioned earlier that ribosomes found in the nucleus were particularly evident in the nucleolar region (Frenster et al., 1960). There have been many reports of active amino acid uptake into nucleolar nonhistone protein (see Perry, 1969), but although it is likely that ribosomal proteins become associated with rRNA in the nucleolus, there is no firm evidence that the proteins are made there, and also (see Section 5.C) some reason to assume that such proteins are able to enter the nucleus from the cytoplasmic sites of synthesis.

Functions for the nucleolus beyond that of ribosome assembly are occasionally proposed. One of the most recent is that of Harris (1970), who has suggested from his work on heterokaryons of chick erythrocytes with various mammalian cell lines that "the flow of genetic information from the nucleus [of these erythrocytes] to the cytoplasm appears to be coupled to the flow of ribosomes . . . [which] in turn [is] directly dependent upon the synthetic activities of the nucleolus." In other systems, however, the nucleolus cannot be involved in this way as there is good evidence that new RNA can be detected in the cytoplasm either before nucleoli appear or, in the enucleolate mutant of *Xenopus*, under conditions in which the nucleolus is congenitally absent.

C. The Nuclear Membrane

The characteristics of the nuclear membrane—a double layered membrane with "pores" (Fig. 5)—became evident following the first description of the pores by Callan and Tomlin (1950) in amphibian oocyte nuclei. The pores are believed to occupy 5–15% of the surface area of the nucleus (Watson, 1955), although up to 25% has also been suggested (Gall, 1964) .There are normally approximately 35–65 pores per square micron of surface (Franke, 1966). Each membrane is approximately 10 nm thick and has a 10- to 30-nm perinuclear space between the two membranes. Each individual membrane is triple-layered, similar to the postulated structure of the unit membrane (Robertson, 1959). The outer nuclear membrane can sometimes be seen merging into the membrane of the ER (Watson, 1955, 1959). The pores may be up to 100 nm in diameter and contain electron-dense material in their centers which may project into either the cytosol or the nucleoplasm (Afzelius, 1955; Watson, 1955, 1959). The nature and organization of this electron-dense material in animal and plant (Roberts and Northcote, 1970) cells is currently receiving much attention. Electron microscopy has also shown that the inner surface of the nuclear membrane of *Amoeba* and some vertebrate cells is associated with a further electron-dense region, the fibrous lamina (Fawcett, 1966a), which extends from the membrane back into the nucleoplasm (Fig. 6).

There is little doubt that this complex structure of the nuclear membrane is associated with permeability characteristics very different from those of plasma or mitochondrial membranes. They have particularly been studied with respect to cations, amino acids, and protein or ribonucleoprotein. Even with aqueous isolation procedures, thymus nuclei have been found to have a Na^+/K^+ ratio very much higher than in the cytosol (Itoh and Schwartz, 1956). This has also been clearly established for nuclei from other tissues by isolation in nonaqueous media (Langendorf et al., 1961). Similarly, there is a relatively high chloride concentration which would neutralize up to 40% of the intracellular cations (Langendorf et al., 1966). Intranuclear sodium equilibrates very rapidly with extracellular ^{22}Na (Siebert et al., 1965), and direct traverse to the nucleus from the extracellular space has been suggested (Siebert

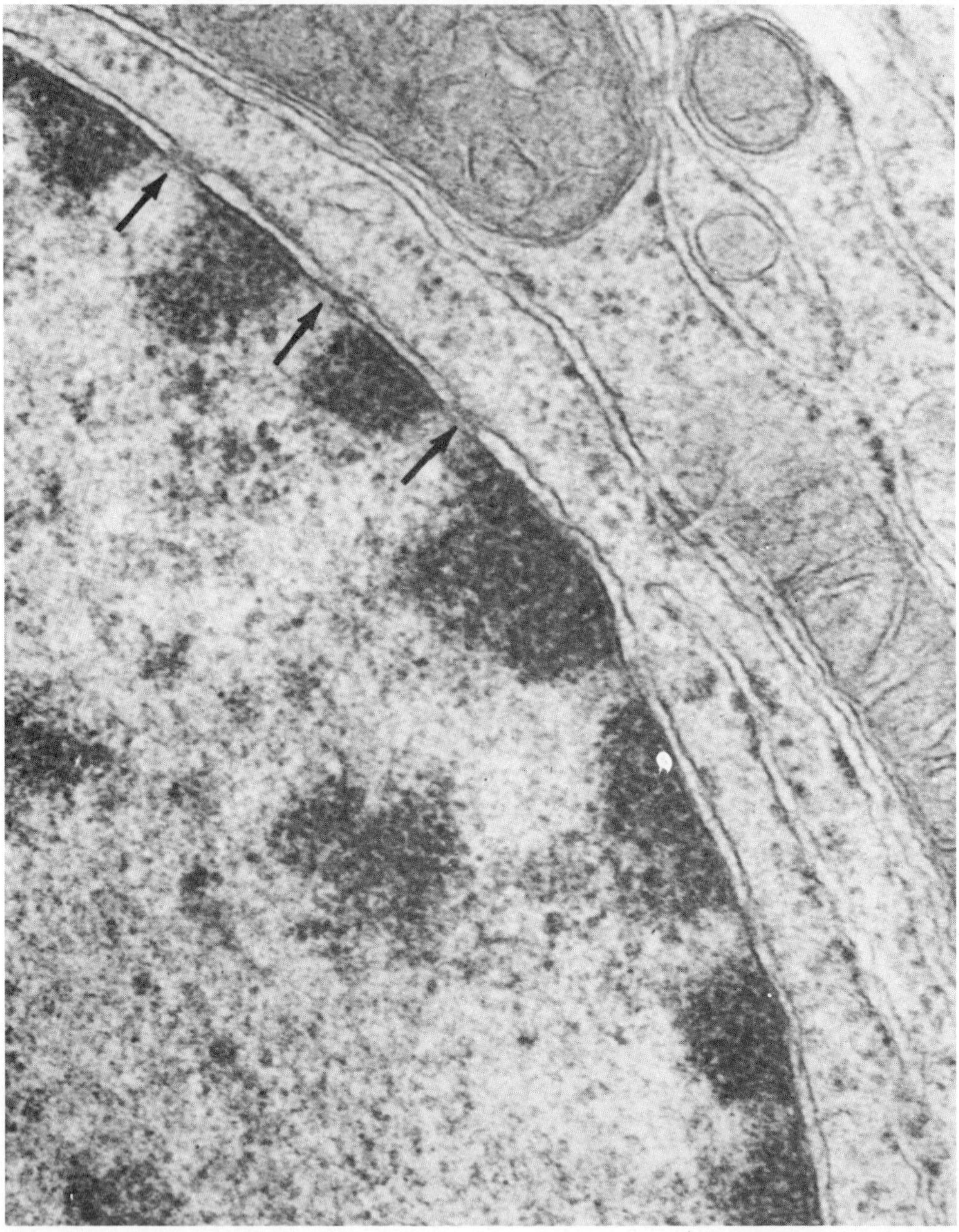

Figure 5. Nuclear pores (from Fawcett, 1966a). A small area of the nucleus and adjacent cytoplasm from an intestinal epithelial cell of *Amphiuna tridactylum*. The fibrous lamina appears to continue across the nuclear pore but is somewhat thinner and less dense in this region. (from Fawcett, 1966a.)

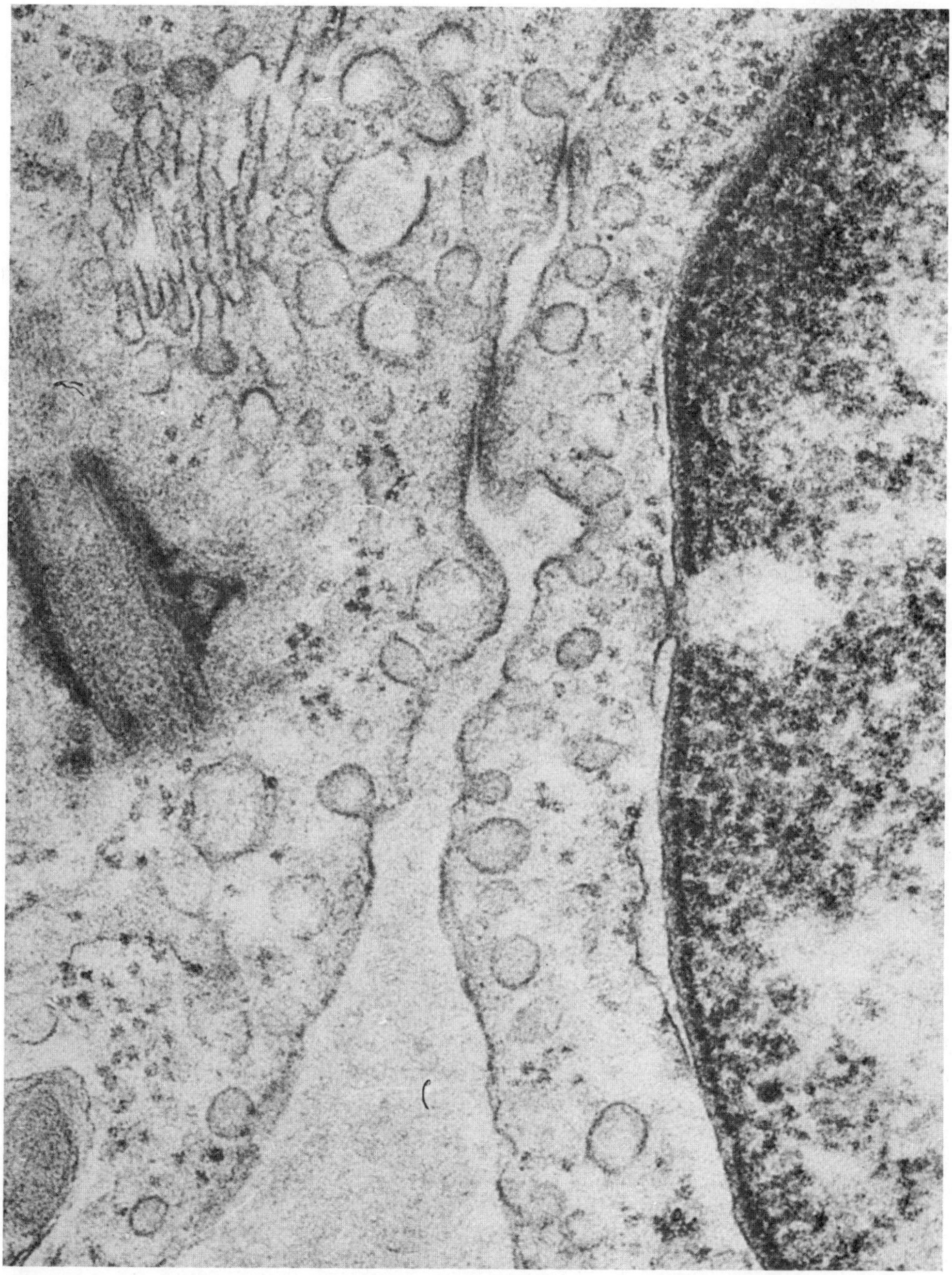

Figure 6. Detail of a nuclear pore and fibrous lamina. The fibrous lamina extends as a thin layer of lower density interposed between the dense chromatin and the dark line representing the inner membrane of the nuclear envelope. (From Fawcett, 1966a.)

et al., 1965). Measurements of ^{22}Na and ^{42}K uptake (Naora et al., 1962) in frog ooctyes and ^{32}P inorganic phosphate uptake in thymocytes (Ord and Stocken, 1961) indicated that in these cells the ions are probably taken up rapidly from the cytosol. The dependence of amino acid uptake by isolated thymus nuclei on Na^+ (Breitman and Webster, 1959; Allfrey et al., 1961) has been discussed earlier; the uptake requires ATP and might be assumed to involve a sodium pump. Amino acid uptake by chick heart nuclei (Klein et al., 1968) was also found to require ATP, Na^+, K^+, Mg^{2+}, and Ca^{2+} and to be inhibited by ouabain, but the ATPase in liver nuclear membranes was not ouabain sensitive nor was it Na^+ or K^+ activated (Delektorskaya and Perevoshchikova, 1969). This and the earlier data of Siebert and his colleagues were interpreted to suggest that in liver nuclei sodium uptake does not occur by active transport across the membrane. The differential retention of Na^+ within the nucleus is still unexplained.

Another approach to the study of ion permeability in the nucleus has been that of Loewenstein (1964), who directly measured conductivity across the nuclear membrane in interphase salivary gland cells of *Drosophila* and *Chironomus* larvae. Strong barriers to the free diffusion of ions were found, in contrast to the absence of such a barrier across the membranes of oocyte nuclei. The marked difference in properties could not be ascribed to differences observable in the electron microscope in size, frequency, or distribution of the pores between the different membranes (Wiener et al., 1965).

Detailed studies of the properties of the pores at different stages of development and of the cell cycle are being made. Merriam (1962) reported that the pores of immature frog oocytes had annuli with more electron-dense material and more central globules than the mature oocytes. In macronuclei from *Tetrahymena*, differences in density of annular material are found between cells in the stationary and logarithmic phases of growth (Wunderlick and Franke, 1968). The ability of the pores to allow the passage of inert material has been studied by Feldherr (1962, 1965). Polyvinylpyrrolidone-coated gold particles were injected into the cytoplasm of large *Amoeba* and the time and route for their penetration into nuclei studied. By 1–10 min after injection the particles could be seen in the center of the nuclear pores. When *Amoeba* and oocytes were compared (Feldherr, 1969), the uptake was much faster in the former; this could not be correlated with the area of pore surface. In *Amoeba proteus* the entry of the gold particles was rapid immediately after cytokinesis and decreased during late interphase, although the pore area again was unaffected. Feldherr (1968) suggested that the decreased rate of penetration in interphase might be linked to the reconstruction of the fibrous lamina which occurs more slowly after mitosis than does the reformation of the nuclear membrane.

Nuclei isolated in aqueous media show both loss of proteins which are retained in nonaqueous isolation procedures and also a capacity to take up protein from their environment (Anderson, 1953). The presence of ribonucleoprotein in the pores can be seen in *Chironomus* salivary gland nuclei with well-developed puffs (Stevens and Swift, 1966; Figs. 7 and 8), indicating that transit of macromolecules from the nucleus can occur through these regions. At present, it is not known how the nature of the annular material differs in pores of differing permeability and, although control of nuclear function by alterations in permeability at different stages of cell development can be postulated, no evidence is yet available as to how this may be effected. The

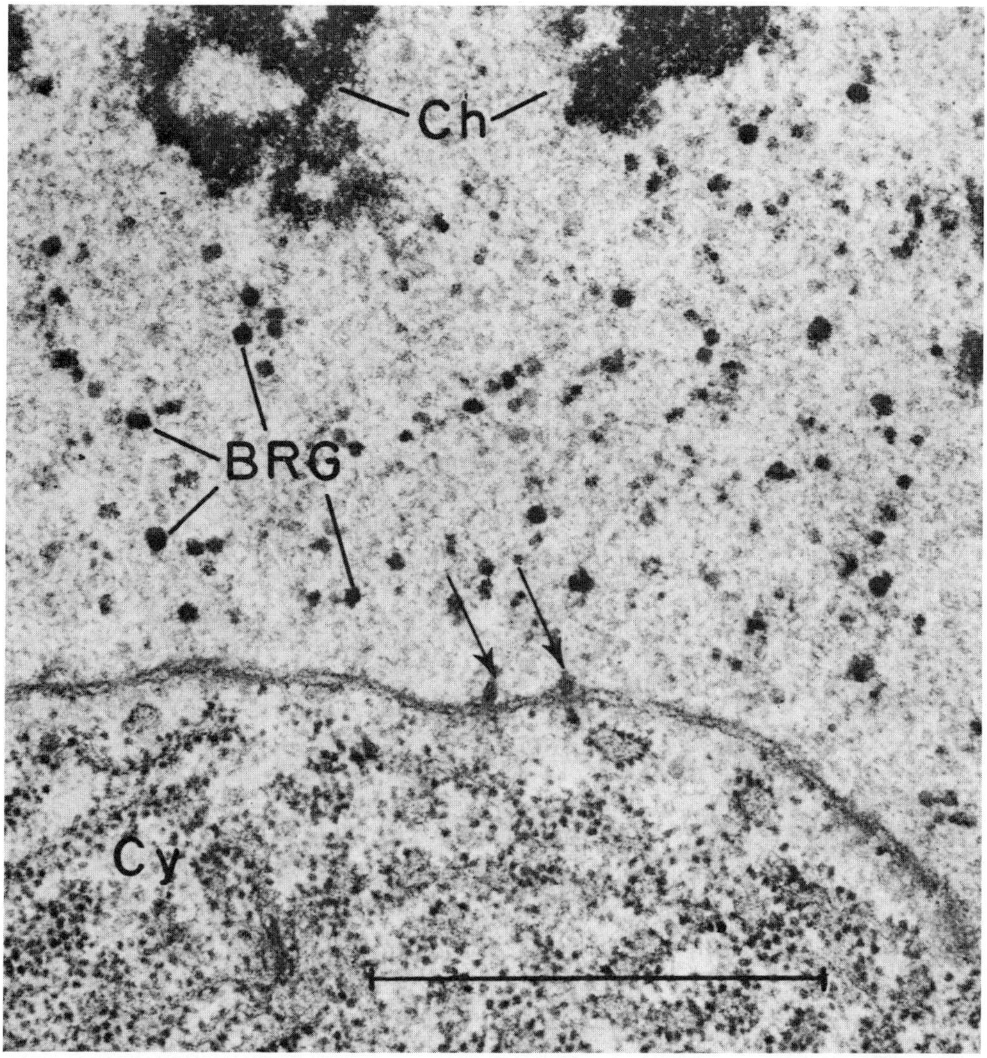

Figure 7. Nucleus from salivary gland of *C. thummi*. Region near nuclear envelope. BRG, Balbiani ring granules; Ch, chromosomes; Cy, cytoplasm. At arrows two granules are seen in intimate association with nuclear envelope. Scale = 1 μ. (From Stevens and Swift, 1966a.)

significance of the channels into the lumen of the ER is also uncertain. It seems possible that where the pores are functionally patent ions might move into the nucleus through these regions; where transit through the pores is minimal, penetration across the membrane may be important.

One further feature of nuclear membranes may be mentioned; from studies on microorganisms it is known that DNA replication starts from a membrane site (see Lark, 1969). This has also been reported for eukaryotic cells (Comings and Kakefuda, 1968; Alfert and Das, 1969). The structure and operation of the site(s) are not yet evident.

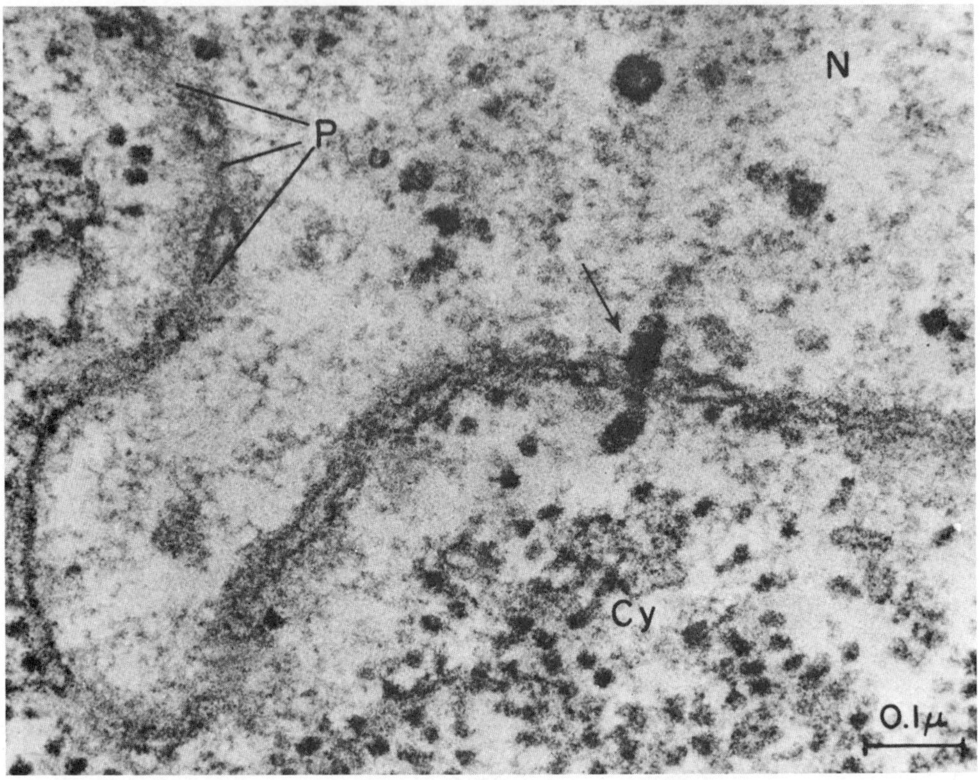

Figure 8. Ribonucleoprotein in pore *Chironomus* nucleus. N, Nucleus; Cy, cytoplasm; P, pore. Arrow indicates Balbiani ring granule transformed into rodlike shape in the center of a pore. (From Stevens and Swift, 1966.)

6. CONCLUSIONS

The picture of the nucleus that emerges from these discussions is of an integrated structure whose components are slowly being disentangled and their chemical composition established. Much more obscure is an understanding of the physical and chemical interactions of the macromolecules and the part these interactions play in the functioning of the nucleus. The sensitivity even of the fully differentiated nucleus to cytoplasmic (Gurdon, 1967; Harris, 1967) and extracellular stimuli is firmly established, and some of the controlling factors may perhaps be predicted. If we exclude the possibility of minor, but significant, reversible changes in DNA, the variables are the availability of metabolites, the nonhistone proteins, the histones, the intranuclear ion concentrations, and the permeability of the nuclear membranes. A major uncertainty is the means by which selectivity in response is obtained.

Current thinking is heavily influenced by the prokaryotic cell model, and the nonhistone proteins are popular candidates for repressors. Even in the case of the repressor of the *lac* operon, however, the molecular interactions leading to its selective binding on DNA have not emerged. Analogous reasoning also suggests that there may be eukaryotic parallels to σ factors (see also Crippa and Toccini-Valenti, 1970). Merriam (1969) and others (see Erhan et al., 1970) showed that cytoplasmic proteins enter transplanted nuclei prior to the initiation of RNA or DNA synthesis. Their target within the nucleus has not yet been demonstrated.

Specificity of histone association to DNA has been discussed; a parallelism may be found here between the appearance of histones particularly associated with terminally differentiated cells and isoenzyme complements found in other systems. The ease with which the microstructure of many histones can be altered by various enzyme changes which could be influenced or promoted from the cytoplasm has already been mentioned. Although some of the changes occur concomitantly with alterations in information flow from the nucleus, causality has not yet been established. It is suggested that alterations in microstructure of histones set up a permissive state to allow new information transfer to take place, and that specific derepression is effected by nonhistone proteins. Changes in regenerating liver suggest that many enzymes associated with division in parenchymal cells may be coordinately linked; conceivably, nonhistone protein might cover the operon for such enzymes and histones be associated with the dependent cistrons. Our almost total ignorance of the genetic strategy of higher organisms precludes further speculation in this field.

Swelling is one of the first changes detected in transplanted nuclei; localized increases in ionic strength could certainly cause dissociation of nucleoprotein, but again a selective response is more difficult to envisage. An increased intranuclear ionic strength during S phase could contribute to the weakened DNA–histone interactions that may be expected during this period, but as far as we are aware no changes in nuclear ion binding have been described during the cell cycle. Generalized alterations in nuclear permeability also might allow increased RNA or DNA synthesis in situations in which nuclear concentrations of metabolites are rate limiting. Again, selectivity is not obvious, but a role for both hormones and cytoplasmic factors affecting membrane synthesis can be postulated.

ACKNOWLEDGMENTS

We are grateful to Drs. D. W. Fawcett, R. Maggio, G. E. Palade, P. Siekewitz, H. Swift, B. J. Stevens, and H. S. Shapiro, and to the Rockefeller University Press, W. B. Saunders and Company, the Wistar Press, and The Chemical Rubber Company for permission to reproduce previously published data. We also thank Dr. C. A. Pasternack and Miss M. E. Cross for their useful comments, and the Cancer Research Campaign for research assistance.

REFERENCES

Adams, R. L. P., and J. G. Lindsay (1969). Changes in the activity of nuclear deoxyribonucleic acid polymerase in synchronised mouse cells. *Biochem. J.*, **114,** 57P.

Adler, A. J., B. Schaffhausen, T. A. Langan, and G. D. Fasman (1971). Altered conformational effects of phosphorylated lysine-rich histone F_1 in F_1-DNA complexes. Circular dichroism and immunological studies. *Biochemistry*, **10,** 909.

Afzelius, B. (1955). The ultrastructure of the nuclear membrane in the sea urchin oocyte as studied with the electron microscope. *Exptl. Cell Res.*, **8,** 147.

Alfert, M., and N. K. Das (1969). Evidence for control of the rate of nuclear DNA synthesis by the nuclear membrane in eukaryotic cells. *Proc. Natl. Acad. Sci., U.S.*, **63,** 123.

Allfrey, V. G. (1959). In *The Cell* (J. Brachet and A. E. Mirsky, eds.), Vol. I, p. 193, North-Holland, Amsterdam.

Allfrey, V. G. (1970). Changes in chromosomal proteins at times of gene activation. *Federation Proc.*, **29**, 1447.

Allfrey, V. G., M. M. Daly, and A. E. Mirsky (1955). Some observations on protein metabolism in chromosomes of non-dividing cells. *J. Gen. Physiol.*, **38**, 415.

Allfrey, V. G., A. E. Mirsky, and S. Osawa (1957). Protein synthesis in isolated cell nuclei. *J. Gen. Physiol.*, **40**, 451.

Allfrey, V. G., R. Meudt, J. W. Hopkins, and A. E. Mirsky (1961). Sodium-dependent "transport" reactions in the cell nucleus and their role in protein and nucleic acid synthesis. *Proc. Natl. Acad. Sci., U.S.*, **47**, 907.

Allfrey, V. G., R. Faulkner, and A. E. Mirsky (1964a). Acetylation and methylation of histones and their possible role in the regulation of RNA synthesis. *Proc. Natl. Acad. Sci., U.S.*, **51**, 786.

Allfrey, V. G., V. C. Littau, and A. E. Mirsky (1964b). Methods for the purification of thymus nuclei and their application to studies of nuclear protein synthesis. *J. Cell. Biol.*, **21**, 213.

Anderson, N. G. (1953). Studies on isolated cell components. *Exptl. Cell Res.*, **5**, 361.

Anderson, N. G. (1968). Preparative particle separation in density gradients. *Quart. Rev. Biophys.*, **1**, 217.

Ando, T., K. Suzuki, S. Watanabe, and S. Inoue (1967). Studies on protamines. *Proc. VIIth Intern. Congr. Biochem. (Tokyo)*, Symposium I, 1.

Asao, T. (1969). Behavior of histones and cytoplasmic basic proteins during embryogenesis of the Japanese newt *Triturus pyrrhogaster*, *Exptl. Cell Res.*, **58**, 243.

Avery, O. T., C. M. Mcleod, and M. McCarty (1944). Induction of transformation by a deoxyribonucleic acid fraction isolated from *Pneumococcus* type III. *J. Exptl. Med.*, **79**, 137.

Bäckström, S. (1965). Basic proteins in sea urchin embryo. *Acta Embryol. Morphol. Exptl.*, **8**, 20.

Baltus, E. (1954). Observations sur le rôle biochimique du nucléole. *Biochim. Biophys. Acta*, **15**, 263.

Beerman, W. (1963). Cytological aspects of information transfer in cellular differentiation. *Am. Zoologist*, **3**, 23.

Behrens, M. (1932). Untersuchung an isolierten Zell- und Gewebsbestandteilen. I. Mitteilung: Isolierung von Zellkernen des Kalbherzmuskels. *Z. Physiol. Chem.*, **209**, 59.

Bekhor, I., G. M. Kung, and J. Bonner (1969). Sequence-specific interactions of DNA and chromosomal protein. *J. Mol. Biol.*, **39**, 351.

Berendes, H. D., and W. Beerman (1969). In *Handbook of Molecular Cytology*, p. 500 (A. Lima-de-Faria, ed.), North Holland Publishing, Amsterdam.

Betel, I. (1967a). The occurrence of cytochromes in nuclei isolated from rat thymus. *Biochem. Biophys. Acta*, **143**, 62.

Betel, I. (1967b). Decreased nucleotide content as the cause of decreased ATP synthesis in thymus nuclei after irradiation. *Intern. J. Radiation Biol.*, **12**, 459.

Betel, I. (1969). The endogenous substrate for nuclear oxidative phosphorylation. *Arch. Biochem. Biophys.*, **134**, 271.

Betel, I., and H. M. Klouwen (1967). Oxidative phosphorylation in nuclei isolated from rat thymus. *Biochim. Biophys. Acta*, **131**, 453.

Birnstiel, M. L., M. I. H. Chipchase, and R. J. Hayes (1962). Incorporation of L-[^{14}C]-leucine by isolated nuclei. *Biochim. Biophys. Acta*, **55**, 728.

Birnstiel, M. L., M. I. H. Chipchase, and W. G. Flamm (1964). On the chemistry and organisation of nucleolar proteins. *Biochim. Biophys. Acta*, **87**, 111.

Birnstiel, M., M. Grunstein, J. Speirs, and W. Hennig (1969). Family of ribosomal genes in *Xenopus laevis*. *Nature*, **223**, 1265.

Bijvoet, P. (1957). Ethanol precipitation analysis of thymus histone. *Biochim. Biophys. Acta*, **25**, 502.

Boivin, A., R. Vendrely, and C. Vendrely (1948). L'acide désoxyribonucleique du noyau cellulaire depositaire des caractéres hereditaires: Arguments d'ordre analytique. *Compt. Rend.*, **226**, 1061.

Bollum, F. J. (1960). Calf thymus polymerase. *J. Biol. Chem.*, **235**, 2399.

Bollum, F. J., and V. R. Potter (1958). Incorporation of thymidine into deoxyribonucleic acid by enzymes from rat tissues. *J. Biol. Chem.*, **233,** 472.

Borun, T. W., M. D. Scharff, and E. Robbins (1967). Rapidly labeled, polyribosome-associated RNA having the properties of histone messenger. *Proc. Natl. Acad. Sci., U.S.*, **58,** 1977.

Boublik, M., E. M. Bradbury, and C. Crane-Robinson (1970). An investigation of the conformational changes of histone F1 and F2a1 by proton magnetic resonance spectroscopy. *European J. Biochem.*, **14,** 486.

Brachet, J. (1940). La détection histochimique des acides pentosnucleiques. *Compt. Rend. Soc. Biol.*, **133,** 88.

Brachet, J. (1942). La localisation des acides pentosnucleiques dans les tissus animaux et les oeufs d'amphibiens en voie de développement. *Arch. Biol.* (*Liège*), **53,** 207.

Brachet, J. (1957). *Biochemical Cytology*. Academic, New York.

Bradbury, E. M. (1969). Conformations of nucleohistones and histones. *Biochem. J.*, **114,** 49P.

Bradbury, E. M., C. Crane-Robinson, D. M. P. Phillips, E. W. Johns, and K. Murray (1965). Conformational investigations of histones. *Nature,* **205,** 1315.

Bradbury, E. M., C. Crane-Robinson, H. Goldman, H. W. E. Rattle, and R. M. Stephens (1967). Spectroscopic studies of the conformations of histones and protamine. *J. Mol. Biol.*, **29,** 507.

Breitman, T. R., and G. C. Webster (1959). Monovalent cations and the incorporation of metabolites by isolated thymus nuclei. *Nature,* **184,** 637.

Britten, R. J., and D. E. Kohne (1969a). In *Handbook of Molecular Cytology* (A. Lima-de-Faria, ed.), p. 21. North Holland, Amsterdam.

Britten, R. J., and D. E. Kohne (1969b). In *Handbook of Molecular Cytology* (A. Lima-de-Faria, ed.), p. 37. North Holland, Amsterdam.

Brown, D. D. (1967). The genes for ribosomal RNA and their transcription during amphibian development, *Curr. Top. Develop. Biol.*, **2,** 48.

Brown, D. D., and J. B. Gurdon (1964). Absence of ribosomal RNA synthesis in the anucleolate mutant of *Xenopus laevis*. *Proc. Natl. Acad. Sci., U.S.*, **51,** 139.

Buckingham, R. H., and L. A. Stocken (1970). Histone F1 from rat thymus. Subfractionation and incorporation of ^{32}P phosphate *in vitro*. *Biochem. J.*, **117,** 509.

Burzio, L., and S. S. Koide (1972). *In vitro* effect of NAD on DNA synthesis in isolated nuclei from regenerating rat liver and Novikoff hepatoma. FEBS Letts., **20,** 29.

Busch, H. (1965). *Histones and Other Nuclear Proteins*. Academic, New York.

Busch, H. (1968). *Comprehensive Biochemistry* (M. Florkin and E. H. Stotz, eds.), **23,** 39. Elsevier, Amsterdam.

Bustin, M., and R. D. Cole (1968). Species and organ specificity in very lysine-rich histones. *J. Biol. Chem.*, **243,** 4500.

Bustin, M., S. C. Rall, R. H. Stellwagen, and R. D. Cole (1969). Histone structure: Asymmetric distribution of lysine residues in lysine-rich histone. *Science,* **163,** 391.

Cairns, J. (1963). The bacterial chromosome and its manner of replication as seen by autoradiography. *J. Mol. Biol.*, **6,** 208.

Callan, H. G. (1963). The nature of lampbrush chromosomes. *Intern. Rev. Cytol.*, **15,** 1.

Callan, H. G. (1967). The organisation of genetic units in chromosomes. *J. Cell Sci.*, **2,** 1.

Callan, H. G. (1969). In *Handbook of Molecular Cytology* (A. Lima-de-Faria, ed.), p. 540. North Holland, Amsterdam.

Callan, H. G., and S. G. Tomlin (1950). Experimental studies on amphibian oocyte nuclei. *Proc. Roy. Soc.* (*London*), **B137,** 366.

Caspersson, T., and J. Schultz (1939). Pentose nucleotides in the cytoplasm of growing tissues. *Nature,* **143,** 602, 609.

Chambon, P., M. Ramuz, and J. Doly (1965). Relation between soluble DNA-dependent RNA polymerase and "aggregate" RNA polymerase. *Biochem. Biophys. Res. Commun.*, **21,** 156.

Champagne, M., and A. Mazen (1966). Calf thymus and chick erythrocytes nucleoproteins. *Nature,* **209,** 1131.

Chauveau, J., Y. Moulé, and Ch. Rouiller (1956). Isolation of pure and unaltered liver nuclei. Morphology and biochemical composition. *Exptl. Cell Res.*, **11**, 317.

Chesterton, C. J., and P. H. W. Butterworth (1971). A new form of mammalian DNA-dependent RNA polymerase and its relationship to the known forms of the enzyme. *FEBS Letters*, **12**, 301.

Chipchase, M. I. H., and M. L. Birnstiel (1963). On the nature of nucleolar RNA. *Proc. Natl. Acad. Sci., U.S.*, **50**, 1101.

Clever, U. (1961). Genaktivitäten in den Riesenchromosomen von *Chironomus tentans* und ihre Beziehungen zur Entwicklung. I. Genaktivierung durch Ecdyson. *Chromosoma*, **12**, 607.

Clever, U. (1967). In *The Control of Nuclear Activity*, p. 161 (L. Goldstein, ed.). Prentice Hall, Englewood Cliffs, New Jersey.

Clever, U. (1968). Regulation of chromosome function. *Advan. Genet.*, **2**, 11.

Cohen, P., and C. Kidson (1968). Conformational analysis of DNA-poly-L-lysine complexes by optical rotatory dispersion. *J. Mol. Biol.*, **35**, 241.

Comings, D. E., and T. Kakefuda (1968). Initiation of DNA replication at the nuclear membrane. *J. Mol. Biol.*, **33**, 225.

Conover, T. E. (1970a). Low temperature spectrum of respiratory components in calf thymus nuclei. *Arch. Biochem. Biophys.*, **136**, 541.

Conover, T. E. (1970b). Cytochrome oxidase activity in calf thymus nuclei. *Arch. Biochem. Biophys.*, **136**, 551.

Crampton, C. F., S. Moore, and W. H. Stein (1955). Chromatographic fractionation of calf thymus histone. *J. Biol. Chem.*, **215**, 787.

Crampton, C. F., S. Moore, and W. H. Stein (1957). Comparative studies on chromatographically purified histones. *J. Biol. Chem.*, **225**, 363.

Crawley, J. C. W. and H. Harris (1963). The fine structure of isolated HeLa cell nuclei. *Exptl. Cell Res.*, **31**, 70.

Creasey, W. A., and L. A. Stocken (1959). The effect of ionizing radiation on nuclear phosphorylation in the radiosensitive tissues of the rat. *Biochem. J.*, **72**, 519.

Crick, F. (1971). General model for the chromosomes of higher organisms. *Nature*, **234**, 25.

Crippa, M., and G. P. Tocchini-Valentini (1970). Performance of a bacterial RNA polymerase factor in an amphibian oocyte. *Nature*, **226**, 1243.

Cross, M. E., and M. G. Ord (1970). Changes in the phosphorylation and thiol content of histones in phytohaemagglutinin-stimulated lymphocytes. *Biochem. J.*, **118**, 191.

Dabeva, M. D., and R. G. Tsanev (1966). Extraction of rapidly labeled undegraded ribonucleic acid from isolated rat liver nuclei. *Anal. Biochem.*, **17**, 390.

Daly, M. M., and A. E. Mirsky (1955). Histones with high lysine content. *J. Gen. Physiol.*, **38**, 405.

Dancheva, K. I. (1969). Incorporation of ^{32}P into the nucleotides of thymus nuclei incubated with ribose and glucose. *Biokhymia*, (*Engl. Transl.*), **34**, 791.

Davidson, J. N. (1969). *The Biochemistry of the Nucleic Acids*, 6th edition. Methuen, London.

Davidson, J. N., R. M. S. Smellie, H. M. Keir, and A. H. McArdle (1958). Synthesis of deoxyribonucleic acid in extracts of mammalian cells. *Nature*, **182**, 589.

DeLange, R. J., and E. L. Smith (1971). Histones: Structure and Function. *Annu. Rev. Biochem.*, **40**, 279.

Delektorskaya, L. N., and K. A. Perevoshchikova (1969). Adenosine triphosphatase activity of nuclear membrane and nuclei of rat liver and of Zajdela hepatoma in rats and Ehrlich's ascites carcinoma in mice. *Biokhymia*, **34**, 161.

Dingman, C. W., and A. C. Peacock (1968). Analytical studies on nuclear ribonucleic acid using polyacrylamide gel electrophoresis. *Biochemistry*, **7**, 659.

Dixon, G. H., and M. Smith (1968). Nucleic acids and protamine in salmon testes. *Progr. Nucleic Acid Res. Mol. Biol.*, **8**, 9.

Dixon, G. H., K. Marushige, V. Ling, M. Sung, and D. T. Wigle (1969). Phosphorylation of histones in maturing trout testis. *Federation Proc.*, **28**, Abstr. no. 1898.

Dounce, A. L., and R. Ickowicz (1969). The effect of the method of isolation of cell nuclei on the ratios of nuclear proteins to DNA. *Arch. Biochem. Biophys.*, **131**, 359.

Dounce, A. L., and R. Umaña (1962). The proteases of isolated cell nuclei. *Biochemistry*, **1,** 811.

Dounce, A. L., F. Seaman, and M. Mackay (1966). Effect of the pH of isolation on the quantities and properties of histones extractable from calf thymus and rat liver nuclei. *Arch. Biochem. Biophys.*, **117,** 550.

Duerksen, J. D., and B. J. McCarthy (1971). Distribution of DNA sequences in fractionated chromatin. *Biochemistry*, **10,** 1471.

Edström, J. E. (1968). *Masters, slaves and evolution. Nature*, **220,** 1196.

Edström, J. E., B. Daneholt, E. Egyházi, B. Lambert, and U. Ringborg (1969).Formation and processing of ribonucleic acid in subnuclear compartments of *Chironomus tentans. Biochem. J.*, **114,** 51P.

Edwards, L. J., and L. S. Hnilica (1968). The specificity of histones in nucleated erythrocytes. *Experientia*, **24,** 228.

Erhan, S., S. Reisher, E. A. Franko, S. A. Kamath, and R. J. Rutman (1970). Evidence for "the wedge" initiation of DNA replication. *Nature*, **225,** 340.

Fambrough, D. M. (1969). In *Handbook of Molecular Cytology* (A. Lima-de-Faria, ed.), p. 437. North Holland, Amsterdam.

Fawcett, D. W. (1966a). On the occurrence of a fibrous lamina on the inner aspect of the nuclear envelope in certain cells of vertebrates. *Am. J. Anat.*, **119,** 129.

Fawcett, D. W. (1966b). *An Atlas of Fine Structure*. Saunders, Philadelphia.

Feldherr, C. M. (1962). The nuclear annuli as pathways for nucleocytoplasmic exchanges. *J. Cell Biol.*, **14,** 65.

Feldherr, C. M. (1965). The effect of the electron opaque pore material on exchanges through the nuclear annuli. *J. Cell Biol.*, **25,** 43.

Feldherr, C. M. (1968). Nucleocytoplasmic exchanges during early interphase. *J. Cell Biol.*, **39,** 49.

Feldherr, C. M. (1969). A comparative study of nucleocytoplasmic interactions. *J. Cell Biol.*, **42,** 841.

Fisher, H. W., and H. Harris (1962). The isolation of nuclei from animal cells in culture. *Proc. Roy. Soc. (London)*, **B156,** 521.

Fitzgerald, P. J., W. H. Marsh, M. G. Ord, and L. A. Stocken (1970). Histone changes during degeneration and regeneration of the pancreas. *Biochem. J.*, **117,** 711.

Flemming, W. (1882). *Zellsubstanz, Kern und Zellteilung*. Leipzig. Verlag von F. C. W. Vogel.

Forer, A. (1969). Chromosome movements during cell division, In *Handbook of Molecular Cytology* (A. Lima-de-Faria, ed.), p. 553. North Holland, Amsterdam.

Franke, W. W. (1966). Isolated nuclear membrane. *J. Cell. Biol.*, **31,** 619.

Freed, J. J., and L. Mezger-Freed (1970). Stable haploid cultured cell lines from frog embryos. *Proc. Natl. Acad. Sci., U.S.*, **65,** 337.

Frenster, J. H. (1965). Nuclear polyanions as de-repressors of synthesis of ribonucleic acid. *Nature*, **206,** 680.

Frenster, J. H. (1969). In *Handbook of Molecular Cytology*, (A. Lima-de-Faria, ed.), p. 251. North Holland, Amsterdam.

Frenster, J. H., V. G. Allfrey, and A. E. Mirsky (1960). Metabolism and morphology of ribonucleoprotein particles from the cell nucleus of lymphocytes. *Proc. Natl. Acad. Sci., U.S.*, **46,** 432.

Frenster, J. H., V. G. Allfrey, and A. E. Mirsky (1963). Repressed and active chromatin isolated from interphase lymphocytes. *Proc. Natl. Acad. Sci., U.S.*, **50,** 1026.

Friedman, D. L. (1970). DNA polymerase from HeLa cell nuclei levels of activity during a synchronised cell cycle. *Biochem. Biophys. Res. Commun.*, **39,** 100.

Gall, J. G. (1963). *Chromosomes and Differentiation in Cytodifferentiation and Macromolecular Synthesis*, (M. Locke, ed.), p. 119. Academic, New York.

Gall, J. G. (1964). *Electron microscopy of the nuclear envelope*. Protoplasmatologia, V, No. 2, p. 4, Springer, Vienna.

Gallwitz, D. (1970). Extraction and partial purification of two histone-specific transacetylases from rat liver nuclei. *Biochem. Biophys. Res. Commun.*, **40,** 236.

Gallwitz, D., and G. C. Mueller (1969). Histone synthesis *in vitro* on HeLa cell microsomes. *J. Biol. Chem.*, **244,** 5947.

Georgiev, G. P. (1967). The nature and biosynthesis of nuclear ribonucleic acids. *Progr. Nucleic Acid Res. Mol. Biol.*, **6,** 259.

Gershey, E. L., G. Vidali, and V. G. Allfrey (1968). Chemical studies on histone acetylation. *J. Biol. Chem.*, **243,** 5018.

Gilbert, I. G. F., and J. M. Radley (1964). A procedure for the isolation of cell nuclei for trace-metal studies. *Biochim. Biophys. Acta*, **82,** 618.

Gross, S. R. (1969). Genetic regulatory mechanisms in the fungi. *Advan. Genet.*, **3,** 395.

Gurdon, J. B. (1964). The transplantation of living cell nuclei. *Advan. Morphogenesis*, **4,** 1.

Gurdon, J. B. (1967). On the origin and persistence of a cytoplasmic state inducing nuclear DNA synthesis in frogs' eggs. *Proc. Natl. Acad. Sci., U.S.*, **58,** 545.

Gurley, L. R., and R. A. Walters (1971). Response of histone turnover and phosphorylation to X-irradiation. *Biochemistry*, **10,** 1588.

Haines, M. E., I. R. Johnston, A. P. Mathias, and D. Ridge (1969). The synthesis of nicotinamide-adenine dinucleotide and poly(adenosine diphosphate ribose) in various classes of rat liver nuclei. *Biochem. J.*, **115,** 881.

Hancock, R., and H. Amos (1968). Nuclear binding of exogenous histones by L cells at low pH. *J. Cell Biol.*, **36,** cl.

Harris, H. (1967). The reactivation of the red cell nucleus. *J. Cell Sci.*, **2,** 231.

Harris, H. (1968). *Nucleus and Cytoplasm*. Clarendon, Oxford.

Harris, H. (1970). *Cell Fusion*. Clarendon, Oxford.

Hennig, W., and P. M. B. Walker (1970). Variations in the DNA from two rodent families. *Nature*, **225,** 915.

Henson, P., and I. O. Walker (1970). The partial dissociation of nucleohistone by salts. Hydrodynamic and denaturation studies. *European J. Biochem.*, **14,** 345.

Heyden, H. W. V., and H. G. Zachau (1971). Characterization of RNA in fractions of calf thymus chromatin. *Biochim. Biophys. Acta*, **232,** 651.

Hnilica, L. S. (1967). Proteins of the cell nucleus. *Prog. Nucleic Acid Res. Mol. Biol.*, **7,** 25.

Hodes, M. E., C. G. Palmer, and D. S. Williams (1958). Possible use of synthetic surface-active agents for the preparation of tumour cell nuclei. *Nature*, **182,** 529.

Hogeboom, G. H., and W. C. Schneider (1952). The synthesis of diphosphopyridine nucleotide by the liver cell nuclei. *J. Biol. Chem.*, **197,** 611.

Hogeboom, G. H., W. C. Schneider, and G. E. Palade (1948). Isolation of intact mitochondria from rat liver: Some biochemical properties of mitochondria and submicroscopic particulate material. *J. Biol. Chem.*, **172,** 619.

Hogeboom, G. H., and W. C. Schneider (1952). The synthesis of diphosphopyridine nucleotide by the liver cell nuclei. *J. Biol. Chem.*, **197,** 611.

Hogeboom, G. H., W. C. Schneider, and M. J. Striebich (1952). On the isolation and biochemical properties of liver cell nuclei. *J. Biol. Chem.*, **196,** 111.

Howk, R., and T. Y. Wang (1970). DNA polymerase from rat liver chromosomal proteins. *European J. Biochem.*, **13,** 455.

Huang, R. C. C., and J. M. Bonner (1962). Histone, a suppressor of chromosomal RNA synthesis. *Proc. Natl. Acad. Sci., U.S.*, **48,** 1216.

Huang, R. C. C., and J. M. Bonner (1965). Histone-bound RNA, a component of native nucleohistone. *Proc. Natl. Acad. Sci., U.S.*, **54,** 960.

Huang, R. C. C., and P. C. Huang (1969). Effect of protein-bound RNA associated with chick embryo chromatin on template specificity of the chromatin. *J. Mol. Biol.*, **39,** 365.

Huberman, J. A., and G. Attardi (1967). Studies of fractionated HeLa cell metaphase chromosomes. *J. Mol. Biol.*, **29,** 487.

Huberman, J. A., and A. D. Riggs (1966). Autoradiography of chromosomal DNA fibers from Chinese hamster cells. *Proc. Natl. Acad. Sci., U.S.*, **55,** 599.

Huberman, J. A., and A. D. Riggs (1968). On the mechanism of DNA replication in mammalian chromosomes. *J. Mol. Biol.*, **32,** 327.

Hughes, A. (1959). *A History of Cytology.* Abelard-Schuman, New York.

Hurlbert, R. B., M. C. Liau, and A. Orengo (1964). Synthesis of RNA in isolated nuclei and nucleoli of rat liver. *Federation Proc.*, **23,** Abstr. 525.

Ingles, C. J., and G. H. Dixon (1967). Phosphorylation of protamine during spermatogenesis in trout testis. *Proc. Natl. Acad. Sci., U.S.*, **58,** 1011.

Itoh, S., and I. L. Schwartz (1956). Sodium and potassium content of isolated nuclei. *Nature*, **178,** 494.

Itoh, S., and I. L. Schwartz (1957). Sodium and potassium distribution in isolated thymus nuclei. *Am. J. Physiol.*, **188,** 490.

Iwai, K., K. Ishikawa, and H. Hayashi (1970). Amino acid sequence of slightly lysine-rich histone. *Nature*, **226,** 1056.

Jergil, B., M. Sung, and G. H. Dixon (1970). Species and tissue-specific patterns of phosphorylation of very lysine-rich histones. *J. Biol. Chem.*, **245,** 5867.

Johns, E. W. (1964). Preparative methods for histone fractions from calf thymus. *Biochem. J.*, **92,** 55.

Johns, E. W. (1967). A method for the selective extraction of histone fractions F2a1 and F2a2 from calf thymus deoxyribonucleoprotein at pH 7. *Biochem. J.*, **105,** 611.

Johns, E. W., and J. A. V. Butler (1962). Further fractionation of histones from calf thymus. *Biochem. J.*, **82,** 15.

Johns, E. W., and J. A. V. Butler (1964). Specificity of the interactions between histones and DNA. *Nature*, **204,** 853.

Johns, E. W., and S. Forrester (1969). The binding of extra acidic proteins to deoxyribonucleoprotein during the preparation of nuclear proteins. *European J. Biochem.*, **8,** 547.

Johns, E. W., and T. A. Hoare (1970). Histones and gene control. *Nature*, **226,** 650.

Johns, E. W., D. M. P. Phillips, P. Simson, and J. A. V. Butler (1960). Improved fractionation of arginine-rich histones from calf thymus. *Biochem. J.*, **77,** 631.

Johnston, I. R., A. P. Mathias, F. Pennington, and D. Ridge (1968). The fractionation of nuclei from mammalian cells by zonal centrifugation. *Biochem. J.*, **109,** 127.

Jones, K. W. (1970). Chromosomal and nuclear location of mouse satellite DNA in individual cells. *Nature*, **225,** 912.

Keir, H. M. (1965). DNA polymerases from mammalian cells. *Progr. Nucleic Acid Res. Mol. Biol.*, **4,** 81.

Keir, H. M., R. M. S. Smellie, and G. Siebert (1962). Intracellular location of DNA nucleotidyltransferase. *Nature*, **196,** 752.

Keyl, H. G. (1965). Duplikationen von Untereinheiten der Chromosomalen DNA während der Evolution von *Chironomus thummi*. *Chromosoma*, **17,** 139.

Kinkade, J. M., and R. D. Cole (1966a). The resolution of four lysine-rich histones derived from calf thymus. *J. Biol. Chem.*, **241,** 5790.

Kinkade, J. M., and R. D. Cole (1966b). A structural comparison of different lysine-rich histones of calf thymus. *J. Biol. Chem.*, **241,** 5798.

Klein, R. L., C. R. Horton, and A. Thureson-Klein (1968). Evidence for an amino-acid transport system in nuclei isolated from embryonic heart. *European J. Biochem.*, **6,** 514.

Kleinschmidt, A. K. (1967). Nukleinsäuren. *Naturwissenschaften*, **54,** 417.

Kleinsmith, L., V. G. Allfrey, and A. E. Mirsky (1966). Phosphoprotein metabolism in isolated lymphocyte nuclei. *Proc. Natl. Acad. Sci., U.S.*, **55,** 1182.

Kleinsmith, L. J., J. Heidema, and A. Carroll (1970). Specific binding of rat liver nuclear proteins to DNA. *Nature*, **226,** 1025.

Klouwen, H. M., and I. Betel (1963). Radiosensitivity of nuclear ATP synthesis. *Intern. J. Radiation Biol.*, **6,** 441.

Klouwen, H. M., I. Betel, A. W. M. Appelman, and C. Arts (1965). Synthesis of adenosine 5′-triphosphate by isolated rat liver nuclei. *Biochim. Biophys. Acta*, **97,** 152.

Klyszejko-Stefanowicz, L. and Y. Khouvine (1968). Noyaux de pancréas de rat isolés dans le saccharose. Étude chimique et physico-chimique des histones. *Bull. Soc. Chim. Biol.*, **50,** 1245.

Kodama, R. M., and H. Tedeschi (1963). An electron microscopy study of calf thymus nuclear preparations isolated in sucrose solutions. *J. Cell. Biol.*, **18,** 541.

Krakow, J. S., H. O. Kammen, and F. S. Canellakis (1961). The incorporation of ribonucleotides into terminal positions of deoxyribonucleic acid. *Biochim. Biophys. Acta,* **53,** 52.

La Cour, L. F. (1966). In *Chromosomes Today* (C. D. Darlington and K. R. Lewis, eds.), p. 150. Oliver and Boyd Ltd. Edinburgh and London.

Langan, T. A. (1967). A phosphoprotein preparation from liver nuclei and its effect on the inhibition of RNA synthesis by histones. *B.B.A. Library Series* **10,** 233.

Langan, T. A. (1968). In *Some Regulatory Mechanisms for Protein Synthesis in Mammalian Cells* p. 101 (A. San Pietro, M. Lamborg, and F. T. Kenney, eds.) Academic, New York.

Langan, T. A. (1969). Phosphorylation of liver histone following the administration of glucagon and insulin. *Federation Proc.,* **28,** 600.

Langan, T. A., and L. Smith (1966). The effect of a phosphoprotein preparation from liver nuclei on the inhibition of RNA synthesis by histones. *Federation Proc.,* **25,** Abstr. 778.

Langan, T. A., and L. K. Smith (1967). Phosphorylation of histones and protamines by a specific protein kinase from liver. *Federation Proc.,* **26,** 603.

Langan, T. A., S. C. Rall, and R. D. Cole (1971). Variation in primary structure at a phosphorylation site in lysine-rich histones. *J. Biol. Chem.,* **246,** 1942.

Langendorf, H., G. Siebert, I. Lorenz, R. Hannover, and R. Beyer (1961). Kationenverteilung im Zellkern und Cytoplasma der Rattenleber. *Biochem. Z.,* **335,** 273.

Langendorf, H., G. Siebert, K. Kesselring, and R. Hannover (1966). High nucleo-cytoplasmic concentration gradient of chloride in rat liver. *Nature,* **209,** 1130.

Lark, K. G. (1969). Initiation and control of DNA synthesis. *Ann. Rev. Biochem.,* **38,** 569.

Lett, J. T., I. Caldwell, C. J. Dean, and P. Alexander (1967). Rejoining of X-ray induced breaks in the DNA of leukemia cells. *Nature,* **214,** 790.

Lezzi, M. (1965). Die Wirkung von DNAase auf isolierte Polytänchromosomen. *Exptl. Cell Res.,* **39,** 289.

Lindahl, T., and G. M. Edelman (1968). Polynucleotide ligase from myeloid and lymphoid tissues. *Proc. Natl. Acad. Sci., U.S.,* **61,** 680.

Littau, V. C., V. G. Allfrey, J. H. Frenster, and A. E. Mirsky (1964). Active and inactive regions of nuclear chromatin as revealed by electron microscope autoradiography. *Proc. Natl. Acad. Sci., U.S.,* **52,** 93.

Littau, V. C., C. J. Burdick, V. G. Allfrey, and A. E. Mirsky (1965). The role of histones in the maintenance of chromatin structure. *Proc. Natl. Acad. Sci., U.S.,* **54,** 1204.

Loeb, J. F., and C. Creuzet (1969). Electrophoretic comparison of acidic proteins of chromatin from different animal tissues. *FEBS Letters,* **5,** 37.

Loewenstein, V. R. (1964). Permeability of the nuclear membrane as determined with electrical methods. Protoplasmatologia V, No. 2, p. 26. Springer, Vienna.

McCarthy, B. J. (1969). In *Handbook of Molecular Cytology* (A. Lima-de-Faria, ed.), p. 3. North Holland, Amsterdam.

McEwen, B. S., V. G. Allfrey, and A. E. Mirsky (1963a). Comparison of nuclear and mitochondrial phosphorylation in the thymus. *J. Biol. Chem.,* **238,** 758.

McEwen, B. S., V. G. Allfrey, and A. E. Mirsky (1963b). Pathways of aerobic carbohydrate catabolism in thymus nuclei. *J. Biol. Chem.,* **238,** 2571.

McEwen, B. S., V. G. Allfrey, and A. E. Mirsky (1963c). Participation of glycolysis and the citric acid cycle in nuclear ATP synthesis in the thymus. *J. Biol. Chem.,* **238,** 2579.

Maggio, R., P. Siekevitz, and G. E. Palade (1963a). Isolation and chemical characterization of a nuclear fraction from guinea pig liver. *J. Cell Biol.,* **18,** 267.

Maggio, R., P. Siekevitz, and G. E. Palade (1963b). Isolation and chemical characterization of nucleolar and nucleoplasmic subfractions. *J. Cell. Biol.,* **18,** 293.

Maio, J. J., and C. L. Schildkraut (1967). Isolated mammalian metaphase chromosomes. *J. Mol. Biol.,* **24,** 29.

Mantsavinos, R., and E. S. Canellakis (1959). Studies on the biosynthesis of deoxyribonucleic acid by soluble mammalian enzymes. *J. Biol. Chem.*, **234**, 628.

Marushige, K., V. Ling, and G. H. Dixon (1969). Phosphorylation of chromosomal basic proteins in maturing trout testis. *J. Biol. Chem.*, **244**, 5953.

Marzluff, W. F., and K. S. McCarty (1970). Two classes of histone acetylation in developing mouse mammary gland. *J. Biol. Chem.*, **245**, 5635.

Maul, G. G., and T. H. Hamilton (1967). The intranuclear localization of two DNA-dependent RNA polymerase activities. *Proc. Natl. Acad. Sci., U.S.*, **57**, 1371.

Mazia, D. (1961). In *The Cell* (J. Brachet and A. E. Mirsky, eds.), Vol. 3, p. 77. Academic, New York.

Merriam, R. W. (1962). Some dynamic aspects of the nuclear envelope. *J. Cell Biol.*, **12**, 79.

Merriam, R. W. (1969). Movement of cytoplasmic proteins into nuclei induced to enlarge and initiate DNA or RNA synthesis. *J. Cell Sci.*, **5**, 333.

Miescher, R. (1897). *Die Histochemischen und Physiologischen Arbeiten*. Vogel, Leipzig.

Miller, G. L., and B. R. Beatty (1969). In *Handbook of Molecular Cytology* (A. Lima-de-Faria, ed.), p. 605. North Holland, Amsterdam.

Mirsky, A. E., and H. Ris (1949). Variable and constant components of chromosomes. *Nature*, **163**, 666.

Mitchell, H. K. (1967). Biochemical aspects of *Drosophila*. *Advan. Genet.*, **1**, 185.

Monneron, A., and W. Bernard (1969). Fine structural organisation of the interphase nucleus in some mammalian cells. *Ultrastr. Res.*, **27**, 266.

Morgan, T. H. (1919). *The Physical Bases of Heredity*. J. B. Lippincott Co., London and Philadelphia.

Muramatsu, M., K. Smetana, and H. Busch (1963). Quantitative aspects of isolation of nucleoli of the Walker carcinosarcoma and liver of the rat. *Cancer Res.*, **23**, 510.

Muramatsu, M., J. L. Hodnett, W. J. Steele, and H. Busch (1966). Synthesis of 28S RNA in nucleolus. *Biochim. Biophys. Acta*, **123**, 116.

Murray, K. (1964). The occurrence of ε-*N*-methyl lysine in histones. *Biochemistry*, **3**, 10.

Murray, K. (1966). The acid extraction of histones from calf thymus deoxynucleoprotein. *J. Mol. Biol.*, **15**, 409.

Murray, K., G. Vidali, and J. M. Neelin (1968). The stepwise removal of histones from chicken erythrocyte nucleoprotein. *Biochem. J.*, **107**, 207.

Naora, H., H. Naora, M. Izawa, V. G. Allfrey, and A. E. Mirsky (1962). Some observations on differences in composition between the nucleus and cytoplasm of the frog oocyte. *Proc. Natl. Acad. Sci., U.S.*, **48**, 853.

Nooij, E. H. de, and H. G. K. Westenbrink (1962). Isolation of a homogeneous lysine-rich histone from calf thymus. *Biochim. Biophys. Acta*, **62**, 608.

Novello, F., and F. Stirpe (1970). Simultaneous assay of RNA polymerase I in nuclei isolated from resting and growing rat liver with the use of α-amanitin. *FEBS Letters*, **8**, 57.

O'Brien, B. R. A. (1964). A rapid method for the isolation and collection of nuclei from whole cell suspensions. *J. Cell Biol.*, **20**, 525.

Ohba, Y. (1966). Structure of nucleohistone. Hydrodynamic behaviour. *Biochim. Biophys. Acta*, **123**, 76.

Ohlenbusch, H. H., B. M. Olivera, D. Tuan, and N. Davidson (1967). Selective dissociation of histones from calf thymus nucleoprotein. *J. Mol. Biol.*, **25**, 299.

Ord, M. G., and L. A. Stocken (1961). Acid soluble phosphate metabolism in nuclei from rat thymus gland. *Biochem. J.*, **81**, 1.

Ord, M. G., and L. A. Stocken (1962a). Phosphate transfer reactions by nuclei from rat thymus gland. *Biochem. J.*, **84**, 593.

Ord, M. G., and L. A. Stocken (1962b). The effects of 200 r X-irradiation *in vivo* on phosphate transfer reactions in nuclei from rat thymus gland. *Biochem. J.*, **84**, 600.

Ord, M. G., and L. A. Stocken (1966). Metabolic properties of histones from rat liver and thymus gland. *Biochem. J.*, **98**, 888.

Ord, M. G., and L. A. Stocken (1967). Changes in the phosphorylation of histones during liver regeneration. *Biochem. J.*, **103**, 5P.

Ord, M. G., and L. A. Stocken (1968). Variations in the phosphate content and thiol/disulphide ratio in histones during the cell cycle. *Biochem. J.*, **107,** 403.

Ord, M. G., and L. A. Stocken (1969). Further studies on the phosphorylation and the thiol/disulphide ratio of histones in growth and development. *Biochem. J.*, **112,** 81.

Ord, M. G., and L. A. Stocken (1970). Changes in the proportion of thiol to disulphide in acid-soluble nuclear proteins of *Echinus* during the first cell cycle. *Biochem. J.*, **116,** 415.

Osawa, S., V. G. Allfrey, and A. E. Mirsky (1957). Mononucleotides of the cell nucleus. *J. Gen. Physiol.*, **40,** 491.

Paik, W. K., and S. Kim (1969). Enzymic methylation of histones. *Arch. Biochem. Biophys.*, **134,** 632.

Panyim, S., and R. Chalkley (1969). A new histone found only in mammalian tissues with little cell division. *Biochem. Biophys. Res. Commun.*, **37,** 1042.

Paoletti, R. A., and R. C. C. Huang (1969). Characterisation of sea urchin sperm chromatin and its basic proteins. *Biochemistry*, **8,** 1615.

Pardon, J. F., M. H. F. Wilkins, and B. M. Richards (1967). Superhelical model for nucleohistone. *Nature*, **215,** 508.

Patel, G., V. Patel, T. Y. Wang, and C. R. Zobel (1968). Studies of the nuclear residual proteins. *Arch. Biochem. Biophys.*, **128,** 654.

Paul, J., and R. S. Gilmour (1968). Organ-specific restriction on transcription in mammalian chromatin. *J. Mol. Biol.*, **34,** 305.

Pawse, A., M. G. Ord, and L. A. Stocken (1969). Enzymic phosphorylation of histone F1 in regenerating rat liver. *Biochem. J.*, **114,** 54P.

Pawse, A. R., M. G. Ord, and L. A. Stocken (1971). Histone kinase and cell division. *Biochem. J.*, **122,** 713.

Pelc, S. R. (1968). Turnover of DNA and function. *Nature*, **219,** 162.

Pelling, C. (1966). A replicative and synthetic chromosomal unit—The modern concept of the chromomere. *Proc. Roy. Soc.* (*London*), **B164,** 279.

Penman, S., I. Smith, and E. Holtzman (1966). Ribosomal RNA synthesis and processing in a particulate site in the HeLa cell nucleus. *Science*, **154,** 786.

Perry, R. P. (1969). In *Handbook of Molecular Cytology* (A. Lima-de-Faria, ed.), p. 620. North Holland, Amsterdam.

Perry, R. P., A. Hell, and M. Errera (1961). The role of the nucleolus in RNA and protein synthesis. *Biochim. Biophys. Acta*, **49,** 47.

Phillips, D. M. P. (1971). ed. Histones and Nucleohistones. Plenum Press, London.

Phillips, D. M. P. (1963). The presence of acetyl groups in histones. *Biochem. J.*, **87,** 258.

Phillips, D. M. P., and E. W. Johns (1965). A fractionation of the histones of group F2A from calf thymus. *Biochem. J.*, **94,** 127.

Platz, R. D., V. M. Kish, and L. J. Kleinsmith (1970). Tissue specificity of non-histone chromatin phosphoproteins. *FEBS Letters*, **12,** 38.

Plaut, W. (1963). On the replicative organisation of DNA in the polytene chromosome of *Drosophila melanogaster*. *J. Mol. Biol.*, **7,** 632.

Pogo, B. G. T., A. O. Pogo, V. G. Allfrey, and A. E. Mirsky (1968). Changing patterns of histone acetylation and RNA synthesis in regeneration of the liver. *Proc. Natl. Acad. Sci., U.S.*, **59,** 1337.

Rapkine, L. (1931). Sur les processus chimiques au cours de la division cellulaire. *Ann. Physiol. Physico-Chim. Biol.*, **7,** 382.

Rasmussen, P. S., K. Murray, and J. M. Luck (1962). On the complexity of calf thymus histone. *Biochemistry*, **1,** 79.

Reisfeld, R. A., U. J. Lewis, and D. E. William (1962). Disc electrophoresis of basic proteins and peptides on polyacrylamide gels. *Nature*, **195,** 281.

Renwick, J. H. (1969). Progress in mapping human autosomes. *Brit. Med. Bull.*, **25,** 65.

Rhoades, M. M. (1961). In *The Cell* (J. Brachet, J. and A. E. Mirsky, eds.), Vol. 3, p. 1. Academic, New York.

Richardson, C. C. (1969). Enzymes in DNA metabolism. *Ann. Rev. Biochem.*, **38,** 795.

Ris, H. (1957). In *The Chemical Basis of Heredity* (D. McElroy and B. Glass, eds.) p. 23. Johns Hopkins Press, Baltimore.

Ris, H. (1969). In *Handbook of Molecular Cytology* (A. Lima-de-Faria, ed.), p. 221. North Holland, Amsterdam.

Roberts, K., and D. H. Northcote (1970). Structure of the nuclear pore in higher plants. *Nature,* **228,** 385.

Robertson, J. D. (1959). The ultrastructure of cell membranes and their derivatives. *Biochem. Soc. Symp.*, **16,** 3.

Roeder, R. G., and W. J. Rutter (1969). Multiple forms of DNA-dependent RNA polymerase in eukaryotic organisms. *Nature,* **224,** 234.

Roeder, R. E., and W. J. Rutter (1970). Specific nucleolar and nucleoplasmic RNA polymerases. *Proc. Natl. Acad. Sci., U.S.*, **65,** 675.

Roodyn, D. B. (1959). A survey of metabolic studies on isolated mammalian nuclei. *Intern. Rev. Cytol.*, **8,** 279.

Sagdopal, A., and J. Bonner (1970). Proteins of interphase and metaphase chromosomes compared. *Biochim. Biophys. Acta,* **207,** 227.

Sakai, H. (1967). A ribonucleoprotein which catalyses thiol-disulphide exchange in the sea urchin egg. *J. Biol. Chem.*, **242,** 1458.

Samarina, O. P. (1964). The distribution and properties of cytoplasmic deoxyribonucleic acid-like ribonucleic acid (messenger ribonucleic acid). *Biochim. Biophys. Acta,* **91,** 688.

Schildkraut, C. L., and J. J. Maio (1968). Studies on the intranuclear distribution and properties of mouse satellite DNA. *Biochim. Biophys. Acta,* **161,** 76.

Seifart, K. H., and C. E. Sekeris (1969). Extraction and purification of DNA-dependent RNA polymerase from rat liver nuclei. *European J. Biochem.*, **7,** 408.

Shepherd, G. R., and L. R. Gourley (1966). High resolution disc electrophoresis of histones. An improved method. *Anal. Biochem.*, **14,** 356, 364.

Shepherd, G. R., B. J. Noland, and C. N. Roberts (1970). Phosphorus in histones. *Biochim. Biophys. Acta,* **199,** 265.

Shepherd, G. R., J. M. Hardin, and B. J. Noland (1971a). Methylation of lysine residues in histone fractions from synchronised mammalian cells. *Arch. Biochem. Biophys.*, **143,** 1.

Shepherd, G. R., B. J. Noland, and J. M. Hardin (1971b). Histone phosphorylation in synchronised mammalian cell cultures. *Arch. Biochem. Biophys.*, **142,** 299.

Sheridan, W. F., and H. Stern (1967). Histones of meiosis. *Exptl. Cell Res.*, **45,** 323.

Sibatani, A., K. Yamana, K. Kimura, and H. Okagaki (1959). Fractionation with phenol of ribonucleic acid of animal cells. *Biochim. Biophys. Acta,* **33,** 590.

Siebert, G. (1961). Enzyme und Substrate der Glykolyse in Isolierten Zellkernen. *Biochem. Z.*, **334,** 369.

Siebert, G. (1968). In *Comprehensive Biochemistry* (M. Florkin and E. H. Stotz, eds.), Vol. 23, p. 1. Elsevier, Amsterdam.

Siebert, G., and G. B. Humphrey (1965). Enzymology of the nucleus. *Advan. Enzymol.*, **27,** 239.

Siebert, G., H. Langendorf, R. Hannover, D. Nitz-Litzow, B. C. Pressman, and C. Moore (1965). Untersuchungen zur Rolle des Natrium-Stoffwechsels im Zellkern der Rattenleber. *Z. Phys. Chem.*, **343,** 101.

Siebert, G., M. G. Ord, and L. A. Stocken (1971). Histone phosphokinase activity in nuclear and cytoplasmic fractions from normal and regenerating rat livers. *Biochem. J.*, **122,** 721.

Smith, E. L., R. J. De Lange, and J. Bonner (1970). Chemistry and biology of the histones. *Physiol. Rev.*, **50,** 159.

Sober, H. A. (1968). *Handbook of Biochemistry.* The Chemical Rubber Company. Cleveland, Ohio.

Starr, J. L., and B. H. Sells (1969). Methylated ribonucleic acids. *Physiol. Rev.*, **49,** 623.

Stellwagen, R. H., and R. D. Cole (1969). Histone biosynthesis in mammary gland during development and lactation. *J. Biol. Chem.*, **244,** 4878.

Stern, H., and Y. Hotta (1969). In *Handbook of Molecular Cytology* (A. Lima-de-Faria, ed.), p. 520. North Holland, Amsterdam.

Stevely, W. S., and L. A. Stocken (1968). Variations in the phosphate content of histone F1 in normal and irradiated tissues. *Biochem. J.*, **110,** 187.

Stevens, B. J., and H. Swift (1966). RNA transport from nuclei to cytoplasm in *Chironomus* salivary gland. *J. Cell Biol.*, **31,** 55.

Sung, M. T., and G. H. Dixon (1970). Modification of histones during spermiogenesis in trout: A molecular mechanism for altering histone binding to DNA. *Proc. Natl. Acad. Sci., U.S.*, **67,** 1616.

Sung, M. T., G. H. Dixon, and O. Smithies (1971). Phosphorylation and synthesis of histones in regenerating rat liver. *J. Biol. Chem.*, **246,** 1358.

Swift, H. (1965). Molecular morphology of the chromosome *in vitro* (C. J. Dawe, ed.), Tissue Culture Association, Baltimore, Md.

Teng, C. S., C. T. Teng, and V. G. Allfrey (1971). Studies of nuclear acidic proteins. *J. Biol. Chem.*, **246,** 3597.

Tidwell, T., V. G. Allfrey, and A. E. Mirsky (1968). The methylation of histones during regeneration of the liver. *J. Biol. Chem.*, **243,** 707.

Turkington, R. W., and M. Riddle (1969). Hormone-dependent phosphorylation of nuclear proteins during mammary gland differentiation *in vitro*. *J. Biol. Chem.*, **244,** 6040.

Ui, M. (1957). Preparation, fractionation and properties of calf thymus histone. *Biochim. Biophys. Acta*, **25,** 493.

Vidali, G., and J. M. Neelin (1968). The stability of acid-soluble chromosomal proteins from avian erythrocytes. *Can. J. Biochem.*, **46,** 781.

Vincent, W. S. (1952). The isolation and chemical properties of the nucleoli of starfish oocytes. *Proc. Natl. Acad. Sci., U.S.*, **38,** 139.

Vincent, W. S., and E. Baltus (1960). In *The Cell Nucleus* (J. S. Mitchell, ed.), p. 18. Butterworths, London.

Waldeyer, W. (1888). Ueber Karyokinase und ihre Beziehungen zu den Befruchtungsvorgängen. *Arch. Mikroskop. Anat.*, **32,** 1.

Walker, P. M. B. (1971). "Repetitive" DNA in higher organisms, *Prog. Biophys.*, **23,** 145.

Walker, P. M. B., W. G. Flamm, and A. McLaren (1969). In *Handbook of Molecular Cytology* (A. Lima-de-Faria, ed.), p. 52. North Holland, Amsterdam.

Wang, T. Y. (1961). Incorporation of [^{14}C]amino-acids into ribonucleoprotein fraction of isolated thymus cell nuclei. *Biochim. Biophys. Acta*, **49,** 108.

Wang, T. Y. (1962). The chemical composition of nuclear ribosomes of calf thymus. *Arch. Biochem. Biophys.*, **97,** 387.

Wang, T. Y. (1966). Solubilization and characterization of the residual proteins of the cell nucleus. *J. Biol. Chem.*, **241,** 2913.

Wang, T. Y. (1968). Restoration of histone-inhibited DNA-dependent RNA synthesis by acidic chromatin proteins. *Exptl. Cell Res.*, **53,** 288.

Wang, T. Y., and E. W. Johns (1968). Study of the chromatin acidic proteins of rat liver: Heterogeneity and complex formation with histones. *Arch. Biochem. Biophys.*, **124,** 176.

Watson, M. L. (1955). The nuclear envelope. *J. Biophys. Biochem. Cytol.*, **1,** 257.

Watson, M. L. (1959). Further observations on the nuclear envelope of the animal cell. *J. Biophys. Biochem. Cytol.*, **6,** 147.

Weismann, A. (1892). *The Germ Plasma, a Theory of Heredity* (W. N. Parker and H. Rönnfeldt, translators, Walter Scott, Ltd.). London, 1893.

Weiss, S. B. (1960). Enzymatic incorporation of ribonucleoside triphosphates into the interpolynucleotide linkages of ribonucleic acid. *Proc. Natl. Acad. Sci., U.S.*, **46,** 1020.

Widnell, C. C., and J. R. Tata (1966). Studies on the stimulation by ammonium sulphate of the DNA-dependent RNA polymerase of isolated rat liver nuclei. *Biochim. Biophys. Acta*, **123,** 478.

Wiener, J., D. Spiro, and W. R. Loewenstein (1965). Ultrastructure and permeability of nuclear membranes. *J. Cell. Biol.*, **27,** 107.

Wunderlick, F., and W. W. Franke (1968). Structure of macronuclear envelopes of *Tetrahymena pyriformis* in the stationary phase of growth. *J. Cell Biol.*, **38,** 458.

Zimmerman, E. G., and M. R. Lee (1968). Variation in chromosomes of the cotton rat, *Sigmodon hispidus*. *Chromosoma*, **24,** 243.

Zubay, G., and P. Doty (1959). Isolation and properties of deoxyribonucleoprotein particles containing single nucleic acid molecules. *J. Mol. Biol.*, **1,** 1.

6

THE GOLGI COMPLEX

D. H. Northcote

1. INTRODUCTION

An extremely important part of the process of cell and tissue differentiation is the mechanism by means of which chemical communication takes place between the regions within a single cell and also between a cell and its neighbors. In order to understand the Golgi apparatus in relation to the functioning and growth of the cell, it is important to appreciate its role as one of the main synthetic sites and as a major route for intra- and intercellular communication (Northcote, 1971a, b).

In conjunction with the endoplasmic reticulum (ER), membrane-bounded vesicles, and lysosomes, the Golgi apparatus forms an extensive intercommunicating membrane system within the cell. The lumen enclosed by this membrane system is made up by that of the perinuclear space; the ER; the cisternae, vesicles, and tubules of the Golgi complex; the lysosomes; the pinocytotic and autophagic vesicles; and, in plant cells, the vacuole contained by the tonoplast. The region outside the plasmalemma must also be included as part of this intercommunicating transport route. All these spaces are connected either by a functional continuity or by a direct morphological connection, so that together they form a distinct channel which is separated from the rest of the cytoplasm and nucleoplasm of the cell and in which material can be synthesized, modified, and transported. The movement of material can take place either within the cell or from the inside to the outside of the cell and also, to a limited extent, from the exterior back into the vesicles and lysosomes. The Golgi apparatus occupies a key position in the functioning of this membrane system, since not only are specific synthetic functions carried out within the organelle and its associated vesicles, but also, as part of the transport route, it presents an irreversible connection between the perinuclear space and ER on the one hand and the vesicles and plasmalemma on the other (De Duve, 1969). Because it is also a relatively localized region of the diffuse membrane system, it represents a site along the route where synthesis and transport could be controlled. The connection with the nucleus is also probably very important for the transport of material from this central control site during the growth and development of the cell.

2. FORM OF THE GOLGI APPARATUS

The existence of the Golgi apparatus has been doubted since its very first discovery, and the interpretation of images seen with the optical microscope has been constantly discussed and criticized. The Golgi apparatus can be seen in sections of fixed cells stained with metallic stains, or in the living cell by using phase-contrast optics (Golgi, 1898a, b; Dalton and Felix, 1952, 1953; Beams and Kessel, 1968). The organelle varies in form from a compact discrete granule or mass to a more dispersed filamentous reticulum. It is pleomorphic, and its shape as well as its function varies with the metabolic and developmental state of the cell (Elliott and Zieg, 1968; Flickinger, 1969c; Northcote, 1969b). The advent of a thin-sectioning technique for the electron microscope made possible a more detailed and definitive description of an organelle designated as the Golgi apparatus (Mollenhauer and Morré, 1966b; Mollenhauer et al., 1967).

The model of the apparatus is based upon images seen in thin sections and in freeze-etched replicas of plant (Figs. 1–3, 5–10, and 13–16) and animal tissues (Fig. 4), and also upon negatively and positively stained cell fractions prepared by centrifugation from broken cells in order to concentrate the Golgi apparatus free from other cell organelles and membrane material (Cunningham et al., 1966) (Figs. 11–12). It is a characteristic, easily recognizable structure and occurs in nearly all the cells of animals and plants. It consists of a stack of disc-shaped cisternae or saccules (usually about 3 to 12) one above the other (Figs. 1–10); they are all slightly curved, so as to give a concave and a convex face to the whole stack (Figs. 1 and 9). In some cases material can be seen between the cisternae of a stack, and this may be part of a definite intercisternal structure (Figs. 5 and 6). From the edges of the cisternae, an anastomosing network of tubules arises and in places these swell to form various types of vesicles (Figs. 5, 8–10, and 14–16). The cisternae may therefore not be formed of complete sheets of membrane, but the membranes may be fenestrated (Fig. 10) and this fenestration can become more pronounced progressively down the stack of cisternae (from convex to concave face), so that at the convex surface the membranes are complete and at the other surface the cisternal membranes are extensively broken by pores which fragment the cisternae into the network of tubules (Morré and Mollenhauer, 1964; Mollenhauer and Morré, 1966a, b; Mollenhauer et al., 1967; Flickinger, 1969a).

3. FORMATION

The development and formation of the Golgi apparatus may involve a series of different processes such as the synthesis of a stack of cisternae in the absence of a pre-

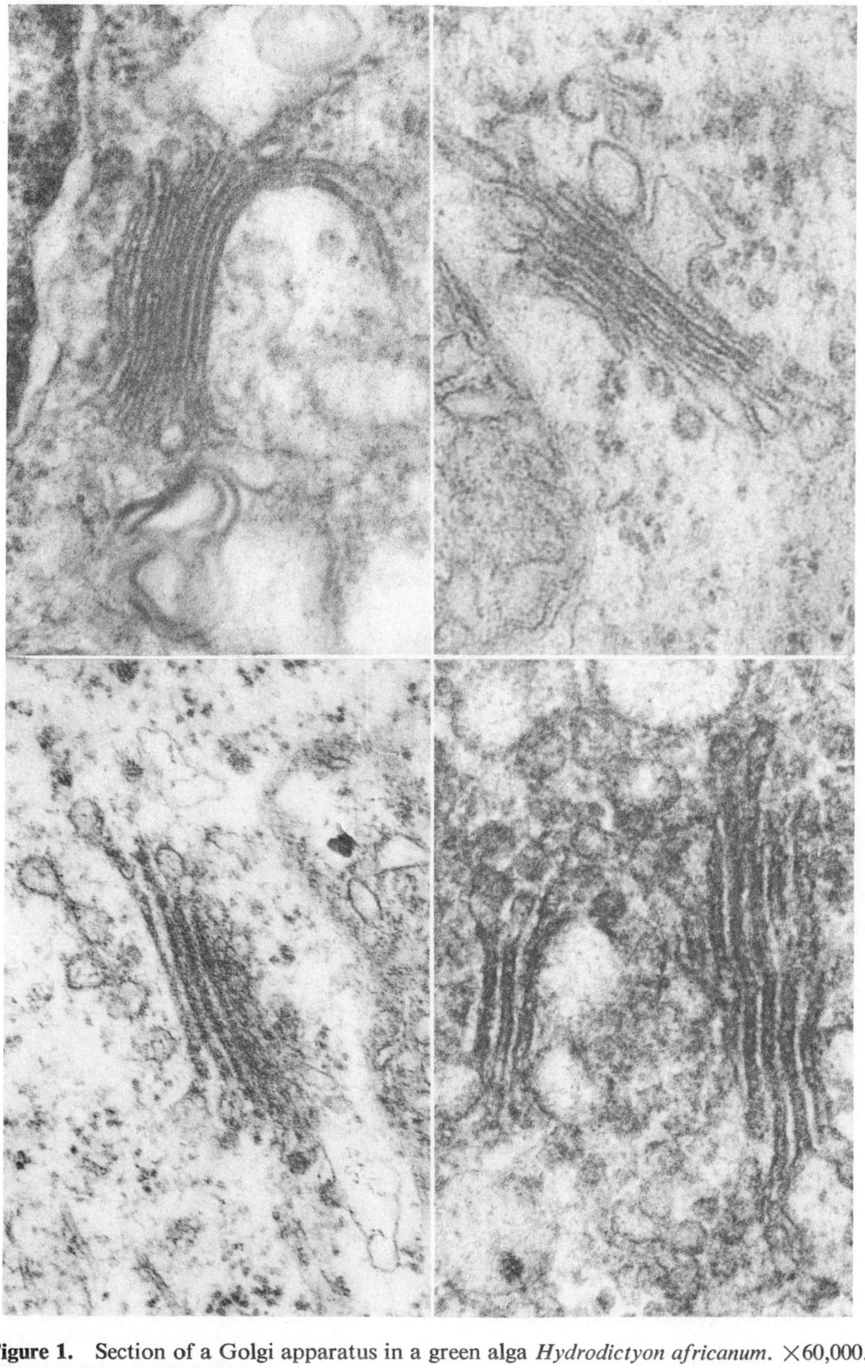

Figure 1. Section of a Golgi apparatus in a green alga *Hydrodictyon africanum*. ×60,000.
Figure 2. Section of a Golgi apparatus in the peripheral cytoplasm of a bean cell (*Phaseolus vulgaris*) in tissue culture. ×68,000.
Figure 3. A Golgi apparatus in the cell plate region of a sycamore cell (*Acer pseudoplatanus*). The cell is in mitosis at telophase. The vesicles at the Golgi cisternae are also found at the developing cell plate. ×53,000.
Figure 4. Golgi apparatus and associated vesicles in a tunicate (*Ciona intestinalis*). ×65,000.

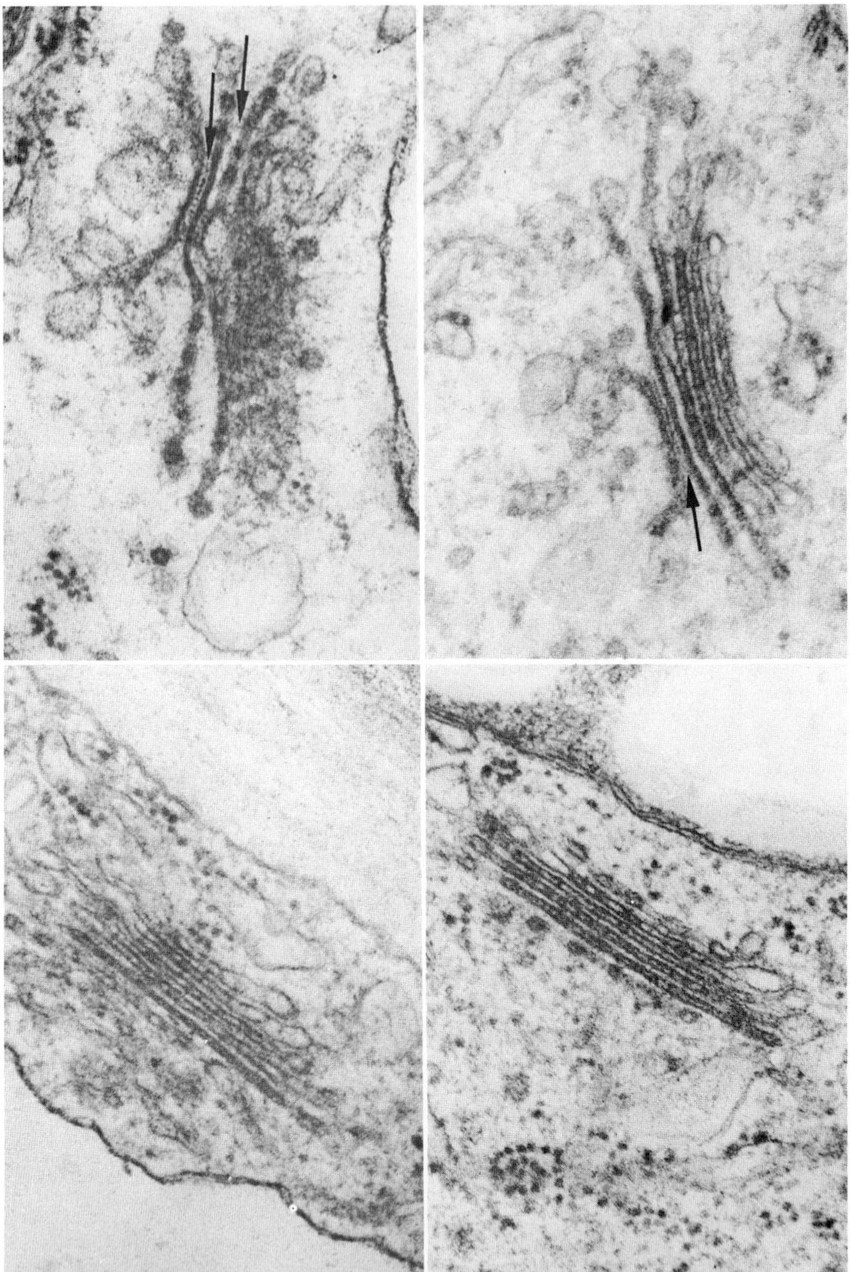

Figures 5–8. Golgi bodies in the cytoplasm of bean cells in tissue culture. Intercisternal material can be seen (arrows) in the complexes shown in Figs. 5 and 6. The relationship of the Golgi apparatus to rough and smooth ER can be seen in Figs. 7 and 8. Fig. 5: ×45,000. Fig. 6: ×65,000. Fig. 7: ×40,000. Fig. 8: ×65,000.

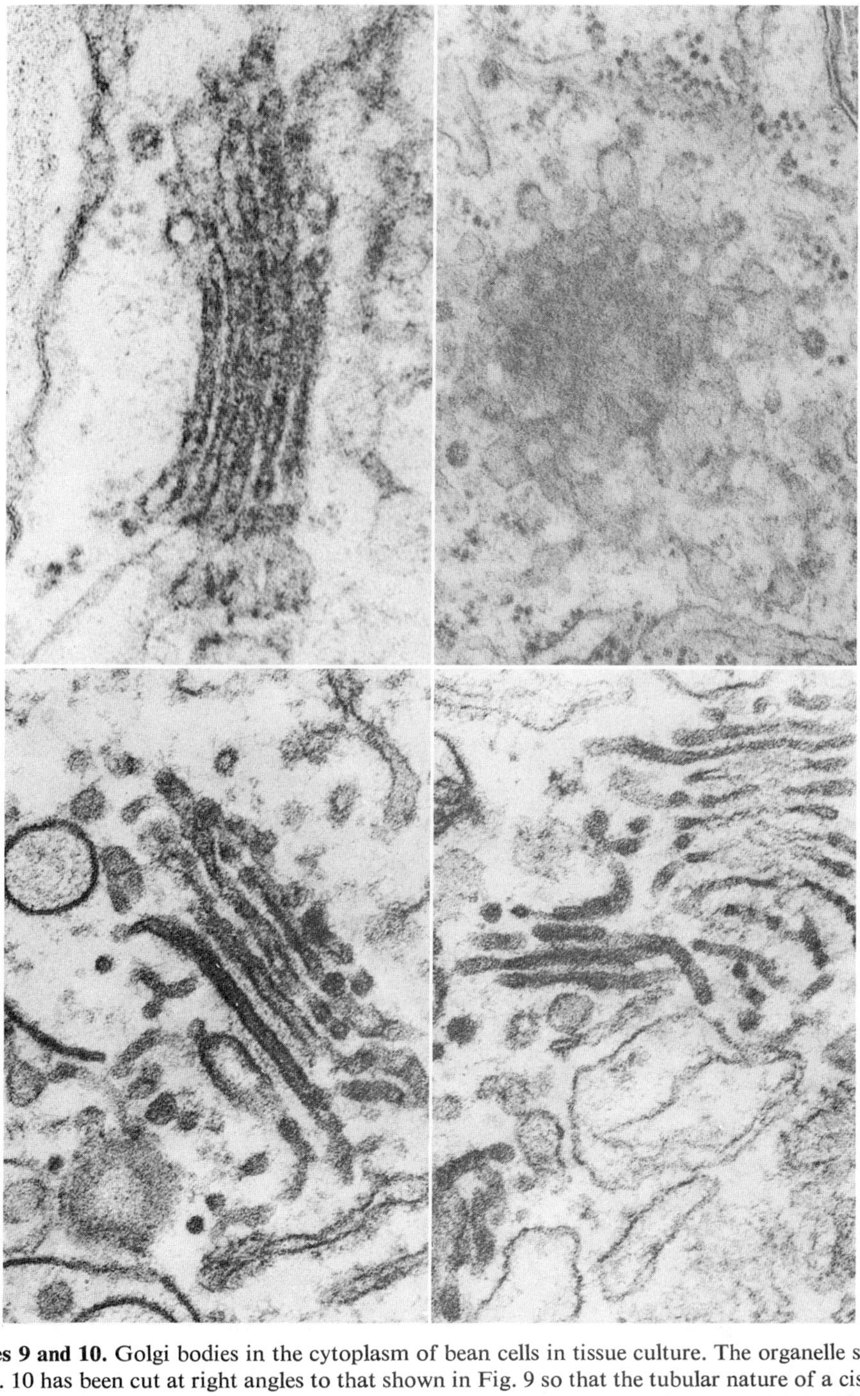

Figures 9 and 10. Golgi bodies in the cytoplasm of bean cells in tissue culture. The organelle shown in Fig. 10 has been cut at right angles to that shown in Fig. 9 so that the tubular nature of a cisterna at its periphery and the connections with the vesicles can be seen. Fig. 9: ×65,000. Fig. 10: ×55,000.

Figures 11 and 12. These micrographs show sections of pellets prepared from broken cells of pea roots. The homogenate of the broken cells was centrifuged to concentrate the Golgi membranes and cisternae. The material was fixed with potassium permanganate. Fig. 11: ×64,000. Fig. 12: ×52,000.

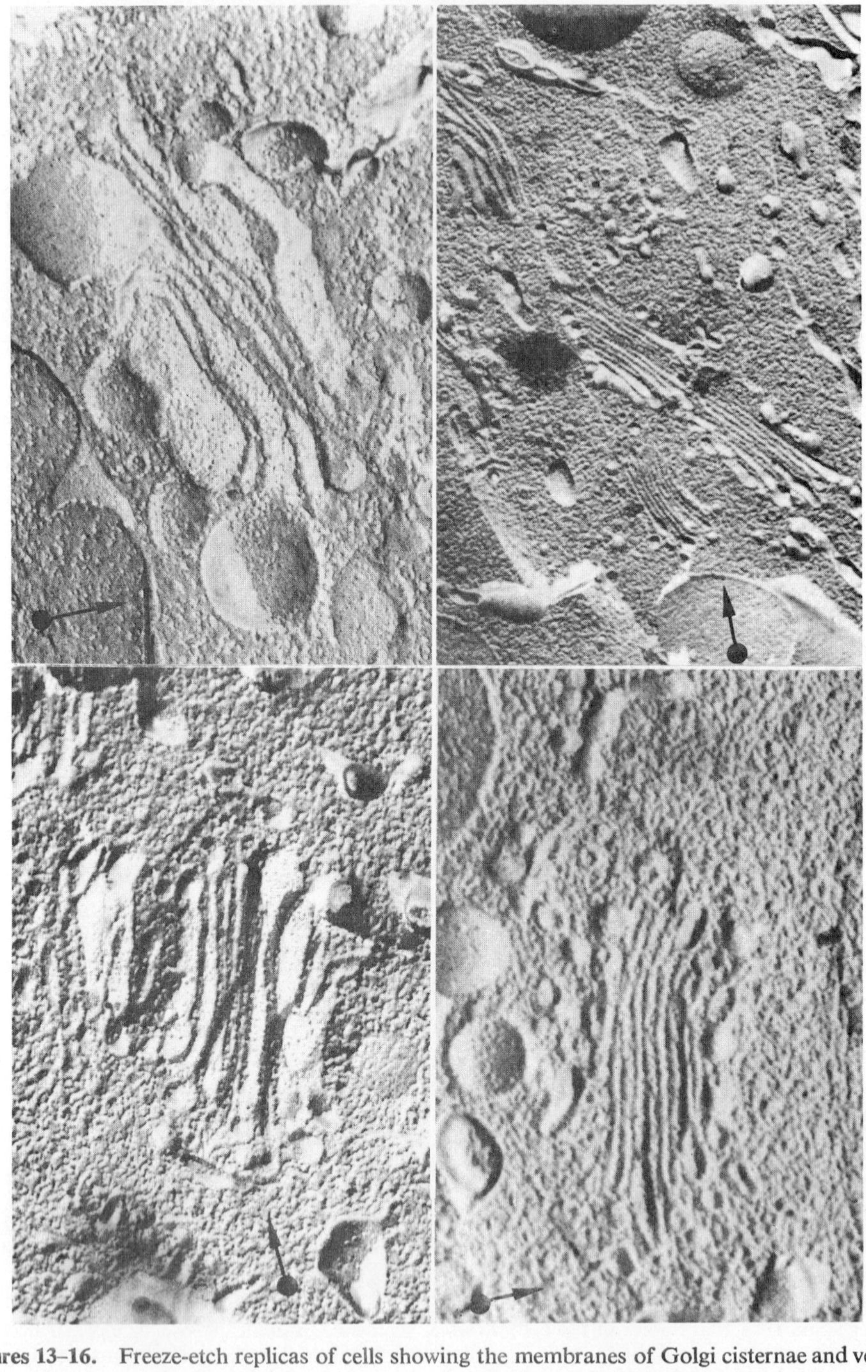

Figures 13–16. Freeze-etch replicas of cells showing the membranes of Golgi cisternae and vesicles. Figure 13 shows a cell of a pea root, Figures 14, 15, and 16 are micrographs of cells from pea leaves. The tissues were not fixed. The large arrow shows the direction in which the platinum was evaporated onto the replicas. Fig. 13: ×53,000. Fig. 14: ×30,000. Fig. 15: ×56,000. Fig. 16: ×61,000.

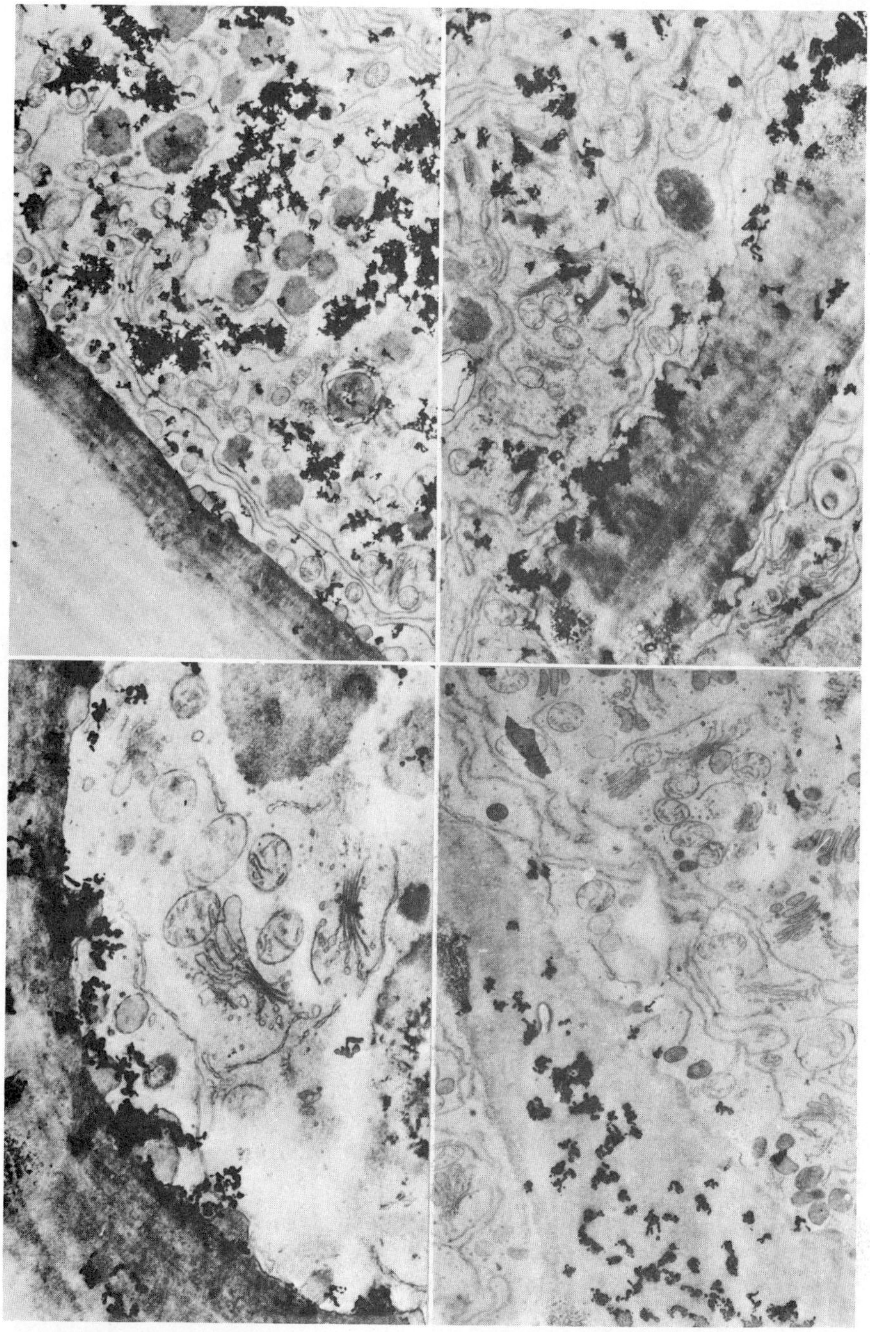

Figures 17–20. Radioautographs. Figure 17 shows a longitudinal section of a root cap cell of wheat exposed to D-[6-^{3}H] glucose (5 mCi /ml) for 10 min before fixation. Accumulation of a labeled derivative (a pectic polysaccharide containing galactose) in the Golgi bodies can be seen. Figures 18, 19, and 20 are micrographs of cells of the root cap of wheat treated similarly to that shown in Figure 17 but incubated for 10 min, 30 min, and 60 min, respectively, in nonradioactive glucose before fixation. The Golgi bodies lose the radioactive material which becomes incorporated across the plasmalemma into the wall. Fig. 17: ×4,500. Fig. 18: ×5,000. Fig. 19: ×11,500. Fig. 20: ×6,000.

existing apparatus within the cell, the alteration in type and number of vesicles formed, increases in the number of stacks, and variations in the number and size of cisternae in each stack and in the amount of tubular and of vesicular regions of the stack (Flickinger, 1969c). All these forms of growth may, however, be part of a single process of membrane conversion and formation, but it is possible that some of the details of the separate stages might be different. For instance, it might be that individual stacks of cisternae can arise from preexisting stacks by division or fragmentation, and this would be a different process from a *de novo* formation of the apparatus. Nevertheless, many cytological, biochemical, and chemical studies can be interpreted by postulating that there is a flow of membrane material from the endoplasmic reticulum via the Golgi apparatus to the plasmalemma. The most likely theory for the formation of the Golgi apparatus is that the cisternal membranes arise from the rough ER, which changes to smooth ER and then becomes the Golgi cisternae, and that these cisternae break down to vesicles which can fuse with and extend the plasmalemma.

The maintenance of the Golgi complex requires the presence of the nucleus, and in its absence or in the presence of actinomycin D, the Golgi bodies decrease in size and eventually are no longer detectable (Flickinger, 1969b). In enucleated amebas small, curved, smooth cisternae are found within 30 min to 1 hr after renucleation, and these aggregate in certain parts of the cytoplasm. A similar progressive alignment of single cisternae to form stacks has been observed in the cells of a ciliate *Tetrahymena* during mating, when Golgi bodies are formed (Elliott and Zieg, 1968). In amebas the Golgi complexes grow both in size and number during the 6–24 hr after renucleation, and during this time the rough ER seems to participate in the process of membrane production. This is shown by the presence of an electron-dense material both in the lumen of the ER and also in the lumen of the cisternae of the Golgi apparatus, as well as by the appearance of a direct continuity between the ER and the Golgi cisternae (Flickinger, 1969b). Grimstone (1959), in an earlier study on the regeneration of the Golgi apparatus in the flagellate *Trichonympha*, also showed a close association between the ER and the Golgi cisternae and, in addition, the Golgi bodies were observed to appear sequentially after the regeneration of the rough ER during refeeding of a starved organism.

The radioautographic evidence of Palade and his co-workers (Caro and Palade, 1964; Jamieson and Palade, 1967a, b) demonstrated that material was transferred from the rough ER by means of small vesicles to the outer cisternae of the Golgi complex, thus although there need be no direct connection, the ER system is in continuity with the Golgi bodies. There have been numerous experimental results that confirm these ideas and also indicate that there are transitional elements of smooth ER between the rough ER and the Golgi cisternae. Since the outermost cisternae of the Golgi apparatus could be constantly formed by the fusion of vesicles derived from the reticulum system, while the inner cisternae in any one stack could become fenestrated and pinch off vesicles, the concept has arisen whereby the Golgi apparatus is thought of as having a newly forming face at one surface and a mature or secreting face at the opposite surface. In colonic goblet cells, Neutra and Leblond (1966a) (see also Brown, 1969) estimated that the entire Golgi apparatus is formed and reformed within 20–40 min and that individual cisternae are released once every 2–4 min.

Chemical analysis of the membranes of the Golgi apparatus has also indicated that they have a composition between that of the ER and of the plasmalemma (Keenan and

Morré, 1970). This has also been suggested as the result of a study of the cytoplasmic membranes of the plant pathogenic fungus *Pythium ultimum* Trow in which the overall membrane thickness, the staining intensity, and the substructural pattern of stain deposition were investigated. It was shown that at one extreme the plasmalemma and the membranes of the vesicles (free in the cytoplasm and attached to the Golgi cisternae at the distal, mature face) stained intensely, were thicker (75 Å), and were clearly unit membranes, while at the other extreme the nuclear envelope and the ER membranes stained faintly, appeared thinner (25–40 Å), and rarely revealed the dark-light-dark pattern of the unit membrane (Grove et al., 1968). The membranes of the Golgi apparatus were differentiated across the stack so that those at the forming face appeared similar to those of the ER and nucleus, whereas those at the mature face resembled the plasmalemma; and membranes of the midregion of the stack were intermediate in their appearance. These results indicated that the membranes were modified within the Golgi stack and that one of the functions of the apparatus is to alter the membranes of the ER so that they resemble that of the plasmalemma. Presumably, this process is irreversible. Thus the Golgi apparatus represents a one-way valve within the system: nuclear envelope → ER → vesicles → Golgi apparatus → vesicles → plasmalemma.

4. THE FUNCTION OF THE GOLGI APPARATUS

A general feature of the Golgi apparatus is that it is concerned with the formation and packaging of material for export from the cell across the plasmalemma by a process of reverse pinocytosis. In this way cell wall material is deposited in plants (Northcote, 1969a, b), and secretory proteins, digestive enzymes (Zeigel and Dalton, 1962; Jamieson and Palade, 1967a, b), and the matrix material of the connective tissue (Neutra and Leblond, 1966b) are transported in animal tissues. There are also many examples of secretions derived from the Golgi apparatus that form internal cell structures such as the acrosome of the spermatocyte and spermatid (Fawcett, 1966), certain yolk substances of oocytes (Kessel, 1966), or the tubular inclusions of endothelial cells (Sengel and Stoebner, 1970). There is now a very large list of materials known to be packaged and transported through the Golgi apparatus and its associated vesicles in animals and plants (Sievers, 1965; Beams and Kessel, 1968; Schnepf, 1969). Many of these substances are polysaccharides, or they are proteins or lipids which acquire carbohydrate moieties to become glycoproteins or glycolipoproteins. The carbohydrate is acquired in the smooth membrane systems of the Golgi apparatus and its vesicles, and it may be that conjugation with carbohydrate is a prerequisite for subsequent transport of the material across the plasmalemma (Eylar, 1965).

By using the electron microscope, substances can sometimes be seen in the ER, in the Golgi apparatus, in the vesicles, and outside the plasmalemma, and often it appears that the same substance occurs at all these sites. An apparent progression of the form of the material in the lumens of these organelles can be seen, which makes a tentative series of events and flow of products within the cell from one part of the membrane system to the other seem very probable (Thiéry, 1969). The work of Manton (1966, 1967) on the scale production of green and brown algae is of particular interest since these scales, which have a characteristic shape, can be seen within the Golgi cisternae and are progressively developed and transported in the vesicles to the outside of the cell. The scales have a dorsoventral asymmetry and are oriented in a definite way within the vesicle. Since the scale continues to develop in this asymmetric way within the vesicle before it is finally transported to the outside of the cell, it appears that the vesicle also has a dorsoventral asymmetry either in the lumen or on the membranes. A physical and chemical investigation of the scales produced in the Golgi apparatus of one of these algae has indicated that they are made up in part of polysaccharide (Green and Jennings, 1967; Brown et al., 1970).

Direct experimental evidence that clearly demonstrates a flow of material and a possible modification of its structure within the membrane system of the cell can be obtained by the use of radioactive markers. The radioactive tracer is supplied as a pulse to the cell so that it is incorporated into the material, and it is then chased through the system by a cold precursor. The passage of the labeled material through the cell can be followed by direct autoradiography of thin sections of the tissue prepared for electron microscopy (Figs. 17–20) (Northcote and Pickett-Heaps, 1966). It can also be followed by determining the kinetics of the incorporation of the radioactivity supplied as a short pulse of the precursor into the tissue, and by preparing from the tissue cell fractions which separate the various parts of the membrane system during the period of the pulse chase experiment. In order to see the characteristics of the transport-synthesis process, several systems which have been investigated by the techniques described above are discussed in more detail.

The intracellular transport of secretory protein in the pancreatic exocrine cells has been studied by Palade and his co-workers (Caro and Palade, 1964; Jamieson and Palade, 1967a, b). They have shown that the digestive enzymes (e.g., α-amylase) or their precursors (e.g., α-chymotrypsinogen), which are synthesized on the ribosomes at the outer surface of the rough ER, are transferred across the reticulum membrane into the lumen by some mechanism unknown at present. Once in the lumen they are transported to elements of smooth ER where they are packed into small vesicles and carried into the forming surface of the Golgi apparatus. In the Golgi body the protein is concentrated into membrane-bounded zymogen granules which arise from the Golgi cisternae. These granules migrate to the cell apex and discharge their contents into the pancreatic ducts where they pass to the intestine. The passage of the material to the outside of the cell is brought about by reverse pinocytosis in which the membrane of the granule fuses with and becomes part of the cell membrane.

A similar series of events has been shown to occur in the goblet cells of the colon of rats by Leblond and his co-workers (Neutra and Leblond, 1966a, b; see also Thiéry, 1969). In this tissue a protein synthesized at the rough ER is passed to the Golgi apparatus where it is conjugated with carbohydrate, and the resultant mucopolysaccharide is transported as mucigen granules to become the mucus of the colon.

A large number of the glycoproteins exported from animal cells have the carbohydrate portion of the molecule joined to the protein moiety by a covalent linkage in which (*N*-acetyl)-D-glucosamine forms an *N*-glycosidic linkage with the amide group of aspartamide [*N*-(β-aspartyl)-β-D-(*N*-acetyl)-glucosaminide linkage]. Glucosamine also occurs in other positions in the carbohydrate prosthetic group. It is of considerable interest therefore to know if the peptide is labeled with a glucosamine radical at the polysome level so that the rest of the carbohydrate can be added onto the tag in the Golgi complex or whether all the carbohydrate is added onto the peptide when it is passed to the smooth membrane part of the system. Radioautographic evidence, cell fractionation, and the use of puromycin to inhibit polypeptide synthesis have indicated that some glucosamine is incorporated into the nascent polypeptide chains, probably at the level of polyribosomes attached to the membrane of the rough ER and that the rest of the carbohydrate, including more glucosamine, is subsequently attached onto the initial glucosamine in the Golgi complex (Molnar et al., 1965; Lawford and Schachter, 1966; Li et al., 1968; Schenkein and Uhr, 1970; Zagury et al., 1970).

In some instances however, especially when the glycoproteins synthesized are subunits of the cell membrane structure, the assembly of the carbohydrate moiety onto the polypeptide chain takes place solely in the smooth membrane part of the system, probably within the Golgi apparatus. The attachment of the carbohydrate to the protein in these cases may not be an *N*-glycosidic bond to aspartamide but an *O*-glycosidic bond to serine or threonine (Eylar, 1965; Hagopian et al., 1968; Bosmann et al., 1969).

The wall of the plant cell is composed of polysaccharides which can be classified into two types: a matrix composed of pectic substances and hemicelluloses, and a fibrillar network of α-cellulose (Northcote, 1969a). The first stage of wall formation is at cytokinesis when a cell plate is developed at the center of the mitotic spindle, and this is progressively enlarged and thickened in an organized manner. The cell plate is contained by a membrane which becomes the new plasmalemma of each daughter cell. The material of the matrix of the cell plate and of the membrane is derived from vesicles, some of which can be seen to be pinched off from the Golgi apparatus and which move to the site of cell plate formation where they fuse together to form both the matrix substance and the cell membrane (Whaley et al., 1966; Roberts and Northcote, 1970). Microfibrils of cellulose are woven into the clear matrix of the cell plate at a later stage (Roberts and Northcote, 1970). The wall is continually enlarged by the deposition of polysaccharide matrix material into it from the Golgi vesicles by reverse pinocytosis. Golgi bodies from the roots of pea-seedlings have been isolated after the roots had been incubated with radioactive glucose, and it has been shown that the Golgi bodies and the associated vesicles contain pectic substances and hemicelluloses of the cell wall (Harris and Northcote, 1971). The type of polysaccharide incorporated into the wall changes during its development, although the method of deposition into the wall does not, so that the mechanism for synthesizing polysaccharides which probably occurs within the Golgi cisternae and vesicles is controlled and changes during the differentiation of the cell (Northcote, 1969a).

The location of cellulose synthesis is at present not clear. Various studies have indicated that the Golgi apparatus and its vesicles form the cellulose microfibrils and export them to the outside of the cell (Ray et al., 1969; Brown et al., 1969, 1970). Brown et al. (1970) investigated the polysaccharide scales found in certain flagellates. They are found within the Golgi complex before export across the plasmalemma by reverse pinocytosis (Manton, 1966). These investigators showed that part of the polysaccharide complex that forms the scale is composed of glucose chains linked by $1 \rightarrow 4\,\beta$ bonds and that it resembles cellulose in other chemical and physical properties. However, autoradiographic evidence suggests that cellulose is formed at the plasmalemma surface (Wooding, 1968) and freeze-etch studies have also indicated that cellulose microfibrils arise from the surface of the plasmalemma at discrete sites which appear on the membrane surface of the freeze-etch replica as particles (Staehelin, 1966; Northcote and Lewis, 1968). Microfibrils can also be seen to form within the matrix of the developing cell plate, and in this case too they appear to radiate from particles within the matrix of the cell plate (Roberts and Northcote, 1970).

These two apparently different hypotheses can be reconciled when it is realized that the Golgi apparatus probably contributes to the formation of the plasmalemma and that therefore the enzymic sites that may have been attached to the Golgi membranes on the inside of the cisternae or vesicles become the sites on the outside of the plasmalemma after the vesicles have fused with the cell membrane. By this means it is also possible that enzymes such as transglycosylases might be transported to the outside of the cell during growth to modify polysaccharides already present in the wall and thus change the physical nature of the wall, hence the growth response of the cell (Stoddart and Northcote, 1967; Rubery and Northcote, 1970).

During the thickening of the cell wall, the material is often deposited in an organized manner at definite localized sites which give rise to spiral or reticulate thickening of the wall. This phenomenon, together with others of a similar nature, such as the localized formation of the cell plate in plant cells, the secretion of protein granules at a definite cell surface, and the formation of the brush border in animal cells, indicates that it is necessary to consider not only the formation of the material and its packaging for transport but also the direction and site of transfer at the right time. This must be an organized and controlled process. During cell plate and wall formation in plant cells, this directed transport can be seen to be mediated partly by the alignment and laying down of microtubules within the cytoplasm. They are found at those positions where the vesicles are incorporated into the cell plate (Esau and Gill, 1965; Pickett-Heaps and Northcote, 1966a) or into the wall (Wooding and Northcote, 1964; Northcote, 1969b).

One of the functions of the apparatus, as already discussed, is that it is part of an internal transport system of the cell, coupled with this function, which involves secretory and digestive processes, are synthetic activities, not the least of which are those concerned with the formation and modification of lipoprotein membranes. The most obvious way in which this is achieved is by the incorporation of a mucopolysaccharide coat which appears inside the vesicles derived from the Golgi cisternae but which, when exported to the plasmalemma, appears on the outside of this membrane. This could imply that not only is the outer surface coat of the membrane derived and placed at the surface of a preexisting and independently formed plasma membrane, but that the whole complex macromolecular architecture of the cell membrane itself—lipoprotein and mucopolysaccharide—is assembled by the Golgi apparatus and its vesicles, and that this is then transferred to the growing points on the surface of the cell where the membrane is being extended (Stockem, 1969), sometimes in a very specialized way, as in the brush borders of the intestinal absorptive cells (Bonneville and Weinstock, 1970).

The plasmalemma of the cell plate of plants arises from the membranes of the vesicles that fuse together to form the plate, and some of these vesicles are formed from the Golgi apparatus so that at this time there is a direct contribution of the membrane developed at the Golgi complex to the new plasmalemma formed at telophase (Roberts and Northcote, 1970). The Golgi apparatus continues to contribute to the extension of the plasmalemma during growth in the area of the cell wall, when material contained within membrane-bounded vesicles derived from the Golgi apparatus is passed into the wall by fusion of the vesicles with the cell membrane (Pickett-Heaps and Northcote, 1966b). During secondary thickening of the wall, however, there is still an extensive incorporation of material from the Golgi apparatus into the wall (Wooding and Northcote, 1964; Northcote and Wooding, 1966, 1968) and a corresponding contribution to the plasmalemma without any very great increase in surface area of the cell membrane. It is possible therefore that membrane material can be continuously formed and aggregated at the ER and Golgi apparatus, and possibly disaggregated and released back to the cytoplasm at the plasmalemma. In HeLa cells, Bosmann et al. (1969) showed that smooth internal membranes (probably derived from the Golgi complex) were the site for the assembly of membrane glycoproteins and glycolipids which form the subunits of the plasma membrane.

The change and modification of membrane material thought to occur at the Golgi complex ensures that material packed in membranes and pinched off from the secretory face of the Golgi apparatus is surrounded by membrane of a nature similar to that of the plasmalemma. Hence pinocytosis at the outer cell membrane is more possible than at the endoplasmic reticulum or other internal cell membrane. Not all material packed and transported from the Golgi apparatus is exported across the cell membrane. In polymorphonuclear leukocytes two distinct types of granules are found. Both are formed from the Golgi cisternae. The larger granules (80 nm) are formed only during the progranulocyte stage, and the material arises from the proximal or concave face of the Golgi apparatus in the form of vacuoles which have an electron-dense core and which aggregate to form the large granule. The smaller granules (50 nm) are formed during the myelocyte stage and arise from the distal or convex face of the Golgi apparatus (Bainton and Farquhar, 1966). These observations show the changing metabolic activity of the Golgi complex during different developmental stages of the cell, and they also indicate a functional compartmentalization within the cisternal stack. In addition, they indicate that different materials can be passed out from the Golgi body, presumably for different functions and possibly wrapped in different, although similar, membranes, which predisposes their fate within the cell. The granular material of the polymorphonuclear leukocyte is utilized intracellularly, and the granules are discharged into phagocytic vacuoles.

5. ENZYMES AND METABOLIC ACTIVITY OF THE GOLGI APPARATUS

The Golgi apparatus seems to bring about both synthesis and modification of the material contained in the lumen of its cisternae and vesicles. In many instances this is accomplished by a formation of oligosaccharide or polysaccharide, so that there must be present in the membrane system enzymes such as nucleoside diphosphate sugar transglycosidases that are necessary for this type of synthesis. In addition, vesicles of the Golgi apparatus possibly form primary lysosomes which contain hydrolytic enzymes (Novikoff, 1963). By fusion with pinocytotic vesicles or with autophagic vesicles, these primary lysosomes give rise to secondary lysosomes or to the digestive vacuoles of animal cells (Cohn and Fedorko, 1969). The hydrolytic enzymes are presumably first made at the rough ER and can pass to the smooth ER, which can either give rise to primary lysosomes directly, or the enzymes can be transferred via vesicles to the Golgi apparatus for further modification and concentration. These hypotheses concerning the role of the Golgi apparatus and ER in the formation of lysosomes are based on cytochemical evidence for the distribution of hydrolytic enzymes such as phosphatases in these membrane systems (Novikoff et al., 1962;

Novikoff et al., 1969; Novikoff et al., 1971). In plant cells some of the digestive function takes place in vacuoles (secondary lysosomes) which may arise from the ER, and they may also receive material from the vesicles of the Golgi apparatus (Matile, 1969).

The Golgi apparatus and its vesicles must therefore carry enzymes of two types: those used for the synthesis of lipoprotein membranes and polysaccharides, and those present because they are being concentrated, possibly modified, and transferred through the membrane system to be used in another part of the cell or organism.In either case the enzyme activity of the apparatus changes not only with the reversible changes that can take place in a cell's metabolism because of biochemical feedback mechanisms or supply of hormones (Smith and Farquhar, 1966; Hopkins, 1969), but also with the slower irreversible changes that are part of the development or differentiation of the cell as the organism grows and matures (Northcote, 1969a). Thus any enzyme activity found to be present in the Golgi apparatus need not be a constant or a general feature of the organelle.

The enzymic associations of the Golgi apparatus region of the cells have been studied by cytochemical techniques adapted to the electron microscope. In many cells it can be shown that acid phosphatases are present together with nucleotide diphosphatase and thiamine pyrophosphatase activity. The thiamine pyrophosphatase may be different from the nucleotide phosphatase or, as in liver microsomes, it may be the same enzyme (Yamazaki and Hayaishi, 1968). It has been suggested that thiamine pyrophosphatase may be a constant marker enzyme of the Golgi apparatus (Shanthaveerappa and Bourne, 1965; Dauwalder et al., 1966). These cytochemical studies have also demonstrated that not all the cisternae in a single stack of the Golgi body carry the same enzyme, but that various activities may be localized in different saccules at the mature or forming surface. It has further been shown that the ER may contain enzymic activities related to those of the vesicles near the Golgi apparatus, which again suggests a functional connection between the membrane systems in the region of the Golgi body (Novikoff et al., 1969).

A direct analysis of enzymic activity can be carried out on purified cell fractions made from broken cells (Fleischer et al., 1969; Cheetham et al., 1970). In Golgi bodies prepared from rat and bovine liver tissue, a high activity of an antimycin-insensitive NADH–cytochrome c reductase and a NADPH–cytochrome c reductase is found. These fractions also concentrate various sugar transferase activities which may be important for glycoprotein synthesis (Fleischer et al., 1969). Thus a UDP-*N*-acetylglucosamine: protein transferase and a UDP-D-galactose: *N*-acetylglucosamine (or: glycoprotein containing a terminal *N*-acetylglucosamine residue; or: glucose) transferase have been found in various fractions rich in Golgi bodies prepared from several tissues (Bosmann et al., 1968; Hagopian et al., 1968; Morré et al., 1969; Ovtracht et al., 1969; Caccam and Eylar, 1970).

Autoradiographic evidence has indicated that pectic polysaccharides and hemicelluloses are synthesized and transported by means of the Golgi apparatus in plant tissues (Northcote and Pickett-Heaps, 1966; Northcote and Wooding, 1966, 1968; Wooding, 1968). The transferases that form these polysaccharides from nucleoside diphosphate sugar compounds have been prepared from plant cells, and they have all been shown to be membrane bound (Hassid, 1969). Kauss and his colleagues (Kauss and Swanson, 1969; Kauss et al., 1969) showed that the enzymes for the polymerization and methylation of polygalacturonic acid (a pectic substance) are present in a localized and membrane-bound structural compartment of the cell and that it is a complex unit which allows organized cooperation of several enzymes necessary to build up a heteropolysaccharide such as the pectic material. This unit might well be the Golgi apparatus and its associated vesicles.

During polysaccharide synthesis in higher plants, it has recently been suggested that one of the initial steps is a transfer of a glycosyl unit from a nucleoside diphosphate sugar to a noncarbohydrate acceptor which may be a lipid component (Kauss, 1969; Villemez, 1970). A subsequent transglycosylation reaction of the saccharide carried on the lipid or lipoprotein carrier to a carbohydrate acceptor gives the complete polysaccharide (Northcote, 1969a). The polysaccharide synthetases are isolated as membrane-bound enzymes, and they are probably located within the Golgi complex. It is possible therefore that the lipoprotein membrane of the Golgi body and its vesicles take part as acceptor substances during the glycosyl transfer processes that assemble the polysaccharide molecules.

REFERENCES

Bainton, D. F., and M. G. Farquhar (1966). Origin of granules in polymorphonuclear leukocytes. Two types derived from opposite faces of the Golgi complex in developing granulocytes. *J. Cell Biol.*, **28**, 277.

Beams, H. W., and R. G. Kessel (1968). The Golgi apparatus: Structure and function. *Intern. Rev. Cytol.*, **23**, 209.

Bonneville, M. A., and M. Weinstock (1970). Brush border development in the internal absorptive cells of *Xenopus* during metamorphosis. *J. Cell Biol.*, **44**, 151.

Bosmann, H. B., A. Hagopian, and E. H. Eylar (1968). Glycoprotein biosynthesis: The characterization of two glycoprotein fucosyl transferases in HeLa cells. *Arch. Biochem. Biophys.*, **128**, 470.

Bosmann, H. B., A. Hagopian, and E. H. Eylar (1969). Cellular membranes: The biosynthesis of glycoprotein and glycolipid in HeLa cell membranes. *Arch. Biochem. Biophys.*, **130**, 573.

Brown, R. M. (1969). Observations on the relationship of the Golgi apparatus to wall formation in the marine Chrysophycean alga *Pleurochrysis scherffelii* Pringsheim. *J. Cell Biol.*, **41**, 109.

Brown, R. M., W. W. Franke, H. Kleinig, H. Falk, and P. Sitte (1969). Cellulosic wall component produced by the Golgi apparatus of *Pleurochrysis scherffelii*. *Science*, **166**, 894.

Brown, R. M., W. W. Franke, H. Kleinig, H. Falk, and P. Sitte (1970). Scale formation in Chrysophycean algae. I. Cellulosic and noncellulosic wall components made by the Golgi apparatus. *J. Cell Biol.*, **45**, 246 .

Caccam, J. F., and E. H. Eylar (1970). Glycoprotein biosynthesis: Purification and characterization of a glycoprotein: Galactosyl transferase from Ehrlich ascites tumor cell membranes. *Arch. Biochem. Biophys.*, **137**, 315.

Caro, L. G., and G. E. Palade (1964). Protein synthesis, storage, and discharge in the pancreatic exocrine cell. An autoradiographic study. *J. Cell Biol.*, **20**, 473.

Cheetham, R. D., D. J. Morré, and W. N. Yunghans (1970). Isolation of a Golgi apparatus-rich fraction from rat liver. Enzymatic characterization and comparison with other cell fractions. *J. Cell Biol.*, **44**, 492.

Cohn, Z. A., and M. Fedorko (1969). In *Lysosomes in Biology and Pathology* (J. T. Dingle and H. B. Fell, eds.), Vol. 1, p. 43. North Holland, Amsterdam.

Cunningham, W. P., D. J. Morré, and H. H. Mollenhauer (1966). Structure of isolated plant Golgi apparatus revealed by negative staining. *J. Cell Biol.*, **28**, 169.

Dalton, A. J., and M. Felix (1952). "Lipochondria" and the Golgi substance in epithelial cells of the epididymis. *Nature*, **170**, 541.

Dalton, A. J., and M. Felix (1953). Studies on the Golgi substances of the epithelial cells of the epididymis and duodenum of the mouse. *Am. J. Anat.*, **92**, 277.

Dauwalder, M., J. E. Kephart, and W. G. Whaley (1966). Phosphatases and the Golgi apparatus in differentiating cells. *J. Cell Biol.*, **31,** 25A.

De Duve, C. (1969). In *Lysosomes in Biology and Pathology* (J. T. Dingle and H. B. Fell, eds.), Vol. 1, p. 3. North Holland, Amsterdam.

Elliott, A. M., and R. G. Zieg (1968). A Golgi apparatus associated with mating in *Tetrahymena pyriformis*. *J. Cell Biol.*, **36,** 391.

Esau, K., and R. H. Gill (1965). Observations on cytokinesis. *Planta*, **67,** 168.

Eylar, E. H. (1965). On the biological role of glycoproteins. *J. Theoret. Biol.*, **10,** 89.

Fawcett, D. W. (1966). In *An Atlas of Fine Structure. The Cell, Its Organelles and Inclusions*, p. 128. Saunders, Philadelphia.

Fleischer, B., S. Fleischer, and H. Ozawa (1969). Isolation and characterization of Golgi membranes from bovine liver. *J. Cell Biol.*, **43,** 59.

Flickinger, C. J. (1969a). Fenestrated cisternae in the Golgi apparatus of the epididymis. *Anat. Record*, **163,** 39.

Flickinger, C. J. (1969b). The development of Golgi complexes and their dependence upon the nucleus in amoebae. *J. Cell Biol.*, **43,** 250.

Flickinger, C. J. (1969c). The pattern of growth of the Golgi complex during the foetal and postnatal development of the rat epididymis. *J. Ultrastruct. Res.*, **27,** 344.

Golgi, G. (1898a). Sur la structure des cellules nerveuses. *Arch. Ital. Biol.*, **30,** 60.

Golgi, G. (1898b). Sur la structure des cellules nerveuses des ganglions spinaux. *Arch. Ital. Biol.*, **30,** 278.

Green, J. C., and D. H. Jennings (1967). A physical and chemical investigation of the scales produced by the Golgi apparatus within and found on the surface of the cells of *Chrysochromulina chiton* Parke et Manton. *J. Exptl. Botan.*, **18,** 359.

Grimstone, A. V. (1959). Cytoplasmic membranes and the nuclear membrane in the flagellate *Trichonympha*. *J. Biophys. Biochem. Cytol.*, **6,** 369.

Grove, S. N., C. E. Bracker, and D. J. Morré (1968). Cytomembrane differentiation in the endoplasmic reticulum–Golgi apparatus–vesicle complex. *Science*, **161,** 171.

Hagopian, A., H. B. Bosmann, and E. H. Eylar (1968). Glycoprotein biosynthesis: The localization of polypeptidyl: *N*-acetylgalactosaminyl, Collagen: Glucosyl, and Glycoprotein: Galactosyl transferases in HeLa cell membrane fractions. *Arch. Biochem. Biophys.*, **128,** 387.

Harris, P. J., and D. H. Northcote (1971). Polysaccharide formation in plant Golgi bodies. *Biochim. Biophys. Acta*, **237,** 56.

Hassid, W. Z. (1969). Biosynthesis of oligosaccharides and polysaccharides in plants. Mechanisms of enzymic synthesis of complex plant carbohydrates are reviewed. *Science*, **165,** 137.

Hopkins, C. R. (1969). The fine structural localization of acid phosphatase in the prolactin cell of the teleost pituitary following the stimulation and inhibition of secretory activity. *Tissue Cell*, **1,** 653.

Jamieson, J. D., and G. E. Palade (1967a). Intracellular transport of secretory proteins in the pancreatic exocrine cell. I. Role of the peripheral elements of the Golgi complex. *J. Cell Biol.*, **34,** 577.

Jamieson, J. D., and G. E. Palade (1967b). Intracellular transport of secretory proteins in the pancreatic exocrine cell. II. Transport to condensing vacuoles and zymogen granules. *J. Cell Biol.*, **34,** 597.

Kauss, H. (1969). A plant mannosyl-lipid acting in reversible transfer of mannose. *FEBS Letters*, **5,** 81.

Kauss, H., and A. L. Swanson (1969). Cooperation of enzymes responsible for polymerisation and methylation in pectin biosynthesis. *Z. Naturforsch.*, **24,** 28.

Kauss, H., A. L. Swanson, R. Arnold, and W. Odzuck (1969). Biosynthesis of pectic substances. Localisation of enzymes and products in a lipid-membrane complex. *Biochem. Biophys. Acta*, **192,** 55.

Keenan, T. W., and D. J. Morré (1970). Phospholipid class and fatty acid composition of Golgi apparatus isolated from rat liver and comparison with other cell fractions. *Biochemistry*, **9,** 19.

Kessel, R. G. (1966). Some observations on the ultrastructure of the oocyte of *Thyone briareus* with special reference to the relationship of the Golgi complex and endoplasmic reticulum in the formation of yolk. *J. Ultrastruct. Res.*, **16,** 305.

Lawford, G. R., and H. Schachter (1966). Biosynthesis of glycoprotein by liver. The incorporation *in vivo* of ^{14}C-glucosamine into protein-bound hexosamine and sialic acid of rat liver subcellular fractions. *J. Biol. Chem.*, **241,** 5408.

Li, Y., Y. Li, and M. R. Shetlar (1968). Isolation of glycopeptides from rat liver microsomes involved in the biosynthesis of plasma glycoprotein. *J. Biol. Chem.*, **243,** 656.

Manton, I. (1966). Observations on scale production in *Prymnesium parvum*. *J. Cell Sci.*, **1,** 375.

Manton, I. (1967). Further observations on scale formation in *Chrysochromulina chiton*. *J. Cell Sci.*, **2,** 411.

Matile, Ph. (1969). In *Lysosomes in Biology and Pathology* (J. T. Dingle and H. B. Fell, eds.), p. 406. North-Holland, Amsterdam.

Mollenhauer, H. H., and D. J. Morré (1966a). Tubular connections between dictyosomes and forming secretory vesicles in plant Golgi apparatus. *J. Cell Biol.*, **29,** 373.

Mollenhauer, H. H., and D. J. Morré (1966b). Golgi apparatus and plant secretion. *Ann. Rev. Plant Physiol.*, **17,** 27.

Mollenhauer, H. H., D. J. Morré, and L. Bergmann (1967). Homology of form in plant and animal Golgi apparatus. *Anat. Record.*, **158,** 313.

Molnar, J., G. B. Robinson, and R. J. Winzler (1965). Biosynthesis of glycoproteins. IV. The subcellular sites of incorporation of glucosamine-1-^{14}C into glycoprotein in rat liver. *J. Biol. Chem.*, **240,** 1885.

Morré, D. J., and H. H. Mollenhauer (1964). Isolation of the Golgi apparatus from plant cells. *J. Cell Biol.*, **23,** 295.

Morré, D. J., L. L. Merlin, and T. W. Keenan (1969). Localization of glycosyl transferase activities in a Golgi apparatus-rich fraction isolated from rat liver. *Biochem. Biophys. Res. Commun.*, **37,** 813.

Neutra, M., and C. P. Leblond (1966a). Synthesis of the carbohydrate of mucus in the Golgi complex as shown by electron microscope radioautography of goblet cells from rats injected with glucose-H^3. *J. Cell Biol.*, **30,** 119.

Neutra, M., and C. P. Leblond (1966b). Radioautographic comparison of the uptake of galactose-H^3 and glucose-H^3 in the Golgi bodies of various cells secreting glycoproteins or mucopolysaccharides. *J. Cell Biol.*, **30,** 137.

Northcote, D. H. (1969a). The synthesis and metabolic control of polysaccharides and lignin during differentiation of plant cells. *Essays Biochem.*, **5,** 89.

Northcote, D. H. (1969b). Fine structure of cytoplasm in relation to synthesis and secretion in plant cells. *Proc. Roy. Soc.* (*London*), **B173,** 21.

Northcote, D. H. (1971a). The Golgi apparatus. *Endeavour,* **30,** 26.

Northcote, D. H. (1971b). Organisation of structure, synthesis and transport within the plant during cell division and growth. *Symp. Soc. Exptl. Biol.*, **25,** 51.

Northcote, D. H., and D. R. Lewis (1968). Freeze-etched surfaces of membranes and organelles in the cells of pea root tips. *J. Cell Sci.*, **3,** 199.

Northcote, D. H., and J. D. Pickett-Heaps (1966). A function of the Golgi apparatus in polysaccharide synthesis and transport in the root-cap cells of wheat. *Biochem. J.*, **98,** 159.

Northcote, D. H., and F. B. P. Wooding (1966). Development of sieve tubes in *Acer pseudoplatanus*. *Proc. Roy. Soc.* (*London*), **B163,** 524.

Northcote, D. H., and F. B. P. Wooding (1968). The structure and function of phoem tissue. *Sci. Prog.* (*Oxford*), **56,** 35.

Novikoff, A. B. (1963). In *Lysosomes* (A. V. S. de Reuck and M. P. Cameron, eds.), p. 36. Churchill, London.

Novikoff, A. B., E. Essner, S. Goldfischer, and M. Heus (1962). Nucleosidephosphatase activities of cytomembranes. *Symp. Intern. Soc. Cell Biol.*, **1,** 149.

Novikoff, A. B., E. Essner, and N. Quintana (1969). Golgi apparatus and lysosomes. *Federation Proc.*, **23,** 1010.

Novikoff, P. M., A. B. Novikoff, N. Quintana and J-J. Houw (1971). Golgi apparatus, GERL, and lysosomes of neurons in rat dorsal root ganglia, studied by thick section and thin section cytochemistry. *J. Cell Biol.*, **50,** 859.

Ovtracht, L., D. J. Morré, and L. M. Merlin (1969). Isolement de l'appareil de Golgi d'un glande secretrice de mucopolysaccharides chez l'escargot. *J. Microscopie*, **8**, 989.

Pickett-Heaps, J. D., and D. H. Northcote (1966a). Cell division in the formation of the stomatal complex of the young leaves of wheat. *J. Cell Sci.*, **1**, 121.

Pickett-Heaps, J. D., and D. H. Northcote (1966b). Relationship of cellular organelles to the formation and development of the plant cell wall. *J. Exptl. Botan.*, **17**, 20.

Ray, P. M., T. L. Shininger, and M. M. Ray (1969). Isolation of β-glucan synthesis particles from plant cells and identification with Golgi membranes. *Proc. Natl. Acad. Sci., U.S.*, **64**, 605.

Roberts, K., and D. H. Northcote (1970). The structure of sycamore callus cells during divisions in a partially synchronized suspension culture. *J. Cell Sci.*, **6**, 299.

Rubery, P. H., and D. H. Northcote (1970). The effect of auxin (2,4-dichlorophenoxyacetic acid) on the synthesis of cell wall polysaccharides in cultured sycamore cells. *Biochem. Biophys. Acta*, **222**, 95.

Schenkein, I., and J. W. Uhr (1970). Immunoglobulin synthesis and secretion. I. Biosynthesis studies of the addition of the carbohydrate moieties. *J. Cell Biol.*, **46**, 42.

Schnepf, E. (1969). Sekretion und Exkretion bei Pflanzen. *Protoplasmatologia*, **8**, 1.

Sengel, A., and P. Stoebner (1970). Golgi origin of tubular inclusions in endothelial cells. *J. Cell Biol.*, **44**, 223.

Shanthaveerappa, T. R., and G. H. Bourne (1965). The thiamine pyrophosphatase technique as an indicator of the morphology of the Golgi apparatus in the neurons. *Cellule Rec. Cytol. Histol.*, **65**, 201.

Sievers, A. (1965). In *Funktionelle und morphologische Organisation der Zelle: Sekretion und Exkretion Funktion des Golgi-Apparates in pflanzlichen und tierischen Zellen*, p. 89. Springer, Berlin.

Smith, R. E., and M. G. Farquhar (1966). Lysosome function in the regulation of the secretory process in cells of the anterior pituitary gland. *J. Cell Biol.*, **31**, 319.

Staehelin, A. (1966). Die Ultrastruktur der Zellwand und des Chloroplasten von *Chlorella*. *Z. Zellforsch. Mikroskop. Anat.*, **74**, 325.

Stockem, W. (1969). Pinocytose und Bewegung von Amöben. III. Die Funktion des Golgiapparates von *Amoeba proteus* und *Chaos chaos*. *Histochemie*, **18**, 217.

Stoddart, R. W., and D. H. Northcote (1967). Metabolic relationships of the isolated fractions of pectic substances of actively growing sycamore cells. *Biochem. J.*, **105**, 45.

Thiéry, J.-P. (1969). Role de l'appareil de Golgi dans la synthese des mucopolysaccharides etude cytochimique. I. Mise en évidence de mucopolysaccharides dans les vésicules de transition entre, l'ergastoplasme et l'appareil de Golgi. *J. Microscopie*, **8**, 689.

Villemez, C. L. (1970). Characterization of intermediates in plant cell wall biosynthesis. *Biochem. Biophys. Res. Commun.*, **40**, 636.

Whaley, W. G., M. Dauwalder, and J. E. Kephart (1966). The Golgi apparatus and an early stage in cell plate formation. *J. Ultrastruct. Res.*, **15**, 169.

Wooding, F. B. P. (1968). Radioautographic and chemical studies of incorporation into sycamore vascular tissue walls. *J. Cell Sci.*, **3**, 71.

Wooding, F. B. P., and D. H. Northcote (1964). The development of the secondary wall of the xylem of *Acer pseudoplatanus*. *J. Cell Biol.*, **23**, 327.

Yamazaki, M., and O. Hayaishi (1968). Allosteric properties of nucleoside diphosphatase and its identity with thiamine pyrophosphatase. *J. Biol. Chem.*, **243**, 2934.

Zagury, D., J. W. Uhr, J. D. Jamieson, and G. E. Palade (1970). Immunoglobulin synthesis and secretion. II. Radioautographic studies of sites of addition of carbohydrate moieties and intracellular transport. *J. Cell Biol.*, **46**, 52.

Zeigel, R. F., and A. J. Dalton (1962). Speculations based on the morphology of the Golgi systems in several types of protein-secreting cells. *J. Cell Biol.*, **15**, 45.

7

LYSOSOMES

Viviane Maggi

1. HISTORY OF LYSOSOMES

The term "lysosome" was introduced by De Duve and his collaborators in 1955 (Appelmans and De Duve, 1955; Appelmans et al., 1955; De Duve et al., 1955) when these investigators suggested the existence in rat liver of a population of organelles limited by a membrane which enclosed several hydrolases. The latter had a pH optimum in the acid range and were not normally active in the cell but could, however, be activated by an action exerted on the lysosomal membrane, either as a physiological stimulus or as a damaging factor. The membrane would become more permeable and permit the entry into the lysosome and into contact with the enzymes of various substrates likely to be digested by the enzymes.

In 1955 this ingenious concept was purely biochemical in that lysosomes had not been described as cytomorphological entities. Although Sheldon et al. in 1955 had found clusters of lead deposits in ultrathin sections of rat intestine incubated at an acid pH in the presence of lead nitrate and glycerophosphate, it was not until 1959 that acid phosphatase-positive granules were first tentatively identified as lysosomes in rat kidney (Holt, 1959) and not until 2 yr later that Novikoff (1961) and Holt and Hicks (1961) described ultrastructure and acid phosphatase content of these organelles in rat liver employing a modification of the cytochemical technique of Gomori (1950) for the demonstration of this lysosomal enzyme. Thus, with the application of the Gomori technique to tissue sections, it had become possible to localize lysosomes in the cell. This fundamental step made it possible to gather a considerable amount of information on the presence of lysosomes in various types of cells in several living organisms, and also to investigate their role in the metabolic processes taking place in the cell under physiological and pathological conditions. At present, the bulk of the information available on lysosomes relies on cytological and cytochemical data; therefore these form the major part of this article.

2. THE "LYSOSOME CONCEPT"

De Duve (1963) envisaged the lysosome as a bag of hydrolytic enzymes, all active at an acid pH and surrounded by a membrane which had the function of protecting the rest of the cytoplasm from the digestive action of the enzymes. Digestion would take place within the lysosome after the substance to be digested had penetrated the organelle. Such substances could include protein molecules, polysaccharides, lipoproteins, nucleic acids, and other compounds of a lower molecular weight. They could originate either from the external, extracellular milieu and represent nutritional or toxic factors, in which latter case the lysosome would act with a detoxicating mechanism; or they could originate from within the cytoplasm of the cell, particularly in cases in which the cell was undergoing autolysis. In the first case the organelles were called "heterophagosomes" or lysosomes. When cell autolysis was involved, the organelles were called "autophagosomes." In this chapter "cytolysome" (Novikoff, 1961) is used instead of autophagosome in order to avoid confusion with the term "phagosome." This will be used to describe the large vacuoles present mainly in primitive organisms such as amoebae, where they have the function of taking up food for the cell (physiological digestive function). "Phagosome" and "phagocytosis" will also be employed to define the uptake of large bodies such as bacteria and viruses by monocytes and polymorphs of the blood, or by fixed and free macrophages of the reticulo-endothelial system (RES). The term "vacuole" will be used to describe bodies resulting from either the enlargement of vesicles or their fusion together, whether the vesicles originate from the cell membrane (endocytic vesicles) or from the Golgi apparatus. They are present in most cells of higher organisms, particularly mammalian kidney. Neither phagosomes nor vacuoles have been found to possess acid hydrolases (Straus, 1967).

Additional definitions of terminology used in this chapter are:

Endocytic vesicle (Fig. 1*a**): Small invagination of the cell wall which forms in contact with macromolecules originating from the external milieu; the invagination eventually closes around the macromolecules; it has not been found to contain acid hydrolase activity.

Vacuole (Fig. 1*b*): This organelle contains constituents originating from the external, extracellular milieu, and which are to be digested or stored; it probably represents fusion of several endocytic vesicles or an enlarged vesicle; it does not contain acid hydrolases.

Golgi apparatus (Fig. 1*c*): System of membranes comprising elongated *saccules*, large *vacuoles*, and small *vesicles*; a relation of the Golgi apparatus to lysosomes appears likely but has not been elucidated completely.

Lysosome (Fig. 2a-h): Organelle limited (usually) by a single membrane and containing at least one acid hydrolase.

Cytolysome (Fig. 3*a–g*): Organelle limited by a single (sometimes double) membrane and containing cytoplasmic residues such as mitochondrial fragments and membranes, and at least one acid hydrolase.

Residual body (Fig. 4a-b): Whorl of myelin figures thought to represent the final stage in the process of lysosome degradation, whether deriving from an heterophagosome or from a cytolysome; it usually does not contain acid hydrolases although sometimes some residual activity can be demonstrated (Fig. 4*b*).

3. ORIGIN OF LYSOSOMES AND OF LYSOSOMAL ENZYMES

In order to absorb a large molecule, a microorganism, or food particles the cell wall forms an invagination which surrounds and finally encloses and isolates the molecule or organism or particles themselves (Metchnikoff, 1892). At this stage the endocytic vesicle, or phagosome does not appear to possess adequate enzyme equipment to digest the absorbed compound. This is achieved when lysosomes fuse with the phagosome or vacuole or endocytic vesicles and make their hydrolases available for degradation of the absorbed substrate. Since the absorbing surface of cells is not always logistically close to sites of lysosomes, the phagosome, vacuole, or vesicle, and lysosome must travel a certain distance within the cell in order to make contact

* In all figures except Fig. 1*a* the scale is equivalent to approximately 0.5 μ. In Fig. 1*a* it is equivalent to 0.1 μ. Figures 1*b*, *e*, *f*; 2*a*, *b–d*, *f*, *h–j*, *m*, *n*; 3*f*, *g*; 4*b*; 6*a–g*; 7*a*; 10; and 11 show tissues fixed in 2% glutaraldehyde in 0.1 *M* cacodylate buffer at pH 7.1–7.4 for 4 hr, rinsed for 18–36 hr in cold (+4°C) 0.1 *M* cacodylate buffer at pH 7.1–7.4 and incubated for the demonstration of acid naphthol AS/B1 phosphatase. Figure 3*d* shows a tissue incubated for naphthol AS-β-δ glucuronidase. All tissues were processed as recommended by Livingston et al. (1969). The tissues shown in Figs. 1; 2*a–d*, *f*, *h–j*, *m*, *n*; 3*f*, *g*; 6*a–d*, *g*; 7*a*; 10; and 11 were stained in alkaline lead citrate; those shown in Figs. 2*b*, *e*; 3*d*; 4*b*; 5; 6*e*, *f*; and 12 were unstained. Figures 2*e* and 5 show tissues fixed as above but incubated for the demonstration of acid β-glycerophosphatase according to Holt and Hicks (1961). All other figures show tissues which were fixed in 2% osmium tetroxide for 1 hr and rinsed, dehydrated, and embedded conventionally. They were block-stained with a saturated solution of uranyl acetate in 70% ethanol, and sections were stained with alkaline lead citrate.

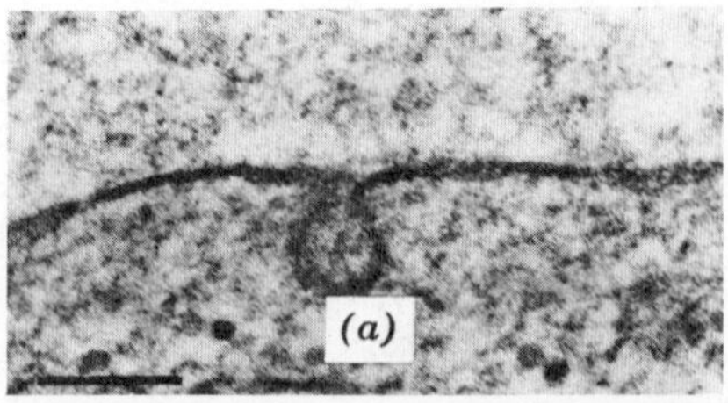

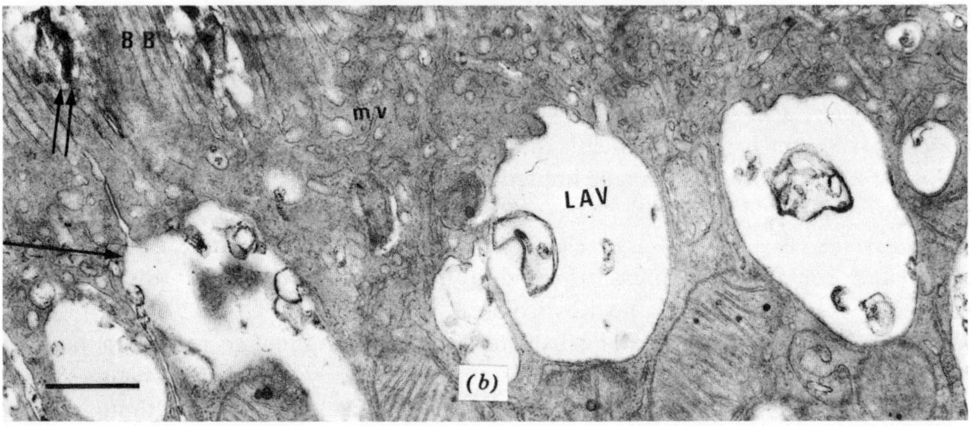

Figure 1. Structures involved in the formation of lysosomes. (*a*) Enolocytic vesicle of neutrophil leukocyte. (*b*) Large apical vacuoles (LAV) of mouse kidney proximal convoluted tubules. The connection between LAV and micropinocytic vesicles (mv) invaginating from the microvilli of the brush border (BB) is demonstrated in this photograph (arrow). Through these connections large molecules enter the LAV. In this micrograph these structures contain membrane residues which are unlikely to derive from the tubule lumen and are more likely to be destined to be eliminated through the microvilli, as suggested by the presence of residues in the brush border itself (double arrow). Thus these structures have a dual function of absorption and defecation. (*c*) Cell from a mouse fibrosarcoma. Golgi bodies (G), multivesicular bodies (arrow), and lysosomes (L) are present in the same area. (*d*) Budding off smooth ET containing electron-dense material and very similar to a lysosome. It is suggested that this may represent one way by which lysosomes are formed directly from the ER. Human colon mucosa. (*e*) Seminal vesicle of mouse treated with oestradiol for 6 months. Strands of acid naphthol phosphatase-positive ER membranes are closely apposed to a mitochondrion (M). It is suggested that this is the first step toward the formation of a cytolysome. These organelles are found frequently under these experimental conditions. (*f*) Empty-looking vacuole limited by a single membrane joined to a thick membrane showing a clear periodicity and surrounding homogeneous dense material. This structure was found close to the brush border of a mouse proximal convoluted tubule where usually both large apical vacuoles (phagosomes) and lysosomes are found. It is suggested that the structure shown here represents a stage of fusion of these two organelles. The type of membrane appears rather unusual. (g) Proximal convoluted tubule cell from a 1-wk-old mouse. Two prominent Golgi bodies (G) with numerous Golgi vesicles (V) and lysosomelike organelles (L). All phases between rough and smooth ER membranes can be seen (arrows) and together with the Golgi bodies form the GERL. MB, Microbody. (From Maggi and Oddy, 1968; courtesy of Chapman and Hall, London.)

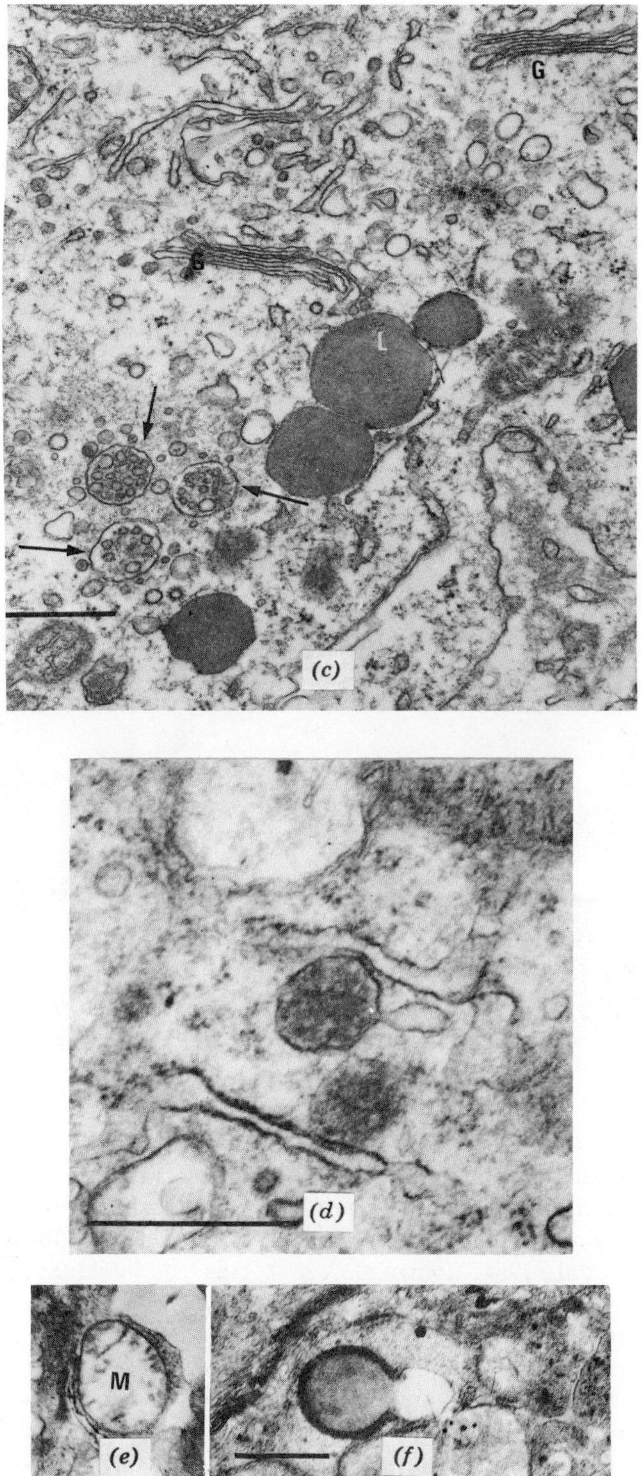

Figure 1. continued

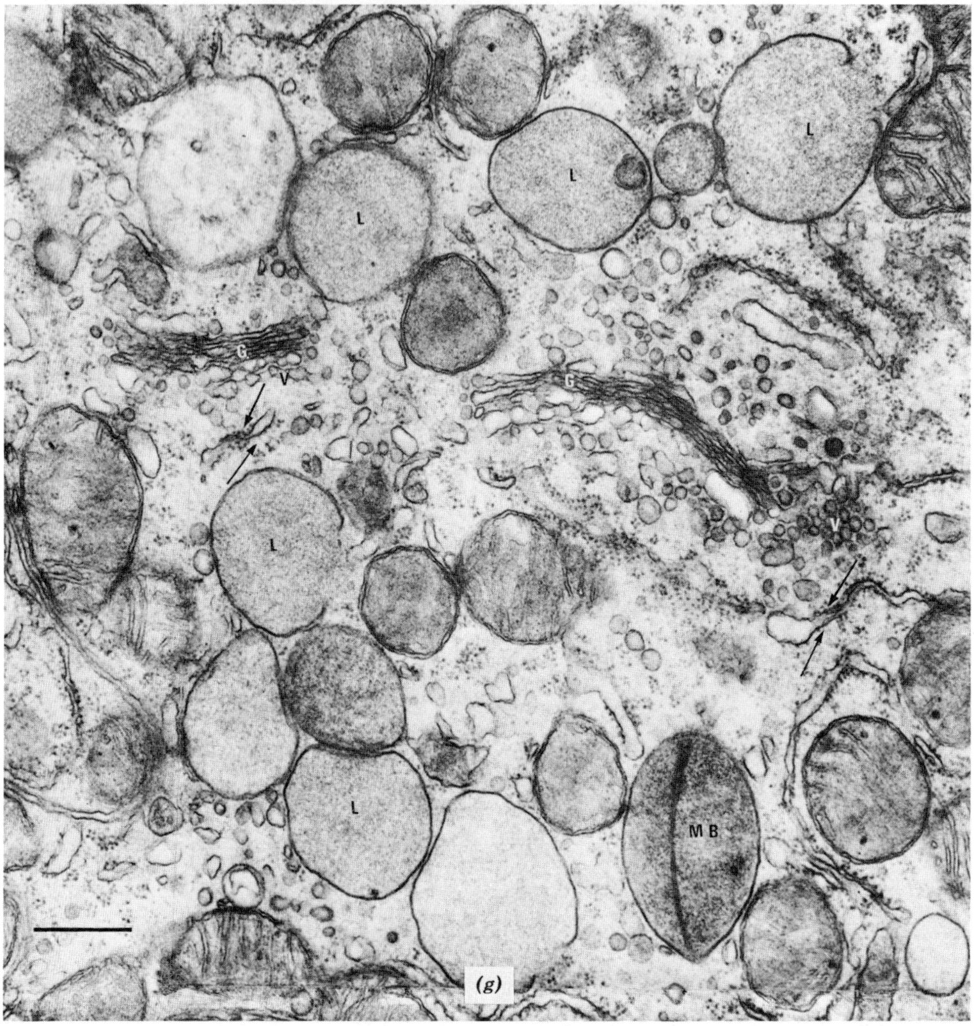

Figure 1. continued

In several studies it was found that it is the vacuolar structure that travels within the cell (Miller and Palade, 1964; Straus, 1964; Maunsbach, 1966a, b) and that the absorbed material enters in contact with the Golgi apparatus for a certain period of time (Novikoff et al., 1964; Loni et al., 1967). This led Novikoff to suggest that the Golgi apparatus was primarily involved in the formation of lysosomes and that the Golgi vesicles were possible carriers of acid hydrolases, the latter in turn originating from the ribosomes of the rough endoplasmic reticulum (ER). In support of this hypothesis, Novikoff et al. (1964) noted the similarity between the Golgi vesicles and the vesicles comprising the multivesicular body (Fig. 1*c*), an organelle that contains acid hydrolases. In this hypothesis therefore the multivesicular body would be a young lysosome which would mature to form a dense body or adult lysosome. The chief obstacle to this attractive theory was that Golgi vesicles usually did not appear

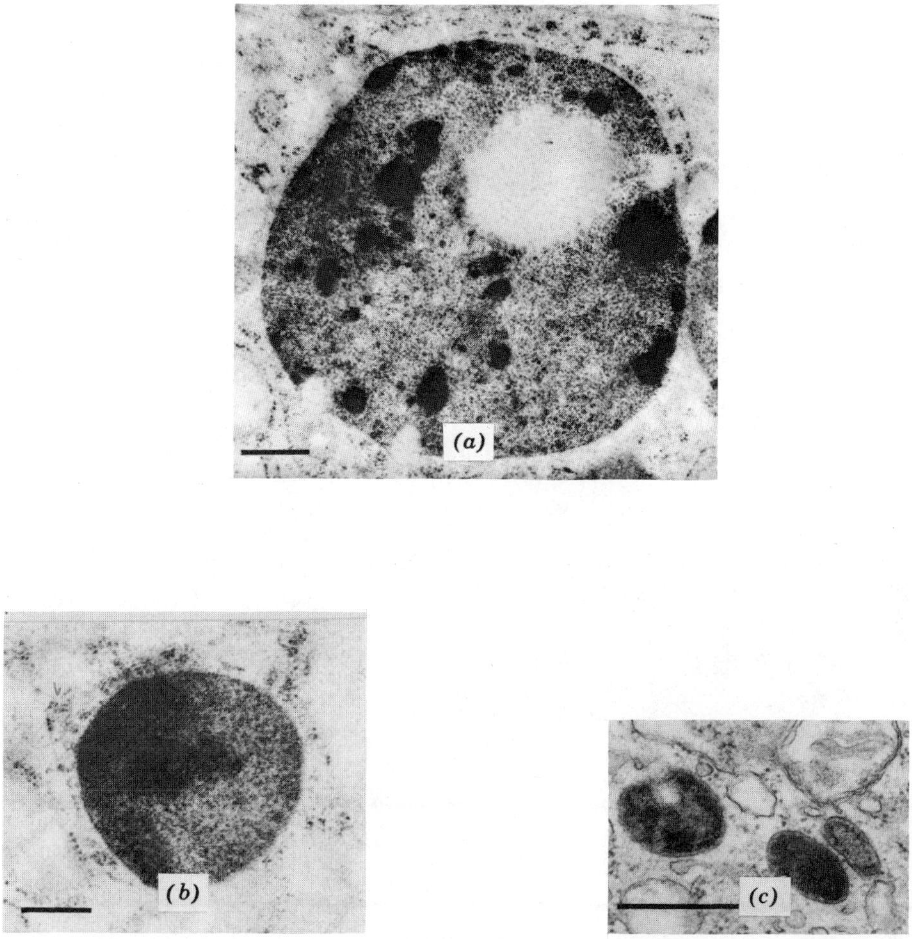

Figure 2. Differences in the cytomorphology of lysosomes. (*a–d*) All bodies illustrated in this series were found in mouse seminal vesicles. The tissues were processed in the same manner, that is, fixed in glutaraldehyde and incubated for the demonstration of acid naphthol AS phosphatase. The end product of enzyme activity is present in all cases. (*e–f*) Lysosomes of rat kidney containing end product of acid β-glycerophosphatase activity (*e*) and end product of acid naphthol AS phosphatase activity (*f*). Fixation was the same and processing was different, but this does not account for the difference in morphology between the two types of organelles. (*g–h*) Lysosomes of mouse kidney processed conventionally (*g*), and incubated for the demonstration of acid naphthol AS phosphatase activity (*h*). Note the difference between the two types of organelles and between (*g*) and (*e*). For other types of mouse kidney lysosomes, see Figs. 1*g* and 7*a*. (*i–k*)Lysosomes of rat heart. The lysosomal nature of (*i*) is not certain but is likely on account of its positivity for acid naphthol AS phosphatase activity; (*i*) and (*j*) were processed according to Livingson et al. (1969), and (*k*) was processed conventionally; the difference between the various types of organelles is not attributable to the difference in techniques, however. (*l*) Membranes of two lysosomes are seen to be in continuity. This may represent either a lysosome in the process of dividing or two lysosomes in the process of fusing. Mouse kidney. (*m*) Dense body containing an unusual inclusion. Mouse seminal vesicle as in (*g*). The dense material in the dense body is naphthol AS–lead phthalocyanin end product of acid phosphatase activity, which is present also in a cisterna of the ER (er). (*n*) Dense body with crystal inclusions in mouse coagulating gland.

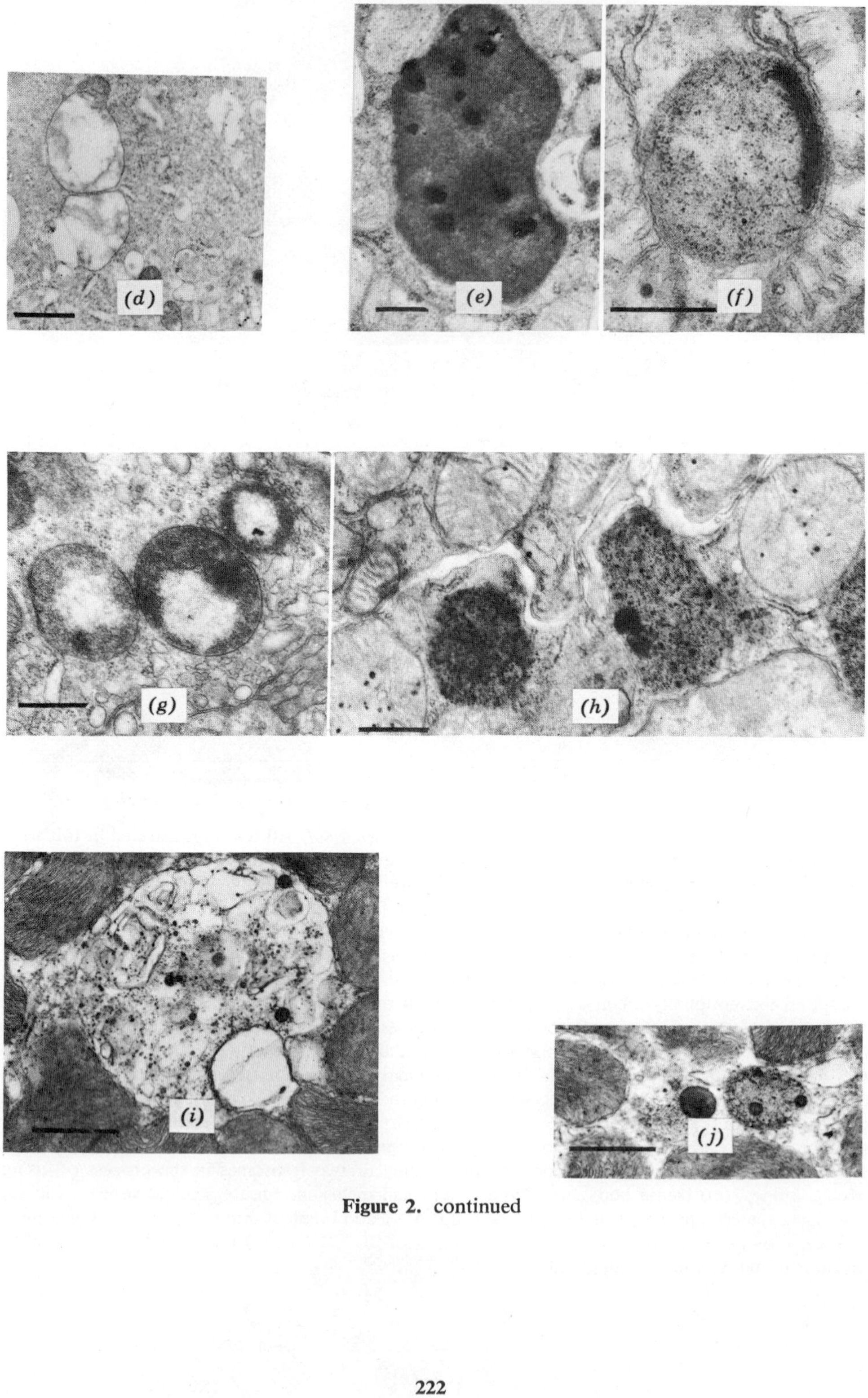

Figure 2. continued

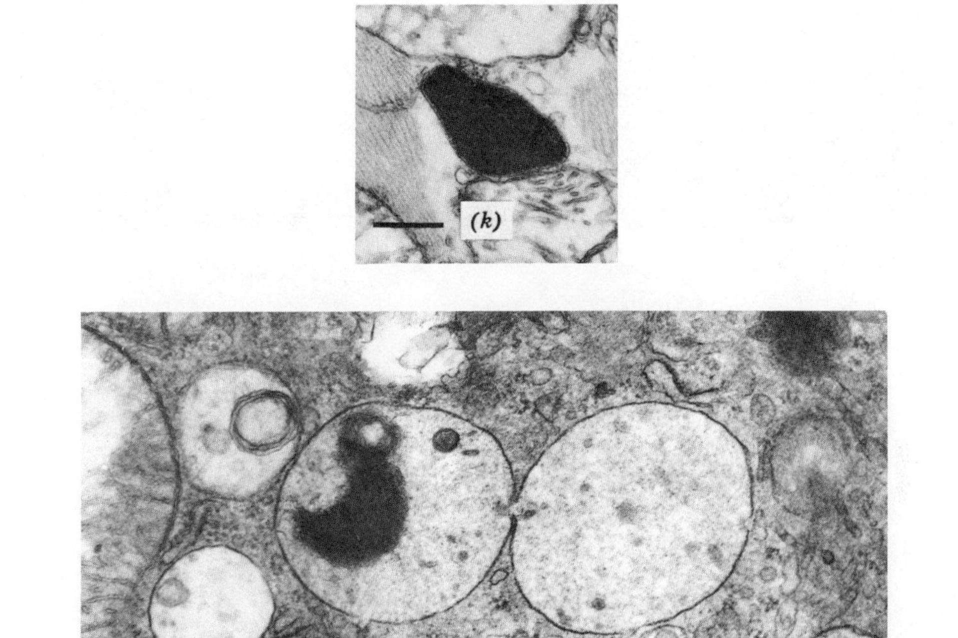

Figure 2. continued

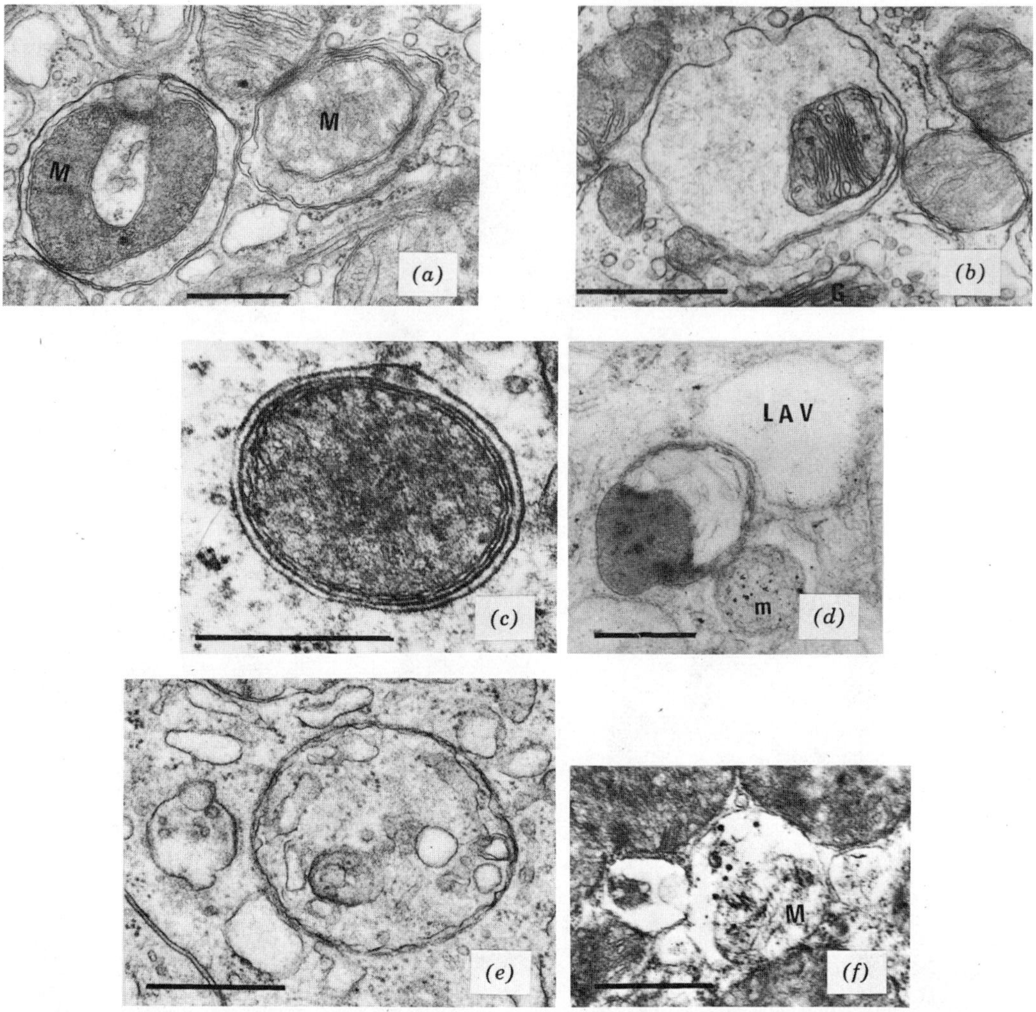

Figure 3. Differences in the cytomorphology of cytolysomes. (*a*) Two cytolysomes found in a cell of kidney proximal convoluted tubule of a mouse deprived of food for 18 hr. In both, the degenerating mitochondria (M) are surrounded by a double membrane. Their density is very different, suggesting different stages of intralysosomal degradation. (From Maggi and Oddy, 1968; courtesy of Chapman and Hall, London.) (*b*) Cytolysome formed by a partly single membrane enclosing a mitochondrial fragment. Cell of kidney proximal convoluted tubule of a 1-wk-old mouse. (*c*) Single membrane enclosing a triple-layered mitochondrial residue. Experimental conditions and tissue are the same as in (*a*). (*d*) Large apical vacuole (LAV) of a normal mouse kidney proximal convoluted tubule closely apposed to a lysosome which in turn contains membrane residues as well as a partly dense β-glucuronidase-positive matrix; the latter is in close contact with a cytolysome containing a mitochondrial residue (m) in which some end product of β-glucuronidase activity can be seen. (*e*) Two cytolysomes ot different sizes but both containing vesicles; the larger body also contains smooth and rough ER fragments, and some free ribosomes. Seminal vesicle of a castrated mouse given estradiol for 3 wk. (*f*) Membrane of the sarcoplasmic reticulum of rat heart enclosing a mitochondrial fragment (M) to form a cytolysome. The dense particles present in the organelle represent the end product of acid naphthol AS phosphatase activity. (*g*) Dense body in seminal vesicle of a mouse given estradiol for 1 wk, containing various cytoplasmic residues.

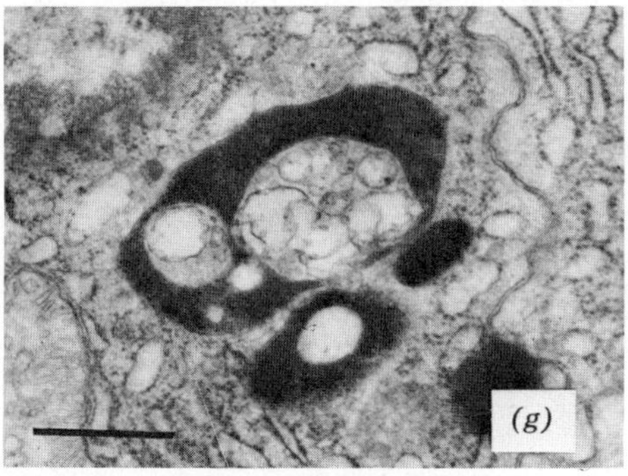

Figure 3. continued

to contain acid phosphatase or other acid hydrolases, and that when such hydrolases were present in the Golgi apparatus they appeared to be located in the vacuoles or in the saccules rather than in the vesicles (Fig. 5). Further observations led Novikoff (Novikoff and Shin, 1964; Holtzman et al., 1967) to modify his theory of the exclusive formation of lysosomes from Golgi vesicles and to suggest that acid hydrolases would be synthesized on the ribosomes of the rough ER, and that they would be passed along the membranes and the cisternae of the smooth ER from which they could be directed along one of the following pathways: (1) the smooth ER would bud off to form a lysosome (Fig. 1*d*), or (2) the smooth ER would wrap itself around cytoplasmic fragments due to be autolyzed (Fig. 1*e*), or (3) the acid hydrolases would be passed on to a specialized part of the ER which is normally present in close apposition to the Golgi saccules and from there they could be transferred to the saccules and/or the vacuoles, and/or to the Golgi vesicles. This complex is known as the GERL (Golgi endoplasmic reticulum lysosomes, Fig. 1*g*). Lysosomes could originate directly from

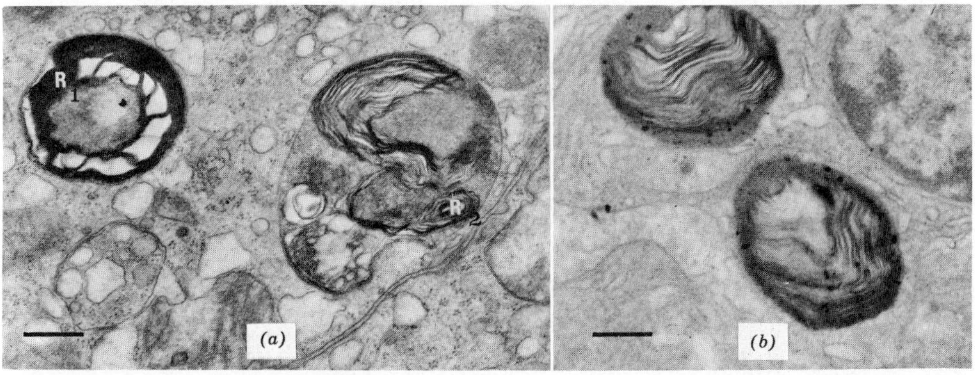

Figure 4. Residual bodies. (*a*) Residual body (R_1) in its final stage of life when it is composed exclusively of lipid membranes. Another body (R_2) appears to still contain some visible cytoplasmic residues which are probably in the process of being transformed in myelin whorls before being extruded in the extracellular milieu. Kidney from mouse starved for 18 hr. (*b*) End product of acid naphthol AS phosphatase activity in two typical residual bodies of the seminal vesicle of a mouse given estradiol for 1 wk. The presence of enzyme activity in these bodies is exceptional.

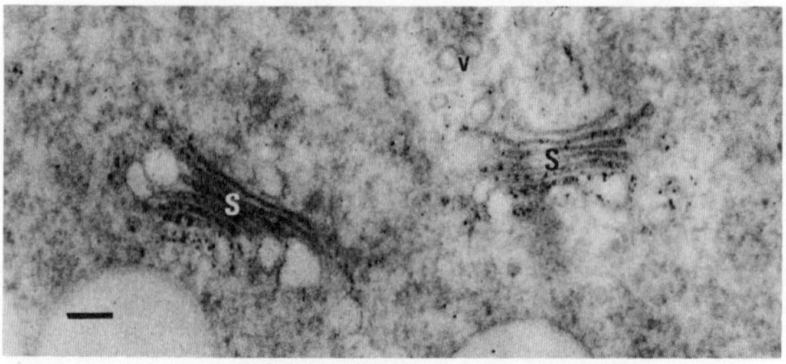

Figure 5. The Golgi apparatus. Two Golgi bodies from a section of *Vicia faba* incubated for the demonstration of acid β-glycerophosphatase activity. Enzyme end product is present in the saccules (*S*) but not in the vesicles (v). (Courtesy of D. James.)

the vacuoles, or multivesicular bodies could originate from Golgi vesicles penetrating the vacuoles.

The main argument against these theories in the past has been that acid hydrolases were not found in the ribosomes or the ER and seldom, and in only certain types of cells, in the Golgi apparatus (for a review see Novikoff et al., 1964). However, the presence of acid hydrolases in subcellular structures other than lysosomes had been suggested by several investigators from studies using optical cytochemical techniques. Thus Rosenbaum and Rolon (1962) described an acid phosphatase in the liver of rabbits which was not localized close to the peribiliary canaliculi as the lysosomal enzyme should be; Maggi et al. (1966) described an acid phosphatase in the basement membrane area of the seminal vesicles and the coagulating gland of the mouse where lysosomes are very scarce. This acid phosphatase had a substrate affinity, a response to inhibitors, and a pH optimum different from the lysosomal enzyme. More recently, and as a result of the progress in electron microscope techniques for the demonstration of acid hydrolases other than acid β-glycerophosphatase (Holt and Hicks, 1966; Livingston et al., 1969, 1970; Smith and Fishman, 1969), it has become possible to describe the presence of these enzymes at the ultrastructural level, that is, in the cisternae of the smooth ER or in the ribosomes of several cells under a variety of experimental conditions. Thus Livingston et al. (1969, 1970), Maggi and Carbonell (1969), and Maggi et al. (1970) described a ribosomal acid phosphatase in the kidney of adult mice and in the ribosomes of the rough ER of seminal vesicles; Livingston et al. (1969) and Gahan and MacLean (1967, 1968) described the presence of esterases and acid phosphatase in the Golgi apparatus and the ER of plants, and Holt and Hicks (1966) found nonspecific esterases in the cisternae of smooth ER of rat kidney. Fishman et al. (1967, 1970) reported the presence of β-glucuronidase in the ER of mammalian liver and kidney; Smith and Fishman (1969) and Maggi et al. (1970) found acid phosphatase in the smooth ER cisternae of rat liver and preputial gland, and in the seminal vesicles of castrated mice, respectively. Furthermore, Maggi (in preparation) found acid naphthol AS phosphatase in the sarcoplasmic reticulum and around lipid droplets of rat heart (Fig. 6*b* and *c*). This enzyme is present on the nuclear membrane of several cells (Fig. 6*d* and *f*). The presence of acid β-glycerophosphatase in the ER of rat kidney adenomas (Seljelid, 1966) and of gingival mucosa (Katchburian and Holt, 1969) has been reported. Biochemical evidence for the

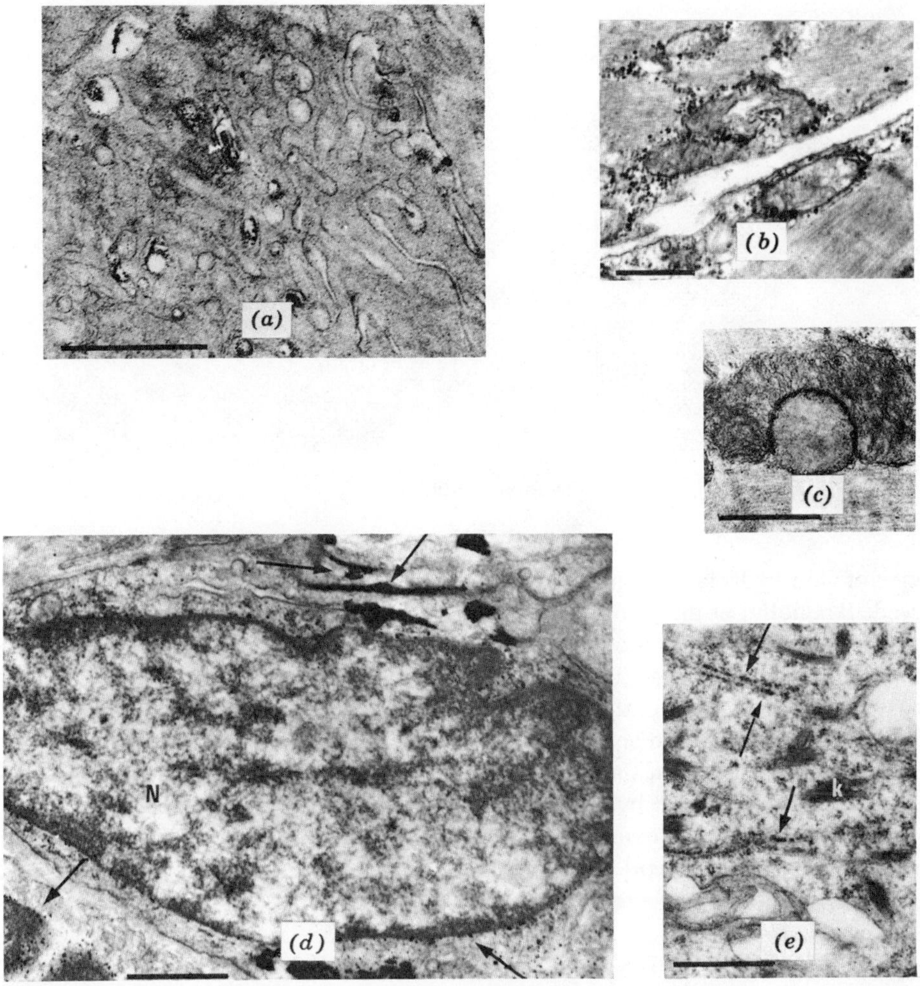

Figure 6. Nonlysosomal localization of acid phosphatase. (*a*) Oblique section of the apical area of a cell of the proximal convoluted tubule of mouse kidney showing the endocytic vesicles containing end product of acid naphthol AS phosphatase reaction. The tissue was incubated in the presence of a lead phthalocyanin derivative, which makes the preservation and cutting of the tissue very difficult (Livingston et al., 1970). (*b*) Section of 2-day-old rat heart showing the deposition of end product of acid naphthol AS phosphatase reaction on the membranes of the sarcoplasmic reticulum closely apposed to mitochondria. Some glycogen granules are also present. Tissue incubated in the presence of sodium fluoride. (*c*) Section of adult rat heart showing the deposition of end product of acid naphthol AS phosphatase activity around a lipid droplet. (*d*) Basal region of acinus of mouse seminal vesicle. Acid naphthol AS phosphatase end product (arrow) is present in the cisternae of the smooth ER and on the nuclear membrane. The animal was castrated and treated with estradiol for 6 wk. Section incubated with sodium molybdate. N, Nucleus. (*e*) Seminal vesicle of mouse castrated and treated with estradiol for 8 wk. Keratinization of the epithelium has taken place. Acid naphthol AS phosphatase end product is present along the remaining strands of rough ER (arrow) but very little can be seen on free ribosomes. k, Keratin. (*f*) Blood cell. The nuclear membrane contains acid naphthol AS phosphatase end product. Incubation with sodium molybdate. (*g*) Epithelial cell of mouse seminal vesicle incubated with sodium molybdate. Acid naphthol AS phosphatase activity is free in the keratinized cytoplasm. The animal was castrated and treated with estradiol for 6 wk. N, Nucleus.

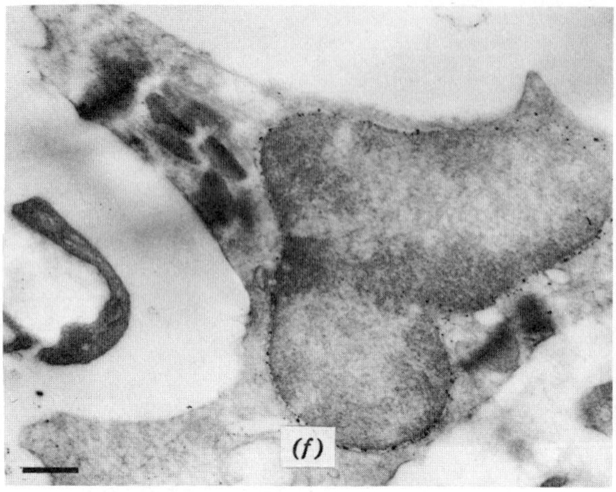

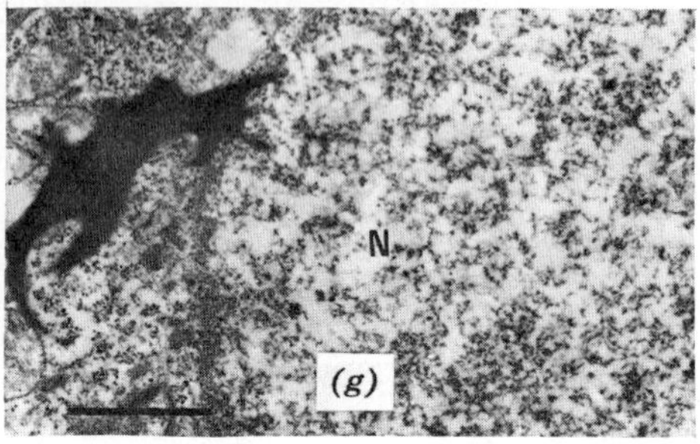

Figure 6. continued

presence of acid hydrolases in nonlysosomal fractions of several tissues has also been observed. Baggiolini et al. (1969), Moore and Angeletti (1961), Maggi (1965a, 1968, 1969), Neil and Horner (1964a, b), Nelson (1966), and Zeja and Spitznagel (1971 a,b) all showed that different acid phosphatases with different kinetic properties and response to inhibitors exist in the cells of several tissues. Their behavior under experimental conditions such as in spontaneous and transplanted tumors or tissue culture cells is different from that of the lysosomal enzyme (Maggi and Wyatt, unpublished). It is possible that these acid hydrolases are the precursors of the lysosomal enzymes. This subject has been discussed by Maggi (1968) and by Fishman and his group (Fishman et al., 1970; Ide and Fishman, 1970).

4. PHYSIOLOGY OF LYSOSOMES IN CELLS OF HIGHER ANIMALS

A. Phagocytosis, Pinocytosis, Endocytosis—Intralysosomal Digestion

When a cell function is that of absorbing large molecules (for example, cells of the kidney's proximal convoluted tubule, intestinal cells), this is performed through a mechanism called endocytosis. At least in amoeba this process is affected by the isoelectric point of the compound (protein) to be ingested (Chapman Andresen, 1962), and there is no reason to suppose that the same phenomenon does not occur in mammalian cells. The cell membrane forms an invagination and closes around the molecule, thus forming intracytoplasmic vesicles or vacuoles of differing sizes. At this early stage the vacuole or vesicle does not appear to have acid hydrolases and cannot be classified as a lysosome; however, it possesses enzymes of the cell membrane, such as adenosine triphosphatase and alkaline phosphatase; the role of these enzymes in the degradation of the large molecules after endocytosis is unknown, but it is possible that they begin the process of digestion.*

* Preliminary results from our laboratory indicate that some hydrolytic activity can be found in endocytic vesicles of the cells of kidney proximal convoluted tubules (Fig. 6*a*). These data require further investigation.

Subsequently, the phagosome or vacuole fuses with the lysosome already present in the cell, or with the Golgi vesicles. This process is very rapid and seldom observed when only morphological techniques are used. It had been elegantly demonstrated, however, by Straus (1964), who injected rats with an enzyme protein, horseradish peroxidase, and observed the enzyme histochemically as being taken up by the cells of the kidney proximal convoluted tubule by an endocytotic mechanism. This histochemical reaction for peroxidase yields a blue color. Straus (1964) demonstrated the fusion of the acid phosphatase-containing lysosome with the horseradish peroxidase-containing apical vacuole by reacting the kidney sections for acid phosphatase using a technique that yields a red color. Thirty minutes after the injection of horseradish peroxidase, the latter, as detected by the presence of blue color, was present in vacuoles near the brush border of the kidney cells; the acid phosphatase reaction, as detected by the presence of red color, was present near the basement membrane of the same cells (this is the main site of location of lysosomes in rat kidney cells). Later the blue droplets migrated toward the base of the cell nearer to the red (lysosomes) droplets, and after 1–3 days the two distinct colors could no longer be resolved; large purple (red plus blue) drops could be observed at the base of the cells.

This experiment has been recently repeated by Graf et al. (1971) who used electron microscopy cytochemical techniques, and, with the aid of colored filters superimposed on a series of black and white positive electron micrographs, succeeded in reproducing the end-product of horse radish peroxidase reaction in a vacuole as a yellow image, and that of acid β-glycerophosphatase in a lysosome as a green image. The membranes of the two organelles were clearly seen to fuse at one point of contact while most of the end-product of the two enzyme reactions was each in its respective organelle.

Other investigators have published micrographs suggesting the fusion of the two organelles (Maunsbach, 1966b; Daems et al., 1969; see also Fig. 1*f*). However, most investigators have shown the two successive stages—absorption of a colored or of a radioactive protein by endocytosis and presence of the color or of the radioactivity in lysosomes—but have not observed the actual moment of fusion, probably owing to the rapidity with which the fusion takes place.

B. Exocytosis of Acid Hydrolases and Extracellular Digestion

Acid hydrolases are found in extracellular fluids such as bile or serum in normal conditions, and their origin from lysosomes has been postulated rather than proved. If it is accepted that the presence of acid hydrolases in the cell sap outside the lysosomes represents a damaging or even a lethal factor for the cell, theories must be elaborated as to how these enzymes can be secreted by the cell into the extracellular environment without ever coming into contact with the cell sap. One such theory has been put forward by De Duve and Wattiaux (1966) and by Dingle et al. (1967), according to whom the lysosomal membrane could fuse with the plasma membrane and the content be extruded into the extracellular milieu. Several experimental conditions have been described in which an increased synthesis of acid hydrolases in the cells is accompanied by an increased secretion of enzymes. These have included the administration of excesses of vitamin A to bone rudiments in culture (Fell and Dingle, 1963), of toxic amounts of sucrose to the same material (Glauert et al., 1969), and of parathyroid hormone causing bone reabsorption (Vaes, 1965). In these experiments the synthesis and the secretion of acid hydrolases were measured, and in some cases considerable vesiculation of the cytoplasm was observed. However, the identification of the cytoplasmic vesicles by cytochemical means was not performed, nor was the fusion of the vesicles with the cell membrane convincingly demonstrated. Since it must now be accepted that not all acid hydrolases are entirely lysosomal, it is possible that some of these enzymes are secreted by mechanisms not involving membrane fusion.

The ideal organ for a study of the secretion of acid hydrolases appears to be the prostate in which secretion of acid phosphatase is part of the physiological role of the epithelial cells. These cells show a large number of secretion droplets which are packed with acid phosphatase-containing material, a well-developed Golgi apparatus in which acid phosphatase activity does not appear to be present (Brandes, 1965), and lysosomes throughout the cell. The extracellular secretion contains acid phosphatase. It appears unlikely that the secretion droplets can be classified as lysosomes, since these structures have no role in the intracellular digestion of macromolecules or in autophagy. Yet they contain an acid hydrolase which fulfils all the criteria laid down for lysosomal enzymes, and although the mechanism of secretion of this enzyme has not been investigated, it appears likely that secretion occurs through exocytosis. If the function of this acid hydrolase could be elucidated, it would provide a very important insight into the mechanism and function of extracellular secretion of acid hydrolases in general.

C. Lysosomes and Hormones

1. Protein Hormones. (*a*) *Thyroglobulin.* A well-documented role of lysosomes in hormone secretion involves the thyroid organelles. In this organ the epithelium absorbs thyroglobulin by endocytosis and forms colloid droplets, the hormonal activity of which varies greatly. Wollman (1969) showed that thyroid lysosomes fuse with the colloidal droplets containing the glycoprotein thyroglobulin and thus supply the enzyme equipment needed to free the active form of the hormones from their protein support. Since this process of fusion is not synchronous, the level of activity in each droplet varies according to whether one or several lysosomes have fused with it, and according to the stage reached by lysosomal activity. Although this is not the only mechanism of thyroid hormones production, it is certainly of the uttermost importance in the physiology of the thyroid gland.

(*b*) *Other Pituitary Hormones.* The role of lysosomes in disposing of hormones present *in excess* in the anterior pituitary gland has been elucidated by Farquhar (1969). One of the most interesting findings by this investigator was that the secretion droplets containing either somatotrophic or mammotrophic or thyrotrophic hormones were not disposed of by an autophagocytotic mechanism with formation of cytolysomes but were engulfed within lysosomes proper where the droplets lost their limiting membrane but maintained their other morphological characteristics. This indicated a fusion of the membranes of the secretion droplets with those of the lysosomes rather than a wrapping of the ER membranes around the droplets. Under conditions of complete functional normality of the gland, however, the pituitary lysosomes do not appear to play a role in the process of hormone secretion, which appears to occur by a mechanism of direct fusion of the membrane limiting the secretion droplet with the cell plasma membrane, with consequent extrusion of the hormones into the blood.

(*c*) *Insulin.* Lysosomes do not appear to play a part in disposing of excess insulin (Maggi et al., 1969), another peptide hormone of low molecular weight. This finding offers a good explanation for the pharmacological effects of the hormone, since if lysosomes could digest insulin presumably there would not be any hypoglycemic effect by this substance. Since most peptide hormones appear to act on their target organs even when administered in very small doses and even in other species than those from which they have been obtained, it is interesting to speculate that lysosomes probably have no part in disposing of these either; however, no studies appear to exist on the subject of the fate of peptide hormones administered to animals

2. Steroid Hormones. These hormones are certainly capable of acting on lysosomes, and particularly on the lysosomal membrane, either by stabilizing or labilizing it. Most of the studies dealing with this aspect of the question have been experimental, however, and have tended to use nonphysiological doses of the various steroids tested, as well as nonphysiological conditions; this is particularly true of the studies performed *in vitro*. There appear to be very few studies on the behavior of lysosomes in human or in animal tissues, for example, under conditions of hormonal inbalance, except those of Brandes et al. (1962) on the lysosomes of rat prostate, and of Maggi et al. (1970) on the lysosomal and nonlysosomal acid phosphatases of castrated mice treated with estradiol. In both cases the investigators showed very severe and specific changes in these organelles.

Lysosomes of the cells of the adrenal have no role in the secretion of steroid hormones (Rhodin, 1971). However, the stabilizing effect of the adrenal steroids cortisone and hydrocortisone, although not of corticosterone, on the lysosomal membrane is well-documented (Weissmann, 1969) and explains at least in part the antiinflammatory effects of these hormones. This point is discussed further (see Section 9).

D. Lysosomes and Mitosis

The time required to visualize the end product of a histochemical reaction for an acid hydrolase has been assumed to represent the time of "latency" (Bitensky, 1962) described first as a biochemical concept. This time varies depending on the tissue, on the techniques of preparation, on the substrate used, and on the enzyme investigated. However, in strictly comparable conditions, during mitosis in HeLa and L cells in culture, this time is considerably shortened. This has been taken to indicate an increase in the fragility of the lysosomal membrane during cell division (Maggi, 1965b). Although Allison and Mallucci (1964a) have claimed that this increased fragility may result in a final rupture of the membrane with enzyme release into the cell sap, Maggi (1965b) has denied this on the grounds that if the tissues are incubated for the shortest possible time to enable the visualization of the end product of the enzyme reaction, then the end product is always particulate in appearance, whether this be in mitotic or in interphase cells. The variables here are: (1) the length of time required to visualize the end product, this being shorter during mitosis; and (2) the number of particles found after all the lysosomes have reacted, regardless of time, the number being the same in mitotic cells and in interphase cells. It is of course possible to create an artifactual "diffusion" of the enzyme in the cell sap by prolonging the incubation time. As demonstrated by Appelmans and de Duve (1955) and by Gianetto and de Duve biochemically (1955), and by several other investigators who applied biochemical techniques to histochemical preparations (Bitensky, 1962; Gahan, 1965), prolonged incubation at an acid pH damages the lysosomal membrane and causes a release of enzymes or of end product into the cell sap; this damaging action of the pH is naturally more pronounced when the membrane is more fragile, as during mitosis.

That lysosomes are involved in cell division also appears to be likely from the work of Robbins and Gonatas (1964a) and of Holt (personal communication). These investigators found that at prophase, during the dissolution of the cell membrane, the lysosomes are disposed around the nucleus; this position represents a change from that of the cell in interphase, when the lysosomes are disposed at the two poles of the cell in the case of the HeLa cells, and near the peribiliary canaliculi in the case of the hepatocytes of regenerating liver. The lysosomes remain as a circle around the chromosomes during metaphase and gather at each pole of the equatorial plate during anaphase and telophase. Maggi (1965b) noted that at this stage, and before cell division with the formation of two nuclei, the particles thought to be lysosomes were equally divided at each pole of the equatorial plate and the sum of the two groups of particles was equal to the total number of lysosomes present in the cell during interphase. This observation shows that immediately after division each cell must possess only half its final number of lysosomes and that these must be formed *de novo* during interphase, before the next division is initiated.

However, Gahan et al. (1971) have claimed that while this localization of lysosomes is typical of single cells during division, it is not found in differentiated tissues such as the epithelium of the digestive tract, the testes, and tissues of plants, in which cell division occurs as a physiological mechanism of cell replacement. These observations appear to invalidate the hypothesis that the localization of lysosomes around the chromatin material during division may play a role in the digestion of the cell membrane, as was suggested by Maggi (1965b); or at least it invalidates the extrapolation of the data obtained on cells in culture to other types of living cells.

E. Lysosomes and Differentiation

The role of lysosomes in the differentiation of an organ appears to be related to their digestive and autolytic properties. For example, these organelles appear to be responsible for the digestion of the proteins forming the Müllerian ducts of the chick embryo when it is differentiating into a male. However, no such digestion occurs when the embryo develops into a female, in which case the Müllerian ducts eventually form the oviducts (Scheib, 1963), a process controlled by hormonal secretion. During the process of differentiation of the tadpole into a frog, lysosomes play an important role in operating the autolysis of the tadpole tail (Weber, 1963).

In mammals, however, the role of these organelles in the process of differentiation of the whole organism from embryo to a fully developed animal has been little investigated. It is unlikely that phenomena similar to those described in birds and frogs occur here since processes of autodigestion within the embryo are not as relevant to embryonic development. The mammalian embryonic tissues are oriented toward synthesizing large molecules from small unit components supplied by the mother through the placenta, a process in which the lysosomes of the visceral layer of the yolk sac but not the embryo's lysosomes play a part (Beck and Lloyd, 1968).

The distribution of these organelles in tissues that have not differentiated is different from that of the equivalent tissue once differentiation has taken place. Clark (1957) demonstrated the presence of large "protein" droplets in the kidney proximal convoluted tubule cells of newborn mice; these droplets were distributed at random within the cells and were very numerous. In the adult mouse kidney, such organelles had greatly decreased in number and were localized in close proximity to the brush border. Maggi and Carbonell (1969) demonstrated these droplets to be lysosomes in that they contained acid phosphatase. These investigators also described similar differences in several tissues of young and adult mice. No difference in the distribution of lysosomes was found between adult (6–9 months old) and old (2- and 3-yr old) animals.

These investigators also found that one specific difference between nondifferentiated and differentiated tissues concerned a fluoride and molybdate-resistant acid phosphatase which is not lysosomal (Livingston et al., 1969; Maggi and Carbonell, 1969; Maggi et al., 1970). This enzyme was absent in nondifferentiated cells, and appeared in mouse tissues between 3 and 4 weeks after birth. Since the enzyme localization, at least in the kidney, appeared to be on the ribosomes of the rough ER, its absence or scarcity in mouse kidney cells may be attributable to the scarcity of these subcellular structures in these cells. The possibility that this ribosomal enzyme be a precursor of the, or of one of the lysosomal acid phosphatase(s) has been discussed in Section 3.

F. Cytolysomes in Adult Cells

As defined earlier, cytolysomes appear to be deputed to intracellular segregation and digestion of cytoplasmic components of the cell itself. In pathological conditions in which autolysis occurs at a more-or-less accelerated rate, these organelles are commonly found (see Section 9). They are also found under physiological conditions in which a high turnover of cell components would be expected, such as in embryonic tissues (Larsson et al., 1969), or in tissues undergoing involution (Wellings and Deome, 1963), although always to a very limited extent (Helminen and Ericsson, 1968a–d).

5. PHYSIOPATHOLOGY: LYSOSOMES AS A SYSTEM OF CELL DEFENSE

The role of lysosomes in response to factors damaging the cell, whether the damage be reversible or irreversible, has been studied by several workers. Two hypotheses, for which investigators provided evidence of a mainly circumstantial nature, have been advanced. One was that the damaging agent would first enter the lysosomes, where it would cause rupturing of the lysosomal membrane, hence would be released together with the intralysosomal acid hydrolases to cause cell damage and death through enzymic digestion of the cell constituents. If the cell were able to deal with this process and overcome it, the damage would be temporary and reversible.

The other hypothesis is that the damaging agent acts directly and primarily on each target cell constituent in a differentiated manner, depending mainly on the damaging factor itself. For example, anoxia would be expected to damage chiefly mitochondria, as the site of oxygen-linked respiratory processes. Maleic hydrazide would damage the chromosomes (McLeish, 1953), while an overloading of lysosomes with dextran (Daems and van Rijssel, 1961) or Triton WR-1339 (Novikoff and Essner, 1962) would damage these organelles primarily. Under these circumstances the role of lysosomes would be that of a "second line of defense," and they would be involved in "cleaning up" the irreversibly damaged organelles, in which case the irreversibility of the damage would only concern single organelles and not necessarily the whole cell. It is clear that if too many organelles of a single type, or several types of organelles in a cell, are damaged, the lysosomes may become overloaded and the cell would die as a consequence of a collapse of the second line of defense. If the lysosomes were able to deal with the effects of the toxic factor, then the damage would be reversible.

That *all* factors leading to cell damage and/or death act through a direct action on lysosomes is a theoretical concept that is difficult to accept. However, that cell death can be a consequence of a direct action on the lysosomal membrane, with consequent release of lysosomal enzymes *into* the cell sap (intracellular release of acid hydrolases) and hydrolysis of intracellular components, has been claimed by Allison et al. (1966). These investigators introduced suspensions of silica particles into a cell culture medium. The particles were found to be present in phagosomes and were then assumed to have been taken up by lysosomes (although no cytochemical identification of the latter was attempted at an ultrastructural level). At a later stage some of the particles were found free in the cell sap. A release of acid hydrolases into the cell sap was deduced not from ultrastructural cytochemical investigations but from histochemical studies at the optical microscope level only. Some of the cells were incubated in a medium for the demonstration of β-glycerophosphatase activity, a typical lysosomal enzyme. The incubation was prolonged for 60 min, and no shorter incubation periods were used. After such a time interval, the acid β-glycerophosphatase activity appeared in a diffuse form in the cells, whereas in untreated control cells incubated for the same length of time, the reaction was particulate. Allison et al. thus assumed that the enzyme had diffused out of the lysosomes, as a consequence of the presence of the silica particles, and was present in the cell sap where it caused cell death.

According to the same investigators (Allison and Mallucci, 1965), viruses could act similarly to silica particles and kill cells through their rupturing action on the lysosomal membrane rather than through a direct toxic action on cellular structures. Apart from the existing evidence that most viruses are not found in lysosomes but reproduce freely in the cell sap (Dales, 1969), the histochemical data of Allison and Mallucci (1965) are easier to explain if the data of Bitensky (1962) and Maggi (1965b) are remembered. These investigators showed that a prolonged incubation of cells in an acid incubation medium such as that used for the histochemical demonstration of acid phosphatases can result in an increased fragility of the lysosomal membrane. The question of the occurrence of artifacts during the use of histochemical methods has been mentioned previously (see Section 4, D) and is discussed in more detail elsewhere (Section 10). Moreover, although Allison and Mallucci (1965) had postulated an intracellular release of acid β-glycerophosphatase from the damaged lysosomes, from the biochemical data published by these investigators in a separate paper concerning the same experiments (Mallucci and Allison, 1965) it appears that under their experimental conditions net synthesis of lysosomal enzymes had occurred, which was reflected in an increase of both the "bound" (lysosomal) and the "free" (nonlysosomal) enzyme forms, rather than an increase in the free enzyme form at the expense of the bound form.

Some damaging factors, usually of a reversible nature, may act on the cell by altering the lysosomal membranes. Such factors are, for example, anesthetics, as demonstrated by Bitensky and Gahan (1962) and Al Azzawi and Stoward (1970). These investigators administered ether or chloroform or nitrogen gas to rats. Pieces of liver and kidney were frozen and sections of unfixed material were incubated for lengths of time just sufficient to obtain reaction end product for acid β-glycerophosphatase. Under these conditions, while in normal controls the reaction end product appeared as black granules after a 20-min incubation, in anesthetized animals the reaction end product appeared as brownish diffuse deposits after only a 5-min incubation. These workers interpreted these results as evidence for the presence of acid hydrolases of lysosomal origin in the cell sap. When the animals were allowed to recover from the anesthetics, their livers and kidneys behaved as those of the controls. Since no electron microscopy or quantitative biochemical investigation was performed under these conditions, it is difficult to assess the results of the experimentation in full. It is just possible that the "diffuse" reaction observed by Bitensky and Gahan (1962) was a result of the *de novo* formation of very small lysosomes, not resolvable with the optical microscope, as shown by Seljelid (1966). It is also possible that ultrastructural changes leading to an increased synthesis of the enzyme, such as those observed by Maggi and Oddy (1968) during starvation (see Section 9), were responsible for these findings.

An increase in the free activity of acid hydrolases at the expenses of the bound activity was observed by Slater et al. (1963) and Greenbaum et al. (1965) during involution of the mammary gland following suppression of lactation both in normal animals and in rats with dimethylbenzanthracene-induced mammary carcinomas. The same investigators found that carbon tetrachloride and other toxins had little effect on the release of some lysosomal hydrolases (Slater and Greenbaum, 1965). Dianzani (1963) and Beaufay and de Duve (1959), on the contrary, found a remarkable increase in the activity of acid hydrolases after carbon tetrachloride poisoning and in ischemic liver. In these cases, however, the primary damage seemed to occur in mitochondria, and lysosomes were affected subsequently. From the observations of Dingle (1961), Fell and Dingle (1963), and Dingle et al. (1969) on the effect of vitamin A or sucrose on embryo cartilage in organ culture, it is evident that there was no intracellular release of acid hydrolases but that a net synthesis of enzymes occurred in the cells, probably through an activation of the Golgi system, as demonstrated by Glauert et al. (1969). The enzymes were secreted, probably through a mechanism of exocytosis, into the tissue extracellular matrix where they started digesting the extracellular material; this in turn was reabsorbed by different cells in the same tissue, for the breaking-down process to be terminated through a further lysosomal digestion. At no stage did intracellular release of enzymes into the cell sap occur.

It must also be emphasized that in studies in which biochemical techniques are used there is always the possibility that an increase in the free activity of the tissue cells may mean merely an increased susceptibility of the lysosomes to the rather crude homogenization and fractionation procedures. The fragility of the lysosomal membranes varies greatly from tissue to tissue and from species to species; it is to be expected therefore that factors harmful to the cell may exert a labilizing action on the lysosomal membrane without causing its true *in vivo* rupture until the tissue is subjected to homogenization. There is also the possibility that an increase in free activity does not reflect a leakage of enzymes from the lysosomes but an increased activity or an increased synthesis of nonlysosomal acid hydrolases which on subfractionation separate from sucrose homogenates in the supernatant. So far, rupture of the lysosomal membrane as a consequence of a direct damaging factor has been observed *in vitro* and *in vivo* as a consequence of the action of streptolysin S (Hirsch et al., 1963; Weissmann et al., 1963; 1964; Zucker-Franklin and Hirsch, 1964; Zucker-Franklin, 1965) and after administration of staphylococcal leucocydin to leukocytes in culture (Gladstone and van Heyningen, 1957). These two bacterial toxins disrupt the lysosomal membrane of leukocyte lysosomes, and cell death follows. Hirsch and Cohn (1964) have noted, however, that their data cannot be taken as final conclusive evidence of a direct causal effect of lysosomal rupture on cell death.

The presence of acid phosphatase activity in the cell sap concomitantly with cell death was observed by Gahan and Maple (1965) in plant cells in which lignification had taken place. These investigators, however, were unable to determine whether the appearance of "diffuse" (nonlysosomal?) reaction end product preceded or followed the cell death; since ultrastructural observations were not made, the results of these studies are difficult to interpret.

Ultrastructural studies of cells in which damage to organelles has occurred have not consistently indicated the primary role of lysosomes. In some instances the primary response of the cell to toxic factors seems to be a hyperproduction of membranes with formation of vesicles. For example, Glauert et al. (1969) described the vesiculation of cells from embryonic skeletal tissue cultured in the presence of sucrose. In this case no enzyme cytochemical studies were performed. Maggi and Oddy (1968) described the complete vesiculation of some cells of the proximal convoluted tubule of mouse kidney after short-term starvation; after prolonged starvation, fatty degeneration followed. Maggi and Oddy (1968) found that a large number of vesicles were acid phosphatase-positive and suggested that they might represent Golgi vesicles. Maggi et al. (1970) found the same hyperproduction of acid phosphatase-positive Golgi vesicles in the seminal vesicles of mice that had been castrated and treated with estradiol. While in the first set of experiments there was no evidence for the hyperproduction of lysosomes, in the second case the cells contained an exceedingly high number of such subcellular organelles which were all acid phosphatase-positive. No evidence for an intracellular release of free enzyme was found, although signs of free acid phosphatase activity were present in the cytoplasm after the lysosomes had disappeared as a possible consequence of their overloading by estrogen esters (Fig. 6*d*, *e*, *g*). However, signs of increased enzyme synthesis by hyperplastic Golgi bodies were observed.

In the great majority of cases, a distinctive feature of cell damage, which appears likely to be involved in "cleaning up" the organelles of the damaged cell, are the cytolysomes (Fig. 3). These have been found in all cases in which severe cellular damage has occurred. Novikoff (1963) found acid phosphatase-positive cytolysomes in kidneys of rats whose ureters had been ligated. Orlandi (cited in Gahan, 1967) found cytolysomes in livers of patients treated with chlorpromazine or with androgenlike steroids. These organelles were also found in kidney cells of starved mice (Maggi and Oddy, 1968), in the epithelial and smooth muscle cells of mice castrated and treated with estradiol (Maggi et al., 1970), in hearts of old mice, rats, and hamsters (Maggi, unpublished), in starved *Euglena gracilis* (Brandes and Bertini, 1964; Brandes, 1965), and in long dormant embryonic cells of *Fraxinus* seeds (Villiers, 1967). Cytolysomes were found in skeletal muscle of rats with trichinous myositis (Maeir and Zaiman, 1966) and of patients with heart failure (Wheat, 1965); these last-mentioned findings are of particular interest since lysosomes are very rare in heart (Novikoff, 1961). Although in several cases mentioned above cell death followed, or was concomitant with, the appearance of cytolysomes, in other cases the process appeared to be reversible. Swift and Huban (1964) found cytolysomes in a number of pathological conditions in which the metabolic alterations were reversible. Cytolysomes have also been found in physiological conditions, such as very young kidneys (Larsson et al., 1969; Maggi and Carbonell, 1969). They are infrequent in a type of lesion, namely, cancer cells, in which the life span of subcellular structures such as nuclei, mitochondria, cell membranes, must be very limited. Their mechanism of formation is still debated. Novikoff et al. (1964) have suggested that the membranes of the smooth ER wrap themselves around the cell residues to form a membrane-limited body. Measures of the thickness of the limiting membrane in this body support this viewpoint (Maunsbach, 1966d). Thus the Golgi apparatus would have no role in the formation of these organelles. However, the mechanism with which these organelles, usually acid phosphatase-positive, obtain their enzyme endowment is still obscure. It has been suggested either that the acid phosphatase-producing Golgi vesicles fuse with the cytolysome once it is formed (Swift and Hruban 1964), or that the enzymes derive from the ribosomes attached to the rough ER membranes prior to the transformation of the latter into smooth ER and finally into cytolysomes. This mechanism is made more likely by the recent findings of Smith and Fishman (1969), Holt and Katchburian (1969), Fishman et al. (1970), and Maggi et al. (1970) that the cisternae of the smooth ER contain acid phosphatase.

A third mechanism which would explain the process of digestion of mitochondria has been suggested by Maggi and Oddy (1968). This would involve a direct fusion of the Golgi vesicles or of lysosomes or of any acid hydrolase-containing structure with the mitochondrial membrane. This mechanism was first suggested by Novikoff (1963) who later (Novikoff et al., 1964) attributed it to artifacts of fixation with glutaraldehyde, since in osmium tetroxide-fixed material signs of fusion could not be found. However, since the cell in Figure 7*a* was fixed in glutaraldehyde and that in Figure 7*b* fixed in osmium tetroxide, it is difficult to explain these images on the basis of a fixation artifact. It appears therefore that the hypothesis of a direct fusion of the cell membranes, whether smooth or rough ER, with the mitochondrial membrane cannot be dismissed completely and may represent yet another mechanism of lysosome formation.

6. LYSOSOMES IN OTHER LIVING ORGANISMS

A. Unicellular Organisms

Simple forms of life such as bacteria secrete acid hydrolases in the extracellular milieu where processes of digestion of large molecules take place; these organisms have no lysosomes. Unicellular organisms such as protozoa, cells in culture or nucleated blood cells, however, show particles in their cytoplasm which possess all the characteristics of lysosomes, both morphologically and functionally. It appears therefore that a certain structural organization of the cell is required for the lysosomes to exist, and it seems likely that intracellular membranes, such as the ER, are an essential part of this organization.

Lysosomes in these types of cells appear to perform the same function as lysosomes in tissues of higher animal organisms. They engulf, store, and digest large molecules (Cohn and Hirsch, 1960; Cohn and Benson, 1965a–d), ferritin (Loni et al., 1967), or foreign substances such as silica or asbestos (Allison et al., 1966), and possess acid hydrolase activity (Munro et al., 1964; Robbins and Gonatas, 1964b; Cohn and

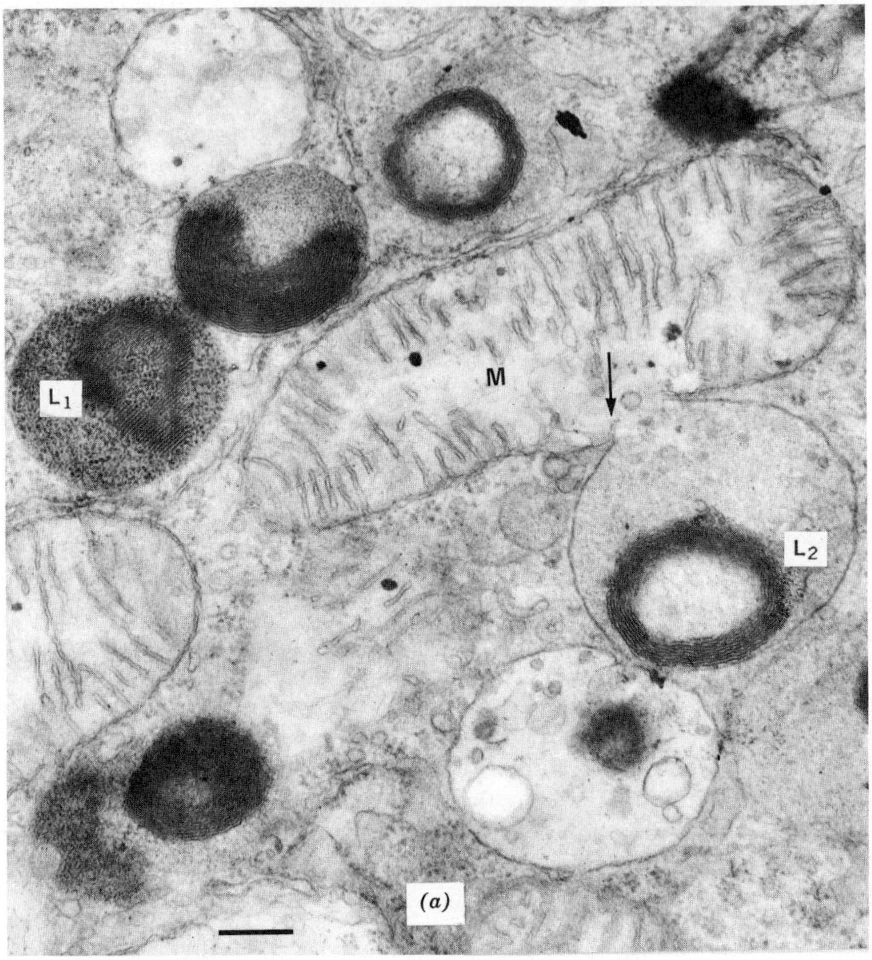

Figure 7. Mitochondrion–lysosome fusion. (*a*) Cell of the kidney proximal convoluted tubule of 2-wk-old mouse showing several bodies with myelin figure-type inclusions. The lysosomal nature of these bodies is demonstrated by the presence of end product of acid phosphatase activity in one of them (L_1). All the bodies are very close to or appear to fuse with each other. The fusion of the mitochondrial (M) membrane with that of L_2 is apparent. A small vesicle can be seen in the area of fusion. Tissue fixed with glutaraldehyde. (b) Smooth muscle cell from seminal vesicle of mouse castrated and treated with estradiol for 5 wk. Tissue fixed in osmium tetroxide. Several membranes, whether smooth (arrow) or rough (double arrow), appear to be fusing with the outer mitochondrial membrane. Note also the close apposition of a number of small vesicles (Golgi vesicles?) to one side of the organelle where the membrane is condensing.

Benson, 1965a–c; Maggi and Riddle, 1965). When subjected to various types of damage, such as starvation, some unicellular organisms develop typical cytolysomes (Brandes and Bertini, 1964; Brandes, 1965).

The role of lysosomes in organisms such as *Plasmodium* or *Amoeba histolytica* is completely unknown but could be important in the pathogenesis of the related diseases malaria and amebiasis. The role of the lysosomes in leukocytes is certainly of primary importance as a defense mechanism against bacterial infections (see Section 9).

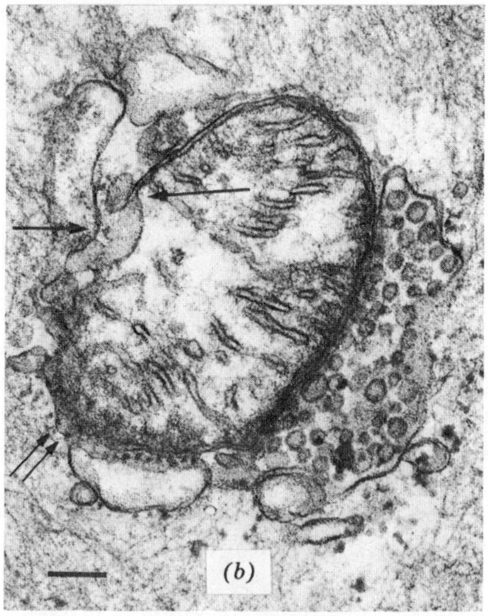

Figure 7. continued

B. Plants

The long debated argument as to whether or not plants possess a lysosomal system similar to that in animal cells appears to have been resolved positively and is reviewed by Gahan (1972). The existence of plant organelles containing acid phosphatase was first described cytochemically by Jensen (1956) in mitochondria of root tip cells, at the same time that De Duve and his coworkers (1955) identified the lysosomes as distinct from mitochondria. A complete description of the lysosomal system in plants had to await the elaboration of special biochemical procedures for the isolation of the very fragile vacuoles of seedlings (Matile et al., 1965). These investigators found that the vacuoles contained a number of acid hydrolases which were similar to the acid hydrolases of animal cells, namely acid β-glycerophosphatase, acid ribonuclease, acid protease, β-amylase, α-naphthyl esterase, and α-glycosidase. The function of these vacuoles, their formation and the mechanism of production of the acid hydrolases, were studied mainly by cytochemical techniques. It now seems likely that in plant cells the vacuoles are formed by one of the two following mechanisms: (1) vesicles pinch off from the SER and fuse together, carrying with them their content in acid hydrolases, or (2) the vacuoles are pre-formed in the cell and are devoid of acid hydrolases until dyctiosomal vesicles fuse with them and provide acid hydrolases. The parallel with animal cells is evident. Vesicles pinching off the SER are found in animal cells (see Fig. 1d). The dyctiosomal vesicles and the plant vacuoles can be equated to the Golgi vesicles and the phagosomes, or vacuoles, in animal cells respectively; thus the fusion of dyctiosomal vesicles with the vacuoles parallels the fusion of the Golgi vesicles with the animal cell vacuoles, thus forming the multivesicular bodies. That both vesicles pinching off the SER and dyctiosomes contain acid hydrolases has been inequivocally demonstrated by Dauwalder et al. (1969), Gahan and McLean (1969) and Poux (1970).

The functions of the vacuolar, or lysosomal, system in plants appear to be multiple. This system stores food such as proteins and starch and digests these compounds when the cell needs metabolites (Matile and Moore, 1968), and controls the release of food reserve material to the cell. It performs the autolysis of structures such as fruiting bodies and of cytoplasmic fragments and residues, or of such material as is disposed of during cell differentiation from the meristematic state to a mature form. Cytolysomes are found in plant cells in conditions of stress such as enforced, prolonged dormancy in seeds of *Fraxinus excelsior* (Filliers, 1967) or in cells of *Cucurbita pepo* (Coulomb and Buvat, 1968). In spite of the great number of data published on the possible involvement of acid hydrolase-containing bodies and of acid hydrolases in cell death (Gahan and Maple, 1966; Sutcliffe and Sexton, 1968; Sexton and Sutcliffe, 1969), the question as to whether the vacuolar system plays a primary or a secondary role in determining cell death has not been settled (Gahan, 1972). Similarly, the question as to whether or not plant cells are capable of taking up large molecules by a mechanism of endocytosis leading to the formation of endocytic vesicles which would subsequently fuse with acid hydrolase-containing vacuoles has not been resolved. So far, although it has been proved beyond any reasonable doubt that intact plant cells can take up such large molecules as RNAse (Brachet, 1955) and bacterial DNA (Stroun et al., 1966; 1967), the mechanism whereby this takes place has not been elucidated. Wheeler and Hanchey (1971) have suggested the possibility that viruses and macromolecules enter plant cells by a mechanism of endocytosis, but have not provided evidence for the subsequent fusion of the endocytic vesicles with the vacuoles. In any case, endocytosis in plants does not appear to be a physiological mechanism of uptake of nutrients. It seems, therefore, that the vacuolar (lysosomal) system of plants is primarily a system for the disposal of stored or waste substances, and may not play such an important part in the physiological uptake of nutrient macromolecules or in processes of cell defence as it does in animal cells.

7. CHEMICAL STRUCTURE OF LYSOSOMES AND LYSOSOMAL MEMBRANES. HETEROGENEITY OF LYSOSOMES

Although it is easy to define a lysosome in general terms (see Section 1), lysosomes in various tissues of different living organisms exhibit an enormous variety of appearances. This depends in part on the tissue and is therefore likely to be related to the different function of each tissue and organ and of different cell types in them (see, for example, lysosomes of hepatocytes, Kupffer cells, bile duct system, endothelial cells, and of gallbladder, all in the liver); it also depends on the various species for the same cell type (see rat and mouse kidney, rat and mouse prostate) and on the functional state of the particular cell at a particular time. Figures 2–4 and 8–11 indicate the kind of variability existing in the morphology of lysosomes, which is likely to be reflected in their different chemical composition.

This composition includes structural components such as proteins, lipids, and polysaccharides as well as acid hydrolases. While the latter are difficult to analyze separately from the structural proteins and represent a specific protein component of the lysosomes, the lysosomal lipids appear to be similar to those present in mitochondria (Guss et al., quoted by Tappel, 1969a). However, Barrett and Dingle (1967) have described in isolated rat kidney lysosomes a lipid component which is strongly

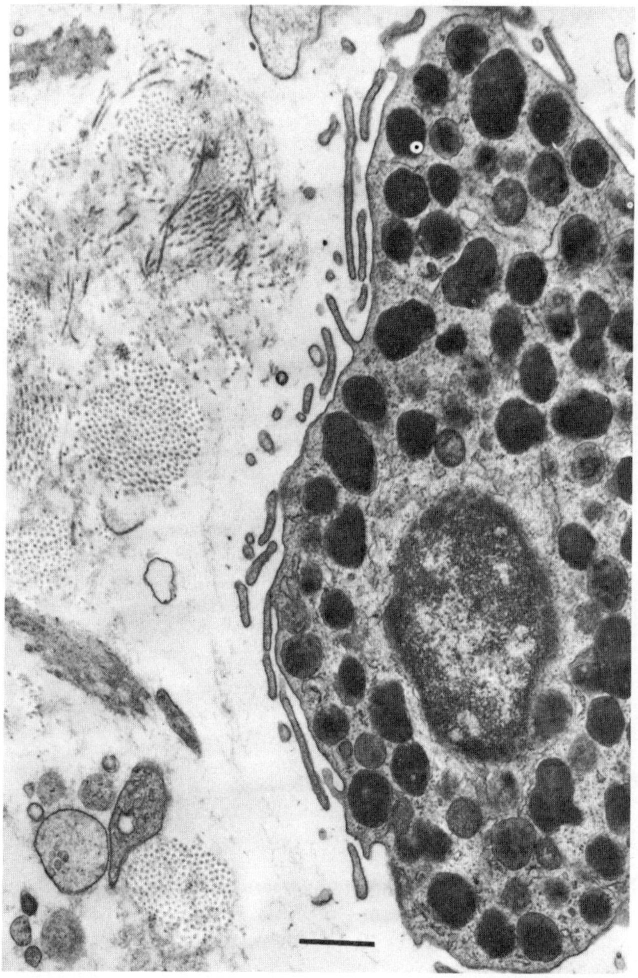

Figure 8. Basophil polymorph cell of mouse blood. The granules of these cells have been demon strated to be lysosomes.

acidic and which is capable of binding lipophilic and cationic molecules. According to these investigators, this substance could represent the specific component that binds toxic substances in order to protect the cell. Whether this lipid is present in lysosomes from other tissues has not been ascertained.

Other peculiarities of the chemical composition of the lysosomes include a high proportion of flavins and a high content of metals, particularly iron (Tappel, 1969a).

The analysis of the lysosomal membrane has shown that it is similar in composition to that of the plasma membrane of erythrocytes and mitochondria (Thinès-Sempoux, 1967), but it appears to contain a high proportion of sphingomyelin (Granda and Posner, 1969).

The next question concerning lysosomes is whether each lysosome contains all acid hydrolases so far found in lysosome fractions obtained by means of ultracentrifugation

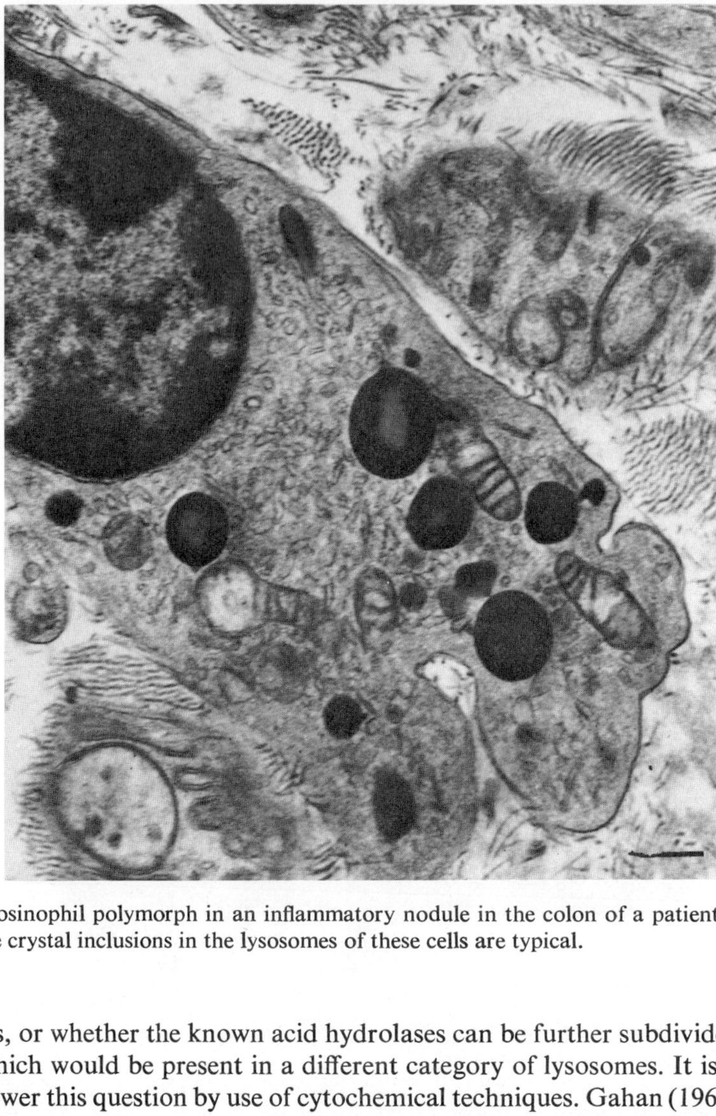

Figure 9. Eosinophil polymorph in an inflammatory nodule in the colon of a patient with Crohn's disease. The crystal inclusions in the lysosomes of these cells are typical.

techniques, or whether the known acid hydrolases can be further subdivided in groups each of which would be present in a different category of lysosomes. It is rather difficult to answer this question by use of cytochemical techniques. Gahan (1968) incubated sections of roots of *Vicia faba* at pH 5.0 in the presence of naphthol AS/B1 phosphate and Fast Red Violet LB first, and subsequently at pH 6.0 in the presence of naphthol AS/D acetate and Fast Blue B. This investigator succeeded in demonstrating that the red granules representing the end-product of acid phosphatase reaction mixed always with the blue granules representing the end-product of acid esterase reaction thus yielding purple granules. The author concluded that these two hydrolases are present in the same population of lysosomes. The difficulties in interpreting such data are obvious, however. In animal cells, lysosomes tend to be rather small, and often at the limit of resolution of most optical microscopes, particularly when cryostat-, rather than paraffin-embedded sections are used (the latter should not be used in enzyme cytochemistry because of the high degree of enzyme inactivation occurring during the processing of the tissue). This means that it is impossible to equate one granule of end-product with one lysosome, and that often one granule, or droplet, can be assumed to

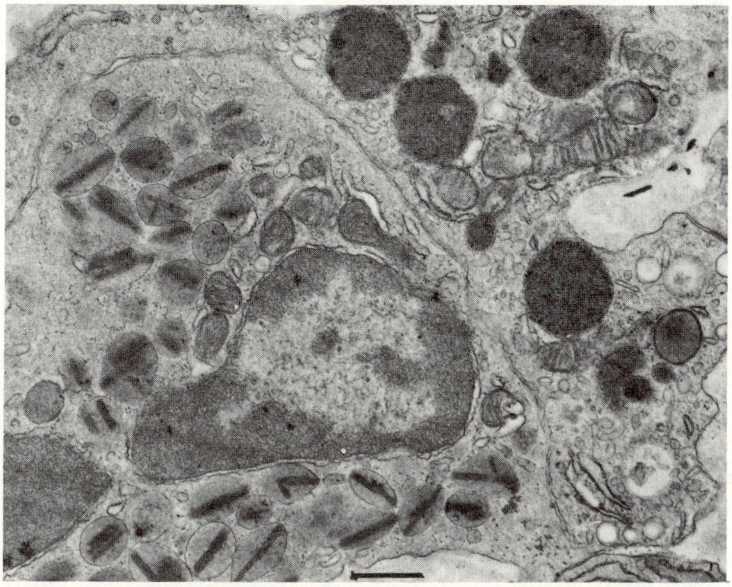

Figure 10. Two cells of a small blood vessel wall of mouse seminal vesicle. Note the difference in the morphology of the granules of these two cell types. Note also that in the cell on the left there is no acid naphthol AS phosphatase end product in the granules, while the other is rich in lysosomal enzyme activity.

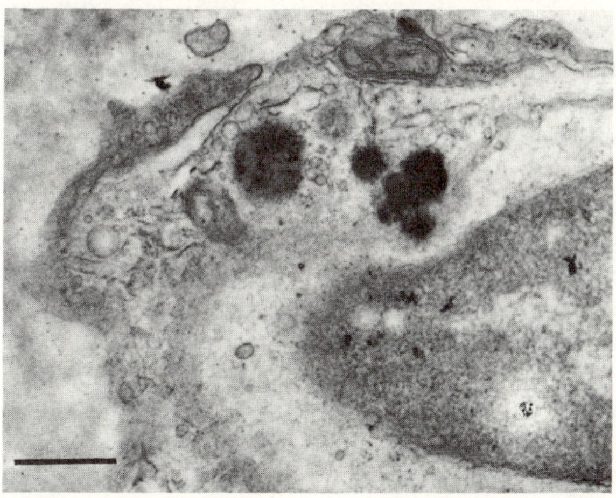

Figure 11. Cell of a capillary blood vessel of mouse seminal vesicle. The lysosomes contain end product of acid naphthol AS phosphatase activity; their morphology is different from that of the lysosomes in Fig. 10.

represent several lysosomes clumped together. When using ultrastructural enzyme cytochemistry, the number of substrates available for the demonstration of acid hydrolases has been restricted sofar to those capable of yielding a lead-containing complex, which makes the distinction between two different types of end-product impossible. It is hoped that the development of new chemical procedures such as those of Seligman et al. (1968), Livingston et al. (1969, 1970), and Smith and Fishman (1969), of new processing techniques such as those of Bernhard and Leduc (1967), Leduc et al. (1967), Appleton (1969), Gahan et al. (1970), and Bernhard and Viron (1971) and possibly the application of biophysical techniques such as x-ray diffraction will make progress in this field possible, particularly since biochemical data such as those of Baggiolini et al. (1969) and of Zeja and Spitznagel (1971a, b) clearly indicate the existence of at least 3 three classes of lysosomes, containing different types of acid hydrolases, in polymorphonuclear leukocytes of rabbit blood and bone marrow.

8. FUNCTION OF LYSOSOMES WITH RESPECT TO THE FUNCTION OF DIFFERENT TISSUES

A precise correlation between the activity of lysosomal enzymes and tissue function(s) is difficult to establish. If it is assumed that lysosomes are involved only in processes of intracellular digestion, whether it be auto- or heterodigestion, it is then hard to understand, for example, why the human prostate contains such enormous quantities of acid phosphatase, a "typical" lysosomal enzyme; this tissue does not appear to be involved in processes of absorption and digestion of nutrients, does not undergo processes of self-destruction, and the enzyme is present in secretion droplets as an integral part of the secretion rather than in lysosomes exclusively. Even if it is assumed that the prostate is an exception in which lysosomes may play a role in secretion, it is clear that lysosomal enzymes vary greatly in quantity according to the species for the same tissue. For example, although rat and mouse kidney have the same function, their content of lysosomes appears to be very different; the rat tissue is very rich, and the mouse tissue rather poor, in these organelles. The same could be said for other tissues such as liver or even striated muscle and heart (Maggi, unpublished observations). The same difference exists among the various animal species concerning the intracellular localization of lysosomes; for example, in rat kidney proximal convoluted tubules, the lysosomes are more numerous in the basal region of the cells, whereas in human, guinea pig, mouse, and hamster proximal convoluted tubules, they are concentrated near the brush border and often in immediate proximity to the large apical vacuoles that function as a possible final station before defecation of end products of intracellular digestion (see Fig. 1*b*). In this respect it seems that the latter localization is more economical for the functioning of the cell, since the fusion between lysosome and vacuole would occur close to the sites of absorption or of excretion. However, within the same species there appears to exist a correlation between the number of lysosomes in a tissue and the function of the latter. Tissues, the function of which is to absorb and digest large molecules, particularly proteins and their complexes, contain a greater number of lysosomes than tissues in which metabolism is geared to the utilization of small molecules. Typical examples of these two types of tissues are the kidney and the heart, respectively; the proximal convoluted tubule cells of the first organ absorb relatively large protein molecules such as insulin (Maggi et al., 1969), horseradish peroxidase (Straus, 1964), ferritin (Maunsbach, 1966b), and homologous (Maunsbach, 1966a) and heterologous (Miller and Palade, 1964) albumins. The heart, however, utilizes only small molecules such as hexoses, short-chain fatty acids, and amino acids. Accordingly, the first tissue is rich in lysosomes while the second is not.

At the same time it is quite possible that the number of lysosomes in each tissue can be correlated to the more-or-less active turnover of cell organelles. If this were true, the number of cytolysomes would be high in a tissue in which the life-span of, for example, mitochondria is very short, and vice versa. Since so little is known about the life-span of subcellular organelles in many tissues, however, this statement is difficult to prove at present. Certainly no correlation can be seen to occur between the life-span of cells of certain tissues and their content in lysosomes. For example, the life-span of an intestinal cell is certainly very short while that of a kidney cell is fairly long; yet both cell types contain a high number of lysosomes with respect to other tissue cells within the same species. Tissues such as heart, however, in which respiratory processes are very active and whose mitochondria would be expected to turn over very rapidly, are not rich in either lysosomes or cytolysomes. Although cytolysomes are occasionally found in otherwise perfectly normal tissues, and particularly in tissues of very young animals or embryonic tissues (Larsson et al., 1969; Maggi and Carbonell, 1969) they seem to be involved more when pathological processes affect the cells. This point is discussed further in Section 9.

9. LYSOSOMES IN PATHOLOGY

Lysosomes are probably involved in all diseases that involve an accelerated turnover of cells or of cell constituents. Thus one would expect to find increased lysosomal activity in numerous conditions, such as bacterial, viral, and parasitic infections, nutritional deficiencies, and intoxications.

A. Infections, Inflammation, Parasitoses

Relationships between infections and lysosomes have been approached experimentally and have yielded much information on the role of lysosomes in general. Weissman et al. (1963, 1964) and Hirsch et al. (1963) showed that streptolysin, a streptococcal toxin, can act both *in vivo* and *in vitro* by disrupting the lysosomal membrane of leukocytes and causing cell death. If the bacteria are injected instead of the toxin, they will be taken into the same cells by phagocytosis; leukocyte granules will then fuse with the bacteria-containing phagosomes and their hydrolase content will proceed to digest the bacterial wall and finally the bacterium itself. In this case the leukocyte remains intact, its cytoplasm and nucleus having been protected from the harmful effects of acid hydrolytic digestion.

It is not clear in these instances how toxins are taken up by the cells. It could be that they diffuse through the cell membrane into the cell sap and attack the lysosomal membrane from the outside; or it is possible that they are taken up by endocytosis and that they disrupt the lysosomal membrane from the inside after fusion of the lysosome with the endocytic vacuole. At present, these hypotheses are pure speculation and await further experimental evidence. Another instance in which bacteria are taken up by macrophages by a phagocytotic mechanism and apparently digested by lysosomes is Whipple's disease (Trier et al., 1965).

A primary role in the occurrence of rheumatoid arthritis has been attributed to the lysosomes of the synovial membrane and particularly to the lysosomal proteases cathepsin B and D (Fell and Thomas, 1960; Thomas et al., 1960; Dingle, 1963, 1966; Thomas, 1964; Weiss and Dingle, 1964; Dingle et al., 1971). This would involve a mechanism of digestion of the subsynovial cartilage matrix similar to that occurring during vitamin A deficiency (Roels et al., 1965) or hyperdosage (Fell and Dingle, 1963).

Lysosomes of polymorphonuclear leukocytes might be involved in the inflammatory changes characteristic of gout (Rajan, 1968), as well as in inflammation processes generally, although there is so far no clear-cut evidence that a release of lysosomal enzymes from polymorphs in the surrounding tissue is responsible for the inflammatory response of the latter. Support for this, however, appears to come from the finding of a cationic protein which is contained in leukocyte lysosomes and has an inflammatory action in rat mesentery (Janoff and Zweifach, 1964), and of a lysosomal mucoprotein protease (Burke et al., 1964) which increases vascular permeability.

De novo formation of lysosomes can apparently be induced in striated muscle by the parasite responsible for *trichinous myositis*; in this condition a large number of acid phosphatase-positive bodies is formed in the muscle cells, which are normally very poor in lysosomes (Maeir and Zaiman, 1966). Other parasitic infestations do not appear to have been studied with respect to the behavior of lysosomes. Owing to the large size of most parasites, it is to be expected that lysosomes would be involved in dealing with the reactive changes caused by the parasites in the tissue rather than in digesting the parasite itself.

Although it has been demonstrated that most viruses infect cells by a mechanism of endocytosis or phagocytosis these virus particles are later found free either in the cytoplasm or in the nucleus where they reproduce. Since viruses are particularly large nucleic acid–protein complexes, this special behavior of the cells toward them seems to merit particular attention.

B. Nutritional and Other Metabolic Defects

The role of lysosomes during starvation has been studied to a certain extent. During prolonged deprivation of food, rat liver becomes very fatty and many typical cytolysomes are found containing residues of numerous cell organelles, particularly mitochondria (Novikoff et al., 1964; Swift and Hruban, 1964). After starvation for only 18 hr, mouse kidney also becomes very fatty, and many cells of the proximal convoluted tubule of the kidney become filled with acid phosphatase-positive vesicles (Maggi and Oddy, 1968). It has been suggested that these vesicles belong to the hyperplastic Golgi bodies which are also found in these cells to contain acid β-glycerophosphatase. According to Novikoff et al. (1964), this increased production of Golgi structures should inevitably lead to the increased formation of lysosomes and/or cytolysomes.

Maggi and Oddy (1968) suggested that the changes observed in the kidney of starved mice might be linked to a state of acidosis and deficiency in the respiratory processes which were deduced from the presence of severe morphological alterations of the mitochondria in the same cells. In this respect, the findings of Wheat (1965) acquire a special significance, since this investigator found an increase in the number of lysosomes in hearts of patients with severe heart failure, in which respiratory processes are severely affected. It is possible that the same mechanism may be involved in the *de novo* production of lysosomes in the cells of the papilla of hamster kidney, where they are normally absent during K^+ deficiency (Morrison et al., 1963; Morrison and Panner, 1964).

The same mechanism may also be involved in the marked increase in the activity of lysosomal β-glycerophosphatase, which has been found by Maggi (1971) in hearts of strains of hamsters affected by congenital myocardiopathy. Both cytochemical and ultrastructural data, although preliminary, appear to confirm that in the myocardium of these animals the number and the activity of the lysosomes have increased considerably (Maggi and Hart, in preparation).

Other interesting instances of abnormal behavior of lysosomes (either an increase in their number or in activity, or a high number of cytolysomes) are reported by Gabrielescu (1970), who described the *de novo* formation of dense bodies in the neurons of rats and squirrels subjected to stresses. Vitamin deficiencies, particularly of vitamins A and E, and vitamin hyperdosages can also lead to an increase in the permeability of the lysosomal membrane (Fell and Dingle, 1936; Roels, 1969). It is unfortunate that most of the available data refer to purely experimental conditions, and that no investigation has been conducted on human beings affected by the factors mentioned above, that is, starvation or nutritional deficiencies, hypovitaminoses, or intoxications by substances such as pesticides, riot gases, and others.

C. Endocrinological Diseases

The same is true of another interesting field of pathology, that of endocrinological diseases. While numerous investigations have demonstrated the labilizing or stabilizing action of corticosteroids, estrogens, androgens, and thyrotropin (De Duve et al., 1961; Scheib, 1963; Wollman et al., 1964; Bangham et al., 1965; Weissman, 1965a, b; Weissman et al., 1965; Maggi et al., 1970), no data have been reported on the behavior of lysosomes and of lysosomal enzymes in clinical conditions involving the above-mentioned hormones; yet these clinical conditions are very numerous and include the adrenogenital syndrome, cryptorchidism, thyrotoxycosis, hypersecreting ovarian adenomas and carcinomas, Cushing's and Addison's syndromes, and many others.

D. Inborn Diseases

In all the conditions described above, lysosomes are not primarily involved but only intervene as a secondary factor of reaction to the primary injury. Some clinical conditions have been described, however, in which the disease has been claimed to affect lysosomes primarily. All these are inborn conditions in which lysosomal enzyme activities are somewhat impaired owing to a genetic defect. The most classical example of these, glycogenosis II, was first described by Baudhuin et al. (1964), and by Hers (1965), and completely investigated cytochemically, biochemically, and morphologically. The disease is characterized by an intralysosomal accumulation of glycogen in the liver, apparently due to a lack of the lysosomal acid hydrolase α-glycosidase. However, other types of glycogenosis, in which the excess glycogen is present in the cell sap, do not appear to involve lysosomes.

The intralysosomal accumulation of various kinds of mucopolysaccharides in diseases such as Tay-Sachs's syndrome, Fucosidosis, Hurler's syndrome (McKusick, 1966), is also unlikely to involve lysosomes as a primary factor, since none of these substances is a natural substrate for any of the known lysosomal enzymes. This was evident in a case of Sanfilippo disease, where the accumulation of the sulfated mucopolysaccharides heparitin sulfate and chondroitin sulfate was not lysosomal but occurred within the striated muscle fibers, and was accompanied by a wastage of the myofilaments (Maggi et al., in preparation), whereas intralysosomal storage of these substances took place in cells of the RES. This is true also in a number of congenital gangliosidoses or lipidoses such as the Gaucher, the Niemann-Pick, the Wolman disease or familial lipidosis, and others. Thus for example in a child affected by the familial lipidosis most tissues were very rich in triglycerides and cholesterol esters: an ultrastructural analysis of these tissues showed signs of remarkable enhancement in the synthesis of these two lipid types in hepatocytes and cells of the adrenal glands with storage in SER vesicles. In very few of these cells lysosomes were filled with lipid and cholesterol esters, whereas most lysosomes were normal (Fig. 12). However, the lysosomes of the cells of the RES were grossly abnormal (Fig. 13). On the basis of these data (Maggi et al., in preparation) it was suggested that this disease is not due to a deficiency in a lysosomal acid lipase, as was hypothesized by Patrick ahd Lane (1971) but is more likely to be due to a disturbance in the synthetic pathway of triglycerides and/or cholesterol esters, leading to an accumulation of these substances in the SER and the lysosomes of the RES as a secondary phenomenon.

It seems therefore that further evidence must be awaited before a final and comprehensive classification of inborn primary lysosomal diseases can be made; and that in the meantime some restriction must be exerted in defining any storage disease with abnormal lysosomal granules as primarily lysosomal. Thus far, only one disease has been thoroughly characterized—biochemically, cytologically, and cytochemically—as being lysosomal, namely glycogenosis II. In another disease, Fabry's disease, a genetically-induced deficiency of a lysosomal acid hydrolase (galactosyl-galactosyl-glucosylceramide: galactosyl-hydrolase) appears to be the cause for the accumulation of the cerebroside in brain's lysosomes (Mapes et al., 1970).

E. Teratogenesis

A most interesting hypothesis has been advanced concerning the possible role of lysosomes in teratogenesis. In 1948, Gillman et al. demonstrated that trypan blue is a teratogenic agent in rats. Subsequently, in pregnant rats this substance was found to be taken up by the lysosomes of the yolk sac (Lloyd et al., 1968) where apparently it remains without ever reaching the embryonic tissues and where it inhibits the activity of lysosomal enzymes (Davies et al., 1969). Since proteins injected into pregnant rats are endocytosed by the lysosomes of the yolk sac where they are digested (Beck and Lloyd, 1968; Williams et al., 1969), it was suggested that lysosomal degradation of proteins is a prerequisite for nourishment of the fetus, at least at an early stage in embryonic development, and at least for certain specific proteins, if not for homologous maternal serum proteins or maternal antibodies. Trypan blue would therefore inhibit such degradation of nutrient maternal proteins, and teratogenesis would be the result of an indirect, and possibly specific, shortage of certain products

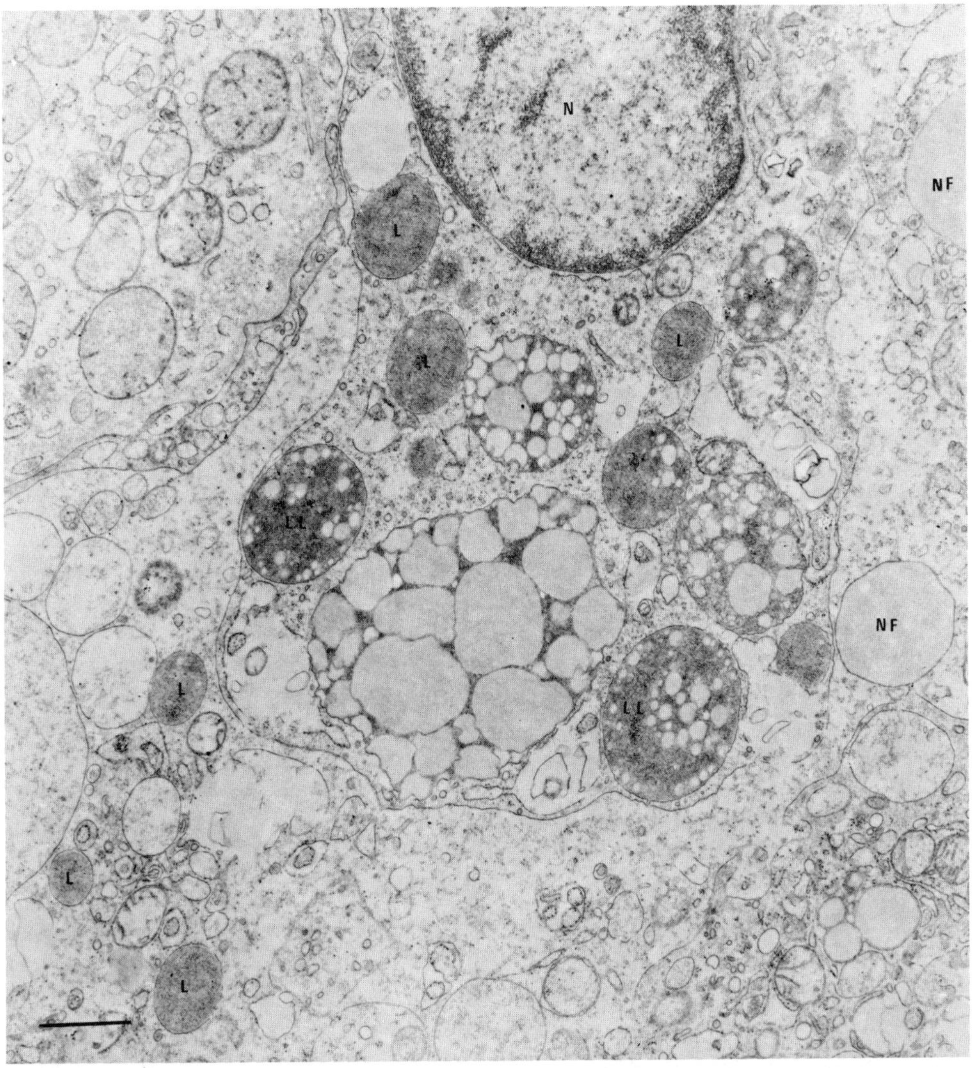

Figure 12. Kupffer cell in the middle of the field, containing a number of lysosomes filled with small vesicles of variable electrondensity (LL). Other lysosomes (L) are normal. Hepatocytes surround the Kupffer cell and one of them contains normal lysosomes (L). The RER is completely disrupted; the SER is dilated into a great number of vesicles of variable size, some of which contain lipid material (NF = neutral fats). N = Nucleus.

of protein degradation, that is, amino acids. Further evidence for the early, and irreversible, action of trypan blue on the yolk sac also derives from the data of Critchfield and Daniel (1965), who found the substance to act in quails to which it had been administered days before ovulation. Some of the offspring of these animals were consistently abnormal.

There is no need to emphasize the importance of these data, which could be relevant to the interpretation of teratogenesis in human embryos, particularly concerning chemical (thalidomide, defoliants) or infective (Rubella virus) teratogenesis.

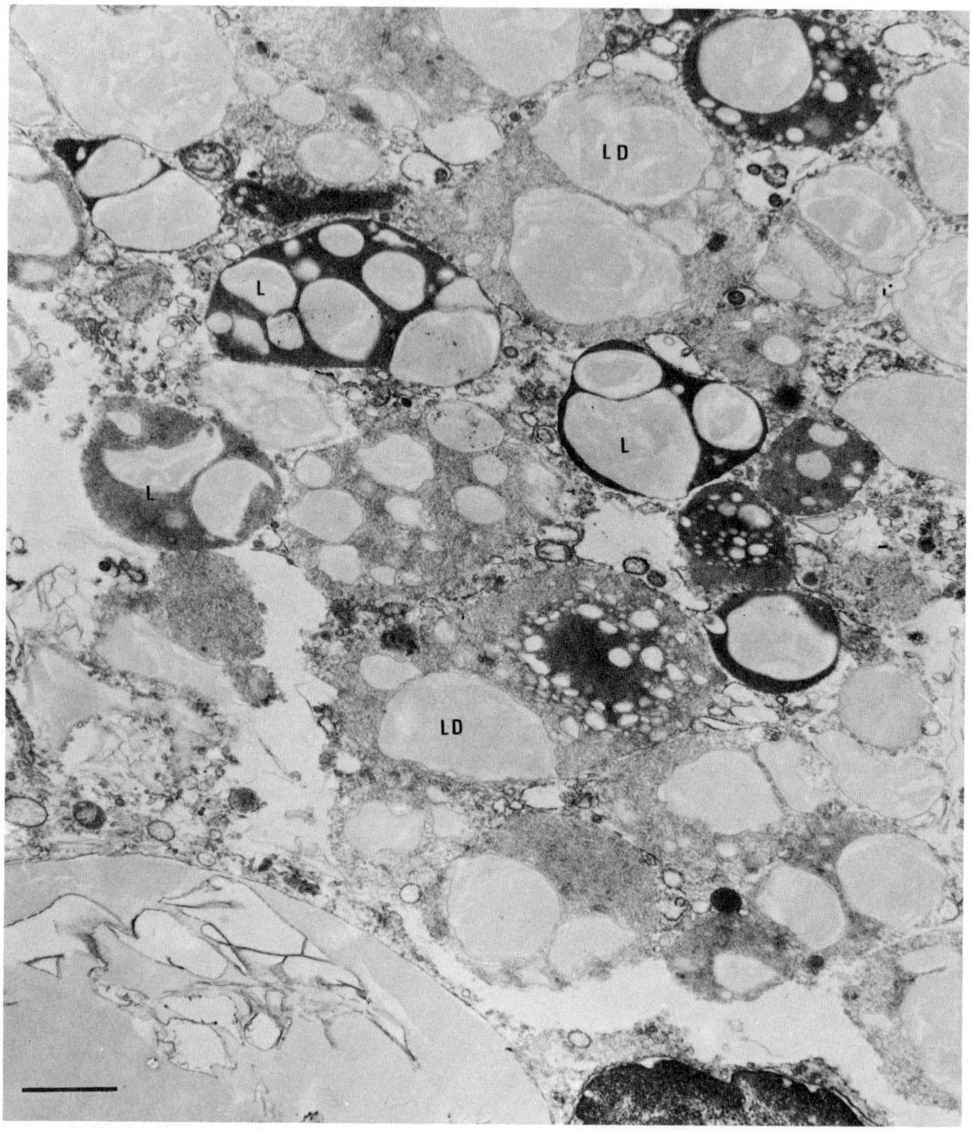

Figure 13. Part of a macrophage from a lymph node of the same child as in Fig. 12. Lysosomes (L) are filled with vesicles, most of which appear to contain the same type of crystalline material as present in the cytoplasm outside the lysosomes (LD). It has been assumed (Maggi et al., in preparation) that the lipid material present in LD is produced in excess by the SER of liver and adrenals, and is deposited in most other tissues where it is phagocytosed by lysosomes of fixed and free macrophages.

F. Cancer

Surprisingly little work has been done on this subject, and it is still not clear whether lysosomes have a special role in cancer cells. Sylvèn and Malmgren (1957) and Sylvèn and Bois-Svensson (1964) suggested that lysosomal cathepsins might be responsible for the autolysis of the normal tissue by the cancer cells preceding the invasion. Burstone (1962) showed considerable "leucine aminopeptidase" activity in the cells of the invading edge of some tumors. Apart from the uncertainties about the lysosomal nature of the latter enzyme (Emmelot et al., 1968; Barrett and Poole, 1969), many morphological, cytochemical, and biochemical data (Wattiaux et al., 1969) appear to contradict the concept that lysosomes play a specific part in cancer cells.

An examination of a great number of ultrathin sections from a large number of tumors (Franks and Rowlatt, personal communication) failed to reveal a specific pattern of lysosome number, distribution, and morphology with respect to either tumor type, cell type, or phase of cell or nuclear division. A series of investigations on acid hydrolase behavior in both human and experimental tumors (Maggi and Wyatt, unpublished) has confirmed the data of Franks and Rowlatt and has shown no

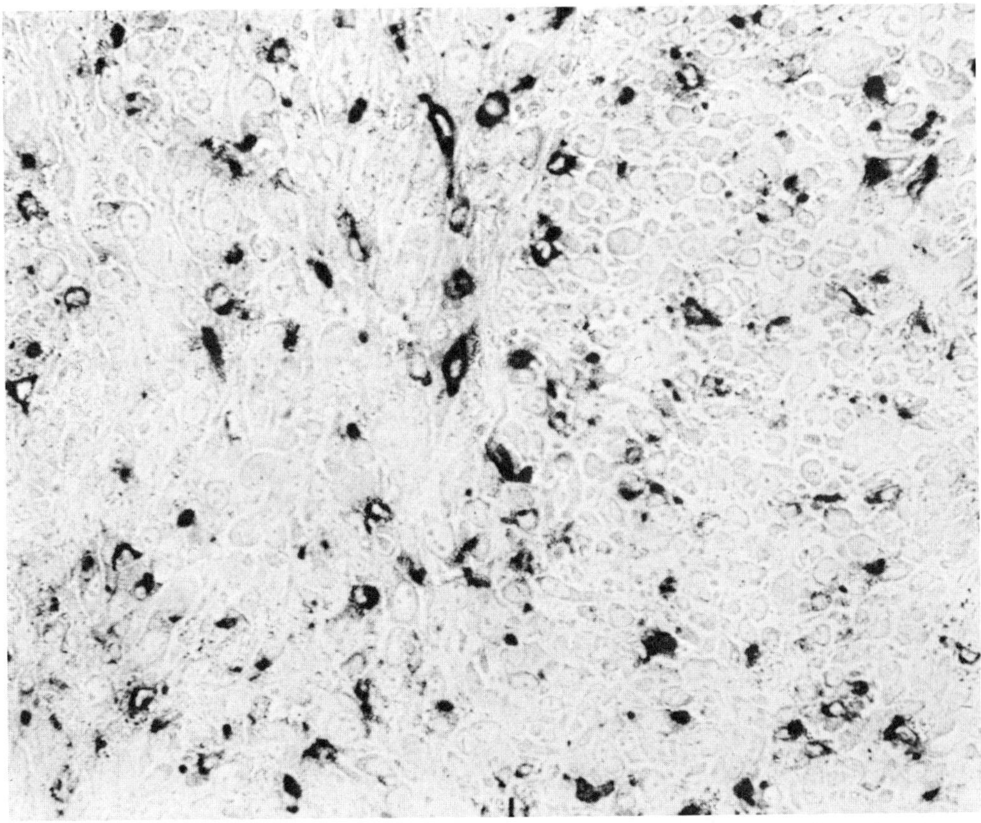

Figure 14. Mouse fibrosarcoma. Tissue fixed in formalin. Section incubated 5 min at 37°C for the demonstration of naphthol AS acid phosphatase using fast red violet LB as coupler. When glycerophosphate was used as substrate, no reaction end product was found even when the incubation was prolonged for 60 min. When their ultrastructure was examined, these cells contained very few lysosomes.

correlation between enzyme behavior and type of tumor, both with respect to its tissue origin and its malignancy. It was also apparent in these studies that no correlation existed between enzyme activity and number of lysosomes, and it was often possible to find tumor cells with considerable acid hydrolytic activity which contained very few lysosomes, and vice versa. In some of these tissues the behavior of lysosomal enzymes such as acid phosphatase was investigated by means of polyacrylamide gel electrophoresis; this revealed considerable discrepancies between the lysosomal β-glycerophosphate phosphatase activity and other acid phosphatases, such as *p*-nitrophenylphosphatase, β-naphthylphosphatase, and naphthol AS phosphatases. For example, the tumor shown in Figure 14 had very high naphthol AS phosphatase activity but very little β-glycerophosphatase activity; in this tumor only a small β-glycerophosphatase band was present electrophoretically, but bands with different localizations were present when the gel was incubated with naphthol AS phosphate as substrate. In other cases nonspecific esterases, thought to be lysosomal enzymes, were very active, while the number of lysosomes in the tumor cells was very low. On the whole, although the question of the relationship between lysosomes and cancer is still debated, and Allison (1969) and Allison and Paton (1969) have suggested a primary role in carcinogenesis for these organelles, it appears to us that the data available at present do not support the hypothesis that lysosomes play an important or, even less, a specific role in the onset of cancer or in the metabolism of cancer cells.

10. METHODOLOGY AND METHODOLOGICAL WARNINGS

This section provides a brief outline and some basic references for those readers who plan to undertake studies involving lysosomes. In particular, it is intended to give some warnings as to the main difficulties experienced both in our laboratory and by other investigators in dealing with some of the techniques concerning these organelles.

A. Biochemistry

The main biochemical criterion for the definition of a tissue preparation as containing lysosomes is that such preparation must possess enzyme activity in a latent form. Latency in turn can be defined as the lapse of time between the reaching of the final equilibrium by the various components of the enzyme reaction mixture (substrate, enzyme, cofactors) and the beginning of enzyme activity; in practical terms any hydrolytic enzyme present in lysosomes does not begin to split its substrate immediately after addition of the latter, but only after a lapse of time. This has been interpreted as being attributable to the presence of an intact lysosomal membrane which would act at first as a barrier between the extralysosomal substrate and the intralysosomal enzyme. This latency must be assessed in conditions of full protection of the lysosomal membrane from any osmotically or otherwise damaging agent. This is usually achieved by using special precautions during the homogenization of the tissue, which is performed routinely in sucrose at different concentrations, since lysosomes from various tissues of various species have different sensitivities to different sucrose molarities. Since prolonged incubation at an acid pH damages the lysosomal membrane (Gianetto and De Duve, 1955; Beaufay and De Duve, 1959), and since the pH optimum of lysosomal enzymes lies within an acid range, the incubation aiming to assess the existence of latency in a subcellular fraction cannot be prolonged for too long. In turn, this means that high levels of activity must exist in the fraction and that it is very difficult to assess the existence of latency in a fraction if the enzyme activity is very low and can only be detected with a prolonged incubation. A typical case is the mammalian heart, in which experiments with lysosomal-mitochondrial pellets suspended in 0.32 *M* sucrose or in distilled water or frozen and thawed six times or treated with a detergent with a view to disrupting the lysosomal membrane did not yield any detectable activity when the pellets were incubated for 10 min with β-glycerophosphate as a substrate at a variety of pH values between 4.0 and 5.8. When incubated for 60 min however, all preparations possessed the same activity which was of a very low level. It was clear in this case that the level of activity in the membrane-damaged preparations was too low to be detected after only 10 min, but that after 60 min the membranes of the "intact" preparations had been damaged to the same extent as those of "damaged" preparations, thus making access of the enzyme to the substrate possible. In this case latency could not be detected. Its existence was proved by a further experiment in which a pellet much richer in lysosomes was obtained by pooling a greater number of hearts.

Both whole homogenates and their subcellular fractions can be assayed for the presence of enzyme latency. The difference between the various types of preparation depends on the level of sophistication of the centrifugation procedures used and thus on the level of lysosome purity achieved. Several techniques have been published for the separation of reasonably pure lysosomal pellets from numerous mammalian and plant tissues (Appelmans et al., 1955; De Duve et al., 1955; Trouet, 1964; Maunsbach, 1966c; Leighton et al., 1968; Romeo et al., 1968; Matile, 1969). Depending on the cell type, on the composition of the tissue, on the fragility of the lysosomes in each tissue (which may vary greatly from tissue to tissue and for the same tissue from species to species), on the heterogeneity of the various lysosome types (which varies according to cell type and to the functional state of the cell at the time of investigation), and on the richness of lysosomes in each tissue, these techniques may be fairly elaborate. As a rule, they involve the use of density gradient centrifugation. For many tissues the separation of a pure lysosomal fraction has not been described. Even in the best of circumstances, that is, when a tissue is fairly homogeneous and its lysosomes osmotically resistant and numerous, a certain degree of damage to the lysosomal membranes, with ensuing leakage of the enzymes into the supernatant, can always be expected and has been reported by all the authors mentioned above. For a generally valid and scientifically sound approach to the problem of subcellular fractionation, the reader is referred to the excellent article by Roodyn (1965).

The study of the various lysosomal enzymes has been the object of long, detailed, and careful studies [Barrett (1967) for cathepsins; Fishman (1967) for β-glucuronidase; Roy (1960) for aryl sulfatases; for a review, see Tappel (1969a)]. In this context methodological warnings are particularly important. When dealing with enzymes possessing a nondefined substrate specificity, such as cathepsins, aryl sulfatases, or acid phosphatases, the choice of the substrate must be made carefully. Although it has been claimed (Tappel, 1969b) that *p*-nitrophenylphosphate can be used as a substrate for assaying lysosomal acid phosphatase, other investigators have demonstrated that this compound is split preferentially by nonlysosomal acid phosphatases with kinetic properties different from the lysosomal enzyme (Neil and Horner, 1964; Maggi, 1969). Lysosomes contain several acid phosphatases (Barrett, personal communication), of which one is only partially inhibited by sodium fluoride (Maggi et al., 1970) and splits naphthol AS and *p*-nitrophenylphosphates. This latter enzyme does not exhibit latency (Maggi, in preparation). The classic lysosomal acid phosphatase splits β-glycerophosphate, and only this substrate must be used to study lysosomes and compare data with those accepted in the literature.

This type of study involves the determination of inorganic phosphorus liberated by the enzyme. Several methods are available for this purpose; however, care must be taken in the choice of the reducing agent, since it has been found in our laboratory that, for example, ascorbic acid as used in the method of Chen et al. (1956) interferes both with sucrose (usually present in the incubation medium when latency is investigated) and with sodium fluoride, to inhibit the development of the blue color of reduced phosphate–molybdate complex, thus yielding false negative values.

B. Histo- and Cytochemistry

The conditions of choice for the study of lysosomes and of their enzymes in tissue sections have been described by Gahan (1967), and there is very little that can be added here. For an organelle in a tissue section to be defined as a lysosome with certainty, it is indispensable that this organelle contain at least one acid hydrolase. That any one cytochemically demonstrable acid hydrolase is present in a lysosome is quasi-impossible to determine when using optical microscopy techniques alone. These methods at best yield a series of granules the morphological nature of which it is difficult to assess except in very few cases in which the organelles are large enough to be resolved by means of phase-contrast microscopy, for example, Chinese hamster fibroblasts in culture (Munro et al., 1964), HeLa cells (Robbins and Gonatas, 1964b), and rat kidney (Maunsbach, 1966d). In other cases the presence of particles in tissue cells after they have been incubated for the demonstration of an acid hydrolase is no more evidence of the lysosomal nature of the particles than the presence of a diffuse stain is evidence for the presence of nonlysosomal enzyme.

For an organelle to be defined as a lysosome, it must be seen with an electron microscope as having a membrane—usually single—and as containing end product of an hydrolytic enzyme. The content of these organelles can vary, as can be seen in Figures 2, 3, and 8–11, but it does not define a lysosome. For example, there is no morphological way to distinguish between a kidney lysosome and a kidney peroxisome, particularly since the latter very seldom contains the otherwise typical crystals (Fig. 1*g*). To achieve this, it is necessary to ascertain which of these otherwise very similar organelles contains peroxidase and which contains acid hydrolases. This is usually a much simpler way than the very precise measurements of comparative membrane thicknesses as carried out in highly specialized laboratories by very skilled workers (Maunsbach, 1966d).

Having made these provisos, light microscopy histochemical techniques are certainly very useful as an indication of the presence, the localization, and the activity of lysosomes and their enzymes in a tissue. A few precautions must be taken here too. The choice between fixed and nonfixed tissues has been the subject of argument among cytochemists. While aldehyde fixation undoubtedly offers a much better preservation of tissue structures and thus allows a much more precise localization of sites of lysosomal activity, some hydrolases are selectively and completely inhibited by certain fixatives such as formaldehyde (Maggi et al., 1966), and others are partially inhibited (Holt, 1959) so that when only a low amount of enzyme is present the use of fixatives may make the enzyme undetectable. In unfixed cryostat sections there may occur a loss of enzyme through diffusion, although in our laboratory this has happened only when freezing of the unfixed tissue was performed incorrectly (Maggi and Riddle, 1965). The use of fixed tissues becomes imperative when ultrastructural studies are undertaken. The choice of fixatives here is very limited, glutaraldehyde usually giving a better preservation than formaldehyde, although the latter yields excellent results when employed according to Holt's formula (1958). The final choice depends on the possible different sensitivity of the enzyme under investigation to the various aldehydes (this should be checked at optical microscopy level first) and on the importance of good tissue preservation.

The choice of substrates is limited by the necessity to produce an insoluble and visible end product of the enzyme reaction under investigation exactly at the site where the enzyme is present in the cell. A further and very important factor when naphthol AS monoesters are used as substrates (Burstone, 1958; Lojda et al., 1964) is the choice of method of coupling—simultaneous or postcoupling—and the choice of the diazotate. It is essential that the rate of coupling of the naphthol AS molecule with the azo dye be very rapid, lest the naphthol AS molecule move away from its site of precipitation before it has coupled. This is particularly important if azo dyes such as lead phthalocyanin (Tice and Barnett, 1965; Livingston et al., 1969, 1970), hexazonium pararosaniline (Lojda, 1958; Davis and Ornstein, 1959; Barka, 1960; Benes et al., 1961) or indigo dyes (Holt and Hicks, 1966) are used to visualize sites of end product deposition at the electron microscope level. The conditions of coupling, such as pH optimum for each type of diazotate and naphthol, should therefore be tested and possible signs of diffusion of the end product within the tissues or cells, as well as outside the sections, should be investigated. Another possibility often overlooked is that the end product may be soluble in the dehydration and embedding media used in electron microscopy; thus it was found (Livingston et al., 1969) that the lead phthalocyanin diazotate–naphthol AS/BI complex produced in the incubated sections was partially soluble in ethanol even after osmication, although it was insoluble when tested as a pure compound in the test tube and that the hexazonium pararosaniline–naphthol AS/BI complex was also soluble under the same conditions. While the lead phthalocyanin complex was not soluble in water-soluble Durcupan or Araldite, the hexazonium pararosaniline complex was.

When β-naphthylamide is the final product of an enzyme reaction, its complexes with some diazotates such as fast blue B or fast garnet GBC are soluble in water-soluble or in ethanol-soluble mounting media such as polyvinylpyrrrolidone (Kaplow and Burstone, 1963) or Euparal, although not in ethanol or in water. In this instance the sections must be dehydrated to xylene and mounted in a xylene-soluble medium (XAM, Canada balsam), in order to avoid solubilization. The complexes of the same azo dyes with naphthol AS molecules are completely soluble in xylene but are insoluble in water-soluble or in ethanol-soluble media, or in ethanol or water. They can therefore either be mounted in polyvinylpyrrolidone, Farrants's medium, glycerol, (in which case fading of the stain can be expected within a few weeks) or dehydrated to absolute ethanol, with permanence of the stain for at least 1 year.

ACKNOWLEDGMENTS

I am greatly indebted to my husband, P. B. Gahan, for patiently discussing this article with me. I wish to express my gratitude to Miss Maura F. Oddy and Mr. G. C. Grenoak without whose help the experiments described here could not have been performed; Mr. G. E. Leach and his staff for their assistance in the processing of the photographs, all of which except one are part of the author's collection; and last but not least, Miss Anne Houlton, who was so patient in her typing of the manuscript.

REFERENCES

Al-Azzawi, H. T., and P. J. Stoward (1970). The effects of various forms of euthanasia on the histochemically demonstrable activities of lysosomal enzymes. *Roy. Microscop. Soc. Proc.*, **5,** 13.

Allison, A. C. (1969). In *Lysosomes in Biology and Pathology*. (J. T. Dingle and H. B. Fell, eds.), Vol. 2, pp. 178–206. North Holland Publ. Comp., Amsterdam.

Allison, A. C., and L. Mallucci (1964a). Lysosomes in dividing cells, with special reference to lymphocytes. *Lancet*, **2,** 1371.

Allison, A. C., and L. Mallucci (1964b). Uptake of hydrocarbon carcinogens by lysosomes. *Nature*, **203,** 1024.

Allison, A. C., and L. Mallucci (1965). Histochemical studies of lysosomes and lysosomal enzymes in virus-infected cell cultures. *J. Exptl. Med.*, **121,** 463.

Allison, A. C., and G. R. Paton (1969). Lysosomes, chromosomes and cancer. *Biochem. J.*, **115,** 31P.

Allison, A. C., J. S. Harington, and M. Birbeck (1966). An examination of the cytotoxic effects of silica on macrophages. *J. Exptl. Med.*, **124,** 141.

Appelmans, F., and C. De Duve (1955). Tissue fractionation studies. 3. Further observations on the binding of acid phosphatase by rat liver particles. *Biochem. J.*, **59,** 426.

Appelmans, F., R. Wattiaux, and C. De Duve (1955). Tissue fractionation studies. 5. The association of acid phosphatase with a special class of cytoplasmic granules in rat liver. *Biochem. J.*, **59,** 438.

Appleton, T. C. (1969). In *Autoradiography of Diffusible Substances* (L. J. Roth and W. E. Stumpf, eds.), pp. 304–319. Academic Press, New York.

Baggiolini, J., J. G. Hirsch and C. De Duve (1969). Resolution of granules from rabbit heterophil leukocytes into distinct populations by zonal sedimentation. *J. Cell Biol.*, **40,** 529.

Bangham, A. D., M. M. Standish, and G. Weissman (1965). The action of steroids and streptolysin S on the permeability of phospholipid structures to cations. *J. Mol. Biol.*, **13,** 253.

Barka, T. (1960). A simple azo-dye method for histochemical demonstration of acid phosphatase. *Nature*, **187,** 248.

Barrett, A. J. (1967). Lysosomal acid proteinase of rabbit liver. *Biochem. J.*, **104,** 601.

Barrett, A. J., and J. T. Dingle (1967). A lysosomal component capable of binding cations and a carcinogen. *Biochem. J.*, **105,** 20P.

Barrett, A. J., and A. R. Poole (1969). Unsuitability of leucine naphthylamide for the histochemical demonstration of lysosomal proteolytic activity. *Nature*, **224,** 279.

Baudhuin, P., H. G. Hers, and H. Loeb (1964). An electron microscopic and biochemical study of type II glycogenosis. *Lab. Invest.*, **13,** 1139.

Beaufay, H., and C. De Duve (1959). Tissue fractionation studies. 9. Enzymic release of bound hydrolases. *Biochem. J.*, **73,** 604.

Beck, F., and J. B. Lloyd (1968). Methods for the study of embryotrophic nutrition. *Lab. Animals*, **2,** 157.

Benes, K., Z. Lojda, and B. Horavka (1961). A contribution to the histochemical demonstration of some hydrolytic and oxidative enzymes in plants. *Histochemie*, **2,** 313.

Bernhard, W., and E. H. Leduc (1967). Ultrathin frozen sections. I. Methods and ultrastructural preservation. *J. Cell Biol.*, **34,** 757.

Bernhardt, W., and A. Viron (1971). Improved techniques for the preparation of ultrathin frozen sections. *J. Cell Biol.*, **49,** 731.

Bitensky, L. (1962). The demonstration of lysosomes by the controlled-temperature freezing method. *Quart. J. Microscop. Sci.*, **103,** 205.

Bitensky, L., and P. B. Gahan (1962). Reversible activation of lysosomes in normal cells. *Biochem. J.*, **84,** 13P.

Brachet, J. (1954). Effects of ribonuclease on the metabolism of living root-tip cells. *Nature*, **174,** 876.

Brachet, J. (1955). Action of ribonuclease and ribonucleic acid on living amoebae. *Nature*, **175,** 851.

Brandes, D. (1965). Observations on the apparent mode of formation of "pure" lysosomes. *J. Ultrastruct. Res.*, **12,** 63.

Brandes, D., and F. Bertini (1964). Role of Golgi apparatus in the formation of cytolysomes.*Exptl. Cell Res.*, **35,** 194.

Brandes, D., F. Györkey, and D. P. Groth (1962). Fine structural and histochemical study of the effect of castration on the rat prostatic complex. I. The coagulating gland. *Lab. Invest.*, **11,** 339.

Burke, J. S., T. Uriuhara, D. R. L. MacMorine, and H. Z. Movat (1964). A permeability factor released from phagocytosing pmn-leukocytes and its inhibition by protease inhibitors. *Life Sci.*, **3,** 1505.

Burstone, M. S. (1958). Histochemical demonstration of acid phosphatases with naphthol AS phosphates. *J. Natl. Cancer Inst.*, **21,** 523.

Burstone, M. S. (1962). In *Enzyme Histochemistry and Its Application in the Study of Neoplasms*, pp. 411–415, (M. S. Burstone, ed.), Academic Press, New York.

Chapman, Andresen C. (1962). Studies on pinocytosis in Amoebae. *Compt. Rend. Trav. Lab. Carlsberg*, **33,** 73.

Chen, P. S., T. Y. Toribara, and H. Warner (1956). Microdetermination of phosphorus. *Anal. Chem.*, **28,** 1756.

Clark, S. L. (1957). Cellular differentiation in the kidneys of new born mice studied with the electron microscope. *J. Biophys. Biochem. Cytol.*, **3,** 349.

Cohn, Z. A., and B. Benson (1965). The differentiation of mononuclear phagocytes. Morphology, cytochemistry and biochemistry. *J. Exptl. Med.*, **121,** 153.

Cohn, Z. A., and B. Benson (1965b). The *in vitro* differentiation of mononuclear phagocytes. I. The influence of inhibitors and the results of autoradiography. *J. Exptl. Med.*, **121,** 279.

Cohn, Z. A., and B. Benson (1965c). The *in vitro* differentiation of mononuclear phagocytes. II. The influence of serum on granule formation, hydrolase production and pinocytosis. *J. Exptl. Med.*, **121,** 835.

Cohn, Z. A., and B. Benson (1965d). The *in vitro* differentiation of mononuclear phagocytes. III. The reversibility of granule and hydrolytic enzyme formation and the turnover of granule constituents. *J. Exptl. Med.*, **122,** 455.

Cohn, Z. A., and J. G. Hirsch (1960). Isolation and properties of specific cytoplasmic granules of rabbit polymorphonuclear leucocytes. *J. Exptl. Med.*, **112,** 983.

Coulomb, C., and R. Buvat (1968). Processus de dégénerescence cytoplasmique partielle dans les cellules de jeunes racines de Cucurbita pepo. *Compt. Rend, Séances Acad., Sci. Paris*, **267,** 843.

Critchfield, C., and J. C. Daniel (1965). Teratogenic effects of trypan blue on the Coturnix quail when injected into the mother. *Growth*, **29,** 301.

Daems, W. T., and T. G. van Rijssel (1961). The fine structure of the peribiliary dense bodies in mouse liver tissue. *J. Ultrastruct. Res.*, **5,** 263.

Daems, W. Th., E. Wisse, and P. Brederoo (1969). In *Lysosomes in Biology and Pathology* (J. T. Dingle and H. B. Fell, eds.), Vol. 1, pp. 64–112. North Holland Publ. Comp., Amsterdam.

Dales, S. (1969). In *Lysosomes in Biology and Pathology* (J. T. Dingle and H. B. Fell, eds.), pp. 69–86· North Holland Publ. Comp., Amsterdam.

Dauwalder, M., W. G. Whaley, and J. E. Kephard (1969). Phosphatases and differentiation in Golgi apparatus. *J. Cell Sci.*, **4,** 455.

Davies, M., J. B. Lloyd, and F. Beck (1969). Protein digestion in isolated lysosomes inhibited by intralysosomal trypan blue. *Science*, **163,** 1454.

Davis, B. J., and L. Ornstein (1959). High resolution with a new diazo reagent "hexazonium pararosaniline." *J. Histochem. Cytochem.*, **7,** 297.

De Duve, C. (1963). In *Lysosomes* (A. V. S. de Reuck and M. P. Cameron, eds.), pp. 1–35. Churchill, London.

De Duve, C., and R. Wattiaux (1966). Function of lysosomes. *Ann. Rev. Physiol.*, **28,** 435.

De Duve, C., B. C. Pressman, R. Gianetto, R. Wattiaux, and F. Appelmans (1955). Tissue fractionation studies. 6. Intracellular distribution patterns of enzymes in rat-liver tissue. *Biochem. J.*, **60,** 604.

De Duve, C., R. Wattiaux, and M. Wibo (1961). Effects of fat-soluble compounds on lysosomes *in vitro*. *Biochem. Pharmacol.*, **9,** 97.

Dianzani, M. U. (1963). In Ciba Found. Symp. on *Lysosomes* (A. V. S. de Reuck and M. P. Cameron, eds.), pp. 335–352. Churchill, London.

Dingle, J. T. (1961). Studies on mode of action of excess of vitamin A. 3. Release of bound protease by action of vitamin A. *Biochem. J.*, **79,** 509.

Dingle, J. T. (1963). In *Lysosomes* (A. V. S. de Reuck and M. P. Cameron, eds.), pp. 384–398. Churchill, London.

Dingle, J. T. (1966). In *Modern Trends in Rheumatology* (A. G. S. Hill, ed.), pp. 110–120. Butterworths, London.

Dingle, J. T., A. J. Barrett, and P. D. Weston (1971). Cathepsin D. Characteristics of immunoinhibition and the confirmation of a role in cartilage breakdown. *Biochem. J.*, **123,** 1.

Dingle, J. T., H. B. Fell, and R. R. A. Coombs (1967). The breakdown of embryonic (chick) cartilage and bone cultivated in the presence of complement-sufficient antiserum. 2. Biochemical changes and the role of the lysosomal system. *Intern. Arch. Allergy*, **31,** 283.

Dingle, J. T., H. B. Fell, and A. M. Glauert (1969). Endocytosis of sugars in embryonic skeletal tissues in organ culture. IV. Lysosomal and other biochemical effects. General discussion. *J. Cell Sci.*, **4,** 139.

Emmelot, P., A. Visser, and E. L. Benedetti (1968). Studies on plasma membranes. VII. A leucyl-β-naphthylamidase-containing repeating unit on the surface of isolated liver and hepatoma plasma membranes. *Biochim. Biophys. Acta*, **150,** 364.

Farquhar, M. G. (1969). In *Lysosomes in Biology and Pathology* (J. T. Dingle and H. B. Fell, eds.), pp. 462–482. North Holland Publ. Comp. Amsterdam.

Fell, H. B., and G. T. Dingle (1963). Studies on the mode of action of excess of vitamin A. Lysosomal protease and the degradation of cartilage matrix. *Biochem. J.*, **87,** 403.

Fell, H. B., and L. Thomas (1960). Comparison of effects of papain and vitamin A on cartilage. II. Effects on organ cultures of embryonic skeletal tissue. *J. Exptl. Med.*, **111,** 719.

Fishman, W. H. (1967). Determination of β-glucuronidases. *Methods Biochem. Anal.*, **15,** 77.

Fishman, W. H., S. S. Goldman, and R. De Lellis (1967). Dual localization of β-glucuronidase in endoplasmic reticulum and in lysosomes. *Nature*, **213,** 457.

Fishman, W. H., H. Ide, and R. Rufo (1970). Dual localization of acid hydrolases in endoplasmic reticulum and in lysosomes. I. β-Glucuronidase staining reactions and cytochemical studies on kidney in androgen-stimulated mice. *Histochemie*, **20,** 287.

Gabrielescu, E. (1970). The lability of lysosomes during the response of neurons to stress. *Histochem. J.*, **2,** 123.

Gahan, P. B. (1965). Reversible activation of lysosomes in rat liver. *J. Histochem. Cytochem.*, **13,** 334.

Gahan, P. B. (1967). Histochemistry of lysosomes. *Intern. Rev. Cytol.*, **21,** 1.

Gahan, P. B. (1968). In *Plant Cell Organelles* (J. Pridham, ed.), pp. 228–238. Academic Press, New York.

Gahan, P. B. (1972). In *Lysosomes in Biology and Pathology*, North Holland Publ. Comp., Amsterdam, Vol. III in the press.

Gahan, P. B., G. C. Greenoak and D. James (1970). Preparation of ultrathin frozen sections of plant tissues for electron microscopy. *Histochemie*, **24,** 230.

Gahan, P. B., and J. McLean (1967). Acid phosphatases in roots of *Vicia faba*. *Biochem. J.*, **102,** 47.

Gahan, P. B., and J. McLean (1968). Hydrolase activity in dividing and differentiating plant cells. *Abstr. Intern. Congr. Histochem. Cytochem.* Springer, Verlag, New York, p. 67.

Gahan, P. B., and J. McLean (1969). Subcellular localization and possible functions of acid β-glycerophosphatases and naphthol esterases in plant cells. *Planta*, **89,** 126.

Gahan, P. B., V. Maggi, and H. Boro (1971). Hydrolase activity in mitotic cells. *Proc. Can. Fed. Biol. Soc. 14th Ann. Meeting*, Toronto, n° 95.

Gahan, P. B., and A. J. Maple (1965). The behaviour of lysosome-like particles during cell differentiation. *J. Exptl. Botany*, **17,** 151.

Gianetto, R., and C. De Duve (1955). Tissue fractionation studies. 4. Comparative study of the binding of acid phosphatase, β-glucuronidase and cathepsin by rat-liver particles. *Biochem. J.*, **59,** 433.

Gillman, J., C. Gilbert, T. Gillman, and I. Spence (1948). A preliminary report on hydrocephalus, spinabifida and other congenital anomalies in the rat produced by trypan blue. *S. African J. Med. Sci.*, **13,** 47.

Gladstone, G. P., and E. van Heyningen (1957). Staphylococcal leucocidins. *Brit. J. Exptl. Pathol.*, **38,** 123.

Glauert, A. M., H. B. Fell, and J. T. Dingle (1969). Endocytosis of sugars in embryonic skeletal tissues in organ cultures. II. Effect of sucrose on cellular fine structure. *J. Cell Sci.*, **4,** 105.

Gomori, G. (1950). An improved histochemical technique for acid phosphatase. *Stain Technol.*, **25,** 81.

Graf, J., D. Kerjaschki, and H. Hörandner (1971). Simultaneous demonstration of exogenous horseradish peroxidase and acid phosphatase activities in phagolysosomes. *J. Histochem. Cytochem.*, **19,** 569–572.

Granda, J. L., and A. S. Posner (1969). The physical, chemical and biological aspects of thin films. III. Lysosomal membrane of rabbit polymorphonuclear leukocytes: Chemical composition and partial characterization of its protein moiety. *Trans. N.Y. Acad. Sci.*, **31,** 879.

Greenbaum, A. L., T. F. Slater, and D. Y. Wang (1965). Lysosomal enzyme changes in enforced mammary-gland involution. *Biochem. J.*, **97,** 518.

Helminen, H. J., and J. L. E. Ericsson (1968a). Studies on mammary gland involution. I. On the ultrastructure of the lactating mammary gland. *J. Ultrastruct. Res.*, **25,** 193.

Helminen, H. J., and J. L. E. Ericsson (1968b). Studies on mammary gland involution. II. Ultrastructural evidence for auto- and heterophagocytosis. *J. Ultrastruct. Res.*, **25,** 214.

Helminen, H. J., and J. L. E. Ericsson (1968c). Studies on mammary gland involution. III. Alterations outside auto- and heterophagocytic pathways for cytoplasmic degradation. *J. Ultrastruct. Res.*, **25,** 228.

Helminen, H. J., J. L. E. Ericsson, and S. Orrenius (1968). Studies on mammary gland involution. IV. Histochemical and biochemical observations on alterations in lysosomes and lysosomal enzymes. *J. Ultrastruct. Res.*, **25,** 240.

Hers, H. G. (1965). Inborn lysosomal diseases. *Gastroenterology*, **48,** 625.

Hers, H. G., and F. van Hoof (1969). Lysosomes and mucopolysaccharidoses. *Biochem. J.*, **115,** 34P.

Hirsch, J. G., and Z. A. Cohn (1964). Digestive and autolytic functions of lysosomes in phagocytic cells. *Federation Proc.*, **23,** 1023.

Hirsch, J. G., A. W. Bernheimer, and G. Weissmann (1963). Motion picture study of the toxic action of streptolysins on leukocytes. *J. Exptl. Med.*, **118,** 223.

Holt, S. J. (1958). In *General Cytochemical Techniques* (J. F. Danielli, ed.), Vol. 1, pp. 375–399. Academic Press, New York.

Holt, S. J. (1959). Factors governing the validity of staining methods for enzymes, and their bearing upon the Gomori acid phosphatase technique. *Exptl. Cell Res., Suppl.*, **27,** 1.

Holt, S. J., and R. M. Hicks (1961). The localization of acid phosphatase in rat liver cells as revealed by combined cytochemical staining and electron microscopy. *J. Biophys. Biochem. Cytol.*, **11,** 47.

Holt, S. J., and R. M. Hicks (1966). The importance of osmiophilia in the production of stable azoindoxyl complexes of high contrast for combined enzyme cytochemistry and electron microscopy. *J. Cell Biol.*, **29,** 361.

Holtzman, E., A. B. Novikoff, and H. Villaverde (1967). Lysosomes and GERL in normal and chromatolytic neurons of the rat ganglion nodosum. *J. Cell Biol.*, **33,** 419.

Ide, H., and W. H. Fishman (1970). Dual localization of β-glucuronidase and acid phosphatase in lysosomes and microsomes. II. Membrane-associated enzymes. *Histochemie*, **20,** 300.

Janoff, A., and B. W. Zweifach (1964). Production of inflammatory changes in the microcirculation by cationic proteins extracted from lysosomes. *J. Exptl. Med.*, **120,** 747.

Jensen, W. A. (1956). The cytochemical localization of acid phosphatase in root tip cells. *Am. J. Bot.*, **43,** 50.

Kaplow, L. S., and M. S. Burstone (1963). Acid-buffered acetone as a fixative for enzyme cytochemistry. *Nature*, **200,** 690.

Katchburian, E., and S. J. Holt (1969). Role of lysosomes in amelogenesis. *Nature*, **223,** 1367.

Lake, B. D., and A. D. Patrick (1970). Wolman's disease: Deficiency of E-600-resistant acid esterase activity with storage of lipids in lysosomes. *J. Pediatrics*, **76,** 262.

Larsson, L., A. B. Maunsbach, L. Saxen, and J. Wartiovaara (1969). Lysosomes in developing kidney tubule cells *in vitro*. *J. Ultrastruct. Res.*, **29,** 570.

Leduc, E. H., W. Bernhard, S. J. Holt, and J. P. Tranzer (1967). Ultrathin frozen sections. II. Demonstration of enzymic activity. *J. Cell Biol.*, **34,** 773.

Leighton, F., B. Poole, H. Beaufay, P. Baudhuin, J. W. Coffey, S. Fowler, and C. De Duve (1968). The large-scale separation of peroxisomes, mitochondria and lysosomes from the livers of rats injected with Triton WR 1339. *J. Cell Biol.*, **37,** 482.

Livingston, D. C., M. M. Coombs, L. M. Franks, V. Maggi, and P. B. Gahan (1969). A lead phthalocyanin method for the demonstration of acid hydrolases in plant and animal tissues. *Histochemie*, **18,** 48.

Livingston, D. C., S. W. Fisher, G. E. Greenoak, and V. Maggi (1970). A new procedure for the ultrastructural localization of acid phosphatases. *Histochemie*, **24,** 159.

Lloyd, J. B., F. Beck, A. Griffiths, and L. M. Parry (1968). In *Interaction of Drugs and Subcellular Components in Animal Cells* (P. N. Campbell, ed.), pp. 171–200. Churchill, London.

Lojda, Z. (1958). The azo-coupling reactions in the histochemical detection of enzymes. *Stat. Zdrav. Nakl.* 1958. Cited in Benes et al. (1961).

Lojda, Z., B. Vecerek, and H. Pelichova (1964). Some remarks concerning the histochemical detection of acid phosphatase by azo-coupling reactions. *Histochemie*, **3,** 428.

Loni, M. C., M. Borgers, and J. Hugon (1967). Uptake of ferritin particles by ATP-stimulated HeLa cells. *Z. Zellforsch. Mikroscop. Anat.*, **76,** 525.

McKusick, V. A. (1966). In *Heritable Disorders of Connective Tissue*, pp. 325–399. The C. V. Mosby Company, St. Louis.

McLeish, J. (1953). The action of maleic hydrazide in *Vicia*. *Heredity, Suppl.*, **6,** 125.

Maeir, D. M., and H. Zaiman (1966). The development of lysosomes in rat skeletal muscle in trichinous myositis. *J. Histochem. Cytochem.*, **14,** 396.

Maggi, V. (1965a). Biochemical evidence for the presence of two acid phosphatases in the coagulating gland of the mouse. *Biochem. J.*, **96,** 12P.

Maggi, V. (1965b). A study of lysosomal acid phosphatase during mitosis in HeLa cells. *J. Roy. Microscop. Soc.*, **85,** 291.

Maggi, V. (1968). Lysosomal and non-lysosomal localization of acid hydrolases in animal cells. *Biochem. J.*, **111,** 25P.

Maggi, V. (1969). Acid phosphatases in rat heart. *Biochem. J.*, **115,** 51P.

Maggi, V. (1971). Distribution of acid phosphatases in the myocardium of hamsters with congenital heart failure. *Proc. Can. Fed. Biol. Soc. 14th Ann. Meeting*, Toronto, n° 321.

Maggi, V., and A. W. Carbonell (1969). Lysosomes and acid phosphatases during growth and differentiation in mice; A light and electron microscope study. *Histochem. J.*, **1,** 383.

Maggi, V., and M. F. Oddy (1968). Effect of short-term starvation on the cells of the proximal convoluted tubule of the mouse: A cytological and cytochemical study. *Histochem. J.*, **1,** 78.

Maggi, V., and P. N. Riddle (1965). Histochemistry of tissue culture cells: A study of the effects of freezing and of some fixatives. *J. Histochem. Cytochem.*, **13,** 310.

Maggi, V., L. M. Franks, and A. W. Carbonell (1966). Histochemical study of a number of acid hydrolases, including a new acid phosphatase, in various tissues of men and mice. *Histochemie*, **6,** 305.

Maggi, V., L. M. Franks, P. D. Wilson, and A. W. Carbonell (1969). Localization of insulin in mouse tissues using fluorescence microscopy and light microscope and high resolution autoradiography. *Diabetologia*, **5,** 67.

Maggi, V., A. W. Steggles, and P. B. Gahan (1970). Effects of castration and of oestradiol treatment on the sub-cellular structure of the mouse seminal vesicles with particular regard to the function of the Golgi apparatus and the lysosomes. *Histochem. J.*, **2,** 381.

Mallucci, L., and A. C. Allison (1965). Lysosomal enzymes in cells infected with cytopathic and non-cytopathic viruses. *J. Exptl. Med.*, **121,** 477.

Mapes, C. A., R. L. Anderson, and C. C. Sweeley (1970). Galactosyl galactosyl glucosyl ceramide: Galactosyl hydrolase in normal human plasma and its absence in patients with Fabry's disease. *FEBS Letters*, **7,** 180.

Matile, Ph., J. P. Balz, E. Semadeni, and M. Gost (1965). Isolation of spherosomes with lysosome characteristics from seedlings. *Z. Naturforsch.*, **20b,** 693.

Matile, Ph., and H. Moore (1968). Vacuolation in origin and development of the lysosomal apparatus in root-tip cells. *Planta*, **80,** 159.

Maunsbach, A. B. (1966a). Absorption of I^{125}-labelled homologous albumin by rat kidney proximal convoluted tubule cells. A study of microperfused single proximal tubules by electron microscopic autoradiography and histochemistry. *J. Ultrastruct. Res.*, **15,** 197.

Maunsbach, A. B. (1966b). Absorption of ferritin by rat kidney proximal tubule cells. *J. Ultrastruct. Res.*, **16,** 1.

Maunsbach, A. B. (1966c). Isolation and purification of acid phosphatase-containing autofluorescent granules from homogenates of rat kidney cortex. *J. Ultrastruct. Res.*, **16,** 13.

Maunsbach, A. B. (1966d). Observations on the ultrastructure and acid phosphatase activity of the cytoplasmic bodies in rat kidney proximal tubule cells. *J. Ultrastruct. Res.*, **16,** 197.

Metchnikoff, E. (1892). *Leçon sur la Pathologie comparée de L'Inflammation*. Paris, Masson et Cie.

Miller, F., and G. Palade (1964). Lytic activities in renal protein absorption droplets: Electron microscopical cytochemical study. *J. Cell Biol.*, **23,** 519.

Moore, B. W., and P. V. Angeletti (1961). Chromatographic heterogeneity of some enzymes in normal tissues and tumors. *Ann. N.Y. Acad. Sci.*, **94,** 659.

Morrison, A. B., and B. J. Panner (1964). Lysosome induction in experimental potassium deficiency. *Am. J. Pathol.*, **45,** 295.

Morrison, A. B., B. Y. Panner, and G. Gasic (1963). Lysosomes in renal papillae of rats: Formation induced by potassium-deficient diet. *Science*, **142,** 1066.

Munro, T. R., M. R. Daniel, and J. T. Dingle (1964). Lysosomes in Chinese hamster fibroblasts in culture. *Exptl. Cell Res.*, **35,** 515.

Neil, M. W., and M. W. Horner (1964a). Studies on acid hydrolases in adult and foetal tissues. Acid *p*-nitrophenyl phosphate phosphohydrolases of adult guinea pig liver. *Biochem. J.*, **92,** 217.

Neil, M. W., and M. W. Horner (1964b). Studies on acid hydrolases in adult and foetal tissues. 2. Acid phenyl-phosphomonoesterases of adult mouse liver. *Biochem. J.*, **93,** 220.

Nelson, B. D. (1966). Rat liver acid phosphatase: Differences in lysosomal and cytoplasmic forms. *Proc. Soc. Exptl. Biol. Med.*, **121,** 998.

Novikoff, A. B. (1961). In *The Cell* (J. Brachet and A. E. Mirsky, eds.), Vol. II, pp. 424–488. Academic Press, New York.

Novikoff, A. B. (1963). In *Lysosomes* (A. V. S. de Reuck and M. P. Cameron, eds.), p. 39. Churchill, London.

Novikoff, A. B., and E. Essner (1962). Cytolysomes and mitochondrial degeneration. *J. Cell Biol.*, **15,** 140.

Novikoff, A. B., and W. Y. Shin (1964). The endoplasmic reticulum in the Golgi zone and its relation to microbodies, Golgi apparatus and autophagic vacuoles in rat liver cells. *J. Microscopie*, **3,** 187.

Novikoff, A. B., E. Essner, and N. Quintana (1964). Golgi apparatus and lysosomes. *Federation Proc.*, **23,** 1010.

Poux, N. (1970). Localisation d'activités enzymatiques dans le meristème radiculaire de Cucumis sativus L. III. Activité phosphatasique acide. *J. Microscopie*, **9**, 407.

Rajan, K. T. (1966). Lysosomes and gout. *Nature*, **210**, 959.

Robbins, E., and N. K. Gonatas (1964a). The ultrastructure of a mammalian cell during the mitotic cycle. *J. Cell Biol.*, **21**, 429.

Robbins, E., and N. K. Gonatas (1964b). Histochemical and ultrastructural studies on HeLa cell cultures exposed to spindle inhibitors with special reference to the interphase cell. *J. Histochem. Cytochem.*, **12**, 704.

Roels, O. A. (1969). In *Lysosomes in Biology and Pathology* (J. T. Dingle and H. B. Fell, eds.), pp. 254–275. North Holland Publ. Comp., Amsterdam.

Rhodin, J. A. (1971). The ultrastructure of the adrenal cortex of the rat under normal and experimental conditions. *J. Ultrastruct. Res.*, **34**, 23.

Roodyn, D. B. (1965). The classification and partial tabulation of enzymes studies on subcellular fractions isolated by differential centrifuging. *Intern. Rev. Cytol.*, **18**, 99.

Rosenbaum, R. M., and C. I. Rolon (1962). Species variability and the substrate specificity of intracellular acid phosphatases: A comparison of the lead-salt and azo-dye methods. *Histochemie*, **3**, 1.

Roy, A. B. (1960). The synthesis and hydrolysis of sulfate esters. *Advan. Enzymol.*, **22**, 205.

Scheib, D. (1963). In *Lysosomes* (A. V. S. de Reuck and M. P. Cameron, eds.), pp. 264–277. Churchill, London.

Seligman, A. M., H. L. Wasserkrug, C. Deb, and J. Hanker (1968). Osmium-containing compounds with multiple basic or acidic groups as stains for ultrastructure. *J. Histochem. Cytochem.*, **16**, 87.

Seljelid, R. (1966). An electron microscopic study of the formation of cytosomes in a rat kidney adenoma. *J. Ultrastruct. Res.*, **16**, 569.

Sexton, R., and J. F. Sutcliffe (1969). The distribution of β-glycerophosphatase in young roots of Pisum sativum. *Ann. Botany (London)*, **33**, 407.

Sheldon, H., H. Zetterquist, and D. Brandes (1955). Histochemical reactions for electron microscopy: Acid phosphatase. *Exptl. Cell Res.*, **9**, 592.

Slater, T. F., and A. L. Greenbaum (1965). Changes in lysosomal enzymes in acute experimental liver injury. *Biochem. J.*, **96**, 484.

Slater, T. F., A. L. Greenbaum, and D. Y. Wang (1963). In *Lysosomes* (A. V. S. de Reuck and M. P. Cameron, eds.), pp. 311–334. Churchill, London.

Smith, R. E., and W. H. Fishman (1969). *p*-(Acetoxymercuric)aniline diazotate, a reagent for visualizing the naphthol AS-B1 product of acid hydrolase action at the level of the light and electron microscope. *J. Histochem. Cytochem.*, **17**, 1.

Straus, W. (1964). Occurrence of phagosomes and phagolysosomes in different segments of the nephron in relation to the reabsorption, transport, digestion, and extrusion of intravenously injected horse radish peroxidase. *J. Cell Biol.*, **21**, 295.

Straus, W. (1967). In *Enzyme Cytology* (D. B. Roodyn, ed.), pp. 239–319. Academic, New York.

Stroun, M., P. Anker, P. Charles, and L. Ledoux (1966). Fate of bacterial deoxyribonucleic acid in *Lycopersicum esculentum*. *Nature*, **212**, 397.

Stroun, M., P. Anker, and L. Ledoux (1967). Translocation of DNA of bacterial origin in *Lycopersicum esculentum* by ultracentrifugation in caesium chloride gradient. *Nature*, **215**, 975.

Sutcliffe, J. F., and R. Sexton (1968). β-glycerophosphatase and lateral root development. *Nature*, **217**, 1285.

Swift, H., and Z. Hruban (1964). Focal degradation as a biological process. *Federation Proc.*, **23**, 1026.

Sylven, B., and I. Bois-Svensson (1964). Studies on the histochemical "leucine aminopeptidase" reaction. IV. Chemical and histochemical characterization of the intracellular and stromal LNA reactions in solid tumor transplants. *Histochemie*, **4**, 135.

Sylven, B., and H. Malmgren (1957). The histological distribution of proteinase and peptidase activity in solid tumour transplants. *Acta Radiol., Suppl.*, **154.**

Tappel, A. L. (1969a). In *Lysosomes in Biology and Pathology* (H. T. Dingle and H. B. Fell, eds.), pp. 207–244. North Holland Publ. Comp., Amsterdam.

Tappel, A. L. (1969b). In *Lysosomes in Biology and Pathology* (H. T. Dingle and H. B. Fell, eds.), pp.

547–554. North Holland Publ. Comp., Amsterdam.

Thinès-Sempoux, D. (1967). Chemical similarities between the lysosome and plasma membrane. *Biochem. J.*, **105,** 2p0P.

Thomas, L. (1964). *Injury, Inflammation and Immunity.* Williams and Wilkins, Baltimore.

Thomas, L., R. T. McCluskey, J. L. Potter, and G. Weissmann (1960). Comparison of effects of papain and vitamin A on cartilage. I. Effects in rabbits. *J. Exptl. Med.*, **111,** 705.

Tice, L. W., and R. J. Barnett (1965). Diazophthalocyanins as reagents for fine structural cytochemistry. *J. Cell Biol.*, **25,** 23.

Trier, J. S., P. C. Phelps, S. Eidelman, and C. E. Rubin (1965). Whipple's disease: Light and electron microscope correlation of jejunal mucosal histology with antibiotic treatment and clinical status. *Gastroenterology*, **48,** 684.

Trouet, A. (1964). Immunisation de lapins par des lysosomes hépatiques de rats traités au Triton WR 1339. *Arch. Intern. Physiol. Biochim.*, **72,** 698.

Vaes, G. (1965). Excretion of acid and of lysosomal hydrolytic enzymes during bone resorption induced in tissue culture by parathyroid extract. *Exptl. Cell Res.*, **39,** 470.

Villiers, T. A. (1967). Cytolysomes in long-dormant plant embryo cells. *Nature*, **214,** 1356.

Wardrop, A. B. (1968). Occurrence of structures with lysosome-like function in plant cells. *Nature*, **218,** 968.

Wattiaux, R., S. Wattiaux, de Coninck, J. M. Van Dijck, and H. P. Morris (1969). Lysosomes in hepatomas. *Biochem. J.*, **115,** 52P.

Weber, R. (1963). In *Lysosomes* (A. V. S. de Reuck and M. P. Cameron, eds.), pp. 282–300. Churchill, London.

Weiss, L., and J. T. Dingle (1964). Lysosomal activation in relation to connective tissue disease. *Ann. Rheumatic Diseases*, **23,** 57.

Weissmann, G. (1965a). Lysosomes. *New Engl. J. Med.*, **273,** 1084.

Weissmann, G. (1965b). Lysosomes. *New Engl. J. Med.*, **273,** 1143.

Weissmann, G. (1969). In *Lysosomes in Biology and Pathology* (J. T. Dingle and H. B. Fell, eds.), Vol. 1, pp. 276–298. North Holland Publ. Comp., Amsterdam.

Weissmann, G., H. Keiser, and A. W. Bernheimer (1963). Studies on lysosomes. III. The effects of streptolysins O and S on the release of acid hydrolases from a granular fraction of rabbit liver. *J. Exptl. Med.*, **118,** 205.

Weissman, G., B. Becher, and L. Thomas (1964). Studies on lysosomes. The effects of streptolysins and other hemolytic agents on isolated leucocyte granules. *J. Cell Biol.*, **22,** 115.

Wellings, S. R., and K. B. Deome (1963). Electron microscopy of milk secretion in the mammary gland of the C3H/Crgl mouse. III. Cytomorphology of the involuting gland. *J. Natl. Cancer Inst.*, **30,** 241.

Wheat, M. W. (1965). Ultrastructure, autoradiography and lysosomes. Studies in myocardium. *J. Mt. Sinai Hosp.*, **32,** 107.

Wheeler, H., and P. Hanchey (1971). Pinocytosis and membrane dilation in uranyl-treated plant roots. *Science*, **171,** 68.

Williams, K. E., J. B. Lloyd, and F. Beck (1969). The digestive capacity of rat yolk-sac in organ culture. *Biochem. J.*, **115,** 66P.

Wollman, S. H. (1969). In *Lysosomes in Biology and Pathology* (J. T. Dingle and N. B. Fell, eds.), pp. 483–512. North Holland Publ. Comp., Amsterdam.

Wollman, S. H., S. S. Spicer, and M. S. Burstone (1964). Localization of esterase and acid phosphatase in granules and colloid droplets in rat thyroid epithelium. *J. Cell Biol.*, **21,** 191.

Zeja, H. I., and J. K. Spitznagel (1971a). Characterization of cationic protein-bearing granules of polymorphonuclear leukocytes. *Lab. Invest.*, **24,** 229.

Zeja, H. I., and J. K. Spitznagel (1971b). Isolation of polymorphonuclear leukocyte granules from rabbit bone marrow. *Lab. Invest.*, **24,** 237.

Zucker-Franklin, D. (1965). Electron microscope study of the degranulation of polymorphonuclear leukocytes following treatment with streptolysin. *Am. J. Pathol.*, **47,** 419.

Zucker-Franklin, D., and J. G. Hirsch (1964). Electron microscope studies of the degranulation of rabbit peritoneal leukocytes during phagocytosis. *J. Exptl. Med.*, **120,** 569.

8

NUCLEOCYTOPLASMIC INTERACTIONS

J. W. Watts

1. INTRODUCTION

Any attempt to consider systematically nucleocytoplasmic interactions in eukaryotic organisms is soon embarrassed by the shortage of information in many important areas of the subject. A study of some of the reviews of the subject shows that progress has been limited to certain restricted fields and that no grand synthesis is yet possible (Brachet, 1961; Hämmerling, 1963; Gurdon and Woodland, 1968; Harris, 1968; de Terra, 1969). Inevitably, then, many of the questions considered here cannot at present be answered.

Earlier work on the genetics of higher plants and animals demonstrated convincingly the essential role of the chromosomes in inheritance, so that at least in the long term the nucleus is known to be the prime controller of cellular activity. More recent work, concentrating on the genetics and molecular biology of microorganisms, has concerned itself with the way in which information relating to molecular structure is contained in the genome and is transcribed and translated for use in protein biosynthesis. The messenger hypothesis and the genetic code are now almost household phrases (Woese, 1967; Geiduschek and Haselkorn, 1969). Unfortunately, the emphasis placed on microorganisms means that the findings are only of limited value in clarifying the problems of nucleocytoplasmic interactions. Virtually all the recent studies of gene action have been made with bacteria, mainly *Escherichia coli;* in other words, with organisms that lack the distinctive nucleocytoplasmic organization of higher cells.

It is intrinsically probable that many of the fundamental mechanisms of biosynthesis elucidated with microbial systems also operate in higher cells; for example, those concerned with the synthesis of DNA, the transcription of DNA, and the translation of RNA, the genetic code itself, all play a part in eukaryotic cells. It seems certain, however, that there are mechanisms at work in nucleocytoplasmic interactions that are not found in bacteria, if only because they are not necessary. The evidence available does not encourage the view that the mechanisms for translocation of molecules in bacteria are adequate for eukaryotic cells. Even the messenger theory, which provides a model of the way in which information is passed from the genome to the rest of the cell, requires considerable modification when applied to higher cells. The assumption that there are common mechanisms of cellular control operating in all biological systems may contain some truth, but it is not very helpful when extrapolating from pro- to eukaryotes, since it is precisely in the area of nucleocytoplasmic interactions that real qualitative differences exist between the two classes.

Cell diameters range from about 1 μ up to many centimeters; the problems associated with controlling events over intracellular distances that differ by perhaps three or four orders of magnitude may be expected to require a variety of mechanisms, no single one of which is satisfactory for more than a restricted range of conditions. The organization of the nucleus into nucleolus and chromatin, with all the associated apparatus for mitosis and meiosis, separated from the cytoplasm by a double-layered membrane system, presumably represents a solution to the problems associated with an increasing cell size and complexity qualitatively different from that found in the bacteria. It requires little imagination to see that, although the eukaryotic form of organization must have adaptive value in relation to increasingly complex cellular activities, it also introduces new problems. The most obvious of these are the increased intracellular distances and the presence of a membrane system separating the genome, with its essential store of information coding for cellular growth and development, from the main sites of protein synthesis in the cytoplasm. The relatively simple mechanisms of control and interaction currently thought to operate in bacteria could not be expected to function effectively in this situation.

The balanced growth of a cell must require close coordination of nuclear and cytoplasmic activities. Inevitably, then, there is a continuous exchange of information in both directions across the nuclear membrane. It is with the nature of these exchanges and interactions between nucleus and cytoplasm that this chapter is concerned.

2. MODELS OF NUCLEOCYTOPLASMIC INTERACTIONS

A. The Bacterial or Prokaryotic Model

Although the nucleocytoplasmic interactions of eukaryotic cells are our main concern, a brief summary of the way in which the genome interacts with and regulates the rest of the bacterial cell will illustrate the basic mechanisms of interaction and at the same time emphasize the great differences between the two classes of organism.

The bacterial genome is believed to consist of a circular, virtually naked strand of DNA, in contrast to that of the eukaryotic cell in which the DNA is associated with proteins, largely the basic histones, in the chromosomal structures. The base sequences of the DNA are transcribed into primary gene products, various classes of RNA, which are then used to direct and regulate cell metabolism. Part of the genome codes for the structures of the ribosomal RNA (rRNA) molecules, including transfer RNA (tRNA) and other small molecules, but the bulk of the genome is concerned with the production of messenger RNA (mRNA). Regulation of mRNA synthesis is obtained by means of the operon unit (Jacob and Monod, 1961). This often takes the form of a polycistronic section of DNA which contains a sequence of several genes which are activated or repressed together. A polycistronic sequence usually specifies the structure of a series of enzymes which form a biosynthetic pathway. The *lac* operon, for example, contains the genes that determine the structure of the enzymes involved in the metabolism of lactose (Geiduschek and Haselkorn, 1969). The structural genes are activated by the operator gene which is at the end of the group of genes at which transcription begins. The operator gene is in turn controlled by a regulator gene which may be in quite a different part of the genome. The regulator gene is thought to respond to the levels of metabolic products in the cell, and the repressor itself is probably a protein which binds to the operator and prevents the initiation of synthesis of mRNA on the other structural genes. mRNA is usually very short-lived, so that synthesis of an enzyme ceases within a few minutes of repression of messenger formation. The actual translocation of mRNA from DNA to ribosomes presents no problems since the ribosomes can come into direct contact with the genome. It appears that mRNA begins to attach to the ribosome while it is still being synthesized on the DNA (Geiduschek and Haselkorn, 1969).

The mechanisms of interaction and regulation in bacteria are thus relatively simple. Translocation of molecules is largely achieved by diffusion over very short distances, and control is accomplished through the binding of molecules to specific sites, with a great deal of emphasis placed on the lability of mRNA.

B. The Eukaryotic Model

All the larger and more complex organisms, both protozoan and metazoan, have the eukaryotic type of organization. Nuclear membranes effectively isolate nuclear and cytoplasmic material but, at the same time, paradoxically, constitute the link between nucleus and cytoplasm. The genome itself is organized into chromosomes in which the DNA is closely associated with protein, while the ribosomes are almost entirely confined to the cytoplasm. The simple bacterial model for interaction therefore cannot be applied without considerable modification.

The intact nuclear membrane is studded with pores which are presumably a major route by which particles and macromolecules move between nucleus and cytoplasm (Fig. 1). They are plainly of some importance in nucleocytoplasmic exchanges, but their precise structure and function remain uncertain. What is clear is that the pores are not open holes in the membrane, since electron microscopy has shown that they have an elaborate annular structure with a central plug. Conductivity measurements also show that there is a barrier to the free flow of electrolytes between nucleus and cytoplasm (Loewenstein, 1965). There is some indication that the number of pores per unit area of membrane is related to the metabolic activity of the cell; the actual

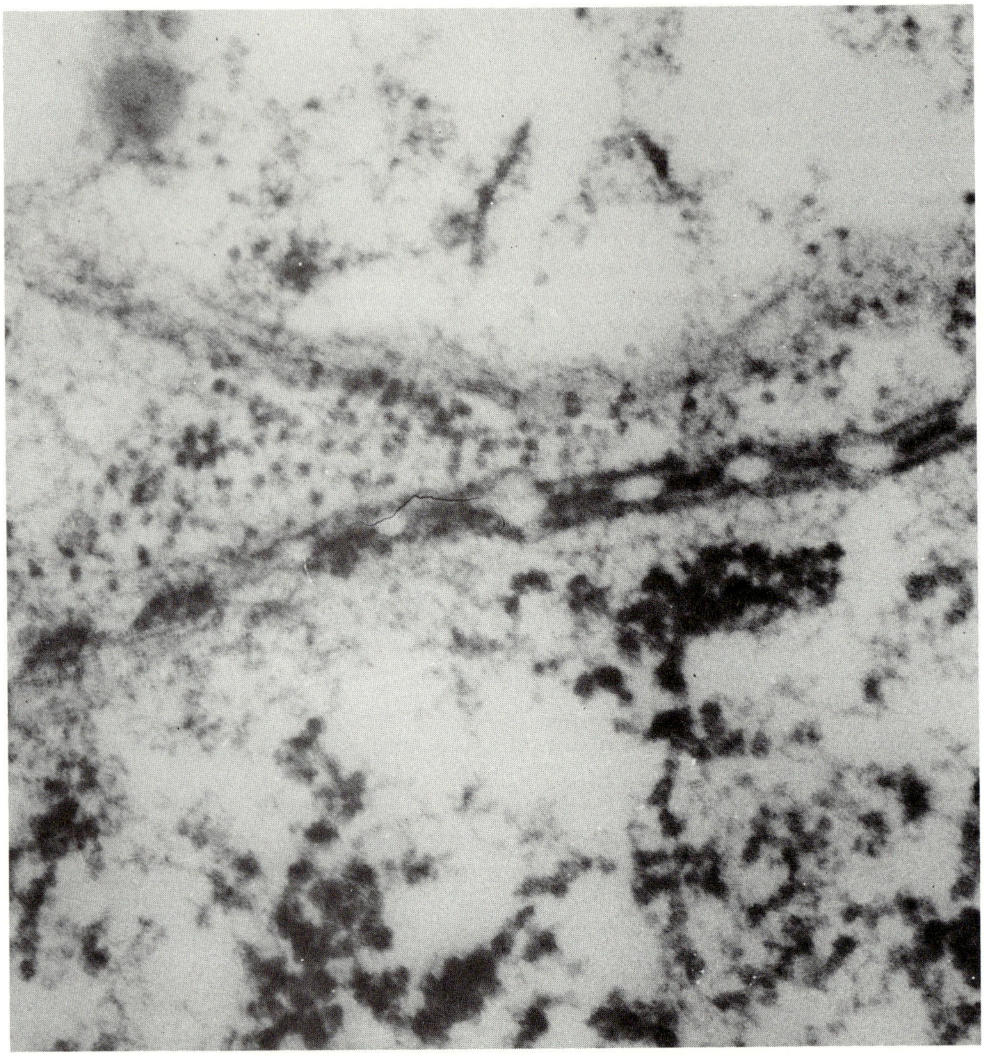

Figure 1. Movement of material from nucleus to cytoplasm through the nuclear pores. A transverse section through the nuclear pores in pollen mother cells at pachytene in *Fritillaria lanceolata*. The polarization of materials, apparently passing through the pores is a relatively uncommon event, judged by the frequency with which it is observed. The nucleus is at the top. × 136,500. (Courtesy of L. F. LaCour.)

density of pores can vary greatly not only between different cell types, but even in the same cell at different times in the cell cycle.

It is thus necessary, in view of these differences in intracellular organization, to decide what features of the bacterial model can be retained or modified and what new elements must be introduced to account for the known facts of nucleocytoplasmic interactions in eukaryotic cells. The similarity of organization of eukaryotic cells suggests that there are common mechanisms for the exchange of information between nucleus and cytoplasm, but it is probable that there are considerable changes in emphasis on different aspects of this mechanism from cell to cell. Superimposed upon the common elements of organization are many variations; for example, in many cells the nuclear membrane fragments during mitosis, reassembling at late telophase, but this is not a universal rule. The syncytial myxomycetes, the true slime molds, retain their nuclear membrane during mitosis (Gray and Alexopoulos, 1968). It follows that while in some cells the disappearance of the membrane during mitosis may play an important part in the exchange of materials between nucleus and cytoplasm, it cannot be a universal mechanism.

Any model of nucleocytoplasmic interactions in eukaryotic cells must take into account those features not found in bacterial cells. Therefore it is well to consider some of the evidence that is available, first at the descriptive and then at the molecular level.

3. THE CHARACTERISTICS OF NUCLEOCYTOPLASMIC INTERACTIONS

A. Nucleus-Determined Effects

1. Evidence from Genetics. The many studies of genetics of higher plants and animals have shown conclusively that the main location of genetic material is in the nucleus, and specifically in the chromosomes. There is little doubt that the fundamental mechanism of heredity is essentially the same as that in microorganisms. The information specifying the structure of the cell proteins resides in the DNA of the genome in the form of specific sequences of pyrimidine and purine bases.

The extent of nuclear dominance in inheritance is most readily appreciated when searching for non-Mendelian types of inheritance. Cytoplasmic inheritance is an important factor in genetics, but nowhere does it occur on the same scale as Mendelian (i.e., nuclear) inheritance. It is one of the problems of nucleocytoplasmic interactions that, in the face of this nuclear dominance, the cytoplasm is still able to play a decisive role in the development of the cell.

2. Evidence from Nuclear Transplants. Several excellent studies have utilized microsurgical techniques to enucleate cells or transfer nuclei from one cell to another to determine which functions are critically dependent on the nucleus. While it is true that anucleate cells are always capable of some limited metabolism, they are very restricted both in functions and life-span when compared with nucleated cells. Not surprisingly, the nucleus is essential. The results obtained by Hämmerling (1963) and others, using *Acetabularia* as an experimental organism, probably give the best illustration of the extent and some of the subtleties of nuclear control.

Acetabularia is a genus of giant marine algae with several unusual features which are useful in the study of nucleocytoplasmic interactions. The more commonly used species such as *Acetabularia mediterranea* grow in the course of 6 months from a zygote, which is a few microns in diameter, to a uninucleate plant approximately 5 cm in length and 0.3–0.4 mm in width. Throughout its life cycle the single diploid nucleus remains in the rhizoid at the lower end of the stalk. When the alga approaches maturity, the other extremity of the stalk differentiates to form an umbrellalike cap which has a morphology characteristic of the species. After cap formation the nucleus divides to produce a second generative nucleus which then undergoes a series of mitotic divisions to produce many more nuclei. These are transferred by cytoplasmic streaming to the cap, where cysts are formed; further mitoses occur within the cysts, followed by meiosis to produce gametes. New plants are formed by the fusion of gametes of suitable mating types to produce the diploid zygote.

The size and shape of the plant lend themselves to a variety of techniques. Enucleation can be performed simply by cutting off the rhizoid with a pair of scissors. At a slightly more elaborate level, the nucleus can be removed from the rhizoid, washed in sucrose, and placed in another plant with a minimum of cytoplasmic contamination. Interspecific grafts can be made very easily by pushing the cut ends of stems together so that the smaller end enters the larger. Grafts of this type are usually successful and can also be used to produce multinucleate plants.

The following points, which have been established by studies of nuclear transfers with *Acetabularia*, demonstrate that the nucleus plays an essential role in determining the activity of the cytoplasm. They also show, however, that the situation is more complex than the simple bacterial model suggests.

1. The shape of the cap is determined by the nucleus. If the cap is cut off, a new one, characteristic of the species, is regenerated. If the nucleus of an *A. mediterranea* plant is replaced with one from an *Acetabularia crenullata* plant and the cap then cut off, the regenerated cap is of the *A. crenullata* form; the instructions for the new cap have thus come from the nucleus, and the presence of a large excess of *A. mediterranea* cytoplasm has had no apparent effect.

2. The information necessary for cap formation is normally present in the cytoplasm long before differentiation begins to produce a cap. For example, if a young, capless plant is enucleated, it continues to grow and ultimately produces a cap of normal appearance. The new cap often develops prematurely, suggesting that while there is information for cap formation there is a deficiency of instructions for normal growth. The information specifying cap morphology is not uniformly distributed throughout the stalk but shows a gradient from tip to rhizoid, so that different sections of the stalk show varying potential for cap development; the regions nearest the rhizoid have the least power of differentiation.

3. The information for cap formation present in the cytoplasm of the young plant is responsible for initiating complex biochemical changes many weeks after enucleation (Keck, 1969). Biochemical studies have shown that weeks after enucleation new species of protein and cap-specific polysaccharides make their appearance in the cytoplasm. It is therefore necessary to think in terms of some type of "masked" information if conventional ideas about the transfer of information from nucleus to cytoplasm are to be invoked. It has been suggested that the new enzymes induced are actually the result of minor modifications in enzymes already present in the cytoplasm (Keck, 1969). If this were the case, the need for masked mRNA would be removed, although the precision with which a nuclear-determined process is performed long after enucleation still requires explanation.

Acetabularia is, however, an unusual organism, and not all cells show the same ability to continue growth and development without a nucleus. There is always an ultimate dependence on the presence of a nucleus, but enucleation usually results in a very rapid loss of most cytoplasmic activities. Anucleate *Amoeba*, for example, lose the power of controlled movement and can no longer ingest food; these functions are rapidly restored when a nucleus is placed in the cell (Brachet, 1961). The great value of the *Acetabularia* experiments is the way in which they show that the influence of the nucleus is not simply a short-term-influence effect but may continue long after the nucleus has ceased to function. They also direct attention to the precision with which the cytoplasm controls and manipulates nuclear-derived information, a fact that is obscured in many situations and which serves to emphasize that the cell must be visualized as a series of interacting parts.

B. Cytoplasm-Determined Effects

Largely because of the emphasis placed on genetics that stresses nuclear characteristics, there is a general inclination to relegate cytoplasm to a purely subsidiary role in cell growth. The cell is, however, an interacting system of parts, none of which can function satisfactorily in isolation, and it is unreal to ask which is more important, cytoplasm or nucleus. It is in fact limitations in experimental techniques that result in the unbalanced development of a subject, and only in recent times have the effects of cytoplasm become the subject of detailed investigation.

1. Evidence from Syncytial Organisms. A simple demonstration that the cytoplasm plays an important part in the control of cell development and, specifically, nuclear activity, is provided by the vegetative diploid plasmodium of the true slime molds. The true slime molds, the myxomycetes, have a life cycle that alternates between the uninucleate, haploid, ameboid form and the syncytial, multinucleate, diploid form. The nuclei in the diploid plasmodium share a common cytoplasm which exhibits very rapid streaming; the velocity of cytoplasmic particles can be as high as 1.4 mm/sec. These high rates of streaming actually occur in localized channels reminiscent of a circulatory system. The cytoplasm and its contents are thus rapidly and continually circulated and mixed, thus overcoming any tendency for gradients of cytoplasmic components to form. A feature of the myxomycetes is the retention of the nuclear membrane during mitosis; the whole apparatus of mitosis, including the spindle, is within a greatly expanded nuclear membrane. There is thus at all times a clear separation between nuclear and cytoplasmic material. It seems probable that this is an advantage, if not a necessity, because of the shearing forces associated with rapid cytoplasmic streaming (Gray and Alexopoulos, 1968).

A marked feature of the plasmodium is the very high degree of synchrony of nuclear events; sharing a common cytoplasm, the nuclei also share a common pattern of development. This is very evident at mitosis, when nuclear synchrony is so good that in plasmodia with diameters up to 5 cm mitosis occurs virtually simultaneously in over 99% of the nuclei. The extent of variation between nuclei in a vigorously growing plasmodium is 1–2 min in a mean generation time of about 10 hr. This observation in itself must mean that it is the cytoplasm that controls nuclear division. This view is supported by experiments in which plasmodia are fused. It is a normal feature of the slime molds that when two plasmodia of suitable types come into contact they rapidly fuse to form a single plasmodium, the components of which are mixed by cytoplasmic streaming. As a general rule, whatever the respective stages in interphase of the original plasmodia, interactions occur in the composite plasmodium which cause a rapid resynchronization of cell chemistry. This is apparent at the first division after fusion, when the nuclei are found to be synchronized. For example, when one of the original plasmodia is in early interphase, synthesizing DNA, and the other is approaching mitosis, after fusion the nuclei of the more advanced plasmodium are restrained from dividing until the other class of nuclei have completed DNA synthesis and reached the same stage. Only then does mitosis occur (Rusch et al., 1966). This simple experiment, which avoids all the problems of trauma associated with cell surgery, shows that the cytoplasm has a decisive role in determining some aspects of nuclear activity. There must be a movement of information from the cytoplasm to the nucleus, but whether this is in the form of specific messengerlike molecules or is a relatively nonspecific effect, for example, the relative amounts of nuclear and cytoplasmic material, cannot be determined by these experiments.

There is another feature of this fusion experiment that is of some interest. Nuclei that have already completed DNA synthesis are not normally induced to replicate their genome again. The cytoplasmic signals are only effective and dominant when the nuclei are competent to obey. Resynchronization is therefore not an example of cytoplasm overriding the nucleus but the result of a true cooperation between nuclei and cytoplasm.

A rather different picture is presented by the parasitic protozoans of the genus *Opalina* (Nitova and Sukhanova, 1964). These organisms also have a multinuclear syncytial organization, and the nuclei appear to be of one type. There is, however, a wide variation in nuclear activity, including both RNA and DNA synthesis, within a single organism. This absence of synchrony is almost certainly related to the presence in the cytoplasm of well-defined gradients of material, which result in the differential activation of nuclei. This is a situation reminiscent of the existence in some eggs of morphogenetic gradients which determine the early stages of embryogenesis.

2. Evidence from Nuclear Transplants. Among the earliest experiments designed to investigate the ability of cytoplasm to modify nuclear behavior are those that involve nuclear transplantation. This type of approach is sometimes criticized on the grounds that the process is quite unnatural and that the trauma produced by surgery may lead to abnormal responses. Despite this problem, and it is a very real one, the overall picture obtained with nuclear transfers is consistent with the results of other studies, and the results are so impressive that no examination of nucleocytoplasmic interactions can ignore them. The most dramatic results have been obtained with nuclear transfers into amphibian eggs (Gurdon and Woodland, 1968), but a great deal of work has also been done with *Amoeba*, *Stentor*, *Tetrahymena*, and other organisms (Brachet, 1961; Hämmerling, 1963; de Terra, 1969), all of which supports the conclusions reached with amphibian cells.

It is widely assumed that in most organisms the somatic cells possess identical genomes, although a certain degree of polyploidy or polyteny may be observed in some tissues. During differentiation therefore the organization of the nucleus and cytoplasm are modified to ensure that only certain genes are active.

The fertilized egg is, in higher animals, the only cell normally capable of growing and dividing to produce a new organism. It has proved possible, however, to transfer the nucleus of a somatic cell into a mature amphibian egg with remarkable consequences. A variety of somatic cells has been used to supply nuclei, and in all cases a common pattern of development is observed. The final stage in the development of the amphibian egg is the breakdown of the germinal vesicle; somatic nuclei transferred into the egg at this stage normally increase in size, and if they possess a nucleolus it becomes very diffuse or disappears. The transplanted nucleus then begins to synthesize DNA. The process of differentiation normally appears to be essentially irreversible for many cell types, but in the egg cytoplasm even nuclei derived from nonmultiplying end cells from the brain begin to synthesize DNA. The process of reorganization of nuclear activity continues in favorable circumstances and may even result in the development of a more-or-less normal embryo. Since nuclei from a variety of cells, both dividing and nondividing, from both immature and adult tissues, have been shown capable of reactivation (or, rather, reorganization) to permit DNA, RNA, and protein synthesis, and subsequent division to permit embryonic development, it is clear that the egg cytoplasm possesses remarkable properties. These changes are dependent, however, on the condition of the egg cytoplasm, and do not occur until the germinal vesicle has broken down. Nuclei transplanted into the oocyte prior to this stage do not synthesize DNA; indeed, if DNA synthesis normally occurs in the transferred nucleus, it is soon prevented by the egg cytoplasm.

The cytoplasm of the mature egg thus possesses the ability to impose a new pattern of activity on the nucleus and even to reverse the process of nuclear differentiation. Since the immature egg exhibits similar properties, it can be concluded that cytoplasm is able to reorganize nuclear activity with great precision. The cytoplasmic factors involved in initiating DNA synthesis are not species-specific since nuclei from the somatic tissues of the mouse also begin to synthesize DNA when implanted in frog egg cytoplasm (Graham et al., 1966).

3. Evidence from Cell Hybridization. Another technique used in recent years to investigate nucleocytoplasmic interactions is the fusion of different cell types to produce multinucleate heterokaryons. It is possible to fuse many types of vertebrate cells from both cold- and warm-blooded animals by mixing them in the presence of inactivated virus, usually Sendai virus. The virus causes an unknown change in the character of the cell walls so that the membranes fuse to produce cytoplasmic bridges between the cells. After a relatively short time, these bridges expand so that there is complete fusion of the cells. In this way heterokaryons can be prepared from the cells of many vertebrate species. The number of nuclei in each heterokaryon can also be regulated by controlling the amount of added virus (Harris, 1970).

The results obtained with heterokaryons are qualitatively similar to those obtained with nuclear transplants but, because of the simplicity and versatility of the technique, it has been possible to investigate some features of nucleocytoplasmic interactions in much greater detail. The earliest studies used heterokaryons formed by fusion of cells from man and mouse (HeLa and Ehrlich ascites tumor cells) (Harris and Watkins, 1965). Both these cell lines are derived from cancerous tissue; the HeLa cell is an established line which has been maintained in tissue culture for many years; the Ehrlich cell does not grow *in vitro* and is maintained by growth in the peritoneal cavity of the mouse. When these two cell types are fused, the Ehrlich cell nuclei begin to synthesize DNA in the heterokaryon and ultimately divide. Thus once more the familiar pattern of cytoplasmically determined nuclear behavior is observed. Another type of heterokaryon used is formed from HeLa cells and rabbit macrophages. The macrophage from the peritoneal cavity of the rabbit is a nondividing end cell which synthesizes RNA but not DNA. Within the heterokaryon, however, the macrophage nuclei begin to synthesize DNA, so that once more the HeLa cytoplasm is able to activate nuclear processes.

Heterokaryons normally contain cytoplasm derived from both types of cells, but this is not the case with heterokaryons formed from HeLa and hen red cells (Harris, 1970). The red cells of birds contain nuclei which have highly pyknotic (heterochromatic) chromatin and are essentially inactive. A feature of heterokaryon formation between HeLa and hen red cells is that the cytoplasm of the red cell is not incorporated to any appreciable extent. It is therefore possible to consider the heterokaryon as nuclei from the two cells in HeLa cytoplasm. The sequence of events observed after fusion provides an example of nucleocytoplasmic interaction similar in many ways to nuclear transfers into amphibian eggs, but one that has proved susceptible to much more detailed analysis.

Shortly after fusion of the HeLa and hen red cells the small hen nucleus begins to swell and the chromatin becomes less condensed. At a later stage the red cell nucleus begins to synthesize RNA and DNA. This is as far as the experiment can normally be taken because the HeLa cell nuclei begin to divide within the heterokaryon, and this causes disintegration of the red cell nuclei which are induced to divide prematurely. If, however, the HeLa cell nuclei are inactivated before fusion, for example, by irradiation with ultraviolet light, so that they are no longer capable of division, premature disintegration of the red cell nuclei is avoided. The process of activation of the nuclei can then be followed to a later stage. Continued nuclear reorganization now results in the appearance of a nucleolus, and simultaneously new rRNA begins to appear in the cytoplasm and proteins characteristic of the hen cell are synthesized. The process of reactivation of the red cell nucleus is thus virtually complete, but only to the stage found in the original HeLa cells.

An analogous experiment with heterokaryons formed by fusion of rabbit macrophages with hen red cells results in the activation of the red cell nucleus to a stage similar to that of the macrophage nucleus; there is synthesis of RNA but not of DNA. These relationships are summarized in Figure 2.

These experiments thus confirm and extend the results obtained by other techniques. The environment of the nucleus, the cytoplasm, clearly controls and regulates the biosynthetic activities of the nucleus, hence plays an important part in differentiation. To some extent at least, the process of differentiation must be reversible, and the ability to synthesize RNA and DNA consistently seems to be dominant over the suppression of these functions. The work with heterokaryons does not demonstrate that only cytoplasmic factors are responsible for nuclear activation, since two classes of nuclei are present and one is activated only as far as the level of the other. The

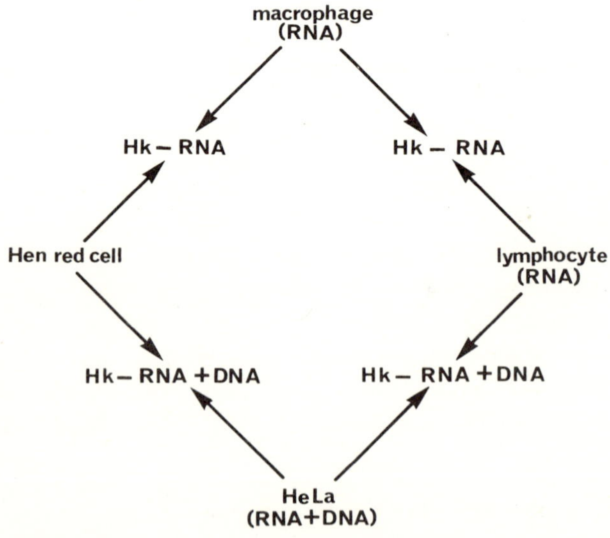

Figure 2. Metabolic relationships between heterokaryons and their parent cells. The figure shows that nucleic acid metabolism is determined by the parent cell with the greater biosynthetic ability.

factors could therefore be nuclear in origin. What is clear, however, is that the signals are mediated by the cytoplasm.

Perhaps the most remarkable feature of the fusion studies is the demonstration that the factors responsible for nuclear reorganization are not species-specific; heterokaryons have been made between human, mouse, rabbit, rat, hen, and frog cells, and in all cases the same general pattern has been observed. A certain amount of caution must be exercised when considering this aspect of cell fusion since at some stages the nucleus itself must be involved. This is shown by the appearance of nucleus-determined molecules such as rRNA and proteins after a nucleolus has formed. The initiation of RNA or DNA synthesis also shows that there is nuclear participation at an early stage.

The position could be summarized as follows. The nucleus ultimately is the source of the basic information that determines a cell's potentials. The realization of those potentials is, however, the result of complex and often very subtle interactions between nucleus and cytoplasm. In general, once the nucleus has passed to the cytoplasm instructions which initiate a particular sequence of events, it ceases to be able to control the subsequent events, except perhaps to exert a moderating influence. This situation is probably most clearly evident in the amphibian egg in which the highly specialized cytoplasm is able to modify established patterns of differentiation and redirect nuclear development to the pathways of embryogenesis.

4. THE MOLECULAR BASIS OF NUCLEOCYTOPLASMIC INTERACTIONS

It is now necessary to consider the question how interactions between the nucleus and cytoplasm are mediated and regulated. This entails an examination of the way in which molecules with an informational content pass between the parts of the cell. It is possible to visualize a variety of molecules that might fulfil these functions.

1. The most obvious class includes the large molecules such as RNA and proteins which could contain specific items of information. The best known examples are the different types of mRNA which contain sequences of purine and pyrimidine bases specifying the amino acid structure of a particular protein. Although there is relatively little evidence in favor of the idea, it is probable that highly specific proteins are involved in the regulation of sections of the genome (Britten and Davidson, 1969; Crick, 1971).
2. Large molecules, particularly of proteins, could contain relatively broad classes of information regulating the general direction of cell metabolism. There is a rapid and large-scale exchange of material, including several classes of protein, between nucleus and cytoplasm, and it is hard to avoid the conclusion that they play some part in nucleocytoplasmic interactions (Goldstein and Prescott, 1967). A possible example of this class is DNA polymerase (Gurdon and Woodland, 1968).
3. Small molecules and ions may have profound effects on cell structures and be responsible for initiating certain processes. For example, divalent cations such as magnesium and calcium affect the state of aggregation of chromatin. Small molecules such as cyclic AMP and the steroid hormones are known to influence a range of biosynthetic activity (Jost and Rickenberg, 1971).

The fact that the cytoplasmic factors that promote nuclear reorganization are not species-specific suggests that at least some of the molecules involved in nucleocytoplasmic interactions do not have either very elaborate structures or high specificity. Obviously many signals, such as those coding for protein structure, are highly specific, so that there is every likelihood that all these classes of molecules may be involved.

A. Regulation by the Nucleus

1. The Synthesis and Translocation of mRNA. The synthesis of RNA in the nucleus and its translocation into the cytoplasm have been investigated in some detail (Darnell, 1968); most of these studies have been concerned with the mechanisms by which the genome regulates cellular events, and these have been visualized broadly in terms of protein biosynthesis. The information specifying the primary structure of a protein is transcribed from the DNA into RNA molecules which then move into the cytoplasm where they program the synthesis of the specific protein by the ribosomes. A tiny fraction of the cell DNA specifies the structures of rRNA and RNA, the remainder is assumed to specify the structure of the cell proteins and to fulfil other, unknown functions. One of the problems inherent in this view concerns the amount of DNA, only a small fraction of which should be necessary to code for all the cell's proteins. There thus remains a question as to what the bulk of the DNA does. The most reasonable answer is that it participates in the regulation of cell development at a qualitatively different level. Some speculative models have in fact been proposed, but in the absence of experimental evidence we are left with the uncomfortable situation in which there is far too much DNA with no known function (Britten and Davidson, 1969; Crick, 1971).

Several classes of RNA are involved in the synthesis of protein, a process that is predominantly cytoplasmic.

1. Over 60% of cell RNA is usually found in the ribosomes and forms two classes with sedimentation constants around 18 and 28S. Within each of these groups, the molecules are usually presumed to be identical but of unknown function.
2. tRNA and other species with sedimentation constants around 4–5S comprise this class. The best understood of these are the tRNA species concerned in the activation of amino acids for use in protein synthesis. The roles of other species such as the 5S RNA associated with ribosomes are not known.
3. mRNA is, from the point of view of nucleocytoplasmic interactions, perhaps the most important class of all. As a species of RNA, it is only poorly defined and cannot really be said to have been unequivocally identified in eukaryotic cells. A fraction of RNA with a sedimentation constant of approximately 9S has been isolated from reticulocyte ribosomes; it apparently contains the base sequences that code for the protein chains of hemoglobin, but some workers consider that it is strictly a product of partial degradation of rRNA (Hadjiolov, 1967).
4. Nuclear RNA resembles rRNA in containing two classes of molecules with sedimentation constants 16–18 and 28S. Various roles have been attributed to the rather special fractions of nuclear RNA which become labeled immediately after exposure of cells to radioactive precursors of RNA, for example, rRNA precursor and mRNA. No well-defined function has, however, been ascribed to the bulk of the nuclear RNA which in many proliferating cells, for example, HeLa and the cells of plant root meristems, may account for 10–20% of the total cell RNA.

The nuclear origin of virtually all RNA has been demonstrated by the technique of hybridizing DNA and RNA. This has also shown that some fractions of RNA are restricted to the nucleus (Shearer and McCarthy, 1967). The work of Melli and Bishop (1969) has, however, questioned the precise significance of DNA–RNA hybridization, and it is not clear how much weight can be attached to these observations.

When a eukaryotic cell is grown in medium containing a radioactive precursor of RNA, for example uridine, the observed pattern of incorporation of label into RNA is one that is common to virtually all cells synthesizing RNA. Its general features may be summarized as follows.

1. Initially, the most heavily labeled area is the nucleus. When the precursor is uridine, the most heavily labeled intranuclear site is the nucleolus, where this organelle is present. Other precursors, for example, adenosine, may give the reverse pattern of intranuclear labeling, with the chromatin being more rapidly labeled than the nucleolus. The rate of incorporation into the nucleus is virtually maximal from the beginning and only tails off when the exogenous supply of label fails or the specific radioactivity of nuclear RNA approaches that of the precursor. When examined by ultracentrifugation, the rapidly labeled nuclear RNA has at first a highly polydisperse sedimentation pattern, often characterized by a broad peak at approximately 45S. At this early stage there is already a small but significant labeling of rRNA in the cytoplasm, and some labeling of tRNA.
2. As the period of labeling increases, the pattern of labeling of nuclear RNA progressively changes until after several hours it corresponds closely with the sedimentation pattern of the bulk of the nuclear RNA with its two peaks at 18 and 28S.
3. The rate at which radioactivity appears in rRNA increases during the first 2–4 hr until it too reaches a constant rate. During the first few minutes of labeling, the lighter 18S fraction incorporates radioactivity faster than the 28S fraction, but this difference largely disappears after about 2 hr.
4. If the labeled cell is transferred to medium containing unlabeled precursors of RNA before appreciable labeling of cytoplasmic RNA has occurred, there is a rapid fall in the amount of label in nuclear RNA and, simultaneously, radioactive RNA appears in the cytoplasm.

These general observations have been widely construed as revealing a precursor–product relationship between nuclear and cytoplasmic RNA. The bulk of the cytoplasmic RNA presumably originates in the nucleus, but there is in fact no equivalence between loss from the nucleus and gain by the cytoplasm. It has been increasingly recognized in recent years that no real confidence can be placed in apparent precursor–product relationships of this sort because there is a large-scale turnover of RNA within the nucleus. This is explicit in the scheme shown in Figure 3 which includes details of a proposed sequence by which rRNA is synthesized and matured in the nucleolus. This scheme proposes that approximately 50% of the 45S RNA molecule produced by the ribosomal cistrons turns over. There is much evidence showing that there is extensive turnover of nuclear RNA (Harris, 1968; Darnell, 1968).

The existence of nuclear RNA turnover means that the arguments for the nuclear origin of rRNA based on kinetic studies are of doubtful validity. Thus, although there is agreement that translocation of RNA is a reality, firm evidence for the process is difficult to obtain. The best support for translocation now comes in fact from experiments in which the nucleolus is either inactivated by irradiation with an ultraviolet microbeam or activated by formation of heterokaryons. These indicate that the

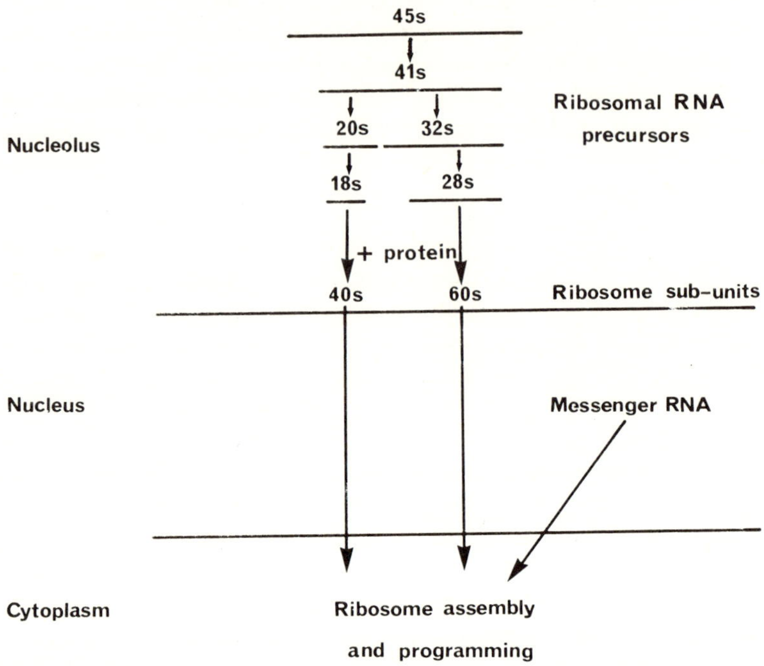

Figure 3. Conventional scheme for translocation of mRNA and rRNA from nucleus to cytoplasm. (After Darnell, 1968; Weinberg and Penman, 1970.)

presence of an intact nucleolus is essential for the translocation of all species of nuclear RNA from nucleus to cytoplasm (Harris, 1970).

Since mRNA is probably the key to those aspects of cytoplasmic activity regulated from the nucleus, it is useful to examine briefly the current situation. Much has been written demonstrating that the original experiments with bacteria did not in fact demonstrate the formation of mRNA (Harris, 1968), and the subsequent search for mRNA in eukaryotic cells has not been very successful. The existence of RNA species equivalent to mRNA is not seriously questioned; the outstanding problems are concerned rather with the form of the message and the manner of its translocation into the cytoplasm. Two models have been proposed; the first, based on the bacterial model, has already been outlined. Studies of bacterial systems fit quite well a simple scheme in which mRNA is released from the DNA and simultaneously attaches to the ribosome where it is translated into protein (Geiduschek and Haselkorn, 1969). The polysomes are then the ribosomal apparatus actually in the process of moving along the strand of mRNA. In order to apply this scheme to eukaryotic cells, it is usually assumed that the mRNA detaches from the DNA and then diffuses out of the nucleus until it engages with a free ribosome in the cytoplasm (Fig. 3). The presence of RNase activities in both nucleus and cytoplasm introduces serious but not insuperable problems, and the messenger strand is in fact widely assumed to translocate within a protective coat of protein (Hadjiolov, 1967). The most doubtful feature of the scheme is the rather haphazard character of diffusion over relatively large distances, a process that hardly seems designed to allow the degree of control that the cell undoubtedly exerts over protein synthesis. Some kind of active and directional process of translocation therefore seems desirable. It is possible, however, that the situation is at once simpler and more precise than the bacterial model suggests.

When the messenger hypothesis was first proposed, one feature of the messenger molecule that was stressed was its short life (Jacob and Monod, 1961). The rates of synthesis of proteins were presumably determined by the rate at which messenger was made, and protein synthesis ceased soon after suppression of the relevant gene. Many studies of protein synthesis have shown that mRNA can have very great stability. The classic example is of course the hemoglobin messenger in reticulocytes, which continues to code for protein synthesis for several days after the nucleus has been extruded. Evidence has also accumulated that control of protein synthesis can occur at the level of translation as well as of transcription. These facts have some relevance to the design of a model of nucleocytoplasmic interactions because they remove the necessity for a rapid and continuous flow of mRNA from the nucleus. It now becomes possible to envisage a situation in which preprogrammed ribosomes are activated or repressed as well as reprogrammed.

An alternative approach to the problem of translocation of messenger has been discussed by several workers in recent years. It assumes that the mRNA strand is an integral part of the ribosome (Hadjiolov, 1967; Harris, 1970) (Fig. 4). The reasons for such a radically different approach are found in several observations which are not easily accommodated by the simple bacterial model. Some of the principal difficulties are as follows.

1. Searches for mRNA have been relatively unsuccessful. The work with reticulocytes has demonstrated that a messengerlike RNA is present in the ribosomes (Laycock and Hunts, 1969; Holder and Lingrel, 1970). Some workers do not, however, obtain this 9S component and conclude that it is the result of slight degradation of rRNA which can be prevented by better control of RNase activity. If this criticism is accepted

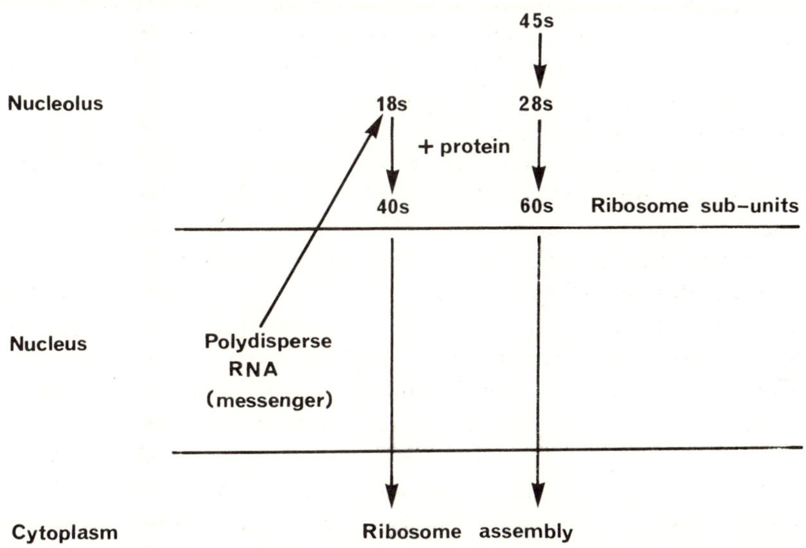

Figure 4. Alternative scheme for translocation of mRNA and rRNA from nucleus to cytoplasm. In this model mRNA becomes an integral part of the ribosome subunit before translocation into the cytoplasm.

it would appear that the messenger component is readily split off from the rest of the rRNA (Hadjiolov, 1967).

2. mRNA cannot only have a long life in the cytoplasm, as in the reticulocyte, but can also be held in an inactive condition. For example, during cap regeneration in *Acetabularia*, there is an induction of nucleus-determined enzymes several weeks after enucleation (Keck, 1969). In the case of the amphibian egg, large numbers of ribosomes are made during growth of the oocyte, but they do not synthesize proteins until fertilization, when they are rapidly activated (Nemer, 1967). The evidence suggests that the mRNA needed for this new protein synthesis was already present in the cytoplasm long before fertilization and is unmasked by a factor, possibly a protease, which is released into the cytoplasm by fertilization. Control of protein synthesis therefore appears to occur at the level of translation as well as transcription.

3. The work on heterokaryons indicates that the synthesis of new species of protein cannot begin without the concomitant synthesis of new rRNA. There is, for example, a large-scale synthesis of nuclear RNA of the kind often called "messenger" on the chromatin of hen nuclei in HeLa–hen red cell heterokaryons. It is not, however, until a nucleolus is formed that any ribosomal or other RNA from the hen nucleus appears in the cytoplasm; and it is at this stage that hen-specific proteins begin to be synthesized. Either mRNA is also processed independently, or it is incorporated into the ribosomal subunit, in the nucleolus. If the bacterial model were correct, synthesis of hen-specific proteins would be expected to occur at a very early stage since mRNA would be able to diffuse out into the cytoplasm and reprogram HeLa ribosomes.

4. All the available evidence shows that the nucleolus is essential for the translocation of rRNA from nucleus to cytoplasm. The usual view is that 18 and 28S RNA are synthesized coordinately as part of a polycistronic RNA molecule (Fig. 3). The two species do not, however, appear coordinately in the cytoplasm, and there is evidence that they are not necessarily synthesized coordinately. Very low concentrations (0.05 μg/ml) of actinomycin D almost entirely suppress incorporation of labeled uridine into nucleolar RNA, but a low level of incorporation continues into chromosomal RNA and cytoplasmic rRNA (Bramwell, 1970). The label appearing in the cytoplasm is in 18S RNA, while that in the chromosomal RNA is in polydisperse RNA. Both these fractions have similar base compositions, and it is a relatively simple matter reversibly to convert the polydisperse RNA to an 18S type by suitable heat treatment. It seems possible therefore that the 18S rRNA originates on the chromatin, rather than being synthesized coordinately with 28S in the nucleolus.

5. If the ribosome contains an 18S RNA that determines the function of the ribosome, that is, the structure of the protein it can synthesize, an extensive reconstruction of the ribosome will be necessary in order to reprogram it. Provided there is control at the translational level, and there is excellent evidence for this, reconstruction would, however, be essential only when there is a reorganization of protein synthesis on a large scale, for example, during differentiation. That reconstruction of this type takes place is suggested by results obtained with the cellular slime mold *Dictyosteleum discoideum* (Mizukami and Iwabuchi, 1970). When the vegetative amebas are labeled with uridine, the usual pattern of incorporation into rRNA is observed; the lighter fraction labels initially much faster than the heavier, but within 2 hr the specific radioactivities of the two species of RNA are similar. The cells can now be transferred to a medium containing an excess of labeled uridine and induced to move into the migratory phase which culminates in the production of a sporangia-

like body. Approximately 11 hr after transfer to unlabeled medium, the process of differentiation is well advanced, and when the distribution of radioactivity in rRNA is examined, the pattern of labeling has been greatly modified; the specific activity of the lighter fraction is now only about half that of the heavy molecules. This can readily be understood only if there has been an extensive and preferential turnover of the lighter fraction of rRNA.

A rather similar phenomenon is observed with *Physarum polycephalum*. When a plasmodium with labeled RNA is fused with an unlabeled one that is at a different stage in the cell cycle, there is a rapid change in the relative levels of labeling of the two types of rRNA, and evidence suggests that this is attributable to turnover of the lighter fraction of RNA.

If the 18S rRNA contains the messenger, many of these observations and the problems associated with translocation of messenger are simply resolved. Both Hadjiolov (1967) and Harris (1970) have argued in support of a model of this general form. There is, however, a considerable amount of evidence to support the more conventional view of the origin and translocation of mRNA and rRNA, so that it would again be unwise to prejudge the issue. It is conceivable that a combination of both models may be at work.

2. Regulation at the Level of Translation. The ability of the cell to control protein synthesis at the level of the ribosome has already been mentioned. The precise nature of this control is, however, uncertain. Activation of translation in the amphibian egg immediately after fertilization has been considered to result from unmasking of messenger by a protease (Nemer, 1967). Tomkins et al. (1969) investigated induction of the enzyme tyrosine aminotransferase in a cell line derived from rat hepatoma and concluded that synthesis of the enzyme is controlled at both transcription and translation levels. The inducer is an adrenal steroid which increases the level of enzyme about 10-fold above the basal, uninduced level; the cell can, however, only be stimulated to increase enzyme production over about two-thirds of the cell cycle. One of the more interesting features of this system is the effect of actinomycin D which actually "superinduces" enzyme formation. These investigators suggest that the nucleus regulates synthesis in two ways. There is, first, the conventional control at the level of mRNA synthesis, and, second, synthesis of another RNA species at a regulatory gene which promotes the formation of a repressor, probably a protein. The repressor is presumed to inhibit the translation of messenger by the ribosomes and at the same time promote the degradation of messenger. Induction is then envisaged as repression of the production of repressor.

A scheme of this type is clearly of great interest, but much more information will be necessary before it can be accepted with real confidence.

3. Other Nucleus-Determined Molecules Regulating Cytoplasmic Activity. The overwhelming importance of RNA in nucleocytoplasmic interactions to some extent reflects current interest in nucleic acids. Protein must be involved in translocation of RNA into the cytoplasm since it provides a protective covering. Goldstein and Prescott (1967) have also demonstrated a movement of protein between nucleus and cytoplasm. Little is known about the function of these molecules in the regulation of cytoplasmic activity. Other types of molecules may also be involved but, again, virtually nothing is known about their relevance to cytoplasmic regulation.

B. Regulation by the Cytoplasm

The experiments discussed earlier show that the cytoplasm very largely determines the future activities of the nucleus, hence of the cell. In the absence of a clear understanding of the underlying mechanisms, there is always a temptation to attach undue importance to the results. Again, we need to remind ourselves that the special place of the nucleus in cellular organization is unequivocally established; it contains the bulk of the genome. Whatever information may come from the cytoplasm to affect nuclear behavior, the presumption must be that if it is in the form of RNA or protein molecules the information originally specifying these originated in the nucleus. What is really under consideration is the ability of the cytoplasm to manipulate information and to ensure a controlled release of certain classes of information from the nucleus. It appears to be the role of the cytoplasm to convert the one-dimensional form of the original information, that is, the base sequences in DNA, into the four-dimensional structure that is the cell in space and time. It is possible for the relevant structural information to be coded into the form of linear sequences, but it is hard to see how this can be converted into the necessary four-dimensional array without a suitable organization to select and use the information. In a sense therefore the cytoplasm is not simply the expression of the genome, or at least a part of the genome; it possesses by virtue of its continuing, if changing, structure and function, a structural content not strictly contained in the genome (Sonneborn, 1970). That is, nucleus and cytoplasm are indispensable to one another, possessing complementary functions.

1. Regulation by Relatively Nonspecific Molecules. Experiments in which nuclei are transferred into foreign cytoplasm, either by transplantation or cell fusion, show conclusively:

1. The cytoplasm determines whether or not a nucleus synthesizes RNA and DNA.
2. Ability to synthesize nucleic acids is dominant over inability.
3. At least some of the cytoplasmic factors conditioning the nucleus are not species-specific.

Since nucleic acids have a high degree of structural specificity, it is likely that the cytoplasmic factors responsible for nuclear reorganization do not belong to this class of macromolecules. It seems intrinsically more probable that the initiating factors are the presence and availability of enzymes or small molecules which are widely distributed among life forms. The latter might include inorganic ions, small cofactors, and molecules such as the riboside triphosphates which play a key role in the supply of energy and which largely originate in the cytoplasm.

There is some indication that the process of nuclear reorganization has two stages: initiation, which is determined by relatively unspecific factors; and development, which is largely nuclear controlled. This scheme is supported by the following evidence.

1. Reorganization of the nucleus, for example, the hen nucleus in a HeLa–hen red cell heterokaryon, is rapidly initiated by the foreign cytoplasm, as judged by enlargement of the nucleus and the start of RNA synthesis. There is no evidence for translocation of RNA from the hen nucleus to the cytoplasm at this stage, nor is there any synthesis of new hen proteins. The stage of reorganization that produces a nucleolus, rRNA, and protein synthesis, takes several days more. The obvious conclusion must be that the early stage is a general reactivation of the chromatin mediated by cyto-

plasmic factors to which the hen nucleus is unable to contribute; the later stage is connected with a sequence of nuclear events which enables the nucleus to modify the cytoplasm.

2. The cytoplasm of a cell capable of RNA and DNA synthesis is dominant as far as these two activities are concerned, but this is not generally true of other synthetic activities. For example, immature hen red cells which are synthesizing hemoglobin can be fused with HeLa cells to give a heterokaryon which includes cytoplasm from both cells (Harris, 1970). The ability to synthesize hemoglobin is at first stimulated in the heterokaryon, but within 24 hr begins to decline and is soon irreversibly lost. In contrast, when A_9 cells, which lack the enzyme inosinic pyrophosphorylase, are fused with hen red cells, the heterokaryons, as expected, are unable to incorporate hypoxanthine into the nucleic acids. At the time when a nucleolus makes its appearance in the hen nuclei, and hen-specific proteins begin to be made, the heterokaryon also acquires the ability to incorporate hypoxanthine, and enzymic analysis shows that inosinic acid pyrophosphorylase is now being synthesized. In this latter case, then, a gene has been derepressed, and the process appears to be a nucleus-determined one; in the former case a nucleus-determined process, hemoglobin synthesis, has been rapidly suppressed and is not observed again.

A sufficient number of cases of unpredictable changes in cell activities has been reported to make it evident that, while the final pattern of activation at the level of RNA and DNA synthesis is predictable, this is not true of other nuclear activities.

The most obvious initial effect in nuclear activation is an increase in nuclear size and expansion of the chromatin. This is observed with both heterokaryons and nuclear transplants into amphibian eggs. Effects reminiscent of these are observed with isolated nuclei *in vitro*. It is commonly observed that cations such as Ca^{2+} and Mg^{2+} have marked effects on nuclear structure; nuclei isolated in isotonic sucrose solutions normally burst unless low concentrations (1–3 mM) of these cations are added to the isolation medium. Even lower concentrations of heavy metal ions are effective. Divalent cations visibly affect the state of aggregation of the chromatin; increasing concentrations cause the chromatin to condense and become very granular and pyknotic, while addition of chelating agents such as EDTA has the reverse effect. The magnitude of the effect varies with the organism; for example, the chromatin in interphase nuclei of *P. polycephalum* is always rather granular, and when nuclei are isolated in medium containing 2–3 mM calcium it becomes so condensed that it resembles prochromosomes, that is, chromatin which at the beginning of prophase is condensing into visible chromosomes. These visible changes are also accompanied by considerable changes in the metabolism of RNA in isolated nuclei.

The evidence in no way demonstrates that cations are responsible for initiating reorganization in heterokaryons, but it shows that molecules of this type can have profound effects on a nucleus. To suppose that gross variations in ionic strengths in the cytoplasm are responsible for the effects would be rather naive. The nuclear membrane is a complex structure which does not normally allow free movement of ions (Loewenstein, 1965). Further, any large variations in the concentrations of metal ions would be expected to have an adverse effect on cytoplasmic structures such as the ribosomes which are very sensitive to magnesium concentration. Despite these obstacles there remain various observations on the effects of salt concentrations on embryogenesis. One of the better known of these is the effect of lithium chloride and sodium bicarbonate on differentiation in amphibian embryos. These chemicals profoundly modify the normal pattern of differentiation and therefore provide a clear indication that small ions are capable of participating in nucleocytoplasmic interactions (Flickinger et al., 1970).

There is presumably a range of small molecules capable of influencing nuclear behavior. It does not necessarily follow that they all exert their influence directly on the nucleus, but alterations in one part of the cell must affect the rest. The various plant and animal hormones are examples of small, relatively nonspecies-specific molecules that cause far-reaching effects on the pattern of behavior of cells. Some of these molecules may operate at the level of transcription of DNA, others may affect translation of messenger, while others may directly influence enzymic activity. It often appears that the small molecule combines with a much larger molecule, for example, of protein, and this combination interacts with the relevant cell structures to produce its effect (Tomkins et al., 1969; Jost and Rickenberg, 1971). This in turn suggests that the specificity of action of nonspecies-specific factors may depend on the combination of a small cytoplasmic agent with a molecule in the nucleus to provide a molecule capable of manifesting high specificity.

2. Regulation by Proteins. The absence of species specificity discussed above should not be allowed to obscure the fact that much species specificity does exist. The cytoplasm of a specialized cell such as the egg certainly contains a great many highly specific factors which control nuclear development in a characteristic way for a considerable number of divisions. Embryogenesis can hardly be a case in which simple considerations of the concentrations of small molecules are adequate to explain the complex sequence of events that results in the development of a viable embryo. This is indicated by the fact that interspecific nuclear transfers into amphibian eggs fail to develop satisfactorily after a few divisions (Gurdon, 1962). Similarly, the cytoplasm of the egg is itself highly organized and often cannot be disturbed, for example, by centrifugation, without destroying the ability of the egg to develop. Without pressing the point further, it is apparent that in addition to factors lacking high species specificity there are those that are highly specific.

The obvious mediators of species-specific regulation of nuclear activity are the nucleic acids and proteins. Whenever a nonradioactive plasmodium of *P. polycephalum* is fused with one that has either labeled nucleic acids or proteins, there is a rapid movement of labeled RNA or protein into the unlabeled nuclei. The results obtained with RNA are rather equivocal because of the difficulty of ensuring that intracellular pools of labeled precursors have been reduced to insignificant levels; even if this could be ensured, there still remains the intracellular turnover of labeled RNA which is continually returning labeled precursors to the intracellular pool. It is therefore not entirely certain that the appearance of radioactivity in the unlabeled nuclei represents translocation of molecules rather than *de novo* synthesis of RNA. This is not the case with proteins, however, since pool effects can be eliminated. There is clear evidence for a relatively large-scale exchange of proteins between nucleus and cytoplasm.

Experiments with heterokaryons and nuclear transplantation show the same phenomenon, and it is tempting to suppose that it plays an important part in nucleocytoplasmic interactions. Goldstein and Prescott (1967) studied the translocation of proteins in *Amoeba proteus*. These investigators used a series of nuclear transfers as shown in Figure 5. The nucleus N1 from a cell that had been given radioactive food to label its proteins was transplanted into an unlabeled cell with cytoplasm C2 and nucleus N2. At this stage there was an appreciable movement of label out of N1 and a

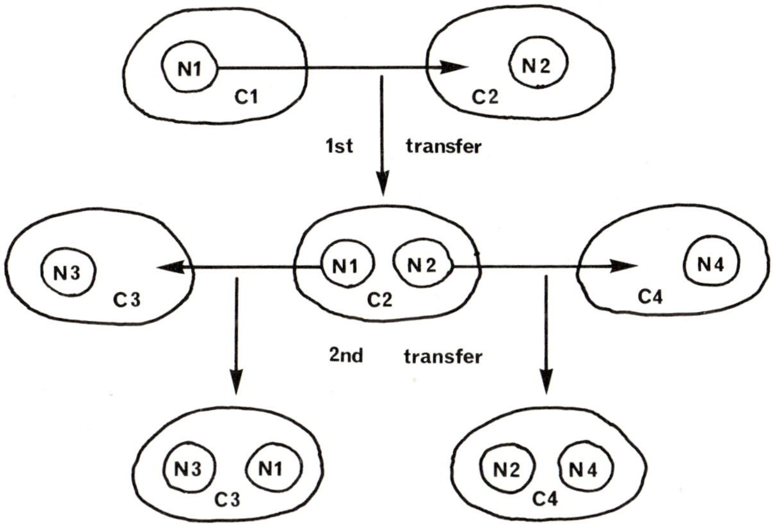

Figure 5. The experiment of Goldstein and Prescott (1967) which demonstrates movement of proteins between nucleus and cytoplasm in *A. proteus*. See text for details.

corresponding appearance of labeled proteins in the cytoplasm and in nucleus N2. The counts in the two nuclei were in the ratio of approximately 6:1. The nuclei N1 and N2 were then transplanted into two more unlabeled cells, and the redistribution of label again measured. The ratio of the counts in the cell containing N1 was about 11:1 in favor of N1, while it was about 1:1 in the cell containing N2. The experiments thus showed that there are two classes of proteins that presumably fulfil different roles. Further experiments showed that the fraction of protein that remained in nucleus N1 in fact also migrated very slowly into the cytoplasm. Thus virtually all the nuclear proteins appeared to turn over, and two classes moved between nucleus and cytoplasm, one very rapidly, the other much more slowly. Calculations showed that the actual amounts of the rapidly migrating proteins in nucleus and cytoplasm were approximately the same, but bcause of the small size of the nucleus, approximately 2% of the cell volume, the concentration in the nucleus was about 50-fold greater than in the cytoplasm.

Relatively little is known about these migrating proteins, but DNA polymerase has been suggested to be in this class, and to be involved in the induction of DNA synthesis (Gurdon and Woodland, 1968). For example, in amphibian eggs the enzyme is found in the cytoplasm until the nucleus begins to synthesize DNA. It seems certain that a migration of this kind occurs in heterokaryons in which DNA synthesis is induced, since the nuclei that are induced do not appear competent to make the enzyme themselves until nuclear reorganization is well advanced. It is unlikely, however, that this is a sufficient cause of induction, since lymphocytes can be induced to make DNA polymerase and DNA and even divide under the stimulus of phytohemagglutinin. The effect soon ceases however, so that the mere presence of DNA polymerase is not sufficient to maintain the process (Loeb et al., 1969).

5. CONCLUSION

It is not possible at present to construct a coherent model of nucleocytoplasmic relationships, largely because of an almost complete absence of data in several important areas. The current ideas about RNA synthesis, the mRNA theory, and protein biosynthesis, provide a framework which accounts reasonably well for some of the nucleus-determined effects, but the mechanism by which the correct part of the genome is selected for use or repression is not understood. A much better understanding of the structure of the chromosome seems to be one of the keys to this problem, since we are presumably considering the ways in which the chromosome is expanded to expose selected areas of the genome which are then activated or repressed. Some of the bonds holding the chromatin in a condensed form appear sensitive to the concentrations of quite unspecific molecules, but others respond only to highly specific stimuli.

The problem of differential repression of regions in the genome is something of a mystery. Repression by inhibitor molecules, perhaps proteins, similar to that observed in bacteria is probably responsible for regulating individual genes. Repression in this case would then occur by feedback of secondary gene products or by the action of regulator genes producing suppressor molecules. Britten and Davidson (1969) have discussed the problem of activation and repression of different groups of genes entirely in terms of intranuclear phenomena. Their scheme is designed to account for the wide range of combinations of gene activities observed in the cells of higher animals and is therefore much more complex than the bacterial model. They suggest that groups of genes are activated by RNA molecules produced by regulator genes; the activator RNA may or may not operate by producing a protein molecule which is the actual molecule interacting with the structural gene. The essential feature of their model, which distinguishes it from that applied to bacteria, is the assumption that the regulator genes exist in batteries which correspond to the genetic activity of a particular type of differentiated cell. Each of these batteries of regulator genes is in turn switched on and off by a sensor gene which responds to stimuli directing the course of differentiation. Any particular regulator gene may be duplicated many times, but each one is associated with a different combination of other regulators.

The model described by Britten and Davidson (1969) is interesting because it provides a function for that part of the genome that cannot be accounted for in terms of structural genes. Further, unlike most theories of repression, it appears to offer a function for the nuclear RNA that never appears in the cytoplasm. The sensor gene in turn offers a means by which extranuclear signals could modify the pattern of cell development.

The genome of the somatic cells of higher animals contains many genes that are replicated during the cell cycle but never used. Taking the simple view that expansion of the chromatin exposes many genes that are not required for cell growth, why do they remain repressed? One answer seems to be that the cells must be associated during ontogeny with a suitable tissue for this to occur, so that extracellular influences may in the end be decisive. Certainly, embryogenesis seems to support this view. The utilization of those genes not essential to cell growth would thus ultimately depend on the specialized structure of the original egg cytoplasm. Once more, then, we find that it is the interactions of nucleus and cytoplasm that determine the behavior of both.

REFERENCES

Brachet, J. (1961). In *The Cell* (J. Brachet and A. E. Mirsky, eds.), Vol. II, pp. 771–841. Academic, New York.

Bramwell, M. E. (1970). Intranuclear accumulation of RNA resembling the smaller ribosomal RNA component. *J. Cell Sci.*, **6,** 53.

Britten, R. J., and E. H. Davidson (1969). Gene regulation for higher cells: A theory. *Science*, **165,** 349.

Crick, F. (1971). General model for the chromosomes of higher organism. *Nature*, **234,** 25.

Darnell, J. E. (1968). Ribonucleic acids from animal cells. *Bacteriol. Rev.*, **32,** 262.

De Terra, N. (1969). Cytoplasmic control over the nuclear events of cell reproduction. *Intern. Rev. Cytol.*, **25,** 1.

Flickinger, R. A., D. M. Kohl, M. R. Lauth, and P. J. Stambrook (1970). Effect of rate and number of cell divisions on RNA synthesis. *Biochim. Biophys. Acta*, **209,** 260.

Geiduschek E. P., and R. Haselkorn (1969). Messenger RNA. *Ann. Rev. Biochem.*, **38,** 647.

Goldstein, L., and D. M. Prescott (1967). The fundamental characteristics of the rapidly migrating proteins and the slow turnover proteins of the *Amoeba proteus* nucleus. *J. Cell Biol.*, **33,** 637.

Graham, C. F., K. Arms, and J. B. Gurdon (1966). The induction of DNA synthesis by frog egg cytoplasm. *Develop. Biol.*, **14,** 349.

Gray, W. D., and C. J. Alexopoulos (1968). *Biology of the Myxomycetes*. Ronald, New York.

Gurdon, J. B. (1962). The transplantation of nuclei between two species of *Xenopus*. *Develop. Biol.*, **5,** 68.

Gurdon, J. B., and H. R. Woodland (1968). The cytoplasmic control of nuclear activity in animal development. *Biol. Rev.*, **43,** 233.

Hadjiolov, A. A. (1967). Ribonucleic acids and information transfer in animal cells. *Prog. Nucleic Acid Res. Mol. Biol.*, **7,** 196.

Hämmerling, J. (1963). Nucleo-cytoplasmic interactions in *Acetabularia* and other cells. *Ann. Rev. Plant Physiol.*, **14,** 65.

Harris, H. (1968). *Nucleus and Cytoplasm*. Clarendon, Oxford.

Harris, H. (1970). *Cell Fusion*. Clarendon, Oxford.

Harris, H., and J. F. Watkins (1965). Hybrid cells derived from mouse and man: Artificial heterokaryons of mammalian cells from different species. *Nature*, **205,** 640.

Holder, J. W., and J. B. Lingrel (1970). The localization of the hemoglobin mRNA on the 40S ribosomal subunit of rabbit reticulocyte ribosomes. *Biochim. Biophys. Acta*, **204,** 210.

Jacob, F., and J. Monod (1961). Genetic regulatory mechanisms in the synthesis of proteins. *J. Mol. Biol.*, **3,** 318.

Jost, J-P., and H. V. Rickenberg (1971). Cyclic AMP. *Ann. Rev. Biochem.*, **40,** 741.

Keck, K. (1969). Metabolism of enucleated cells. *Intern. Rev. Cytol.*, **26,** 191.

Laycock, D. G., and J. A. Hunt (1969). Synthesis of rabbit globin by a bacterial cell free system. *Nature*, **221,** 1118.

Loeb, L. A., S. S. Agarwal, and A. M. Woodside (1969). Induction of DNA polymerase in human lymphocytes by phytohemagglutinin. *Proc. Natl. Acad. Sci., U.S.*, **61,** 827.

Loewenstein, W. R. (1965). Permeability of a nuclear membrane: Changes during normal development and changes induced by growth hormone. *Science*, **150,** 909.

Melli, M., and J. O. Bishop (1969). Hybridization between rat liver DNA and complementary RNA. *J. Mol. Biol.*, **40,** 117.

Nemer, M. (1967). Transfer of genetic information during embryogenesis. *Prog. Nucleic Acid Res. Mol. Biol.*, **7,** 243.

Nitova, V. K., and K. M. Sukhanova (1964). Synthesis of nucleic acids in *Opalina ranarum*, Ehrbg. *Nature*, **204,** 459.

Rusch, H. P., W. Sachsenmaier, K. Behrens, and V. Gruter (1966). Synchronization of mitosis by the fusion of the plasmodia of *Physarum polycephalum*. *J. Cell Biol.*, **31,** 204.

Shearer, R. W., and B. J. McCarthy (1967). Evidence for RNA molecules restricted to the cell nucleus. *Biochemistry*, **6,** 283.

Sonneborn, T. M. (1970). Gene action in development. *Proc. Roy. Soc.*, B, **176,** 347.

Tomkins, G. M., T. D. Gelehrter, D. Granner, D. Martin, H. H. Samuels, and E. B. Thompson (1969). Control of specific gene expression in higher organisms. *Science*, **166,** 1474.

Weinberg, R. A., and S. Penman (1970). Processing of 45S nucleolar RNA. *J. Mol. Biol.*, **47,** 169.

Woese, C. R. (1967). *The Genetic Code*. Harper, New York.

9

NUCLEIC ACID BIOSYNTHESIS AND INTERACTIONS

R. H. Burdon

1. THE BIOSYNTHESIS OF DNA

A. General Considerations

During the cell division cycle the genome, comprising a unique collection of DNA molecules, is precisely replicated. One of these newly manufactured collections is then distributed into each of the daughter cells. In the cells of higher organisms, such as plants and mammals, the whole genome is replicated not at the actual phase of division, or mitosis, but at a particular time during the interphase or resting phase of the cell division cycle. This period, from one-third to one-half of the cycle time, has been designated the S (synthetic) phase. It is preceded and followed by other well-defined periods of the interphase, designated G_1 and G_2, during which no DNA synthesis occurs. This contrasts with the situation in a bacterium such as *Escherichia coli* in which, under normal growth conditions, 90–97% of cell cycle time is spent in DNA replication, there being no periods that can be considered equivalent to the G_1 and G_2 phases.

Although the mammalian cell genome is replicated completely during the S phase, the replication process does not proceed uniformly. DNA molecules are replicated at defined locations within the nucleus as well as at defined times within the S phase itself. Moreover, the replication rate varies according to both the location and time (see Cleaver, 1967). From Chapter 5 on the cell nucleus, it is clear that the DNA molecules comprising the mammalian genome are organized as complexes with basic and acidic proteins to form chromosomes which condense and become clearly visible at the time of mitosis. However, during interphase, there is much less structural detail to be seen either by the light or electron microscope. Nevertheless, two different states of DNA have been recognized, namely, euchromatin and heterochromatin. The first of these is an active site of RNA manufacture and consists of loosely coiled fibers. Heterochromatin, however, is far less active as far as RNA manufacture is concerned and from the morphological point of view comprises tightly coiled fibers. What then is known regarding the replication of the DNA contained in these two states? Considerable insight into this type of problem has come through the use of autoradiography at the microscopic level, employing tracer levels of the DNA nucleoside thymidine labeled with tritium (^3H), to indicate the intracellular sites and levels of DNA synthesis. The precise mechanism whereby thymidine taken up by cells exposed for suitable times to the tracer is discussed later in this chapter. However, in principle, the thymidine is converted within the cell to thymidine 5′-triphosphate which then serves as a substrate for the enzyme DNA polymerase in the manufacture of new DNA. A property of the newly synthesized DNA in contrast to the triphosphate substrate is its insolubility in the acid media used to fix the cell for subsequent microscopic examination. Thus it is a relatively simple matter to detect the intracellular location of the radioactive, newly synthesized, acid-fixed DNA by coating the fixed cell preparation with suitable photographic emulsion and storing in the absence of light for a given amount of time. The distribution of radioactive atoms within the nuclear structure is then detected by examining the pattern of developed silver grains that the emitted radiation produces in the overlying photographic emulsion. By this means it was clear that DNA of euchromatin was replicated early in S phase whereas that in heterochromatin was replicated late. Isolation and autoradiographic examination of DNA molecules from cells exposed to [^3H]thymidine indicated a rate of replication of the order of approximately 1 μ/min. This is too slow, however, for replication to be completed within the time of S phase by means of a single replication site per chromosome. Indeed, several hundred sites may be required on each chromosome through S phase. Additional autoradiographic experiments (see Cleaver, 1967) now show that replication does occur at discrete positions along each chromosome and moreover at several positions simultaneously. In general, chromosome replication is asynchronous. Homologous chromosomes may sometimes have differing replication patterns, and certain chromosomes have regions that consistently replicate late in the S phase. For instance, late replication of the X chromosome in the XX females of animals is observed. In addition to these complexities, examination of the genome of the mouse, for example, at the molecular level using techniques of ultracentrifugation, showed that 10–12% of the DNA was readily separable and chemically distinct from the remainder (Walker, 1971), and for this reason it has been called "satellite" DNA. Ostensibly, it is made up of tandem repeats of similar six nucleotide sequences and does not appear to play a role in coding for proteins and RNA. Recently, by making use of a novel technique of intracellular molecular hybridization, Jones (1969) not only localized satellite DNA in the heterochromatin fraction of interphase nuclei but found it to be concentrated in the centromere regions of mitotic chromosomes. Thus in short it is clear that a complete understanding of the molecular mechanism of DNA replication as it occurs in higher organisms must take all these above-mentioned complexities into account. Not surprisingly perhaps, as evident later, progress in this direction has been slow.

By comparison, however, the situation in bacteria appears less complex, at least superficially. DNA replication in bacteria has been the subject of intensive study in recent years and has yielded a valuable insight into the molecular processes involved, which has been of great value to investigators confronted with the problems existing in more complex organisms.

B. The Replication of the Bacterial Chromosome

1. General. From autoradiographic studies at the molecular level, and from genetic studies, the chromosome of the bacterium *E. coli* is known to be a single circular DNA molecule. What are the features of a DNA molecule? It is now fairly well established that DNA molecules are composed of complementary polynucleotide sequences arranged in the form of a double helix around a common axis. Such a helical molecule is maintained in this configuration by the aid of specific hydrogen bonding between the complementary base moieties of nucleotides comprising the two strands. The base adenine (A) can form what is termed a base pair with thymine (T), and guanine (G) can form a base pair with cytosine (C). A and T are therefore said to be complementary base pairs, as are G and C. In the *E. coli* chromosome, since its molecular weight is 3000 million, there are 5 million such pairs. In addition, it should be remembered at this stage that the component strands of the DNA double helix are of the opposite polarity. That is, the 3′ and 5′ positions of the deoxyribose moieties in one strand run in a direction opposite those in the complementary strand. This is illustrated in Figure 1. Furthermore, it should be noted that one end of each strand is now commonly termed the 5′ end whereas the other end of each strand is termed the 3′ end (Fig. 1).

Meselson and Stahl (1958) provided compelling evidence for the conservation of each strand of the original double-helical molecule during the process of replication in *E. coli*. They grew the bacterium in a medium containing ^{15}N (the "heavy" isotope of nitrogen) for several generations and then transferred them to a medium containing ^{14}N ("light" nitrogen); samples of the culture were withdrawn at intervals for several generations. DNA was isolated from these cell samples and its buoyant density determined by ultracentrifugation in gradients of cesium chloride. Initially, the DNA appeared as a single band corresponding to heavy or ^{15}N-labeled nucleic acid, that is, containing ^{15}N in its bases. Later, DNA molecules in a different position containing one-half the level of ^{15}N began to appear, and one generation after culturing in the ^{14}N medium these hybrid molecules (one light ^{14}N strand and one heavy ^{15}N strand) alone were present. After a second generation two bands of DNA were found, one in the position of the hybrid just described and another where the ^{14}N DNA would be found (see Fig. 2). Such observations suggested that during replication each existing molecule is effectively split into two single strands, each of these strands ultimately appearing in a daughter molecule, the other strand of which is the product of new DNA synthesis. Some clues regarding the molecular mechanism of this new synthesis were reported in 1956. Kornberg and his colleagues (see Kornberg, 1961)

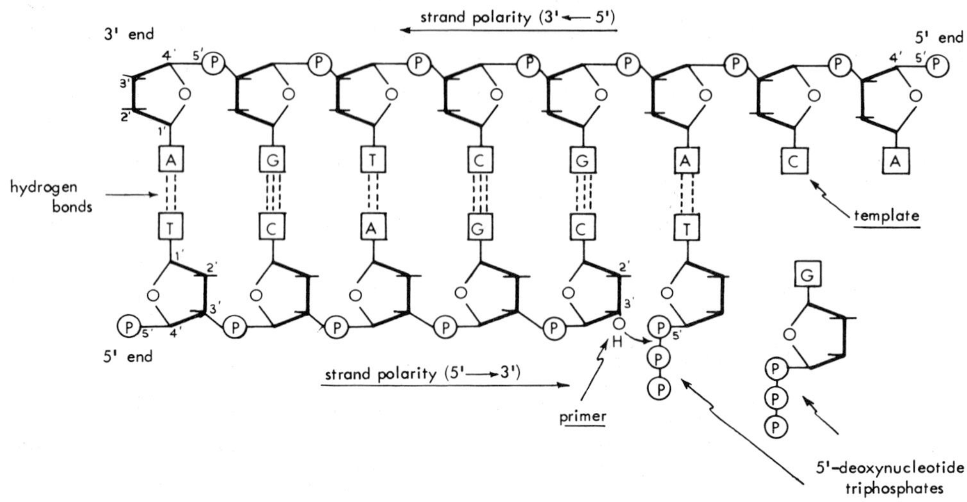

Figure 1. A diagrammatic representation of DNA structure illustrating the opposite polarity of the component strands and the mode of action of *E. coli* DNA polymerase using one strand as template and the other as primer.

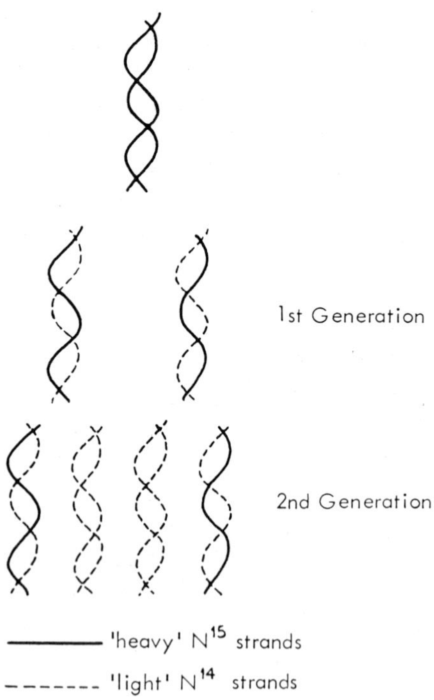

Figure 2. Schematic representation of the results of the Meselson and Stahl experiment.

showed that in cell-free preparations of *E. coli* isotopically labeled thymidine was incorporated into an acid-insoluble material which proved to be DNA. Soon they discovered that the preparation in fact contained an enzyme system which converted the thymidine to thymidine 5′-triphosphate (dTTP). This in turn served as one of the substrates for an enzyme capable of the *in vitro* synthesis of new DNA. In addition, the enzyme required the 5′-triphosphates of deoxyadenosine, deoxycytidine, and deoxyguanosine, as well as magnesium ions and some added DNA. This enzyme, now designated DNA polymerase, has since been detected in higher organisms (see Keir, 1966), but it is the bacterial enzyme that has been purified several thousandfold. With the purified *E. coli* enzyme, 10- to 20-fold increases in the amount of DNA originally present in the reaction mixture can be demonstrated. Moreover, chemical analyses indicated that the nucleotide composition of the newly synthesized DNA mimicked almost precisely that of the original DNA added (Kornberg, 1961). In addition, using an ingenious procedure called "nearest neighbor frequency analysis," Josse et al. (1961) gained some idea of the sequences of nucleotide units in the DNA newly created by the action of the enzyme. The frequency with which any given nucleotide occurs adjacent to any other is a function of the nucleotide sequence. The results of these analyses suggested that the newly synthesized DNA product results from the copying of both strands of the original DNA molecules proffered to the enzyme whatever their source. Thus it appeared that in some manner the original DNA acted as a "template" for the synthesis of new DNA, and the enzymic manufacture of DNA can be envisaged as occurring in three main stages: (1) formation of deoxyribonucleotide 5′-monophosphates (from small precursor molecules, including thymidine); (2) their phosphorylation to the 5′-triphosphate level by appropriate kinase enzymes; and (3) the polymerization of the 5′-triphosphates to DNA by the polymerase in the presence of magnesium ions and template DNA.

2. **The *in Vivo* Situation.** At this stage it is profitable to consider in more detail what is known of the replication process as it occurs *in vivo*. We have already mentioned the classic experiment of Meselson and Stahl. This, together with similar type of experiments, has shown that chromosome replication in bacteria is "semiconservative" and proceeds in a sequential manner in which order is preserved between one cycle of replication and the next. Some insight into the actual processes that might be involved was gained from the elegant autoradiographic studies of Cairns (1963) involving uptake of [^{3}H]thymine. He devised an approach that allowed him to examine DNA molecules literally caught in the act of duplication. Labeled thymine is incorporated into the replicating DNA of the intact bacteria, presumably after phosphorylation and action of DNA polymerase. The cells are very gently lysed so that the DNA molecules can be withdrawn intact and then placed on special membranes coated with photographic emulsion sensitive to the β particles emitted by the isotope. Not only does the chromosome comprise two polynucleotide chains, which separate at the time of duplication, as anticipated, but the circular molecule is duplicated at a single locus that traverses its entire length. Two polynucleotide strands become four at this point and from its Y-shaped appearance it has come to be called the replicating "fork." However, the replicating chromosome as a whole is not simply a Y-shaped entity. The 3′ ends of the Y are apparently joined. The ends of the daughter molecules are joined to each other and to the far end of the parent molecule. In short, the chromosome remains circular while being replicated (Fig. 3*A*). Similar replicating structures have been observed with other bacterial DNAs and also with some virus

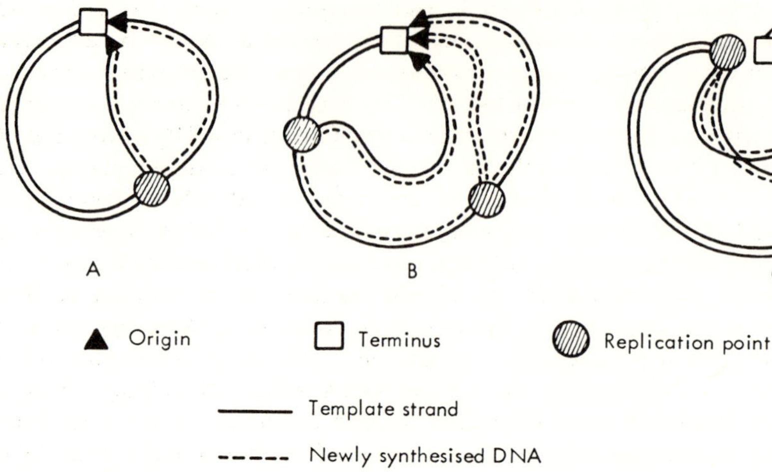

Figure 3. Replication of the cyclic molecule of DNA in the *E. coli* cell according to the basic mechanism proposed by Cairns: (*A*) with one replication fork; (*B*) with two replication forks; (*C*) with one replication fork but positioned at a membrane site close to the origin as envisaged by Sueoka and Quinn (1968). Newly synthesized DNA is indicated by the dashed line.

DNAs. The replication fork from Cairns' studies appears to travel around the chromosome at a fairly constant rate despite bacterial growth rate. About 50 min is required to traverse the whole chromosome, which corresponds to a synthetic rate of one-half million nucleotide pairs per minute. At very fast cell growth rates (generation time of approximately 20 min), however, new replicating forks appear to be initiated (Pierucci and Helmstetter, 1969), leading to the multiforked replication situation illustrated in Figure 3*B*.

In 1961, on the basis of nutritional experiments with various auxotrophs, Maaloe and Hanawalt (1961) postulated that the bacterial chromosome originates at a specific site and proceeds sequentially to the terminus, but that synthesis of protein is required before a new round of replication can commence. A more elaborate proposal was put forward 2 yr later by Jacob et al. (1963). A genetic element replicated in its entirety, such as the bacterial chromosome, was designated a replicating unit, or "replicon" by these investigators. Initiation of replication in their view was brought about by the specific interaction of an initiator substance with a unique region of the replicon. Furthermore, they proposed that in bacteria the replicating unit was attached to a component of the cell membrane to assure segregation of the newly formed genetic material. Indeed, their electron microscope studies supported this view that there was some interaction with membranous material.

The unique region of the chromosome from which sequential replication is initiated is referred to as the replicative "origin," and a cycle of replication is completed when the chromosome "terminus" has been duplicated. This of course poses the question as to whether or not in the bacterial chromosome there is any discontinuity, chemical or otherwise, between the proposed origin and terminus. When protein synthesis is arrested by starving the bacterial cultures of amino acids, for instance, replication proceeds for a while and then is halted, presumably at the terminus–origin region. When the amino acids are replenished, DNA synthesis commences. If, however, the starvation and replenishment of amino acids are repeated, it was found by Pritchard and Lark (1964), using various heavy isotopes to label the newly synthesized DNA, that it was always this hypothetical terminus–origin that is replicated first when DNA synthesis commences after its arrest resulting from inhibition of protein synthesis. Further evidence from the study of the replication of genetic markers indicated that in most strains of *E. coli* replication proceeds from a common origin (close to the xylose marker) and in a clockwise direction in respect to the genetic map of the organism (Ward and Glaser, 1970). Recently, genetic data (Masters and Broda, 1971) do indicate, however, that replication can on some occasions occur in both directions.

On a superficial structural basis, it is clear that for most of the replication cycle the origin must exist in a duplicated form, whereas the terminus does not (Fig. 3*A*). However, this is not the only structural problem. As is well known, the two strands that comprise the DNA molecule are wound around each other in a helical configuration, one complete turn for every 10 nucleotide pairs. Thus separation of the strands during replication, similar to the strands of a rope, must involve the rotation of the parent molecule with respect to the daughter molecules. A bacterium dividing approximately every 50 min would then have to unwind 500,000 turns in that time, that is, 10,000 revolutions per minute. However, a firmly closed circular molecule cannot be unwound. Thus to replicate and rotate such a molecule, it is probably necessary to break one (or even both strands). Indeed, experiments carried out in the study of the replication of circular helical bacteriophage DNAs indicate this to be necessary. Nevertheless, such a scission of the bacterial chromosome would require some endonuclease-type enzyme specific for a certain nucleotide sequence. It is known that certain cytosine and adenine residues already in DNA nucleotide sequences can be methylated by specific methylase enzymes to yield methyladenine and methylcytosine. When this is prevented, initiation of replication is also prevented (Lark, 1968). Perhaps at the origin it is a special sequence containing one of these modified bases that is recognized by a key scission enzyme.

In addition to indicating this scission step, the bacteriophage experiments also implicated attachment of the origin of the replicon to a membrane component as a necessary part of the initiation process, and according to a study with protein synthesis inhibitors this probably accounts for at least a part of the protein synthetic requirement for initiation. Although replicating DNA is complexed to the cell membrane of *E. coli*, there are no data demonstrating specific attachment; certain genetic loci of the *Bacillus subtilis* chromosomes remain preferentially attached during replication, and radioactive label representing the terminus–origin can also be shown to remain attached (Sueoka and Quinn, 1968). However, more interestingly, the most recently made DNA is also attached to membrane components and can be solubilized with detergents and lipase (Ganesan, 1967; Smith and Hanawalt, 1967). This implies that the replication fork itself may be attached to the cell membrane. A modification of the Cairns' model by Sueoka and Quinn (1968) takes this into account (Fig. 3*C*).

What molecular events then occur at this fork? In 1968, Okazaki and his colleagues (Okazaki et al., 1968) exposed cultures of *E. coli* to labeled thymine for 15 sec at 20°C and then carefully isolated the DNA. Surprisingly, they found the label incorporated during this time into short discrete pieces of DNA approximately 1000 nucleotides long rather than into the very large DNA molecule that constituted the chromosome. Nevertheless, after a long exposure to the labeled thymidine, the label did occur in high-molecular-weight DNA. More elaborate "chase"-type experiments showed that the short, single-stranded pieces probably served in some way as precursors of the larger DNA molecule. These investigators advanced the theory that DNA synthesis during the replication process is a discontinuous process, but just how the fragments produced were subsequently incorporated into the large DNA chromosome was not clear. However, there was known to be present in *E. coli* a ligase-type enzyme capable of joining two segments of an interrupted strand in a duplex (Richardson, 1969). This enzyme requires NAD as coenzyme and effects the joining of a single-stranded break in a DNA helix with the release of AMP (Fig. 4). The question was then asked whether or not such an enzyme might be responsible for the integration of the small fragments into large DNA molecules. Reduction of the *E. coli* NAD level to 10% of its normal level can be achieved in nicotinamide-requiring strains and leads to a situation in which all newly synthesized DNA is as small pieces. Integration of these pieces into normal-sized DNA molecules occurred when the NAD concentration was restored to normal (Nozawa and Mizino, 1969). Furthermore, a temperature-sensitive mutant of *E. coli* is known (Pauling and Hamm, 1969), which has a much reduced ligase activity,

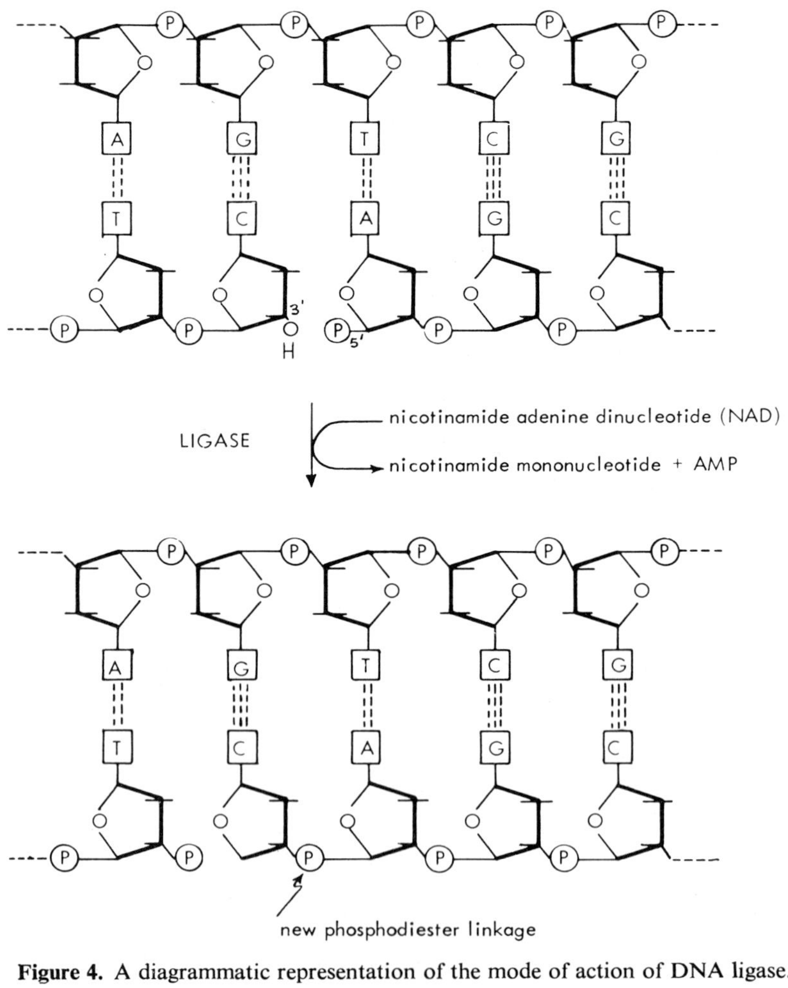

Figure 4. A diagrammatic representation of the mode of action of DNA ligase.

and when this mutant is grown at the restrictive temperature, DNA fragments accumulate. Work with temperature-sensitive mutants of bacteriophage known to be defective in ligase also supports Okazaki's hypothesis, similar accumulations of fragments being observed at the restrictive temperature. However, when ligase activity is restored by growth at the permissive temperature the fragments can be "chased" into very large molecules. Recent results of Okazaki and his colleagues now indicate, first, that the synthesis of *all* the short single-stranded fragments occurs in one direction only, namely, from the 5′ end to the 3′ end (Okazaki and Okazaki, 1969) and, second, that half of the fragments have a sequence complementary to sequences in one of the DNA strands of the double-stranded bacteriophage genome, whereas the remainder have sequences complementary to the other DNA strand of the genome (Sugimoto et al., 1969).

While supporting the general idea that an initial step in replication is the discontinuous copying of both strands, the fact that the direction of fragment synthesis is always 5′ → 3′ is puzzling, in view of (1) the antiparallel nature of the strands being copied, and (2) the apparent unidirectional travel of the replication fork. The actual situation at the fork has been variously pictured. One suggestion of Okazaki is illustrated in Figure 5 in which short stretches of DNA, synthesized by some enzyme in the 5′ → 3′ direction, are complementary to each of the parent DNA strands. The actual process may in reality oscillate rapidly from strand to strand, giving the

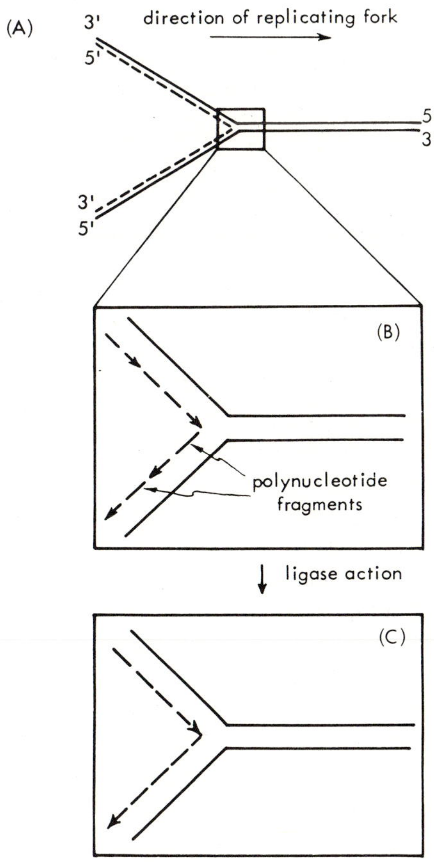

Figure 5. Schematic representation of events taking place at the replication fork. (*B*) and (*C*) indicate in more detail the processes occurring in the box indicated in (*A*). Newly synthesized DNA is indicated by the dashed line, and the arrowhead indicates the direction of synthesis, that is 5′→ 3′.

appearance of simultaneous replication of both strands. Okazaki then suggested that the fragments are connected together, and to the growing polynucleotide chain, by the ligase-catalyzed formation of phosphodiester links. Thus a reasonable role for the participation of ligases in the replication process is offered, but despite this the general enzymology of the replication process is not clear.

3. DNA Polymerase. The enzyme DNA polymerase, some features of which have already been described, was not unreasonably held by many to be a good candidate for the job of producing the fragments described by Okazaki and his colleagues. Since its discovery by Kornberg and his collaborators, the enzyme has been intensively studied. There are about 400 molecules of this enzyme in the *E. coli* bacterium, and each molecule is known to be a single polypeptide chain with a molecular weight of 109,000 (Jovin et al., 1969). If they were spherical they would each have a diameter of 65 Å which is three times the diameter of the DNA helix. The ability of purified molecules of this enzyme to initiate DNA synthesis in the test tube has been carefully studied. Initiation requires a 3′-hydroxyl-terminated *primer* DNA strand as well as a *template* strand. This concept is illustrated in Figure 1. A primer can be regarded as a polymer required in a reaction leading to the synthesis of new polymer by a process of elongation or extension of the original chain. However, a template is a single strand of DNA which serves to determine the nature and sequence of the nucleotides comprising the extension to the primer. In this process the nucleotides comprising the new portion of the polymer are aligned by complementary base pairing with those in the template. The purified polymerase apparently has binding sites for primer and template as well as *one* for all the four 5′-triphosphate substrates (Eglund et al., 1969). New DNA synthesis is then achieved by the alignment process mentioned above, the actual mechanism whereby chain growth proceeds being a nucleophilic attack by the 3′-hydroxyl at the end of the *primer* on the pyrophosphate-activated nucleotide residue aligned in the adjacent position to form a 3′-to-5′ phosphodiester bridge (Figs. 1 and 6*a*). The process of enzymic synthesis occurs in one direction only, that is, 5′ → 3′. Problems arise, however, when naturally occurring double-helical DNA molecules are proffered to the enzyme for replication. How is synthesis initiated? The enzyme could add nucleotides to the 3′ end, but unfortunately there would be no corresponding base in the template (Fig. 6*b*). Another problem with intact circular DNA molecules would simply be the complete lack of 3′ termini. In fact, when the experiment is carried out with purified polymerase and linear DNA helices, the synthetic rate of new polymer is disappointingly low (100 times less than that *in vivo*). Moreover, the product is a "branched" structure. It is postulated (Schildkraut et al., 1969) that the branching occurs as a result of the enzyme leaving the original template strand and beginning to copy the complementary strand albeit in the correct 5′ → 3′ direction. Possibly, initiation is achieved at a single-strand break (see Fig. 6*c*). Such a break may be present in a DNA double helix either naturally, or is produced inadvertently during its isolation, but at present no enzymic activity capable of carrying out such an endonucleolytic scission can be found in association with the purified polymerase. Nevertheless, for extensive synthesis of DNA starting at this break, one of the original strands would have to be displaced (see Fig. 6*c*). Moreover, in view of the reported properties of the enzyme, it is difficult to envisage the situation whereby discontinuous synthesis occurs on both strands at the fork as pictured in Figure 5 since an obvious difficulty is the initiation of synthesis in the absence of a free 3′-hydroxyl. However, this might be preceded by the formation of a branch-type situation (see Fig. 6*c*) with subsequent enzymic cleavage of the newly synthesized DNA at a point where the new chain spans the region between the two chains of the fork, creating a new 3′-hydroxyl-terminated primer. In summary, it thus appears that to date test tube conditions have not yet been found under which the purified DNA polymerase actually replicates faithfully a double-helical DNA molecule to produce two daughter molecules. It could be reasonably argued, however, that such conditions require not only DNA polymerase but also appropriate ligase and scission enzymes as the whole process is complex and requires the regulated contribution of several enzymes. Nevertheless, it is difficult at present to assign unequivocally DNA polymerase a role in DNA replication.

A feature of the purified enzyme that has recently puzzled investigators is its apparent ability to act as an exonuclease, in as much as it can cleave DNA single strands one nucleotide unit at a time in the 5′ → 3′ direction. Moreover, under certain conditions the enzyme "leapfrogs" a few units at the 5′ end of a single-stranded molecule and cuts in, to yield a fragment several units long (Kelly et al., 1969). For several years it has been known that ultraviolet radiation can have lethal effects on cells by causing the formation of chemical bonds between adjacent pyrimidine nucleotides in the DNA to form a dimer, preventing subsequent replication. Nevertheless, this damage can normally be repaired *in vivo*. One means is by the process known as *dark reactivation*. This is thought to involve the enzymic excision of the dimers followed by the repair of the gap produced (Howard-Flanders, 1968). First, a dimer, such as that between two thymine bases, introduces a local distortion into a DNA double helix and it appears that an endonuclease specific for this region makes a single-strand break close to the dimer on its 5′ side. Such a nuclease has been found in certain ultraviolet-resistant bacterial strains. Local denaturation now occurs since the dimer itself cannot participate in hydrogen bonding. However, according to Kelly et al. (1969), the duplex structure can be restored by the DNA polymerase itself, first by adding nucleotides covalently to the 3′-hydroxyl terminus using the intact strand as template until it reaches the opposite side of the gap at which the nuclease activity of

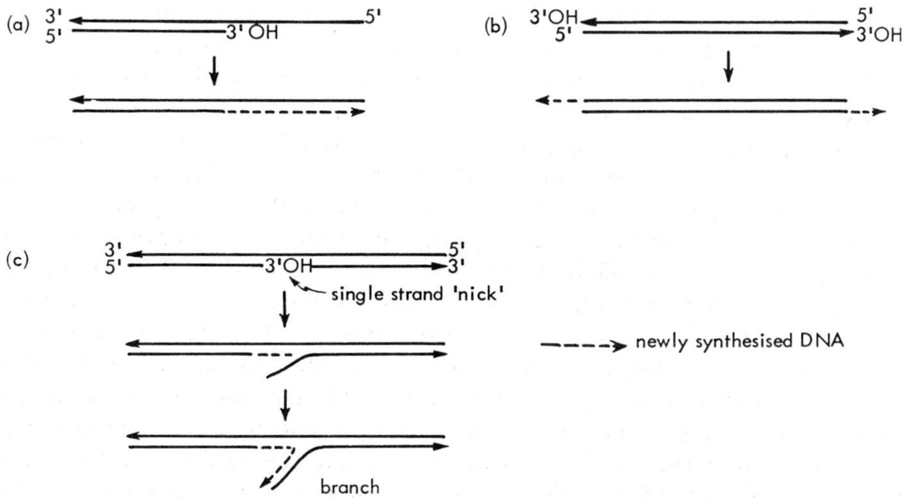

Figure 6. Possible modes of DNA polymerase action (see text for explanation).

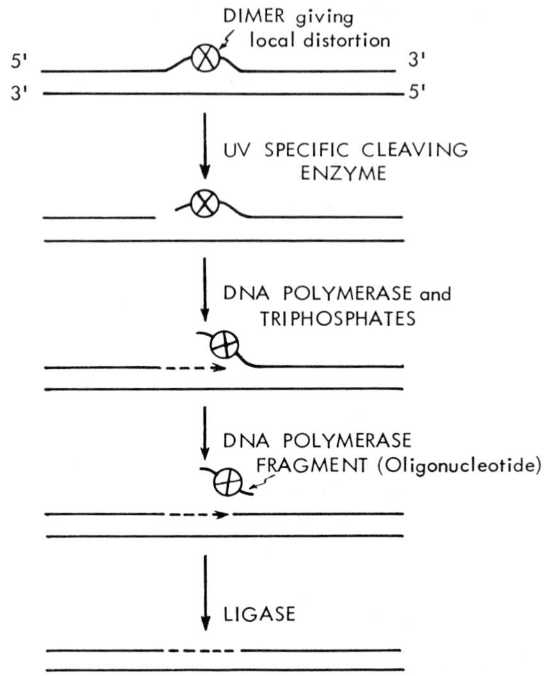

Figure 7. Hypothetical scheme for participation of DNA polymerase in the repair of irradiated DNA. (After Kelly et al., 1969.)

the polymerase comes into action and removes the damaged region as an oligonucleotide fragment. A ligase can then close the gap (see Fig. 7).

Thus some circumstantial evidence is now accumulating suggesting that inside the cell the repair of DNA could be the chief if not the sole occupation of DNA polymerase. Some additional support for this view might be taken from a recent observation of De Lucia and Cairns (1969). They found a mutant of *E. coli* that possessed almost negligible levels of soluble DNA polymerase. This mutant multiplies normally but, significantly, it has acquired an increased sensitivity to ultraviolet light. Both this sensitivity and the lack of soluble polymerase appear to be the result of a single mutation step which, according to Gross and Gross (1969), is an *amber* mutation and turns out to be in the structural gene for the soluble DNA polymerase (Kelley and Whitfield, 1971). On this basis it now seems that recovery of *E. coli* from the effects of ultraviolet radiation is partly the responsibility of the soluble DNA polymerase. Thus despite considerable intensive enzymology, the question of what enzyme, or enzymes, are actually responsible for DNA replication remained a mystery. Attention has now turned to the nonsoluble fractions of *E. coli* homogenates for possible clues. For instance, membrane fractions from certain bacteria have been known for some time to be capable of supporting the *in vitro* synthesis of DNA (Ganesan, 1967), and work of Smith et al. (1970) and Knippers and Stratling (1970) indicates that a fraction can be isolated from the mutant containing the DNA chromosome and membrane components, which is capable of DNA synthesis at a rate compatible with that detected *in vivo* (i.e., about 500,000 nucleotide pairs per minute). Kornberg and Gefter (1970) have now managed to solubilize a DNA-synthesizing activity. Although deoxyribonucleoside 5′-triphosphates serve as substrates, this activity can be distinguished from the previously described soluble DNA polymerase on the basis of its sensitivity to high ionic strength and to *p*-chloromercuribenzoate. However, although it is insensitive to antiserum directed against the "repair" DNA polymerase, the authors cannot rule out the possibility that the activity is composed of, in part, an altered form of the soluble "repair" DNA polymerase. As yet there is no precise idea of the mechanism whereby DNA replication could be achieved by these crude systems. Membrane structures can certainly be envisaged which might be complexes of a DNA polymerase-like entity, together with necessary nucleases and ligases whose concerted action is required in the replicative process. After all, the actual replication fork, as already mentioned, is also membrane-associated.

Replication of DNA on the other hand might well involve synthesis both from 5′ →3′ and 3′ → 5′ and be catalyzed by some enzyme very different from the well-studied soluble "repair" DNA polymerase. A recent radical hypothesis of Werner (1971) proposed on the basis of kinetics of isotope labeling, implies that whereas the "repair" activity utilizes the deoxyribonucleotide 5′-triphosphate, the "replicative" activity is quite distinct and the substrates are some modified forms of the nucleoside monophosphates.

4. **Models of Replication.** From the previous discussion it may be premature to advance any general picture of the process whereby the bacterial chromosome is replicated. Nevertheless, several hypothetical models have been put forward. One, however, not mentioned so far and which has provoked intense debate is that of Gilbert and Dressler (1968), the so-called "rolling circle" model (Fig. 8). Reasonably, they propose synthesis to begin by opening one strand of the original circle at a specific site by introducing a single-strand break displaying a 3′-hydroxyl and 5′-phosphoryl end group. The 5′ end is transferred to a specific membrane site. Synthesis then proceeds by elongation of the 3′ end and uses the intact strand as *template*. The old "positive" strand is peeled off the circular template as a new "negative" strand is created. As the growing point continues around the circle, daughter molecules are peeled off endlessly. It is an asymmetric process, resulting in long DNA molecules containing several genomes' worth of information, as is sometimes detected in virus-infected cells. In fact, some of the original attractions of this model lie in this, together with the fact that a large number of replicons exist naturally as circles, for example bacterial chromosomes, bacteriophage chromosomes, episomal DNA, mitochondrial DNA, and chloroplast DNA. Moreover, where the genome of a bacteriophage was

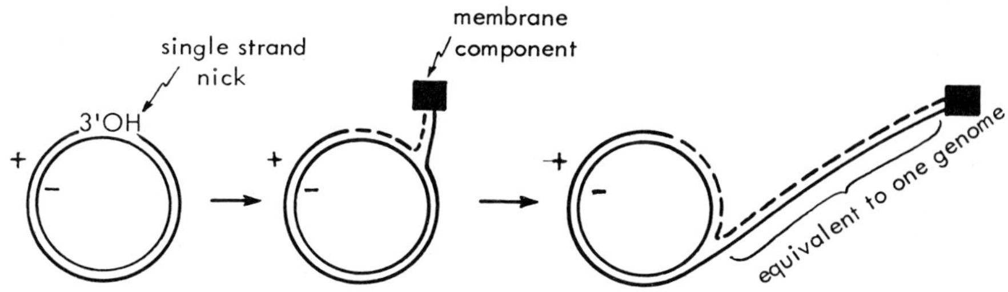

Figure 8. The rolling circle model for DNA replication. The symbols + and − indicate the individual DNA strands of the circular duplex, + being complementary in base sequence to −.

known not to be circular, there was nonetheless considerable evidence for circular intermediates to be involved in its replication. Certainly, circular daughter molecules similar in size to the parental molecules would obviously have to involve cleavage of one genome's worth of DNA followed by some recombination event. Whereas the process as originally envisaged by Cairns (Fig. 3) for chromosomal replication is a symmetrical process, Gilbert and Dresslers' is asymmetric, and while it will probably explain the replication of certain specific bacteriophage genomes, recent evidence from studies of segregation after multiforked replication of the *E. coli* genome suggests Cairns' scheme to be more likely in *E. coli* (Caro, 1970).

C. Replication in Higher Organisms

1. Studies at the Enzymic Level. Activity similar in certain aspects to that of the *E. coli* polymerase has been detected in soluble preparations obtained from mammalian cells (Smellie et al., 1960; Keir, 1965). Thus from that point of view it could be said that such eukaryotic cells are equipped with some enzymic machinery for the manufacture of deoxynucleotide polymers. Attempts to purify the enzyme have in general been less satisfactory than for the bacterial enzyme. The enzyme activity seems to be more labile. Greene and Korn (1970) nevertheless have recently achieved 400-fold purification of a DNA polymerase from cultured human cells. The enzyme, similar to that from *E. coli*, appears to require a 3′-hydroxyl terminus for initiation of synthesis which proceeds in the $5' \rightarrow 3'$ direction. A variety of templates can be copied *in vitro* to yield products that accurately reflect the base composition of the template used. Nevertheless, in view of the problems in assigning an intracellular role to the *E. coli* polymerase, a similar dilemma exists in the eukaryote. For instance, early attempts to detect DNA polymerase in mammalian cells revealed that the bulk of the activity responsible for incorporating deoxyribonucleotide 5′-triphosphates into DNA-like polymers was readily soluble, and when cell fractionations were carried out the activity seemed apparently predominantly located in cytoplasmic fractions (Keir, 1965). This seemed paradoxical in view of the autoradiographic evidence which clearly indicated a nuclear site for DNA replication (Cleaver, 1967). Of course, many nuclear enzymes are known to be readily solubilized from nuclei during the course of cell fractionation techniques involving the use of aqueous media. Another dilemma lay in the fact that the DNA polymerase-type activity readily detectable in soluble cell fractions operated 2 to 50 times less effectively on double-helical DNAs than on single-stranded DNAs. Doubts were expressed regarding the possible role of such enzyme activity in the process of replicating the nuclear DNA of the eukaryote. In view of these issues, nuclei have been examined more carefully for enzyme activity capable of supporting the synthesis of DNA. Lindsay and Adams (1968) showed in exponentially growing mouse cells that a substantial level of synthetic activity was nevertheless to be found in the nuclear fraction, and what is more it seems to be effective toward double-helical DNAs. Howk and Wang (1969) partially purified an enzyme with similar properties from the nonhistone chromosomal proteins of rat liver. A new type of question can now be asked. Does the nuclear activity represent the bound complex involved in the actual replication process and the soluble, apparently cytoplasmic, polymerase represent molecules of enzyme not actively engaged in the replication process? After all, only approximately 20% of an exponentially growing cell population are in that phase of the cell cycle in which

DNA synthesis is actually taking place (the S phase). There have been reports (Gold and Helleiner, 1964; Littlefield et al., 1963) of changes in the DNA polymerase activity in cultured mouse cells as they pass through the cell division cycle. The enzyme activity in the soluble cell fraction fell slightly at the time the cells were in the S phase, whereas that bound in the nuclear fraction rose somewhat, suggesting but not proving some migration of enzyme molecules to the nucleus at that time.

In the same context an interesting situation exists in sea urchin embryos. Loeb and his colleagues found a DNA polymerase enzyme operating on double-helical DNAs to be strongly associated with nuclei at the hatching stage (Loeb et al., 1969). In fact, the enzyme was subsequently found to be bound to chromatin (Loeb, 1970). Then during development up to the hatching stage, they found a progressive increase in the activity of this polymerase activity strongly associated with nuclear components and a concomitant decrease in the soluble fraction (Fansler and Loeb, 1969). However, whether this represented an actual molecular migration is not yet clear.

In regard to the mammalian cell, Mueller (1969) has provided some recent evidence to suggest that the activity of the nuclear bound enzyme is dependent on five separable cytoplasmic proteins, distinct from the soluble DNA polymerase. These active principles are manufactured by the cell during transit through S phase and may well play a role in the sequential synthesis of DNA already described. Moreover, they could be possible subunits of the bound enzyme complex that may well be involved in the controlled replication process occurring within the interphase nucleus.

2. In Vivo Studies. Studies on cultured mammalian cells and livers of rats exposed for various periods to [^{3}H]thymidine indicate the possibility, similar to that in bacterial systems, whereby replication of DNA in chromosomes proceeds by the production of short fragments approximately 1000 nucleotides long which are then apparently linked into chains of very high molecular weight (Tsukada et al., 1968; Schandl and Taylor, 1969). However, although ligases are known to be present in cells of higher organisms (Lindahl and Edelman, 1968), the actual means whereby this linkage is achieved in mammalian cells is unknown at present.

The involvement of the nuclear membrane in DNA replication in mammalian cells has also been invoked. Electron autoradiography of HeLa cells after short exposures to [^{3}H]thymidine indicates the newly synthesized DNA to be predominantly located close to the nuclear membrane (Comings and Kakeguda, 1968). Moreover, use of chemical techniques, such as extraction with phenol or chloroform to isolate the DNA, indicate that the new DNA in cells so labeled is attached to some cellular component which, according to Friedman and Mueller (1959), may consist of lipopolysaccharide material.

Immediately following this synthesis of DNA, certain cytosine residues of mammalian DNA are then methylated to form 5-methylcytosine residues (Burdon and Adams, 1969) and, as noted for bacteria, this may well serve to indicate points of chain scission by sequence-specific cleaving enzymes (endonucleases), which would be a prelude to subsequent DNA replication in the next cycle of cell division. Thus, in conclusion, it is tempting to envisage in higher organisms a nuclear membrane-bound replicative enzyme complex comprising polymerase, ligase, and perhaps nucleases specific for certain modified sequences all of whose various activities are programmed in some manner, perhaps by additional cytoplasmic protein subunits, to replicate the eukaryote DNA in the complex manner observed.

3. A General Scheme for Eukaryote DNA Replication. At the chromosomal level Taylor et al. demonstrated in 1957 that the incorporation of [^{3}H]thymidine into chromosomal DNA at S phase could be detected in all chromosomes. Moreover, the label is equally distributed between the two chromatids of each chromosome. When the cells were placed in a medium containing unlabeled thymidine and a second round of chromosome duplication allowed to take place, one set of chromosomes was labeled and the other not. This suggested, at least at a superficial level, that in higher organisms chromosomal replication is a semiconservative process. Proof at the molecular level of a situation similar to that discovered by Meselson and Stahl (1958) in *E. coli* was provided in 1960 by Djordjenic and Szybalski, who used bromouracil as a density label for DNA in growing cultures of human cells (HeLa). However, just how this is achieved is not yet understood. Although the *E. coli* chromosome can be envisaged as a single replicon which, under normal growth conditions, is replicated at a single forklike growing point, the situation in the mammalian cell must be more complex since the problem of several independent replicating units has yet to be resolved. Huberman and Riggs (1968) have applied Cairns' technique of visualizing the chromosomes in the actual process of replication by autoradiography. Surprisingly, their data showed that the long fiber of chromosomal DNA is apparently made up of many tandemly joined sections, in *each* of which DNA is replicated at a forklike growing point. In Chinese hamster cells most of these sections are probably less than 30 μ long and the rate of DNA replication in the region of 1–2 μ/min. (*E. coli* DNA is replicated at 30 μ/min.). Perhaps more unexpected was the finding that replication seemed to proceed in opposite directions at adjacent growing points. Furthermore, these adjacent growing points appeared to initiate replication at the same time. This is illustrated in Figure 9. Now an X chromosome contains DNA about 51,000 μ long which would correspond to 800 replicons if each were 60 μ, that is, perhaps approximately 20,000 to 100,000 per mammalian cell. However, further work will be required before this can be stated with certainty, but it certainly illustrates the enormity of the problem.

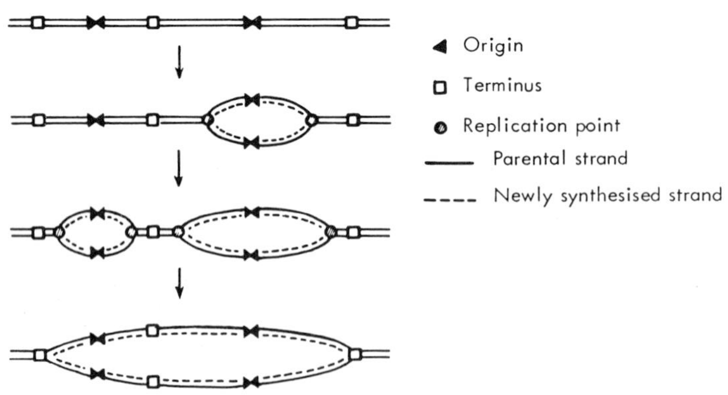

Figure 9. A representation of the bidirectional model for DNA replication in mammalian cells. (Adapted from Riggs and Huberman, 1968.)

2. THE BIOSYNTHESIS OF RNA

A. General Considerations

When techniques for handling the very large DNA molecules encountered in the chromosomes of higher organisms become more refined, such DNA may prove not to be a uniform collection of polymers but many different molecules of various defined lengths. This is certainly the case for RNA from both eukaryotes and prokaryotes. Historically, the first type of RNA to be characterized was transfer RNA (tRNA) now established as the carrier of amino acids in protein synthesis (see Hoagland, 1960). There is at least one tRNA for every amino acid, and to date the nucleotide sequences of approximately 15 individual tRNA species have been established and can be arranged in the well-known cloverleaf tRNA structure (see Levitt, 1969).

Soon after, messenger RNA (mRNA) was demonstrated in bacteria (Brenner et al., 1961; Gros et al., 1961). This species is now recognized as the means whereby genetic information is transported from the DNA genome to the protein synthetic sites on the ribosomes of the cytoplasm. However, tRNA accounts for 10–12% of the RNA of a cell and mRNA for only about 1–2%. By far the most abundant species are the RNAs found as integral parts of cytoplasmic ribosomes themselves. Upon extraction of this ribosomal RNA (rRNA) and analysis by ultracentrifugation, three species are revealed. These were subsequently identified for convenience by their sedimentation coefficients. Thus the single-stranded, high-molecular-weight RNA from the smaller ribosomal subunit of mammalian cells is designated 18S (16S from bacterial smaller ribosomal subunits), whereas the RNAs from the larger ribosomal subunits are termed 28S (23S in bacteria) and 5S. The latter, similar to tRNA is relatively small. It has only 110 nucleotides, but although its sequence is known (Brownlee et al., 1967) no secondary structure has yet been elucidated. The 28S and the previously

Table 1. Comparison of Composition and Size of RNAs from Various Sources

	G + C content (%)			Molecular weight of rRNAs ($\times 10^{-6}$) and apparent sedimentation coefficient		
Source	Genome	RNA	tRNA	From large subunit	From small subunit	First ribosomal precursor RNA
Human (HeLa)	41[f]	63[e]	54[e]	1.75 (28S)[a]	0.70 (18S)[a]	4.4 (45S)[b]
Amphibian (*Xenopus*)	41	58[b]	68[c]	1.54 (28S)[a]	0.69 (18S)[a]	2.5 (35S)[b]
Yeast	36[f]	47[f]	55[f]	1.30 (26S)[a]	0.72 (17S)[a]	2.8 (42S)[d]
Escherichia coli	50[f]	54[f]	60[f]	1.07 (23S)[a]	0.56 (16S)[a]	—
Micrococcus lysodeikticus	72[f]	58[f]	56[f]	~1.07 (23S)	~0.56 (16S)	—

[a] Loening (1968).
[b] Loening et al. (1969).
[c] Brown and Littna (1966).
[d] Retel and Planta (1970).
[e] Darnell (1968).
[f] Mahler and Cordes (1966).

mentioned 18S species, however, have molecular weights of 1.9 million and 0.7 million, respectively. The sedimentation coefficients and molecular sizes of several species are shown in Table 1. Generally speaking, a manifestation of the evolutionary processes is increased rRNA size (Loening, 1968). Over the last decade, other less abundant RNA species have been discovered, as discussed in Section 2B, and our understanding of their features has increased, but the situation is complex and their various roles within the cell are not yet clearly appreciated.

B. The Patterns of Cellular RNA Synthesis

1. rRNA and tRNA. While virtually all the DNA of the eukaryote (except 1% in mitochondria) is confined to the nucleus, the bulk of the cellular RNA is located in the cytoplasm. Thus the question of the role of the nucleus in its biosynthesis arose. Just as [^{3}H]thymidine had been used to detect DNA synthesis, autoradiographic experiments involving the use of radioactively labeled ribonucleosides were designed to follow RNA synthesis. This, together with cell fractionation techniques involving differential centrifugation of homogenates of cells exposed to the labeled nucleoside for various times, indicated the nucleus and not the cytoplasm to be the site of synthesis. Studies at the enzymic level indicated the presence of an enzyme exclusive to the nucleus which was not only capable of synthesizing new RNA from ribonucleoside-5′ triphosphate units but which also apparently took its directions for this synthesis from a DNA template (Weiss, 1960, Burdon and Smellie, 1962). Further support for the concept that the composition and nucleotide sequence of cytoplasmic RNA was under the control of nuclear DNA was obtained from molecular hydridization experiments in which it was shown that there were sequences of DNA precisely complementary to the nucleotide sequences of rRNA and tRNA. In bacteria those complementary to rRNA amounted to approximately 0.1–0.2% of the genome (Yankofsky and Spiegelman, 1962), while those complementary to tRNA to approximately 0.02% (Goodman and Rich, 1962). The word complementary is used here in the sense previously discussed in connection with DNA replication. A guanine (G) in a DNA strand base pairs with a cytosine (C) in an RNA strand, a C in a DNA strand with a G in an RNA strand, a thymine (T) in a DNA strand with an adenine (A) in an RNA strand and an A in a DNA strand with a uracil (U) in an RNA strand. There is no T in RNA it is remembered. As detailed in Section 2C, it was soon shown that the product of the nuclear RNA-synthesizing enzyme, now known as DNA-dependent RNA polymerase, has a nucleotide sequence complementary to *one* of the strands of the DNA used as template (Shenkin and Burdon, 1966). This process, now known as "transcription," is discussed in Section 2C (the molecular mechanism is shown in Fig. 13). As mentioned earlier, the euchromatin fraction is more active in this respect than the heterochromatin. These fractions, however, contain relatively little RNA, the bulk of nuclear RNA being located in the nucleolus, and for this reason some researchers initially felt that the nucleolus might store the RNA transcribed from chromatin sites. Nevertheless, the nucleolus seemed very active itself in the manufacture of RNA (see Prescott, 1964; Perry, 1967). In fact, molecular hybridization experiments demonstrated that the part of the DNA genome (approximately 0.1%) containing the nucleotide sequences which are precisely complementary to the sequences found in cytoplasmic rRNA could be demonstrated in the nucleolus of certain eukaryotes (Brown and Dawid, 1968). In *Xenopus*, for instance, this nucleolar DNA can be shown to contain equimolar amounts of sequences complementary to 28 and 18S RNAs (Birnsteil et al., 1968).

We now know that the nucleolus is the actual site of rRNA biosynthesis. Our present understanding of the molecular events occurring there that give rise to the RNA owes much to the intensive efforts of Darnell and Penman and their respective colleagues during the last few years. In 1962, Scherrer and Darnell demonstrated the existence of relatively short-lived species of RNA in cultures of human cells exposed to ^{3}H-labeled uridine. This RNA occurred only in small amounts (about 1% of the total) and was distinguishable from rRNA or tRNA. It sedimented at 45S, and its molecular weight is now known to be about 4.5 million. Although the site of its synthesis was soon found by cell fractionation techniques to be the nucleolus (Penman et al., 1966), a more interesting feature was discovered. Shortly after its synthesis, this long, single-stranded molecule undergoes a series of molecular "tailoring" events in the nucleolus whereby its molecular dimensions are progressively reduced to give eventually rise to the 28 and 18S rRNA species found in cytoplasmic ribosomes. Detailed studies on the kinetics and mechanism of the various nucleolar cleavage events were undertaken (see Darnell, 1968), and by 1970 Weinberg and Penman were able to formulate the present-day concept of the scissions involved in the processing of the 45S molecule in the nucleolus, and now referred to as the 45S ribosomal precursor RNA. This is illustrated in Figure 10. Ribosome manufacture itself is now known to take place in the nucleolus (Perry, 1967). Particulate precursors of cytoplasmic ribosomes ("nascent ribosomal particles") comprising the ribosomal precursor RNA species of the various sizes just described, plus newly synthesized ribosomal proteins and the small RNA component designated 5S, have been isolated from

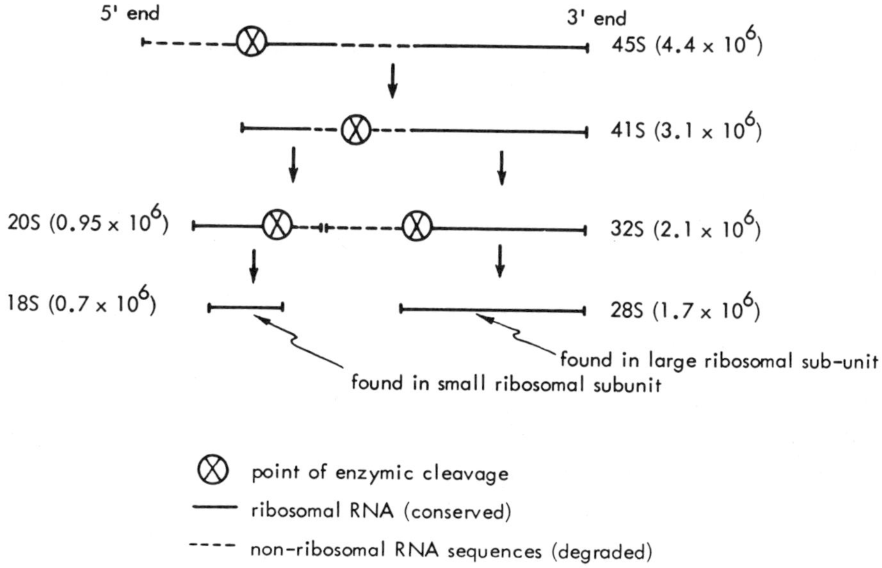

Figure 10. A schematic illustration of the possible events occurring in the processing of the 45S rRNA precursor in mammalian cell nucleoli. The actual linear sequence of ribosomal and nonribosomal sections in the 45S molecule are only hypothetical, but it is thought that the 28S region is nearest the 3′ end. Approximate molecular weights of each species are indicated in parentheses. (After Weinberg and Penman, 1970.)

nucleoli (Warner and Soiero, 1967). The complex 45S processing steps appear to take place not just within the nucleolus but within the nascent ribosomal particles themselves.

One of the intriguing features of the processing as a whole was revealed by analysis of nucleotide composition and by competition molecular hybridization (Jeanteur et al., 1968). About half of the 45S ribosomal precursor RNA molecule does not in fact contain nucleotide sequences destined to appear in the mature ribosomes of the cytoplasm. Apparently, these are lost in the processing, presumably as a result of degradation to nucleotides (Fig. 10). What kinds of molecular mechanisms control the apparently specific cleavage steps? Are specific endonuclease enzymes directed to certain nucleotide sequences by some type of chemical signal? Shortly after synthesis certain nucleotide sequences of the 45S molecule are known to be altered by virtue of methylation of base moieties of certain nucleotides (Burdon, 1966; H. Greenberg and Penman, 1966). Some typical methylated bases are shown in Figure 11*A*, and their formation in intact newly synthesized 45S RNA molecules is thought to be catalyzed by sequence-specific methylase enzymes of the nucleolus (Culp and Brown, 1970). Our fundamental understanding of methylases comes from the work of Borek and his colleagues (see Borek and Srinivasan, 1966). In short, RNA methylases catalyze the transfer of intact methyl groups from *S*-adenosyl-L-methionine to specific bases in certain sequences of nucleotides in *intact* RNA molecules as shown in Figure 11*B*. These methylated nucleotides could then serve to protect certain sequences from cleavage by nuclease enzymes or serve to indicate the sequences to be removed or cleaved from a polymer by nuclease action. Interestingly, it is the sequences that are nonmethylated that are lost during the processing of the 45S precursor of rRNA (Jeanteur et al., 1968). Although reptiles, insects, and plants also appear to process a high-molecular-weight rRNA precursor, it is smaller than the 45S molecule of mammalian cells. Analysis shows it to contain less nonribosomal RNA and to have a molecular weight in the region of 2.5 million (J. R. Greenberg, 1969; Loening et al., 1969).

Despite its presence as part of the larger ribosomal subunit, the precise origin of the 5S RNA species is not yet clear. It cannot arise during the processing of the 45S RNA precursor for two reasons. Hybridization data (Brown and Dawid, 1968) show that the DNA of the nucleolus that gives rise to the 45S RNA does not contain any sequences complementary to 5S RNA molecules. In addition, metabolic inhibitors capable of blocking 45S RNA manufacture have no effect on 5S synthesis (see Burdon, 1970).

Something is now known of animal cell tRNA biosynthesis. That part of the genome with sequences complementary to tRNA molecules (about 0.01% of the whole genome) is located in nonnucleolar regions of the nucleus (see Burdon, 1970). Similar to rRNA, it appears to arise as the result of processing a precursor (Burdon et al., 1967). Whereas tRNA, like rRNA contains methylated bases, the tRNA precursor is deficient in these. In addition, the precursor is devoid of pseudouridine (Smillie and Burdon, 1970), a nucleoside common to all tRNAs. Burdon and Clason (1969) found the precursor to be longer than tRNA (by 20 to 30 nucleotides) but it is processed in the cell cytoplasm to yield tRNA. This involves (1) enzymic removal of the extra nucleotides (Smillie and Burdon, 1970), and (2) methylation of certain nucleotides in specific tRNA sequences by methylase enzymes located in the soluble fraction of the cytoplasm (Burdon et al., 1967; Burdon, 1970). A scheme illustrating the general features of rRNA and tRNA synthesis in mammalian cells is shown in Fig. 12.

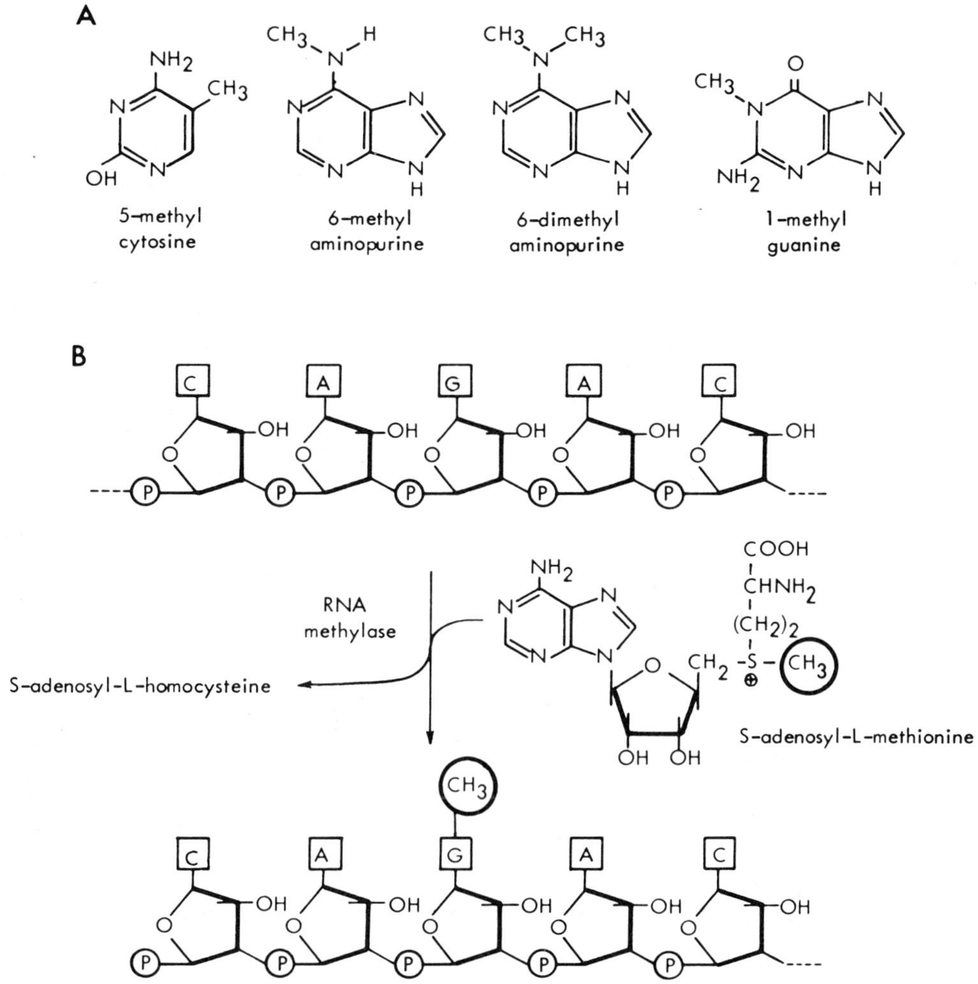

Figure 11. (*A*) Some methylated bases to be found in naturally occurring nucleic acids. (*B*) Diagrammatic representation of RNA methylase action.

The situation with regard to the synthesis of tRNA is different in prokaryotes. To date there has been little evidence for a high-molecular-weight precursor for the 23 and 18S rRNAs in bacteria with extensive nonribosomal sequences. These RNAs appear instead to arise from precursors which are only very slightly larger than the final product (Adesnick and Levinthal, 1969). This of course raises the question as to the role of the apparently extra nonribosomal sequences present within the precursor found in eukaryotes. On the other hand tRNA precursors similar to those in eukaryotes have been detected. A manifestation of evolution seems to be complexity with regard to rRNA biosynthesis, but in turn this may well be a manifestation of more elaborate control mechanisms operating in the production of the major cellular RNA species to meet various requirements (Burdon, 1970).

2. mRNA. One of the distinguishing features of rRNA and the tRNAs is their relatively high content of nucleotides containing the bases guanine and cytosine. In

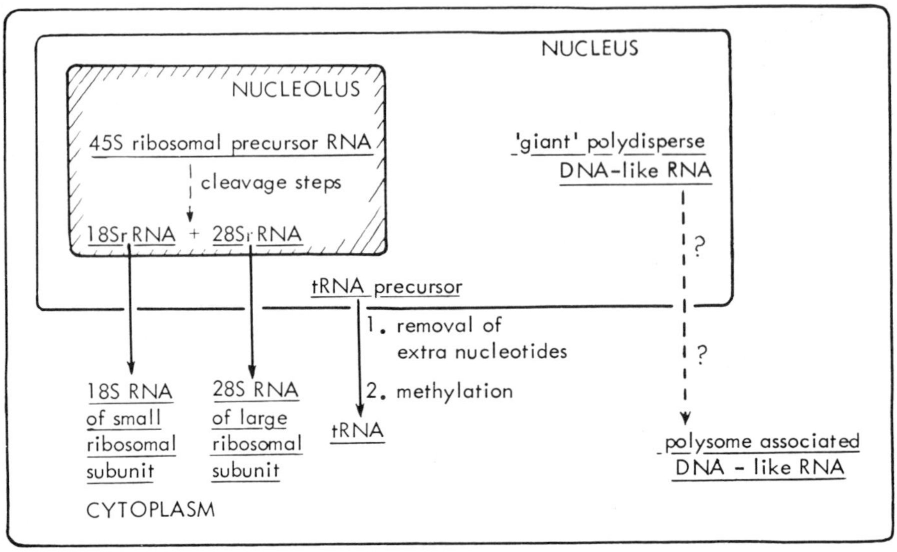

Figure 12. A representation of the intracellular sites of synthesis and processing of rRNA, tRNA, and DNA-like RNA in mammalian cells.

rRNA of human cells in culture (HeLa), these account for about 63% of the total base content of rRNA and for about 54% of the bases in tRNA (Darnell, 1968). However, the guanosine and cytosine content of rRNA and tRNA in most cases remains high and close to these figures, despite large species variation in the base composition of their DNA genome (see Table 1). In 1956, Volkin and Astrachan determined the nucleotide composition of a new class of relatively unstable RNA molecules rapidly synthesized in *E. coli* after infection with T2 bacteriophage. They found that their base or nucleotide composition resembled the composition of the bacteriophage DNA rather than the *E. coli* genome. Hall and Spiegelman (1961) soon demonstrated these new RNA molecules to be complementary in nucleotide sequence to the T2 DNA rather than the *E. coli* DNA, and Brenner et al. (1961) showed that they attached themselves to preexisting ribosomes to form polysomes. Thus the new "T2-specific RNA" has acted in the manner expected of a "messenger" from the bacteriophage DNA to the ribosomes. The existence of a similar unstable class of RNA molecules in uninfected *E. coli* was also demonstrated at about the same time (Gros et al., 1961). On a short exposure (a few seconds) to a labeled ribonucleotide, the newly synthesized RNA molecules were heterogeneous in molecular size, but in the main were found to sediment on ultracentrifugation between tRNA and the 16S rRNA of *E. coli*. Their average nucleotide composition corresponded closely to that of *E. coli* DNA (Hayashi and Spiegelman, 1961) and they associated with ribosomes to form polysomes (Gros et al., 1961). Recent experiments indicate this RNA species to have a decay time of 3 or 4 min (Salser et al., 1968).

In mammalian cells, after exposure to radioactively labeled nucleoside for about 1% of the cell generation time, the 45S rRNA precursor in the nucleolus as well as the precursor to tRNA become labeled. However, in addition the presence of newly synthesized high-molecular-weight RNA molecules is noticeable in nonnucleolar sites of the nucleus (Soiero et al., 1966; Houssais and Attardi, 1966). They are heterogeneous with respect to molecular weight, ranging from 2 to 10 million but, most interestingly, their nucleotide composition closely resembles that of the mammalian genome (i.e., guanine and cytosine residues accounting for about 42% of the total base content). For this reason they have been termed giant, heterodisperse, DNA-like RNA molecules. However, another class of DNA-like RNA molecules can be found, after short exposures to labeled ribonucleosides, to be associated with polysomes (Latham and Darnell, 1965). On removal from the ribosomes themselves they were, similar to the nuclear species, shown to be heterogeneous in size. However, they were much smaller than the nuclear species, their molecular weight being on average about 0.6 million. Nevertheless, their association with microsomes or polysomes and their DNA-like composition initally suggested that they might represent mRNA of mammalian cells. However, if this is so, the question that arises is how they actually reach the cytoplasm to direct protein synthesis. The problem is made more complex by the fact that no molecules of size and composition similar to the polysome-associated DNA-like RNA can be detected in the nucleus. What, *if any*, is their relationship to the giant, heterodisperse, DNA-like RNA of the nucleus? This is difficult to assess because the giant molecules prove to be very unstable and are almost completely degraded *within* the nucleus itself (Houssais and Attardi, 1966). What then is the function of these giant molecules? At present, the following possibilities are being considered: (1) most of each giant molecule is degraded but a very small portion could serve as mRNA and be exported to the cytoplasm; (2) many extra copies of the giant molecules are made but only a very few are used to generate mRNA; and (3) most of the giant species represent types of molecule never used even in part, but a small fraction of the total number of giant molecules might be used either wholly or in part.

3. Nuclear Low-Molecular-Weight RNAs. Careful analysis of nuclear RNA from mammalian cells by the sensitive technique of polyacrylamide gel electrophoresis has revealed the presence of several distinct monodisperse low-molecular-weight RNAs (Prestayko and Busch, 1968; Weinberg and Penman, 1968). While exclusive to the nucleus, they are distributed between chromatin and nucleoli. Their size ranges from 100 to 180 nucleotide residues, and some of their nucleotides are methylated. In each nucleus there are only about 10^5 molecules, and from labeling experiments they seem to be very stable metabolically; it has been suggested that these molecules might be important in the regulation of genetic expression of chromosomes. However, to date their function remains an unknown. Also, in the nucleus of higher organisms is a small RNA (40 to 60 nucleotides in length) with an unusually high content of dihydropyrimidine bases. It is *covalently* bound to chromosomal protein and can form hybrids with double-helical DNA (Bekhor et al., 1969a). For this reason it has been implicated in directing the information of certain chromosomal proteins with the genome of higher organisms to dictate the regions available for transcription by the DNA-dependent RNA polymerase. The characteristics of synthesis and stability of this RNA are as yet unknown.

C. Enzymic Synthesis of RNA

1. Early Studies on RNA Polymerase in Mammalian Cells. As mentioned briefly, synthesis of RNA in normal cells is brought about at the enzymic level by DNA-dependent RNA polymerase. The enzyme catalyzes the formation of single-stranded RNA molecules whose sequence is dictated by a DNA template.

In 1960, Weiss reported that a particulate, or "aggregate," preparation from rat liver nuclei was capable of incorporating ribonucleotide units derived from the 5′-triphosphates of adenosine, guanosine, cytidine, and uridine into an acid-insoluble, nondialyzable material which was characterized as RNA, since, for instance, it was not attacked by deoxyribonuclease but was digested by ribonuclease, snake venom diesterase, and by alkali. Optimal synthesis proceeded only in the presence of all *four* ribonucleoside 5′-triphosphates and required either magnesium or manganese ions. Prior incubation of the crude preparation with a small amount of deoxyribonuclease abolished its activity. Soon after this several attempts were made to solubilize this activity in other mammalian tissues but served only to underline the fact that DNA-dependent RNA polymerase was very tenaciously bound to the chromatin fraction of the nucleus (Burdon and Smellie, 1962), and any enzymic activity readily made soluble was capable only of adding nucleotide units to the end of existing polyribonucleotides acting as primers rather than catalyzing the extensive synthesis of new RNA under the direction of a DNA template (Burdon and Smellie, 1961).

2. The RNA Polymerase of Bacteria. Because of the problems involving the apparent insolubility of RNA polymerase of mammalian systems, considerable progress in the early 1960s was made toward an understanding of its basic mechanism in studies on RNA polymerases that were more readily isolated and purified from bacterial sources (see Hurwitz and August, 1963). In fact the most purified preparations now come from *E. coli*. The bacterial enzymes were shown to require DNA as a template, manganese or magnesium ions, and the four ribonucleoside 5′-triphosphates. Careful analysis of the product has shown it to consist of single-stranded polyribonucleotides of high molecular weight containing all four ribonucleoside residues linked by 3′:5′-phosphodiester bonds. DNA from a wide variety of sources can be utilized as template, and the salient feature of these experiments was that the nucleotide composition of the RNA product was determined by that of the DNA used. This is true even though the amount of RNA formed in the test tube may greatly exceed the amount of DNA added initially. Moreover, when the DNA is reisolated after the reaction has been stopped, it appears unaltered. For instance, its buoyant density in cesium chloride gradients is unaffected and if template DNA with ability to "transform" bacteria is used it retains this activity (see Hurwitz and August, 1963). It appears therefore that the DNA used as template behaves only catalytically in the process of RNA synthesis. When double-helical DNA is used as template, the newly formed RNA can be found in the reaction medium free of any template DNA. In contrast, when single-stranded DNA is used, the first product is a hybrid molecule of RNA, hydrogen-bonded to the template DNA. As the reaction proceeds, free RNA is released while new RNA molecules are synthesized on this template (Sinsheimer and Lawrence, 1964). The nearest neighbor frequency analyses carried out on the RNA produced supported the view that the sequence of nucleotides in the product is complementary with that in the DNA used as template. Also, each new strand is synthesized in the 5′-to-3′ direction and with a polarity opposite that of the DNA strand from which it is derived (Hurwitz and August, 1963).

Further convincing evidence that the RNA product was complementary in nucleotide sequence to the DNA template comes from molecular hybridization studies

(Geiduschek et al., 1961), but together with the other studies raised the question as to whether or not the enzyme used both strands of the DNA template or just one. Genetic data (Champe and Benzer, 1962) suggested that *in vivo* only one of the strands of the DNA molecule acts as template for the synthesis of functional RNA. This concept was certainly supported from studies on bacteria infected with certain bacteriophages whose double-helical DNA genomes could be separated into two distinguishable single strands of widely differing base composition by virtue of their buoyant density differences in cesium chloride gradients. The individual strands of the bacteriophage genome could then be tested for their ability to form specific molecular hybrids with the naturally occurring mRNA isolated from cells infected with these bacteriophages. The result of such experiments clearly demonstrated that only one strand of the DNA forms a hybrid with the natural messenger (see Bautz, 1966). With the purified RNA polymerase it appears that whether or not one DNA strand is used as template (asymmetric transcription) or both are used simultaneously (symmetric transcription) depends somewhat on the condition of the DNA used and the ionic conditions (Hayashi et al., 1964). DNA molecules that are degraded and in whch single-stranded regions are exposed tend to be transcribed symmetrically, whereas intact circular helical DNAs are transcribd asymmetrically. A general idea of the molecular mechanism of transcription is shown in Figure 13. Another question that arises when considering RNA polymerase as the enzyme responsible for *in vivo* RNA synthesis is the rate at which it can make RNA in a test tube. By increasing the ionic strength of the RNA polymerase reaction mixtures, the rate at which RNA chains are manufactured is increased. At 0.2 M potassium chloride T4 bacteriophage DNA was found by So et al. (1967) to direct RNA molecules to grow at a rate of more than 20 per second which is close to the value of 28 per second detected *in vivo* for T4 DNA

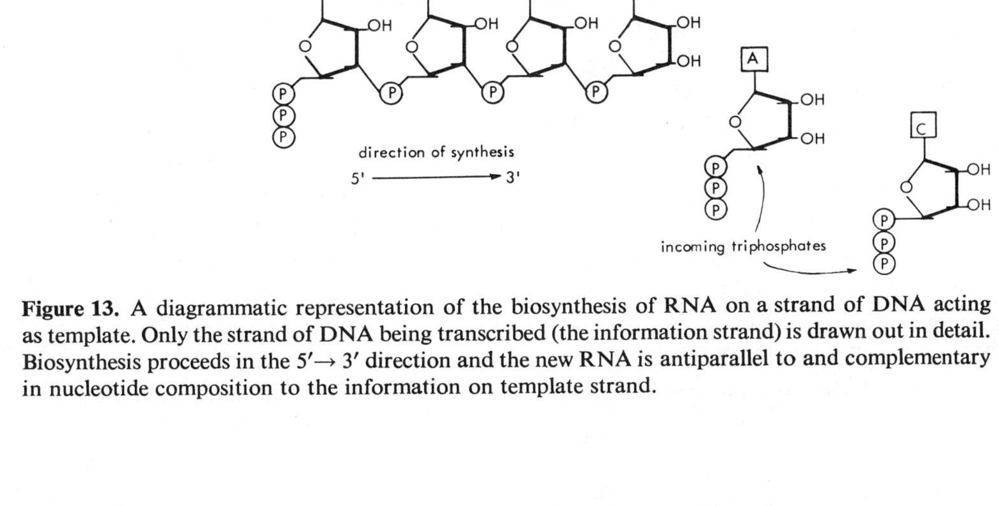

Figure 13. A diagrammatic representation of the biosynthesis of RNA on a strand of DNA acting as template. Only the strand of DNA being transcribed (the information strand) is drawn out in detail. Biosynthesis proceeds in the 5′→ 3′ direction and the new RNA is antiparallel to and complementary in nucleotide composition to the information on template strand.

transcription (Bremer and Yuan, 1968). This high concentration of ions appears to aid in the release of the RNA product from the template, which otherwise tends to inhibit the formation of new molecules, and may in fact be close to the potassium chloride concentration within *E. coli* itself (see Richardson, 1969).

What then are the factors responsible for initiation of asymmetric transcription of a DNA molecule into complementary RNA? There is evidence that not only does chain growth occur by the addition of new nucleotides to the 3′-hydroxyl end of nascent RNA chains but that the terminal nucleotide still retains its three 5′-phosphate groups and is more often than not a purine-containing nucleotide (Maitra and Hurwitz, 1965). In general, there is a preference for GTP, but this depends very much on the DNA template used. This could be a clue to the necessary first step in the initiation of synthesis, that is, the interaction between the enzyme and the template. The binding of RNA polymerase to DNA has been studied extensively (see Richardson, 1969), and it can occur rapidly and reversibly in the absence of the four ribonucleoside triphosphates. The association is sensitive to ionic strength, does not require magnesium ions, and from quantitative studies and electron microscope analyses seems to have a specificity for certain sites. On the average there is about one site per 1000 or 2000 nucleotides (Richardson, 1969). It is possible that RNA polymerase can recognize particular nucleotide sequences, but it is not yet clear how this is achieved. Perhaps sequences altered by methylation of certain bases are required. Since the enzyme binds to intact circular DNAs, other alternatives such as single-strand breaks or areas where there is a collapse of the normally helical structure are unlikely to be involved. However, certain DNAs are known to contain regions rich in pyrimidine nucleotides (Szybalski et al., 1966), and since the first nucleotide unit in the RNA produced by the enzyme is usually of the purine-containing variety, it may develop that such pyrimidine-rich sequences in DNA strands are responsible for the initial binding of the enzyme. Despite this, there may well be distinct differences between sites that bind the enzyme and those where the initiation of RNA synthesis takes place. After all initiation, if the transcription is from one strand of the DNA only, involves not just recognition of a particular DNA site but also the local unwinding of the DNA helix at that locus to expose a strand for use as template (see Fig. 13). Whether or not RNA polymerase can destabilize a region of DNA helix in this way as a result of binding to it is not yet known and will almost certainly involve complex physical studies. In addition, there is a problem involving the binding of the first purine nucleoside 5′-triphosphate to the enzyme–DNA complex followed by the condensation of the second nucleoside triphosphate to the first. Any, or all, of these could be rate limiting but serve to illustrate the complexity of the problem.

During the last year further insight into the mode of action of RNA polymerase has been obtained as a result of improved procedures employed for purifying the enzyme from bacteria such as *E. coli*. These methods have yielded a molecule that can be dissociated into several subunits after treatment with 8 *M* urea or 0.1% sodium dodecyl sulfate. These subunits can be separated on polyacrylamide gel electrophoresis and are: two α chains of molecular weight 39,000, one β chain of molecular weight 155,000, one β' chain of molecular weight 165,000, and one σ chain of molecular weight 90,000 to 95,000 ($\alpha_2\beta\beta'\sigma$). Fractionation of the intact RNA polymerase on phosphocellulose columns, however, yielded a "core" enzyme comprising the two α chains complexed with the β and β' chains ($\alpha_2\beta\beta'$) with a combined molecular weight of 400,000 and the separated σ chain (Burgess et al., 1969). This core enzyme from

E. coli retains the ability to catalyze the synthesis of RNA chains under the direction of a DNA template *in vitro* quite efficiently when alien DNA from calf thymus is provided, but this ability is greatly reduced when the DNA comes from *E. coli* or from bacteriophage T4. Nevertheless, in the latter case at least, the addition of the σ-protein chain restores the ability to manufacture RNA molecules to normal levels. Later, Burgess and Travers (1969) demonstrated that the σ protein (sigma factor) is required for the initiation of RNA synthesis. After initiation the sigma factor is released from the RNA polymerase core–sigma factor complex and from the DNA template and may be reused by another polymerase core molecule to initiate the synthesis of a new RNA chain (Fig. 14). An obvious possible use of this type of mechanism is for selecting genes at which transcription is to take place and perhaps causing a local unwinding of the helix at that locus. An analysis of the RNA molecules synthesized in a test tube with T4 bacteriophage DNA as template showed that addition of the sigma factor specifically stimulates the core enzyme into transcribing those genes of the bacteriophage usually expressed *in vivo* during the first minute after infection. In the absence of the sigma factor, RNA synthesis is initiated at random sites along the whole genome (Bautz et al., 1969). A further insight into the complexities of the initiation process may come from the use of the antibiotic rifampycin. Actinomycin D has been known for several years to inhibit DNA-dependent RNA polymerases of both animal and bacterial origin by forming a complex with guanine residues of the DNA used as template. In this complex the peptide chains of the antibiotic appear to be packed into the small groove of the DNA helix (Reich, and Goldberg, 1964). Rifampycin, however, inhibits only the RNA polymerase of bacteria such as *E. coli*. It does so by preventing initiation of synthesis rather than by preventing binding of the polymerase. This is achieved by virtue of forming some specific hydrogen-bonded complex with a component of the core enzyme (probably a β subunit) rather than with the sigma factor (Wehrli and Staehelin, 1970). It is now possible to dissociate the core enzyme into subunits with various levels of lithium chloride and then to reconstitute the complex (Sethi and Zillig, 1970), as follows.

$$\alpha_2\beta\beta' \rightleftharpoons \alpha + \alpha\beta\beta' \rightleftharpoons 2\alpha + \beta\beta'$$

Binding studies with isolated $\beta\beta'$ support the idea that β' is the binding subunit.

The synthesis of a functional RNA molecule clearly requires the enzyme RNA polymerase not only to initiate at an appropriate site on the DNA template but also terminate transcription at an appropriate site. Usually it was found that RNA synthesis *in vitro* with the purified system described above did not end at unique sites but

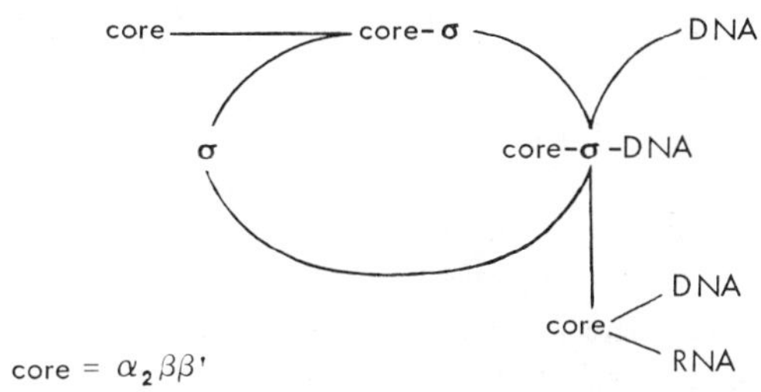

Figure 14. The sigma cycle.

instead terminated at random to produce RNA molecules of various and often clearly excessive lengths. Furthermore, the RNA molecules were not always released from the DNA and enzyme complex described above. However, Roberts reported in 1969 the occurrence of a protein in *E. coli* that caused specific termination and release of RNA synthesized *in vitro*. This termination protein, which is named the ρ (rho) factor (ρ for release), was discovered during a search for factors in a crude extract of *E. coli* that might increase the overall accuracy of RNA synthesis. The ρ factor depresses RNA synthesis from λ bacteriophage DNA as template by terminating the manufacture of RNA molecules at specific sites thereon, the product in fact comprising several discrete RNA species which could correspond to some of the naturally occurring mRNAs synthesized from DNA early in the infectious cycle.

3. RNA Polymerase of Higher Organisms. As mentioned earlier, the mammalian DNA-dependent RNA polymerase activity was first detected tightly bound to the deoxynucleoprotein component of chromatin and seems to transcribe only one strand of the DNA therein (Shenkin and Burdon, 1966). Careful examination, however, of ionic requirements by Widnell and Tata (1964) suggested the possibility of two forms or types of RNA polymerase, since the RNA-synthesizing properties of the crude nuclear preparation varied depending on whether the assay system contained magnesium ions on one hand, or manganese ions together with a high concentration of ammonium sulfate on the other. However, it could not really be determined whether this differential effect was caused by the presence of different species of RNA polymerase in mammalian nuclei or by a salt effect which might have altered the action of inhibitory substances associated with the DNA or the polymerase within the chromatin. Examination of the nucleotide composition of products produced by the crude preparation indicated that under conditions of high ionic strength with ammonium sulfate the product was quite DNA-like in composition, whereas without ammonium sulfate its composition was rather similar to that of rRNA (Widnell and Tata, 1966).

In the last year techniques have improved such that the bulk of RNA polymerase activity can now be solubilized from the nuclei of mammalian cells. In support of the original ideas regarding multiple forms or activities of RNA polymerase, two different forms of the enzyme have been separated from rat liver nuclei (Liao et al., 1969; Roeder and Rutter, 1969) and from calf thymus nuclei (Kedinger et al., 1970). These polymerases, for convenience presently designated I and II, differ in their properties and intranuclear origin. I seems to be present in the nucleolus and II in the nonnucleolar area of the nucleus (Roeder and Rutter, 1970). It is tempting to assume that polymerase I might be responsible for the synthesis of 45S ribosomal precursor RNA in the nucleolus and polymerase II for the synthesis of the giant, heterodisperse, DNA-like RNA and perhaps tRNA precursors in the nonnucleolar regions of the chromatin. A solution to this problem could possibly come from the use of the toxin α-amanitin from *Amanita phalloides* on intact nuclei (Novello and Stirpe, 1970). Kedinger et al. (1970) found that this toxin binds to, and inhibits, the action of partially purified polymerase II without affecting polymerase I.

Although these various forms of RNA polymerase can be detected with slightly differing properties, it is not yet known how complex they are. For instance, they may well be compounded, as is the *E. coli* RNA polymerase, of several subunits, but also they might include various sigma factors or batteries of sigma factors which would influence their transcriptive properties. Indeed, their properties might be important when related to differentiation in eukaryotes. Much evidence now indicates that transcription of the DNA genome is restricted in differentiated cells. RNA in differentiated tissues of an animal is made from only about 5–10% of the DNA present (see McCarthy and Hoyer, 1964). Such a restriction could be achieved quite simply at the enzymic level by various combinations of sigma factors and ρ (terminator) proteins. Alternatively, there is the view that certain parts of the genome are "masked" by histones in conjunction with other molecules such as acidic chromosomal proteins and/or small RNA molecules (Bekhor et al., 1969b; Gilmour and Paul, 1969). However, the situation is such that at present there is no conclusive proof that either mechanism is operating to the exclusion of the other in this context.

3. DNA AND RNA AS TEMPLATES

The previous sections have illustrated DNA functioning as a template (1) in the DNA synthetic processes responsible for the maintenance and replication of the genome, and (2) in the RNA synthetic mechanisms required both in the initial step in the general scheme whereby hereditary information is transmitted from the genome to the protein synthetic sites (mRNA) and in the production of the means whereby this information can be used at these sites (tRNA and rRNA). In general, it can be said that while our knowledge of these processes has increased considerably over the last decade we are only now beginning to comprehend the enormity of the molecular complexity involved. The enzymes participating in these processes appear to do so as complexes, possibly with many other proteins, and any future understanding of the molecular events taking place will depend heavily on our physicochemical knowledge of protein–protein interactions as well as those between protein and nucleic acid. Moreover, the possible contribution of lipid and carbohydrate moieties of membranous structures may well have to be taken into account.

The function of DNA as the template for both DNA and RNA synthesis, as well as the interactions of these nucleic acids after their biosynthesis, has been dealt with in detail here because of their fundamental importance to the biology of the normal cells of living organisms. However, no detailed consideration in this chapter has been given to the nucleic acid syntheses encountered in infected cells. Nevertheless, it should be briefly mentioned that study of cells infected with certain RNA-containing viruses (or bacteriophages) has revealed other mechanisms of RNA and DNA synthesis which depend on RNA rather than DNA functioning as template. For instance, when a small RNA bacteriophage (or virus) infects a cell, the phage RNA acts as a messenger and programs the cell's protein synthetic apparatus to manufacture a specific enzyme capable of catalyzing the replication of that RNA. Although RNA is being used as template, the substrates are still the four ribonucleoside 5′-triphosphates and the direction of synthesis is in the 5′-to-3′ direction. Basically, the process is presently envisaged (see August, 1969) as involving, first, the synthesis of a strand of RNA complementary in sequence to the original bacteriophage RNA to form a double-stranded intermediate (the replicative intermediate). This new strand then acts as template on which a new strand of nucleotide sequence identical to the viral RNA is synthesized which then displaces the original bacteriophage strand. The process (Fig. 15) is then repeated many times so that a batch of new bacteriophage-type strands is produced. The enzyme involved in this process seems now, similar to the other polymerases, to be complex. It is large (molecular weight 150,000) and can be

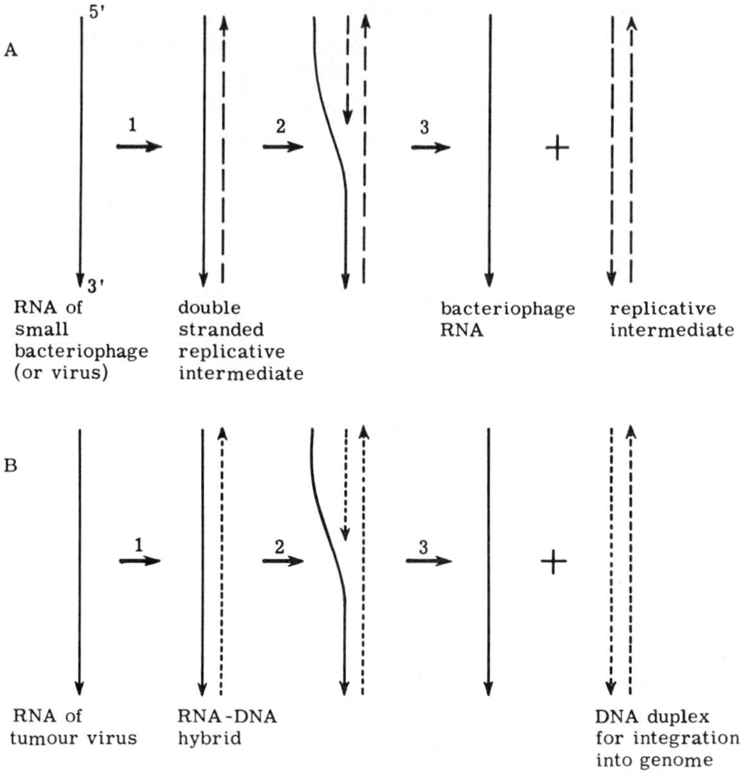

Figure 15. Schematic representation of (*A*) RNA-directed synthesis of RNA as it occurs in the replication of bacteriophage or viral RNA, and (*B*) RNA-directed synthesis of DNA as it might occur in cells infected with a large RNA tumor virus. Original RNA strand ——→; newly synthesized RNA — — →; newly synthesized DNA - - - - - - - →.

separated into light and heavy components, and to participate in the RNA replication process it requires the assistance of two other protein factors which are normally resident in the host bacterium. These, similar to the sigma factor already described, are involved somehow in correct initiation of RNA synthesis.

Perhaps more intriguing than RNA-directed RNA synthesis are the recent findings of Temin and Mizutani (1970) and of Baltimore (1970) that all large oncogenic RNA viruses so far examined have located within them RNA-dependent DNA polymerase activity. This was the outcome of a purposeful search for a possible explanation of a fact known for several years that inhibitors of DNA synthesis can prevent the successful infection of animal cells by such oncogenic viruses. At present, the mechanism whereby DNA synthesis using RNA as template is achieved is not understood. Perhaps there is an intermediate RNA–DNA hybrid. The original RNA strand might then be displaced with the formation of a DNA duplex which could then be integrated into the genome of the host cell (Fig. 15). This integration of tumour virus DNA into a host chromosome could well be brought about by enzymes existing in the cell before it is infected. It appears however that at least some of the viruses carry their own integration machinery (Mizutani et al., 1971). Rous sarcoma virus particles contain DNA endonuclease activity which can cut long DNA chains into shorter pieces as well as a

DNA exonuclease activity which can clip nucleotide units one by one from the end of a chain. Also, the virus particles contain DNA ligase activity. The nucleases thus could cut some section of the DNA of a host cell chromosome and trim away the gap. Into the gap might be inserted the viral DNA which could then be sealed into place by the action of the ligase. Whatever the mechanism, the basic discovery that DNA can actually be synthesized from an RNA template may well have a profound impact on cell biology.

REFERENCES

Adesnick, M., and C. Levinthal (1969). Synthesis and maturation of ribosomal RNA in *E. coli*. *J. Mol. Biol.*, **46,** 281.

August, J. T. (1969). Mechanism of synthesis of bacteriophage RNA. *Nature*, **222,** 121.

Baltimore, D. (1970). RNA-dependent DNA polymerase in virions of RNA tumour viruses. *Nature*, **226,** 1209.

Bautz, E. K. F. (1966). In *Molecular Genetics* (J. H. Taylor, ed.), Part II, p. 213. Academic, New York.

Bautz, E. K. F., F. A. Bautz, and J. J. Dunn (1969). *E. coli* sigma factor: A positive control element in phage T4 development, *Nature*, **223,** 1022.

Bekhor, I., J. Bonner, and G. K. Dahmus (1969a). Hybridisation of chromosomal RNA to native RNA. *Proc. Natl. Acad. Sci., U.S.*, **62,** 271.

Bekhor, I., G. M. Kung, and J. Bonner (1969b). Sequence specific interaction of DNA and chromosomal protein, *J. Mol. Biol.*, **39,** 351.

Birnstiel, M., J. Spiers, I. Purdon, K. Jones, and U. E. Loening (1968). Properties and composition of the isolated DNA satellite of *Xenopus laevis*, *Nature*, **219,** 454.

Borek, E., and P. R. Srinivasan (1966). The methylation of nucleic acids. *Ann. Rev. Biochem.*, **34,** 275.

Bremer, H., and D. Yuan (1968). Chain growth rate of messenger RNA in *E. coli* infected with bacteriophage T4. *J. Mol. Biol.*, **39,** 527.

Brenner, S., F. Jacob, and M. Meselson (1961). An unstable intermediate carrying information from genes to ribosomes for protein synthesis. *Nature*, **190,** 576.

Brown, D. D., and I. B. Dawid (1968). Specific gene amplification in oocytes. *Science*, **160,** 272.

Brown, D. D., and E. Littna (1966). Synthesis and accumulation of low molecular weight RNA during embryogenesis of *Xenopus laevis*. *J. Mol. Biol.*, **20,** 95.

Brownlee, G. G., F. Sanger, and B. G. Barrell (1967). Nucleotide sequence of 5S ribosomal RNA from *E. coli*. *Nature*, **215,** 735.

Burdon, R. H. (1966). Methylation of nucleic acids in Krebs 2 ascites tumour cells. *Nature*, **210,** 797.

Burdon, R. H. (1970). RNA maturation in animal cells. *Prog. Nucleic Acid Res. Mol. Biol.*, **00,** 33.

Burdon, R. H., and R. L. P. Adams (1969). The *in vivo* methylation of DNA in mouse fibroblasts. *Biochim. Biophys. Acta*, **174,** 332.

Burdon, R. H., and A. E. Clason (1969). Intracellular location and molecular characteristics of tumour cell transfer RNA precursors. *J. Mol. Biol.*, **39,** 113.

Burdon, R. H., and R. M. S. Smellie (1961). Incorporation of UTP into RNA by enzyme fractions from Ehrlich ascites carcinoma cells. *Biochim. Biophys. Acta*, **51,** 153.

Burdon, R. H., and R. M. S. Smellie (1962). RNA- and DNA-dependent RNA polymerases of mammalian cells, *Biochim. Biophys. Acta*, **61,** 633.

Burdon, R. H., B. T. Martin, and B. M. Lal (1967). Synthesis of low molecular weight RNA in tumour cells. *J. Mol. Biol.*, **28,** 357.

Burgess, R. R., A. A. Travers, J. J. Dunn, and E. K. F. Bautz (1969). Factor stimulating transcription by RNA polymerase. *Nature*, **221,** 43.

Cairns, J. (1963). The bacterial chromosome and its manner of replication as seen by autoradiography. *J. Mol. Biol.*, **6,** 208.

Caro, L. G. (1970). Chromosome replication in *E. coli*. III. Segregation of chromosomal strands in multiforked replication. *J. Mol. Biol.*, **48,** 329.

Champe, S. P., and S. Benzer (1962). Reversal of mutant phenotypes by 5-fluorouracil: An approach to nucleotide sequences in messenger RNA. *Proc. Natl. Acad. Sci., U.S.*, **48,** 532.

Cleaver, J. E. (1967). In *Thymidine Metabolism and Cell Kinetics* (A. Neuberger and E. L. Tatum, eds.), North-Holland Research Monographs, Frontiers of Biology, Vol. 6. North-Holland, Amsterdam.

Comings, D. E., and T. Kakefuda (1968). Initiation of DNA replication at the nuclear membrane in human cells, *J. Mol. Biol.*, **33,** 225.

Culp, L. A., and G. M. Brown (1970). RNA methylases from rat liver nucleoli. *Arch. Biochem. Biophys.*, **137,** 222.

Darnell, J. E. (1968). RNAs from animal cells. *Bacteriol. Rev.*, **32,** 262.

Djordjenic, B., and W. Szybalski (1960). Genetics of human cell lines. III. Incorporation of 5-bromo and 5-iododeoxyuridine into the DNA of human cells and its effect on radiation sensitivity. *J. Exptl. Med.*, **112,** 509.

Eglund, P. T., J. A. Huberman, T. M. Jovin, and A. Kornberg (1969). Enzymatic synthesis of DNA. XXX. Binding of triphosphates to DNA polymerase. *J. Biol. Chem.*, **244,** 3038.

Fansler, B., and L. A. Loeb (1969). Sea urchin nuclear DNA polymerase. II. Changing localisation during early development. *Exptl. Cell Res.*, **57,** 305.

Friedman, D. L., and G. C. Mueller (1969). Studies on the nature of replicating DNA of HeLa cells. *Biochim. Biophys. Acta*, **174,** 253.

Ganesan, A. T. (1967). In *Organisational Biosynthesis* (H. J. Vogel, J. O. Lampen, and V. Bryson, eds.), p. 19. Academic, New York.

Geiduschek, E. P., T. Nakamoto, and S. B. Weiss (1961). The enzymatic synthesis of RNA: Complementary interaction with DNA. *Proc. Natl. Acad. Sci., U.S.*, **47,** 1405.

Gilbert, W., and D. Dressler (1968). DNA replication: The rolling circle model. *Cold Spring Harbor Symp. Quant. Biol.*, **33,** 473.

Gilmour, R. S., and J. Paul (1969). RNA transcribed from reconstituted nucleoprotein is similar to natural RNA. *J. Mol. Biol.*, **40,** 137.

Gold, M., and C. W. Helleiner (1964). DNA polymerase in L cells. I. Properties of the enzyme and its activity in synchronised cell cultures. *Biochim. Biophys. Acta*, **80,** 193.

Goodman, H. M., and A. Rich (1962). Formation of a DNA-soluble RNA hybrid and its relation to the origin, evolution and degeneracy of soluble RNA. *Proc. Natl. Acad. Sci., U.S.*, **48,** 2101.

Greenberg, H., and S. Penman (1966). Methylation and processing of ribosomal RNA in HeLa cells. *J. Mol. Biol.*, **21,** 527.

Greenberg, J. R. (1969). Synthesis and properties of ribosomal RNA in *Drosophila*. *J. Mol. Biol.*, **46,** 85.

Greene, R., and D. Korn (1970). Partial purification and characterisation of DNA polymerase from KB cells. *J. Biol. Chem.*, **245,** 254.

Gros, F., W. Gilbert, H. Hiatt, C. Kurland, R. W. Risebrough, and J. D. Watson (1961). Unstable RNA revealed by pulse labelling of *E. coli*. *Nature*, **190,** 581.

Gross, J., and M. Gross (1969). Genetic analysis of an *E. coli* strain with a mutation affecting DNA polymerase. *Nature*, **224,** 1166.

Hall, B. D., and S. Spiegelman (1961). Sequence complementarity of T2 DNA and T2-specific RNA. *Proc. Natl. Acad. Sci., U.S.*, **47,** 137.

Hayashi, M., and S. Spiegelman (1961). The selective synthesis of informational RNA in bacteria. *Proc. Natl. Acad. Sci., U.S.*, **47,** 1564.

Hayashi, M., M. N. Hayashi, and S. Spiegelman (1964). DNA circularity and the mechanism of strand selection in the generation of genetic messages. *Proc. Natl. Acad. Sci., U.S.*, **51,** 351.

Hoagland, M. B. (1960). In *The Nucleic Acids* (E. Chargaff and J. N. Davidson, eds.), Vol. 3, p. 349. Academic, New York.

Houssais, J. F., and G. Attardi (1966). High molecular weight non-ribosomal type nuclear RNA and cytoplasmic messenger RNA in HeLa cells. *Proc. Natl. Acad. Sci., U.S.*, **56,** 616.

Howard-Flanders, P. (1968). DNA repair. *Ann. Rev. Biochem.*, **37,** 175.

Howk, R., and T. Y. Wang (1969). DNA polymerase from rat liver chromosomal proteins. I. Partial purification and general characteristics. *Arch. Biochem. Biophys.*, **133,** 238.

Huberman, J. A., and A. D. Riggs (1968). On the mechanism of DNA replication in mammalian chromosomes. *J. Mol. Biol.*, **32,** 327.

Hurwitz, J., and J. T. August (1963). The role of DNA in RNA synthesis. *Prog. Nucleic Acid Res. Mol. Biol.*, **1,** 59.

Jacob, F., S. Brenner, and F. Cuzin (1963). On the regulation of DNA replication in bacteria. *Cold Spring Harbor Symp. Quant. Biol.*, **28,** 329.

Jeanteur, P., F. Amaldi, and G. Attardi (1968). Partial sequence of ribosomal RNA from HeLa cells. II. Evidence for non-ribosomal type sequences in 45S and 32S ribosomal RNA precursors. *J. Mol. Biol.*, **33,** 757.

Jones, K. W. (1970). Chromosomal and nuclear location of mouse satellite DNA in individual cells. *Nature*, **225,** 912.

Josse, J., A. D. Kaiser, and A. Kornberg (1961). Enzymatic synthesis of DNA. VIII. Frequencies of nearest neighbor base sequences in DNA. *J .Biol. Chem.*, **236,** 864.

Jovin, T. M., P. T. Englund, and L. L. Bertsch (1969). Enzymatic synthesis of DNA. XXVI. Physical and chemical studies of a homogeneous DNA polymerase. *J. Biol. Chem.*, **244,** 2996.

Kedinger, C., M. Gniazdowski, J. C. Mandel, Jr., F. Gissinger, and P. Chambon (1970). α-Amanitin: A specific inhibitor of one of two DNA-dependent RNA polymerase activities from calf thymus. *Biochem. Biophys. Res. Commun.*, **38,** 165.

Keir, H. M. (1965). Mammalian DNA polymerases. *Prog. Nucleic Acid Res. Mol. Biol.*, **4,** 82.

Kelly, R. B., M. R. Atkinson, J. E. Huberman, and A. Kornberg (1969). Excision of thymine dimers and other mismatched sequences by DNA polymerase of *E. coli*. *Nature*, **224,** 495.

Kelley, W. S., and H. J. Whitfield, Jun. (1971). Purification of an altered DNA polymerase from an *E. coli* strain with a *pol* mutation. *Nature*, **230,** 33.

Knippers, R., and W. Stratling (1970). The DNA replicating capacity of isolated *E. coli* cell wall–membrane complexes. *J. Mol. Biol.*, **46,** 281.

Konberg, T., and M. L. Gefter (1970). DNA synthesis in cell-free extracts of a DNA polymerase defective mutant. *Biochem. Biophys. Res. Commun.*, **40,** 1348.

Kornberg, A. (1961). *Enzymatic Synthesis of DNA*, Ciba Lectures in Microbial Biochemistry. Wiley, New York.

Lark, C. (1968). Effect of the methionine analogs ethionine and nor-leucine on DNA synthesis in *E. coli* 15T$^-$. *J. Mol. Biol.*, **31,** 401.

Lark, K. G. (1969). Initiation and control of DNA synthesis. *Ann. Rev. Biochem.*, **38,** 569.

Latham, H., and J. E. Darnell (1965). Distribution of mRNA in the cytoplasmic polyribosomes of the HeLa cell. *J. Mol. Biol.*, **14,** 1.

Levitt, M. (1969). Detailed molecular model for transfer RNA. *Nature*, **224,** 759.

Liao, S., D. Sagher, A. H. Lin, and S. Fang (1969). Magnesium and manganese specific forms of soluble liver RNA polymerase. *Nature*, **223,** 297.

Lindahl, T., and G. M. Edelman (1968). Polynucleotide ligase from myeloid and lymphoid tissues. *Proc. Natl. Acad. Sci., U.S.*, **61,** 680.

Lindsay, J. G., and R. L. P. Adams (1968). DNA nucleotidyl-transferase in nuclei and supernatant fractions of mouse fibroblasts. *Biochem. J.*, **108,** 43P.

Littlefield, J. W., A. P. McGovern, and K. B. Margeson (1963). Changes in the distribution of polymerase activity during DNA synthesis in moust fibroblasts. *Proc. Natl. Acad. Sci., U.S.*, **49,** 102.

Loeb, L. A. (1970). Molecular association of DNA polymerase with chromatin in sea urchin embryos. *Nature*, **226,** 298.

Loeb, L. A., B. Fansler, R. Williams, and D. Mazia (1969). Sea urchin nuclear DNA polymerase. I. Localisation in nuclei during rapid DNA synthesis. *Exptl. Cell Res.*, **57,** 298.

Loening, U. E. (1968). Molecular weight of ribosomal RNA in relation to evolution. *J. Mol. Biol.*, **38,** 355.

Loening, U. E., K. Jones, and M. L. Birnstiel (1969). Properties of ribosomal RNA precursor in *Xenopus laevis*: Comparison to the precursor in mammals and in plants. *J. Mol. Biol.*, **45,** 353.

Lucia, P. de, and J. Cairns (1969). Isolation of an *E. coli* strain with a mutation affecting DNA polymerase. *Nature*, **224,** 1164.

Maaloe, O., and P. C. Hanawalt (1961). Thymidine deficiency and the normal DNA replication cycle. *J. Mol. Biol.*, **3,** 144.

McCarthy, B. J., and B. H. Hoyer (1964). Identity of DNA and diversity of mRNA molecules in normal mouse tissues. *Proc. Natl. Acad. Sci., U.S.*, **52,** 915.

Mahler, H. R., and E. H. Cordes (1966). *Biological Chemistry*, p. 176. Harper, New York.

Maitra, U., and J. Hurwitz (1965). The role of DNA in RNA synthesis. IX. Nucleoside triphosphate termini in RNA polymerase products. *Proc. Natl. Acad. Sci., U.S.*, **54,** 815.

Masters, M., and P. Broda (1971). Evidence for the bidirectional replication of the *E. coli* chromosome. *Nature New Biol.*, **232,** 137.

Meselson, M., and F. W. Stahl (1958). The replication of DNA in *E. coli*. *Proc. Natl. Acad. Sci., U.S.*, **44,** 671.

Mizutani, S., H. M. Temin, M. Kodama, and R. T. Wells (1971). DNA ligase and exonuclease activities in virions of Rous sarcoma virus. *Nature New Biol.*, **230,** 233.

Mueller, G. C. (1969). Biochemical events in the animal cell cycle. *Federation Proc.*, **28,** 1780.

Novello, F., and F. Stirpe (1970). Simultaneous assay of RNA polymerase I and II in nuclei isolated from resting and growing rat liver with the use of α-amanitin. *FEBS Letters*, **8,** 57.

Nozawa, R., and D. Mizino (1969). Replication and properties of DNA in nicotinamide adenine dinucleotide deficiency of *E. coli* cells. *Proc. Natl. Acad. Sci., U.S.*, **63,** 904.

Okazaki, T., and R. Okazaki (1969). Mechanism of DNA chain growth. IV. Direction of synthesis of T4 short DNA chains as revealed by exonucleolytic degradation. *Proc. Natl. Acad. Sci., U.S.*, **64,** 1243.

Okazaki, R., T. Okazaki, K. Sakabe, K. Suzimoto, and A. Sugino (1968). Mechanism of DNA chain growth. I. Possible discontinuity and unusual secondary structure of newly synthesized chains. *Proc. Natl. Acad. Sci., U.S.*, **59,** 598.

Pauling, C., and L. Hamm (1969). Properties of a DNA ligase mutant of *E. coli*. II. Intermediates in DNA replication. *Biochem. Biophys. Res. Commun.*, **37,** 1015.

Penman, S., I. Smith, and E. Holtzman (1966). Ribosomal RNA synthesis and processing in a particulate site in the HeLa cell nucleus. *Science*, **154,** 789.

Perry, R. D. (1967). The nucleolus and the synthesis of ribosomes. *Prog. Nucleic Acid Res. Mol. Biol.*, **6,** 219.

Pierucci, O., and C. E. Helmstetter (1969). Chromosome replication, protein synthesis and cell division in *E. coli*. *Federation Proc.*, **28,** 1755.

Prescott, D. M. (1964). Cellular sites of RNA synthesis, *Prog. Nucleic Acid Res. Mol. Biol.*, **3,** 33.

Prestayko, A. W., and H. Busch (1968). Low molecular weight RNA of the chromatin fraction from Novikoff hepatoma and rat liver nuclei. *Biochim. Biophys. Acta*, **169,** 339.

Pritchard, R. H., and K. G. Lark (1964). Induction of replication by thymine starvation at the chromosome origin in *E. coli*. *J. Mol. Biol.*, **9,** 288.

Reich, E., and I. H. Goldberg (1964). Actinomycin and nucleic acid function. *Prog. Nucleic Acid Res. Mol. Biol.*, **3,** 183.

Retel, J., and R. J. Planta (1970). Non-conservative processing of ribosomal precursor RNA in yeast. *Biochim. Biophys. Acta*, **34,** 286.

Richardson, C. C. (1969). Enzymes in DNA metabolism. *Ann. Rev. Biochem.*, **38,** 795.

Richardson, J. P. (1969). RNA polymerase and the control of RNA synthesis. *Prog. Nucleic Acid Res. Mol. Biol.*, **9,** 75.

Roberts, J. W. (1969). Termination factor for RNA synthesis. *Nature*, **224,** 1168.

Roeder, R. G., and W. J. Rutter (1969). Multiple forms of DNA-dependent RNA polymerase in eukaryotic organisms. *Nature*, **224,** 234.

Roeder, R. G., and W. J. Rutter (1970). Specific nucleolar and nucleoplasmic RNA polymerases. *Proc. Natl. Acad. Sci., U.S.*, **65,** 675.

Salser, W., J. Janin, and C. Levinthal (1968). Measurement of the unstable RNA in exponentially growing cultures of *Bacillus subtilis* and *E. coli*. *J. Mol. Biol.*, **31,** 237.

Schandl, E. K., and J. H. Taylor (1969). Early events in the replication and integration of DNA into mammalian chromosomes. *Biochem. Biophys. Res. Commun.*, **34,** 291.

Scherrer, K., and J. E. Darnell (1962). Sedimentation characteristics of rapidly labelled RNA from HeLa cells. *Biochem. Biophys. Res. Commun.*, **7,** 486.

Schildkraut, C. L., C. C. Richardson, and A. Kornberg (1964). Enzymic synthesis of DNA. XVII. Some unusual physical properties of the product primed by native DNA templates. *J. Mol. Biol.*, **9,** 24.

Sethi, V. S., and W. Zillig (1970). Dissociation and reconstitution of active DNA-dependent RNA polymerase from *E. coli*. *FEBS Letters*, **8,** 236.

Shenkin, A., and R. H. Burdon (1966). Asymmetric transcription of DNA by DNA dependent-RNA polymerase of Krebs 2 ascites tumour cells. *Biochem. J.*, **98,** 5C.

Sinsheimer, R. L., and M. Lawrence (1964). *In vitro* synthesis and properties of a ΦX DNA-RNA hybrid. *J. Mol. Biol.*, **8,** 289.

Smellie, R. M. S., E. D. Gray, H. M. Keir, J. Richards, D. Bell, and J. N. Davidson (1960). Studies on the biosynthesis of DNA by extracts of mammalian cells. III. Net synthesis of polynucleotides. *Biochim. Biophys. Acta*, **37,** 243.

Smillie, E. J., and R. H. Burdon (1970). The enzymic conversion of transfer RNA precursor to 4S RNA. *Biochim. Biophys. Acta*, **213,** 248.

Smith, D. W., H. E. Schaller, and F. J. Bonhoeffer (1970). DNA synthesis *in vitro*. *Nature*, **226,** 711.

Smith, D. W., and P. C. Hanawalt (1967). Properties of the growing point region in the bacterial chromosome. *Biochem. Biophys. Acta*, **149,** 519.

So, A. G., E. W. Davie, R. Epstein, and A. Tissières (1967). Effects of cations on DNA-dependent RNA polymerase, *Proc. Natl. Acad. Sci., U.S.*, **58,** 1739.

Soiero, R., H. C. Birnboim, and J. E. Darnell (1966). Rapidly labelled HeLa cell nuclear RNA. II. Base composition and cellular localization of a heterogeneous RNA fraction. *J. Mol. Biol.*, **19,** 362.

Sueoka, N., and W. G. Quinn (1968). Membrane attachment of the chromosome replication origin in *B. subtilis*. *Cold Spring Harbor Symp. Quant. Biol.*, **33,** 695.

Sugimoto, K., T. Okazaki, Y. Imae, and R. Okazaki (1969). Mechanism of DNA chain growth. III. Equal annealing of T4 nascent short DNA chains with the separated complementary strands of phage DNA. *Proc. Natl. Acad. Sci., U.S.*, **63,** 1343.

Szybalski, W., H. Kubinski, and P. Sheldrick (1966). Pyrimidine clusters on the transcribing strand of DNA and their possible role in the initiation of RNA synthesis. *Cold Spring Harbor Symp. Quant. Biol.*, **31,** 123.

Taylor, J. H., P. S. Woods, and W. L. Hughes (1967). The organisation and duplication of chromosomes as revealed by autoradiographic studies using tritium labelled thymidine. *Proc. Natl.Acad. Sci., U.S.*, **43,** 122.

Temin, H. M., and S. Mizutani (1970). RNA-dependent DNA polymerase in virions of Rous sarcoma virus. *Nature*, **226,** 1213.

Travers, A. A., and R. R. Burgess (1969). Cyclic re-use of RNA polymerase sigma factor. *Nature*, **222,** 537.

Tsukada, K., T. Moriyama, W. E. Lynch, and I. Lieberman (1968). Polydeoxynucleotide intermediates in DNA replication in regenerating liver. *Nature*, **220,** 162.

Volkin, E., and L. Astrachan (1956). Intracellular distribution of labelled RNA after phage infection of *E. coli*. *Virology*, **2,** 433.

Walker, P. M. B. (1971). Origin of satellite DNA. *Nature*, **229,** 306.

Ward, C. B., and D. A. Glaser (1970). Origin and direction of DNA synthesis in *E. coli* B/r. *Proc. Natl. Acad. Sci., U.S.*, **62,** 881.

Warner, J. R., and R. Soiero (1967). Nascent ribosomes from HeLa cells. *Proc. Natl. Acad. Sci., U.S.*, **58**, 1984.

Werner, R. (1971). Nature of DNA precursors. *Nature New Biol.*, **233**, 99.

Wehrli, W., and M. Staehlin (1970). In *RNA Polymerase and Transcription* (L. Silvestri, ed.), Lepetit Colloquia on Biology and Medicine, Vol. 1, p. 65. Wiley, New York.

Weinberg, R. A., and S. Penman (1968). Small molecular weight monodisperse nuclear RNA. *J. Mol. Biol.*, **38,** 289.

Weinberg, R. A., and S. Penman (1970). Processing of 45S nucleolar RNA. *J. Mol. Biol.*, **47,** 169.

Weiss, S. B. (1960). Enzymatic incorporation of ribonucleoside triphosphates into the interpolynucleotide linkages of RNA. *Proc. Natl. Acad. Sci., U.S.*, **46,** 1020.

Widnell, C. C., and J. R. Tata (1966). Studies on the stimulation by ammonium sulphate of the DNA-dependent RNA polymerase of isolated rat liver nuclei. *Biochim. Biophys. Acta*, **123,** 478.

Widnell, C. C., and J. R. Tata (1969). Evidence for two DNA- dependent RNA polymerase activities in isolated rat liver nuclei. *Biochim. Biophys. Acta*, **87,** 531.

Yankofsky, S. A., and S. Spiegelman (1962). The identification of the ribosomal RNA cistron by sequence complementarity. II. Saturation of and competitive interaction at the RNA cistron. *Proc. Natl. Acad. Sci., U.S.*, **48,** 1466.

10

PROTEIN SYNTHESIS IN ANIMAL CELLS

B. E. H. Maden

1. INTRODUCTION

In Chapter 9 the nucleic acids and their role in the storage and transfer of genetic information are discussed. This chapter concerns the translation of this information into the functional proteins upon which cellular activity depends.

Until recently, the aspects of protein synthesis that received the greatest attention were the genetic code and the mechanism of the translation process. Of particular interest to cell biologists, however, are several more specialized aspects of protein synthesis in eukaryotic cells, such as the intracellular organization of protein synthesis, the synthesis of specific proteins by differentiated and cultured cells, and protein synthesis in virus-infected cells. Also, the control of protein synthesis in higher organisms is far from clearly understood.

This chapter opens with a general account of the components and mechanism of protein synthesis as revealed largely from bacterial systems. Subsequent sections survey special aspects of protein synthesis in animal cells, particularly in the areas mentioned above. A more complete account of the mechanism of protein synthesis in bacteria is given in a review by Lengyel and Söll (1969). Volume 31 of the *Cold Spring Harbor Symposium on Quantitative Biology* (1966) deals in detail with the genetic code.

2. GENERAL ASPECTS OF PROTEIN SYNTHESIS

A. Gene Structure and Protein Structure

The linear topologies of gene and polypeptide carry the obvious implication that the two structures are colinear. Proof of colinearity of gene and protein was first obtained from an analysis of mutants of the head protein of bacteriophage T4 (Sarabhai et al., 1964) and the A protein of tryptophan synthetase of *Escherichia coli* (Yanofski et al., 1964).

The genetic code relating gene to protein is triplet in nature (a triplet of nucleotides in DNA codes for a single amino acid in a polypeptide), nonoverlapping (a point mutation affects a single amino acid), and degenerate (in general, more than one alternative triplet specifies each amino acid). The triplet nature of the code was first deduced from an analysis of "frameshift" mutations in the *rII* region of the genetic map of bacteriophage T4 (Crick et al., 1961). Detailed codon assignments were made mainly from *in vitro* studies on protein synthesis directed by synthetic polynucleotides of defined sequence (see *Cold Spring Harbor Symposium on Quantitative Biology*, 1966). Some of these codon assignments are referred to in subsequent sections. The code is shown in Figure 1.

B. The Components of Protein Synthesis

Protein synthesis requires the prior transcription of genetic information into messenger RNA (mRNA). Translation of the genetic message then involves the sequential interaction of aminoacyl tRNA with mRNA on ribosomes and requires

1st ↓ 2nd→	U	C	A	G	3rd
U	Phe	Ser	Tyr	Cys	U
	Phe	Ser	Tyr	Cys	C
	Leu	Ser	C.T.	C.T.	A
	Leu	Ser	C.T.	Trp	G
C	Leu	Pro	His	Arg	U
	Leu	Pro	His	Arg	C
	Leu	Pro	Gln	Arg	A
	Leu	Pro	Gln	Arg	G
A	Ile	Thr	Asn	Ser	U
	Ile	Thr	Asn	Ser	C
	Ile	Thr	Lys	Arg	A
	Met (fMet)	Thr	Lys	Arg	G
G	Val	Ala	Asp	Gly	U
	Val	Ala	Asp	Gly	C
	Val	Ala	Glu	Gly	A
	Val } fMet }	Ala	Glu	Gly	G

Figure 1. The genetic code, as deduced mainly from studies with *E. coli*. It is likely that the code is universal. Parentheses indicate chain initiation and chain termination. Note that AUG and GUG (initiating codons) are read *internally* as met and val, respectively.

in addition various supernatant protein factors, GTP, and inorganic cations. Properties of these components of the protein-synthesizing apparatus are now summarized. Their mode of interaction in protein synthesis is outlined in the following section.

1. Ribosomes. It has been known for many years that ribosomes are the subcellular particles on which protein synthesis takes place. Consequently, much effort is directed toward the analysis of ribosome structure. At present, however, it is only possible to correlate structure with function in rather general terms. We therefore take the prerogative of giving a rather abbreviated summary of ribosome structure. New details are constantly appearing in the literature, and much of the fairly recent work is described in Volume 34 of the *Cold Spring Harbor Symposium for Quantitative Biology* (1969).

Ribosomes from all sources are built on the same plan. They are roughly ellipsoidal ribonucleoprotein particles of molecular weight ranging from 2.5×10^6 in bacteria to approximately 4×10^6 in mammals. They consist of two subunits, of which the larger is 2 to 2.5 times the size of the smaller. The two subunits associate together and function as an integrated unit in protein synthesis. They dissociate between "rounds" of protein synthesis *in vivo* and can be made to dissociate *in vitro* by exposure to low Mg^{2+} concentration, high concentration of monovalent cations, or EDTA. (Subunits prepared by EDTA treatment are usually inactive in subsequent *in vitro* protein synthesis.) The relative S values of the monomeric ribosome and the subunits are: in bacteria, 70, 50, and 30S; in higher organisms, 80, 60, and 40S.

Somewhat more than half of the mass of each ribosomal subunit consists of RNA, the remainder being made up of protein. The small subunit consists of a single RNA molecule of $0.55–0.75 \times 10^6$ daltons (16–18S, the larger values being for mammalian ribosomes) and several proteins. In *E. coli* these proteins comprise 20 distinct polypeptide species. The majority of the polypeptides are present to the extent of 1 per 30S ribosome; a few are present in fractional amounts. The large subunit contains one large ($1.1–1.75 \times 10^6$ daltons; 23–28S) and one small (5S) RNA molecule and a further complex group of proteins (approximately 36 polypeptides in *E. coli*). Definitive information is not yet available on the numbers and molar ratios of animal cell ribosomal proteins, but partial fractionation of the proteins on acrylamide or starch gels indicates a high degree of complexity.

At present there is only indirect evidence relating to the three-dimensional arrangement of rRNA and proteins in the ribosome. Physicochemical data suggest that rRNA consists of alternating single- and double-helical regions (Cox, 1966), and that much of the surface of the ribosome consists of RNA (Cotter et al., 1967), presumably facilitating interaction of rRNA with the other RNA components of the protein-synthesizing system.

Not surprisingly, detailed functions of the RNA and numerous protein components of the ribosome remain unknown, although spectacular progress is being made in *E. coli* with studies on the *in vitro* reassembly of the 30S ribosome from its RNA and protein components (Mizushima and Nomura, 1970, and references therein). Evidence is given in Section 2C as to the overall functions of the two ribosomal subunits in protein synthesis.

2. mRNA. Whereas ribosomes are the "molecular machines" (Spirin, 1969) upon which polypeptide assembly takes place, the *information* for protein synthesis derives from a separate RNA species that "programs" these machines, mRNA. The historical background of the discovery of mRNA has been amply documented elsewhere and need not be dealt with here (see, for example, Brenner, 1961; Perutz, 1962; Burdon, this volume). There are, however, several points in connection with mRNA that may be mentioned at this point.

1. mRNA is characterized by its reversible attachment to ribosomes, its heterogeneous size, DNA-like base composition, and complementarity to a considerable fraction of the genome in DNA–RNA hybridization experiments (at least in bacteria; see Section III.A for animal cells). rRNA, by contrast, is homogeneous in size, of distinctive base composition, and complementary to a very restricted segment of the genome.

2. Usually several ribosomes are found in association with a single molecule of mRNA, the resulting aggregate being termed a polyribosome or *polysome*. In radioactive labeling experiments nascent protein is located almost exclusively on polysomes and not on single ribosomes or subunits.

3. In general, mRNA is more rapidly labeled *in vivo* and has a shorter half-life than ribosomes which are metabolically stable. However, the metabolic stability of mRNA varies widely in different tissues. The very rapid (1–2 min) turnover originally attributed to mRNA applies only to bacteria and bacteriophages. The average half-life for mRNA in HeLa cells is of the order of 3 hr, and in reticulocytes is as long as that of the ribosomes themselves (a few days).

4. Protein synthesis *in vitro* can be directed by the addition of synthetic polyribonucleotides or by viral RNA. So far, bacterial systems have proved more responsive than animal cell systems to the addition of exogenous mRNA, probably for reasons connected with the difficulty of obtaining efficient *initiation* of polypeptide synthesis in animal cell systems. In such experiments the nature of polypeptide products is determined by the nature of the added template RNA. As already mentioned, the use of synthetic polynucleotides of defined sequence has been instrumental in solving the genetic code.

5. With the exception of certain viral RNAs, it has until recently been impossible to obtain homogeneous preparations of natural mRNA for a single protein or limited group of proteins. By the use of newer techniques such as purification of RNA by acrylamide gel electrophoresis, coupled with the proper selection of biological material, it has now become possible to obtain pure or partially pure preparations of globin mRNA (Labrie, 1969; Laycock and Hunt, 1969; Williamson et al., 1970), and histone mRNA (Borun et al., 1967). These developments will undoubtedly facilitate the study of *in vitro* synthesis of specific proteins and will also open the way to sequencing individual mRNA species.

3. tRNA. The other central component of protein synthesis is transfer RNA (tRNA). tRNA serves a triple role in protein synthesis. It functions in amino acid activation, acts as adaptor in mRNA-directed amino acid specification, and binds the growing polypeptide to the ribosome.

Before incorporation into protein the amino acids are first enzymatically esterified to tRNA. There is one or more species of tRNA specific for each amino acid. Esterification of an amino acid to its cognate tRNA occurs at the expense of hydrolysis of ATP to AMP and inorganic pyrophosphate. The ester bond of aminoacyl tRNA possesses a free energy of hydrolysis equivalent to that of a pyrophosphate bond in ATP, which is sufficient to drive subsequent peptide bond formation strongly in the forward direction. The activation reaction is itself rendered effectively irreversible by hydrolysis of inorganic pyrophosphate to orthophosphate in the cell.

Selection of the correct amino acid in protein synthesis is achieved by base pairing between a triplet of nucleotides in mRNA (the codon) and a complementary triplet of nucleotides on tRNA (the anticodon). Crick (1966) has proposed a series of rules whereby a single tRNA species can interact with two or three alternative codons which differ only in the third base (the "wobble" hypothesis). The wobble hypothesis explains the pattern of degeneracy of the genetic code shown in Figure 1 and makes predictions as to the nature of the third base in the anticodon. In every case in which tRNA sequence information is available (see next paragraph), the structure of the anticodon is consistent with the corresponding codon assignments and the wobble hypothesis.

Because tRNA molecules are relatively small (75 to 80 nucleotides long), many different tRNA species from various sources have recently been sequenced (for a compilation of references see Lengyel and Söll, 1969). Each tRNA whose sequence is known can be arranged in the form of a common secondary structure, the cloverleaf pattern (Fig. 2). It is therefore highly probable that some derivative of this basic form represents the actual three-dimensional structure of tRNA (for a plausible detailed model see Levitt, 1969). In every case the cloverleaf structure leaves exposed

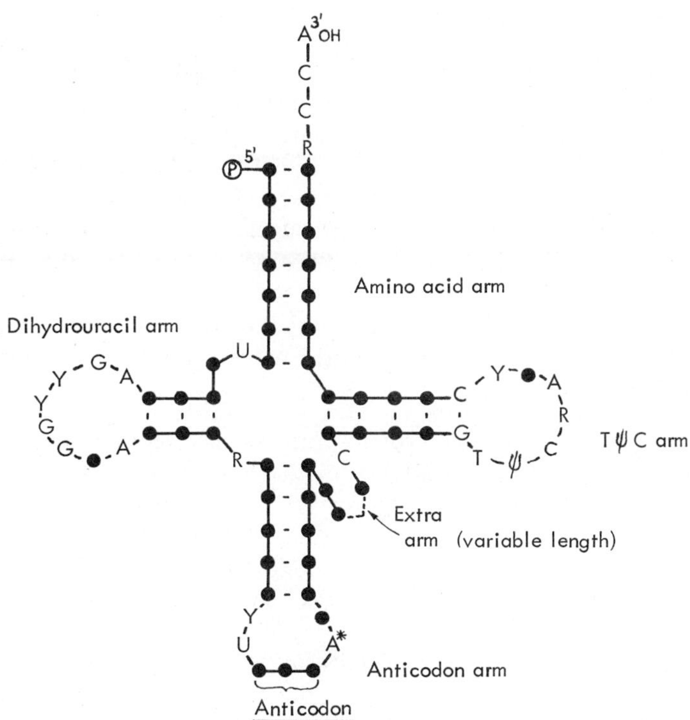

Figure 2. A generalized version of the cloverleaf model of tRNA. ●, Any base; -, hydrogen bonding between bases; R, purine; Y, pyrimidine; T, ribothymidine; ψ, pseudouridine; A* modified adenine usually found adjacent to anticodon. In a proposed three-dimensional structure, Levitt (1969) envisages the amino acid arm stacked on the TψC arm, and the dihydrouracil arm stacked on the anticodon arm to form two "double" double helices. (Adapted from Levitt, 1969).

a triplet of nucleotides complementary to the corresponding codon (the presumed anticodon) at the end of the central arm of the cloverleaf. It is also worth noting that the two other arms always contain common features (TψC and dihydrouracil, respectively), suggesting that these loops may be involved in common functions of all tRNA molecules, such as ribosome binding.

The rapid accumulation of knowledge on the structures of individual tRNA species stems not only from the development of RNA sequencing methods but also from techniques for obtaining fairly pure preparations of individual tRNA species. To the original rather cumbersome techniques of column chromatography and countercurrent distribution have now been added two elegant methods. One of these depends upon the selective retention of charged tRNA species, chemically modified in the aminoacyl moiety, on columns of benzoylated DEAE-cellulose (Gillam et al., 1967). The other important development is the coupling of the gene for a specific tRNA, the tyrosine suppressor tRNA, to the bacteriophage ϕ80. Infection of a bacterial cell with this modified phage then results in production of large quantities of tyrosine suppressor tRNA (which may be selectively labeled with ^{32}P (J. D. Smith et al., 1966; Goodman et al., 1968)). (A brief account of suppression is given in Section 2.C.5.) By selecting for mutants that have lost the suppressor function and obtaining tRNA preparations by the above technique, it has become possible to study the effects of single base changes on the different functions of tRNA in protein synthesis, namely amino acid activation, binding of tRNA to the mRNA–ribosome complex, and incorporation of the amino acid into protein (Abelson et al., 1970).

4. **Supernatant Protein Factors and GTP.** A cell-free system containing only ribosomes, mRNA, and aminoacyl tRNA shows little activity in protein synthesis. When the ribosomes are first washed by centrifugation through solutions of high salt concentration, no activity is obtained at all. Activity may be restored to the system by addition of postribosomal supernatant protein, GTP, and a GTP-regenerating system. Prior fractionation of the postribosomal supernatant reveals that several distinct protein factors are involved in protein synthesis. Of these no less than three are concerned with the initial formation of the mRNA–tRNA–ribosome complex (chain initiation), two with polypeptide chain growth, and two with chain termination. The way in which these factors are involved in the overall process of protein synthesis is described in Section 2.C.

5. **Inorganic Cations.** *In vitro* protein synthesis also requires suitable concentrations of monovalent cation (K^+ or NH_4^+) and divalent cation (usually Mg^{2+}, although this is wholly or partially replaceable by Mn^{2+}, Ca^{2+}, or spermidine).

Certain interactions in protein synthesis occur in the absence of monovalent cations [e.g., mRNA binding to ribosomes (Moore, 1966)], but K^+ or NH_4^+ are essential for other steps in the overall process, for example, peptide bond formation (Maden and Monro, 1968).

With respect to divalent cations, a suitable Mg^{2+} concentration is essential for maintaining not only the efficiency but also the *fidelity* of translation *in vitro*. When the Mg^{2+} concentration is too low, no protein synthesis occurs. Over a rather sharp concentration range (3–5 mM), the system is rapidly activated. At higher concentrations there is only slight loss of efficiency of incorporation, but incorrect amino acids are incorporated with appreciable frequency (e.g., leucine with polyuridylic acid, which normally codes only for phenylalanine). The first "mistake" to manifest itself at slightly elevated Mg^{2+} concentration (10 mM) is nonspecific *initiation* of protein synthesis. It was because of this effect that it was possible to study both the genetic code and the mechanism of polypeptide chain extension in bacterial cell-free systems before the specific role of fmet tRNA in the initiation of protein synthesis became known (see Section 2.C.4).

C. Polypeptide Chain Assembly

1. Sequential Assembly and Direction of Translation. Polypeptide assembly proceeds sequentially from the N′-terminal to the C′-terminal end. This was first established by Dintzis (1961) for the globin chains of hemoglobin. The principle upon which Dintzis' experiments were based is shown in Figure 3. If reticulocytes are briefly labeled with amino acids, a partition of label will occur between complete and nascent polypeptide chains. If assembly is sequential, there will result, after short labeling periods, a gradient of radioactivity along the *completed* chains such that the end synthesized *last* will contain the most radiaoctivity. In the experiments complete chains were separated from nascent (ribosome-bound) chains by differential centrifugation. α- and β-globin chains were separated by column chromatography and

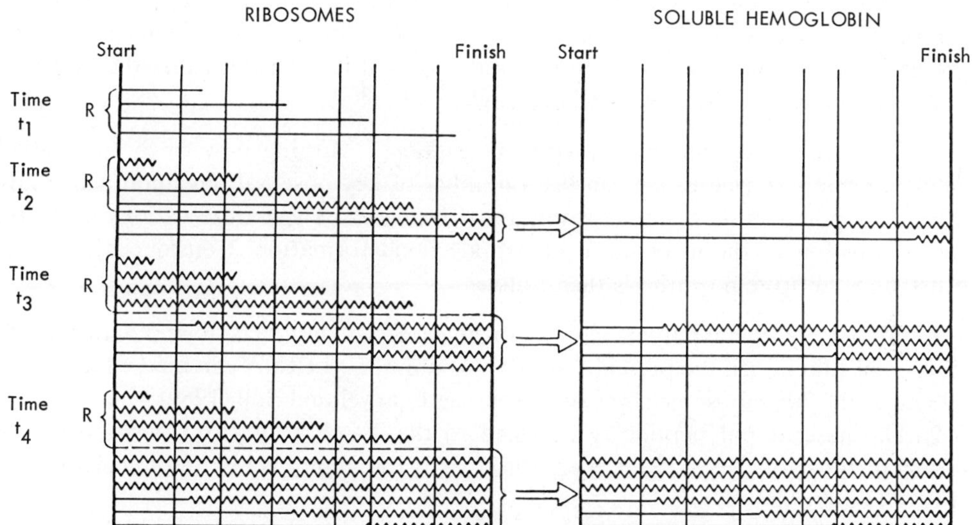

Figure 3. Sequential polypeptide assembly; predicted distribution of radioactivity between peptides of complete and nascent protein chains after labeling for short periods with radioactive amino acids. Labeled amino acids are added at time t_1. Straight lines represent unlabeled protein synthesized before t_1; zigzag lines represent labeled protein synthesized after t_1. After a labeling period, which is short compared with the time required for polypeptide chain assembly, there is a uniform distribution of radioactivity between the peptides of nascent protein, but in completed chains only peptides near the end assembled *last* are labeled ($t_1 - t_2$). After somewhat longer labeling periods ($t_1 - t_4$), there are gradients of radioactivity of opposite polarity between the peptides of nascent and complete polypeptide chains. After still longer labeling periods (not shown), the distribution of label between peptides in nascent chains remains the same, but in complete chains labeling of the peptides approaches uniformity. (From Dintzis, 1961.)

cleaved into specific peptides with trypsin. After short labeling periods a steep radioactivity gradient was clearly evident across each chain, the C′-terminal peptide containing the most radioactivity.

The distribution of label between complete and incomplete polypeptide chains in these experiments permitted estimation of another important parameter, polypeptide chain growth rate. This was calculated to be approximately 2 min at 37°C for globin chains (approximately 150 amino acids long). Subsequent careful analysis of the distribution of label between tryptic peptides of *nascent* chains has indicated that the rate of polypeptide chain growth is uniform along the length of the chains (Hunt et al., 1968). In *E. coli* the polypeptide chain growth rate is considerably faster than the figure given above for reticulocytes (McQuillen et al., 1959).

Sequential assembly from the N′-terminal end of the polypeptide has been confirmed by studies with cell-free systems. Stepwise synthesis of various di-, tri- and oligopeptides from the N′-terminal end has been shown, using either polynucleotides of partly defined sequence or MS2 or related phage RNAs as template (Bretscher, 1968a; Haenni and Lucas-Lenard, 1968; Erbe et al., 1969).

Use of cell-free systems programmed by synthetic polynucleotides with distinguishing codons at the 5′-end has also permitted the demonstration that the 5′-end of mRNA corresponds to the N′-terminal end of the polypeptide (M. A. Smith et al., 1966). This has been confirmed *in vivo* by correlating amino acid replacements in a frameshift mutation of bacteriophage T4 lysozyme with codon assignments for the respective amino acids; correlation is possible only if it is assumed that the direction of translation is $5' \rightarrow 3' = N' \rightarrow C'$ (Terzaghi et al., 1966).

2. The Ribosome and Peptide Bond Formation. Turning now to the individual steps in protein synthesis, we can list a number of key observations upon which any model for polypeptide chain assembly must be based. These observations lead first to a discussion of the mechanism of peptide bond formation. Consideration of the other steps in protein synthesis then follows.

1. mRNA (synthetic polynucleotides or phage RNA) interacts *in vitro* with the 30S ribosomal subunit. In the presence of mRNA, aminoacyl tRNA can bind either to the 30S or to the 70S ribosome (for references, see Lengyel and Söll, 1969).

2. The nascent polypeptide is esterified to the 2′- (3′-) hydroxyl position of the terminal ribose of tRNA (Bretscher, 1963), as is the amino acid in aminoacyl tRNA (Fig. 4*A*).

3. When ribosomes that have been active in *in vitro* protein synthesis are dissociated into subunits by exposure to low Mg^{2+} concentration, the nascent polypeptide tRNA complex remains associated with the 50S subunit (Gilbert, 1963).

4. When the nascent polypeptide–tRNA–50S ribosome complex is incubated with puromycin (an analog of the aminoacyladenosine end of aminoacyl tRNA (Fig. 4*B*), the nascent polypeptide is transferred from tRNA into peptide linkage with puromycin (J. D. Smith et al., 1965). This reaction does not require supernatant factors or GTP (Monro et al., 1967) and is not inhibited by the compound, guanylyl methylene diphosphonate (GMP–PCP, Fig. 5), a competitive inhibitor of GTP-dependent reactions in protein synthesis (Hershey and Monro, 1966; see below).

5. Polyribosomes isolated after *in vivo* labeling experiments contain, on the average, *two* tRNA molecules per ribosome (Warner and Rich, 1964).

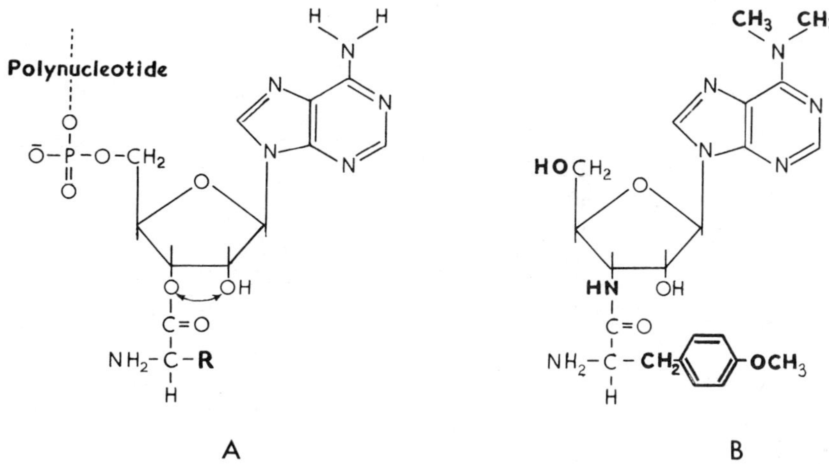

Figure 4. The structures of the 3′-terminal end of aminoacyl tRNA (*A*), and puromycin (*B*). Differences between the two structures are indicated in bold type.

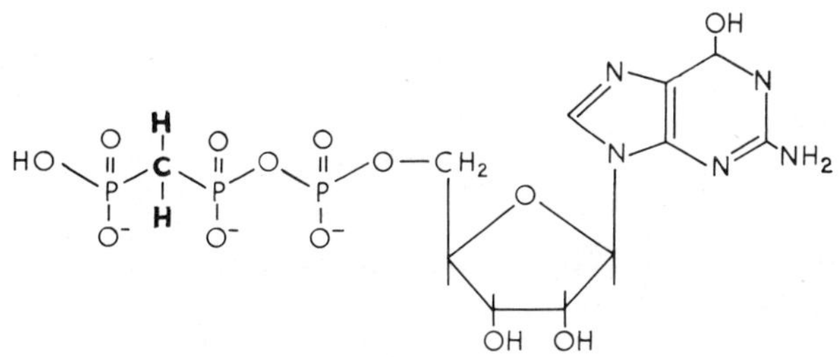

Figure 5. Guanylyl methylene diphosphonate.

From these observations it is possible to construct the ribosomal model for peptide bond formation shown in Figure 6. The model shows two parallel ribosomal binding sites for tRNA, each bridging the two ribosomal subunits. Codon–anticodon interaction takes place on the 30S subunit. Peptide bond formation takes place on the 50S subunit and results in transfer of nascent polypeptide from one tRNA to the next aminoacyl tRNA.

Evidence that peptide bond formation in *E. coli* is in fact catalyzed by the 50S ribosome itself (presumably some protein component thereof) is discussed in detail by Maden et al. (1968). Since the reaction does not require GTP, it probably occurs through simple group transfer via the nucleophilic attack mechanism shown in Figure 7. Peptide bond formation is therefore commonly termed *peptidyl transfer*.

With the above model in mind, we now turn to the remaining questions of polypeptide chain assembly: (1) What are the functions of supernatant factors and GTP? (2) How is polypeptide chain synthesis initiated? (3) How is the completed polypeptide released from its ester linkage with tRNA?

3. Translocation, Supernatant Factors, and GTP. In both bacterial and animal cell systems, two distinct supernatant factors and GTP are required for the overall

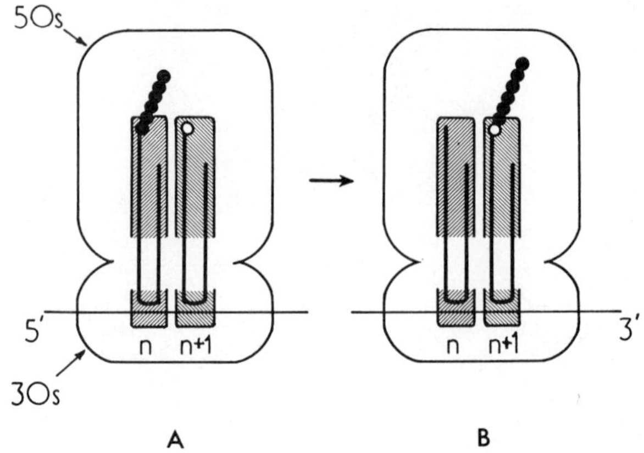

Figure 6. Model for peptidyl transfer showing the active complex before transfer (*A*) and after transfer (*B*).

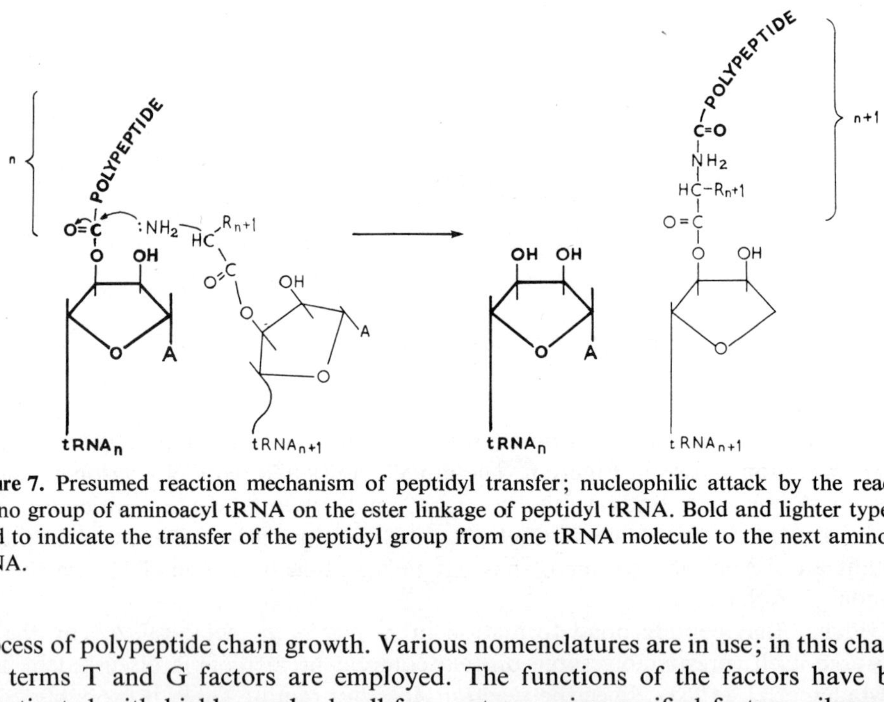

Figure 7. Presumed reaction mechanism of peptidyl transfer; nucleophilic attack by the reactive amino group of aminoacyl tRNA on the ester linkage of peptidyl tRNA. Bold and lighter type are used to indicate the transfer of the peptidyl group from one tRNA molecule to the next aminoacyl tRNA.

process of polypeptide chain growth. Various nomenclatures are in use; in this chapter the terms T and G factors are employed. The functions of the factors have been investigated with highly resolved cell-free systems using purified factors, ribosomes, synthetic messengers, aminoacyl tRNA, puromycin, GTP, and the GTP analog GMP–PCP. This compound, which was mentioned above (Fig. 5), is nonhydrolyzable between the β- and γ-phosphate groups and permits identification of the steps in protein synthesis at which GTP is hydrolyzed. Reactivity toward puromycin defines which of the two putative sites on the ribosome peptidyl tRNA (or aminoacyl tRNA) occupies. If peptidyl (aminoacyl) tRNA reacts with puromycin, it is presumed to occupy the peptidyl or P site of the 50S ribosome (Fig. 8); if not, it is presumed to occupy the aminoacyl or A site.

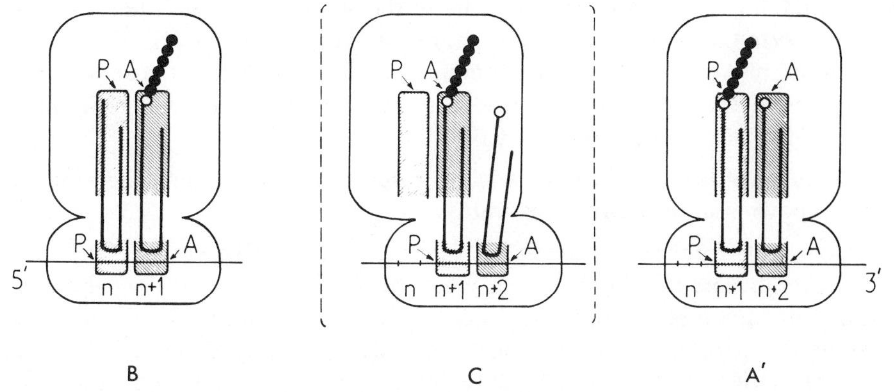

Figure 8. Model for translocation. (*B*) After peptidyl transfer (as in Fig. 6*B*); (*C*) proposed intermediate in translocation; (*A'*) Translocation complete (as in Fig. 6*A*). (From Bretscher, 1968.)

Experiments with supernatant factors have been performed by several groups, in particular by Haenni and Lucas-Lenard (1968) and Erbe et al. (1969) for bacterial systems, and Skogerson and Moldave (1968) for rat liver. As already indicated, the supernatant factors and GTP are not required for peptide bond formation itself. Their overall function in polypeptide chain growth is twofold: (1) to *translocate* peptidyl tRNA from the A site, which it occupies *after* peptidyl transfer, to the P site before the next peptidyl transfer reaction; this step automatically brings the next codon on mRNA into the A site; and (2) to bind the next incoming aminoacyl tRNA to the A site in a conformation that will permit peptidyl transfer.

The experimental data are consistent with a model in which G factor plus GTP mediate step (1), and T factor plus GTP mediate step (2), GTP being hydrolyzed at both steps. In its simplest form the model suggests that the ribosome moves as a single unit relative to the mRNA–peptidyl tRNA complex during translocation. While this may be true, momentary detachment of the whole ribosome from the mRNA–peptidyl tRNA complex could cause problems. Partly to overcome such difficulties, Bretscher (1968b) proposed the model for translocation shown in Figure 8. Here translocation is proposed to take place in two steps, the 30S subunit moving first to form a "hybrid" intermediate (Fig. 8*C*). This model, although hypothetical, is attractive for several reasons:

1. During translocation the peptidyl tRNA complex detaches from only one part of the ribosome at a time, the other part serving as an "anchor."

2. Binding of the incoming aminoacyl tRNA occurs first at the A site of the 30S subunit alone. Thus accidental access of an unspecified tRNA to the A site of the 50S ribosome is prevented, and a possible source of errors in protein synthesis is eliminated.

3. The model could explain a paradox connected with initiation of polypeptide synthesis. There is evidence suggesting that the initiating tRNA is bound first in the A site of the 50S ribosome. However, the unique degeneracy pattern for initiation ($^{A}_{G}$UG), affecting the *first* base of the codon, suggests that codon–anticodon recognition for this tRNA may occur at a site other than the A site of the 30S ribosome. The ribosomal conformation shown in Figure 8*C* would overcome this difficulty by permitting

binding of the initiating tRNA at the P site of the 30S ribosome and the A site of the 50S ribosome (initiation is considered in more detail in Section 2.C.4).

4. Separate movement of the two subunits during each cycle of reactions leading to peptide bond formation would provide a plausible explanation for the universal existence of two separable subunits of the ribosome.

In summary, although the details of translocation are still unclear, the model described above is an attractive possibility.

4. **Polypeptide Chain Initiation in Bacterial and Mammalian Systems.** It has been clear for several years that in bacteria a special mechanism initiates polypeptide chain synthesis. This involves the ribosomal binding of fmet tRNA under the direction of the codons AUG or GUG. Either of these codons, at the beginning of a cistron, initiates translation in the correct phase by defining the frame of reading of that cistron (Fig. 9). The mechanism therefore permits translation to commence from points other than the 5′-end of the messenger. Thus reinitiation can occur in polycistronic messengers, and in fact fmet is found at the N′-terminal end of all three nascent proteins of MS2 bacteriophage (Vinuela et al., 1967). The special tRNA species involved in initiation inserts methionine *only* into N′-terminal positions, regardless of the presence or absence of the formyl group. Moreover, AUG is read internally as met (by a different met tRNA species) and GUG internally as val.

The order of formation of the initiation complex *in vitro* is: (1) the 30S ribosome binds to mRNA; (2) fmet tRNA binds to the 30S ribosome–mRNA complex; (3) the 50S ribosome binds to the above complex.

Each of these interactions is mediated by a separate initiation supernatant factor. The fmet tRNA appears to be bound initially at the A site of the 50S ribosome at low Mg^{2+} concentration and undergoes GTP-dependent translocation prior to binding of the second aminoacyl tRNA to the ribosome. When the latter has been bound, peptide bond formation takes place, and subsequent polypeptide chain growth occurs as already described.

These steps in initiation were identified at low (3–5 m*M*) Mg^{2+} concentration. At this Mg^{2+} concentration protein synthesis occurs *only* in the presence of fmet tRNA and the appropriate codon in mRNA. At a higher (10 m*M*) Mg^{2+} concentration, initiation takes place nonspecifically with any synthetic mRNA (such as polyuridylic acid) and with random selection of frame of reading.

Escherichia coli possesses an enzyme system which cleaves either the formyl group, or fmet, from the N′-terminal end of many proteins (Adams, 1968). Thus many completed proteins no longer bear N′-terminal methionine or formylmethionine. A more detailed review of initiation in *E. coli* is given by Lengyel and Söll (1969).

Figure 9. Phasing effect of AUG near 5′-end of synthetic polynucleotides. With random selection of frame of reading, any of the three polypeptides indicated could be formed. In fact at low Mg^{2+}–concentration only fmet phe $_{(n)}$ is formed (bold type).

Whether or not animal cells possess a special mechanism for initiating polypeptide synthesis was in question for a long time. However, strong evidence is now accumulating that *methionine* (not formylmethionine) is involved in initiation of mammalian protein synthesis. As in bacteria, a special met tRNA species permits incorporation of methionine specifically into N′-terminal positions of polypeptides (Brown and Smith, 1970; A. E. Smith and Marcker, 1970). At the time of writing, various laboratories have obtained evidence that methionine is the N′-terminal amino acid in nascent rabbit globin (Jackson and Hunter, 1970) and nascent trout protamine (Wigle and Dixon, 1970). The N′-terminal methionine is cleaved from the chains early during polypeptide assembly, which accounts for failure to detect this methionine in the past. Thus a special species of met (or fmet) tRNA seems to play a wide if not universal role in initiation of polypeptide synthesis. Whether *reinitiation* occurs (in polycistronic messengers) in mammalian systems is an interesting question which is discussed in Section 3.C.

5. **Polypeptide Chain Termination.** At the end of polypeptide chain growth, the completed polypeptide is still attached to tRNA. Chain termination is defined as the release of polypeptide from tRNA. The signal for release is contained in mRNA, and mutants exist in microorganisms that bring about premature release of polypeptide from tRNA. These mutants arise from base changes within a codon specifying an amino acid, producing a codon with any one of the constitutions: UAG, UAA, or UGA. It can be shown that these triplets, incorporated into synthetic messengers, bring about chain termination in a cell-free system. Messengers lacking these codons do not specify termination.

It might be thought that a codon for chain termination would be read in the same way as any other codon, that is, by a tRNA or related molecule. However, all attempts to isolate a chain-terminating RNA species have failed. There exists a mutant of bacteriophage MS2 which causes premature chain termination in the head protein at the position of the seventh amino acid. In a cell-free system, release from tRNA occurs immediately after stepwise formation of a hexapeptide from six highly purified aminoacyl tRNA species. No other species of tRNA is required for release (Bretscher, 1968a).

It has in fact been found that two supernatant protein factors are involved in chain termination. One of these reads UAG and UAA; the other reads UAA and UGA. The mechanism whereby these factors bring about cleavage of the peptidyl tRNA bond is still unkhown. For further details and references, the reader is referred to Capecchi and Klein (1970) and Beaudet and Caskey (1970).

All information on chain-terminating mutants has been obtained from microbial systems in which the otherwise lethal effects of the mutations can be suppressed by "suppressor" strains of bacteria. These strains permit limited expression of the wild-type phenotype by allowing translation to proceed beyond the chain-terminating codon (Stretton and Brenner, 1965). The efficiency of propagation is characteristic of the suppressor strain employed. Propagation is brought about by insertion of an amino acid at the site of termination. One suppressor strain of *E. coli* inserts tyrosine in this way. Insertion of tyrosine results from the presence in the cell of an extra tyrosine tRNA species with a mutation in the anticodon such that it recognizes UAG rather than the normal tyrosine codons $\mathrm{UA}^{\mathrm{U}}_{\mathrm{C}}$ (Goodman et al., 1968). It is probable that tRNA mutations of this type will provide a general explanation of this kind of suppression.

Because of the key role of suppression in the study of chain termination in microbial systems, little progress has been made in this area in mammalian systems. However, on the basis of the model for polypeptide chain growth already considered, the conclusion seems inescapable that here too some special signal in mRNA must be required to bring about release of the completed polypeptide from tRNA.

3. PROTEIN SYNTHESIS IN ANIMAL CELLS

A. Organization of Protein Synthesis

We turn now to some aspects of protein synthesis of particular interest to the cell biologist. First to be considered is the intracellular organization of protein synthesis in animal cells. One of the principal features of eukaryotic cells is their compartmentalization into nucleus and cytoplasm. In this section evidence is reviewed which indicates that the cytoplasm is the site of nearly all if not all cellular protein synthesis. Problems of information transfer from the nucleus are examined, the significance of the endoplasmic reticulum (ER) is discussed, and mitochondrial protein synthesis is briefly mentioned.

1. The Cytoplasm and Protein Synthesis. At least 99% of the ribosomes in liver and HeLa cells are situated in the cytoplasm. These ribosomes are mostly in the form of polysomes and can be shown to bear nascent protein in pulse-labeling experiments. Reticulocyte ribosomes are also by definition cytoplasmic, since this cell no longer possesses a nucleus. Some of the ribosomes in HeLa and liver cells are bound to the outer nuclear membrane which is continuous with the ER. Demonstration of their cytoplasmic identity requires treatment of a crude nuclear fraction with mild detergents (Blobel and Potter, 1966; Penman, 1966). Failure to do this in the past may have led to confusion in attempts to determine the intracellular distribution of ribosomes in many cell types. In fact, it now appears to be true of eukaryotic cells in general that all functional ribosomes are located in the cytoplasm.

On the basis of this evidence, it is likely that all or nearly all cellular protein synthesis takes place in the cytoplasm. There is in fact strong evidence that in HeLa cells the histones, characteristic nuclear proteins, are synthesized in the cytoplasm and thereafter rapidly migrate to the nucleus (Robbins and Borun, 1967). The same is true for protamine synthesis in trout testis (Ling et al., 1969). Very recently, evidence has been obtained in several laboratories that ribosomal proteins are also synthesized in the cytoplasm, whereupon they migrate rapidly into the nucleolus for ribosome assembly.

The role of the nucleolus in ribosome assembly has recently been reviewed in detail by Maden (1971). In HeLa and liver cells, a very small number of ribosomal particles is found in the *nucleoplasm*, even after detergent treatment of nuclei (Vaughan et al., 1967; Sadowski and Howden, 1968). In HeLa cells these nucleoplasmic ribosomal particles are unequally represented with respect to the two subunits (60S > 40S) and display the metabolic characteristics of newly formed particles in transit from nucleolus to cytoplasm (Vaughan et al., 1967). It is therefore unlikely that they are involved in protein synthesis within the nucleus.

It should be mentioned that Gallwitz and Mueller (1969) and Zimmerman et al. (1969) have obtained evidence for low levels of protein synthesis in isolated nuclear and nucleolar preparations from HeLa cells. There is agreement that this can account for, at most, not more than 1–2% of the total protein synthesis of the cell, and the nature of the amino acid-incorporating system remains unclear.

2. Information Transfer from the Nucleus. The cytoplasmic location of protein synthesis raises the obvious question how mRNA is transferred from genes to ribosomes in eukaryotic cells. This is a problem that is specific to higher organisms. In bacteria there is no morphological barrier between genes and ribosomes, and there are several lines of evidence indicating that translation is closely coupled to transcription, ribosomes attaching to the 5′-end of nascent mRNA and commencing translation while mRNA synthesis is still proceeding toward the 3′-end.

Whatever the significance of the partition of genes from ribosomes in eukaryotic cells, it has been generally believed that mRNA is transcribed in some form within the nucleus and subsequently finds its way out to the cytoplasm to attach to ribosomes.

It is well known that there exists in the nucleoplasm of eukaryotic cells a high-molecular-weight, heterogeneously sedimenting, metabolically unstable DNA-like RNA (Harris and Watts, 1962; Attardi et al., 1966; Scherrer et al., 1966; Soeiro et al., 1966; Warner et al., 1966; Penman et al., 1968; Willems et al., 1969). Any discussion of the origin of mRNA in eukaryotic cells is bound to center upon its possible relationship to this heterogeneous nucleoplasmic RNA (hnRNA). Evidence on this subject is now summarized.

1. hnRNA and mRNA share common overall properties of DNA-like base composition and metabolic instability. hnRNA appears to be of very high molecular weight ($2–10 \times 10^6$, as determined from sedimentation velocity before and after denaturation in dimethyl sulfoxide). Moreover, recent calculations relating the rate of synthesis of hnRNA to the quantity and half-life of cytoplasmic mRNA indicate that over a short period of time far more hnRNA is synthesized than could possibly be quantitatively converted to mRNA. The remainder must be degraded within the nucleus (Soeiro et al., 1968; see also Harris and Watts, 1962). The function of the large amount of hnRNA that turns over within the nucleus is unknown, but the possibility exists that hnRNA synthesis represents a special mechanism for mRNA production in which only a small part of the total amount of hnRNA synthesized is actually used as messenger, the remainder being discarded.

Cytoplasmic mRNA is of considerably lower molecular weight (approximate range $2–5 \times 10^5$) than hnRNA. Therefore, if mRNA is derived from hnRNA, the process must involve some type of specific cleavage mechanism, for which alternative models have been proposed (Soeiro and Darnell, 1970). It has not so far proved possible either to establish or to rule out a precursor–product relationship between the two RNA species in HeLa cells by kinetic experiments. (However, see item 4 below.)

2. The chemical relationship between hnRNA and mRNA has therefore been investigated by Soeiro and Darnell (1970) by DNA–RNA hybridization. Competition between the two RNA species for the same sites on DNA was observed but, as these investigators noted, interpretation of the results is difficult for the following reasons. First, despite the use of what appeared to be "saturating" amounts of RNA, only a few percent of the DNA were hybridized. Second, and relatedly, it has been shown that in such hybridization experiments, using animal cell DNA, only certain reiterated sequences in DNA normally take part in the hybridization reaction (Melli and Bishop, 1969; Pagoulatos and Darnell, 1970a). This indicates that the hnRNA and

mRNA sequences shown to compete with each other represented reiterated material of unknown function, rather than unique sequences carrying information for synthesis of specific proteins. The question of whether the unhybridized (nonreiterated) hnRNA gives rise to the presumed *informational* content of mRNA remained unanswered. In addition, a new problem was raised: What is the nature of the reiterated (readily hybridizable) material in mRNA? Is it concerned with some common function of mRNA molecules, such as initiation of protein synthesis, or does it have some other significance? Other investigators have also tackled the problem described in this paragraph, but no definite answer has been obtained.

3. To attempt to identify informational sequences in hnRNA and mRNA, these RNA species from SV40-transformed cells have been hybridized to SV40 DNA (Lindberg and Darnell, 1970). SV40-specific sequences were found both in hnRNA and in mRNA. This shows that hnRNA and mRNA share common informational sequences and therefore strongly suggests, but still does not conclusively prove, that mRNA is derived from hnRNA.

4. Following unilateral nephrectomy there is a *decrease* in the amount of hnRNA in the cell nuclei of the remaining kidney. This is followd by an increased appearance of label in cytoplasmic mRNA, which certainly suggests a precursor–product relationship between the two species (Willems et al., 1969).

Could mRNA arise in any way other than from hnRNA? One possibility is that mRNA might be synthesized on DNA adjacent to the nuclear membrane, from which it would pass directly into the cytoplasm and attach to ribosomes. There is at present no direct evidence in favor of this suggestion.

In summary, the origin of mRNA in eukaryotic cells remains an outstanding unsolved problem. At present, continued attempts to develop techniques for identifying informational sequences in hnRNA and mRNA, together with further kinetic analyses, seem to offer the best hope of obtaining a solution. An answer to the question regarding the origin of mRNA, and particularly the role of hnRNA in the process, is crucial for an understanding of the control of gene expression in eukaryotic cells.

3. The Endoplasmic Reticulum: Free and Membrane-Bound Ribosomes. Most cell types possess a more-or-less well-developed ER. On electron microscopy the ER can be resolved into rough and smooth components. The rough and smooth ER are continuous in terms of the membrane structures of which they are composed but are distinguished by the fact that the rough ER possesses large numbers of attached ribosomes, which are bound to the membranes by their large (60S) subunits (Sabatini et al., 1966).

The actual percentage of the total cellular ribosomes bound to membranes varies greatly for different cells, and in general reflects the amount of ER in the respective cell type. Pancreatic exocrine cells possess a very well-developed ER and large numbers of ribosomes in the cell are attached to membranes. Liver cells also have a well-developed ER with many attached ribosomes. Kidney proximal tubule cells possess a less well-developed ER and HeLa cells have relatively little ER; only 15% of HeLa cell ribosomes are attached to membranes. Reticulocytes possess no ER, and the great majority of the ribosomes in the cell are free (a small number are attached to the cell membranes).

In general, a well-developed ER (both rough and smooth) is found in those cells that synthesize considerable amounts of proteins for secretion. The obvious inference has been drawn that membrane-bound ribosomes synthesize proteins for secretion and free ribosomes synthesize proteins that remain in the cell. Glycosylation of certain proteins is also known to take place in the ER after completion of polypeptide chain synthesis.

Autoradiographic electron microscope studies have shown that in pancreatic exocrine cells the bulk of protein synthesis takes place in the region of the rough ER. Subsequently, radioactive protein can be shown to migrate down the smooth ER and into the Golgi apparatus, the final reservoir from which zymogens are secreted from the cell (Caro and Palade, 1964; Jamieson and Palade, 1967). Transport of protein down the smooth ER is not dependent upon continued protein synthesis but is energy dependent (Jamieson and Palade, 1968).

By using liver it is possible to separate free from membrane-bound polyribosomes by differential centrifugation (Blobel and Potter, 1967). On the basis of experiments utilizing such a separation, it has been concluced that the free polysomes in liver synthesize ferritin, a protein retained by the cell, whereas membrane-bound ribosomes synthesize albumin (Hicks et al., 1969; Redman, 1969; Takagi and Ogata, 1969) and glycoproteins (Hallinan et al., 1968), proteins that are secreted from the cell. Glycoprotein synthesis has also been reported to take place on the ER in kidney (Priestley et al., 1969). All these findings therefore support the putative role of free and membrane-bound ribosomes in the synthesis of cellular and secretory proteins (or glycoproteins), respectively.

It is of interest that the ribosome is attached to the ER by the 60S subunit. As already noted, it is on the large ribosomal subunit that polypeptide chain growth takes place. After completion of synthesis the newly formed polypeptide is therefore in a position to be channeled directly into the ER in preparation for glycosylation and/or secretion.

One final point should be mentioned in this section. The problem of information transfer from the nucleus has been discussed. If it is true that proteins for secretion are synthesized only on membrane-bound ribosomes and intracellular proteins on free ribosomes, a mechanism must exist for accurately directing the appropriate mRNA molecules to their respective sites of translation. There is at present no indication as to how this might be achieved.

4. **Mitochondrial Protein Synthesis.** Following earlier reports for *Neurospora*, several workers have recently reported a separate protein-synthesizing system in animal cell mitochondria (Perlman and Penman, 1970; Swanson and Dawid, 1970; Attardi and Ojala, 1971).

Mitochondrial ribosomes are smaller than cytoplasmic ribosomes (60S monomer, 45 and 30S subunits) and contain 16–20S and 12S RNA which are complementary to mitochondrial DNA. Synthesis of the mitochondrial ribosomes is sensitive to ethidium bromide (specific for supercoiled mitochondrial DNA) and less sensitive to actinomycin than the nucleolar synthesis of cytoplasmic ribosomes.

The mitochondrial protein-synthesizing system utilizes fmet tRNA (A. E. Smith and Marcker, 1968), is inhibited by chloramphenicol (as is bacterial protein synthesis), and is insensitive to cycloheximide, a potent inhibitor of cytoplasmic protein synthesis.

The size of the mitochondrial genome is such that it could not code for more than a few of the many proteins found in mitochondria, and mitochondrial protein synthesis accounts for less than 1% of total cellular protein synthesis. Which mitochondrial proteins are synthesized by the mitochondria themselves and which by the cytoplasm is an unsolved question.

B. Protein Synthesis in Differentiated and Cultured Animal Cells: Special Topics

Protein synthesis has been intensively studied in several different animal cell types with different aims in view. For example, the reticulocyte has been widely employed for studies on the mechanism of protein synthesis. Antibody synthesis by plasma cells constitutes a special kind of cellular differentiation. Cultured cells have been used to investigate questions concerning general patterns of cellular metabolism, including protein synthesis during the cell cycle. A line of hepatoma cells is being used to study the synthesis of an inducible enzyme, tyrosine transaminase. Each of these topics is of some general interest, and brief discussions now follow.

1. The Reticulocyte and Hemoglobin Synthesis. The fact that the reticulocyte is devoted almost exclusively to the synthesis of a single protein, hemoglobin, has been widely exploited in studies on the basic mechanism of protein synthesis. Some important findings from the reticulocyte system have already been noted, but a few have not been mentioned. The findings from the reticulocyte system are therefore summarized in this section, references being given for points not previously mentioned:

1. Proof of N′-to-C′ terminal direction of polypeptide chain assembly and demonstration that growth rate is uniform along the length of the globin chains.
2. Discovery of the polyribosome as the active aggregate in protein synthesis (Warner et al., 1962). [Polysomes were discovered at about the same time in liver cells (Wettstein et al., 1963).]
3. Demonstration that there are two tRNA molecules per active ribosome in protein synthesis.
4. Early characterization of supernatant factors involved in animal cell polypeptide chain extension (Hardesty et al., 1963) and more recent characterization of initiation factors in animal cell protein synthesis (Prichard et al., 1970).
5. Demonstration of the role of methionine in initiation of protein synthesis in animal cells.
6. Isolation of the mRNAs for globin.

To the above can be added recent studies on the control of globin synthesis. Heme appears to play a crucial role in initiating the synthesis of globin and in coordinating the relative amounts of synthesis of the α and β chains (Tavill et al., 1968; Adamson et al., 1969).

2. Antibody Formation. Another important example of cells involved largely in the synthesis of a single class of proteins occurs in antibody formation. Antibodies comprise a very large number of different but related proteins, the immunoglobulins, which are produced by plasma cells in response to stimulation by specific antigens.

Immunoglobulins consist of two identical light and two identical heavy polypeptide chains, linked by S—S bonds. Each chain possesses a "constant" region which is similar in most immunoglobulins, and a "variable" region which is different for different immunoglobulins and is believed to contain the antigen-combining site. The variable regions are located toward the N′-terminal end of the light and heavy chains.

The synthesis of specific antibodies in response to antigenic stimulation represents a special type of differentiation, and it is for this reason that the immunological response has aroused so much interest among biologists. While it is not possible to review the field of immunology here, it is worth mentioning certain findings in connection with immunoglobulin formation. Much of the relevant experimental work has been done with myeloma cells, tumor cells producing large amounts of single immunoglobins.

There is good evidence that any particular immunocyte (plasma cell) synthesizes only one type of antibody molecule. The mechanism of antibody assembly is also reasonably well understood. It has been shown that the variable and constant regions of immunoglobulin chains are synthesized together as single polypeptides and not joined after separate synthesis. Light and heavy chains are produced on different classes of polysome (small and large). The completed light chain interacts with the incomplete heavy chain on the "heavy" polysomes. Glycosylation commences on ribosome-bound polypeptides and is completed in the smooth ER prior to secretion of the completed immunoglobulin.

Much experimental data substantiating this brief outline of antibody structure and formation can be found in the review by Milstein and Pink (1970) and in Volume 32 of the *Cold Spring Harbor Symposium for Quantitative Biology* (1967). The latter also contains discussions of the difficult problems concerning the origin of diversity of antibody structure and the cellular events involved in the immune response.

3. Protein Synthesis and the Cell Cycle in Cultured Mammalian Cells. Cell culture provides homogeneous populations of rapidly dividing cells. These offer a convenient system for analyzing general patterns of cellular metabolism, such as the regulation of protein synthesis during the cell cycle.

In cultured cells the overall rate of protein synthesis is fairly uniform throughout interphase. During mitosis this level falls to about 30% of the interphase level. The diminution in protein synthesis that occurs during mitosis in HeLa cells is not attributable to disappearance of mRNA. In actinomycin-treated cells mRNA formed before mitosis is still used for polysome formation and protein synthesis in the daughter cells after mitosis (Hodge et al., 1969). However, the decrease in protein synthesis during mitosis *is* accompanied by polysome disaggregation. The rate-limiting step in protein synthesis during mitosis is attachment of ribosomes to mRNA; polypeptide chain growth proceeds at the same rate in mitotic as in interphase cells (Fan and Penman, 1970).

It is not clear whether inhibition of initiation during mitosis affects the synthesis of all proteins equally or whether the synthesis of certain specific proteins continues undiminished. It has, however, been reported that the induced synthesis of tyrosine transaminase in rat hepatoma cells (see next section) continues undiminished during mitosis (D. Martin et al., 1969a).

Another important question regarding the cell cycle concerns the nature of the proteins synthesized at various stages in the cycle. It has been clearly shown that histone synthesis is coupled to DNA synthesis. Histones are formed only during the S period, and treatments that inhibit DNA synthesis (e.g., cytosine arabinoside) also inhibit histone synthesis (Robbins and Borun, 1967). During the S phase a small discrete mRNA species can be isolated from small polysomes on which histones are synthesized. This mRNA species is not present during G_1 or G_2 and disappears during treatment of cells with cytosine arabinoside (Borun et al., 1967) (Fig. 10). This mRNA therefore has the characteristics expected of histone mRNA. Its synthesis is blocked by

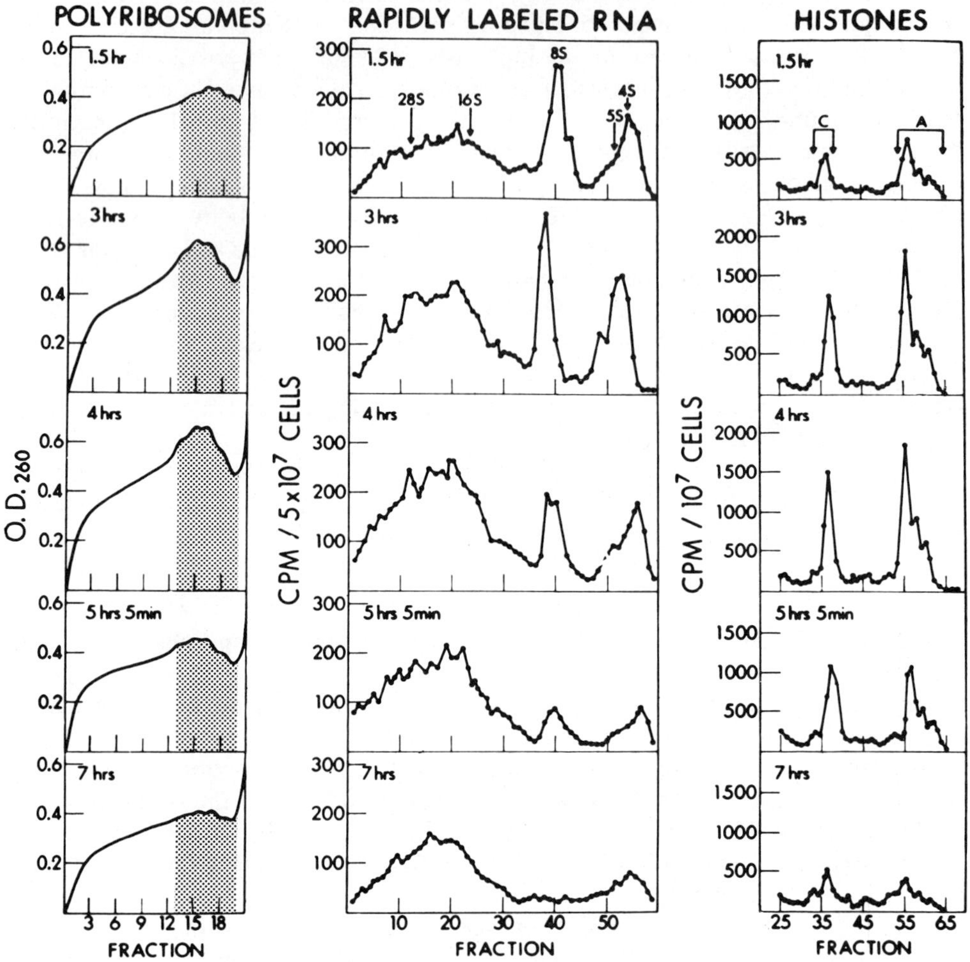

Figure 10. Histone mRNA and histone synthesis in synchronized HeLa cells. The times represent periods after release from block of DNA synthesis. Left-hand column: polysome profiles showing region of small polysomes (shaded) which includes histone-specific polysomes during S phase. Middle column: pulse-labeled RNA from the shaded region of the polysome profiles, analyzed on acrylamide gels. The 8S peak is the putative histone mRNA. Right-hand column; pulse-labeled, acid-soluble proteins analyzed on acrylamide gels. Only the rapidly migrating proteins are shown. The two prominent peaks are histones. (From Borun et al., 1967.)

actinomycin, indicating that control of histone synthesis is exercised at the level of mRNA production.

It may be asked whether or not there is any *general* pattern of mRNA specificity at different periods in the cell cycle. mRNA and hnRNA from HeLa cells in the G_1, S, and G_2 phases have been compared by competition hybridization. So far no differences have been detected either between mRNA or between hnRNA produced during the three phases of the cell cycle (Pagoulatos and Darnell, 1970b). However, the findings were subject to the reservation already mentioned for DNA–RNA hybridization experiments with mammalian cell material, namely, that it is only the

RNA complementary to reiterated DNA sequences that hybridizes efficiently under conditions so far employed.

4. Synthesis of an Induced Enzyme in Cultured Hepatoma Cells—Tyrosine Transaminase. Cellular differentiation leads to synthesis of more-or-less fixed spectra of characteristic proteins in different cell types. The events that bring about this largely irreversible process may well be different from those responsible for the well-known phenomena of induced enzyme synthesis in bacteria. Superimposed upon the basic framework of cellular differentiation, however, are several instances of inducible enzyme synthesis in specialized cells. One of these is the synthesis of tyrosine transaminase, one of the enzymes of gluconeogenesis.

Tomkins and his collaborators (Tomkins et al., 1966) studied tyrosine transaminase activity in a cultured line of hepatoma cells that retain the capacity to produce this enzyme. Enzyme activity is increased at least 10-fold above basal levels by exposure of the cells to dexamethasone (a synthetic glucocorticoid). Experiments with specific antiserum have shown that this increase in activity represents synthesis of new enzyme rather than activation of enzyme preexisting in an inactive form. Induced enzyme synthesis is blocked by simultaneous treatment of the cells with actinomycin, indicating that new RNA synthesis is required for enzyme formation. Subsequent treatment with actinomycin is progressively less effective in inhibiting enzyme synthesis.

There is kinetic evidence for the existence of a second protein which regulates tyrosine transaminase activity by degrading the enzyme with a half-life of a few hours. This evidence derives mainly from studies with inhibitors of RNA and protein synthesis on levels of tyrosine transaminase activity in cells in which induced enzyme synthesis is already in progress.

Finally, although enzyme synthesis continues throughout the cell cycle, including mitosis, in cells that have been induced, induction itself occurs only during late G_1 and S phases. During the remainder of the cycle, synthesis is constitutive (D. Martin et al., 1969b).

C. Protein Synthesis in Virus-Infected Cells

Advances in protein synthesis in bacteria have frequently stemmed from experiments with bacteriophages. Similarly, the study of protein synthesis in animal cells is now benefiting from experiments with animal viruses. Of the several virus–cell systems undergoing intensive study, two are of special interest in connection with specific aspects of protein synthesis. These are poliovirus and Sindbis virus.

1. Poliovirus. Poliovirus is a small spherical virus which contains a single-stranded RNA molecule of molecular weight 2×10^6 (6000 nucleotides) together with four distinct species of polypeptide. The virus replicates in HeLa cells, and radioactive polypeptides from virus grown in cells in the presence of labeled amino acids can be separated on acrylamide gels (Fig. 11).

Shortly after infection of HeLa cells with poliovirus, there is a rapid breakdown of host cell polysomes to single ribosomes, with concomitant inhibition of protein synthesis (Summers et al., 1965). This effect is not a result of degradation of host cell mRNA but appears to be attributable to failure of reattachment of ribosomes to mRNA. The effect does not require viral RNA replication, since it occurs in the presence of 3 mM guanidine which inhibits replication. It does, however, require viral RNA penetration into the cell; attachment of the virus to the cell membrane is not sufficient (Penman and Summers, 1965). Later during infection large virus-specific polysomes accumulate in the infected cell cytoplasm (Penman et al., 1963). Together these findings indicate that a viral product, which becomes available early during infection, selectively inhibits the interaction of ribosomes with host cell mRNA, while permitting interaction with viral RNA. How this selectivity of attachment is achieved is obviously of considerable interest as regards the control of translation of specific mRNA species. Further characterization of the effect is proceeding in at least two laboratories.

The second point of interest concerns the mechanism of translation of viral RNA. It was mentioned above that the mature virion contains four types of polypeptide. Infected cells, in which all residual host-specified protein synthesis has been suppressed by a variety of treatments, contain a much larger number of virus-specific polypeptides (Summers et al., 1965) (Fig. 12). One or more of the "noncapsid" proteins found in the infected cell are presumably connected with virus replication. However, several of the proteins migrate slowly in SDS acrylamide gels, implying that they are large. The sum of the molecular weights of the proteins, estimated from their mobilities on gels, is incompatible with their being separately coded by an RNA molecule of molecular weight 2×10^6 (Maizel and Summers, 1968).

A possible resolution to the paradox was suggested by "pulse-chase" experiments which indicated that some of the larger polypeptides behave as kinetic precursors of the smaller ones (Summers and Maizel, 1968) (Fig. 12).

Jacobson and Baltimore (1968) pursued the question of precursor–product relationships between the polypeptides in polio-infected cells by allowing infection to proceed in the presence of several amino acid analogs, such as *p*-fluorophenylalanine, that are incorporated into proteins. It was expected that if the smaller polypeptides were in fact derived by some type of enzymic cleavage mechanism from larger precursors the

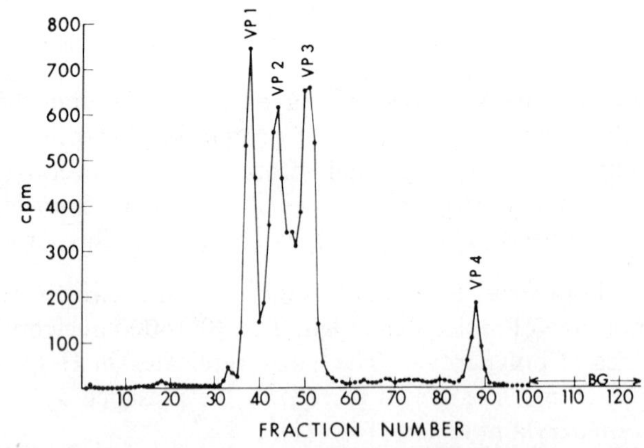

Figures 11 and 12. Acrylamide gel patterns of radioactive proteins from purified poliovirus (Fig. 11) and from viral specific proteins in infected HeLa cell cytoplasm (Fig. 12). In (Fig. 12) infected cells were labeled with amino acids for 3 min and then chased with a large excess of unlabeled amino acids for 2 min (*A*), 10 min (*B*), and 45 min (*C*). VP indicates proteins found in purified poliovirus. Numerals designate proteins found only in the infected cell. Note redistribution of radioactivity from large proteins (left) to smaller proteins (right) during the cold amino acid chase. (From Summers et al. (1965) and Summers and Maizel (1968).

presence of unnatural amino acid residues in the latter would inhibit cleavage. This proved to be the case. Moreover, these investigators identified a previously undetected, unstable polypeptide of very high molecular weight which they suggested was a precursor to all the remaining polypeptides, including intermediates and final products.

Evidence for specific cleavage during the formation of proteins of several other enteroviruses has also been reported (Holland and Kiehn, 1968).

Jacobson and Baltimore explain these findings as follows. They suggest that polycistronic messengers do not exist in mammalian cells. Animal RNA viruses are

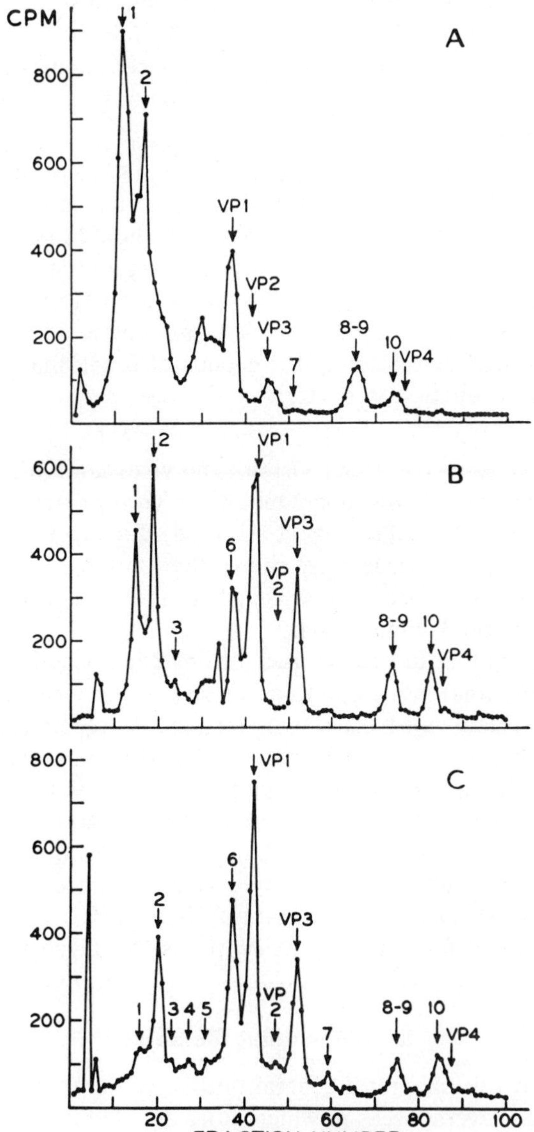

Figure 12.

therefore faced with a special problem in translation of their polycistronic RNA. Translation as a single polypeptide, followed by specific cleavage, is the suggested mechanism whereby the enteroviruses overcome this problem. Possible solutions for other, larger RNA viruses are also suggested by these workers.

Independent evidence suggesting that polycistronic messengers may not exist in animal cells is also cited by Jacobson and Baltimore:

1. In several cell types from which polysomes for specific proteins have been isolated, the size of the polysome is proportional to the size of the polypeptide being synthesized. This suggests that large polysomes do not synthesize multiple small polypeptides.

2. In several animal cell types, there is a linear relationship between polysome size and nascent polypeptide size, as judged by *in vivo* labeling experiments (Kuff and Roberts, 1967). This could not be the case if there were a significant degree of internal reinitiation, or of cleavage of nascent chains, and is not true of bacterial polysomes in which internal reinitiation *does* take place.

Since it is now clear that there is a specific mechanism for initiation of animal cell protein synthesis, which commences translation in the correct phase (Brown and Smith, 1970), it is not clear why internal reinitiation should not be possible. Nevertheless, the suggestion that polycistronic messengers do not normally exist in animal cells is an interesting one with respect to control mechanisms in protein synthesis. It raises the question whether coordinate gene expression in animal cells ever occurs by formation of single mRNA molecules for groups of functionally related proteins, as is the case for operon function in bacteria.

2. Sindbis Virus. Sindbis virus is another small RNA virus which contains only two polypeptide species in the intact virion. The virus contains a nucleocapsid core surrounded by a lipoprotein envelope. One of the viral proteins is found only in the core, the other in the envelope (Strauss et al., 1968). Electron micrograph studies of a similar virus, Semliki Forest virus, have shown that only the core is present inside the infected cell; the envelope becomes attached as the core "buds through" the cell membrane (Acheson and Tamm, 1967).

These findings imply that in such viruses the envelope protein is integrated into the cell membrane before final maturation and release of the virus. This in turn suggests that the envelope protein must bear some structural resemblance to cellular membrane protein(s). Further suggestive evidence that this is true derives from the fact that the envelope protein of Sindbis virus is a glycoprotein (Strauss et al., 1970) (Fig. 13), and that glycosylation is probably brought about by host cell enzymes (Burge and Strauss, 1970). This implies that specific sequences in the viral envelope protein resemble certain host protein amino acid sequences so as to serve as substrates for glycosylation by host cell enzymes. Thus *Sindbis* virus and related viruses may offer model systems for studying the synthesis and glycosylation of single membranelike proteins.

D. Concluding Remarks

In the first part of this chapter a general outline of the mechanism of protein synthesis was given, and several areas in which more detailed knowledge is accumulating were mentioned. In the second part of the chapter certain specific aspects of protein synthesis in animal cells were discussed.

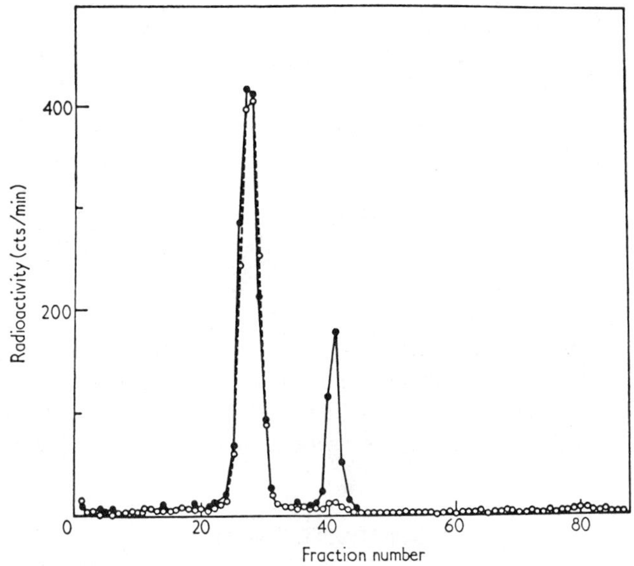

Figure 13. Radioactive proteins from purified double-labeled Sindbis virus analyzed by acrylamide gel electrophoresis. Left, envelope protein; right, nucleocapsid protein; ●, amino acid label. ○, glucosamine label. (From Strauss et al., 1970.)

With respect to overall mechanism, animal cell protein synthesis is similar in most major respects to bacterial protein synthesis. Animal cell ribosomes are larger than bacterial ones and are sensitive to a different range of antibiotics, but the roles of ribosomal subunits and supernatant factors are apparently identical in animal and bacterial cells. Moreover, a special met tRNA is involved in initiation of animal cell protein synthesis, the only difference from bacteria being that methionine is not formylated.

One possible major difference between protein synthesis in bacterial and animal cells is the apparent absence of polycistronic messengers in the latter (Section 3.C). If it is true that there are no polycistronic messengers in animal cells, the initiating met tRNA is presumably used only once per mRNA, to commence translation in the correct frame of reading at or near the 5′-end.

Other differences between bacterial and animal cells are to be found in the control of protein synthesis. In *E. coli* and related organisms, there is a well-substantiated model for the control of induced enzyme synthesis involving the reversible binding of a specific protein repressor molecule to an operator region on DNA. While this may not be the only mechanism responsible for regulating the synthesis of specific proteins in *E. coli*, it is clear that it plays a major role.

In animal cells little is known of the way in which differential gene activity is brought about. Basic features of differentiation, such as irreversibility, are quite different from the phenomena of induced enzyme synthesis in bacteria. As pointed out in Section 3.B, induced enzyme synthesis, when it occurs in animal cells, appears to be superimposed on the basic pattern of cellular differentiation. Histone synthesis, which also occurs during only part of a cell's life, involves cyclical control of mRNA production rather than inducible control in response to a clearly defined external agent. Moreover, the whole question of the nature of the steps linking gene derepression to mRNA production in animal cells is confused by the problem of the precise role of hnRNA in this process.

In addition to mechanisms regulating gene expression at the level of mRNA production in animal cells, there are several well-documented examples of control at the level of mRNA translation. Examples already mentioned include inhibition of initiation of protein synthesis which occurs during mitosis and in poliovirus-infected HeLa cells. Further examples are the unfertilized ovum, in which ribosomes are also inhibited from attaching to and translating existing mRNA [by a trypsin-sensitive substance (Maggio et al., 1968)], and the decreased activity of 50S ribosomal subunits isolated from the muscle of diabetic rats (T. E. Martin and Wool, 1968).

Thus several potential control mechanisms in protein synthesis can be recognized in animal cells. It is likely that this area, which has not been dealt with at length in this chapter, is one in which important developments will occur in the next few years.

ACKNOWLEDGMENT

I wish to thank Professor J. N. Davidson and Professor R. M. S. Smellie for their support and comments during the writing of this chapter.

REFERENCES

Abelson, J. N., M. L. Gefter, L. Barnett, A. Landy, R. L. Russell, and J. D. Smith. (1970). Mutant tyrosine transfer ribonucleic acids. *J. Mol. Biol.*, **47,** 15.

Acheson, N. H., and I. Tamm (1967). Replication of Semliki Forest virus: An electron microscopic study. *Virology*, **32,** 128.

Adams, J. M. (1968). On the release of the formyl group from nascent protein. *J. Mol. Biol.*, **33,** 571.

Adamson, S. D., E. Herbert, and S. F. Kemp (1969). Effects of hemin and other porphyrins on protein synthesis in a reticulocyte lysate cell-free system. *J. Mol. Biol.*, **42,** 247.

Attardi, G., and D. Ojala (1971). Mitochondrial miniribosomes in HeLa cells. *Nature, New Biology*, **229,** 133.

Attardi, G., H. Parnas, M.-I. H. Hwang, and B. Attardi (1966). Giant size rapidly labelled nuclear ribonucleic acid and cytoplasmic messenger ribonucleic acid in immature duck erythrocytes. *J. Mol. Biol.*, **20,** 145.

Beaudet, A. L., and C. T. Caskey (1970). Release factor translation of RNA phage terminator codons. *Nature*, **227,** 38.

Blobel, G., and V. R. Potter (1966). Nuclei from rat liver: Isolation method that combines purity with high yield. *Science*, **154,** 1662. (See last two paragraphs.)

Blobel, G., and V. R. Potter (1967). Studies on free and membrane-bound ribosomes in rat liver. I. Distribution as related to total cellular RNA. *J. Mol. Biol.*, **26,** 279.

Borun, T. W., M. D. Scharff, and E. Robbins (1967). Rapidly labelled, polyribosome-associated RNA having the properties of histone messenger. *Proc. Natl. Acad. Sci., U.S.*, **58,** 1977.

Brenner, S. (1961). RNA, ribosomes, and protein synthesis. *Cold Spring Harbor Symp. Quant. Biol.*, **26,** 101.

Bretscher, M. S. (1963). The chemical nature of the sRNA-polypeptide complex. *J. Mol. Biol.*, **7,** 446.

Bretscher, M. S. (1968a). Polypeptide chain termination: An active process. *J. Mol. Biol.*, **34,** 131.

Bretscher, M. S. (1968b). Translocation in protein synthesis: A hybrid structure model. *Nature*, **218,** 675.

Brown, J. C., and A. E. Smith (1970). Initiator codons in eukaryotes. *Nature*, **226,** 610.

Burge, B. W., and J. H. Strauss (1970). Glycopeptides of the membrane protein of Sindbis virus. *J. Mol. Biol.*, **47,** 449.

Capecchi, M. R., and H. A. Klein (1970). Release factors mediating termination of complete proteins. *Nature*, **226,** 1029.

Caro, L. G., and G. E. Palade (1964). Protein synthesis, storage and discharge in the pancreatic exocrine cell. An autoradiographic study. *J. Cell Biol.*, **20,** 473.

Cotter, R. I., P. McPhie, and W. B. Gratzer (1967). Internal organization of the ribosome. *Nature*, **216,** 864.

Cox, R. A. (1966). The secondary structure of ribosomal ribonucleic acid in solution. *Biochem. J.*, **98,** 841.

Crick, F. H. C. (1966). Codon-anticodon pairing: The wobble hypothesis. *J. Mol. Biol.*, **19,** 548.

Crick, F. H. C., L. Barnett, S. Brenner, and R. J. Watts-Tobin (1961). General nature of the genetic code for proteins. *Nature*, **192,** 1227.

Dintzis, H. M. (1961). Assembly of the peptide chains of hemoglobin. *Proc. Natl. Acad. Sci., U.S.*, **47,** 247.

Erbe, R. W., M. M. Nau, and P. Leder (1969). Translation and translocation of defined RNA messengers. *J. Mol. Biol.*, **39,** 441.

Fan, F., and S. Penman (1970). Regulation of protein synthesis in mammalian cells. II. Inhibition of protein synthesis at the level of initiation during mitosis. *J. Mol. Biol.*, **50,** 655.

Gallwitz, D., and G. C. Mueller (1969). Protein synthesis in nuclei isolated from HeLa cells. *European J. Biochem.*, **9,** 431.

Gilbert, W. (1963). Polypeptide synthesis in *Escherichia coli*. II. The polypeptide chain and sRNA. *J. Mol. Biol.*, **6,** 389.

Gillam, I., S. Millward, D. Blew, M. von Tigerstrom, E. Wimmer, and G. M. Tener (1967). The separation of soluble ribonucleic acids on benzoylated diethylaminoethylcellulose. *Biochemistry*, **6,** 3043.

Goodman, H. M., J. Abelson, A. Landy, S. Brenner, and J. D. Smith. (1968). Amber suppression: A nucleotide change in the anti-codon of a tyrosine transfer RNA. *Nature*, **217,** 1019.

Haenni, A.-L., and J. Lucas-Lenard (1968). Stepwise synthesis of a tripeptide. *Proc. Natl. Acad. Sci., U.S.*, **61,** 1363.

Halinan, T., C. N. Murty, and J. H. Grant (1968). The exclusive function of reticulum-bound ribosomes in glycoprotein biosynthesis. *Life Sci.*, **7,** Part 2, 225.

Hardesty, B., R. Arlinghaus, J. Shaeffer, and R. Schweet (1963). Haemoglobin and polyphenylalanine synthesis with rabbit reticulocyte ribosomes. *Cold Spring Harbor Symp. Quant. Biol.*, **28,** 215.

Harris, H., and J. W. Watts (1962). The relationship between nuclear and cytoplasmic ribonucleic acid. *Proc. Roy. Soc., (London)*, **B156,** 109.

Hershey, J. W. B., and R. E. Monro (1966). A competitive inhibitor of the GTP reaction in protein synthesis. *J. Mol. Biol.*, **18,** 68.

Hicks, S., J. W. Drysdale, and H. N. Monro (1969). Preferential synthesis of ferritin and albumin by different populations of liver polysomes. *Science*, **164,** 584.

Hodge, L. D., E. Robbins, and M. D. Scharff (1969). Persistence of messenger RNA through mitosis in HeLa cells. *J. Cell Biol.*, **40,** 497.

Holland, J. J., and D. E. Kiehn (1968). Specific cleavage of viral proteins as steps in the synthesis and maturation of enteroviruses. *Proc. Natl. Acad. Sci., U.S.*, **60,** 1015.

Hunt, T., T. Hunter, and A. Munro (1968). Control of haemoglobin synthesis: Distribution of ribosomes on the messenger RNA for α and β chains. *J. Mol. Biol.*, **36,** 31.

Jacobson, M. F., and D. Baltimore (1968). Polypeptide cleavages in the formation of poliovirus proteins. *Proc. Natl. Acad. Sci., U.S.*, **61,** 77.

Jackson, R., and T. Hunter (1970). Role of methionine in initiation of haemoglobin synthesis. *Nature*, **227,** 672.

Jamieson, J. D., and G. E. Palade (1967). Intracellular transport of secretory proteins in the pancreatic exocrine cell. I. Role of the peripheral elements of the Golgi-complex. II. Transport to condensing vacuoles and zymogen granules. *J. Cell Biol.*, **34,** 577, 597.

Jamieson, J. D., and G. E. Palade (1968). Intracellular transport of secretory proteins in the pancreatic exocrine cell. III. Dissociation of intracellular transport from protein synthesis. IV. Metabolic requirements. *J. Cell Biol.*, **39,** 580, 589.

Kuff, E. L., and N. E. Roberts (1967). *In vivo* labelling patterns of free polyribosomes: Relationship to tape theory of messenger ribonucleic acid function. *J. Mol. Biol.*, **26,** 211.

Labrie, F. (1969). Isolation of an RNA with the properties of haemoglobin messenger. *Nature*, **221,** 1217.

Laycock, D. G., and J. A. Hunt (1969). Synthesis of rabbit globin by a bacterial cell-free system. *Nature*, **221,** 1118.

Lengyel, P., and D. Söll (1969). Mechanism of protein biosynthesis. *Bacteriol Rev.*, **33,** 264.

Levitt, M. (1969). Detailed molecular model for transfer RNA. *Nature*, **224,** 759.

Lindberg, U., and J. E. Darnell (1970). SV40-specific RNA in the nucleus and polyribosomes of transformed cells. *Proc. Natl. Acad. Sci., U.S.*, **65,** 1089.

Ling, V., J. R. Trevithick, and G. H. Dixon (1969). The biosynthesis of protamine in trout testis. I. Intracellular site of synthesis. *Can. J. Biochem.*, **47,** 51.

McQuillen, K., R. B. Roberts, and R. J. Britten (1959). Synthesis of nascent protein by ribosomes in *Escherichia coli*. *Proc. Natl. Acad. Sci., U.S.*, **45,** 1437.

Maden, B. E. H. (1971). The structure and formation of ribosomes in animal cells. *Progr. Biophys. Mol. Biol.*, **22,** 127.

Maden, B. E. H., and R. E. Monro (1968). Ribosome-catalyzed peptidyl transfer: Effects of cations and pH value. *European J. Biochem.*, **6,** 309.

Maden, B. E. H., R. R. Traut, and R. E. Monro (1968). Ribosome-catalyzed peptidyl transfer: The polyphenylalanine system. *J. Mol. Biol.*, **35,** 333.

Maggio, R., M. L. Vitorelli, I. Caffarelli-Mormino, and A. Monroy (1968). Dissociation of ribosomes of unfertilized eggs and embryos of sea urchin. *J. Mol. Biol.*, **31,** 621.

Maizel, J. V., and D. F. Summers (1968). Evidence for differences in size and composition of the poliovirus-specific polypeptides in infected HeLa cells. *Virology*, **36,** 48.

Martin, D., G. Tomkins, and D. Cranner (1969a). Synthesis and induction of tyrosine aminotransferase in synchronized hepatoma cells in culture. *Proc. Natl. Acad. Sci., U.S.*, **62,** 248.

Martin, D., G. Tomkins, and M. Bresler (1969b). Control of specific gene expression examined in synchronized mammalian cells. *Proc. Natl. Acad. Sci., U.S.*, **63,** 842.

Martin, T. E., and I. G. Wool (1968). Formation of active hybrids from subunits of muscle ribosomes from normal and diabetic rats. *Proc. Natl. Acad. Sci., U.S.*, **60,** 569.

Melli, M., and J. O. Bishop (1969). Hybridization between rat liver DNA and complementary RNA. *J. Mol. Biol.*, **40,** 117.

Milstein, C., and J. R. L. Pink (1970). The structure and evolution of immunoglobulins. *Progr. Biophys. Mol. Biol.*, **21,** 211.

Mizushima, S., and M. Nomura (1970). Assembly mapping of ribosomal proteins from *E. coli*. *Nature*, **226,** 1214.

Monro, R. E., B. E. H. Maden, and R. R. Traut (1967). In *Genetic Elements, Properties and Functions*. Symposium of the Federation of European Biochemical Societies, Warsaw, April, 1966 (D. Shugar, ed.), pp. 179–202. Academic, New York.

Moore, P. B. (1966). Polynucleotide attachment to ribosomes. *J. Mol. Biol.*, **18,** 8.

Pagoulatos, G. N., and J. E. Darnell (1970a). Fractionation of hnRNA: Rates of hybridization and chromosomal distribution of reiterated sequences. *J. Mol. Biol.*, **54,** 517.

Pagoulatos, G. N., and J. E. Darnell (1970b). A comparison of the heterogeneous nuclear RNA of HeLa cells in different periods of the cell growth cycle. *J. Cell Biol.*, **44,** 476.

Penman, S. (1966). RNA metabolism in the HeLa cell nucleus. *J. Mol. Biol.*, **17,** 117.

Penman, S., and D. F. Summers (1965). Effects on host cell metabolism following synchronous infection with poliovirus. *Virology*, **27,** 614.

Penman, S., K. Scherrer, Y. Becker, and J. E. Darnell (1963). Polyribosomes in normal and poliovirus-infected HeLa cells and their relationship to messenger RNA. *Proc. Natl. Acad. Sci., U.S.*, **49,** 654.

Penman, S., C. Vesco, and M. Penman (1968). Localization and kinetics of formation of nuclear heterodisperse RNA, cytoplasmic heterodisperse RNA and polyribosome-associated messenger RNA in HeLa cells. *J. Mol. Biol.*, **34,** 49.

Perlman, S., and S. Penman (1970). Protein-synthesizing structure associated with mitochondria. *Nature*, **227,** 133.

Perutz, M. F. (1962). *Proteins and Nucleic Acids*, pp. 113–122. Elsevier, Amsterdam.

Prichard, P. M., J. M. Gilbert, D. A. Shafritz, and W. F. Anderson (1970). Factors for the initiation of haemoglobin synthesis by rabbit reticulocyte ribosomes. *Nature*, **226,** 511.

Priestley, G. C., M. L. Pryn, and R. A. Malt (1969). Glycoprotein synthesis by membrane-bound ribosomes and smooth ribosomes in kidney. *Biochim. Biophys. Acta*, **190,** 154.

Redman, C. M. (1969). Biosynthesis of serum proteins and ferritin by free and attached ribosomes of rat liver. *J. Biol. Chem.*, **244,** 4308.

Robbins, E., and T. W. Borun (1967). The cytoplasmic synthesis of histones in HeLa cells and its temporal relationship to DNA replication. *Proc. Natl. Acad. Sci., U.S.*, **57,** 409.

Sabatini, D. D., Y. Tashiro, and G. E. Palade (1966). On the attachment of ribosomes to microsomal membranes. *J. Mol. Biol.*, **19,** 503.

Sadowski, P. D., and J. A. Howden (1968). Isolation of two distinct classes of polysome from a nuclear fraction of rat liver. *J. Cell Biol.*, **37,** 163.

Sarabhai, A. S., A. O. W. Stretton, S. Brenner, and A. Bolle (1964). Co-linearity of the gene with the polypeptide chain. *Nature*, **201,** 13.

Scherrer, K., L. Marcaud, F. Zajdela, I. M. London, and F. Gros (1966). Patterns of RNA metabolism in a differentiated cell: A rapidly labelled unstable 60S RNA with messenger properties in duck erythroblasts. *Proc. Natl. Acad. Sci., U.S.*, **56,** 1571.

Skogerson, L., and K. Moldave (1968). Evidence for aminoacyl-tRNA binding, peptide bond synthesis and translocase activities in the aminoacyl-transfer reaction. *Arch. Biochem. Biophys.*, **125,** 497.

Smith, A. E., and K. Marcker (1968). *N*-Formylmethionyl transfer RNA in mitochondria from yeast and rat liver. *J. Mol. Biol.*, **38,** 241.

Smith, A. E., and K. Marcker (1970). Cytoplasmic methionine transfer RNAs from eukaryotes. *Nature*, **226,** 607.

Smith, J. D., R. R. Traut, G. M. Blackburn, and R. E. Monro (1965). Action of puromycin in polyadenylic acid-directed polylysine synthesis. *J. Mol. Biol.*, **13,** 617.

Smith, J. D., J. N. Abelson, B. F. C. Clark, H. M. Goodman, and S. Brenner (1966). Studies on *amber* suppressor tRNA. *Cold Spring Harbor Symp. Quant. Biol.*, **31,** 479.

Smith, M. A., M. Salas, W. M. Stanley, J. R. Wahba, and S. Ochoa (1966). Direction of reading of the genetic message. II. *Proc. Natl. Acad. Sci., U.S.*, **55,** 141.

Soeiro, R., and J. E. Darnell (1970). A comparison between heterogeneous nuclear RNA and polysomal messenger RNA in HeLa cells by DNA-RNA hybridization. *J. Cell Biol.*, **44,** 467.

Soeiro, R., H. C. Birnboim, and J. E. Darnell (1966). Rapidly labelled HeLa cell nuclear RNA. II. Base composition and cellular localization of a heterogeneous RNA fraction. *J. Mol. Biol.*, **19,** 362.

Soeiro, R., M. H. Vaughan, J. R. Warner, and J. E. Darnell (1968). The turnover of nuclear DNA-like RNA in HeLa cells. *J. Cell Biol.*, **39,** 112.

Spirin, A. S. (1969). Structure of the ribosome. *Progr. Biophys. Mol. Biol.*, **19,** Part 1, 133.

Strauss, J. H., B. W. Burge, E. R. Pfefferkorn, and J. E. Darnell (1968). Identification of the membrane protein and "core" protein of Sindbis virus. *Proc. Natl. Acad. Sci., U.S.*, **59,** 533.

Strauss, J. H., B. W. Burge, and J. E. Darnell (1970). Carbohydrate content of the membrane protein of Sindbis virus. *J. Mol. Biol.*, **47,** 437.

Stretton, A. O. W., and S. Brenner (1965). Molecular consequences of the *amber* mutation and its suppression. *J. Mol. Biol.*, **12,** 456.

Summers, D. F., and J. V. Maizel (1968). Evidence for large precursor proteins in poliovirus formation. *Proc. Natl. Acad. Sci., U.S.*, **59,** 966.

Summers, D. F., J. V. Maizel, and J. E. Darnell (1965). Evidence for virus-specific non-capsid proteins in poliovirus-infected HeLa cells. *Proc. Natl. Acad. Sci., U.S.*, **54,** 505.

Swanson, R. F., and I. B. Dawid (1970). The mitochondrial ribosome of *Xenopus laevis*. *Proc. Natl. Acad. Sci., U.S.*, **66,** 117.

Takagi, M., and K. Ogata (1968). Direct evidence for albumin biosynthesis by membrane bound polysome in rat liver. *Biochem. Biophys. Res. Commun.*, **33,** 55.

Tavill, A. S., A. I. Grayzel, I. M. London, M. K. Williams, and G. A. Vanderhoff (1968). The role of heme in the synthesis and assembly of hemoglobin. *J. Biol. Chem.*, **243,** 4987.

Terzaghi, E., Y. Okada, G. Streissinger, J. Emrich, M. Inouye, and A. Tsugita (1966). Change of sequence of amino acids in phage T4 lysozyme by acridine-induced mutations. *Proc. Natl. Acad. Sci., U.S.*, **56,** 500.

Tomkins, G. M., E. B. Thomson, S. Hayashi, T. Gelehrter, D. Granner, and B. Peterkofsky (1966). Tyrosine transaminase induction in mammalian cells in tissue culture. *Cold Spring Harbor Symp. Quant. Biol.*, **31,** 349.

Vaughan, M. H., J. R. Warner, and J. E. Darnell (1967). Ribosome precursor particles in the HeLa cell nucleus. *J. Mol. Biol.*, **25,** 235.

Vinuela, E., M. Salas, and S. Ochoa (1967). Translation of the genetic message. III. Formylmethionine as initiator of proteins programmed by polycistronic messenger RNA. *Proc. Natl., Acad. Sci., U.S.*, **57,** 729.

Warner, J. R., and A. Rich (1964). The number of soluble RNA molecules on reticulocyte polyribosomes. *Proc. Natl. Acad. Sci., U.S.*, **51,** 1134.

Warner, J. R., P. Knoff, and A. Rich (1963). A multiple ribosomal structure in protein synthesis. *Proc. Natl. Acad. Sci., U.S.*, **49,** 122.

Warner, J. R., R. Soeiro, H. C. Birnboim, and J. E. Darnell (1966). Rapidly labelled HeLa cell nuclear RNA. I. Identification by zone sedimentation of a heterogeneous fraction separate from ribosomal precursor RNA. *J. Mol. Biol.*, **19,** 349.

Wettstein, F. O., T. Staehelin, and H. Noll (1963). Ribosomal aggregate engaged in protein synthesis: Characterization of the ergosome. *Nature*, **197,** 430.

Wigle, D. T., and G. H. Dixon (1970). Transient incorporation of methionine at the N-terminus of protamine newly synthesized in trout testis cells. *Nature*, **227,** 676.

Willems, M., H. A. Musilova, and R. A. Malt (1969). Giant nucleoplasmic RNA in the switch-on of compensating renal growth. *Proc. Natl. Acad. Sci., U.S.*, **62,** 1189.

Williamson, R., M. Morrison, and J. Paul (1970). DNA-RNA hybridization of 9S messenger RNA for mouse globin. *Biochem. Biophys. Res. Commun.*, **40,** 740.

Yanofski, C., B. C. Carlton, J. R. Guest, D. R. Helinski, and V. Henning (1964). On the colinearity of gene structure and protein structure. *Proc. Natl. Acad. Sci., U.S.*, **51,** 266.

Zimmerman, E. F., J. Hackney, P. Nelson, and I. M. Anas (1969). Protein synthesis in isolated nuclei and nucleoli of HeLa cells. *Biochemistry*, **8,** 2636.

PART 3

CELL GROWTH AND DIFFERENTIATION

11

THE CONTROL OF CELL GROWTH

E. C. Pollard

1. THE CHARACTERISTICS OF GROWTH

The surface and volume relationships of a cell set limits on its size, just as such relationships determine the size of a drop of liquid. In the case of the liquid drop, the latent heat of condensation must provide the energy to produce the surface. Too small a drop does not have sufficient volume to create the surface and so a drop cannot be smaller than a certain radius. The cell also has a surface–volume relationship. It must use its surface to take in metabolites which are then used to synthesize organelles and macromolecules inside the cell. These units represent a more ordered form of matter than the metabolites and so they are of lower entropy, which must be provided by a correspondingly greater increase in entropy outside. The entropy exchange process takes place at the surface, and if the ratio of the surface area to the volume of organized material becomes small, as is the case for a very large cell, the rate of entropy exchange perforce becomes less and the cell grows more slowly. Such a large but slow growing cell loses in competition for metabolites in the medium in comparison to cells that have divided and so gained a higher proportion of surface. The smaller cells use material to create low entropy order faster and so they win in the process. Thus the living cell tends to divide to increase the rate of metabolism while the drop tends to become larger since its source of energy is its size and it has a different process of entropy exchange.

Since a cell has an overall compulsion to divide, definite limitations of shape can be seen. It is by no means clear that a cell about to divide will choose the bland form of a spherical drop. Rather the very process of division automatically demands a zone of division which should appear to be natural, and this suggests some difference in the cell at the point where division can take place. Observation of dividing cells confirms this apparently planned process, and quite clearly special mechanisms for effecting division have evolved and are readily apparent.

These entropy and growth considerations are worth following numerically to see if they suggest any reasonable predictions. Bayne-Jones and Rhees (1962) measured the heat generated by a bacterial culture, and from this the entropy content per cell was found. It is -1.3×10^{-12} cal/(mole) (degree) per cell. The bacterium *Escherichia coli* has a volume of 2×10^{-12} cm and a surface area of 10^{-7} cm^2. It takes approximately 1300 sec to divide, and so the entropy exchange per second per square centimeter per second is -10^{-8} cal/(degree) (sec) (cm^2). The entropy per unit volume is 0.67 cal/(degree) (cm^3). If we assume these figures hold for all cells, as a first approximation, we can determine the time t required for a cell of radius r to double its negative entropy content. For a spherical cell this is:

$$\frac{4}{3}\pi r^3 \times 0.67 = 4\pi r^2 \times 10^{-8}\, t$$

$$t = 2.2 \times 10^7\, r$$

If we assume that the time of division is 20 hr, the above relation suggests that the radius of the cell would be 32 μ, not an unreasonable figure for the size of a mammalian cell. Thus the process of growth does conform to very broad surface and volume relationships, although not as simply as calculated above because, for example, the shape of the cell need not be spherical. It could instead have much more surface in relation to volume. Quite clearly, such deviations from spherical shape are an advantage, and if possible the cell adopts them.

It is useful to examine some of the statistics concerning cell growth. Most of the information is derived from experiments involving bacterial cells, but good data are becoming available for eukaryotic cells also. Bacteria, such as *E. coli*, grow at 1.5×10^{-7} cm/sec in linear dimensions. This is 15 Å/sec, a very impressive figure. The volume growth is modest, 1.5×10^{-15} cm/sec. This is 3.8×10^{-16} g dry weight per second or, in terms of molecular weight, 1.6×10^7 daltons/sec—1000 small protein molecules per second or 100,000 metabolite-type molecules per second. Twenty-three ribosomes are formed per second; 3000 base pairs per second are assembled into DNA at each growing point for DNA; 50 stretches of messenger RNA (mRNA) are assembled each second. All of these figures are indicative of the rapidity of chemical reaction in the cell. A most startling additional thought is the great precision of the processes involved. Cells such as HeLa cells grow linearly less rapidly but exceed bacteria in most other ways, as would be expected for larger cells. HeLa cells grow at 1.3 Å/sec, increase their volume by 1.3×10^{-14} cm^3/sec, and their mass by 1.3×10^{-15} g or 4.6×10^8 daltons/sec.

A. Exponential Growth

One of the chief features of growth often mentioned is its exponential character. This is a consequence of the process of division, for each time a cell divides into two cells, unless the daughter cells are not equivalent to the parent, they grow as rapidly as the parent and so the overall process is doubled. The rate of growth of number, or mass or volume, then proceeds as 2^g, where g is the number of generations; this is automatically also representable as a power of 10 or of e, the base of natural logarithms. It is important to realize that the actual cellular processes need not be exponential at all, and in fact many are not. In mammalian cells DNA synthesis is actually stopped for a considerable part of the division time; in bacteria there may be only one active growing point at a time, meaning that only one rate of synthesis of DNA takes place, a linear and not exponential process. Thus division and not growth is responsible for the exponential character of a cell culture. However, something very similar to division can occur within the cell, and this can foreshadow the exponential character of growth even though separate unit cells have not appeared in the culture. This is strikingly seen in bacterial DNA and probably also in organelles in general.

B. Multiplication of Intracellular Sites of Synthesis

Bacterial DNA requires two major functioning units for its synthesis: an initiator, and the polymerase or DNA growing point. In a cell in which the DNA exactly matches the envelope, there is one initiator and one DNA growing point which serve to synthesize DNA until the whole chromosome is complete, following which there is division and a new initiator and growing point start up in each new cell. Such conditions can be found, but with difficulty, and it is much more common to find a new initiator and polymerase at work on the DNA before the chromosome is anywhere near completion. This process of speeding up the chromosome growth is one of the responses made by the cell to a favorable medium and is a factor in speeding up cell growth. The result is that the envelope now actually carries a different package of DNA, for when it divides the segregation involves a unit that is more than one simple genome of DNA but is a genome plus the additional part of the genome being synthesized. We thus see one very important feature of the growth of living cells, namely, a separation of the growth of different components and an ability to adapt to and tolerate different contents while remaining the same kind of cell. The existence of this phenomenon also calls attention to the presence of a separate mechanism for initiating each division process and for providing regulation of a cell that has divided abnormally so that it can return to a functional cell before the next division. The control of growth is thus needed for more than the determination of the kind of cell that develops from the genome; it is also needed to keep a cell the same as it was before division. With this general introduction we can now consider the molecular character of cellular activities.

2. THE MOLECULAR CHARACTER OF CELLULAR ACTIVITIES

The physical form of a cell, its rate of growth, its special functions, and its relationships to other cells are all primarily controlled by molecular interactions. These words are not generalizations without precise meaning; rather it is their extreme accuracy that makes the living system what it is. One of the great paradoxes of Nature, and there are many, is that the incredibly beautiful form of the living cell is produced in response to a closed and relatively small molecular pattern which is almost invisible and of a structure that bears no resemblance to the finished product, the cell. While this molecular pattern is of great interest to geneticists and molecular biologists, it is not our primary concern. Our interest is in the actual way in which this pattern becomes that of the cell.

Very often this description is made in terms of the "central dogma" of molecular biology which is familiar to all. A modern useful summary is found in Volume 35 of the *Cold Spring Harbor Symposium on Quantitative Biology* (1971). While we shall have occasion to use the terminology of the molecular biologists quite often, it is still not our main preoccupation, for the processes of control can be discussed more generally than the processes of molecular biology, and by doing so we can see what characteristics to expect.

The very first inkling of the molecular character of cellular operations appeared in Schrödinger's monograph, *What Is Life*? He made it clear that if perfection of form is to be obtained on a very small scale then it must occur in response to the precise molecular patterns of quantum mechanics and can not be statistical in nature. He also clarified the secondary requirement, namely, if perfection of form results in any process entropy must decrease and such a decrease in the cell must be accompanied by an increase somewhere else.

All of life is dominated by the action of the covalent bond. This bond occurs because it is possible for electrons of two different atoms to share orbitals. When this possibility exists and the atoms are close enough together, the operation of quantum mechanics causes a rearrangement of the electrical charge. One such rearrangement favors a strong attraction between the atoms, and this strong attraction is called the covalent bond. The attraction occurs only if the atoms are very close and the extent of this distance is generally speaking, approximately 1 Å. It is usual to think of the strength of the bond in terms of the energy required to separate the two atoms, this being approximately 2 eV or 46,000 cal/mole. Of almost equal importance is the force between the atoms; for example, at half the distance needed to separate them completely, it is 5×10^{-4} dyne. This may seem to be a small number, but it must be remembered that it applies to *one* atom, and if it were scaled up to the force per gram, for example, of water, it would be equal to the weight of the *whole earth*. It is these immense forces that are at work in determining the form of the molecules of life, and it is therefore not surprising, when seen in this perspective, that the form can be firm and definite, even though controlled by a molecular pattern.

The rearrangement of electrical charge that is the basis for the attraction of the covalent bond is by no means always the same. Two major side effects exist. One is that the rearrangment can be so drastic that one whole electronic charge is carried by one atom and the other has one too few. These species are called *ions*. The second is that even when such great asymmetry does not occur there can be considerable variation in the density of charge within an atom. Notably, hydrogen can be almost bare of negative charge on one side; oxygen can have an excess of negative charge; and the combination of the two, in water, can have one end positive and the other negative so that a permanent dipole of considerable magnitude exists. Of almost equal interest is the case of no side effects. If the redistribution of charge results in a completely spherical symmetrical form for the electrical charge, the nearest approach to electrical neutrality is found. However, this neutral molecule *occupies space*. So space must be found for it.

What has just been discussed adds up to two very important cômbinations. The first is between an ion and water. The enormous electrical field of an unneutralized electrical charge has an associated energy, and this energy must be provided if the ion is to be free. Most of the time it is not free; it is in very close combination with one of the other signs and this reduces the extent of the electrical field almost to zero. One way in which the ion can be free is to associate itself with molecules of water. Because water molecules are dipoles, they turn along the field of the ion and a small number bind quite tightly to it. This reduces the extent of the field but not enough to make the ion stable. For a monovalent ion a second tier of water molecules is needed and possibly even a third. Thus a monovalent ion carries with it about 30 associated water molecules, and a divalent ion about 150. Also, these molecules are arranged radially as suggested in Figure 1. This arrangement of water molecules is in contrast to another which we are about to describe, and to which we refer as "polar" water. The radial orientation is not produced in any gentle way; it is strongly coerced by the intense field of the ion. An ion in the midst of water is therefore a very active agent, as it causes a strong effect on many water molecules.

If there is no strongly polarizing agent, the electrical fields of the dipoles themselves become the most potent agencies present and the water molecules arrange themselves in conformity with these fields. Just how this is done is still an open question, and it is likely that there are several ways. All of them, however, involve polar molecules which form some kind of chain. Since this association is three-dimensional, it is more of the nature of a cage than a chain, and since this arrangement is also similar to that observed in a "clathrate," we call it clathrate water. Such structures are usually simple in concept, except that the center is not easily filled, since the center is not part of the chain.

The second combination of interest is that of a completely symmetrical molecule, such as methane, or a nearly symmetrical molecule, such as the higher hydrocarbons, with water. A chainlike structure for water is the only one possible, since the symmetrical structure has very little or no electric field and as a result the chain forms easily with the hydrocarbon in the center of the three-dimensional structure. Thus although the hydrocarbon has no electric field, it can occupy space in such a way that the fields of the polar water molecules are able to interact more closely and as a result the electrical field energy is lower. This surprising result has been found to be most important and has great application in the interpretation of the way in which a long-chain molecule, containing amino acids, can fold. This kind of interaction is called "hydrophobic," and sometimes is dignified by the word "bond," which is inappropriate since the action is emphatically not a process involving pairs but is much more concerned with several molecules, possibly even engaged in avoiding each other.

With even this very brief description of the factors at work in the development of form from molecular structure, we can see two major agencies. If the folding of a protein chain depends on the relation between water and the hydrophobic side chains, the presence of ions, which coercively orient water into the wrong pattern in relation to hydrophobic groups, will exert a strong influence. We see one means of control developing: the presence of ions in the vicinity of protein molecules.

Two other secondary consequences of the nature of atoms must be discussed. The first is the consequence of the asymmetry charge on a hydrogen atom when its electron

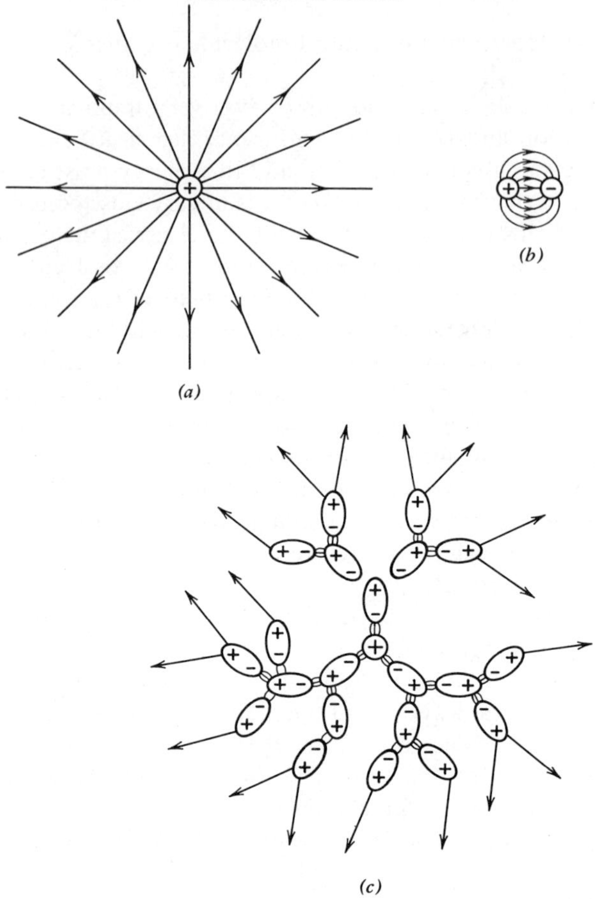

Figure 1. The lines of force radiating from a singly charged ion are shown in (*a*). This electric field has an energy that depends on the square of the field strength and on the extent of the field. For such an ion it is approximately 3 ev, or 69,000 cal/mole. In (*b*) the plus and minus pair is shown. Although the field is intense, it is enormously reduced in extent and so the combination takes far less energy than the separated charges. This occurs in crystals. In (*c*) the appearance of a hydrated ion is shown. Polar water is attracted very strongly to the ion itself, with a second relatively tight tier of water molecules and a third loosely ordered tier outside. The field is considerably reduced, and the ion is able to exist "free" in solution. Notice that the water is arranged along the lines of force in a manner we called "polar" water.

is shared, an asymmetry that makes one side of the atom quite positive. This, together with the tendency for oxygen and nitrogen atoms to gain negative charge, makes it possible for a strong attraction to occur between a hydrogen in one part of a molecule and an oxygen or nitrogen in another. This attraction is called the hydrogen bond and is very familiar in all discussions of molecular biology. For our purposes two very important aspects of hydrogen bonds require mention. The first is that the H bond operates over the greatest distance of any bond normally encountered in biology. Because it is really a primary electrostatic interaction, it has a full range of electrostatic forces. The hydrogen bond thus can serve a very useful function for first *recognition*, to be followed by a consolidation in some other way. The second aspect is that the bond depends on the relative asymmetry of charge in the hydrogen-oxygen pair and this may in turn depend on the other atomic combinations in the vicinity of the bond. Thus there is a measure of control possible on the strength of the bond. This proves to be of great importance and almost certainly is needed to explain enzyme action.

The last molecular interaction to be discussed is the remarkable interaction between transient dipoles and induced dipoles. The resulting forces are often called London forces, dispersion forces, or van der Waals forces. Because in any atom the electron does not revolve around a nucleus in a defined path but must be described in terms of probability, it follows that at any one time the separation of the electron from the nucleus must be such that there exists a transient dipole. This exerts an electrical field, and this field may be of considerable strength and last long enough to be very influential. The influence is to produce an induced dipole in any nearby atom; the nature of the law of electrostatic force demands that such an induced dipole is attracted to the causative transient dipole. If the distances involved are small, the attraction is considerable; this interaction is at work in the process of liquefaction of an inert gas such as argon. The importance of this kind of interaction is that it is both attractive and additive, so that every atom attracts every other. This additivity has been shown by Hamaker of the school of Verwey and Overbeek (1948) to be responsible for considerable binding between large molecules. It is of importance in our context, because it means that regardless of the nature of two molecules there is some attraction between them attributable to this effect, and therefore there can exist a loose combination within which the two can actually sample orientations to see whether or not a specific combination exists. This process makes collisions far more effective than they would be if, in order for a specific combination to take effect, they had to take place with the correct orientation.

This type of force varies very sharply with distance. Because this is true, any kind of shape that prevents close contact diminishes the force between macromolecules, and any kind of surface fit makes a very strong attachment possible. This force is at work in the assembly of the protein components of viruses.

We are now in a position to see, broadly, how living systems operate. Two very fundamental operations must take place; energy must be transferred, and the synthesis of new molecules must occur. Both of these involve the rearrangement of covalent bonds, the first by utilizing a case in which the complexity of the atomic species permits alternative arrangements of covalent bonds, as for example in the combination of three phosphates or of two and one (the familiar ATP, ADP, and P), and the second by the assembly of the component molecules (sugars, amino acids, bases) into a covalently bonded chain constructed to order by a template. Neither of these processes can occur wholly randomly with any efficiency, and so biological systems have evolved specific *active surface elements*, each of which is remarkably chosen for the process it is designed to facilitate and expedite. These surface elements can be the active sites of enzymes, the recognition points for a transfer RNA, the region of active transport of an ion, and the surface combination areas for the components of a virus head or tail. These are probably only a few, but clearly they afford excellent opportunity for control—the subject of this chapter.

It is not necessary to describe any one of these elements in detail initially. The kind of process involved is as follows. A protein chain, having a certain grouping of amino acid residues, some of which have a hydrocarbon character and are hydrophobic, others of which are capable of having an ionized $-NH_3^+$ or $—COO^-$ group, reacts to aqueous surroundings in a rather definite, *but not unique*, way. It is likely that a protein molecule with, for example, 150 residues, has five or six very stable ways of folding and that these can be influenced by hydrogen ion concentration or by ionic strength. In the cell the local environment of these protein molecules can be varied considerably, so that one or another of the configurations can be found and the molecule may not be in the same form throughout the cell.

The specially folded protein chain is not the construct of an organic chemistry laboratory; rather it is the product of evolution. Thus it may have, and probably does have, a part of its surface that can utilize one of the secondary forces just described to bind a particular molecular arrangement, for example, a molecule of urea or a particular pair of bases on a chain of nucleic acid. The binding holds the molecular arrangement—the substrate—and may release it unchanged. If, however, further events are possible, such as the introduction of stored energy derived from the alternative forms of the covalent bond and a collision with another molecule such as water, the substrate is released in chemically changed form and the protein will have then acted as a specific catalyst. The cell uses such catalysts in an absolutely major way and the chemistry that results is rapid, accurate, and most economical of the needed material. Evolution has indeed provided an amazing method of achieving varied chemical reactions, including those needed for synthesis and growth.

Clearly, these reactions can be influenced in many ways, some of which are: alteration in the folding of the enzyme; blocking, or expediting, the transfer of the configurational energy (group potential) to the active site of the enzyme; reducing the concentration of substrate; introducing competitors that wastefully occupy the active site; reducing or increasing the concentration of enzyme; segregating, by some means of compartmentation, one of the needed factors in the reactive process. Although all these need consideration, space does not permit full discussion.

3. CONFIGURATION CHANGES OF A PROTEIN MOLECULE

The configuration of a protein molecule—the way in which it folds—is clearly dependent on at least four factors. Four are enough to show the character of the processes of control.

The first is the location of charged groups on the molecule. An $—NH_2$ group, which acquires an extra hydrogen ion to become $—NH_3^+$, is influenced by two major factors. On the outside of the chain, where polar water molecules are found, this positive charge results in local orientation of the water into polar water. On the inside, where there is relatively little water, the ion has an electric field which interacts with other ions. Since primary electrostatic forces are strong, this is a firm influence. Hydrogen ion concentration, which influences the number of such charged residues, is a strong agent in determining secondary structure.

The second factor is temperature. Temperature influences the motion of water molecules, and its influence may extend to the kind of cage structure assumed by the water molecules. An increase in the mobility of water molecules has an effect on the viscosity of the solvent—the cytoplasm in the cell. An increase in thermal motion may make possible a configuration that is adopted, but which includes some water molecules not free to rotate, to change to one in which the water is released. Thus temperature is a factor, and it is to be expected that enzymes undergo changes in activity as a function of temperature. This may sometimes include sudden loss of activity at low temperatures.

The third factor is the presence of ions, particularly divalent ions. As already noted, each such ion carries with it approximately 30 (monovalent) or 150 (divalent) molecules of water held in the polar form. An ion so close to a protein molecule that it can form some kind of binding with it can release water molecules and in so doing also generates an electric field where the water has left. Both the free water and the free electric field are strong agencies; the former may increase the hydration of part of the protein chain, thus diminishing the influence of the hydrophobic groups; and the latter may repel other similarly charged ions and so cause the secondary structure to assume a new stable form.

The fourth, and in many ways the most interesting factor, is the combination (through secondary forces) of small molecules with a protein molecule. One such combination involves the substrate, and it is now known that the conformation of lysozyme, for example, is changed when the substrate tetrasaccharide is in place. The mere formation of a complex with an "effector" molecule may compel the protein molecule to take a second form. Sometimes several combination sites for effectors exist, and the change in shape of the protein occurs only when four or five such effectors are attached. The attachment by secondary forces is never as strong as for a covalent bond, and so there is generally an equilibrium between molecules that are attached and those that are dissociated. This equilibrium can clearly be influenced by temperature and possibly other physical agents including pressure and pH. The use of this means of modifying the shape, hence the action, of a protein molecule is certainly one of the most potent methods of controlling growth. The subject therefore occupies a considerable amount of our attention. It is quite clear that each protein has *several* possible conformations and that the attachment of small molecules can be used to develop or restrict the formation of any one. Thus control can be exercised in both ways—positively or in an inhibitory manner.

In examining the four methods of influencing the conformation of a protein molecule, it is seen that they can be useful to a cell under different circumstances. Temperature, for example, cannot be maintained differentially in a cell. The scale of a cell is so small that the surface-to-volume ratio in any one element is so great that equalization of temperature takes place in milliseconds even in as large an organelle as the nucleus. Therefore the consequence of temperature changes must be thought of as determining the active temperature range of the cell and not as one used to develop differential action within the cell. In many ways this is true of pH. A compartmentation that can segregate hydrogen ions is not easy to develop, so that one usually accepts the idea that the whole cell has the same pH. This need not be true for well-developed organelles such as mitochondria or chloroplasts or the nucleus. These could conceivably maintain local hydrogen ion concentrations.

Sodium, potassium, and divalent ions *can* be segregated in specialized concentrations, and the use of changes in concentration of these ions to influence enzymic processes in the cell is well known. However, since ions act primarily through the action of electrical fields and changes in the character of water, they are not adapted to produce specific change in any particular enzyme. Thus they are considered more as in the class of pH or temperature, though the existence of a variety of ions makes their action rather more varied.

When we come to the molecules that can bind specifically to protein or other molecules, a wholly new dimension in the potency of control emerges. First, the binding site can be the only one of its kind in the cell. This means that only this enzyme, and no other, can be affected by the binding. Thus a single particular process can be turned on or off. Second, the number of binding sites on the macromolecule can be greater than one, and this means that the change in conformation of the enzyme requires simultaneous occupancy of, for example, four sites. The kinetics of change of conformation are now very different from the case in which a single attachment is sufficient; a higher concentration of effector molecule is necessary and, accordingly, the change can be more abrupt when it takes place. In addition to these two very important factors, the development of sufficient concentration of effector molecules is something that can be influenced, for example, by competitive binding or by varying the rate of formation of the molecules. Thus it is not at all surprising that attention is being given to the nature of effector molecules and the kind of system they influence. Specific examples are given later. For the present we need to consider the process of random collision and specific selection as a general phenomenon because this underlies the action of enzymes, repressors, inhibitors, and even macromolecular synthesis.

4. THE SPECIFIC SELECTION OF A MOLECULE FROM RANDOM BOMBARDMENT

Diffusion, and the equivalent term Brownian movement, represent the only way in which the approach and combination of molecules are initiated. It is a highly random process, very fast on a small scale, and very slow on a large scale. It is important to have some idea of how it is possible for collisions to occur and for specific combinations to develop in this wholly active and random world.

It is first necessary to estimate the rate at which collisions take place at a particular point on a molecular surface. This problem was first dealt with by Smoluchowski (1916). A modification of his theory for collisions with small regions on a molecular surface leads to the following result. The rate of collision per second ϕ is given by

$$\phi = 2\pi DRC_\infty \tag{1}$$

where D is the diffusion constant in centimeters per second, R is the effective radius for collision, and C_∞ is the concentration in molecules per milliliter of the colliding molecules. For example, a small molecule may have a diffusion coefficient, even in the viscous milieu of the cell, of 10^{-6} cm^2/sec. The collision radius is the sum of the radius of the colliding molecule and the effective radius of the surface element involved in the collision. For our calculation we can assume 10 Å or 10^{-7} cm. A concentration of 1000 molecules in a cell the size of a bacterium gives 10^{15} per milliliter, about 1 μM. The resulting collision rate is then found to be 600 per second, not a number that is limiting in any way. However, there are two important additional factors to consider. The first is concerned with the number of wrong molecules, molecules that also collide with the surface but that are unwanted, and the fraction of time that they prevent the specific surface from being able to accept the correct molecule. The second is the question of the aspect of the colliding molecule. Does it have to present a particular aspect, or can it search for, and establish, the right surface of contact?

These two considerations are interrelated. If the pool of small molecules in a bacterial cell is 3% of the dry mass, there are roughly 10^8 such molecules per cell and the resulting concentration is 100,000 times higher than the above calculation; thus there will be a total of 60,000,000 collisions per second involving all kinds of molecules, of which only 600 are of the right kind. The surface is thus constantly bombarded and if any kind of attachment takes place the surface may become useless. It is this type of action that renders inhibitors so important.

If the collision must be made with a favorable aspect, so that only one exact presentation of the molecule is effective, the number of effective collisions is reduced so much that the surface can hardly be useful. The method that must be operative uses one of the secondary forces described above: the transient dipole-induced dipole force, often called London force or van der Waals force. This is nonspecific and always attractive. It guarantees that when a collision, even with the wrong aspect, is made, there will be a loose attraction of the type to permit the molecule to rotate. Rotational Brownian movement of a small molecule is extremely fast, and the time required to sample the surface of the colliding molecule is approximately not much more than a microsecond. Thus we see that the many wrong collisions involve the interesting molecular process of a momentary attachment, very rapid rotation, and a moving away *unless* the specific forces that cause the two surfaces to fit are present. If they are, the result is a sharp, strong binding, and the specific molecular process begins. This is the process designated "random collision and specific selection" (Pollard, 1961, 1963, 1968), and it undoubtedly describes the major aspects of the molecular biochemistry of the cell with respect to collision.

Before illustrating the way in which this process operates, it is worthwhile to look at it in a different but equivalent way. Brownian movement and diffusion are both the same process, but a formal description of the two leads to different equations. We have used the formalism of diffusion; an equivalent approach involving random migration can be used. In Figure 2 a molecule is presumably moving with the random, aimless motion of Brownian movement, or of thermal agitation if that is preferred terminology. It can be seen that it sweeps out a collision volume, and that it does so by a set of random jumps as shown. If we assume that the average length of a random jump is $\bar{r}$, we can see that the volume swept out v is related to the number of jumps N and the area of the swath S (the collision cross section) by the relation

$$v = N\bar{r}S$$

There also exists the relation deduced by Einstein, namely,

$$Dt = \tfrac{1}{6}N\bar{r}^2 \qquad (2)$$

from which we find

$$v = 6DtS/\bar{r} \qquad (3)$$

This is a very useful result, for it gives us an idea of how long it will take any one particular molecule to make contact with every element in the volume of the cell, and it shows why the scale of the cell is so small.

The formula is deficient in confidence in one term, the average distance of a jump. For a small molecule a distance of 1 Å can be assumed. It cannot very well be much more, and if it is very much less the volume must be calculated in a different way. By using this number, we see that the volume swept out by collisions in 1 sec is 6×10^{-12} cm^3 for a molecule of radius 6 Å. Thus any small molecule completely makes contact with every element in a bacterium in $\tfrac{1}{6}$ sec. It would take 4 sec to do the same in the nucleus of a mammalian cell and 1300 sec to traverse the whole cytoplasmic region of a

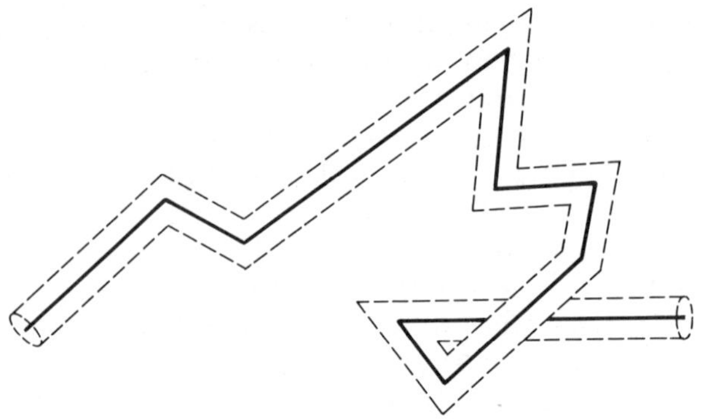

Figure 2. The random wandering of a molecule by Brownian-movement and the collision volume that it sweeps out. The dotted "cylinder" of cross section S represents the volume in which collisions occur. The average length of each jump is $\bar{r}$.

human cell. The bacterium responds quickly to such a molecule, the nucleus more sluggishly; it is safe to say that a whole cell would not respond in any sensible way at all to one molecule released for the purpose of a control mechanism. This very simple calculation makes it clear why the cell uses organelles, why there are membranes within the cell, as, for example, in the endoplasmic reticulum, and why the nucleus must be separate from the cytoplasm. One consequence of the development of organelles and compartmentation is the parallel development of concentration differences in the cell, and one very important consideration is a quantitative estimate of the character of concentration differences. This topic is now discussed.

On a small scale diffusion is very rapid. A representative small molecule traverses, by diffusion, a distance of 23 μ in 1 sec. Thus it is unrealistic to presume that a concentration gradient can be maintained in a cell in the face of this rapid equalization unless some other factor is present. Two factors are of importance. The first is, in a sense, trivial; it is the intervention of membranes that confine components. The second factor is more related to life itself; it is the constant generation and consumption of molecules that can take place at a speed that matches the equalizing speed of diffusion and which, while the cell or organism is alive, can maintain concentration gradients and so utilize the consequences of their existence.

Once again, a very simple numerical analysis shows the dimensions of this process. Suppose that a synthetic process causes the incorporation of 100 molecules per second out of the cell milieu into a macromolecular structure—RNA, for example. If the area over which this reaction takes place is 100 Å^2, or 10^{-12} cm^2, then the application of the diffusion equation gives

$$\frac{dn}{dt} = DA\frac{dc}{dx}$$

where n = number of molecules removed per second, x = distance, A = area, c = concentration, and D = diffusion coefficient. That is to say,

$$10^2 = 10^{-6} \times 10^{-12}\frac{dc}{dx}; \quad \frac{dc}{dx} = 10^{20}$$

This means that the concentration gradient needed to sustain the loss of 100 molecules per second is 10^{20} per milliliter per centimeter and suggests that in a cell 1 μ long such a rate of synthesis might very well maintain a concentration of 10,000 molecules at one end and none at the other. This is an extreme number, and presumes one-dimensional diffusion, which is not the case, but it serves to illustrate that a metabolizing and synthesizing system similar to a cell can produce concentration gradients even within its own method of operation. This means that when a point of synthesis starts up, the local concentrations of the metabolites involved are appreciably altered. Thus if in a cell the supply of deoxytriphosphates is limiting, and experimental studies indicate that this is often true, the formation of a new initiator and DNA growing point will change the concentrations available for synthesis and this will in turn elicit a response—either a slowdown in the rate of DNA synthesis at each point or, more likely, a change in the inhibition of the formation of the triphosphates by feedback from concentration levels near the enzyme systems that produce them. The living cell therefore should not be thought of as having a bland concentration distribution of all its internal components; rather it has very considerable local modifications of the average, and these certainly play a part in control.

Before leaving this topic, we note that the results of applying Eq. 3 to molecules of various size are of interest. Since the cross section term varies with the square of the radius of the molecule, while the diffusion constant varies inversely with the radius, the net volume swept out by a large molecule is greater than that of a small one, provided the distance of each jump is the same, a most unlikely situation. One other factor enters, which is the difference between the viscosity of a great mixture of components and that of water. For small molecules there is no viscosity, but for large molecules the interstices between the various components of the cytoplasm are so small that the viscosity increases and the diffusion constant decreases. Thus it is a fair approximation to assume that the volume swept out is nearly the same regardless of the size of the molecule. Perhaps it is for this reason that the bacterial cell seems to have no special location for the ribosomes but relies on normal collision behavior, as for small molecules.

5. GROWTH BY SELF-ASSEMBLY

There are two major methods by which growth is achieved. One, which at first seemed mysterious, is the process of self-assembly, now no longer hard to understand. The second, the enzymic addition to a cell wall, for example, is by no means a solved problem. First we consider self-assembly.

Many good examples of this process exist. The best studied is the assembly of various components of bacteriophages in which powerful genetic methods have shown the exact character of the process. Several stages involve no enzymic mediation and are consequently genuine self-assembly. The assembly of a microtubule is another example. The bacterial pili are yet another. These remarkable appendages can be removed by high shear, but are replaced even in the absence of protein synthesis, indicating that the cell has a supply of the components available for replacement and that self-assembly can be initiated when the pili are removed. Tobacco mosaic virus is another well-studied system, and ribosomes are rapidly becoming understood in the same kind of detail.

The process is *par excellence* an example of random collision and specific selection. Take, for example, the assembly of a phage tail. A representative assembly unit might be a segment of tail 100 Å across and 500 Å long. This corresponds to roughly 100 subunits, and simplicity almost certainly requires that the genome that codes for these units cannot be of the length necessary to code for 100 different units. The most economical and likely assumption is that they are all identical, as has indeed often been found. They are, however, evolved protein molecules and so have a special shape. A trivial example of a special shape, far too simple to have any biological reality, is shown in Figure 3. It can be seen that these molecules can stack together and that by random collision, together with the rapidity of rotational Brownian movement, they meet, sample all orientations, and stick in exactly the correct way *and no other*. The result is a steady accretion of stacks, which is part of the necessary process of assembly. However, this kind of shape is really very characterless. One has only to introduce a little asymmetry, plus the use of the third dimension, and the structure

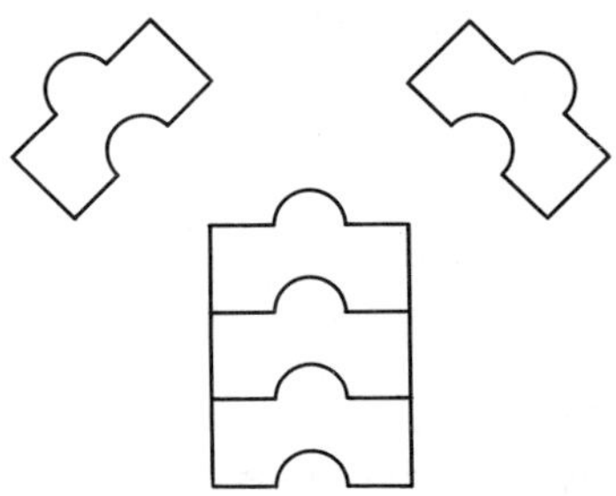

(*a*)

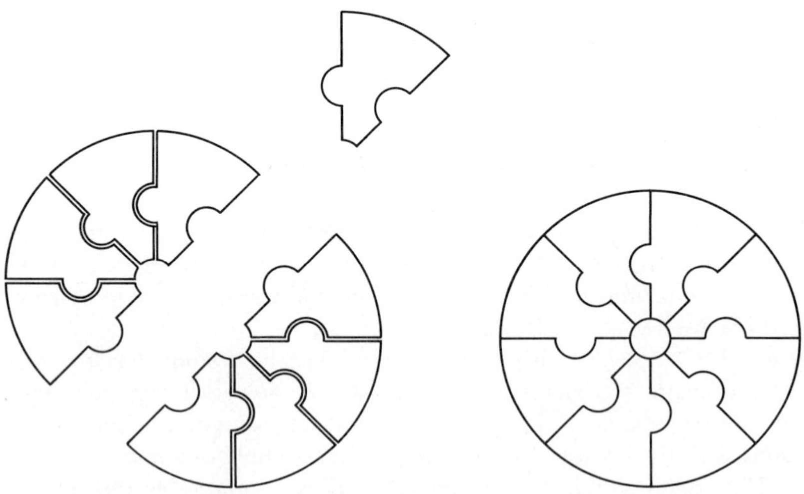

(*b*)

Figure 3. The self-assembly of related shapes. If the close contact is such that binding occurs by London forces, then collision will sample all aspects until the correct one—followed by binding—occurs. The result is shown in (*a*) to be a stack of units; (*b*) shows how a special shape for each unit can lead to a limiting shape—in this case a circular cylinder with a central hole.

that now assembles itself is much more full of character. A helically formed microtubule is easy to design. Furthermore, the process of sticking when the specific surface elements are close together need not be valid for only one part of the two surfaces. They can stick at the side, for example, and then a dimer can develop which is related to other dimers or even to a monomer. It is not our forte to engage in this kind of biological building block exercise, nor is it appropriate in this discussion, but it is abundantly clear that the rapidity of collisions, the surface sampling by rotation, and the specific attraction for the correct conformation can be used to assemble elaborate structures. It is no real mystery why these should be limited in size or length, for if the individual units have a special symmetry, after a certain number of units have been put in place, the whole structure can be closed and no specific attachment surface be exposed so that the structure ceases to grow. A simple case illustrating the principle is shown in Figure 3. Here a cylinder is formed out of several units and it is seen that when two units that can complete the cylinder approach they firmly combine and there is no more growth. The shapes of biology are much more beautiful and subtle, but they have nonetheless evolved to be interrelated, and so structures that are also subtle and have the form of the living system assemble and complete themselves.

It is worthwhile to examine this process numerically. We can start with Eq. 1. If, in a phage-infected cell, the transcription of 1000 subunits has taken place and 100 are needed to complete a unit, the rate of collision is given by the product of diffusion constant, concentration, and effective radius times 2π. The diffusion constant of protein-sized molecules in the cell medium has been determined by Lehman and Pollard (1965) and is 10^{-7} cm^2/sec, and we can assume a radius for effective collision of 10 Å. The concentration is $10^3/10^{-12}$, the denominator being the effective volume of a bacterium. This is 10^{15} subunits per milliliter. The rate of collision is then

$$2\pi \times 10^{-7} \times 10^{-7} \times 10^{15}$$

and so each subunit becomes incorporated into a structure at about 60 per second—amply fast. However, we must consider two more related factors. If the accretion takes place with no dissociation, then nearly complete units will be formed that are very unfavorable in shape for the final completion and this will be very wasteful of the subunits. Thus some dissociation must occur, and this means that the overall time will be longer. With a free energy of binding of five times the energy of thermal agitation, the equilibrium constant for dissociation would be about 1%, so that an incomplete set of 99 subunits would be as likely as not to be dissociated. If the kind of process envisaged in Figure 3 is assumed to occur, the last binding process involves *two* surfaces attaching, with 10 times the energy of thermal agitation and a dissociation constant 100-fold smaller so the stability of the final product would be sufficient for the next stage. Thus the process of self-assembly, in a space in the cell in which the concentration of subunits is quite modest (micromolar), should not take more than ½ min, about the time observed in phage assembly. Incidentally, the effect of temperature on this process might be very revealing in terms of change in cellular viscocity and in dissociation.

One very interesting question that arises concerns the part played by this self-assembly process in the formation of a mitochondrion or a chloroplast. Both these organelles have part of their structure coded for by their own DNA, with special means of translation of the mRNA, and part of the structure determined by the chromosomal genetics of the cell. They must therefore generate some kind of skeleton structure that can divide, and on this structure the separately made protein components must be assembled. It would be very rash to attempt the whole description of this process here, because much has to be established before that can be done; however, it can be noted that in order to keep the concentration of subunit protein molecules as high as those in the virus-infected bacterium discussed above, very high rates of translation are needed. Also, the possibility that partially complete mitochondria should be observed, and actually are not, suggests that the assembly of new mitochondrial material must be taking place while the new skeletal frame is being made. This is a very interesting example of a two-fold growth process and one that offers great challenge.

Jehle (1970) has made the very interesting suggestion that the development of a whole organ occurs partly by this same means. The contact between surfaces need not be made by collisions, and for really large cells this is not a reasonable suggestion. However, the developing organ is made of growing cells which therefore have surfaces that move over one another as they grow. If the surfaces have several contact points with the exact shape, then the same consideration that makes it possible for the specific selection of a substrate by an enzyme can occur and the two surfaces can lock together to make a firm whole. This unit is capable of more growth and locking together, and so it is conceivable that an organ can develop. The change in shape of a protein molecule with conditions can also play a part here, for if the concentration of an effector molecule became such that an appreciable number of the contact molecules that cause binding were to change their shape the accretion of other cells would stop. This offers latitude for control of shape, which is briefly discussed later.

6. ENZYMATICALLY CONTROLLED STRUCTURAL GROWTH

Much harder to understand than the process of self-assembly is the growth of the peptidoglycan shell that surrounds bacteria and that forms a model for the growth of plant cell walls. Different from but clearly related to this is the growth of lipoprotein membranes. Speculation on these is possible, and in so speculating an inkling of what must be happening in the case of peptidoglycan can be gained.

The sacculum or cell wall of a bacterium is a single covalently bonded structure. If one stretches the usual meaning of the word "molecule" to include a covalently bonded structure of polysaccharide and peptides, then this "bag" is about the largest molecule encountered anywhere. It is larger than the bacterial chromosome molecule, of 2½ billion molecular weight. In the case of the DNA macromolecule, the mystery of its growth is somewhat lessened by the knowledge that it forms its own template and that this operates at a specific polymerase site which we have called the "growing point." For the peptidoglycan no such special process is available; instead we find ourselves invoking a myriad of separate enzymes that are engaged in breaking open the structure and inserting new elements into it in a very orderly, although apparently physically unrelated, way. In addition to this already physically obscure process is the requirement that the enzyme systems change their behavior just prior to division and initiate the formation of the septum that divides the cell. If one is willing to take the purely qualitative approach of isolating and characterizing the enzymes and substrates that do this remarkable job, then there is no problem. However, even when we know all about these enzymes in detail, the problem of the regulation of the process to produce the exact shape of the sacculum remains, and that is the problem involved with growth. In this instance it is far easier to suggest ways to control growth than it is to offer an explanation of how the growth takes place.

Since the peptidoglycan has a generally accurate but not detailed precise structure (unlike DNA which is exact to the atom) the enzymes that are at work on the synthesis and insertion of the new sections must be rather accurately distributed; however, the requirement of template synthesis, that the process wait if one component is not assembled until it is found and inserted, is not present in cell wall synthesis. Further, the precise structure of the peptidoglycan is determined by the genetics of the cell and so the enzymes must be transcribed and translated from the DNA. The spatial relationship of the newly synthesized enzymes varies during the cell cycle, and so it is not possible to rely on diffusion and local concentrations to deliver these enzymes to the proper location. In view of this impasse, the role of the lipoprotein membrane can be invoked, and while this discussion is speculative, it indicates the kind of thing that must be happening.

Lipids, with phosphate, or any polar endings have a strong advantage from the point of view of developing a sheetlike structure that can grow (or contract). Because lipids are formed from covalently bonded structures with full charge symmetry, they have no reaction to polar water molecules, except to occupy space and to compel the water to adapt. Thus lipids can line up alongside each other in a relatively orderly way, and if one more or less is included, the neighbors will adapt to include it or to take the space made available. Thus a synthesis of lipid near a lipid layer naturally results in the tendency for lipids to be incorporated into the existing layer, with a concomitant lessening of coercive action on water, meaning that the combination into the layer gives a lower free energy.

If in addition to this process there are proteins that are associated with the lipid, either by a combination with the polar end of the phospholipid, or by an association with the nonpolar residues of the protein, the result will be that the protein molecule will be carried into the upper or lower surface of the lipid layer. If such a molecule is also an enzyme that can catalyze the various processes necessary for the formation of the peptidoglycan cell wall, it can be seen how the seemingly impossible task of breaking and inserting into the cell wall can be achieved. A reasonable arrangement, and one that is often observed, is two bimolecular lipid "leaflets" with associated protein molecules above and below. In the space between these two layers the cell wall is located. The lipids secure a special concentration of the unit material needed for assembly of the wall; the enzymes are available because they are attached to the lipid, and the only unknown factor is the actual arrangement of the various enzymes necessary to achieve the breaking open of the network of peptidoglycan and the insertion of new units. Probably not more than five or six associated enzymes are needed, and it is conceivable that they might even be present on the lipid layer randomly, because if we also assume that the lipid is steadily growing, as it is, each enzyme will be carried along the surface of the cell wall by growth and will thus be able to achieve its function in a moderate amount of time.

To recapitulate this speculation, we suppose that long-chain lipids are synthesized close to a lipid surface. They naturally add to and enter the surface. In addition they are presumed to be capable of association with proteins of special sequence, either by a polar association or by the pulling together of hydrocarbonlike residues. The growth is then primarily of the lipid layer; associated with this is the protein. The suggestion is that two such layers confine a section of the cell in which peptidoglycan is made.

If this concept is adopted, it is relatively easy to see how the septum can form. If transcription of the DNA involves an orderly process, then just prior to the completion of the chromosome a transcription and translation of a modified enzyme system, also lipid-associated, can be assumed. This enters the cell in a rather high concentration and joins with the new lipid being synthesized. The two enter the lipid layer where there is no protein to block their entrance, namely, at the new synthesizing point, and there the peptidoglycan is broken into and undergoes insertion of a set of subunits that have the necessary geometry to initiate the growth of the septum across the normal direction. It is probable that the entry point for the new enzymes is not left to chance as suggested above, but that the inner proteins of the inner cytoplasmic membrane have an entry site specially designated for this process.

Because verbal descriptions are sometimes inadequate, Figure 4 is included as an aid. Warning should be given that this is a schematic diagram; it is intended to evoke thought, not to predict the exact form that will one day be seen. The peptidoglycan is shown in between the two membranes. It is necessary for some parts of the cell wall to be in contact with the enzymes, but this has not been shown. Also, the thickness of the

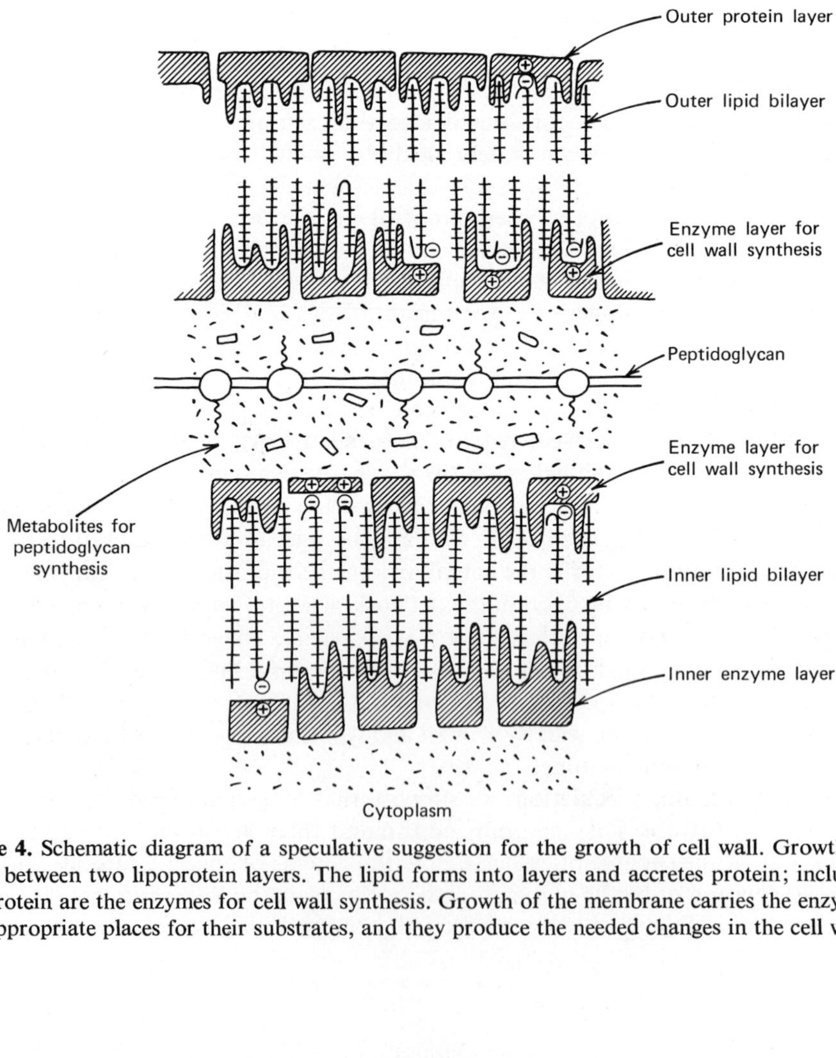

Figure 4. Schematic diagram of a speculative suggestion for the growth of cell wall. Growth takes place between two lipoprotein layers. The lipid forms into layers and accretes protein; included in the protein are the enzymes for cell wall synthesis. Growth of the membrane carries the enzymes to the appropriate places for their substrates, and they produce the needed changes in the cell wall.

cell wall is greater and the structure of the mucopolysaccharide is certainly much thicker. Nevertheless, the general idea can be gathered.

Note that for this part of the cell to grow the major requirement is the synthesis of the lipid molecules at a reasonable rate near the lipid bilayer and in such a way that they can enter the bilayer. As long as they can move by diffusion into this region, the hydrophobic action of the hydrocarbon guarantees their incorporation into the layer. As a few of these accrete, lipid becomes exposed. If there is enough protein with the requisite hydrophobic residues or the right relation to the charge, and if the lipids have polar ends, they will by random collision and specific selection arrive and take their place. Thus growth, to a moderately definite pattern, is achieved by no more than the operation of the protein and lipid synthesis mechanisms. Many variants of this can be devised, such as the transcription of different enzymes to develop longer-chain hydrocarbons, or branched hydrocarbons, when the septum must be produced. These examples are really not appropriate here. The mechanism suggested could work to generate a cell wall, and the shape of the wall could conceivably be under genetic control.

7. THE OVERALL CONTROL OF GROWTH PROCESSES

It is not sufficient to discuss the two methods of growth: that by self-assembly and that by association with a lipid organized membrane. One must also discuss the overall method used by the cell to control growth.

The major element that is controlled is enzyme action. Minor control is exerted on the accretion or dissociation of subunits in organelles, such as the pili. In any event the control involves the determination, by the surroundings, of the shape of a molecule. Since the physical method by which surfaces are brought together is Brownian movement, which cannot be stopped except by interposing membranes or a very large rise in viscosity, both of which are drastic and not easily reversible, the control method used is to permit the collisions which are bound to happen, but prevent the sticking together of the surfaces. Shape alternatives exist; they are not to be thought of as gradual modifications of shape. The forces that cause atoms to combine are very great. When a large number of atoms form a macromolecule, the resulting shape need not be unique, but the various shapes possible are formed in response to these large forces and so do not change gradually from one to another; they jump, or "flip" in a very definite way. Often the word "allosteric" is used to describe this phenomenon; it is useful, but it conveys the idea that some special proteins are allosteric while in reaility any sufficiently large molecule is bound to have several structural forms available, so that it can confidently be predicted that all proteins have several forms. This is amply documented, for example, in the study of the effect of temperature on enzymes.

Because the "flip" causes the macromolecule to assume a completely different configuration, its functioning is also different; it may lose one enzymic function and gain another. In particular, it may be able to undergo self-assembly in one configuration and be quite useless in another. All the means by which these conformational shifts can be achieved are used by the cell; temperature is one, and the range of growth and development of a cell can only occur within the temperature ranges that allow the macromolecules to perform their functions. Above and below this range, the conformations of some vitally needed molecules are different, and thus the cell cannot function. Hydrogen ion concentration and ionic strength are of great importance and can be used for control. These factors enter into environmental relationships and may also be locally created, in the case of doubly charged ions, within the cell, by release from organelles. None can operate on specially selected molecules and so they do not correct a special situation requiring only one enzyme. For this the small effector molecules serve a powerful function.

The effector molecules act at two levels. The simplest is direct binding to the free enzyme. Here one or more molecules can bind, and when the correct number are bound the enzyme changes shape. This action underlies feedback inhibition, in which the product of an enzymic chain acts as the effector molecule and when it is in excess causes a change in shape of one enzyme in the chain and stops further production. The next level is more drastic. Certain sections of the DNA are specially selected for the control of the production of protein. These sections have what amounts to giant effector molecules which can bind to the DNA at the precise region where the polymerase that reads the DNA chain to give the mRNA must start. While these repressors, which are protein macromolecules, are in place, the DNA cannot be transcribed. If one of these repressor molecules receives collisions with attachment of the correct number of effector molecules, termed "inducers," the repressor undergoes a conformational change, no longer binds to the DNA, and the process of transcription is now free to begin. This operation does much more than control an enzyme; it causes the generation of perhaps 10 new enzyme molecules in fair concentration. Thus the cell is truly different after this process has occurred.

Variants of this process can be readily envisaged. The cell can transcribe enzymes that catalyze the formation of effector molecules, and these in turn can derepress or induce. Quite clearly, wide swings in cell behavior can be produced. Indeed, what needs to be further studied is the overall behavior that can include these variations and still retain the identity of the cell.

In bacteria the process of transcription yields mRNA that is destroyed after some use. This seems to obviate the use of control at the level of translation. In eukaryotic cells the mRNA has a long life and it may well be that control is used at this level.

A third level of control has been mentioned; this is the development of new synthetic points for the chromosome. In bacterial cells the mechanism for this is not yet clear as it involves both the transcription of two molecules—the initiator and the DNA polymerase—and the attachment of these to the cell membrane. Thus, in the formation of the lipoprotein membrane, provision must be made for these special sites. It is most likely that the lipids of the membrane are not bland and uniform, but rather that the enzymes that make them are varied and that concentrations of differing lipids are produced in bursts at certain times. The study of these should be very rewarding.

8. CONCLUDING REMARKS

All of the above discussion is of only detailed interest. The absorbing problem is the explanation of how a proliferating set of cells can divide and grow to the shape required by an ear or a kidney and then stop. The explanation is certainly not going to be given here in an offhand manner, and yet it is a good exercise to see whether or not the mechanisms so far proposed could achieve that goal.

First, from the very limited considerations given here, there is no problem with respect to the development of different organs from the same cell. The cell must have, and does have, far more DNA than is necessary for transcription of the enzymes needed for maintenance. In mammalian cells it is even possible that fewer such enzymes are needed than in bacteria, and yet the whole genome is many times larger. If 90% of the DNA is repressed, and differing 10% lengths are expressed, then almost any cell can develop, within reason. Certainly, it seems that the whole problem of differentiation could be covered by this idea.

The second question is more subtle; it concerns the way in which a cell that has turned on the part of the chromosomes that may make it a kidney, and so achieved primary differentiation, can proliferate just to the point that develops the shape of the kidney and then stop. This is not easy to answer, because the very first consideration is whether the definite form of the kidney is "written in" by the chromosome, or whether the determination of the edge is made by the nature of the neighboring medium. At present, the only point of view that seems to be useful is that the primary shape of the organ is indeed encoded in the DNA, but that the encoding may very well involve a response to the nature of the medium surrounding the edge of the organ. In other words, if the cells, as they divide, gradually permit the transcription of enzymes that can undergo the "flip" to a new form, and at the same time the maturing intracellular medium changes so that it becomes more concentrated in the effector molecules that can cause the "flip" to occur, there could then develop a rather sharp boundary of the necessary shape. It is very likely that this is *not* all that is involved. This aspect of growth is still the subject of intense research.

REFERENCES

Cold Spring Harbor Symposium on Quantitative Biology, Vol. 35, (1970). *Translation of Genetic Material.*

Jehle, H. (1970). Bilateral symmetry in the morphogenesis of embryos. *Bull. Am. Phys. Soc., Ser. II*, **15** (11), 1335.

Lehman, R. C., and E. C. Pollard (1965). Diffusion rates in disrupted bacterial cells. *Biophys. J.*, **5**, 109.

Pollard, E. C. (1961). Physical aspects of protein and deoxyribonucleic acid synthesis. *J. Theoret. Biol.*, **1**, 328.

Pollard, E. C. (1963). Collision kinetics applied to phage synthesis, messenger RNA and glucose metabolism. *J. Theoret. Biol.*, **4**, 98.

Pollard, E. C. (1968). In *Formation and Fate of Cell Organelles* (K. B. Warren, ed.), pp. 291–303. Academic, New York.

Schrodinger, E. (1962). *What Is Life?* Cambridge Univ. Press, Cambridge.

Setlow, R. B., and E. C. Pollard (1962). *Molecular Biophysics*. Addison-Wesley, Reading, Massachusetts.

Verwey, E. J. W., and J. Th. G. Overbeek (1948). *Theory of the Stability of Lyophobic Colloids*, p. 100. Elsevier, Amsterdam.

von Smoluchowski, M. (1916). Drei Vorträge über Diffusion, Brownische Molekularbewegung und Koagulation von Kolloidteilchen. I. *Phys. Z.*, **17**, 557.

12

COORDINATION BETWEEN CELL STRUCTURE AND PROTEIN SYNTHESIS DURING GROWTH AND DEVELOPMENT

J. R. Tata

1. INTRODUCTION

Cellular structures in higher organisms are in a dynamic state of turnover and are not static entities. As cells differentiate, grow, and mature, the changing pattern of biosynthetic activity is accompanied by an increasing complexity of intracellular structures. Cell biologists are becoming increasingly aware that many processes of protein synthesis are related to structural elements of the cell, especially membranes of the endoplasmic reticulum (ER). However, relatively little is known about the relationship between the mechanisms that regulate growth and development via a control of RNA and protein synthesis, on the one hand, and the cellular structural requirements for such regulations to be effective, on the other.

There is good agreement among molecular biologists that the rapid modulation of messenger RNA (mRNA) synthesis as established in microbial enzyme synthesis is unlikely to be a key or sole mechanism for regulating protein synthesis in animal cells. It is now accepted that more than one type of mechanism must be involved at both the transcriptional and translational sites, many of which are unique to nucleated cells (see Tomkins et al., 1969). It is in the light of these possible mechanisms of higher organisms that have been recently postulated, and summarized below, that we discuss the question how changes in cellular structure, particularly the ER, are coordinated with those of protein synthesis during growth and development. Much of the experimental work described, in which hormones have been used to initiate or accelerate growth and development, has been performed in our laboratory; however, the phenomena described are common to most late embryonic growth and developmental processes.

2. GENERAL REGULATORY MECHANISMS OF PROTEIN SYNTHESIS IN ANIMAL CELLS

In molecular terms it is quite obvious that growth and development are best explained by a preferential or sequential synthesis of proteins. Numerous sites of control of protein synthesis in animal cells have been proposed in the last 5 yr and this number is bound to increase as more unforeseen phenomena appear. Some of the currently "popular" ones are summarized in Table 1.

Table 1. Some Important Processes Likely To Be Involved in the Regulation of Protein Synthesis in Animal Cells

- DNA synthesis
  - New cell population
  - Gene "amplification"
  - Mitochondrial DNA synthesis
- Gene activation and transcription
  - New genes unmasked
  - Increased rate of transcription
  - Intranuclear turnover of RNA
- mRNA utilization
  - Selective transfer of RNA from nucleus to cytoplasm
  - Cytoplasmic stability and activation of mRNA
- Regulation of translation
  - Polypeptide chain assembly and release
  - Polysome deployment and cell structure; attachment to membranes
- Regulation of degradation of proteins

Evidence that any of the above mechanisms may be rate limiting in protein synthesis *in vivo* is often circumstantial. However, the mechanisms discussed below are important because of their relevance to the involvement of cellular structural elements in the control of protein synthesis as well as cellular growth.

A. Nuclear Restriction of RNA

A large amount of the RNA synthesized in animal cells is ribosomal RNA (rRNA), with the result that much of the rapidly labeled RNA is ribosomal precursor RNA but not mRNA. It has also been established that a substantial amount of RNA synthesized in the nucleus turns over rapidly, has an extremely high molecular weight (60–120S) and is heterodisperse. This RNA may or may not be a precursor of messenger or any other cytoplasmic RNA (Scherrer and Marcaud, 1969; Penman et al., 1970). The function of this giant, labile nuclear RNA is not known, but some preliminary reports suggest that it may be of some importance during the processes of differentiation, growth, or enzyme induction in cells of higher organisms (see Britten and Davidson, 1969; Scherrer and Marcaud, 1969). However, before this can be ascertained it is important to know whether the heterodisperse RNA also known to hybridize very rapidly with DNA is made on unique DNA sequences (as much of mRNA would be expected to be) or on repeating sequences which are now known to exist in most animal cells. Recent DNA–RNA hybridization studies have shown that only a fraction of the species of RNA present in the nucleus is present in the cytoplasm, but the mechanism of this selective restriction of RNA to the nucleus remains unknown (Georgiev, 1966). However, the restriction must involve both a selective nuclease action and a mechanism in the nuclear membrane that somehow monitors the species of RNA transferred from the nucleus to the cytoplasm. It is known from the work of Shearer and McCarthy (1967) that qualitatively different populations of RNA molecules are transferred from the nucleus to the cytoplasm in fetal and adult tissues or at different times during regeneration of liver.

B. Nuclear-cytoplasmic RNA Transfer

The transport of RNA from the nucleus to the cytoplasm is a process that requires the formation and maturation of ribonucleoprotein (RNP) particles. From experiments with inhibitors of RNA and protein synthesis, it has been concluded that continued synthesis of both RNA and protein is essential for the processing and transfer of RNA from the nucleus to the cytoplasm. Two transfer mechanisms have been suggested. One is that mRNA enters the cytoplasm as a complex with subribosomal particles and that both components may then be directly assembled into polysomes (Henshaw et al., 1965; McConkey and Hopkins, 1965). The other poossibility is that free messenger does not complex with ribosomal precursors but that it is released into the cytoplasm as RNP particles, also called "informosomes" (see Spirin, 1969), which are then incorporated into polysomes. Such particles have been detected both in the nucleus and in the cytoplasm (Henshaw, 1968; Samarina et al., 1968). The role of the protein associated with mRNA would be to modulate the activity of the template. It seems that there are three major basic proteins associated with cytoplasmic informosome particles (Samarina et al., 1968; Olsnes, 1971). As regards a technical point concerning detection of informosomes in the cytoplasm, Olsnes (1971) found that one of the three proteins associated with these particles is destroyed by deoxycholate, a detergent commonly used to prepare polysomes. The detergent Triton X-100, however, does not have this deleterious effect, which may explain differences in the stability and activity of polysomes prepared with these two detergents. RNP particles, whether ribosomal or informosomal, must mature within the nucleus, but very little is known about the conditions governing the synthesis and interaction of proteins necessary for the continuous generation of the particles.

C. Structural Requirement for Protein Synthesis

It is well known that free polyribosomes with endogenous or synthetic messenger, RNA, effect the incorporation of amino acids into protein. However, in intact cells it seems that some form of structural or topological deployment of ribosomes is required in the synthesis of proteins. This may involve the attachment of ribosomes to cell membranes, as in lower organisms or reticulocytes (Hauge, 1967; Bulova and Burka, 1970), or to membranes of the ER, as in cells of higher organisms (Hendler, 1968; Campbell, 1970). It is of interest that such membranes are continuously formed and broken down according to the demands for protein synthesis (Siekevitz et al., 1968; Tata, 1968, 1970a). Unfortunately, much of the work done so far on the proliferation of the ER has been carried out on predominantly protein-secreting tissues such as the liver and pancreas. The well-known participation of both the rough and smooth ER (see Palade, 1966) in the process of transport of secretory proteins has therefore tended to mask other possible roles for the attachment of ribosomes to endoplasmic membranes, especially during growth. However, it is important to note that in bacteria and nonsecretory animal cells the attached ribosomes appear to be functionally very active in intact cells (Hendler, 1968; Andrews and Tata, 1968, 1971a, b). A possible role for the nonsecretory aspect of ribosome–membrane attachment during growth and development may be to achieve some form of topological segregation *in vivo*, a possibility that is discussed later (Section 5).

3. PROTEIN SYNTHESIS DURING GROWTH AND DEVELOPMENT

It is not necessary to stress that the protein synthetic machinery of the cell is activated during growth and differentiation. Except for a few situations, such as in the immediate development following fertilization of amphibian or sea urchin eggs, in most instances the actual increase in protein synthetic activity is anticipated by changes in the rate and nature of RNA synthesized in the nucleus and its transfer into the cytoplasm. Because recent work has exploited the growth and developmental properties of many hormones in developing experimental systems for studying developmental changes in higher organisms, much of the research described below is based on studies with such hormones. However, almost identical phenomena have been encountered with apparently nonhormonal growth systems such as regeneration (Tsukada et al., 1968), compensatory hypertrophy (Malt, 1969), or carcinogen- or drug-induced growth (Nebert and Gelboin, 1969). It is well established that the large majority of growth-promoting hormones when administered *in vivo* markedly affect protein synthesis *in vivo* or *in vitro* at the level of the ribosomes of the target cell, but that they fail to influence amino acid incorporation when added *in vitro* to the ribosomes (see Korner, 1965; Tata, 1967a, 1970b). This finding prompted investigations of RNA metabolism and soon led to observations that RNA synthesis in target cells is very sensitive to growth and developmental hormones, and experiments with inhibitors of RNA synthesis revealed that part of the RNA synthesized under hormonal influences is essential for the biological expression of hormonal activity (Tata, 1966). In nonhormone-induced growth and development also, an important involvement of RNA synthesis is observed.

A. RNA Synthesis

RNA synthesis during rapid growth, hormonally induced or otherwise, has been studied in two ways: (1) by following the incorporation of radioactive precursors into nuclear RNA *in vivo* following hormonal stimulation *in vivo;* and (2) by examining the effect of hormones administered *in vivo*, or added *in vitro*, on isolated nuclei or chromatin.

A rapid stimulation of nuclear RNA synthesis in the target cells is now a well-known feature of the administration of growth and developmental hormones to immature or hormone-induced animals or at the onset of rapid growth. The relatively rapid effect on RNA synthesis, anticipating that on protein, has been observed in a variety of hormone-induced growth systems (see Tata, 1966, 1967a, 1970b; Hamilton, 1968). This feature is illustrated in Figure 1 for the fast-acting estrogen, as well as for the slower effects of thyroid hormone. Stimulation of synthesis of rapidly labeled nuclear RNA of the uterus can be observed within minutes after the administration of estrogen and the relatively delayed stimulation by thyroid hormone is still one of its earliest actions.

Evidence for the regulation of RNA synthesis, independently of hormonal effects on precursor pools *in vivo*, can be demonstrated by a stimulated RNA polymerase activity in isolated nuclear preparations, or when chromatin template activity is measured (see Kim and Cohen, 1966; Tata, 1966a; Pogo et al., 1967). As a rule, stimulation of the capacity of isolated nuclei or chromatin to synthesize or to prime the synthesis of RNA *in vitro* is manifested after an increase in nuclear RNA *in vivo* has taken place (Fig. 1). In rapidly growing or maturing tissues, the stimulation of RNA polymerase assayed at low ionic strength was in every instance more rapid and more pronounced than when the polymerase was assayed at high ionic strength or when the isolated chromatin was transcribed by a bacterial polymerase. The significance of this finding lies in the fact that the product of the first reaction is rich in rRNA and that of the other two DNA-like or perhaps mRNA, so that the transcription of rRNA cistrons is more sensitive to the stimulus for growth than that of other cistrons.

As regards the nature of additional RNA, the synthesis of all classes of RNA is accelerated in rapidly growing tissues (Tata, 1966; 1967a; 1970b; Hamilton, 1968), At some stage of hormonal control of specific protein synthesis, as in hormone-induced differentiation (O'Malley et al., 1969) or metamorphosis (Cohen, 1970; Tata. 1970c), mRNA must play a key role. Shearer and McCarthy (1967), in studying the nature of RNA synthesized during mouse liver regeneration, found that the species of RNA synthesized changed very rapidly following hepatectomy. For a brief period at the early stages of regeneration, these workers found by DNA–RNA hybridization that the species of nuclear RNA synthesized resembled more those found in fetal liver than in adult tissue.

With most growth and developmental hormones, at the time of a peak shift in the rate of RNA synthesis, the major effect is on the synthesis of rRNA (see Tata, 1966; 1967a, 1970b; Hamilton, 1968). It is interesting to note that, in most of the observations on the abolition by actinomycin D of the growth-promoting or even metabolic actions of hormones, the antibiotic is very effective at doses too low to block all RNA synthesis. Synthesis of rRNA is inhibited to a greater extent than that of DNA-like RNA by low doses of the antibiotic (Perry and Kelly, 1970). However, at early time intervals after hormone administration or the onset of development, when the overall rate of RNA synthesis is unaffected, it is likely that there may be a subtle qualitative change in the nature of RNA synthesized without much effect on rRNA synthesis. It is important to realize that the nature of highly hybridizable RNA from the nucleus is still in doubt. Much of it may not be mRNA but represent the product of highly repetitive sequences of genes whose function is as yet unknown. Britten and Davidson (1969) have recently suggested that this type of RNA may be of considerable importance in differentiation and postdifferentiation adaptation processes.

Although very little is known about the possible effects of growth stimuli on the selective transfer of RNA from the nucleus to the cytoplasm, an exaggerated acceleration of the rate of rRNA synthesis may perhaps facilitate the transfer of ribosomes and mRNA from the nucleus to the cytoplasm. It may be that the hormone or some other growth factor triggers off a qualitative or quantitative change in the synthesis of RNA but that it does not control the eventual selection and transfer of the different species of RNA made. Dingman et al. (1969) found that in the chick oviduct, estrogen,

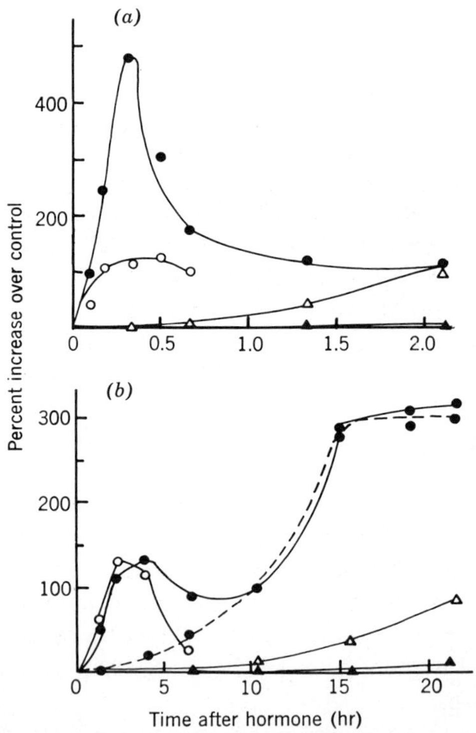

Figure 1a and b. Time course of stimulations of specific activity of nuclear RNA labeled *in vivo* and RNA synthesis by isolated nuclei in (*a*) the uterus of ovariectomized rats after a single injection of estradiol-17β, and (*b*) the liver of a thyroidectomized rat following a single injection of 3,3′,5-triiodothyronine. Labeling *in vivo* was achieved by 2- to 10-min pulses of (*a*) [^{3}H] uridine and (*b*) [^{14}C] orotic acid. ○, Specific activity of UMP; ●, rapidly labeled nuclear RNA (dashed line represents values obtained after correcting for the increase in acid-soluble fraction); △, RNA polymerase assayed at low ionic strength (product mainly RNA); ▲, RNA polymerase assayed at high ionic strength (product mainly DNA-like RNA). Note the differences in the time scales for the two hormones. Curves derived from data of Hamilton (1968) and Tata and Widnell (1966).

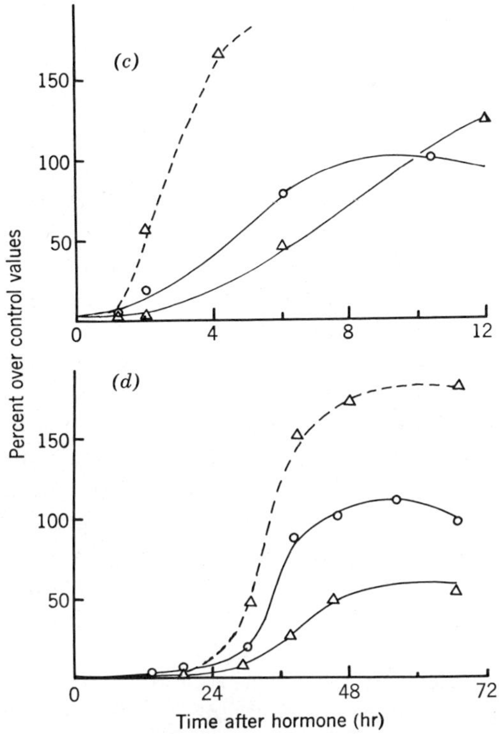

Figure 1c and d. Increase in protein synthetic capacity of cytoplasmic ribosomes coinciding with the appearance of additional polysomes formed as a result of hormone administration. (*c*) Effect of a single injection of estradiol on ovariectomized rat uterus; (*d*) effect of a single injection of triiodothyronine on thyroidectomized rat liver. △, Increase in amount of cytoplasmic ribosomal population (dashed line indicates the increase in radioactivity of the newly labeled polysomes; ○, amino acid incorporation *in vitro* expressed per unit RNA. Curves derived from data of Hamilton (1968), Teng and Hamilton (1967), and Tata and Widnell (1966).

which has a marked effect on the pattern of RNA synthesized in the nucleus (O'Malley et al., 1969), provoked only minor changes in cytoplasmic RNA when it was analyzed by acrylamide gel electrophoresis.

As regards the process of transfer itself, growth hormone was found to accelerate the rate of appearance of 40–50S particles in rat liver just prior to stimulation of protein synthesis (Sells and Takahashi, 1967). These particles were first thought to be precursors of polysomes and to contain mRNA attached to the small ribosomal subunits but it now seems that they may also contain informosomes or messenger RNP particles (Spirin, 1969). Further work is needed to determine whether or not the process of transfer of RNA from the nucleus to the cytoplasm per se is of critical importance during growth and development.

B. Cytoplasmic RNA and Ribosomes

Soon after the enhancement of nuclear RNA synthesis by growth stimulus, a gradual build-up of cytoplasmic RNA and ribosomes occurs. In immature tissues or in tissues of animals deprived of the appropriate hormone, the increase is very rapid and often most impressive. Growth hormone treatment can almost double the population of hepatic polyribosomes in hypophysectomized rats, and androgen and estrogen can cause a four- to eightfold increase in cytoplasmic RNA of their target accessory sexual tissues of castrated rats. To a large extent the accumulation of cytoplasmic RNA is a consequence of enhanced RNA synthesis in the nucleus, but there may also be a concomitant stabilization or reduction in turnover of RNA. For example, Brewer et al. (1969) found that RNase activity of the postmitochondrial fraction of hypophysectomized rat liver is double that in normal animals, and that treatment of hypophysectomized animals with growth hormone reduced it to normal levels. In thyroid hormone-induced metamorphosis of bullfrog tadpoles, the degradation of cytoplasmic ribosomes was found to be selective in that the population accumulating after hormone treatment was more stable; at the same time, however, the breakdown of the "older" ribosomes was accelerated (Tata, 1967b).

Accompanying the appearance of additional newly formed RNA in the cytoplasm following hormonal induction is a gradual change in the polyribosomal profile. In virtually every instance of a growth and developmental system, it has been shown that withdrawal of the hormone or cessation of growth causes a shift to smaller aggregates and an increase in the monomeric ribosomal peak, whereas hormone administration causes a shift in the opposite direction (see Korner, 1965; Tata, 1966, 1967b; Teng and Hamilton, 1970). From the earlier studies on amino acid incorporation in which the response to exogenous synthetic messenger was assayed, it was concluded that the hormone corrected a deficiency of messenger content in the cytoplasm. However, later studies have shown that this does not constitute the principal mode of regulation of cytoplasmic protein synthesis. There is good evidence now to show that hormone deficiency leads to an inherent defect in the capacity of the ribosome itself and that hormone replacement corrects the defect, either directly or through the generation of a new and more active population of ribosomes (see Tata, 1967a).

The relevance of the accumulation or alteration of cytoplasmic RNA to the question of hormonal regulation of protein synthesis becomes most obvious when the timing of changes in protein synthetic activity is taken into account. For many hormones, including androgens, estrogens, thyroid hormones, and growth hormone, or during liver regeneration or kidney hypertrophy, the rate of protein synthesis increased rather abruptly following a relatively long lag period after hormone replacement, hepatectomy, or unilateral nephrectomy (Tata, 1967a; Tsukada et al., 1968; Malt, 1969). The onset of increased protein synthetic activity coincides with the accumulation in the cytoplasm of the hormone-induced newly formed ribosomes. This feature is also illustrated in Figure 1 for the stimulation of growth of the uterus by estrogen and that of liver by triiodothyronine. Similar results were observed with triiodothyronine-induced metamorphosis of bullfrog tadpoles (Tata, 1967b) and hepatic and renal hypertrophy following partial hepatectomy or unilateral nephrectomy (Tsukada et al., 1968; Malt, 1969). It appears therefore that a large part of the effect of the hormone or growth stimulus on protein synthesis is incorporated into the new ribosomes formed after hormone treatment, but it is not certain whether the ribosomes play a passive role as mere transporters of mRNA or whether their intrinsic amino acid incorporation activity is also under hormonal influence. In any case it seems unlikely that hormones control protein synthesis merely by increasing the number of ribosomes, although this would certainly be important for a sustained growth action. Recent work from our laboratory has also underlined the importance of distribution of ribosomes in the cytoplasm, which is discussed in Section 4.

C. Protein Synthesis and Translational Control

Growth and development hormones do not act *in vitro* at the level of polypeptide chain assembly, and from the above discussion it seems that the change in amino acid incorporation activity may be restricted to a fraction of newly formed ribosomes. The concept of translational control has also been recently invoked to account for a few situations in which a hormone brings about a rapid anabolic change or induces specific enzyme synthesis.

Much of the evidence in favor of a direct translational control by growth stimuli is indirect and often based on the differential effect of inhibitors of RNA and protein synthesis. Actinomycin D fails to completely suppress the formation of a rapidly labeled soluble protein in the uterus 30 min after the administration of estrogen. Similarly, the early part of growth hormone effect on amino acid incorporation into protein is insensitive to the antibiotic and therefore independent of RNA synthesis (Clemens and Korner, 1970). Such findings have often been interpreted to mean that the hormone directly controls the synthesis of a single or few specific proteins which have a very short half-life (usually a few minutes) but are synthesized on relatively stable templates. As a consequence of such a preferential synthesis, the tissue may further respond with massive RNA synthesis. However, it is important to establish the identity of such proteins, as the hormonal effect would not be detectable on the basis of an overall incorporation of amino acids into protein.

Although sweeping generalizations are bound to be incorrect, it can be said that the concept of a direct translational control of the existing ribosomes, mRNA, tRNA, factors, and so on, is more applicable to the induction of specific enzymes than to sustained growth and development. Tomkins has invoked translational control in the induction by cortisol or dexamethasone of tyrosine transaminase in cultured human hepatoma cells (Tomkins et al., 1969), which involves the interaction of the hormone with a cytoplasmic translation repressor. There is no doubt that the bacterial model of modulation of mRNA levels is too simple to account for the complexities of highly structured animal and plant cells. However, Kenney and his colleagues have argued that the Tomkins model has been too much based on differential effects of actinomycin D at low and high doses in cultured hepatoma cells and that the situation may be quite different in the induction by cortisol of tyrosine transaminase in intact rat liver (Lee et al., 1970).

What must be considered in discussing translational control is a distinction between a relatively early response of the existing protein-synthesizing apparatus to a growth stimulus and a late and lasting adjustment of a rapidly accumulating population of newly formed ribosomes. There is a likely possibility that the early and late responses may arise from different actions of the hormone. It may also be that one does not necessarily lead to the other. Both these possibilities deserve attention in interpreting sequential phenomena. Some recent work on the control of protein synthesis by growth hormone is most compatible with a dual temporal action. Growth hormone has a very rapid effect on the transport of amino acids, especially into muscle (Kostyo, 1968). An increased availability of amino acids has been shown to increase the activity of ribosomes, according to Clemens and Korner (1970). These investigators have recently shown that growth hormone administration very rapidly increases the ratio of active/inactive ribosomes which also exhibit a concomitantly altered sensitivity to amino acids added in excess. Kostyo (1968) has suggested that the early action that may involve the synthesis of a few specific proteins is not directly related to the slow, overall enhancement of protein anabolism by growth hormone. It is possible, however, that increased transport of amino acids may have a "permissive" action in a later, sustained increase in the rate of protein synthesis. In contrast to this rapid "conditioning" of ribosomes, it is also known that growth hormone enhances the rate of RNA synthesis at a later stage, followed by an almost doubling of the population of ribosomes in the liver of a hypophysectomized rat (Widnell and Tata, 1966). The "new" population of ribosomes shows a higher activity, and at the same time there occurs a rearrangement of the membrane-bound and free polysome fractions, a phenomenon later discussed in detail (Section 4.A, B).

4. THE ROLE OF CELL STRUCTURE IN PROTEIN SYNTHESIS DURING GROWTH

It has been estimated that protein synthesis in cell-free systems has an efficiency less than 1% of that *in vivo*. Thus the disintegration of the structure of the cell must cause the loss of some vital factor(s) for protein synthesis. The importance of cellular architecture in the control of protein synthesis is now being realized (see Hendler, 1968), but it has not yet drawn much attention in problems concerning regulation of protein synthesis during growth and development.

A. Protein Synthesis by Membrane-Bound and Free Polyribosomes of Secretory and Nonsecretory Cells

In almost all animal and plant cells, a certain proportion of ribosomes is attached to cellular membranes or other structural elements, the proportion varying greatly from one type of cell to another (see Fawcett, 1966; Hendler, 1968). Thus far, the major role assigned to this attachment involves the facilitation of protein transport for export, and indeed it is generally accepted that all proteins for secretion are synthesized exclusively on membrane-bound ribosomes or the rough ER (Palade, 1966; Campbell, 1970). It is, however, virtually impossible from these studies to ascribe another role to the ribosome–membrane attachment since they have been exclusively concerned with the vectorial discharge of proteins synthesized on rough ER in predominantly protein-secreting tissues.

That there may be some other function, in addition to secretion, underlying the association between membranes and ribosomes is suggested by the alteration in ribosome distribution on membranes and by the marked proliferation of rough ER during growth and development, as discussed later (see Section 4.C). In both secretory and nonsecretory cells, the proliferation of membrane-bound ribosomes during rapid growth and development is accompanied by the preferential synthesis of intracellular proteins. We have suggested elsewhere that another role for ribosome–membrane attachment may be to achieve topographical segregation of different groups of ribosomes synthesizing different classes of proteins (Tata, 1968b; 1970 a, b). However, as a prerequisite to testing such an idea, it was essential to compare the activity of membrane-bound and free ribosomes of secretory and nonsecretory tissues.

Andrews and Tata (1971a) have recently described methods for the isolation of free and membrane-bound ribosomes from two predominantly nonsecretory tissues of the rat: cerebral cortex and skeletal muscle. Although microsomes and free polyribosomes have been prepared previously from these two tissues, there is no study to date that separates and compares the behavior of membrane-bound and free polyribosomes without the use of detergents. In addition, the protein-synthesizing activity of the bound and free preparations obtained from these tissues was compared with those from liver, for which several preparatory methods have already been described (see Tata, 1969). Particular emphasis was placed on a comparison of the vectorial discharge of nascent protein from membrane-bound ribosomes from liver with the discharge of those obtained from the two nonsecretory tissues. This study was based on the method first described by Redman and Sabatini (1966) for rat liver, by which they showed that much of the newly formed protein, synthesized *in vivo* or *in vitro*, could be discharged by puromycin from membrane-bound ribosomes into the lumen of the microsomal membrane. This fraction of the protein, which is recovered by solubilization with a detergent following puromycin incubation, is thought to be that destined for secretion. Membrane-bound ribosomes from brain and muscle therefore offered an opportunity to determine whether or not proteins made on membrane-bound ribosomes from nonsecretory tissues were handled differently than those from secretory tissues. The results indeed showed that both brain (cerebral cortex) and muscle contain small amounts of membrane-bound ribosomes which differ in a major way from membrane-bound ribosomes from liver in their modes of discharge of newly synthesized polypeptide chains.

In designing the experiments on puromycin-induced release of nascent protein, the main considerations are as follows. If all protein synthesized on membrane-bound ribosomes is transported through the membrane into the vesicle, then a total sequestration of radioactive nascent chains within the particulate fraction should be observed until the membrane is disrupted by deoxycholate. The same pattern of distribution should be observed for membrane-bound ribosomes from secretory and nonsecretory tissues. If, however, membrane-bound ribosomes from secretory and nonsecretory tissues function differently, then a different pattern of distribution of radioactivity between particulate and supernatant fractions should be observed when they are treated successively with puromycin and deoxycholate. On the basis of earlier observations of Redman and Sabatini (1966), only a negligible amount of incorporation occurs into structural proteins of the endoplasmic membrane.

Table 2 shows that after 8 min of incubation puromycin caused the release of a large fraction of radioactivity incorporated into protein *in vitro* by free polysomes of brain (64%), liver (35%), and muscle (46%). The differences in the extent of discharge may reflect tissue-specific differences in the access of puromycin to aminoacyl tRNA binding on the ribosomes. Puromycin releases a substantial part of the nascent protein from free ribosomes of all three tissues directly into the supernatant (Table 3). With membrane-bound ribosomes there is a distinct difference between preparations from the liver on the one hand and those from brain and muscle on the other. Puromycin alone released only 13% of labeled protein from liver preparations, but treatment with puromycin followed by deoxycholate released 55%. In other words, whereas puromycin caused the release of 13% of nascent protein from liver membrane-bound ribosomes directly into the surrounding medium, 18% passed through the membrane to the interior of the vesicle. In contrast, puromycin followed by deoxycholate failed to release more protein than the sum of the individual effects of deoxycholate and puromycin on brain and muscle membrane-bound ribosomes. This difference between membrane-bound ribosomes from secretory and nonsecretory tissues is unlikely to be attributable to the different types of mRNAs associated with them since the same pat-

Table 2. The Effect of Puromycin and Deoxycholate on the Fraction of [^{14}C]Protein Labeled *In Vitro* with [^{14}C]Leucine Retained with the 105,000 × *g* Particulate Fraction[a]

				Total [^{14}C]Protein (%) recovered in 105,000 × *g* sediment after			
Fraction	Tissue	Number of experiments	[^{14}C]Protein range (cpm/mg RNA)	Control	DOC	Puromycin	Puromycin then DOC
Membrane bound	Liver	10	3,200–18,400	83	64	75	37
	Brain	8	1,739–13,960	86	72	37	27
	Muscle	6	1,550–10,400	78	71	36	32
Free ribosomes	Liver	4	9,400–19,000	91	86	59	53
	Brain	8	2,080–14,700	88	83	31	27
	Muscle	6	1,700– 9,350	83	80	47	42

[a] Incubation mixture (1 ml) contained the following amount of free and membrane-bound ribosomes. Liver, 0.5–1.0 mg RNA; brain, 0.1–0.5 mg RNA; muscle, 0.1–0.3 mg RNA. Where indicated, 10^{-3} *M* puromycin was added after 8 min of incubation which was continued for another 8 min. At the end of this period, half the samples were treated with 0.5% (w/v) sodium deoxycholate and the 105,000 × *g* particulate fraction separated from the supernatant. The fraction of [^{14}C]protein recovered in the pellet was then measured. Each value is the mean of 4 to 10 experiments with a S.D. of ±5%. (From Andrews and Tata, 1971a.)

Table 3. The Release of ^{14}C-Protein Labeled *in Vitro* from Membrane-Bound and Free Ribosomes from Liver, Brain, and Muscle, by Treatment with Puromycin and Deoxycholate[a]

		Particulate [^{14}C]protein released (% untreated control) by			
Fraction	Tissue	Deoxycholate a	Puromycin b	Puromycin and deoxycholate c	c − (b + a)
Membrane bound	Liver	24	13	55	+18
	Brain	15	56	70	−1
	Muscle	8	54	60	−2
Free ribosomes	Liver	9	35	44	0
	Brain	7	64	69	−2
	Muscle	4	46	51	+1

[a] In this table the results of the experiments described in Table 4 are expressed as the fraction (percent) of the [^{14}C]protein recovered in untreated control microsomes or ribosomes, that is, released after treatment with 0.05% deoxycholate, 10^{-3} *M* puromycin, or puromycin followed by deoxycholate. (From Andrews and Tata, 1971a.)

tern was obtained with the discharge of [^{14}C]polyphenylalanine synthesized in the presence of polyuridylic acid. When liver and brain proteins were labeled *in vivo*, results similar to those observed with *in vitro* labeling were obtained. Results obtained with prolonged incubation *in vitro* without puromycin (Fig. 2*a* and *b*) suggested that the antibiotic was merely accelerating a normal process and not introducing an abnormal situation. In incubating membrane-bound ribosomes up to 1 hr, rat liver microsomes incorporated more radioactive leucine than did those from brain or muscle, but the fraction of [^{14}C]protein released into the supernatant was much higher for the particles from nonsecretory tissues. This difference was further emphasized when the behavior of free polysomes from liver and brain was found to be identical (compare Fig. 2*a* and *b*); in both instances 25% of newly synthesized polypeptides were released into the supernatant after 60 min of incubation. Figure 3 summarizes the differences in vectorial discharge of proteins synthesized on membrane-bound ribosomes of secretory and nonsecretory tissues. There is no doubt that *in vivo* the major role of attachment of ribosomes to membranes of the ER of predominantly secretory tissues is to facilitate the secretion. We have no direct evidence concerning the function of bound ribosomes in nonsecretory tissues, however, the

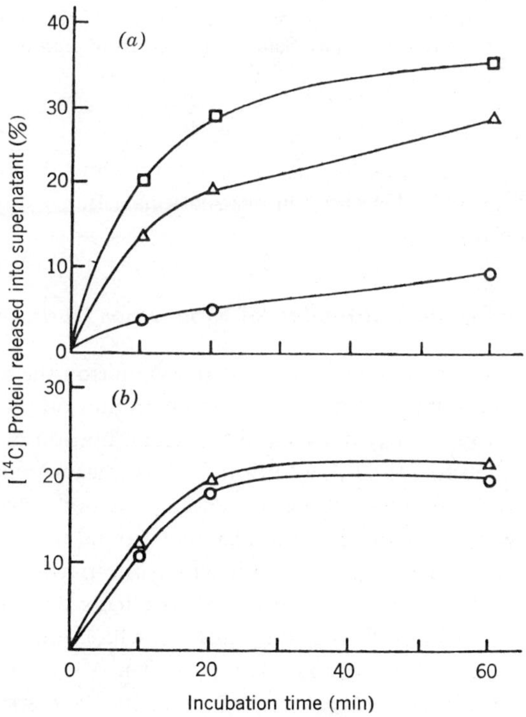

Figure 2. Relationship between time of incubation and the vectorial discharge of nascent protein labeled *in vitro* by membrane-bound ribosomes (*a*) and free ribosomes (*b*) from rat liver (○), brain (△), and muscle (□). The incorporation of [^{14}C] leucine into protein by the different ribosomal preparations was carried out as described for the puromycin-induced discharge experiments of Table 2 except that puromycin or sodium deoxycholate was not used. The fraction of total ^{14}C-labeled protein appearing directly in the 105,000 × *g* supernatant is reported here based on values averaged from two experiments. (From Andrews and Tata, 1971a.)

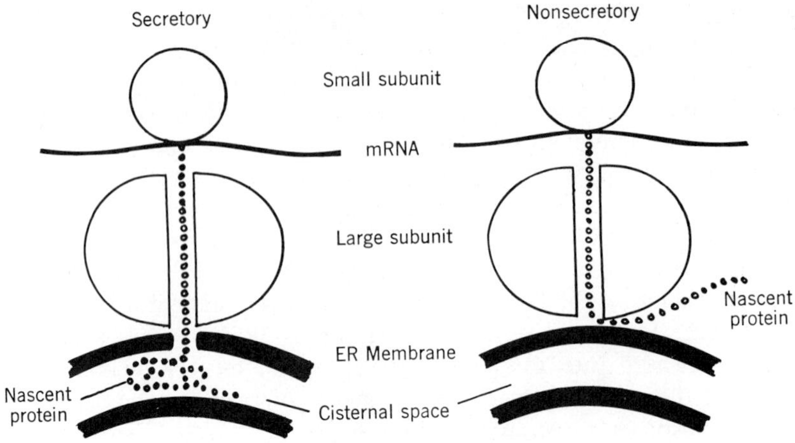

Figure 3. Schematic representation of the difference in vectorial discharge of nascent protein synthesized on membrane-bound ribosomes from secretory and nonsecretory tissues. Only the protein synthesized by secretory cells for export is depicted for these cells. The scheme is based on experiments in which nascent protein was released by puromycin followed by treatment with deoxycholate to determine the extent to which the newly synthesized polypeptides released were discharged directly into the soluble cytoplasmic fraction or into the membrane vesicles. Membrane-bound ribosomes from nonsecretory tissues were thus found to behave in this respect in the same way as free ribosomes from both secretory and nonsecretory tissues. Scheme derived from data presented by Andrews and Tata (1971a).

marked changes in ribosome attachment to membranes and the rapid proliferation of the rough ER during growth, described later (Sections 4.B, C), suggest a nonsecretory function, even in secretory tissues.

B. Cytoplasmic Distribution of Ribosomes During Growth

The importance of structural considerations in evaluating the action of growth and developmental hormones first became obvious from studies on the timing of hormone-induced acceleration of protein synthesis and the accumulation of newly formed RNA in the cytoplasm. At the same time as additional ribosomes were found to accompany the stimulation of protein synthesis it was observed that polysomes from mammalian and amphibian cells were more firmly bound to microsomal membranes in hormonally stimulated tissues as judged by the ease with which the particles were released from membrane complexes with detergents. Both these effects are illustrated in Table 4 for hepatic ribosomes from bullfrog tadpoles in which metamorphosis was precociously induced with triiodothyronine (Tata, 1967b).

A firmer attachment of this type also caused a shift in the distribution of ribosomes from the free polysome fraction to those attached to rough ER membranes as the predominant type. Qualitatively similar, but less pronounced, changes were also observed in the liver of young, thyroidectomized, hypophysectomized, or adrenalectomized rats whose rate of growth was stimulated by growth hormone and thyroid hormone (Tata and Widnell, 1966; Tata and Williams-Ashman, 1967; Cox and Mathias, 1969).

Table 4. Effect of Inducing Metamorphosis on the Relative Distribution of Newly Synthesized RNA in Membrane-Bound and Free Ribosomal Fractions from Liver Mitochondria-Free Supernatants Treated with Different Amounts of Sodium Deoxycholate[a]

Tadpoles[b]	Concentration of sodium deoxycholate (%)	Specific activity (cpm/ϵ_{260}) and distribution (%) of RNA in fractions[c]			
		Membrane-bound ribosomes	Polysomes	Dimers + monomers[d]	50S
Control	0.08	295 (4.7)	227 (21.5)	107 (56.6)	58 (12.0)
	0.16	418 (2.0)	254 (22.8)	92 (53.0)	41 (8.8)
	0.30	— (0)	238 (18.7)	122 (62.5)	73 (14.5)
6 Days after 0.7 μg T_3	0.08	875 (20.5)	430 (33.5)	492 (26.0)	196 (10.5)
	0.16	990 (12.8)	518 (35.1)	631 (32.9)	281 (15.1)
	0.30	1185 (1.5)	550 (25.8)	675 (29.0)	255 (11.9)

[a] Microsomal RNA was labeled by injecting 11.5 μCi of [^{3}H]uridine per tadpole 22.5 hr before death. Equal amounts of mitochondria-free supernatants were treated with different quantities of sodium deoxycholate indicated by their final concentration before sucrose density gradient centrifugation. Fractions of 1 ml were then pooled, after determination of ϵ_{260}, into the four fractions. The membrane-bound ribosomal fraction was the one that did not sediment through the 2 *M* sucrose interface. (From Tata, 1967b.)
[b] T_3, 3,3′,5-triiodothyronine.
[c] Figures in parentheses indicate the percent of total radioactivity delivered on the gradient that was recovered in each of the fractions.
[d] Also includes ribosomal precursor particles smaller than 78S monomers.

In the next series of experiments it was decided to compare the activities *in vitro* of bound and free ribosomes from tissues of animals deprived of or treated with growth-promoting hormones. That it is not necessary to have membrane-attached ribosomes *in vitro* to reflect a hormonal stimulation *in vivo* has been quite well established in numerous studies on ribosomes prepared with detergent treatment (see Korner, 1965; Tata, 1967a; Hamilton, 1968). If, however, hormonal stimulation causes ribosomes to be redistributed in the intact cell between free and membrane-bound fractions, it becomes imperative to prepare submicrosomal fractions *without the use of detergents* and under very mild conditions which cause minimal damage to bound ribosomes. A method based on differential centrifugation was devised, which permitted the separation of microsomes into at least four major fractions: smooth membranes, light rough membranes, heavy rough membranes, and free polysomes (Tata, 1969). The difference between the light and heavy rough microsomes, as revealed by chemical analysis and electron microscopy, is that the ratio of bound ribosomes per unit of membrane is higher in the latter fraction. Table 5 summarizes the results of some experiments on the effect of hypophysectomy and combined treatment with growth hormone and triiodothyronine on the capacity of submicrosomal fractions to incorporate amino acids into protein, both with endogenous and synthetic mRNA (Tata and Williams-Ashman, 1967). When the ability to incorporate [^{14}C]phenylalanine and [^{14}C]isoleucine was measured in the presence and absence of synthetic messengers (polyuridylic acid and a copolymer of uridylic and adenylic acids), it was found that

Table 5. Incorporation of [^{14}C]Phenylalanine and [^{14}C]Isoleucine into Protein by Free Polysomes and Polysomes in Light and Heavy Rough Membrane Fractions of Liver Microsomes from Normal and Hypophysectomized Rats with or without Hormonal Treatment: Response to Poly U and Poly UA[a]

Rats[b]	Polysomes	Amount of RNA (mg/g)	[^{14}C]Phenylalanine incorporated (cpm/10 min/mg RNA) −Poly U	[^{14}C]Phenylalanine incorporated (cpm/10 min/mg RNA) +Poly U	[^{14}C]Isoleucine incorporated (cpm/10 min/mg RNA) −Poly UA	[^{14}C]Isoleucine incorporated (cpm/10 min/mg RNA) +Poly UA
Normal	Free	0.57	13,590	38,090	5,320	6,383
	Light membranes	0.51	18,620	41,840	3,590	5,748
	Heavy membranes	1.62	58,800	128,400	23,800	29,980
Hx	Free	0.44	6,700	30,520	3,120	6,150
	Light membranes	0.46	14,550	44,250	3,560	5,036
	Heavy membranes	0.59	40,100	102,900	18,550	21,190
Hx + HGH + T_3	Free	0.66	26,700	86,300	10,900	20,720
	Light membranes	0.47	22,630	101,130	6,980	11,590
	Heavy membranes	1.40	66,000	144,000	32,100	45,700

[a] Hormonal treatment consisted of three injections of 75 μg of human growth hormone at 21, 47 and 74 hr before death and two injections of triiodothyronine at 47 and 74 hr. Light and heavy rough membrane fractions consist of ribosomes attached to membranes of the ER and differ from one another in the density of ribosome packing. (From Tata and Williams-Ashman, 1967.)

[b] Hx, hypophysectomized; HGH, human growth hormone; T_3, 3,3′,5-triiodothyronine.

hormone treatment corrected the lowered incorporation capacity caused by hypophysectomy in all three types of submicrosomal fractions. The effect of hormone deprivation or replacement was of greater magnitude in the heavy rough membranes than in the light rough membranes. To a large extent the response to synthetic messengers of both the membrane-bound and free ribosomes paralleled that to endogenous messenger, whether growth had been retarded or accelerated. Under the conditions of the rather prolonged hormonal treatment used in these experiments, the population of hepatic ribosomes was 50–60% higher than in the control hypophysectomized rats. Thus there is a distinct possibility that the ribosomes newly formed upon hormone administration were inherently more efficient in handling synthetic or natural messengers. At the same time the relative stimulation by synthetic polynucleotides was slightly higher in the less active preparations, which indicated that a small part of the hormonal effect could result from an increase in the polysomal mRNA content. This type of result raises the possibility that the relatively slow anabolic and developmental actions of hormones may be initially restricted to a small population of newly generated ribosomes rather than modify the activity of all the ribosomes in the cytoplasm.

C. Coordination Between the Proliferation of Microsomal Membrane and Ribosomes and Accelerated Protein Synthesis *in Vivo*

The above findings on the distribution of ribosomes and their amino acid-incorporating activity *in vitro* posed an important question. Are the new ribosomes formed upon growth stimulation merely redistributed upon existing membranes, or do simultaneous alterations in the rate or type of membranes formed also occur? To answer this question the proliferation of membranes of the ER was followed in our laboratory by measuring the incorporation of [^{32}P]phosphate and ^{14}C-labeled choline and glycerol into membrane phospholipids. In order to correlate the effect of hormones or hepatectomy on this process with those on biosynthetic activity of the tissue, studies were made of the rate of incorporation of radioactive amino acids into the nascent proteins of different submicrosomal fractions simultaneously with the accumulation of newly synthesized microsomal RNA. This has now been done in a variety of systems (see Tata, 1970a, b) which include: (1) growth of liver induced by a single injection of growth hormone, triiodothyronine, and testosterone in hypophysectomized, thyroidectomized, and castrated rats, respectively; (2) growth and maturation of the seminal vesicle of the castrated rat when stimulated by testosterone; (3) precocious initiation of enzyme and protein synthesis in bullfrog tadpole liver during thyroid hormone-induced metamorphosis: (4) accelerated synthesis of thyroglobulin or iodoprotein by thyroid tissue under the influence of thyrotrophic hormone (TSH); and (5) regenerating rat liver.

Figure 4 compares the rates of formation of liver microsomal phospholipids with those of microsomal RNA and of nascent protein on microsomes following the combined administration of growth hormone and triiodothyronine to hypophysectomized rats. Since the two hormones have very different lag periods, it was quite simple to detect an enhancement by the hormones of microsomal phospholipid formation temporally coordinated with that of the rates of microsomal RNA and protein synthesis. Thus growth hormone (Fig. 4*a*) stimulated the protein synthetic capacity of hepatic ribosomes several hours earlier than did triiodothyronine (Fig. 4*b*), and the same pattern was followed for the accelerated formation of polysomes (much of the additional microsomal RNA appeared as polysomes). In another type of experiment (Tata, 1970a), hypophysectomized rats were killed at a fixed time corresponding to the peak stimulation for each hormone (i.e., the hormones were not administered together as in Fig. 4). This led to additive stimulations in all three newly made constituents of the rough ER. It should be emphasized that in all these studies of combined treatment the maximal dose of each hormone was used.

Such coordination of proliferation of cellular structures with biosynthetic activity seems to be a feature common to other growth and developmental situations. Good coordination in the onset of accelerated synthesis of phospholipid, RNA, and protein was also observed in microsomes isolated frm the seminal vesicles of castrated rats following growth stimulation with testosterone, as shown in Figure 5. Attempts to obtain mutually uncontaminated submicrosomal fractions from seminal vesicles were unsuccessful. Earlier ultrastructural studies on the depletion of rough ER of the seminal vesicles are also compatible with the above biochemical findings (Szirmai and Van der Linde, 1964). In another system, although ribosomes and membranes were not actually isolated and studied in the same laboratory, a coordinated synthesis of phospholipid and RNA in their respective target tissues may underlie the same process in the case of estrogen (Nishigori and Aizawa, 1968; Teng and Hamilton, 1970). In a growth system not directly involving hormones, as in regenerating rat liver, a simultaneous build-up of rough ER and enhanced protein synthesis are also observed *in vivo* (Tables 6 and 7). Tsukada et al. (1968) and Fex (1970) showed separately that regeneration involves the rapid build-up of ribosomes and membrane phospholipids.

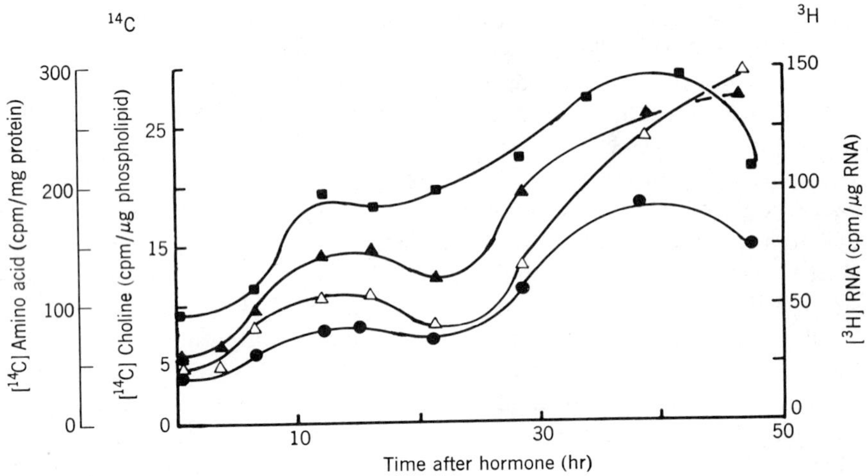

Figure 4. Coordination of the appearance of additional newly formed hepatic microsomal phospholipids (△, ▲) and RNA (●) and the recovery of nascent microsomal protein formed *in vivo* (■) following the administration to hypophysectomized rats of human growth hormone and triiodothyronine together. The rate of formation of microsomal RNA and phospholipids was determined from the incorporation of ^{32}P or [^{14}C]choline, whereas the rate of protein synthesis *in vivo* was derived from the incorporation of ^{14}C-labeled amino acids within 10 min after their administration. The results are expressed as percent increase in specific radioactivities in hormone-treated animals over the values obtained in untreated controls. The first burst of synthetic activity is the response to growth hormone, whereas the second burst is attributable to the slower effect of triiodothyronine. Phospholipid synthesis was measured in both smooth (SM) (△) and rough (RM) (▲) ER. (From Tata, 1970a.)

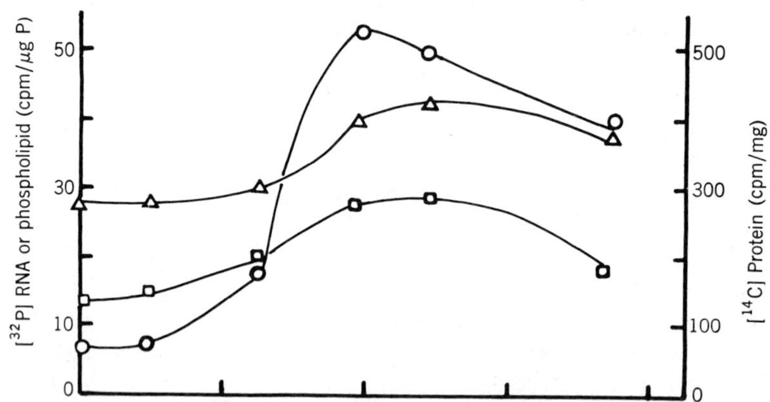

Figure 5. Coordinated acceleration of formation of microsomal phospholipid (△), RNA (○), and protein (□) following the single administration of 250 μg of testosterone propionate to castrated rats. Phospholipid and RNA were both labeled with 95 μc's of [^{32}P]orthophosphate administered 1.5 and 3.5 hr before death, whereas newly synthesized protein was labeled with 15 Ci of ^{14}C-labeled amino acids given 15 min before death. Corrections were made for isotope uptake in each determination. Six castrated rats (160–180 g) were used for each time interval. Seminal vesicles weighed 14 ± 4 mg in control rats (time 0) and 32 ± 6 mg at 35.5 hr after hormone injection. (From Tata, 1970a.)

Table 6. Coordinated Increase in Formation of Rough Microsomal Phospholipid and RNA after Partial Hepatectomy[a]

Time after operation (hr)	Operation	Weight of liver (g)	Specific radioactivity: ^{3}H-Labeled RNA (cpm/mg)	Specific radioactivity: ^{32}P-Labeled phospholipid (cpm/μg of P)
4	Sham	6.8	12,100	270
4	Hepatectomy	1.5	18,480	339
9	Sham	6.9	11,633	465
9	Hepatectomy	1.6	21,070	808
18	Sham	6.7	10,450	366
18	Hepatectomy	1.6	24,600	543
24	Sham	7.0	9,335	297
24	Hepatectomy	1.8	15,870	413

[a] Each group consisted of four sham-operated or partially hepatectomized rats. They were given 52 μCi of [^{3}H]orotic acid or 65 μCi of [^{32}P]orthophosphate 3.2 and 2.0 hr before death. The uptake of the label and acid-soluble fraction of radioactivity in the whole liver homogenate were measured to allow for the correction of specific radioactivities for increases in uptake of label after hepatectomy. (From Tata, 1970a.)

Table 7. Increase in Amino Acid Incorporation into Rough Microsomal Protein after Partial Hepatectomy

Time after operation (hr)	Operation	Weight of liver (g)	Specific radioactivity (cpm/mg of protein)	Protein (mg/mg of liver)
2	Sham	6.4	435	9.2
	Hepatectomy	1.6	398	8.8
4	Sham	6.6	490	8.9
	Hepatectomy	1.9	410	9.3
8	Sham	6.3	385	8.7
	Hepatectomy	1.7	588	9.1
16	Sham	6.3	361	8.8
	Hepatectomy	1.9	639	9.3
24	Sham	6.6	343	8.9
	Hepatectomy	2.0	605	9.6

[a] The experiment was performed in the same way as that in Table 4 except that 8 min before death hepatectomized rats received 4 μCi of a mixture of ^{14}C-labeled amino acids and sham-operated animals were given 10 μCi. Different amounts of radioactivity were administered to the two groups of rats to allow for the different amounts of tissue and corrections were also applied for uptake. (From Tata, 1970a.)

Thus this coordination seems to be in fact a common feature of cells undergoing rapid development in general, irrespective of whether the stimulus is hormonal or not, in which new biochemical functions during this period tend to be concentrated in cytoplasmic structures of increasing complexities (see Dallner et al., 1966; Pollak and Ward, 1967; Siekevitz et al., 1968). The important point to realize in the above discussion is that the accelerated and coordinated assembly of rough ER is not necessarily followed by an enhanced secretion of proteins, but in all cases, including predominantly secretory tissues, there is an accumulation of intracellular components.

An indirect suggestion of some kind of specificity in microsomal structures that proliferate is provided by systems in which the hormone induced the *de novo* synthesis of specific enzymes or proteins. Two examples of this specificity are systems that have been studied in our laboratory: (1) the induction of urea cycle enzymes and serum albumin during triiodothyronine-induced metamorphosis in bullfrog tadpoles, which is largely based on the classic work from the laboratories of Cohen and Frieden (see Cohen, 1970; Frieden and Just, 1970); and (2) the enhancement of thyroglobulin formation by thyroid cells under the influence of TSH (see Freinkel, 1964).

In the latter system TSH causes an almost simultaneous increase in the rate of whole thyroid phospholipid and RNA synthesis *in vivo*. What is particularly interesting is that the onset of this change coincides with the formation of increasing amounts of a thyroglobulin-like protein (Kerkof and Tata, 1967). Since one of the multiple actions of TSH is to stimulate the secretion of thyroglobulin, it could be argued that additional membrane synthesis would be necessary to perform the vectorial discharge of extra iodoprotein. Later studies have confirmed this effect *in vitro* using pig thyroid slices in which it was shown that much of the additional phospholipid, RNA, and iodoprotein formed in response to TSH accumulated in the smooth and rough microsomal fraction (Kerkof and Tata, 1968). The effect of TSH on phospholipid synthesis is particularly striking and has been studied in detail for several years (Freinkel, 1964), but unknown prior to our studies was that this effect was part of the general effect of growth and developmental hormones in enhancing membrane proliferation.

The system of artificial induction of metamorphosis in amphibia has received much attention in our laboratory during the last 6 yr (Tata, 1970c). In Figure 6 are summarized some of our findings on major biosynthetic events that occur during the lag period preceding the appearance of newly formed proteins as a result of administration of triiodothyronine to immature *Rana catesbeiana* tadpoles. It can be seen that there occurred, well within the latent period (5–6 days) for new proteins to be detected, an acceleration in the rate of nuclear and cytoplasmic RNA synthesis. What is of interest are the events that occur between the initial burst of RNA synthesis following triiodothyronine administration and the appearance of newly synthesized proteins. Unlike the mammalian systems mentioned earlier, there occurs during this period a more complex process of breakdown of "old" RNA, alternations in polysome profiles, and a redistribution of ribosomes attached to membranes of the ER. The latter process is accompanied by a simultaneous increase in the synthesis of ribosomes and membranes of the ER as judged by membrane phospholipid synthesis (Fig. 7*a*). It is interesting to note that this simultaneous increase in the formation of the two rough ER components is accompanied by an enhanced rate of protein synthesis (Fig. 7*b*). These biochemical changes are so marked that it is easy to observe structural changes by electron microscopy. Figure 8 shows schematically that during metamorphosis there is a shift in the distribution of ribosomes from around the simple vesicular membrane

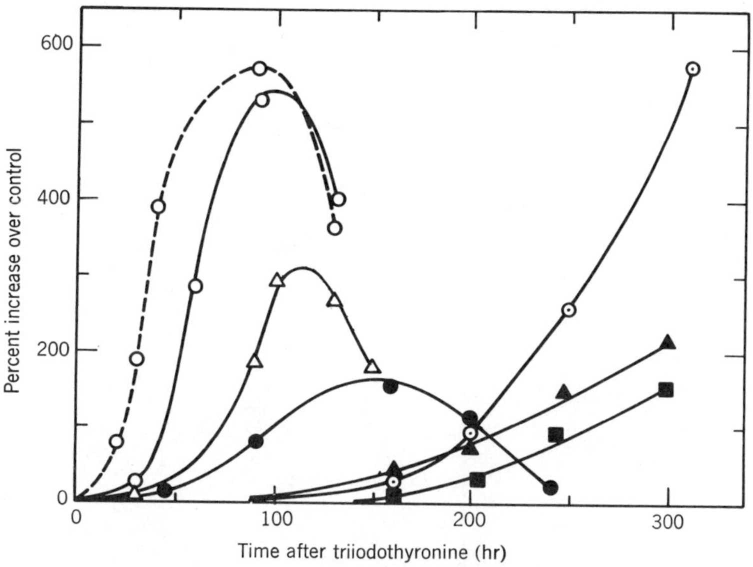

Figure 6. Schematic representation of sequential stimulation of rates of RNA and phospholipid synthesis in relation to the increases in enzymes or protein synthesized upon the precocious induction of metamorphosis in *R. catesbeiana* tadpoles with triiodothyronine. Curves show the rate of rapidly labeled nuclear RNA synthesis (○); specific activity of RNA in cytoplasmic ribosomes (△); amino acid incorporation into nascent microsomal protein *in vivo* (●); urea cycle enzymes (carbamyl phosphate synthetase) (⊙); cytochrome oxidase per milligram mitochondrial protein (▲); appearance of serum albumin in blood (■). The values are expressed as percent increases over those in the noninduced control tadpoles. The decreasing values in curves of RNA and protein synthesis reflect the dilution of specific radioactivity in precursor molecules resulting from the onset of regression of tissues such as the tail and intestine. Dashed line indicates specific activity of nuclear RNA uncorrected for changes in radioactivity in the acid-soluble fraction. (Data from Tata, 1967b.)

structures of the immature larvae to the more complex double lamellar structures more commonly seen in mature tissues. It is interesting that this shift in structural organization of the ER coincides temporally with changes in the formation of microsomal RNA and phospholipids and the synthesis of proteins by polyribosomes. Similar ultrastructural changes have been observed in the ecdysone-induced maturation of insect wing epidermal cells (Greenstein, 1972), in the androgen-dependent growth of male accessory sexual tissues (prostate, seminal vesicles) in rodents (Szirmai and Van der Linde, 1964), in fetal human liver at a stage when new enzymes and functions are acquired (Zamboni, 1965), in the induction of phosvitin by estrogen (Nicholls et al., 1968), and in pea embryo radicles during germination (Yoo, 1970). Topper (1970) has described an interesting phenomenon involving the multihormone-dependent development of the mammary gland in organ culture. In this system corticosteroids cause a marked proliferation of ER, but this in itself does not lead to milk protein synthesis for which prolactin addition is essential.

As regards membrane proliferation, the studies discussed above only describe the rate of formation but not turnover of membrane constituents. It is quite possible that the hormonal stimulus leads to both accelerated formation and retarded degradation of membranes for which the demand is most acute, as for example the rough ER. The

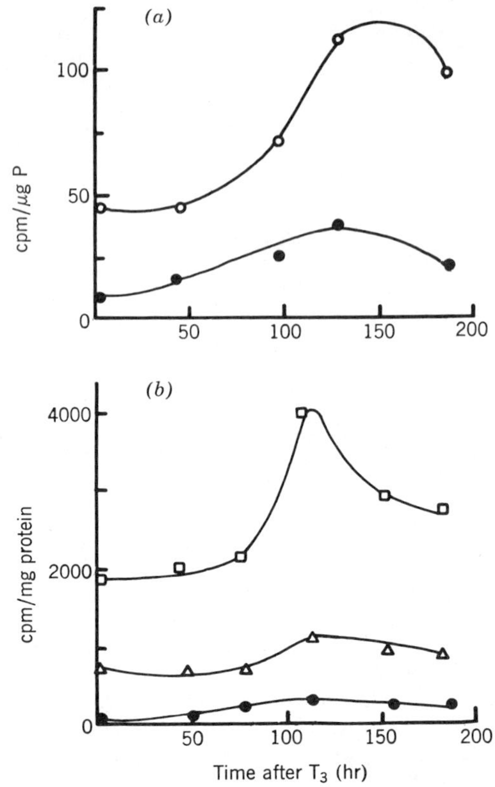

Figure 7. Coordinated proliferation of ER and the increase in protein synthesis *in vivo* in the heavy rough membranes (densely packed ribosomes attached to membranes) during triiodothyronine-induced metamorphosis of young bullfrog tadpoles. (*a*) The appearance of newly synthesized (^{32}P-labeled) heavy rough membrane phospholipids (○) and RNA (●); (*b*) the recovery of labeled nascent proteins formed *in vivo* after a short pulse of radioactive amino acids in heavy rough membranes (□), light rough membranes (△), and free polysomes (●). For details see Tata (1967b).

question of membrane stability is an important one and should be studied along the lines recently laid down by the groups of Palade and Schimke in studying drug-induced proliferation of the ER (Omura et al., 1968; Siekevitz et al., 1968; Arias et al., 1969). What appear to be homogeneous membrane preparations may differ in their composition in a subtle manner so that different enzymes of structural components are continuously inserted and degraded at nonidentical rates. How the phospholipid and protein components of membranes are assembled in an orderly fashion to yield precise structures is not known. It may be that the initial response to a growth stimulus may be to accelerate the assembly or "crystallization" of rigid structures from existing components whose depletion in turn would trigger off the synthesis of additional membrane phospholipids and proteins.

The preferential accumulation of rough ER which, as this chapter has shown, also contains the ribosomal fraction most active in amino acid incorporation *in vivo*, underlines the importance in protein synthesis of the attachment of ribosomes to membranes. As mentioned earlier, the major emphasis on ribosome–membrane

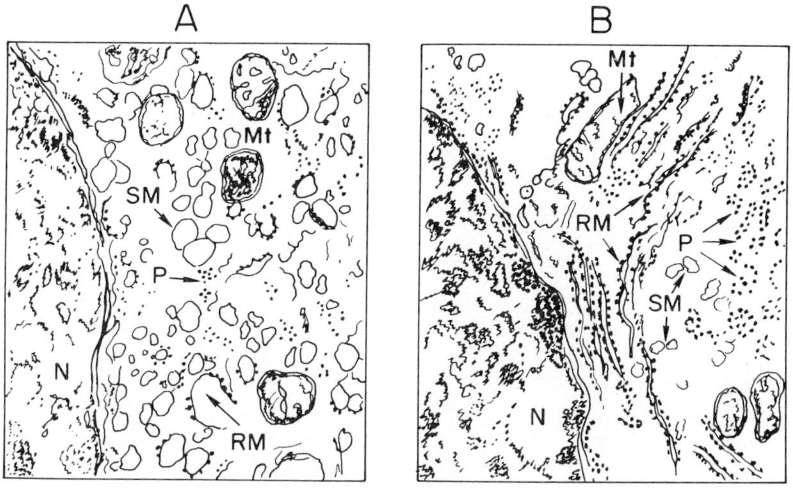

Figure 8. Schematic representation of electron micrographs showing the reorganization of the cytoplasmic membranes and ribosomes during hormone-induced metamorphosis in bullfrog tadpole liver cells. (*A*) Hepatocyte in a premetamorphic showing small groups of ribosomes around simple vesicular structures. (*B*) Similar cell in tadpole 6 days after induction of metamorphosis with triiodothyronine, showing dense ribosomal accumulation often attached to a more differentiated double-lamellar type of reticulum. N, Nucleus; Mt, mitochondrion; SM, smooth membranes; RM, rough membranes; P, typical polysomal chains, rosettes, and spirals. See Tata (1967a, b) for details.

attachment in animal cells has previously been focused on the export of proteins in predominantly secretory cells (Palade, 1966; Siekevitz et al., 1968). However, recent studies from our laboratory on nonsecretory tissues, such as the brain, have shown a different facet of the importance of membrane–ribosome attachment in synthesizing proteins for intracellular destinations during development (Andrews and Tata, 1971b). Figure 9 illustrates the relative protein-synthesizing activity of membrane-bound and free polyribosomes of the postnatal rat cerebral cortex during the first 3 weeks of life. One sees here the well-known rapid decline in amino acid-incorporating activity of brain ribosomes (see Lerner and Johnson, 1970) following birth. The cause of this drop in activity remains unknown, but is unlikely to be largely attributable to changes in levels of mRNA or some soluble factors. The time-related lesion seems to be in the ribosomes themselves. Figure 9 also shows that at birth the membrane-bound ribosomes are considerably more active than the free fraction. On the assumption that about 20% of ribosomes in cerebral cortex are membrane-bound (Andrews and Tata, 1971a) and that the relative activities *in vitro* faithfully reflect *in vivo* function, then over half of protein synthesis at birth may take place on membrane-bound ribosomes. The rapid drop in overall protein synthetic activity following birth is then largely accounted for by the membrane-bound fraction. Thus, although a diminution in protein synthetic rate rather than an enhancement is involved here, as illustrated by all the earlier examples, once again it is evident that membrane attachment of ribosomes plays some important role in the regulation of protein synthesis during growth and maturation.

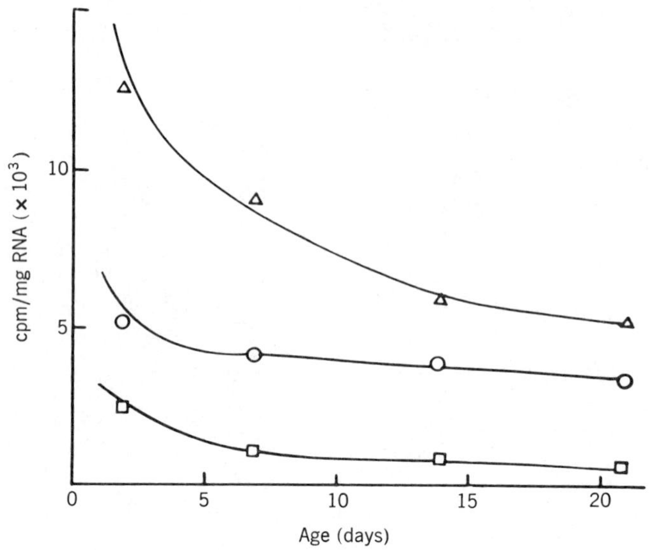

Figure 9. [^{14}C]Leucine incorporation into protein *in vitro* by unfractionated microsomes, free ribosomes, and membrane-bound ribosomes from the brains of normal rats in the neonatal period. The three subfractions were prepared from aliquots of the same homogenates of brain from all four age groups on the same day; 50–100 μg of rRNA were used in 1 ml of incubation suspension and each preparation was incubated in the presence of homologous cell sap. The values given are the means of duplicate determinations. ○, Unfractionated microsomes; △, membrane-bound ribosomes; □, free ribosomes. (From Andrews and Tata, 1971b).

5. SIGNIFICANCE OF RIBOSOMAL ATTACHMENT TO MEMBRANES DURING GROWTH AND DEVELOPMENT: IS THERE A FUNCTION OF RIBOSOMAL SEGREGATION?

From the work described above, it can be concluded that there exists some mechanism in the cell that tightly couples the formation of rough ER in response to an increased demand for protein synthesis following stimulation with a variety of growth and developmental hormones. Perhaps a common fundamental mechanism may also underlie the numerous observations of a marked acceleration of phospholipid synthesis or proliferation of rough ER anticipating or accompanying the initial burst of protein synthesis in apparently a variety of situations of growth and development that are independent of hormones. These include neonatal development of rat liver (Dallner et al., 1966), development of chick embryonic liver (Pollak and Ward, 1967), plasma cell growth following antigenic stimulation (Leduc et al., 1968), maturation of imaginal discs in *Drosophila* larva (Ursprung and Schabtach, 1968), and renal hypertrophy following unilateral nephrectomy (Malt, 1969).

In order to deduce the significance of the coordination between membrane and ribosome proliferation on the one hand, and protein synthesis on the other, it is necessary to know the functions of membrane–ribosome interaction. As mentioned earlier (see Section 4.A), the only well-established function is that of facilitating the export of secretory proteins. However, this function seems to be a most unlikely reason for the proliferation of the rough ER during growth, since even in secretory cells there is relatively little increase in protein secretion during the period of active growth. We have seen earlier that bound ribosomes in nonsecretory and secretory tissues show a major difference in the vectorial discharge of protein newly synthesized on bound ribosomes (Fig. 3) (Andrews and Tata, 1971a). It is quite possible that even in secretory cells the secretory function of rough ER in protein-secreting cells may mask some other function.

A widely held generalization arising from the study of secretion of proteins (see Palade, 1966; P. N. Campbell, 1970) now must be put in proper perspective. It is quite often said that proteins for export are made on bound ribosomes and, conversely, those for intracellular use are made on free ribosomes. Whereas there is no doubt that proteins for secretion or storage are made on the rough ER, there is little reason to believe that some intracellular proteins are not synthesized on bound ribosomes as well. For this reason it is most important that Pitot et al. (1969) have identified, by radioimmunological techniques, bound and not free ribosomes of the liver as the site of synthesis of serine dehydrase, an exclusively intracellular protein. From more superficial studies it also seems that some of the mitochondrial enzymes may be made on microsomes, rather than free ribosomes (Kadenbach, 1970). In our laboratory we are trying to locate urea cycle enzyme on the newly formed rough ER during amphibian metamorphosis (Figs. 8 and 9). Many of the urea cycle enzymes are localized in mitochondria (see Cohen, 1970). It is therefore interesting to note that in the dense perinuclear concentration of rough ER one finds mitochondria of an appearance quite distinct from those in distal cytoplasmic regions (Tata, 1967a, b; Bennett, 1970). A burst in mitochondrial DNA polymerase has also been observed at a time anticipating the increased microsomal proliferation following induction of metamorphosis by thyroid hormone in bullfrog tadpoles (A. M. Campbell et al., 1969). Thyroid hormones also cause a simultaneous increase in mitochondrial and microsomal protein synthesis in rat liver when the two activities are measured separately *in vitro* (Roodyn et al., 1965). Gross (1971) showed, on the basis of a simultaneous acceleration of rat cardiac and liver mitochondrial DNA and protein synthesis induced by triiodothyronine, that there may be a selective turnover of different populations of mitochondria as part of the anabolic action of thyroid hormone. It also seems that microsomal membrane proteins themselves may be made equally on free and bound ribosomes (Ragnotti et al., 1969). Further attempts to identify the site of synthesis of intracellular constituents may indeed reveal a greater involvement of bound ribosomes in the synthesis of intracellular proteins. In this context it would be well worth investigating the proteins synthesized on membrane-bound ribosomes of predominantly nonsecretory tissues.

What other possible functions are served by ribosomal attachment to membranes can only be speculated upon at the moment. Little is known about the factors responsible for attachment of ribosomes to membranes, except that there seems to be some specificity. Ragland et al. (1971) found that ribosomes interact with "stripped" rough membranes but not with smooth membranes, thus suggesting specific sites for ribosomes. Both polyamines (Khawaja and Raina, 1970) and a special type of membrane RNA (Pitot et al., 1969) have been implicated in such sites. According to Pitot et al. (1969), the presence of a special membrane RNA would determine the longevity of translation of mRNA. While a more convincing demonstration of a unique species of membrane RNA has still to be provided, recent reports from Rabin's laboratory (James et al., 1969; Sunshine et al., 1971) are potentially important. It has been shown that the activity of an enzyme in ER termed "rearrangase" which activates a latent RNase is determined by the extent of attachment of ribosomes to membranes. This phenomenon provides one explanation of how ribosomal attachment to membranes would stabilize mRNA, which of course would be an advantage for rapidly growing cells.

The significance of the coordinated membrane–ribosome assembly that we favor, although not based on direct evidence, is that a firm attachment would in effect lead to segregation of different populations of ribosomes. During rapid developmental or adaptive changes, populations of ribosomes formed at different periods would be precoded with different spectra of messengers, so that operationally it would mean topological segregation of ribosomes synthesizing different classes of proteins. There are two important prerequisites for considering such topographical segregation as a likely physiological process: (1) there should be only limited or no exchange between different groups of membrane-bound ribosomes or between bound and free ribosomes, and (2) there are qualitative differences in products or in the composition of products formed by bound and free ribosomes.

As regards the first point, whereas it is technically still not possible to study populations of bound polysomes that would be kept apart *in vivo*, a relatively extensive exchange of ribosomes between bound and free compartments is generally accepted. A critical appraisal of the evidence on which this acceptance is based, however, shows that there is indeed no convincing proof for or against the occurrence of such an exchange. The main evidence for exchange is based on the similarity of the kinetics of labeling and turnover of rat liver bound and free polysomes (Moulé and Delhumeau de Ongay, 1964). However, bound and free ribosomes could well have similar half-lives without the two pools necessarily being exchangeable. Furthermore, all studies on this problem have been carried out in rat liver, a tissue in which over 80% of the ribosomes are bound; in many of these the animals were fasted, which enhances the dissociation of ribosomes from rough membranes (see Tata, 1969). This is not to completely rule out all exchange between bound and free ribosomes of the cell, but it is quite likely that the exchange is minimal and may perhaps be topographically restricted within small regions of the cytoplasm.

There is some evidence for chemical differences between membrane-bound and free ribosomes, and several papers have recently reported on qualitative differences in composition of the identification of different products synthesized on bound and free ribosomes (Table 8). Fridlender and Wettstein (1970) found by polyacrylamide gel electrophoresis of chick embryo ribosomal proteins the absence in the membrane-bound particles of two of the components present in free ribosomes. This result strongly suggests the unlikelihood of an exchange between bound and free ribosomes. An intriguing mechanism has been proposed by Kabat (1970) whereby phosphorylation of some specific ribosomal proteins would render the ribosomes inactive. This process is reversible so that enzymic phosphorylation–dephosphorylation would allow for storage, and perhaps even a topographical translocation, of nonfunctioning ribosomes for periods of sudden protein synthesis demands. Another qualitative difference in ribosomes, reported by Dingman et al. (1970), is in the RNA moiety. These investigators found that removal of the cytoplasmic supernatant from microsomes or the treatment of microsomes with sodium deoxycholate caused a partial degradation of rRNA. By isotopic labeling of RNA, it was found that newly formed RNA and ribosomes were preferentially broken down, suggesting some mechanism in the cytoplasm that allows discrimination between ribosomes formed at different

Table 8. Some Recent Evidence Suggesting Differences between the Composition of or Products Synthesized on Free and Bound Ribosomes

Source of ribosomes	Ribosomal composition or product studies	Difference between free and bound ribosomes	Reference
Rat liver	Composition	Stability of ribosomal and ribosome associated RNA	Dingman, Kakefuda, and Aronow (1970)
	Composition	Dissimilar electrophoretic pattern of RNA	Lechner (1970)
Chicken embryo	Composition	Ribosomal protein patterns not identical	Fridlender and Weinstein (1970)
Escherichia coli	Product formed	β-galatosidase synthesized on membrane-ribosome-DNA complex	Chefurka, Yapo, and Nisman (1970)
Rabbit reticulocytes	Composition	Dissimilar spectrum of ribosomal proteins	Burka and Bulova (1971)
	Product formed	Nascent globin chains in free ribosomes, non-globin proteins on bound ribosomes	Bulova and Burka (1970)

(From Tata, 1971.)

periods of a cell's lifetime. As regards different classes of proteins synthesized by bound and free ribosomes, there is much evidence from microbial or structurally uncomplicated animal cells, such as the reticulocyte in which ribosomes may be attached to the cell membrane or structures other than the ER. Hauge (1967) had suggested the possibility of some primitive form of ER in bacteria and yeast so that different classes of proteins may be synthesized on membrane-bound and unbound ribosomes. Chefurka et al. (1970) have reported that in *Escherichia coli* only the fraction of ribosomes attached to a DNA–membrane complex synthesize β-galactosidase upon enzyme induction. In reticulocytes membrane-bound ribosomes incorporate over 80% of [^{14}C] valine into high-molecular-weight nonglobin proteins, whereas α and β chains of globin are to a large extent synthesized on free ribosomes (Bulova and Burka, 1970).

Thus the idea of topographical segregation of different classes of ribosomes synthesizing different groups of proteins may not be too far from reality. Technical limitations prevent a convincing demonstration of such a segregation *in vivo*. No single set of techniques of cell biology will solve this problem, but a solution is not impossible if immunochemical, histochemical, electron microscope, and isotope labeling methods were to be combined. Until such time, ribosomal segregation will remain an intriguing concept. The advantages of functional segregation of ribosomes would be principally felt in cells undergoing rapid development or adaptation. It would enable such cells to respond (based on selective synthesis of proteins) to hormonal and nonhormonal stimuli without perturbing the synthesis of proteins not initially involved in growth and development.

6. SUMMARY

Figure 10 presents a schematic, and obviously simplified, summary of the main phenomena observed during growth and development and how they may be interrelated. No attempt has been made in this scheme to determine at what sites hormonal or nonhormonal stimuli initially act to trigger off a subsequent chain of events. It is assumed that to a large extent growth and development eventually depend on the induction or preferential synthesis of a predetermined protein pattern. Among the earliest events occurring in target cells following the administration of growth and developmental hormones is a marked alteration in the rate and nature of nuclear RNA synthesis. [Uterine nuclei may respond as early as 2 min after the administration of estrogen to ovariectomized rats (Hamilton, 1968)]. A selective control of mRNA synthesis still remains one of the simplest explanations of specificity of development changes. However, an important technical problem has yet to be resolved, namely, detecting very small changes in what must be a minute fraction of nuclear RNA, especially when a massive increase in the synthesis of rRNA and nonmessenger, DNA-like RNA are provoked.

The extreme sensitivity of ribosomal cistrons may not be a specific property of hormones, but it has been observed in a variety of growth and maturational processes. An exaggerated increase in rRNA synthesis and RNP particles (including informosomes) may be a common feature of nucleated cells in order to facilitate the transport of mRNA from the nucleus to the cytoplasm when a sudden demand is imposed upon its protein-synthesizing machinery. However, before the process of transport must come that of selection of the right species of RNA for utilization in the cytoplasm. Little is known as yet about the role of the extensive synthesis of rapidly turning-over RNA restricted to the nucleus (see Shearer and McCarthy, 1967; Britten and Davidson, 1969). The high degree of nuclear restriction makes it imperative to characterize

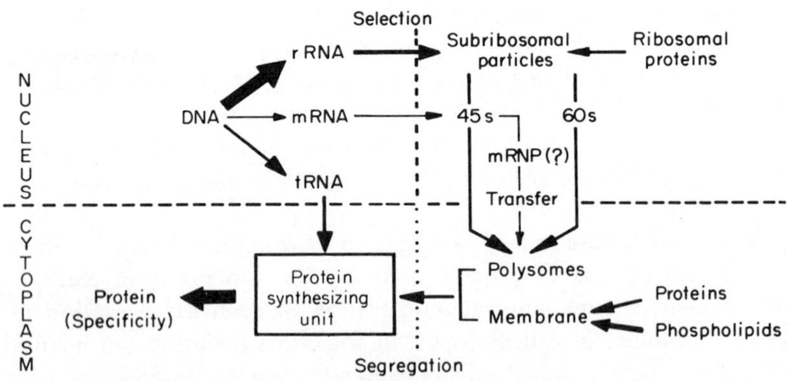

Figure 10. Scheme attempting to integrate some of the processes regulating protein synthesis in nucleated cells and which respond to the growth-promoting stimulus of hormones. It is proposed that the type of growth or development would be determined genetically via mRNA coding for specific proteins, but that its expression requires the generation of ribosomes and mRNA protein particles and that their topographical segregation may be achieved by a firm attachment to membranes. The width of the arrows is roughly related to the magnitude of response of the target cell to growth and developmental hormones with respect to the process studied.

more carefully both the nuclear *and* cytoplasmic RNA formed under the influence of the hormone.

Regardless of the mechanism of selection of RNA, growth stimulation gradually leads to an accumulation of precoded polysomes. Figure 10 indicates that when newly formed RNA accumulates in the cytoplasm there occurs a simultaneous increase in the rate of formation of membranes, especially those to which the ribosomes are attached. There seems to be some mechanism in the cell that coordinates the formation or assembly of membranes of the ER according to the demands for protein synthesis, especially during growth and development. Thus the terminal stages leading to a rather abrupt increase in protein synthesis are complex and include: (1) an increase in overall content of cytoplasmic RNA; (2) a shift toward heavier polysomal aggregates, suggesting a relative increase in messenger content; (3) an elevation of the intrinsic capacity of the ribosome to incorporate amino acids; (4) a shift of polysomal distribution from free to membrane-bound; and (5) a coordinated proliferation of membranes to which the ribosomes are bound.

So far, virtually all the emphasis on ribosome–membrane attachment has been directed toward the function of export of proteins in secretory cells. However, the marked proliferation of the rough ER observed during rapid growth (when relatively more protein is made for intracellular use) in both secretory and nonsecretory cells suggests that the attachment may serve some other function. One such role we have proposed is that the ribosome–membrane association could conceivably restrict the exchange between all the populations of ribosomes and messengers so as to achieve topographical segregation of polysomes. The main advantage of such segregation would be eventually to isolate responses (based on selective synthesis of proteins) to hormonal or nonhormonal stimuli from the orderly synthesis of proteins not involved initially in the growth or developmental response.

REFERENCES

Andrews, T. M., and J. R. Tata (1968). Difference in vectorial release of nascent protein from membrane-bound ribosomes of secretory and non-secretory tissues. *Biochem. Biophys. Res. Commun.*, **32,** 1050.

Andrews, T. M., and J. R. Tata (1971a). Protein synthesis by membrane-bound and free ribosomes of secretory and non-secretory tissues. *Biochem. J.*, **121,** 683.

Andrews, T. M., and J. R. Tata (1971b). Protein synthesis by membrane-bound and free ribosomes of the developing rat cerebral cortex. *Biochem. J.*, **124,** 883.

Arias, I. M., D. Doyle, and R. T. Schimke (1969). Studies on the synthesis and degradation of proteins of the endoplasmic reticulum of rat liver. *J. Biol. Chem.*, **244,** 3303.

Bennett, T. P., and J. S. Glenn (1970). Fine structural changes in liver cells of *Rana catesbeiana* during natural metamorphosis. *Develop. Biol.*, **22,** 535.

Brewer, E. N., L. B. Foster, and B. H. Sells (1969). A possible role for ribonuclease in the regulation of protein synthesis in normal and hypophysectomized rats. *J. Biol. Chem.*, **244,** 1389.

Britten, R. J., and E. M. Davidson (1969). Gene regulation for higher cells: A theory. *Science*, **165,** 349.

Bulova, S. I., and E. R. Burka (1970). Biosynthesis of nonglobin protein by membrane-bound ribosomes in reticulocytes. *J. Biol. Chem.*, **245,** 4907.

Burka, E. R., and S. I. Bulova (1971). Heterogeneity of reticulocyte ribosomes. *Biochem. Biophys. Res. Commun.*, **42,** 801.

Campbell, A. M., M. H. Corrance, J. N. Davidson, and H. M. Keir (1969). The metabolism of DNA in the liver during precocious induction of metamorphosis in *Rana catesbeiana*. *Proc. Roy. Soc.* (*Edinburgh*), **B70,** 295.

Campbell, P. N. (1970). Functions of polyribosomes attached to membranes of animal cells. *FEBS Letters*, **7,** 1.

Chefurka, W., A. Yapo, and B. Nisman (1970). *In vitro* synthesis of β-galactosidase by membrane fractions of *Escherichia coli*. *Can. J. Biochem.*, **48,** 893.

Clemens, M. J., and A. Korner (1970). Amino acid requirement for the growth-hormone stimulation of incorporation of precursors into protein and nucleic acids of liver slices. *Biochem. J.*, **119,** 629.

Cohen, P. P. (1970). Biochemical differentiation during amphibian metamorphosis. *Science*, **168,** 533.

Cox, R. F., and A. P. Mathias (1969). Cytoplasmic effects of cortisol in liver. *Biochem. J.*, **115,** 777.

Dallner, G., P. Siekevitz, and G. E. Palade (1966). Biogenesis of endoplasmic reticulum membranes. I. Structural and chemical differentiation in developing rat hepatocyte. *J. Cell Biol.*, **30,** 73. II. Synthesis of constitutive microsomal enzymes in developing rat hepatocyte. *J. Cell Biol.*, **30,** 97.

Dingman, C. W., A. Aronow, S. L. Bunting, A. C. Peacock, and B. W. O'Malley (1969). Changes in chick oviduct ribonucleic acid following hormonal stimulation. *Biochemistry*, **8,** 489.

Dingman, C. W., T. Kakefuda, and A. Aronow (1970). Rat liver cytoplasmic ribonucleic acids: Some aspects of their localization and isolation. *Biochim. Biophys. Acta*, **224,** 114.

Fawcett, D. W. (1966). The cell: Its organelles and inclusion—An atlas of fine structure. Saunders, Philadelphia.

Fex, G. (1970). Phospholipid metabolism during regeneration of rat liver. *Biochem. J.*, **119,** 743.

Freinkel, N. (1964). In *The Thyroid Gland* (R. Pitt-Rivers and W. R. Trotter, eds.), Vol. 1, pp. 131–162. Butterworths, London.

Fridlender, B. R., and F. O. Wettstein (1970). Differences in the ribosomal protein of free and membrane-bound polysomes of chick embryo cells. *Biochem. Biophys. Res. Commun.*, **39,** 247.

Frieden, E., and J. Just (1970). In *Biochemical Actions of Hormones* (G. Litwack, ed.), pp. 1–52. Academic, New York.

Georgiev, G. P. (1966). The nature and biosynthesis of nuclear ribonucleic acids. *Progr. Nucleic Acid Res. Mol. Biol.*, **6,** 259.

Greenstein, M. E. (1972). The ultrastructure of developing wings in the giant silkmoth, *Hyalophora cecropia*. *J. Morph.*, **136,** 1 and 23.

Gross, N. J. (1971). Control of mitochondrial turnover under the influence of thyroid hormone. *J. Cell Biol.*, **48,** 29.

Hamilton, T. H. (1968). Control by estrogen of genetic transcription and translation. *Science*, **161,** 649.

Hauge, J. G. (1967). In *Symposium on Yeast Protoplasts* (R. Müller, ed.), pp. 273–277. Akademie, Berlin.

Hendler, R. W. (1968). *Protein Biosynthesis and Membrane Biochemistry*. Wiley, New York.

Henshaw, E. C. (1968). Messenger RNA in rat liver polyribosomes: Evidence that it exists as ribonucleoprotein particles. *J. Mol. Biol.*, **36,** 401.

Henshaw, E. C., M. Revel, and H. H. Hiatt (1965). A cytoplasmic particle bearing messenger ribonucleic acid in rat liver. *J. Mol. Biol.*, **14,** 241.

James, D. W., B. R. Rabin, and D. J. Williams. (1969). Role of steroid hormones in the interaction of polysomes with endoplasmic reticulum. *Nature*, **224,** 371.

Kabat, D. (1970). Phosphorylation of ribosomal proteins in rabbit reticulocytes. Characterization and regulatory aspects. *Biochemistry*, **9,** 4160.

Kadenbach, B. (1970). Biosynthesis of cytochrome c. The sites of synthesis of apoprotein and holoenzyme. *European J. Biochem.*, **12,** 392.

Kerkof, P. R., and J. R. Tata (1967). Simultaneous acceleration *in vivo* of the formation of thyroid ribonucleic acid, phospholipid and iodoprotein by thyroid stimulating hormone. *Biochem. Biophys. Res. Commun.*, **28,** 111.

Kerkof, P. R., and J. R. Tata (1969). The subcellular distribution of ^{32}P-labelled phospholipids, ^{32}P-labelled ribonucleic acid and ^{125}I-labelled iodoprotein in pig thyroid slices. *Biochem. J.*, **112,** 729.

Khawaja, J. A., and A. Raina (1970). Effect of spermine and magnesium on the attachment of free ribosomes to endoplasmic reticulum membranes *in vitro*. *Biochem. Biophys. Res. Commun.*, **41,** 512.

Kim, K. H., and P. P. Cohen (1966). Modification of tadpole liver chromatin by thyroxine treatment. *Proc. Natl. Acad. Sci., U.S.*, **55,** 1251.

Korner, A. (1965). Growth hormone control of biosynthesis of protein and ribonucleic acid. *Recent Progr. Hormone Res.*, **21,** 205.

Kostyo, J. L. (1968). Rapid effects of growth hormone on amino acid transport and protein synthesis. *Ann. N.Y. Acad. Sci.*, **148,** 389.

Lechner, M. C. (1970). RNA from free and membrane-bound rat liver ribosomes. *Bull. Soc. Chim. Biol.*, **52,** 1532.

Leduc, E. H., S. Avrameas, and M. Bouteille (1968). Ultrastructural localization of antibody in differentiating plasma cells. *J. Exptl. Med.*, **127,** 109.

Lee, K. I., J. R. Reel, and F. T. Kenney (1970). Regulation of tyrosine α-ketoglutarate transaminase in rat liver. IX. Studies of the mechanisms of hormonal inductions in cultured hepatoma cells. *J. Biol. Chem.*, **245,** 5806.

Lerner, M. P., and T. C. Johnson (1970). Regulation of protein synthesis in developing mouse brain. Alteration in ribosomal activity. *J. Biol. Chem.*, **245,** 1388.

McConkey, E. H., and J. W. Hopkins (1966). Subribosomal patricles and the transport of messenger RNA in HeLa cells. *J. Mol. Biol.*, **4,** 257.

Malt, R. A. (1969). In *Renal Hypertrophy* (M. W. Nowinski and R. J. Goss, eds.), pp. 131–154. Academic, New York.

Moulé, Y., and G. Delhumeau de Ongay (1964). Relations metaboliques entre les ribosomes libres et liés du foie de rat. *Biochem. Biophys. Acta*, **91,** 113.

Nebert, D. W., and H. V. Gelboin (1969). In *Microsomes and Drug Oxidations* (J. R. Gillette, A. H. Conney, G. J. Cosmides, R. W. Estabrook, J. R. Fouts, and G. J. Mannering, eds.), pp. 389–426. Academic, New York.

Nicholls, T. J., B. K. Follett, and P. J. Evennett (1968). The effects of oestrogens and other steroid hormones on the ultrastructure of the liver of *Xenopus laevis* Daudin. *Z. Zellforsch.*, **90,** 19.

Nishigori, H., and Y. Aizawa (1968). Effect of estradiol-17β on the metabolism of phospholipids and their precursors in rat uterus. *Endocrinol. Japon.*, **15,** 209.

Olsnes, S. (1971). Characterization of the complex containing rapidly labelled RNA in EDTA-treated polyribosomes from rat liver. *European J. Biochem.*, **18,** 242.

O'Malley, B. W., W. L. McGuire, P. O. Kohler, and S. G. Korenman (1969). Stuaies on the mechanism of steroid hormone regulation of synthesis of specific proteins. *Recent Progr. Hormone Res.*, **25,** 105.

Omura, T., P. Siekevitz, and G. E. Palade (1967). Turnover of constituents of the endoplasmic reticulum membranes of rat hepatocytes. *J. Biol. Chem.*, **242,** 2389.

Palade, G. E. (1966). Structure and function at the cellular level. *J. Am. Med. Assoc.*, **198,** 815.

Penman, S., M. Rosbash, and M. Penman (1970). Messenger and heterogeneous nuclear RNA in HeLa cells: Differential inhibition by cordycepin. *Proc. Natl. Acad. Sci., U.S.*, **67,** 1878.

Perry, R. R., and D. E. Kelley (1970). Inhibition of RNA synthesis by actinomycin D: Characteristic dose-response of different RNA species. *J. Cell Physiol.*, **76,** 127.

Pitot, H., N. Sladek, W. Ragland, R. K. Murray, G. Mayer, H. D. Soling, and J. P. Jost (1969). In *Microsomes and Drug Oxidations* (J. R. Gillette, A. H. Conney, G. J. Cosmides, R. W. Estabrook, J. R. Fouts, and G. J. Mannering, eds.), pp. 59–79. Academic, New York.

Pogo, A. O., V. C. Littau, V. G. Allfrey, and A. E. Mirsky (1967). Modification of RNA synthesis in nuclei from normal and regenerating liver. *Proc. Natl. Acad. Sci., U.S.*, **57,** 743.

Pollak, J. K., and D. B. Ward (1967). Changes in the chemical composition and the enzymic activities of hepatic microsomes of the chick embryo during development. *Biochem. J.*, **103,** 730.

Ragland, W. L., T. K. Shires, and H. C. Pitot (1971). Polyribosomal attachment to rat liver and hepatoma endoplasmic reticulum *in vitro*. A method for its study. *Biochem. J.*, **121,** 271.

Ragnotti, G., G. R. Lawford, and P. N. Campbell (1969). Biosynthesis of microsomal nicotinamide-adenine dinucleotide phosphate-cytochrome c reductase by membrane bound and free polysomes from rat liver. *Biochem. J.*, **112,** 139.

Redman, C. M., and D. D. Sabatini (1966). Vectorial discharge of peptides released by puromycin from attached ribosomes. *Proc. Natl. Acad. Sci., U.S.*, **56,** 608.

Roodyn, D. B., K. B. Freeman, and J. R. Tata (1965). The stimulation by treatment *in vivo* with tri-iodothyronine of amino acid incorporation into protein by isolated rat liver mitochondria. *Biochem. J.*, **94,** 628.

Samarina, O. P., E. M. Lukanidin, J. Molnar, and G. P. Georgiev (1968). Structural organization of nuclear complexes containing DNA-like RNA. *J. Mol. Biol.*, **33,** 251.

Scherrer, K., and L. Marcaud (1969). Messenger RNA in avian erythroblasts at the transcriptional and translational levels and the problem of regulation in animal cells. *J. Cell Physiol., Suppl.*, **1,** 181.

Sells, B. H., and T. Takahashi (1967). Early changes in liver cytoplasmic RNA of growth hormone-treated rats. *Biochim. Biophys. Acta*, **134,** 69.

Shearer, R. W., and B. J. McCarthy (1967). Evidence for ribonucleic acid molecules restricted to the cell nucleus. *Biochemistry*, **6,** 283.

Siekevitz, P., G. E. Palade, G. Dallner, I. Ohad, and T. Omura (1968). In *Organizational Biosynthesis* (H. J. Vogel, J. O. Lampen, and V. Bryson, eds.), pp. 331–373. Academic, New York.

Spirin, A. S. (1969). Informosomes. *European J. Biochem.*, **10,** 20.

Sunshine, G. H., D. J. Williams, and B. R. Rabin (1971). Role for steroid hormones in the interaction of ribosomes with the endoplasmic membranes of rat liver. *Nature*, **230,** 133.

Szirmai, J. A., and P. C. van der Linde (1965). Effect of castration on the endoplasmic reticulum of the seminal vesicle and other target epithelia in the rat. *J. Ultrastruct. Res.*, **12,** 380.

Tata, J. R. (1966). Hormones and the synthesis and utilization of ribonucleic acids. *Progr. Nucleic Acid Res. Mol. Biol.*, **5,** 191.

Tata, J. R. (1967a). The formation and distribution of ribosomes during hormone induced growth and development. *Biochem. J.*, **104,** 1.

Tata, J. R. (1967b). The formation, distribution and function of ribosomes and microsomal membranes during induced amphibian metamorphosis. *Biochem. J.*, **105,** 783.

Tata, J. R. (1968). Hormonal regulation of growth and protein synthesis. *Nature*, **219,** 331.

Tata, J. R. (1969). In *Subcellular Components* (G. D. Birnie and S. Fox, eds.), pp. 83–104. Butterworths, London.

Tata, J. R. (1970a). Co-ordination between membrane phospholipid synthesis and accelerated biosynthesis of cytoplasmic ribosomes and protein. *Biochem. J.*, **116,** 617.

Tata, J. R. (1970b). In *Biochemical Actions of Hormones* (G. Littwack, ed.), pp. 89–133. Academic, New York.

Tata, J. R. (1970c). In *Control Processes in Multicellular Organisms* (G. E. W. Wolstenholme and J. Knight, eds.), pp. 131–150. Churchill, London.

Tata, J. R. (1971). Ribosomal segregation as a possible function for the attachment of ribosomes to membranes. *Sub-Cell Biochem.*, **1,** 83.

Tata, J. R., and C. C. Widnell (1966). Ribonucleic acid synthesis during the early action of thyroid hormones. *Biochem. J.*, **98,** 604.

Tata, J. R., and H. G. Williams-Ashman (1967). Effects of growth hormone and tri-iodothyronine on amino acid incorporation by microsomal subfractions from rat liver. *European J. Biochem.*, **2,** 366.

Teng, C. S., and T. H. Hamilton (1970). Regulation of polyribosome formation and protein synthesis in the uterus. *Biochem. J.*, **118,** 341.

Tomkins, G. M., T. D. Gelehrter, D. M. Granner, Jr., H. H. Samuels, and E. B. Thompson (1969). Control of specific gene expression in higher organisms. *Science*, **166,** 1474.

Topper, Y. J. (1970). Multiple hormone interactions in the development of mammary gland *in vitro*. *Recent Progr. Hormone Res.*, **26,** 287.

Tsukada, K., T. Moriyama, T. Umeda, and J. Lieberman (1968). Relationship between the ribosomal alteration after partial hepatectomy and the increase in liver protein synthesis *in vivo*. *J. Biol. Chem.*, **243,** 1160.

Ursprung, H., E. Schabtach (1968). The fine structure of the male drosophila genital disk during late larval and early pupal development. *Wilhelm Roux Archiv.*, **160,** 243.

Widnell, C. C., and J. R. Tata (1966). Additive effects of thyroid hormone, growth hormone and testosterone on deoxyribonucleic acid-dependent ribonucleic acid polymerase in rat liver nuclei. *Biochem. J.*, **98,** 621.

Yoo, B. Y. (1970). Ultrastructural changes in cells of pea embryo radicles during germination. *J. Cell Biol.*, **45,** 158.

Zamboni, L. (1965). Electron microscopic studies of blood embryogenesis in humans. I. The ultrastructure of the fetal liver. *J. Ultrastruct. Res.*, **12,** 509.

13

GERM CELLS AND EARLY EMBRYONIC DEVELOPMENT

David Epel

1. INTRODUCTION

An overview of animal development, from coelenterates to mammals, reveals several common patterns. In all cases development is initiated by the fusion of two highly dissimilar cells. One is a large, nonmotile, nutrient-filled cell, the *egg*. The other is a small, motile cell almost devoid of excess cytoplasm, the *sperm*. The fusion of these dissimilar cells—or *fertilization*—"activates" the egg to begin the development of a new individual.

At fertilization the major processes initiated are extracellular surface changes of the egg involved in sperm entry and intracellular cytoplasmic changes which lead to the fusion of sperm and egg nucleus (see Fig. 1). Metabolic events also begin which lead to cycles of DNA synthesis and rapid cell division known as *cleavage*. During cleavage there is no growth, but simply subdivision of the cytoplasm of the egg and the restoration of a more normal cell size (i.e., an increase in the nuclear/cytoplasm ratio). Following the cleavage period there occurs an extensive movement and rearrangement of the cells (*gastrulation*), transforming the previously one-layer-thick embryo into a two- to three-layer-thick embryo (Fig. 1). Typically, there is a decrease in the rate of cell division at this time, and the first evidences of gene transcription become apparent. This transcription in part specifies the proteins required for gastrulation. Transcription later in development specifies the proteins characteristic of the *differentiated* state (e.g., acetylcholinesterase in nerve, myosin in muscle, and melanin in epidermis).

This chapter concerns earliest development, from the formation of the egg, through fertilization and cleavage, to the beginning of gastrulation. We review experimental evidence that the cellular machinery carrying development to gastrulation is laid

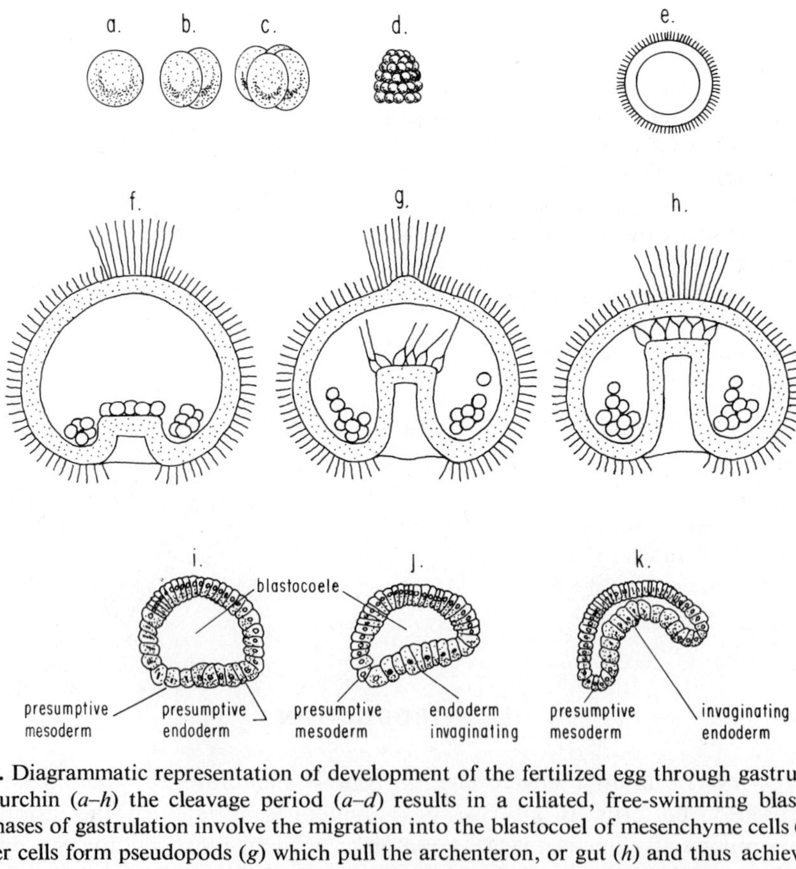

Figure 1. Diagrammatic representation of development of the fertilized egg through gastrulation. In the sea urchin (*a–h*) the cleavage period (*a–d*) results in a ciliated, free-swimming blastula. The initial phases of gastrulation involve the migration into the blastocoel of mesenchyme cells (*f*). These and other cells form pseudopods (*g*) which pull the archenteron, or gut (*h*) and thus achieve separation of the germ layers. Gastrulation of a cephalochordate is shown in cross section in *i–k*. Pseudopodial contraction does not seem to be involved, rather gut invagination involves changes in cell shape.

down during the period of egg formation, and that this machinery is activated at fertilization. We note that cleavage and formation of the blastula do not require parallel messenger RNA (mRNA) synthesis but are dependent on "cytoplasmic factors," that is, that nuclear activity is not required through the blastula stage. To be sure, gene transcription is required, but this is transcription that occurred earlier, during oogenesis.

We also review evidence that embryonic differentiation involves interactions between nuclei that are "equal" and cytoplasms that are "different." During cleavage equivalent cell nuclei end up in these different cell cytoplasms; the resultant interactions between an equivalent nucleus and a differing cytoplasm then yield the responses known as differentiation. The material for these differing cytoplasms is also present in the unfertilized egg, and their unequal distribution is initiated by fertilization. The cleavage pattern, then, insures that the equivalent nuclei end up in cells with differing cytoplasmic environments (Fig. 2).

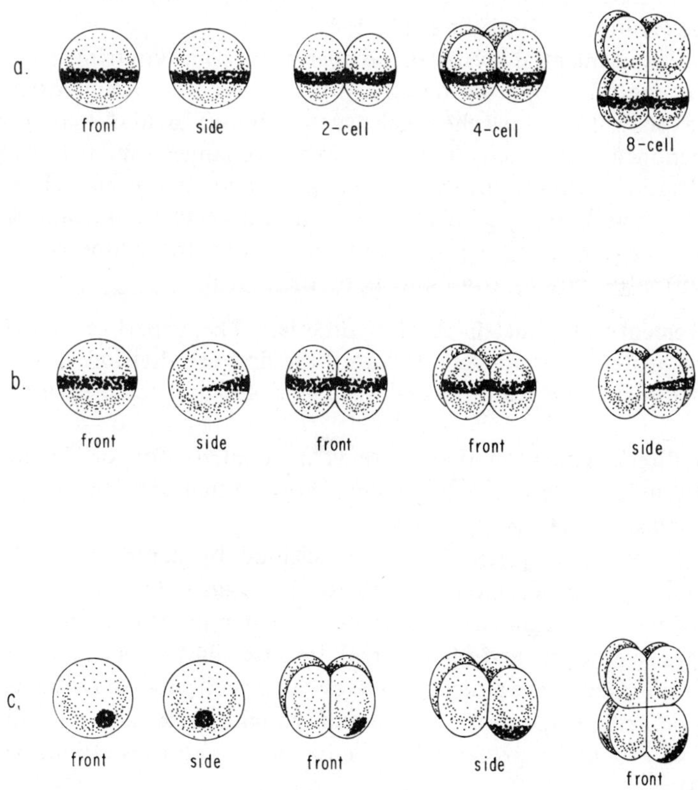

Figure 2. The cytoplasm is differently segregated during cleavage, depending both on the localization of the cytoplasmic inclusion and the particular cleavage pattern. (*a*) Uniform localization below the equator of the egg. Unequal distribution does not occur until the eight-cell stage. (*b*) Partial localization along equator. Equal distribution occurs at the two-cell stage, but unequal distribution takes place at the four-cell stage. (*c*) Highly unequal localization at one pole of the egg. Unequal distribution occurs at the first division (two-cell). Only one of eight blastomeres retains the material at the eight-cell stage.

2. DEVELOPMENTAL INFORMATION PRESENT IN THE OVUM

A large number of experiments indicate that the cellular machinery for early development is already present in the unfertilized egg. This machinery is made under nuclear control during oogenesis; once activated by fertilization, it can proceed with little or no direction from the nucleus. Much of this machinery is concerned with fertilization and cleavage. Some of it, however, is concerned with gastrulation.

A. Information for Fertilization and Cleavage

1. Requirements for Protein Synthesis. Two important experiments clearly show that the changes occurring at fertilization represent activations of preexisting enzymes, as opposed to the *de novo* synthesis of new enzymes. The first experiment shows that the synthesis of new proteins is not required for fusion of sperm and egg nuclei and the first cycle of DNA synthesis. Eggs can be fertilized in the presence of puromycin, an inhibitor of protein synthesis. Although 90% of protein synthesis is inhibited, the eggs develop up to the first cleavage. Chromosome separation takes place, but the egg fails to divide (Hultin, 1961; Wilt et al., 1967).

The second experiment shows that the sperm does not contribute a missing component to the egg that might be required for the activation of preexisting enzymes. Eggs briefly exposed to various chemicals, such as butyric acid or hypertonic seawater, can be parthenogenetically activated. They exhibit changes similar to those following normal fertilization with sperm; they divide, gastrulate, and even yield normal adults (Tyler, 1941). These two experiments, then, indicate that the metabolic response to insemination is strictly an activation, requiring neither the synthesis of new proteins nor the provision of missing components by the sperm.

2. Requirements for Nucleic Acid Synthesis. The experiment with puromycin indicates that proteins are not required until the first cell division. Other experiments indicate that information from the cell nucleus is also not required until the blastula stage.

The most direct indication that there is no contribution of the nucleus to the cleavage pattern is provided by experiments showing that anucleate eggs can "divide" and form a "blastula" (Harvey, 1936).

Unfertilized sea urchin eggs can be enucleated by centrifuging at high speeds against a dense layer or "cushion" of sucrose. The eggs sediment until they reach this cushion and then layer against the sucrose–seawater interface. The centrifugal field causes the eggs to elongate and finally break into two halves, one containing a nucleus and the other anucleate. The anucleate halves can be collected and artificially activated, as with butyric acid. Some of the eggs actually divide in a manner similar to normal division. Some eventually produce a few cilia, similar to those found at the blastula stage (Fig. 3).

Another means of ascertaining the role of gene transcription is to assess the effects of actinomycin D on early development. This drug inhibits DNA-dependent RNA synthesis. Sea urchin eggs can be fertilized and develop through the blastula stage in the presence of this drug (Gross et al., 1964). Development, however, is abnormal in that mRNAs for histones are not made (Kedes et al., 1970), and the chromosomes are correspondingly aberrant (Kiefer et al., 1969). Nevertheless, we can conclude that

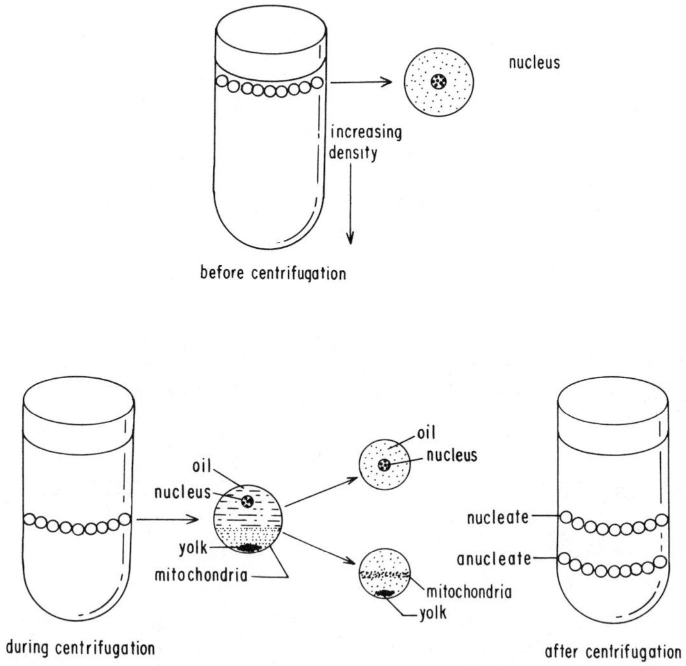

Figure 3. Production of anucleate eggs of the sea urchin. Unfertilized eggs are placed on a sucrose density gradient. During centrifugation the eggs equilibrate at a point in the gradient corresponding to the density of the eggs. The components of the egg then sediment within the egg, and the egg elongates and splits into two halves. The two halves then equilibrate within the tube according to their density.

nuclear direction, at least in terms of the synthesis of mRNAs, is not required for blastula formation.

3. Stable mRNA as One Form of Information. The experiments with actinomycin D indicate that RNA synthesis is not required for cleavage. Conversely, experiments with puromycin indicate a requirement for protein synthesis. This suggests that mRNA molecules might already be present in the unfertilized egg, and that translation of these molecules is required for cleavage. This hypothesis is supported by the direct demonstration of mRNA in eggs (Slater and Spiegelman, 1966) and by the finding that the rate of protein synthesis is the same in (1) artificially activated anucleate eggs, (2) fertilized eggs exposed to actinomycin D, and (3) normally fertilized eggs (Denny and Tyler, 1963; Gross et al., 1964). These experiments, then, suggest that part of earliest development—at least the part requiring protein synthesis—is dependent on mRNA already present in the unfertilized egg (Spirin, 1966).

This mRNA is presumably made earlier, during oogenesis, stored in the egg, and not translated until fertilization. We shall say more about the mechanism of translation later (Section 6A.4). For the moment, the important conclusions from these experiments are that products made during oogenesis specify the types of protein synthesized and the morphological pattern of cleavage.

B. Information for Gastrulation

Gene products are also made during oogenesis that are not required until gastrulation. One of the most interesting examples of this comes from studies of the lethal developmental mutant *o* found in the axolotl (R. R. Humphrey, 1966). *o* is a recessive gene whose action is expressed during gametogenesis but whose effect is not seen until gastrulation. Homozygous females produce eggs whose development is arrested at the gastrula stage. This arrest is independent of the sperm genotype (*OO* or *Oo*). This could result if the lethal allele present in the egg were dominant. However, heterozygous females (*Oo*) crossed with heterozygous (*Oo*) males all produce viable offspring (the homozygous male is sterile). Hence the lethal effect of gene *o* is only apparent in eggs produced by homozygous females, and the lethal effect of this recessive gene cannot be corrected by the sperm genotype. This suggests that the gene *O* is operative during oogenesis, and that the product of this gene is not required until gastrulation.

3. GENE EXPRESSION DURING EARLY DEVELOPMENT

As described below (Section 3.A), differentiation does not involve any irreversible changes in the genome. Except in a few isolated cases, the entire genetic potential of an organism is present in the nucleus of every cell. This is true for a primordial germ cell destined to give rise to a sperm or egg cell; it is equally true for a differentiated cell, such as a muscle or intestine cell.

The critical factor governing nuclear or gene expression is the particular microenvironment around the nucleus. During the cleavage period equivalent nuclei are formed and are partitioned into differing cytoplasms. These differing cytoplasms are formed during oogenesis; their unequal distribution results in part from rearrangements of the cytoplasm and egg surface after fertilization and during cleavage. The differing responses of these nuclei to these different cytoplasms can then account for differentiation.

A. Reversibility of Nuclear Metabolism

The evidence that cell nuclei do not change irreversibly during differentiation comes from a wide variety of experiments. Here we examine three of them.

The first experiment concerns the question whether any irreversible changes occur in the nucleus during development. This question has been approached experimentally by the technique of nuclear transplantation, using the relatively large and easily handled amphibian egg. The method involves removing the nucleus from an unfertilized egg and replacing (transplanting) it with a nucleus from another egg or from a later, more differentiated cell. The transplantation operation also activates the egg, so that fertilization by sperm is unnecessary (Fig. 4).

The results of such experiments indicate that normal tapdoles can develop from an egg with a nucleus transplanted from another unfertilized egg or with a nucleus transplanted from a differentiated cell (Gurdon, 1964; Gurdon and Woodland, 1968). The conclusion from these experiments therefore is that no irreversible changes of the nucleus occur during differentiation.

Some interesting variations on these experiments, elegantly demonstrating the reversibility of nuclear metabolism, have been carried out by Gurdon and his col-

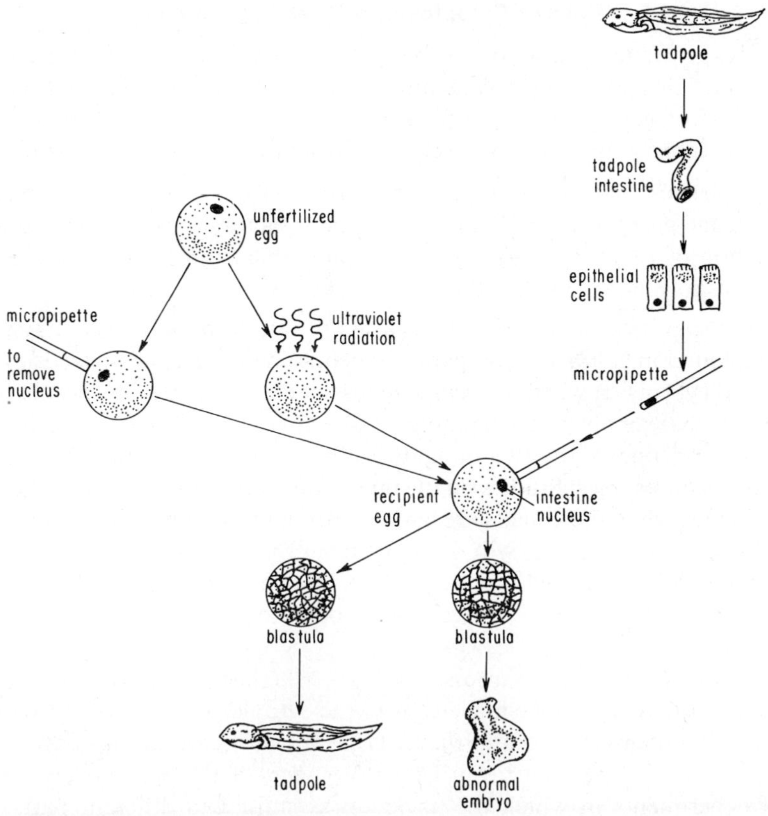

Figure 4. Nuclear transplantation experiments show that nuclei of differentiated cells do not change irreversibly. The nucleus of an unfertilized egg is removed mechanically or by ultraviolet irradiation. Into this egg is injected a nucleus derived from an intestinal cell of a tadpole. The recipient egg is activated, and in a significant number of cases gives rise to a normal tadpole.

laborators. One experiment, which demonstrates the *loss* of a differentiated nuclear function, involved transplantation of a nucleolate nucleus (i.e., containing a nucleolus) from a gastrula stage embryo into an unfertilized egg. Neither the unfertilized egg nucleus nor cleavage stage nuclei contained a nucleolus; the nucleolus of the transplanted nucleus quickly disappeared upon transplantation (Gurdon and Brown, 1965).

The reverse experiment, demonstrating *acquisition* of a function lost during differentiation, can also be shown. Thus, during differentiation of nerve cells, the ability to duplicate DNA and undergo mitosis is lost. Associated with this loss is the absence of the enzyme DNA polymerase. However, if such a polymerase-free nerve cell nucleus is injected into an unfertilized egg, this nucleus acquires the ability to make DNA (Graham et al., 1966). This acquisition of DNA polymerase probably involves the selective transport of the enzyme from the egg cytoplasm into the nucleus (Loeb and Fansler, 1970).

These experiments indicate the reversible nature of nuclear changes during development. Most importantly, they suggest that the environment around the nucleus is critical for these changes. As this environment is the cytoplasm of the egg, the cytoplasm may be the important factor regulating the differential activity of the nucleus.

B. Role of Cytoplasm in Directing the Nucleus

Several experiments point to the cytoplasm as containing information of major developmental import. Here we examine two examples: (1) the cytoplasm necessary for formation of germ cells, as studied in vertebrates and invertebrates, and (2) the cytoplasm necessary for certain mesodermal structures, as studied in the invertebrates.

1. Cytoplasmic Factors Determining Germ Cells. The germinal plasm is the evocative name given to cytoplasmic structures or factors present in eggs and essential for formation of the germ cells. Embryos from which this plasm has been removed develop into normal adults, but they are sterile.

(*a*) *Chromosome Elimination.* The nature of this germinal plasm may vary within the animal kingdom. In *Ascaris*, a parasitic nematode worm, the plasm is evidenced as a specialized cytoplasm which *prevents* the loss of chromosomes in specific cells during cleavage. In these embryos, beginning at the first division, there occurs a loss of chromosomes in one of the two daughter cells. Finally, only one cell retains a complete set of chromosomes. Studies on the fate of this cell show that this is the primordial germ cell. Thus, when the gonads are later formed and rapid division of this primordial germ cell occurs, the gametes will originate from the division products of this cell. As only the primordial germ cells contain these extra chromosomes, it appears that these chromosomes are necessary for gamete production but not for normal somatic differentiation.

The question crucial to our discussion is whether the factors controlling this chromosome loss reside in the nucleus or in the cytoplasm. A series of experiments by Boveri, a brilliant embryologist of the early 1900s, showed that the control of this loss is localized in the cytoplasm toward the yolky end of the egg. Boveri carried out a series of experiments in which the nucleus was shifted to different parts of the egg by subjecting eggs to a centrifugal field. Nuclei distant from the yolky cytoplasm showed the chromosome loss following mitosis. Nuclei in close proximity to this cytoplasm retained their chromosomes. Thus the cytoplasm of the egg, rather than the nucleus, controls the chromosome retention in the germ cell line of *Ascaris* (see Wilson, 1925).

The above experiments seem to be in contradiction to the earlier generalization that differentiation does not involve progressive or irreversible changes in the genome. This exception to the rule has been observed only in some nematodes and a few orders of insects, and is restricted to the separation of somatic cells from germ cells.

This chromosomal behavior, however, suggests that those chromosomes unique to the germ cells are involved in gamete formation. This hypothesis is further supported by studies demonstrating that those chromosomes unique to the germ cells are synthetically active during gametogenesis, at times when the other chromosomes are inactive. In gall midges two cell types can be distinguished on the basis of their chromosome complement. As in *Ascaris*, germ cells retain all chromosomes; somatic cells lose a portion of the chromosome complement. During early oogenesis those chromosomes specific to the germ cells are diffuse within the nucleus; autoradiography shows them to be actively synthesizing RNA. Conversely, the somatic chromosomes are tightly coiled and are surrounded by fibrillar lamellae and do not actively synthesize RNA (Kunz et al., 1970).

(*b*) *Germinal Plasm.* Although chromosome elimination is an exception, it does indicate the special nature of the germ cells as well as the role of the egg cytoplasm in chromosome elimination. Studies on gametes of other organisms also reveal that gamete formation is controlled by cytoplasmic factors, but that these cytoplasmic factors do not affect the chromosomes in so obvious and direct a fashion. In the cases studied—the best studies are with fruit flies and amphibia—there appear to be specific areas of the egg cytoplasm that are partitioned into specific cells during cleavage. These cytoplasmic areas (germinal plasm), which can be recognized by their peculiar cytochemical staining properties, are essential for future germ cell formation. Two types of experiments illustrate this nicely. The first experiment involves studying the effects of the destruction of this germinal plasm. The second experiment attempts to repair this injury by adding (injecting) back cytoplasm containing the putative germinal plasm and again observing the developmental effects.

The first experiment—observation of the effects of germ plasm removal—can be done fairly simply. In eggs of *Drosophila* the germinal plasm is localized at one end of the egg. When this area is damaged, as by cauterization, a normal but sterile adult is produced. Similarly, if this end of the egg is removed by ligation, a sterile adult results. Finally, the same lesion (i.e., sterility) is obtained if the germinal plasm area is irradiated with ultraviolet light. Control experiments involving cauterization, ligation, or ultraviolet irradiations of areas adjacent to the putative germ cell area have no effect and fertile adults develop (reviewed by Tyler, 1955; Smith, 1966).

These experiments indicate that a specific *cytoplasmic* area is involved in the formation of germ cells. If this is the case, one should be able to restore "fertility" to embryos whose germinal plasm has been previously removed.

The testing of this hypothesis has been done with eggs of the common frog *Rana*. In these eggs the germinal plasm is localized in the yolk end of the egg. Ultraviolet irradiation of this region results in sterility, whereas irradiation of the opposite pole of the egg has no effect. The critical experiment, then, was to irradiate the germinal plasm region of eggs and then ascertain whether or not fertility could be restored by injecting back into this egg some cytoplasm from untreated eggs. As predicted, injection of cytoplasm from the germinal plasm area of an unirradiated egg restored fertility; injection of cytoplasm distant from this area had no effect (Smith, 1966).

What is the nature of this specialized cytoplasm? Cytochemical studies indicate it is an area rich in RNA. Electron micrograph studies reveal electron-dense structures (Fig. 5). Surprisingly, these structures are similar in both *Drosphila* and amphibian eggs. (Mahowald, 1962; Mahowald and Hennen, 1971). This similarity is a remarkable example of evolutionary conservation of structure and function and suggests that the control of germ cell differentiation may be similar in widely separated organisms.

2. Cytoplasmic Factors Determining Organ Formation. The second example of the cytoplasmic control of development can be illustrated by experiments on the development of embryos of a mollusc, *Ilynassa*. These embryos possess a very peculiar cleavage pattern. At the first cell division, in addition to two equal daughter cells, a third "cell" is formed at one pole of the egg. However, this cell contains no nucleus and is referred to as the polar lobe. After the division of the two cells, the cytoplasm of the polar lobe flows into one of the cells. At the next division the cell that received the polar lobe material again divides unequally. A polar lobe is formed and again, after the division is completed, the cytoplasm of this lobe flows into one of the cells. The same process is repeated at the third division. The amount of cytoplasm in the polar lobe decreases at each division (Fig. 6).

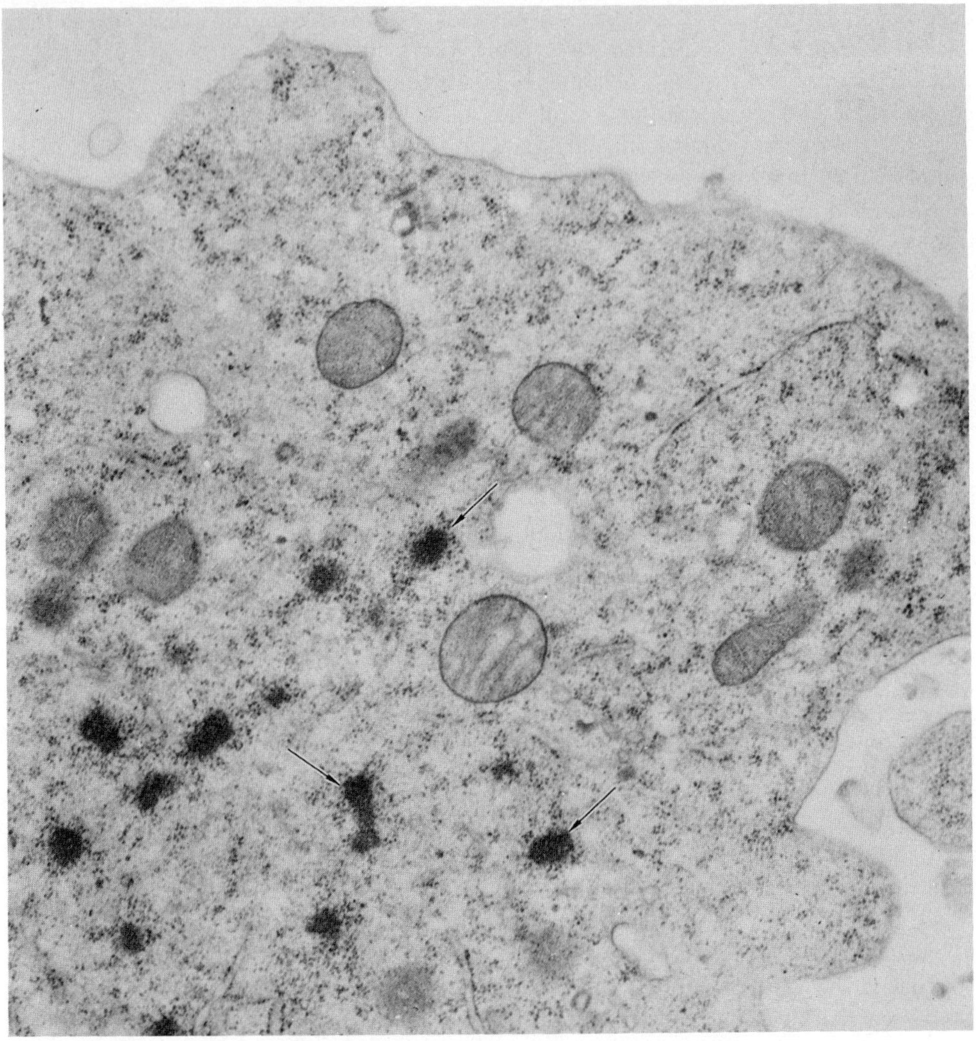

Figure 5. Electron micrograph of the germinal plasm area of *Drosophila melanogaster*. The arrows point to the polar granules characteristic of the germinal plasm. (Courtesy of Dr. A. P. Mahowald.) ×20,000.

The cytoplasm of this anucleate lobe has important developmental significance. The cells into which the polar lobe flows, during normal embryogenesis, give rise to such structures as the heart, muscle of the foot, and the statocysts (gravity-sensing structures). The significant experimental observation, in relation to the role of polar lobe cytoplasm, is that if the polar lobe is removed those structures that would have developed from the cells receiving the lobe material do not appear (Clement, 1962). Thus embryos from which the polar lobe has been removed do not contain heart, foot, or statocysts (Fig. 6).

This experiment indicates that the polar lobe cytoplasm is somehow involved in the differentiation of these organs. The effect of the lobe cytoplasm is probably upon the cell nucleus. At the time of heart and foot differentiation, the amount of RNA synthesis is reduced in embryos lacking the polar lobe (Davidson et al., 1965).

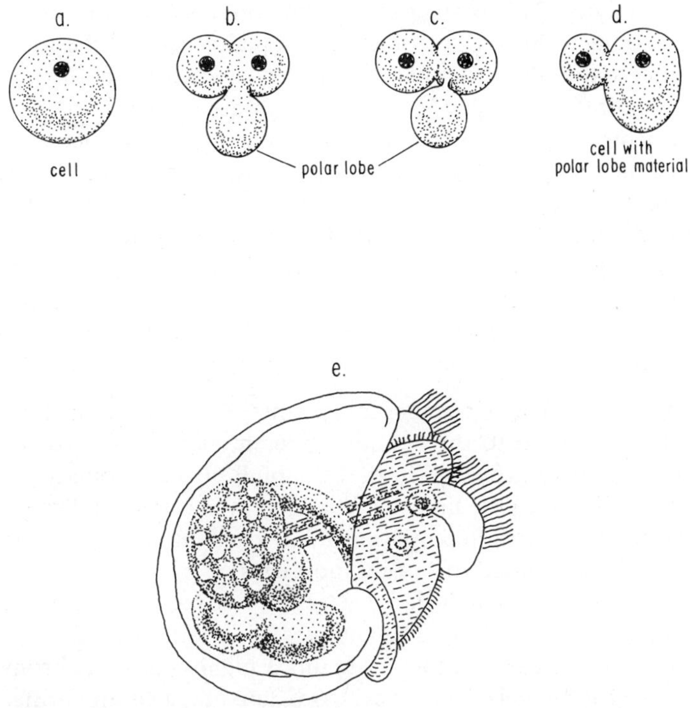

Figure 6. Diagrammatic representation of polar lobe formation in the mollusc *Ilynassa* (*a–d*), and the effects of polar lobe removal on the veligar larva of *Ilynassa*. (*a*) Fertilized egg; (*b*) polar lobe during cleavage; (*c–d*) flow of polar lobe material into one blastomere after cleavage; (*e*) veliger larva of *Ilynassa*. If the polar lobe is removed at the first division, the structures indicated by broken lines are absent in the larva.

4. GAMETOGENESIS

A. Nuclear Activity During Gametogenesis

The preceding discussion indicates that the period of egg formation is a most crucial period of development. We noted that earliest development, at least through the blastula stage in sea urchins, is independent of direct or immediate nuclear control. However, this seeming independence in reality means that the pertinent gene transcription simply occurred at some earlier time, during oogenesis. These gene actions include the synthesis of mRNAs needed for cleavage. They also include the formation of specific cytoplasms, such as the germinal plasm or polar lobe, which later interact with the nucleus during embryogenesis and thus direct development.

The phenomena of oogenesis are thus of major interest to cellular and developmental biologists. In the following discussion, we examine some selected aspects of this important process. First we look at some of the more "dramatic" problems of oogenesis, especially those involved in the synthesis of the bulk material of the unique egg cell. Following this, we examine the phenomena of ovulation, maturation, and fertilization.

1. Oogenesis. As a generalization (to which there are exceptions), it can be stated that the growth and differentiation of the oocyte into the mature egg requires continuous genetic activity. This is suggested by the presence, during the early stages of oogenesis, of extremely high rates of chromosomal RNA synthesis. This synthetic activity is so intense that the chromosomes are rendered visible in the light microscope by the heavy concentrations of RNA and protein associated with the chromosomal fiber. Measurements of rates of RNA synthesis show the oocyte nucleus (or the nucleus of its assistants, the nurse cells) to be one of the most intensive sites of RNA synthesis in the organism. (See review by Davidson, 1968.) Also correlated with this high rate of gene activity is the fact that gene reduction (meiosis) does not take place until after the growth of the eggs has been completed. Thus, the egg is formed with a ploidy of $4n$. Also, as described later (Section 4C.2), many oocytes reduplicate part of the genome so as to amplify the number of genes available for synthesizing RNA. As such, the real ploidy is actually $4n+$.

2. Spermiogenesis. The situation in spermiogenesis is almost completely the opposite. First, as opposed to the egg, the differentiation of the sperm entails what might be called "negative growth," in that the bulk of the cytoplasm is sloughed off or shed from the cell during the final stages of sperm maturation. Second, as opposed to the egg, the final sperm differentiation takes place after meiosis, with a genome corresponding to n (as opposed to $4n+$ in most eggs).

This latter fact suggests that little gene activity is required during the period of sperm formation (spermiogenesis). The analyses of certain meiotic mutants of *Drosophila* show this to be the case. In these mutants a highly unequal chromosome distribution takes place during meiosis. In some cases daughter cells are formed that contain no chromosomes. Nevertheless, these achromosomal cells can undergo spermiogenesis (Lindsley and Grell, 1969). Since protein synthesis occurs during spermiogenesis (Brink, 1968), it appears that stable mRNAs necessary for spermiogenesis are made before the meiotic divisions.

B. Problems of Egg Synthesis

Whether an ostrich egg or a human egg is being discussed, the ovum generally is one of the largest cells produced by an organism. The types of syntheses that must be accomplished by the developing oocyte include those of nutrient material (yolk) as well as large numbers of cellular organelles, such as mitochondria, microtubules, ribosomes, and lysosomes.

As an example of the magnitude of these syntheses, consider the relative number of mitochondria in a normal cell and in an egg. In a typical somatic cell, the amount of nuclear DNA far exceeds that of mitochondrial DNA. Indeed, the amount of nuclear DNA is so much greater that the existence of mitochondrial DNA escaped detection for many years. The reverse situation holds true in the oocyte. Here there are so many mitochondria that the amount of mitochondrial DNA far exceeds the amount of nuclear DNA. Indeed, a puzzling problem in embryology concerns the role of this "cytoplasmic DNA." Recent research has brought to light the fact that a large portion of the cytoplasmic DNA is actually DNA of the mitochondria (Dawid, 1966; Piko et al., 1967).

The magnitude of the egg's synthetic "problems" also varies in relation to the size and numbers of eggs produced, as well as the amount of time available to form the eggs. For example, compare the differences in (1) the human producing one or two 150-μ diameter eggs per month; (2) the domestic hen, producing one 8-cm egg per day; and (3) a short-lived insect, such as the fly, producing several hundred eggs 5 mm long within a few weeks.

The amount of nutritive material required also varies with the organism. In mammals, in which nutrients are provided to the embryo by diffusion from the maternal blood stream through the placenta, the egg is relatively small (e.g., 150 μ in humans; 110 μ in rabbits). In many marine organisms the egg develops rapidly into a motile larva which obtains the nutrients necessary for growth and metamorphosis by feeding on other planktonic organisms. In these organisms the eggs are also relatively small and yolk-free, varying in diameter from 50 to 150 μ. (Organisms having planktonic larvae characteristically produce large numbers of gametes. A sea urchin for instance, can produce 20–100 million eggs per breeding season. One can understand their attraction for the experimental embryologist, since these eggs can be easily spawned from the female and fertilized *in vitro*.) Finally, in many organisms (such as birds, reptiles, fish, and some molluscs), a very large egg is produced which develops directly, without metamorphosis, into a miniature adult.

C. Strategies of Egg Synthesis

As might be expected, the particular strategies employed by different organisms in making eggs varies. Part of this variation is related to the size of the egg and the time available for oogenesis. If a generalization can be made, it is that formation of the egg is generally carried out with the assistance of other cells, that is, that oogenesis is a dependent rather than an independent phenomenon. This dependence is occasionally on extraovarian tissue, such as the liver of birds or the fat body of insects. However, it is usually seen as a dependence on other ovarian cells known as nurse or follicle cells (Telfer, 1965).

1. Synthesis Dependent on Other Cells. As an example of a synthesis that depends on other cells, let us examine yolk formation in chickens and in silkworms (*Cecropia*). In these organisms there is a sex-linked blood protein found only in females and only during periods of egg formation. This blood protein is identical to one of the yolk proteins of the egg; there is excellent evidence that this protein is picked up by the ovary and transferred to the egg (Telfer, 1965).

In the chicken the most thoroughly studied of these proteins are phosvitin, which is made in the liver, and ovalbumin, made in the oviduct. Hormones control the synthesis of both of these proteins. Young roosters, for example, do not ordinarily synthesize phosvitin but, when injected with estrogen, do so. (Greengard et al., 1964).

Chickens and silkworms therefore activate numerous nongametic cells that participate in the synthesis of a major egg constitutent, yolk. Phosvitin is transported to the egg through the blood stream; ovalbumin is added to the egg as it passes through the oviduct.

Perhaps the most interesting example in which oogenesis is dependent on other cells is found in some species of insects. Here growth of the oocyte is almost completely dependent on nurse cells. Nurse cells surround the developing oocyte and are connected to it via cytoplasmic channels. Ribosomes and even mitochondria pass through these channels from the nurse cells to the egg (reviewed by Telfer, 1965).

If the ovary is exposed to radioactive precursors of RNA, the nuclei of the nurse cells become intensely labeled, indicating active RNA synthesis. Later in development part of this label can be found in the oocyte, implying transportation of RNA from nurse cell to egg. The nucleus of the developing oocyte, however, does not become labeled, indicating inactivity of RNA synthesis. Thus in these organisms nurse cells attached directly to the developing oocyte generate constituents of the egg.

2. Synthesis Dependent on Gene Amplification. Above, we emphasized the role of other cells or organs, such as nurse cells or the liver, in providing preformed materials for the oocyte. Such dependence on other cells occurs primarily in insects and birds. In most situations, however, the developing oocyte itself directs part, and occasionally all of its synthetic processes.

As regards protein synthesis, the normal transcription–translation process leading to protein has built-in amplification. Thus several mRNA molecules can be transcribed from each gene. Amplification takes place since each of these mRNA molecules in turn can be translated several times into the corresponding protein.

There can be no amplification in rRNA formation, however, since there is only one step involved (transcription). Perhaps for this reason cells have 200 to 500 duplicate copies of the rRNA gene (Brown and Dawid, 1968), which (by definition) are sufficient for the average cell. However, these extra copies appear to be inadequate for the synthesis of the large number of ribosomes of the oocyte. In some cases, as noted, nurse cells provide ribosomes to the oocyte. In other organisms the oocyte specifically duplicates (or amplifies) the ribosome gene. This phenomenon is referred to as gene amplification. That gene amplification does indeed take place can be shown by a number of elegant techniques. One of these involves the separation of the genes specifying rRNA from the remainder of the genome.

Equilibrium sedimentation of DNA from somatic cells or from sperm cells of the toad *Xenopus laevis* reveals only a single DNA band. The oocyte, however, has three DNA bands of differing density. One corresponds to the somatic DNA and one to mitochondrial DNA. The third band has the density properties expected of the genes specifying rRNA (Brown and Dawid, 1968).

If this third DNA band really corresponds to the DNA specifying rRNA, this DNA should specifically bind to purified preparations of rRNA. This can be tested since the third DNA band can be separated from the other two bands and its binding (i.e., DNA–RNA hybridization) tested. This DNA does bind to rRNA. This evidence suggests that gene amplification for the rRNA gene does occur. In *Xenopus*, calculations of the extent of this amplification indicate a 1000-fold increase in the number of rRNA genes. (As there are 450 copies of the rRNA gene in a normal *Xenopus* cell, there are therefore 450,000 copies of the rRNA genes in oocytes.)

Where are these extra copies located? rRNA synthesis takes place in the nucleolus, and it is therefore reasonable to assume that these extra copies may also reside in the nucleolus. Oocytes often contain multiple nucleoli; in *Xenopus*, the number of nucleoli is estimated at 1000. This correlation suggests that each nucleolus contains a cluster of rRNA genes.

This correlation can be tested, since the gene product (rRNA) can be made radioactive and purified, and since the binding of RNA to its complementary DNA is highly specific. Therefore DNA–RNA hybridization combined with autoradiography can be used to localize the specific gene for rRNA. If radioactive rRNA is applied to toad oocytes under conditions in which nonspecific binding is eliminated, the rRNA binds to the nucleoli. This suggests that the amplified rRNA genes are indeed associated with each of the extra nucleoli (Gall and Pardue, 1969).

5. MEIOSIS AND OVULATION

As described earlier, a difference between egg and sperm formation is that the egg is formed before meiosis, whereas the sperm forms after the completion of meiosis. Here we concentrate on certain cellular events associated with meiosis, ovulation, and the preparation of the egg for fertilization. The oocytes of frogs and starfish have been intensively studied, hence the findings obtained with these organisms are emphasized.

Upon completion of oogenesis in both these animals, the primary oocyte is surrounded by a follicle composed of follicle cells. At ovulation the egg detaches from the follicle and begins the meiotic process preparatory to release into the environment and fertilization. In the starfish ovulation and meiosis, once initiated, are completed in 1 hr. The meiotic divisions occur rapidly and no external coats are added to the egg during its passage through the oviduct. However, in the frog a longer and more complex process occurs. Once initiated, the meiotic divisions require approximately 24 hr. During meiosis the gametes pass through the oviducts where various extracellular jelly layers necessary for fertilization are deposited around the eggs.

The relationship between ovulation, meiosis, and fertilization varies with the organism. In some (e.g., *Urechis*, a wormlike organism belonging to the phylum Echiuroidea), the oocyte is ovulated before meiosis, and meiosis begins after insemination. In others, the egg proceeds to the first (some molluscs) or second (vertebrates) meiotic metaphase, and fertilization induces the completion of meiosis. Sea urchin eggs complete meiosis before ovulation and fertilization.

A. Amphibians

For many years it was known that ovulation in amphibia can be induced by the injection of crude extracts of the pituitary gland. The hormone acts directly on the ovary, since fragments of ovary exposed directly to the extract also commence meiosis. However, the action of the extract is not directly on the eggs but on another ovarian component. For example, if eggs are isolated free of the follicle cells and exposed to pituitary extract, meiosis is not initiated. This suggests that the follicle cells secrete another substance in response to pituitary hormone, and this substance is then inducing meiosis. Evidence is now available that the substance produced by the follicle cells is progesterone (Masui, 1967; Schuetz, 1967). The application of progesterone to eggs previously removed from the follicle results in the initiation of meiosis.

Frog eggs can be fertilized or artificially activated only after the completion of meiosis. Can progesterone-treated eggs be fertilized? This is difficult to answer since these eggs have not passed through the oviduct, hence lack the jelly layer required for fertilization. However, their capability to be artificially activated by a pinprick can be tested. Progesterone-treated eggs can be activated and appear similar to normally ovulated eggs (Smith et al., 1968). It appears, then, that the mechanism of maturation and ovulation in amphibia involves the pituitary-mediated stimulation of progesterone synthesis by the follicle cells surrounding the eggs.

B. Starfish

A mechanism similar in principle but different in detail is involved in meiosis and ovulation of starfish oocytes. Early experiments revealed that a heat-labile substance could be extracted from the radial nerve of starfish, which, when injected into a gravid starfish induced the release of gametes (Chaet, 1963). This neurosubstance has been purified and is a polypeptide of approximately 23 amino acids (Kanatani, 1969).

The target organ for the polypeptide hormone is the ovary. When fragments of ovaries are exposed to the hormone, the eggs in the fragments and those external to but in close proximity to the fragment begin meiosis at the same time. Eggs far removed from the ovarian fragment do not begin meiosis or do so at a much later time. This suggests that the ovary makes a substance in response to the polypeptide hormone, and that this substance induces meiosis and ovulation (Schuetz and Biggers, 1967). Further work showed that a low-molecular-weight, heat-stable substance is released by the wall of the ovary in response to the polypeptide hormone. In a series of brilliant studies made by Kanatani et al. (1969), this substance was identified as 1-methyladenine.

Exposure of isolated eggs or the intact ovary to 1-methyladenine, or injection of this chemical into the female, results in meiosis and ovulation. It appears that 1-methyladenine induces the loss of adhesion among the follicle cells. Once adhesion is lost, the natural turgor of the ovarian wall results in the release of the gametes through the oviducts and into the seawater (Kanatani, 1969).

C. Causal Analysis of Meiosis

What is the mechanism of the hormonal induction of meiosis in amphibians by progesterone and in starfish by 1-methyladenine? Complex sequences of events are initiated by these hormones, including changes in the concentration of cations, water uptake, and the synthesis of protein and RNA. The primary effect of 1-methyladenine is on the surface of the cell. When this compound is microinjected into the oocyte, meiosis is not initiated (Kanatani and Hiramoto, 1970). Although we do not know the exact nature of this effect, some of the secondary effects of the hormone are required for the completion of meiosis or for cleavage. Three of these are the breakdown of the germinal vesicle, the synthesis of RNA, and the synthesis of protein.

1. Role of Germinal Vesicle Constituents. During oogenesis almost all oocytes contain an inordinately large nucleus referred to as the "germinal vesicle." A wide variety of evidence suggests that some components of this vesicle released into the cytoplasm when the germinal vesicle breaks down at the beginning of meiosis are necessary for subsequent development. The analysis of these components has been best studied with amphibian oocytes, and we discuss them here.

(*a*) *Fertilization.* The oocyte can be fertilized or artificially activated only after the completion of meiosis. Are any germinal vesicle components involved in this change? Apparently not. If the germinal vesicle is removed and the oocyte exposed to progesterone, they can become artificially activated (Smith and Ecker, 1969).

(*b*) *Cleavage.* A component released by the germinal vesicle is necessary for cleavage. If the germinal vesicles of progesterone-treated ova are removed and the eggs artificially activated under conditions that should lead to cleavage, the eggs do not divide. Proof that the substance required for division is derived from the germinal vesicle is provided by experiments in which the contents of isolated germinal vesicles are injected back into previously enucleated eggs. These eggs divide when artificially activated (Detlaff et al., 1964; Smith and Ecker, 1969).

(*c*) *DNA Synthesis.* Substances required for DNA synthesis may also be localized in the germinal vesicle. If one injects DNA into oocytes before meiosis, no DNA synthesis can be demonstrated (as by autoradiography). However, if DNA is injected after germinal vesicle breakdown, this DNA is replicated. The ability to synthesize DNA is acquired coincidentally with the rupture of the germinal vesicle (Graham et al., 1966).

(*d*) *Gastrulation.* Finally, there are substances present in the germinal vesicle, which are required for development beyond the gastrula stage. We earlier mentioned the studies on the mutant *o*. The genetic studies indicated that although the product of this gene is made during oogenesis it is not required until gastrulation.

Experimental studies definitively show that this gene product is stored in the germinal vesicle and released into the cytoplasm when the vesicle breaks down. That this product is already present in the egg is shown by the finding that the genetic deficiency of *o/o* eggs can be corrected by injection into these eggs of cytoplasm of normal eggs. Experiments in which different types of normal cytoplasm are injected reveal that the corrective factor is present only in cytoplasm of eggs in which the germinal vesicle has broken down. If germinal vesicles are isolated and injected into mutant eggs, normal development occurs (Briggs and Cassens, 1966).

2. RNA Synthesis. Oocytes of pituitary-stimulated frogs synthesize RNA. Is this RNA necessary for the completion of the meiotic process? When actinomycin D, an inhibitor of RNA synthesis, is injected into eggs, there is no effect on the completion of meiosis. Therefore RNA synthesis is not required for meiosis (Smith and Ecker, 1970).

3. Protein Synthesis. Protein synthesis is initiated by application of progesterone. Is this synthesis required for the completion of meiosis? If puromycin, an inhibitor of protein synthesis, is injected into the eggs, meiosis is arrested (Smith and Ecker, 1970). This suggests, then, that protein synthesis is required for the completion of the meiotic process. However, RNA synthesis is not required. This further suggests that mRNA is already in the oocyte, and that proteins made from this mRNA are necessary for meiosis. As we shall see, there is also mRNA in the oocyte that specifies proteins necessary for cleavage. Translation of this mRNA, however, begins after fertilization.

6. FERTILIZATION

Broadly defined, fertilization concerns the interactions of the haploid sperm and egg resulting in the fusion of these cells and the restoration of the diploid genome. Associated with these interactions are profound metabolic changes which lead to cell division and the formation of a new individual. These metabolic changes are the subject of this section. For the reader interested in the morphological aspects of fertilization, several excellent review articles are available (Colwin and Colwin, 1967; Austin, 1968).

A. Fertilization as an Activation

Earlier, we presented evidence that the responses of the egg to insemination have been previously set up in the egg and are triggered by fertilization. We noted that (1) fertilization changes can occur in the absence of protein synthesis (i.e., eggs can develop up to the first cell division in puromycin), and (2) the sperm does not provide an essential component for activation (i.e., development can be artificially initiated by chemical or mechanical treatments).

What is the nature of these fertilization responses or "activations"? What metabolic and structural changes occur? How do they initiate development of a new individual?

Studies on sea urchin eggs reveal that a large number of structural and enzymic reactions begin after insemination (see reviews by Monroy, 1965; Epel et al., 1969a). The major morphological changes concern the cortical reactions, involving the breakdown of a peripheral layer of granules, 1 μ in diameter, which are located immediately below the plasma membrane. These granules lyse at fertilization, releasing their contents into the space between the egg and the fertilization membrane (strictly speaking, this membrane is not present in the unfertilized egg but is elevated at fertilization through reactions resulting from the cortical granule breakdown). These cortical reactions are involved in the hardening of the fertilization membrane, which prevents extranumerary sperm from entering the egg.

In addition, a diverse series of cytoplasmic reactions are initiated. These involve intracellular changes in the concentration of cations, especially those of sodium, potassium, and calcium. Changes also occur in the concentration of coenzymes such as NAD and NADP, in the rate of respiration, and in the rate of protein and RNA synthesis. Most important is the initiation of DNA synthesis (see Epel et al., 1969a).

An important experiment indicates that there are two parallel, but independent, reaction sequences initiated at fertilization. The first sequence comprises those surface reactions (or cortical reactions) involved in the reception of the sperm. The second, and parallel reaction sequence, involves cytoplasmic changes resulting in the metabolic activation of the egg. The observations suggesting that these cytoplasmic and surface changes are mutually independent come from studies on the effects of hydrostatic pressures on fertilization. Eggs were briefly pressurized during insemination. The eggs were fertilized, but the opening of the cortical granules did not occur. Yet, sperm and egg pronuclei fused and the embryo passed through the cleavage period (Whiteley and Chambers, 1961). These results indicate that the breakdown of the cortical granules is not required for initiating development.

The following discussion concentrates on the cytoplasmic changes beginning after fertilization, especially on the sequence and mechanism of these changes and on their significance in the activation of the egg. The reader should note the similarity of these changes to those occurring after hormonal stimulation of a tissue.

1. Ionic Changes. The earliest changes occurring after fertilization involve alterations in the ionic content of the egg, presumably resulting from permeability changes in the cell membrane. These ionic changes can be measured by direct chemical analysis, or indirectly, for example, by electrophysiological procedures.

If one measures the membrane potential with a microelectrode inserted into an egg, the membrane potential changes at fertilization (Hiramoto, 1959). A recent study shows that this change in membrane potential is one of the earliest detectable changes after fertilization, beginning within 10 sec after addition of sperm to the egg (Stein-

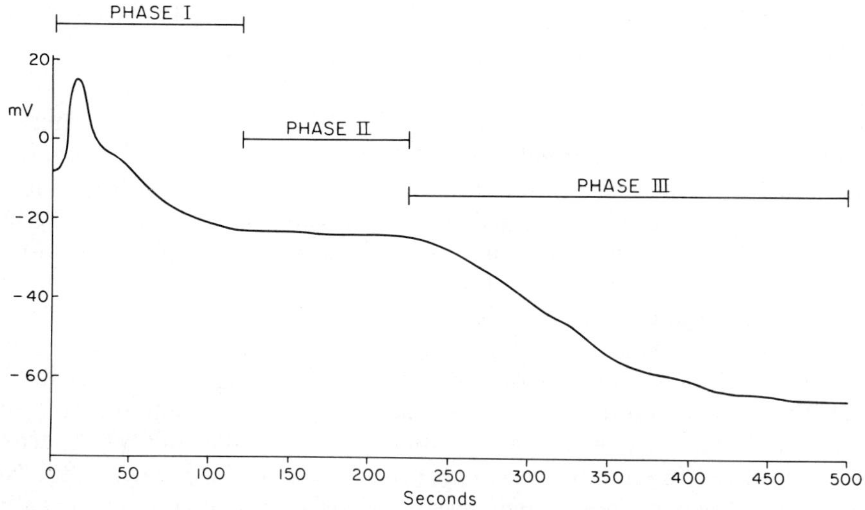

Figure 7. Changes in membrane potential following fertilization of sea urchin eggs. A microelectrode is inserted into the egg and sperm added at zero time. Phase I is the fertilization action potential and corresponds to an influx of Na^+ ion. The "pause" of phase II may be related to cortical granule membrane changes. Phase III corresponds to changes in the membrane relating to K^+ conductance and permeability. (Courtesy of Drs. R. Steinhardt, L. Lundin, and D. Mazia.)

hardt et al., 1971). At fertilization the membrane potential initially becomes more positive. The kinetics of this change are slow as compared to nerve, the rate increase being of the order of seconds rather than milliseconds. Following this initial increase the potential gradually becomes more and more negative (Fig. 7).

Measurement of the membrane potential in seawater of varying ionic composition shows that the potential in the unfertilized egg results from an unequal distribution of chloride ions (rather than potassium, as is the case in nerve, muscle, and presumably most other cells). After insemination the initial increase in potential results from an influx of Na^+ ion. The subsequent slow attainment of a negative potential results from an unequal distribution of K^+ ion across the membrane. Thus the unfertilized egg is different from normal cells in possessing a chloride rather than a potassium potential; after fertilization it becomes more similar to a typical cell in acquiring a potassium-dependent membrane potential (Steinhardt et al., 1971).

Another ionic change involves calcium ion. Although total calcium content does not change after insemination, the free fraction is released (Mazia, 1937). Since changes in calcium ion have been implicated in controlling metabolism (Rasmussen, 1970), this change may be of paramount importance in the metabolic activation of the egg.

2. Coenzyme Changes. The pyridine nucleotide coenzymes NAD and NADP are involved in numerous enzyme-mediated oxidation–reduction reactions. Because of their obligatory role in accepting or donating electrons for these reactions, any changes in concentration or availability of these coenzymes would be expected to have profound metabolic consequences. It is therefore extremely interesting that one of the first cytoplasmic changes occurring after insemination is an interconversion of the pyridine nucleotides involving the phosphorylation of NAD to yield NADP (Epel, 1964a).

This interconversion apparently takes place through the action of the enzyme NAD kinase.

$$\text{NAD} + \text{ATP} \rightarrow \text{NADP} + \text{ADP}$$

In vitro enzyme assays reveal the same activity of this enzyme in unfertilized and fertilized eggs. *In vivo* it is not active until after fertilization. This suggests that the enzyme is activated at fertilization (Epel and Iverson, 1965), and experiments in which unfertilized eggs are incubated in nicotinamide, a precursor of NAD and NADP, prove this to be true. The unfertilized egg converts part of this nicotinamide into NAD, but not NADP or NADPH. Upon fertilization the NAD is phosphorylated to NADP (Krane and Crane, 1963).

Activation of NAD kinase is one of the first changes occurring after insemination. What is the metabolic significance of this change? What effect do the different NADP and NADPH concentrations have on cell metabolism? Examination of the role of NAD and NADP in various metabolic pathways indicate that the NAD coenzymes tend to be involved in oxidative reactions (accepting electrons), whereas the NADP coenzymes tend to be involved in reductive reactions, especially in synthetic pathways (Lowenstein, 1961). The increased NADPH content after insemination may be a critical factor in the beginning of new synthetic pathways (Epel et al., 1969a).

3. Respiration Changes. One of the first investigations of the physiology of the fertilization process was made by Otto Warburg in the early 1900s. Warburg measured the rate of respiration following fertilization of sea urchin eggs and found a five-fold increase in oxygen consumption. Comparative investigations revealed, however, a disconcerting variation in the respiration of eggs after fertilization. In some species respiration increased; in others it decreased; and in still others there was no change. The only generalization that could be made was that in these various organisms there appeared to be a trend toward a similar rate of respiration after fertilization (Whittaker, 1933).

Recently, the respiratory behavior of sea urchin eggs has been reinvestigated using polarographic methods for measuring oxygen consumption. Unfertilized eggs are placed in a closed vessel and the oxygen content of the egg suspension measured with an oxygen electrode. The respiration rate can be calculated from the rate of decrease in O_2 concentration. Such measurements reveal that a transient burst in respiration occurs after fertilization. Within one minute respiration increases 6- to 10-fold, and then decreases to about 4.5 times the unfertilized rate (Epel, 1964b).

When similar polarographic measurements were made with starfish and echiuroid eggs, a surprising result was obtained. The previous manometric measurements with gametes from these organisms indicated that respiration does not increase after fertilization. However, the polarographic measurements reveal a transient burst in respiration (Horwitz, 1965; Paul, 1970). It is not clear whether this burst signifies increased energy utilization or oxidation of some unknown compound. Irrespective of the true situation, such a transient burst may be part of the general response of eggs to fertilization.

4. Protein Synthesis. In sea urchin eggs the rate of protein synthesis increases markedly 5–7 min after insemination (Epel, 1967; Fig. 8). This synthesis is cytoplasmic, as opposed to mitochondrial or nuclear synthesis, since the bulk of this incorporation occurs on cytoplasmic polysomes and the incorporation is inhibited by puromycin (Tyler et al., 1968).

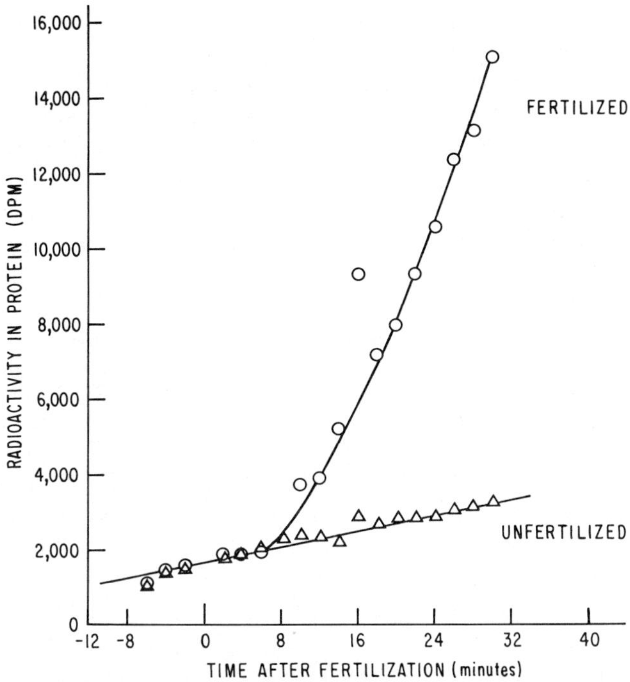

Figure 8. Increased protein synthesis begins 5 min after fertilization of sea urchin eggs. Unfertilized eggs are preloaded with radioactive material and the accumulation of isotope followed as a function of time.

Earlier, we reviewed the evidence that the increase in protein synthesis results from the translation of preexisting or stable mRNA. As noted, the rate of protein synthesis is similar in (1) fertilized eggs (2) enucleated eggs that were artificially activated, and (3) fertilized eggs in which RNA synthesis was inhibited by actinomycin D. The synthesis of mRNA occurs during oogenesis, and at least part of this mRNA remains in the unfertilized egg (Gross et al., 1965).

The observed increase in protein synthesis would result if (1) a greater proportion of the available mRNA were translated, or (2) if the rate of translation increased. Experimental evidence favors the first hypothesis. Exact measurements have shown that (1) the rate of increased protein synthesis is correlated with an increased number of polyribosomes, and (2) the time required to translate the average mRNA molecule is not significantly different between unfertilized and fertilized eggs (T. Humphrey, 1969; Rinaldi and Monroy, 1969).

This translational control of protein synthesis appears to be of common occurrence. It is often observed as an early response to hormonal stimulation. Application of ACTH to the adrenal cortex or adrenal steroid hormones to tissue culture cells results in increased protein synthesis which is not affected by inhibitors of RNA synthesis but is arrested by inhibitors of protein synthesis (Ferguson, 1963; Tomkins et al., 1969).

The mechanism of translational control in response to hormones or insemination is far from clear. Early work on sea urchin eggs showed that *in vitro* protein synthesis increased after a brief exposure of the ribosomes of unfertilized eggs to trypsin (Monroy et al., 1965). Since a protease is also activated at fertilization (Lundblad

1950), one plausible explanation is that the activation of this protease might then lead to *de novo* protein synthesis.

An alternative mechanism, suggested by recent research, emerges from the finding that an inhibitor of protein synthesis initiation, present in unfertilized eggs, is not detectable after fertilization (Metafora et al., 1971). The loss of this inhibitor after fertilization could then result in the activation of protein synthesis. The mechanism of removal or inactivation of this inhibitor is yet unknown. Since the inhibitor in sea urchin eggs is known to be a protein, it may be removed or inactivated by proteolysis. The mechanism of this control is certainly important, and we await its elucidation.

B. General Responses to Fertilization

Although the morphology of fertilization seems to be so similar in the animal kingdom, the metabolic responses of eggs need not be the same. As noted, protein synthesis increases after fertilization of sea urchin eggs. In the frog, however, protein synthesis increases at ovulation but not at fertilization (Smith et al., 1966). Finally, in the eggs of *Urechis* protein synthesis appears to be quite active before meiosis and there is no change at fertilization (Gould, 1969).

Other examples of variation exist. For example, NADP and NADPH levels increase after the fertilization of sea urchin and mollusc eggs (Epel, 1964a; Krane and Crane, 1963). They do not change in *Urechis* eggs (Paul, 1970). Furthermore, a cortical granule enzyme (β-1,3-glucanohydrolase) has been found in sea urchin eggs (Epel et al., 1969b) that is not present in sand dollar eggs (Vacquier, 1971). Both organisms are in the same class (Echinoidea) of the phylum Echinodermata, but in different orders.

Part of this variation may result from the different stages in which the eggs of different organisms are fertilized. As noted, fertilization can occur before meiosis, during meiosis, or after completion of meiosis. These differences, however, do not necessarily account for the metabolic differences. For example, in one case of an NADP-NADPH increase (mollusc), the egg is fertilized before meiosis. In the situation in which no increase occurs (*Urechis*), the egg is also fertilized before meiosis.

Perhaps there is no one "general" metabolic response to fertilization but rather a series of different factors controlling this process. Between different organisms there may be variations in the number of controlling factors. For example, in some organisms the rate of protein synthesis may be limiting. In other species NADP coenzymes may be limiting. In another, both may be limiting.

A general response, if any exist, may be related to changes in the plasma membrane. As noted, the membrane potential change of sea urchin eggs represents a first response to insemination. Are these changes seen in other eggs? Unfortunately, too little comparative data is available to answer this question.

Part of the activation process might follow from the activation of the plasma membrane enzyme, adenyl cyclase, and concomitant formation of 3′-5′-cyclic AMP. Given the ubiquitous response of this enzyme to hormones, and the similarities of fertilization to hormonal activation, could a postfertilization increase in cyclic AMP account for the observed metabolic activation? Such a hypothesis receives support from the finding that adenyl cyclase activity increases about 5 min after insemination (Castaneda and Tyler, 1968). Direct measurements of cyclic AMP, however, reveal no change after insemination (Gray et al., 1970).

Of course, the one general response to fertilization is mitotic cell division. Associated with this is the turning on of DNA synthesis preparatory to the first division. Thus the initiation of DNA synthesis may properly be called a general response to insemination. It is seen in all organisms and begins fairly soon after insemination. Part of this control involves the selective accumulation or uptake of cytoplasmic DNA polymerase by the nucleus (Loeb and Fansler, 1970). This transport of a specific enzyme from the cytoplasm into the nucleus may represent an important part of fertilization.

7. EPILOGUE—PROBLEMS FOR THE FUTURE

The sperm and egg nucleus have fused, DNA replication has occurred, and the embryo is now in the cleavage period. During cleavage the particular pattern of cell division results in equivalent nuclei being placed in unequal or differentiated cytoplasm. The interaction of nuclei with these differentiated cytoplasms later yields cell differentiation.

A major problem for cell and developmental biologists is to describe in molecular terms the nature of these cytoplasms and their interactions with the nucleus. A second problem is determining the mechanism of the highly precise partitioning of these cytoplasms during cleavage. Although segregation of determinants in the polar lobe of molluscs and annelid embryos is easily visualized, such visible evidence of cytoplasmic inequalities is rarely seen in other embryos.

The mechanism of segregation of cytoplasmic determinants is just being approached by experimental scientists. One hypothesis is based on the presence of electric fields across embryos. The resultant potential gradient might then separate different cytoplasmic determinants on the basis of their net electric charge (Jaffee, 1968).

Another favored hypothesis is that the cortex of the egg (i.e., the plasma membrane and the few microns beneath it) may be responsible for segregating cytoplasmic determinants, or the cortex itself may contain these determinants in the form of a blueprint or map for development (Curtis, 1960). Current thought is that part of this map is laid down during oogenesis and that the final plans are more rigidly set after fertilization.

In addition to this major problem of nucleocytoplasmic relationships (see also Chapter 8) there are the many unsolved problems involving gametogenesis and fertilization already noted in this chapter. There is the mechanism of spermiogenesis, with its assembly apparently independent of direct nuclear control. There is the mechanism of specific gene amplification. Does it also occur during terminal differentiation? Or is it restricted to cases in which large numbers of ribosomes must be produced? In cases in which nurse cells are absent, what is the molecular mechanism of synthesis of cell constituents, such as mitochondria or yolk?

How do hormones control meiosis? What membrane effect occurs that intiates ovulation and meiosis? What is the molecular nature of the germinal vesicle constituents required for cleavage and gastrulation?

What are the primary changes in fertilization? If it is the change in membrane potential, how does this in turn trigger subsequent changes, such as NAD kinase activity? Does activation at fertilization involve an unconnected series of separate and independent activations (branched activation sequence)? Or does activation involve one or several dependent activations, with each step dependent on a preceding one (linear activation sequence)?

The answers to these questions are not apparent today. Given the rapid and important progress that has been made in recent years, these answers should soon be known.

REFERENCES

Austin, C. R. (1968). *Ultrastructure of Fertilization*. Holt, New York.

Briggs, R., and G. Cassens (1966). Accumulation in the oocyte nucleus of a gene product essential for embryonic development beyond gastrulation. *Proc. Natl. Acad. Sci., U.S.*, **55,** 1103.

Brink, N. G. (1968). Protein synthesis during spermatogenesis in *Drosophila melanogaster. Mutation Res.*, **5,** 192.

Brown, D. D., and I. Dawid (1968). Specific gene amplification in oocytes. *Science*, **160,** 272.

Castaneda, M., and A. Tyler (1968). Adenine cyclase in plasma membrane preparations of sea urchin eggs and its increase in activity after fertilization. *Biochem. Biophys. Res. Commun.*, **33,** 782.

Chaet, A. B. (1963). A mechanism for obtaining mature gametes from starfish. *Biol. Bull.*, **126,** 8.

Clement, A. C. (1962). Development of *Ilynassa* following removal of the D macromere at successive cleavage stages. *J. Exptl. Zool.*, **149,** 193.

Colwin, L. H., and A L. Colwin (1967). In *Fertilization* (C. B. Metz and A. Monroy, eds.), Vol. 1, pp. 295–368. Academic, New York.

Curtis, A. S. G. (1960) Cortical grafting in *Xenopus laevis. J. Embryol. Exptl., Morphol.*, **8,** 163.

Davidson, E. (1968). *Gene Activity during Early Development*, Academic, New York.

Davidson, E. H., G. W. Haslett, R. S. Ginney, V. G. Allfrey, and A. E. Mirsky (1965). Evidence for prelocalization of cytoplasmic factors affecting gene activation in early embryogenesis. *Proc. Natl. Acad. Sci., U.S.*, **54,** 696.

Dawid, I. (1966). Evidence for the mitochondrial origin of frog egg cytoplasmic DNA. *Proc. Natl. Acad. Sci., U.S.*, **56,** 269.

Deltaff, T. A., L. A. Nikitina, and O. G. Stroeva (1964). The role of the germinal vesicle in oocyte maturation in anurans as revealed by the removal and transplantation of nuclei. *J. Embryol. Exptl. Morphol.*, **12,** 851.

Denny, P. C., and A. Tyler (1963). Activation of protein biosynthesis in non- nucleate fragments of sea urchin eggs. *Biochem. Biophys. Res. Commun.*, **14,** 245.

Epel, D. (1964a). A primary metabolic change of fertilization: Interconversion of pyridine nucleotides. *Biochem. Biophys. Res. Commun.*, **17,** 62.

Epel, D. (1964b). Simultaneous measurements of TPNH formation and respiration following fertilization of the sea urchin egg. *Biochem. Biophys. Res. Commun.*, **17,** 69.

Epel, D. (1967). Protein synthesis in sea urchin eggs: A late response to fertilization. *Proc.Natl. Acad. Sci., U.S.*, **57,** 899.

Epel, D., and R. M. Iverson (1965). In *Control of Energy Metabolism* (B. Chance, R. W. Estabrook, and J. R. Williamson, eds.), pp. 267–272, Academic, New York.

Epel, D., B. C. Pressman, S. Elsaesser, and A. M. Weaver (1969a). In *The Cell Cycle*: *Gene-Enzyme Relationships* (G. Padilla, G. L. Whitson, and I. L. Cameron, eds.), pp. 280–298. Academic, New York.

Epel, D., A. M. Weaver, A. V. Muchmore, and R. T. Schimke (1969b). β-1,3-glucanase of sea urchin eggs: Release from particles at fertilization. *Science*, **163,** 294.

Ferguson, J. J. (1963). Protein synthesis and adrenocorticotropin responsiveness. *J. Biol. Chem.*, **238,** 2754.

Gall, J. G., and M. L. Pardue (1969). Formation and detection of RNA-DNA hybrid molecules in cytological preparations. *Proc. Natl. Acad. Sci., U.S.*, **63,** 378.

Gould, M. C. (1969). A comparison of RNA and protein synthesis in fertilized and unfertilized eggs of *Urechis caupo. Develop. Biol.*, **19,** 482.

Graham, C. F., K. Arms, and J. B. Gurdon (1966). The induction of DNA synthesis by frog egg cytoplasm. *Develop. Biol.*, **14,** 349.

Gray, J. P., J. G. Hardman, T. Bibring, and E. W. Sutherland (1970). High guanyl cyclase activity in sea urchin spermatozoa. *Federation Proc.*, **29,** 608.

Greengard, O., M. Gordon, M. A. Smith, and G. Acs (1964). Studies on the mechanism of diethylstilbestrol-induced formation of phosphoprotein in male chickens. *J. Biol. Chem.*, **239,** 2079.

Gross, P. R., L. I. Malkin, and W. A. Moyer (1964). Templates for the first proteins of embryonic development. *Proc. Natl. Acad. Sci., U.S.*, **51,** 407.

Gross, P. R., L. I. Malkin, and M. Hubbard (1965). Synthesis of RNA during oogenesis in the sea urchin. *J. Mol. Biol.*, **13,** 463.

Gurdon, J. B. (1964). The transplantation of living cell nuclei. *Advan. Morphogenesis*, **4,** 1.

Gurdon, J. B., and D. D. Brown (1965). Cytoplasmic regulation of RNA synthesis and nucleolus formation in developing embryos of *Xenopus laevis. J. Mol. Biol.*, **12,** 27.

Gurdon, J. B., and H. R. Woodland (1968). The cytoplasmic control of nuclear activity in animal development. *Biol. Rev.*, **43,** 233.

Harvey, E. B. (1936). Parthenogenetic merogony or cleavage without nuclei in *Arbacia punctulata. Biol. Bull.*, **71,** 101.

Hiramoto, Y. (1959). Changes in electric properties upon fertilization in the sea urchin egg. *Exptl. Cell Res.*, **16,** 421.

Horwitz, B. A. (1965). Rates of oxygen consumption of fertilized and unfertilized *Asterias*, *Arbacia*, and *Spisula* eggs. *Exptl. Cell Res.*, **38,** 620.

Hultin, T. (1961). The effect of puromycin on protein metabolism and cell division in fertilized sea urchin eggs. *Experientia*, **17,** 410.

Humphrey, T. (1969). Efficiency of translation of messenger RNA before and after fertilization in sea urchins. *Develop. Biol.*, **20,** 435.

Humphrey, R. R. (1966). A recessive factor (o, for ova deficient) determining a complex of abnormalities in the Mexican axolotl (*Ambystoma mexicanum*). *Develop. Biol.*, **13,** 57.

Jaffee, L. (1968). Localization in the developing *Fucus* egg and the general role of localizing currents. *Advan. Morphogenesis*, **7,** 295.

Kanatani, H. (1969). Mechanism of starfish spawning: Action of a neural substance on the isolated ovary. *Gen. Comp. Endocrinol., Suppl.*, **2,** 582.

Kanatani, H., and Y. Hiramota (1970). Site of action of 1-methyladenine in inducing oocyte maturation in starfish. *Exptl. Cell Res.*, **61,** 280.

Kanatani, H., H. Shirai, K. Nakanishi, and T. Kurokawa (1969). Isolation and identification of meiosis inducing substance in starfish *Asterias amurensis. Nature*, **221,** 273.

Kedes, L. H., P. R. Gross, G. Cognetti, and A. L. Hunter (1970). Synthesis of nuclear and chromosomal proteins on light polyribosomes during cleavage. *J. Mol. Biol.*, **45,** 337.

Keifer, B. J., C. F. Entelis, and A. A. Infante (1969). Mitotic abnormalities in sea urchin embryos exposed to dactinomycin. *Proc. Natl. Acad. Sci., U.S.*, **64,** 857.

Krane, S. M., and R. K. Crane (1963). In *Control Mechanisms in Respiration and Fermentation* (B. Wright, ed.), pp.157–166, Ronald, New York.

Kunz, W., H. Trepte, and K. Bier (1970). On the function of the germ line chromosomes in the oogenesis of *Wachtliella persicariae* (Cecidomyiidae). *Chromosoma*, **30,** 180.

Lindsley, D. L., and E. H. Grell (1969). Spermiogenesis without chromosomes in *Drosophila melanogaster. Genetics, Suppl.*, **61,** 69.

Loeb, L. A., and B. Fansler (1970). Intracellular migration of DNA polymerase in early developing sea urchin embryos. *Biochem. Biophys. Acta*, **217,** 50.

Lowenstein, J. M. (1961). Reductions and oxidations in mammalian biosynthesis. *J. Theoret. Biol.*, **1** 98.

Lundblad, G. (1950). Proteolytic activity in sea urchin gametes. *Exptl. Cell Res.*, **1,** 264.

Mahowald, A. P. (1962). Fine structure of pole cells and polar granules in *Drosophila melanogaster. J. Exptl. Zool.*, **151,** 201.

Mahowald, A. P., and S. Hennen (1971). Ultrastructure of the "germ plasm" in eggs and embryos of *Rana pipiens. Develop. Biol.*, **24,** 37.

Masui, D. (1967). Relative roles of the pituitary, follicle cells, and progesterone in the induction of oocyte maturation in *Rana pipiens*. *J. Exptl. Zool.*, **166,** 365.

Mazia, D. (1937). The release of calcium in *Arbacia* eggs upon fertilization. *J. Cellular Comp. Physiol.*, **10,** 291.

Metafora, S , L. Felicetti, and R. Gambino (1971). The mechanism of protein synthesis activation after fertilization of sea urchin eggs. *Proc. Natl. Acad. Sci., U.S.*, **68,** 600.

Monroy, A. (1965). *Chemistry and Physiology of Fertilization*. Holt, New York.

Monroy, A., R. Maggio, and A. M. Rinaldi (1965). Experimentally induced activation of the ribosomes of the unfertilized sea urchin egg. *Proc. Natl. Acad. Sci., U.S.*, **54,** 107.

Paul, M. (1970). Ph.D. Thesis, Stanford University.

Piko, L., A. Tyler, and J. Vinograd (1967). Amount, location, priming capacity, circularity and other properties of cytoplasmic DNA in sea urchin eggs. *Biol. Bull.*, **132,** 68.

Rasmussen, H. (1970). Cell communication, calcium ion and cyclic adenosine monophosphate. *Science*, **170,** 404.

Rinaldi, A. M., and A. Monroy (1969). Polyribosome formation and RNA synthesis in the early post-fertilization stages of the sea urchin egg. *Develop. Biol.*, **19,** 73.

Schuetz, A. W. (1967). Action of hormones on germinal vesicle breakdown in frog (*Rana pipiens*) oocytes. *J. Exptl. Zool.*, **166,** 347.

Schuetz, A. W., and J. D. Biggers (1967). Regulation of germinal vesicle breakdown in starfish oocytes. *Exptl. Cell Res.*, **46,** 624.

Slater, D. W., and S. Spiegelman (1966). An estimation of genetic messages in the unfertilized echinoid egg. *Proc. Natl. Acad. Sci., U.S.*, **56,** 164.

Smith, L. D. (1966). The role of a "germinal plasm" in the formation of primordial germ cells in *Rana pipiens*. *Develop. Biol.*, **14,** 330.

Smith, L. D., and R. E. Ecker (1969). Role of the oocyte nucleus in physiological maturation in *Rana pipiens*. *Develop. Biol.*, **19,** 281.

Smith, L. D., and R. E. Ecker (1970). In *Problems in Biology: RNA in Development* (E. W. Hanley, ed.), pp. 355–380. Univ. of Utah Press, Salt Lake City.

Smith, L. D., R. E. Ecker, and S. Subtelny (1966). The initiation of protein synthesis in eggs of *Rana pipiens*. *Proc. Natl. Acad. Sci., U.S.*, **56,** 1724.

Smith, L. D., R. E. Ecker, and S. Subtelny (1968). *In vitro* induction of physiological maturation in *Rana pipiens* oocytes removed from their ovarian follicles. *Develop. Biol.*, **17,** 627.

Spirin, A. S. (1966). On "masked" forms of messenger RNA in early embryogenesis and in other differentiating systems. *Curr. Top. Develop. Biol.*, **1,** 1.

Steinhardt, R., L. Lundin, and D. Mazia (1971). Bioelectric responses of the echinoderm egg to fertilization. *Proc. Natl. Acad. Sci. U.S.*, **68,** 2426.

Telfer, W. H. (1965). The mechanism and control of yolk formation. *Ann. Rev. Entomol.*, **10,** 161.

Tompkins, G. M., T. D. Gelehrter, D. Granner, D. Martin, H. Samuels, and E. B. Thompson (1969). Control of specific gene expression in higher organisms. *Science*, **166,** 1474.

Tyler, A. (1941). Artificial parthenogenesis. *Biol. Rev.*, **16,** 291.

Tyler, A. (1955). In *Analysis of Development* (B. H. Willier, P. A. Weiss, and V. Hamburger, eds.), pp. 170–212. Saunders, Philadelphia.

Tyler, A., B. S. Tyler, and J. Piatigorsky (1968). Protein synthesis by unfertilized eggs of sea urchins. *Biol. Bull.*, **134,** 209.

Vacquier, V. D. (1971). The appearance of β-1,3-glucanohydrolase activity during the differentiation of the gut of sand dollar plutei. *Develop. Biol.*, **26,** 1.

Whiteley, A. H., and E. L. Chambers (1961). In *Symposium on Germ Cells and Development* (P. S. Ranzi, ed.), pp. 387–401. Inst. Intern. d'Embryologie and Fondazione A. Baselli, Milan.

Whittaker, D. M. (1933). On the rate of oxygen consumption by fertilized and unfertilized eggs. V. Comparison and interpretation. *J. Gen. Physiol.*, **16,** 497.

Wilson, E. G. (1925). *The Cell in Development and Heredity*, 3rd edition, Macmillan, New York.

Wilt, F. H., H. Sakai, and D. Mazia (1967). Synthesis of mitotic apparatus and the first cell division in the sea urchin egg. *J. Mol. Biol.*, **27,** 1.

PART 4

CELLULAR MOTILITY AND AGGREGATION

14

GENERAL FUNCTIONS OF THE CELL SURFACE

A. S. G. Curtis

1. INTRODUCTION

This chapter treats the functions of the plasmalemma. Robertson describes the structure of this organelle in Chapter 1. Most of these functions concern the relation of the cell to its external environment. The main cellular functions in which the cell surface is believed to be involved are discussed below. It should be remembered that at least some contribution is made by internal cell structure in nearly all these categories. In some, indeed, the contribution from the plasmalemma is perhaps small and possibly, although this remains to be established, negligible.

The model of cell surface structure this chapter is based upon is basically that described by Robertson. It should be appreciated that the structure and composition of the surface determine the whole range of surface properties. Conversely, evidence from these surface properties should be used to test various models of the surface. Robertson describes the cell surface as a lipid bilayer with protein and glycosidic components on either side of the lipid. Recent work using techniques such as cell electrophoresis (Mehrishi, 1970), immunochemical techniques (Pardoe and Uhlenbruck, 1971), reaction with lectins (Burger and Goldberg, 1967), and terminal labeling of proteins and glycoproteins (Bretscher, 1971) has revealed much about the nature of the nonlipid components of the cell surface. Bretscher showed that a single glycoprotein molecule may extend through the plasmalemma from one side to the other. The glycosidic components are present as glycolipids and glycoproteins. Because of the hydrophilic nature of the glycosidic groupings they tend to extend into the watery media on either side of the membrane, and this has been confirmed by electrophoresis, immunochemical techniques, and lectin reactions. Although it is clear that part of the protein may be hydrophobic in nature, and thus situated among the lipid components of the membrane, there is no reliable evidence regarding the configuration of this material. Zahler (1970), for example, claims that this protein is in the α-helix conformation, but Curtis (1972) has noted that at present this conclusion cannot be validly deduced from the evidence. In essence the surface probably has a structure of the type shown in Figure 1. The outer half of the plasmalemma can be thought of as being similar to a lawn on which most of the grass (lipid) has been mowed to a constant length, with a number of weeds (glycoproteins, proteins, and glycolipids) standing up above the level of the grass.

In this chapter we see that various different types of cells display very different surface properties. One major question we are trying to answer concerns the nature of the chemical differences that determine these different properties. For example, do glycolipids play a major role in determining the precise adhesiveness of a cell? What differences in cell behavior result from an alteration in the chain length of the acyl components of plasmalemmal phosphatidylcholines? These questions cannot be answered yet, but it is probable that answers of this type might be of considerable medical importance. For example, in a variety of forms of inherited mental deficiency, glycolipid synthesis is deranged. Do the cells of such people have abnormal plasmalemma glycolipids? To what extent do such lipids, if they exist, directly bring about mental deficiency? Do the nerve cells make the wrong connections because of their surface abnormalities? Again, it seems possible (Burger and Goldberg, 1967) that malignant cells have abnormal glycoproteins on their surfaces. A question that requires resolution is whether or not these abnormalities are in any way responsible for the malignant behavior of the cells.

Scientists differentiate to a certain extent between cell surface and plasmalemmal properties. Of course no surface is in reality identical to the mathematical concept. All real surfaces have a depth, that is, a distance over which physical properties change from those of one phase to those of another. In the case of the cell surface, there is a zone of low dielectric constant and high electrical resistance, corresponding to the lipid of the plasmalemma, on one side of the surface. On the other side is the watery medium, surrounding the cell, of high dielectric constant and low electrical resistance. Between the two bulk phases, there is a zone of transition of physical properties. Strictly speaking, one might say that the cell surface occupies this zone of transition and that the bulk of the plasmalemma lies inward of this zone. However, the under-

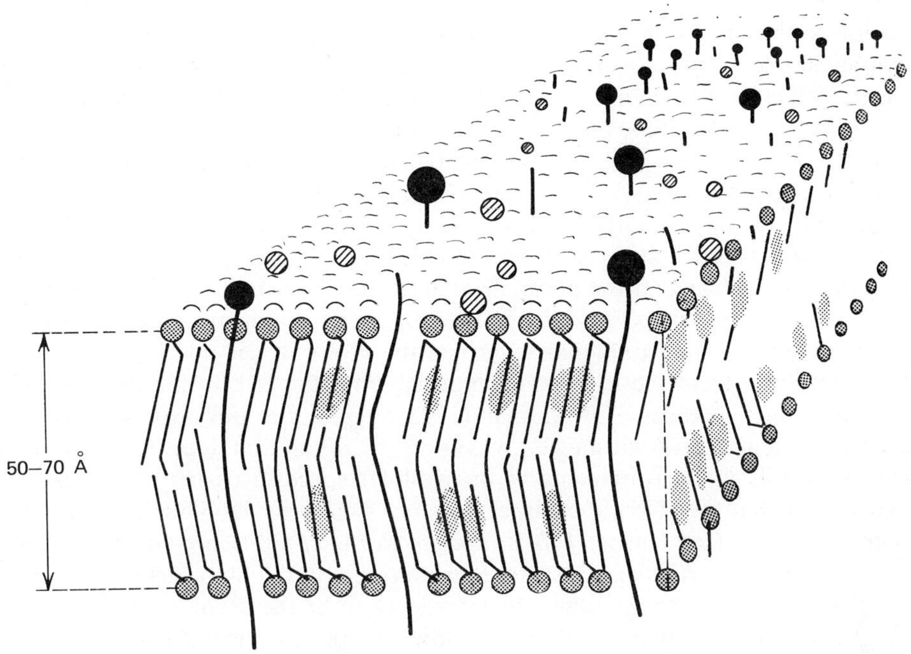

Figure 1. Schematic view of possible cell surface structure. A block of the surface is shown in three-dimensional view with the outer surface of the plasmalemma uppermost. The bilayer of lecithins and sphingomyelins is shown in section as well. The shaded spheres represent the polar head groups of these substances, and the lines emerging from these spheres represent the long-chain fatty acids that form part of these substances. Cholesterol molecules can be seen as shaded irregular ovals clearly apparent along the side. [Note that the tilted nature of the hydrocarbon chains and the highly ordered arrangement of the lipids such that all pairs of hydrocarbon chain have the same orientation and that cholesterols lie in the same plane (the next layer beyind the front one and then alternately) is a feature for which we have little positive evidence.] Protein chains (heavy lines) run through the entire thickness of the surface, protruding a short distance on either side. Some of these chains glycosidic end groupings (solid circles of varying size) on the outer side of the membrane, glycosidic end groups also occur on some of the lipids (hatched circles).

lying lipid phase of the membrane determines many of the features of this zone. For example, if the dielectric constant of the lipid were raised, then the electric potential distribution in the zone of transition would be completely altered. For these reasons it is impossible to make any clear distinction between the terms plasmalemma and cell surface, although the usage of the latter term generally means that attention is being concentrated on the outer part of the plasmalemma. An exactly similar problem applies to the inside of the plasmalemma where it abuts on the cytoplasm, but this surface has received very little attention. Reactions and changes at the inner side of the plasmalemma may affect conditions on the outer surface, and it is possible for the inner cytoplasm to affect outer surface conditions to some extent.

In biophysical terms the cell surface can be treated as one would treat any other surface, although it is probable that the polar parts of protein, lipid, and glycosidic components are spread through a greater depth than is usually found in surface studies. This deep nature of the cell surface has led to a certain amount of controversy. Electrophoretic studies suggest that the surface has a depth of approximately 10 Å on many cells (see discussion by Curtis, 1967), while electron microscopy has led on occasion to the suggestion that the surface is a very deep structure extending over as much as 200 Å or more. A variety of staining reactions suggests that the surface is covered with glycosidic components and that these extend a considerable distance out from the lipid regions of the plasmalemma to form the structure termed the glycocalyx (Bennett, 1963). A very large amount of biochemical, immunochemical, and other research (see particularly Winzler, 1970) shows that such glycosidic components form a major part of the cell surface. However, most techniques provide no information about the depth of such material. If one accepts the evidence from electron microscopy uncritically, it is clear that the plasmalemma of most if not all cells is covered with a thick layer, the glycocalyx. At the opposite extreme it might be maintained that the zone of staining around cells that is termed the glycocalyx is nothing more than an artifact resulting from the leakage of material from the cell during fixation. Tracer experiments such as those of Brightman (1965) suggest that this criticism may have a considerable amount of substance. A variety of other techniques such as electrophoresis of cells or measurements of intercellular space in tissues also suggest in a tentative way that the glycocalyx either does not exist or is very much thinner than is usually supposed (see Curtis, 1967). Recent work by Bretscher indicates that the protein chain of a particular cell surface glycoprotein in red blood cells runs from one side of the membrane to the other. If this proves to be the usual arrangement of cell surface glycoproteins and all the glycosidic components associated with the cell surface prove to be part of these proteins (or directly bound to lipids as glycolipids), it will be clear that these glycosidic components are so intermingled with the plasmalemma lipid that it will be unnecessary to distinguish them as a separate structure.

Cell surface glycolipids have been fairly well characterized (Sweeley, 1970), and it is of interest to note that abnormalities in these components tend to be associated with a variety of forms of mental deficiency. The glycosidic components of these compounds form short chains frequently terminating in neuraminic acid. If it is determined that this neuraminate forms part of the charge of cells revealed by electrophoresis, this will suggest that the charge lies close to the plasmalemmal lipid and thus that the cell coat has no separate existence.

It is of interest that characterization of the glycosidic components of the cell surface glycoproteins would do much to resolve the question of the existence of a separate cell coat. How far from the cell surface lipid do the glycoproteins protrude? Is there any form of cross-linking between glycoproteins? Is there indeed sufficient material to form anything more than a thin layer of glycosidic material on the outside of the lipid? Answers to these questions should resolve the question of the existence of the cell coat.

Unfortunately, the whole problem has been somewhat confused by the careless use of the term basement membrane as an equivalent of glycocalyx. The classic basement membrane is a massive structure reinforced with collagen fibers, entirely separate from cells. Other basement membranes are found, in the kidney, for example, of rather less distinct structure but clearly separate from the cell surface (see Goel and Jurand, 1971).

Whatever the precise geometry of the glycosidic components of the surface, the basic biophysics of the cell surface are likely to be similar to those of other surfaces. One of the more important biophysical aspects of the cell surface is the nature of the surface double layer. The charged groups of the plasmalemma are likely to comprise phosphate and choline on phospholipids, carboxyl groups on neuraminate, aspartate and glutamate, and amines on arginine, lysine, and possibly glucosamine and galactosamine. Electrophoretic evidence has failed to show that the phospholipid groupings are exposed to the outside of the plasmalemma, but this may represent nothing more than lack of research. (In addition, it is probable that anions such as chloride are physically adsorbed from the medium onto the surface.) The overall charge of all cell surfaces appears to be negative. Measurements by cell electrophoresis show that this charge is sufficiently dense to give rise to a surface potential of from -5 to -40 mV. This implies that there are considerably more anions than cationic groupings revealed at the cell surface. Sialic acid, that is, *N*-acetylneuraminic acid, and other neuraminic acids accounts for much of the charge on most cells that have been studied. The surface potential attracts counterions (i.e., cations) from the medium toward the surface, which tend to reduce the potential. In consequence, the electrostatic potential resulting from the surface charge, in the medium adjacent to the surface declines rapidly with increasing distance from the surface (Fig. 2). The distance over which this potential declines to negligible values is termed the thickness of the double layer. Obviously, if the concentration of counterions is increased by raising the ionic strength of the medium, the electrostatic potential is more rapidly neutralized with increasing distance from the surface. Thus the double-layer thickness decreases with increasing ionic strength. Likewise, counterions of a higher valency lead to a diminu-

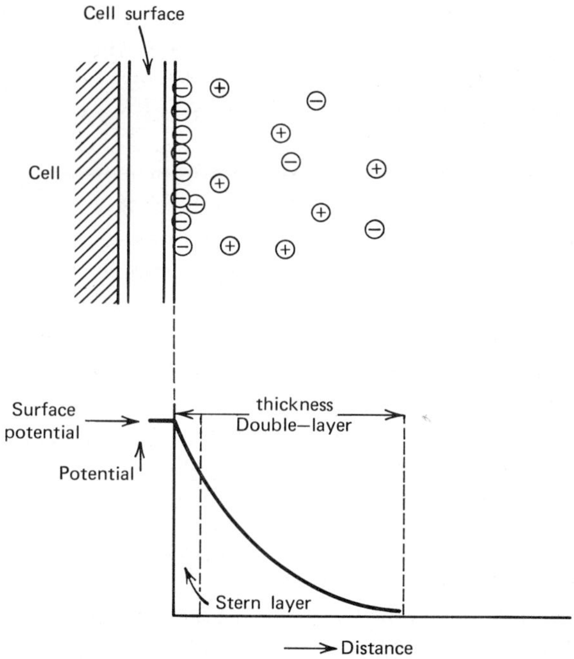

Figure 2. The electrostatic potential in the vicinity of a surface. The upper diagram shows the origin of the potential on a cell surface from the presence of anionic charges on the surface. These charges attract counterions (positive) in the medium. As a result, the electrostatic potential is neutralized over a distance from the surface, the double-layer thickness. The lower diagram shows the relation between distance and potential. If cations are specifically adsorbed to the surface (one is shown in the upper diagram), the potential falls very rapidly initially in the zone termed the Stern layer; conversely, if no adsorption takes place, free cations cannot enter this zone so the potential falls more slowly.

tion of double-layer thickness when compared with the same concentration of counter-ions of lower valency.

The electrical double layer may play an important role in the control of interactions between cells both with each other and with other surfaces. In addition, electrostatic interactions would take place laterally within the surface (see Curtis, 1972), which would act in part to determine its mechanical properties. Finally, electrostatic and other interactions with components of the cell such as microtubules would have an important effect on the cell surface, for in this way an internal stiffening could be provided for the surface or surface contraction could be effected (see Wessels et al., 1971).

The electrostatic interactions would often be repulsive in nature although of course interactions between oppositely charged groupings would lead to attraction. There are a considerable number of forces that may provide the required attraction to counteract the electrostatic repulsions and lead to sufficient cohesion to maintain the structure of the membrane. The relative importance of these various forces is at the moment hard to assess because of lack of experimental work. In consequence, there is some dispute as to whether or not some of the forces of attraction play any relevant role. For this reason, these forces are described and discussed at those points in the text where appropriate experimental work is mentioned.

The biophysics of the cell surface should account for all its functions. However, current knowledge is too incomplete for this to be done. It is possible, nevertheless, to recognize several prime functions of the cell surface which either separately or together determine all the features of cell surface behavior. For instance, cell movement on a solid substrate may represent a secondary function in which the degree of adhesiveness between the cell and substrate and the mechanical properties of the surface modify the basic locomotory mechanism (see Chapter 16). In this way cell surface functions can be divided into primary and secondary classes. One method of classifying cell surface function is in terms of whether one or two surfaces are involved. This scheme is of value in that it distinguishes between biophysically differing situations. Again we can distinguish between those primary and basic functions of the cell surface that can be studied as single biophysical properties and the secondary more derivative properties that result from the cooperation of a variety of cell surface properties. The classification used is:

One surface	Two surfaces
Primary functions	
Mechanical properties: surface pressure; surface viscosity	Mechanical properties: cell adhesion
Electrical properties: transmembrane potential (see Chapter 3)	Electrical properties: electrical coupling of cells (low-resistance pathways)
Secondary functions	
Transmission functions: permeation (see Chapter 3); phagocytosis; pinocytosis	—
Mechanical functions: cell shape; lysis of the cell surface	Mechanical functions: cell movement and more complex contact reactions such as contact inhibition of movement, contact guidance, chemotaxis and kinesis; cell fusion

Owing to limitations of space, only the more salient papers receive specific mention here. The various reviews cited give fuller details.

2. PRIMARY FUNCTIONS OF CELL SURFACE

A. Mechanical Properties of the Cell Surface

Despite the obvious importance of this subject, there has been surprisingly little research on these topics, which has resulted in perhaps still less knowledge of which we can be confident. In part this is because it is difficult to make a mechanical measurement on an object as small as a cell. Another difficulty is that all such measurements involve deformation of the cell to some extent. In practice this means that the mechanical properties of some inner part of the cell make some contribution to the measurement. At the time of writing, our ability to isolate the cell surface in an unaltered fashion is so uncertain that we cannot carry out measurements on it. If we knew the extent of the contribution from internal cell components such as microtubules to the deformability of the whole cell, we might have another method of really measuring cell surface mechanical properties. A further difficulty arises from the considerable depth of the cell surface since its composite nature may lead to complex behavior.

B. Surface Pressure and Surface Tension

Molecules in a membrane surface show a lateral cohesion which prevents them from escaping and mixing with materials on either side of the interface. The greater this cohesion the more close packed are the molecules. Repulsive forces, often of electrostatic nature, oppose the cohesive forces so that when a certain closeness of packing is reached there is a stable balance between forces of cohesion and those of expansion. This balance occurs at a certain pressure, strictly speaking, a two-dimensional pressure in an ideal surface. The cohesive forces tend to make the surface contract spontaneously and this phenomenon is termed *surface tension* (strictly speaking, interfacial tension). Surface tension is related to surface pressure in the following way.

$$\begin{matrix}\text{surface pressure of system} \\ \text{under investigation (mN m}^{-1}\text{)}\end{matrix} = \begin{matrix}\text{surface tension of a reference system} - \\ \text{surface tension of the system (mN m}^{-1}\text{)}\end{matrix}$$

The surface tension in a surface such as that of a cell is in effect a measure of the cohesiveness of the molecules of the surface, low values corresponding to a great cohesion of the surface components as do high surface pressures. The surface pressure of the plasmalemma is probably of the order of 1 mN m^{-1}. This statement has, speaking strictly, rather little meaning because we do not know which part of the surface determines this value and the experimental measurements are suspect (see Curtis, 1967). The usual method for measuring cell surface pressure is to examine the equilibrium shape of the cell, its rate of flattening under its own weight, or the contact angle it makes with a surface of known surface tension. Each of these methods involves deformation of the cell, and if structures such as microfilaments play any role in determining cell shape (see, for example, Wessells et al., 1971) the measurement may be inaccurate.

C. Surface Viscosity

The cell surface shows properties of flow when it is mechanically disturbed. This was suggested by early microdissection experiments. More recently, Frye and Edidin (1970), using the technique of cell fusion, have shown that the antigenic components of two cells slowly intermixed in a manner suggesting that the surfaces of the cells they used are capable of flow. This property of flow can be measured as surface viscosity. Measurements of the surface viscosity of amphibian eggs suggest that the cells have a viscosity of the order of 1×10^5 N/secm^2, and the tissue culture cells examined by Frye and Edidin may have similar values of the order of 1×10^5 N/secm^2. The results from both studies suggest that the plasmalemma is in the liquid state. It would clearly be of great interest to know more about any changes in viscosity that may occur in the cell surface either during development of an animal or plant or during the cell cycle of an individual cell. Changes in viscosity may precede lysis of a cell, and it seems possible that cell fusion (see Section 3.E) requires that the cell surfaces be so fluid that two can flow and fuse together.

Fuller analysis of the surface mechanical properties of red blood cells and of sea urchin eggs has been attempted by Rand and Burton (1964) and by Mitchison and Swann (1955), respectively. The results of these studies are extremely complex, but as yet no general principles concerning the mechanical organization of the cell surface have emerged.

D. Cell Adhesion

The two preceding topics have dealt in brief with the effects of cohesion within the plane of a single surface. Cell adhesion arises from cohesion between two cell surfaces or between one cell surface and another surface, in a direction perpendicular to the plane of each surface. The importance of this phenomenon should be obvious. Cells in our bodies adhere to each other to form tissues, and to intercellular materials such as collagen and bone. The adhesion, for example, of gum cells (epithelia) to the tooth is of importance in the prevention of tooth decay just as the adhesion of bacteria and other components of dental plaque to the tooth may assist that decay. Tissues should adhere to various implanted prosthetic devices, yet blood cells should not adhere to plastic valves placed in the heart. During the development of an animal embryo, extensive cell movements take place in an orderly fashion. Clearly, cell movement requires a certain degree of adhesion between the cells. The cell movements are highly patterned, and eventually cell movement is halted. Does cell adhesion play a role in ordering and terminating these movements? Again, during development cavities and tubes develop in solid masses of tissue. This loss of adhesion between cells, or sometimes between parts of cells, is essential for normal morphogenesis. The malignant spread of tumor cells in the body may represent some deficiency in cell adhesiveness. Finally, it should be remembered that the fusion of cells, in particular cell fusion of zygotes in fertilization, requires the initial adhesion of the cells. The possible role that cell adhesion plays in secondary functions of the surface such as cell movement, cell behavior, phagocytosis, and cell fusion, is considered in Section 3.

There are three main approaches to the investigation of cell adhesion: (1) investigation of the morphology of adhesive contacts of cells; (2) study of the conditions under which adhesion between cells or of cells with noncellular substrates can be broken down, namely, dispersion studies; and (3) examination of the conditions under which adhesions form: it is usual to use cells that have been artificially separated, but in some cases it is possible to use cells that are normally separate during some part of their life cycle, for example, red blood cells or slime mold cells. The process of adhesion of cells is usually referred to as aggregation or reaggregation, although when attachment is being made to an extensive nonliving surface it may be termed settling or attachment.

E. The Morphology of Cell Contacts

The main technique used in such studies is of course electron microscopy. Even when the ever-present possibility of artifacts is allowed for, there is still the problem of demonstrating that a particular type of morphology is associated with a particular adhesive state of the cells. Five main types of cell contact have been recognized (see Farquhar and Palade, 1963; Brightman and Palay, 1963; Goodenough and Revel, 1970). These are:

1. The zonula adherens in which the cell surfaces represented by dark lines of staining are separated by a gap 60–300 Å in width although widths of between 85 and 150 Å are most usual.

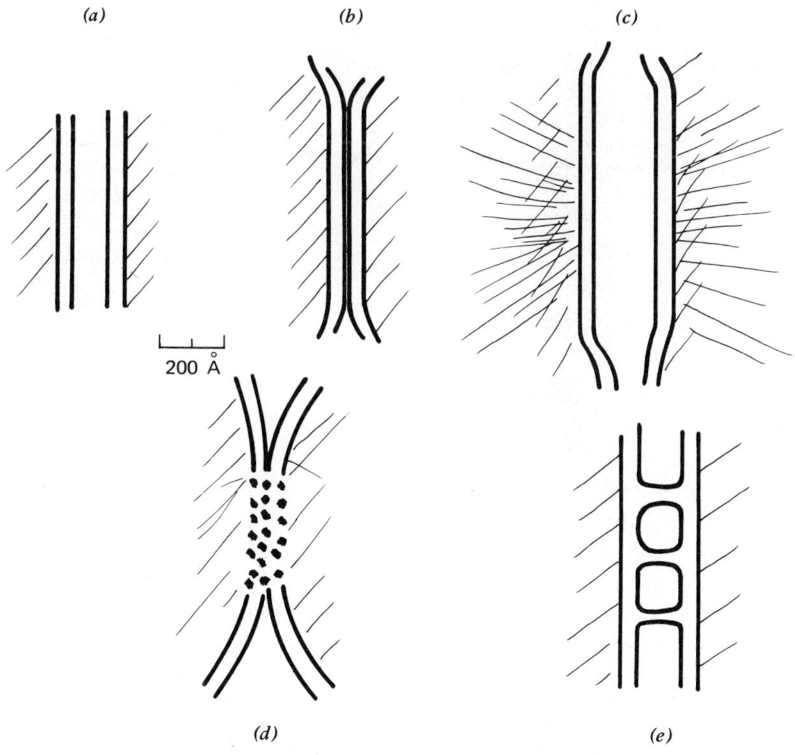

Figure 3. Schematic diagrams showing the structure (in sectional view) of the five main types of cell contact. The plasmalemma is shown as a pair of dark lines separated by a white core, and cytoplasm as hatching. (*a*) The zonula adherens; (*b*) the zonula occludens; (*c*) the macula adherens (with a fibrillar structure in the cytoplasm); staining patterns found in the gap between the two surfaces are not shown; (*d*) the gap junction; and (*e*) the septate desmosome.

2. The zonula occludens in which the cell surfaces come at least within 20 Å of each other. It is not quite certain whether or not there are two classes of zonulae occludentes, one in which there is no gap and another in which a gap of ca. 20 Å is found. Until recently, this type of contact was confused with the gap junction (see item 3), see Figures 3 and 4.

3. The gap junction or nexus (*sensu stricto*). The structure of the cell surfaces of each cell is altered in the zone of contact to form a honeycomb structure freely permeable to tracer molecules; see Figures 3 and 4.

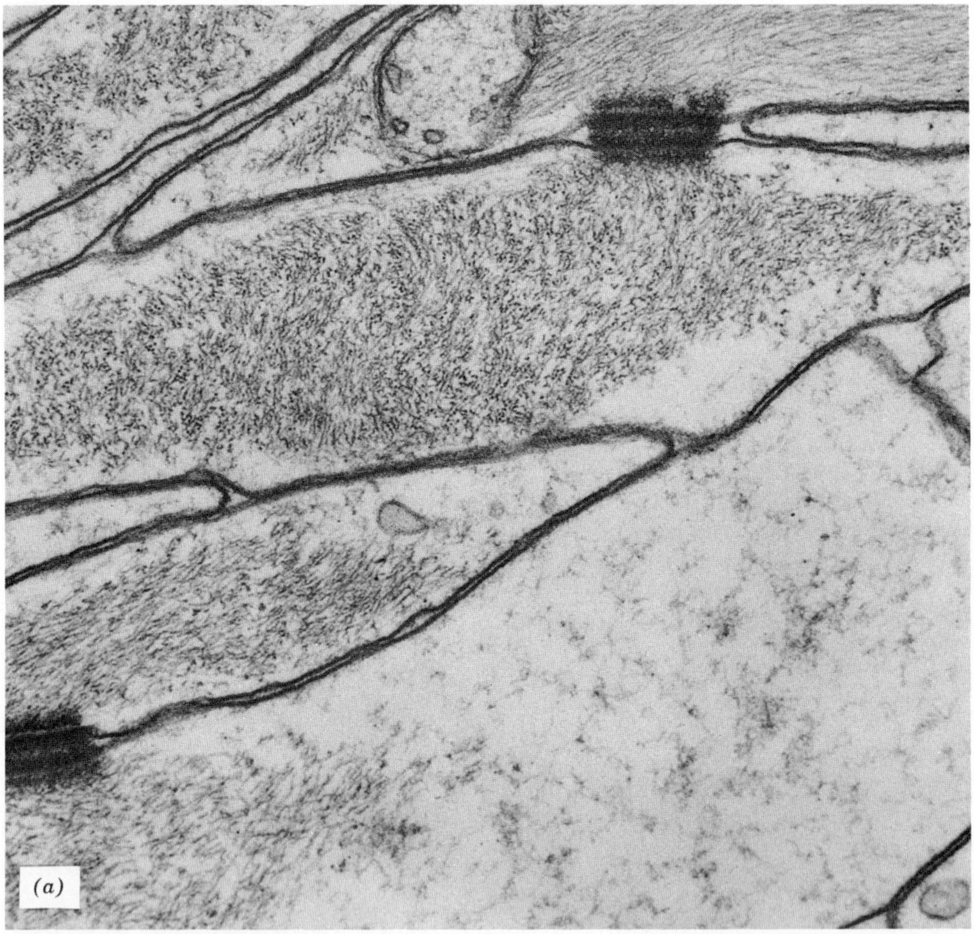

Figure 4. Cell contacts as seen by electron microscopy. (*a*) Zonula adherens contacts between muscle cells. These contacts are interrupted by maculae adherentes (desmosomes) which show dense staining between the plasmalemmas. ×50,000. (Courtesy of D. W. Fawcett.) (*b*) A zonula occludens. Note that the two plasmalemmas lie alongside one another in this type of adhesion. ×94,500. (By kind permission of P. G. Toner and Messrs. Butterworth.) (*c*) and (*d*) Gap junction structure. A freeze-cleaved gap junction is shown in (*c*). The middle arrow points to the particulate lattice structure found in the junction; the pitted zones which start at the left of the two outer arrows are the inner cytoplasmic surface of the junction. (*c*) presents a plan view of the gap junction. In (*d*) the junction is shown in cross section. At × a true cross section is shown but at OBL an oblique view is seen. (*c*) ×190,000 and (*d*) ×80,000. (By kind permission of P. J. Revel and the *Journal of Cell Biology*.)

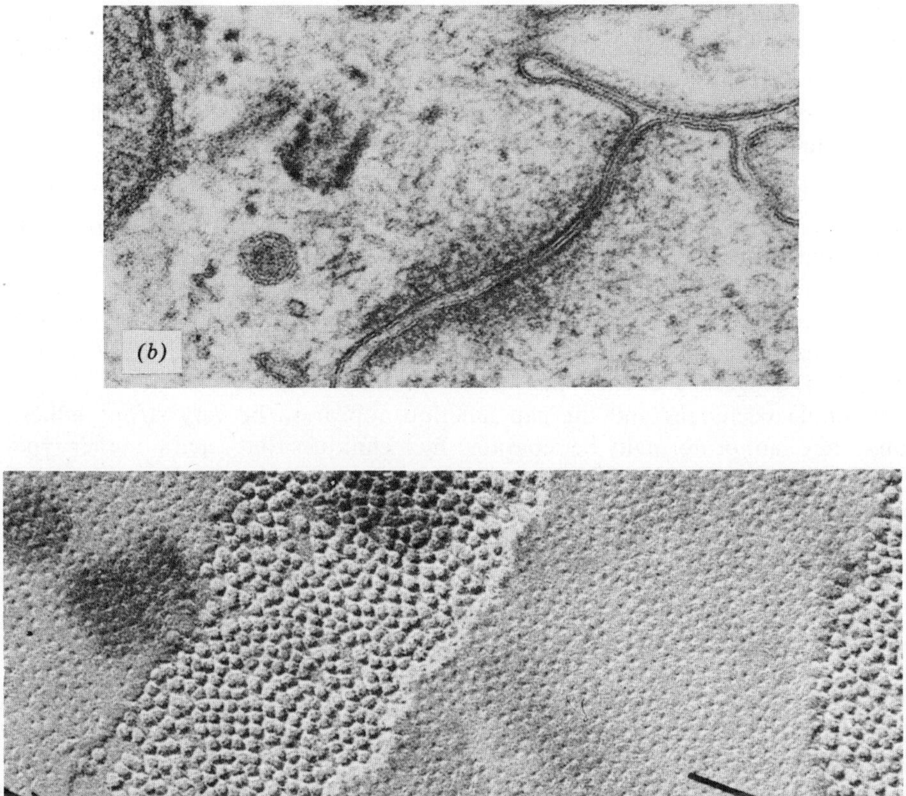

Figure 4. continued

4. The macula adherens or desmosome. This structure is characterized by a dense fibrillar staining in the cytoplasm beneath the plasmalemmae. The dark-staining lines representing the plasmalemma are separated by an appreciable gap. Faint lines of staining may be seen on occasion in the gap between the dark lines of staining.
5. Septate desmosomes, whose structure is still rather unclear.

It should be noted that there may be considerable confusion in the terminology of cell contacts because a very wide range of names have been introduced. A fairly extensive synonomy is given by Curtis (1967).

This classification refers to contacts between cells, both between like and unlike cells. Unfortunately, very few studies have been made of the morphology of contacts between cells and nonliving substrates although the work of Cornell (1969) and of Curtis (1964) suggests that they may have a similar structure.

The zonula occludens and the gap junction appear to be very strong adhesions because they cannot normally be separated by techniques that separate other types of cell contact. It is also clear that the lipid cores of the membranes represented by the dark-staining lines in the zonula occludens are so close that there can be no extensive surface coat or glycocalyx involved in the structure of these adhesions. However, the zonula and macula adherens are two types of contact in which the involvement of the surface coat, if any, presents a problem which is difficult to resolve. One view is that the gap is filled with a cementing material which binds the cells together; presumably, this material is formed from the surface coats on each cell (see Martinez-Palomo, 1970). The main evidence that supports this point of view is that the gap can be stained with certain electron-dense materials. An alternative concept is that the gap is filled with intercellular fluid in life and that no cementing material fills the gap. This may at first seem to be a surprising point of view, but there are good biophysical reasons for the existence of such a system. The most cogent evidence in favor of this idea is that Brightman and others have found that tracer molecules such as ferritin permeate with ease down the gaps between cell surfaces in zonulae adherentes in life, although after fixation the gap is then filled with an impermeable material. Such an experiment suggests that the cementing material is not present, or is present only in very small amounts, during life and that fixation releases sufficient material for the viscosity of the gap to rise to a high value and to provide material that can be stained to appear in an electron micrograph. Evidence is a little less clear concerning the reality of an intercellular material in the gap in the macula adherens, but it seems probable that the same tests should be applied as for the zonula adherens. Thus the question of the nature of the material in the gap immediately raises one concerning the mechanism of cell adhesion, because there are the alternative possibilities that (1) the cell coat cements the surfaces together, or that (2) another system of adhesion is used which permits the existence of a fluid-filled gap between the cells.

F. Dispersion Studies

Cells in animal tissues can be separated from one another or from nonliving substrates by use of the following conditions either separately or in combination: high pH values, generally above pH 8.0; low ionic strength; removal of divalent cations (for example, by the use of chelating agents such as EDTA or EGTA); use of enzymes, in particular trypsin. All these conditions are potentially capable of damaging the cells. In the majority of cases, treatment with reagents to produce such conditions is insufficient of itself to separate the cells. A certain degree of mechanical force is required to complete dispersion (dissociation). In a few instances Brownian motion or the density of the cells themselves provides the force to pull the cells apart after chemical dispersal has weakened the adhesions, but usually mechanical dispersion, for example, by the use of pipetting, is necessary. A few tissues, such as those of certain sponges, can be separated into single cells by the application of mechanical dispersion without prior chemical treatment, but in the majority of cases such treatment induces extensive cell damage.

It is presumed that the main effect of these dispersing agents is to diminish the adhesiveness of the cells to a point at which gentle mechanical action can separate the cells. If this presumption is correct, the nature of the agents that are effective will allow us to make deductions about the nature of cell adhesion. There is, however, the possibility that some or all of these reagents act in another way to effect cell separation. If the cell is damaged at some level internal to the actual components used to effect adhesion, this may result in the cells losing adhesion because the actual adhesive structure has become detached from the cell. If this happens, the results of experiments on cell dispersion will reveal little about the mechanism of cell adhesion. Even if the main action of these reagents is on the actual mechanism of adhesion, there may be additional side effects on other cellular structures, which may affect any subsequent attempt to reaggregate the cells. All these possibilities bear on two questions of importance. The first is whether or not the mechanism of adhesion revealed by the process of disaggregation is identical with that shown in the reaggregation of cells. The second question is whether or not any lack of identity found is attributable to artifact, and if so whether the artifact lies in the process of dispersion or in that of reaggregation. Both these questions can be finally answered only after material on the reaggregation of cells has been considered. However, it should be noted that if the conditions required for cell dispersion are the converse of those required for the formation of adhesions, and if the quantitative measurement of cell adhesion provides answers of the same magnitude both for the establishment as for the break-up of adhesions, then it is possible that there is no change in the adhesive process during dispersion, reaggregation, or in adhesions after they have been established for a long time.

Trypsinization of tissues to produce cell suspensions was introduced by Schieffer-decker (1886) and reintroduced by Rous and Jones (1916). In practice this technique often produces fairly extensive cell lysis although since the damaged cells are usually completely digested the visible cells appear to be of high viability. The lysis of the cells releases DNA which tends to form a jellylike material which may artificially cement cells together. For this reason it is a fairly common practice to include DNAase in the medium for cell dispersal. The action of the enzyme almost certainly lies in its proteolytic nature. Trypsin has been found to release sialoglycopeptides from cell surfaces (see Cook, 1968); it probably damages other cell surface proteins, for example, acyl transferases. Chymotrypsin and pronase have been used on occasion to produce dispersal of tissues, and a certain degree of success has been obtained with papain. Ficin appears to be inactive in affecting cell adhesion. Elastase and collagenase have proved to be most effective in the preparation of cell suspensions from those tissues in which extensive amounts of elastin or collagen are found, but it is uncertain whether these enzymes affect cell adhesion or whether they simply lyse the whole intercellular structure of collagen, leaving almost nothing for the cells to adhere to. Surprisingly, very little systematic testing of the effects of enzymes on cell adhesion has been attempted, and in the majority of cases in which results have been reported no data on enzyme activity or purity have been provided. The conclusions that can be drawn from the actions of enzymes in effecting cell separation are considered later in this section.

The measurement of cell adhesiveness during the dispersion of tissues is fraught with technical and theoretical difficulties. The majority of adult tissues contain appreciable amounts of intercellular materials such as collagen, and many embryonic tissues also contain such substances. If the force required to break a tissue into separate cells is measured, one may be making a measurement of the mechanical strength of the collagen. This problem can be avoided in many instances, but in all we are faced with a set of theoretical problems originating from the fact that there is as yet no general theoretical treatment of the break-up of a multibodied aggregate, whose component particles (cells) are nonrigid bodies of irregular shape. Phillips and Steinberg (1969) have suggested that cell adhesion can be measured for cells in an aggregate by the equilibrium shape assumed by the aggregate in a centrifugal field. This change in shape of the aggregate requires the breaking of old and the establishment of new adhesions. Since we have no idea of the way in which the internal mechanical properties of the cells affect this process, we cannot yet use the technique to measure cell adhesiveness.

Brooks et al. (1967) modified a qualitative method introduced by L. Weiss (1961) to obtain a rough quantitative measurement of cell adhesion in tissues such as liver. They measured the shear force required to break up the tissues into single cells. The magnitude of this force allows the calculation of the energy of adhesion under several assumptions. They concluded that cell adhesions might have energies of between 1 and 10×10^{-8} J m^{-2} of contact.

Cell contacts of the zonula or macula adherens type are easily dispersed by treatment with dispersing agents such as EDTA, whereas zonulae occludentes and gap junctions are unaffected. (Sedar and Forte, 1964; Goodenough and Revel, 1970). This suggests that the two morphologically distinct types of adhesion are indeed also different in mechanism.

G. The Readhesion of Cells (Reaggregation and Aggregation)

If cells in an adhesive state collide with one another they stick to each other in at least a proportion of the collisions. As a consequence, aggregates build up. The collisions may result from the movement of cells on a solid surface, from the swimming of those few cell types that do so, or may be artificially induced by shaking a cell suspension. Gerisch (1960) introduced the method of preparing aggregates by agitating cells in suspension, and Moscona (1961) popularized the technique. Collision takes place in such systems because there are velocity gradients in the medium (i.e., shear); as a consequence cells tend to collide as a faster traveling cell approaches another from behind (see Fig. 5). If the shear rate is too high, however, any pair of cells that has just formed an aggregate will be torn apart by shearing forces. Natural reaggregation, such as that of slime mold cells, often involves cell movement on a solid surface. Unfortunately, such systems are rather less amenable to the experimental study of cell adhesion than those in which a suspension is agitated, because it may be hard to determine whether a treatment that inhibits aggregation does so by affecting cell adhesion or cell movement. However, Beug et al. (1970) have carried out

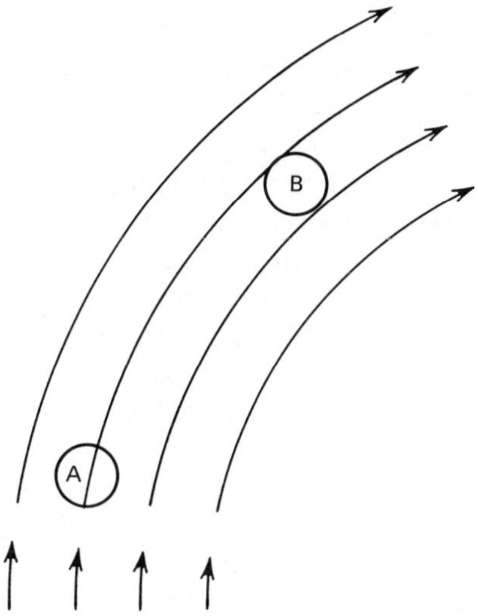

Figure 5. Diagram showing the origin of collisions in a shear flow. The dotted lines show stream lines in the medium, whose flow velocity increases with radius. Consequently, cell A will eventually catch up with cell B and collide because it lies partly on the same stream lines as B but is traveling at a higher velocity.

some remarkable work on the possible relations between surface antigens and aggregation using this system. Aggregation can occur in still unstirred suspensions if the particles (cells) involved are small enough and present at high enough concentration for Brownian motion to induce an appreciable number of collisions. This may take place in platelet aggregation.

Random, variable, and unknown shear is introduced into a cell suspension by shaking it on a gyratory or reciprocating shaker. In practice the size and shape of the flask used to hold the suspension and the depth to which the flask is filled with the suspension may affect the shear rate and thus the collision rate. Obviously, the greater the collision rate the more rapidly aggregates build up, other matters being equal. The collision rate (instantaneous) b_{ij} is given by the relationship: $b_{ij} = (4/3)\ Gn_in_j (r_i + r_j)^3$, where r_i and r_j are the radii of cells i and j present in concentrations n_i and n_j and G is the shear rate. Thus it can be seen that cell size, the population density of cells, and the shear rate all affect the rate of collision. Eventually, the aggregates reach a limiting size. Moscona (1961) suggested that this size provided a measure of the adhesiveness of the cells. A reason for this suggestion is that the shear force across an aggregate increases as the square of the aggregate diameter. This force although tending to bring about collisions, also tends to pull the aggregate to pieces. Eventually, an aggregate should reach a diameter (equilibrium) at which the rate of aggregation is balanced by that of break-off of cells. In consequence, the limiting (equilibrium) size of an aggregate should be determined by the adhesiveness of the cells, with only very adhesive cells forming large aggregates. If this theory is correct, increasing the shear on a group of equilibrium aggregates should reduce their diameter. Since this does not happen (Curtis, in preparation), it is clear that aggregate diameter is not a direct measure of cell adhesiveness.

There are other difficulties in using final aggregate diameter as a measure of cell adhesiveness. Chief among these is that since the aggregates take several hours to reach their equilibrium size, anything that can be deduced from aggregate size about adhesiveness refers to the adhesiveness over a period of hours during which considerable changes may take place. Thus the details of the adhesive process may be obscured.

Curtis and Greaves (1965) introduced the method of using reaggregation kinetics as a measure of cell adhesion. This method examines the rate of formation of adhesions in a suspension. Curtis (1969) developed this method so that the collision efficiency, that is, the probability that an adhesion forms on a collision between cells or aggregates, can be measured. This provides a dimensionless measure of adhesion which can be used to calculate the energy of cell adhesion. There are several advantages in this type of approach. First, the effects of differing shear rates, cell size, cell population density, and medium viscosity can be eliminated so that comparison can be made between different experimental conditions. Second, the kinetics of aggregation can be examined minutely so that any deviation from the kinetics to be expected for a homogenous population of cells of contant adhesiveness can be detected. Changes in adhesion over very small time intervals can be observed.

Such deviations in aggregation kinetics lead to a matter of considerable importance. Several workers have claimed that various chemical reagents prevent aggregation of cells, hence these substances directly inhibit the mechanism of cell adhesion. As a consequence, it has been suggested that cell adhesion requires the synthesis of some compound or compounds. If, however, the cell surface has been damaged during preparation of the suspensions, resynthesis of a normal surface will be necessary before adhesions can form. Consequently, the resynthesis may be the rate-limiting process until a normal surface has been established. Detailed examination of aggregation kinetics (Curtis, 1970; Edwards and Campbell, 1971) suggests that such processes occur. Hence there is the possibility that the discovery that various metabolic inhibitors affect aggregation refers not to the mechanism of cell adhesion but to the resynthesis of the surface.

In practice, aggregation kinetics are followed by visual or electronic counting of the number of cells and aggregates. Light absorption (light-scattering) methods have been introduced for the study of platelet aggregation (Born and Hume, 1967) and also used to follow the aggregation of embryonic cells (Jones, 1965), but so far no method of obtaining the actual aggregation rate from such data has been devised. A certain amount of criticism has been directed at this kinetic method on the grounds that it does not deal with histogenetic aggregation. Histogenetic aggregation is that in which two cell types mixed in suspension yield aggregates that contain the two cell types in a nonrandom, that is, segregated arrangement. It is not yet known at what stage in aggregation this arrangement is established. It has, however, been shown by Elton and Tickle (1971) that aggregates prepared from experiments in which aggregation kinetics were followed show cell segregation.

Methods for studying the adhesion of cells to nonliving substrates are much less well developed. The rate of settling and attachment of cells to such substrates provides a qualitative measure of the process, but as yet no quantitative theory has been developed to take account of other variables in the process such as cell size, relative density, cell shape, and cell deformability, which obviously all affect the process.

In view of the different methods used for the study of cell reaggregation and the possibility that in many cases the findings refer to the "repair" or resynthesis of cell surface structure, it is exceedingly difficult at the moment to present a wholly correct view of the factors that affect cell adhesion.

Ionic conditions affect the readhesion of cells markedly. For example, the adhesion of chick embryonic cells (Curtis, 1963) is markedly pH dependent, so that their adhesiveness increases by a factor of ca. 10^4 on lowering the pH from 8.0 to 7.0. No systematic investigation of the effect of other monovalent cations on adhesion appears to have been made. One difficulty in such work is that the concomitant alteration in osmotic strength may affect cells very extensively; in addition, the large changes in ionic strength required may kill cells. Curtis (1964) showed that high concentrations of Na^+ ions decrease the thickness of the gap between a cell and the glass substrate, presumably strengthening the adhesion between them.

Divalent cations have a considerable effect on cell adhesion. Collins (1966) found that calcium was essential for the adhesion of chick neural retinal cells in a shaker system. Armstrong and Jones (1968) claimed that barium and strontium were ineffective in producing cell adhesion, whereas calcium or magnesium aided adhesion. However, Hornby (1971) found that Ba^{2+} and Sr^{2+} do promote cell adhesion although less markedly than Ca^{2+} or Mg^{2+}. Collins showed that Ca^{2+} reduces the surface charge of cells by about 10%. These results are supported by a large volume of earlier work using less exact techniques which demonstrated that divalent cations increase cell adhesion (reviewed by Curtis, 1967). A few workers, such as Garvin (1968), have been unable to detect any effect of Ca^{2+} on the adhesion of cells. At present, it is unclear as to whether there really are differences between cells or whether technical details explain the discrepancies. It is clear that Stern layer binding of calcium to cells may occur so strongly that it is almost impossible to remove the ion from the cells (Curtis, 1972). This may explain why in some cases addition of calcium did not increase cell adhesiveness. Trivalent cations and other heavy metal ions increase cell adhesiveness (Dan, 1947; Wilkins et al., 1962), but it is probable that the cells are dead.

The effects on cell adhesion of a large number of organic chemicals have been reported. Rather than list the results baldly, we discuss these results in terms of the main theories of cell adhesion they presumably support.

Inhibitors of RNA and protein synthesis and exposure to low temperatures inhibit the reaggregation of cells that have been extensively trypsinized (M. H. Moscona and Moscona, 1966), whereas cells that have not been prepared in this way aggregate in the presence of such inhibitors [Humphreys (1963) for sponge cells; Curtis and Greaves (1965) and Glaeser et al. (1968) for embryonic chick cells; and Guidice (1965) for echinoderm cells or at low temperatures (see Curtis, 1967)]. Moscona has suggested that the dependence of reaggregation on synthesis implies that an adhesive cement is being synthesized. An alternative concept is that trypsinization damages the cell surface so that resynthesis is required before adhesion can take place by another process which is presumably independent of metabolism. Since trypsinization removes surface glycopeptides and may damage the enzymes required for resynthesis of lecithins, glycolipids, and glycoproteins (Fischer et al., 1967), it is clear that resynthesis must occur before a normal cell surface is reattained. It is arguable that chelating agents also damage the surface, and it seems probable that this takes place. However, cells separated with a chelating agent and then exposed to trypsinization show the metabolic requirement typical of cells only exposed to trypsin. Cells separated with trypsin and then exposed to EDTA do not acquire a metabolic independent adhesiveness (Curtis, in preparation). These two facts argue that the metabolic dependence is an artifact induced by trypsinization.

Oppenheimer et al. (1969) found that trypsinized mouse teratoma cells required l-glutamine for the redevelopment of adhesions. Roseman (1970) has developed an interesting hypothesis on the basis of this finding. L-Glutamine is a precursor required in the synthesis of amino sugars. Roseman showed that specific glycosyl transferases occur in the cell surface. These enzymes catalyze the addition of amino sugars to the glycosidic compounds. Roseman suggests that the glycosyl transferase on one cell surface might bind to a glycosidic chain on another. In this way a labile but possibly highly specific attachment might be formed between cells. Roseman notes that Mn^{2+} ions tend to be required for the activity of these enzymes, while Ca^{2+} is often inhibitory. This behavior does not agree well with what is known about the effect of calcium on cell adhesion. The theory suggests that various amino sugar compounds might be specific inhibitors of cell adhesion. The fact that *N*-acetylglucosaminidases and β-galactosidase diminish cell adhesion (Curtis, in preparation) might be expected on Roseman's theory, but it is more difficult to explain why neuraminidase treatment should increase cellular adhesion on the basis of this theory. It also might be expected that the cell surface glycosidic components of one cell could well act as competitive inhibitors of the glycosyl transferases of heterologous cells with the consequence that unlike cells would stick together irreversibly. This does not normally happen. It is of interest that Beug et al. (1970) proved by a univalent antibody technique that slime mold cells carry a glycoprotein on their surfaces associated with cell adhesion. Aggregation-deficient mutants lack this substance.

Nucleotides can affect cell adhesion. Application of ADP increases the adhesiveness of blood platelets markedly, although it appears to have little effect on the adhesiveness of other cells. ATP decreases the adhesiveness of a variety of embryonic cells according to Jones (1966). He has suggested that the ATP produces the contraction of a protein in the cell surface, which in turn affects cell adhesiveness. Although such a mechanism is not impossible, a variety of other mechanisms by which ATP might act have not been disproved. For example, ATP, at the levels used by Jones, is a fairly effective chelating agent for calcium and magnesium.

A considerable number of other compounds have been studied to determine their effect on cell adhesion, but no clear deductions can as yet be made from these results. Curtis (1967) has provided a review of earlier results.

H. Specific Adhesion

Wilson (1970) found that when cell suspensions from two different sponge species were mixed, each single aggregate of the large number that resulted, appeared to be composed of the cells of one species alone. Although a considerable number of mechanisms could have explained this result, Wilson suggested that the cells of each species could adhere only to their own type. In other words, a process of specific adhesion had taken place. When cells from two different tissues of the same species are intermingled, a segregation to different parts of each aggregate takes place at some time during reaggregation or soon afterward. Galtsoff (1925) was probably the first person to make this observation, although the observations of Townes and Holtfreter (1955) on the sorting out of amphibian embryonic cells are also of historical importance. Specific adhesion has been suggested as the mechanism for this sorting out of tissues (histotypic segregation). Steinberg (see Curtis, 1967) separately noted that specific adhesion could not account for the very reproducible patterning of cell position found in such aggregates, for specific adhesion at best would produce nothing more than randomly arranged groups of each cell type.

If specific adhesion takes place even to a small extent, we must construct our theories concerning the mechanism of cell adhesion to take account of it. Thus resolution of the question whether or not specific adhesion occurs is of considerable importance.

Clearly, if specific adhesion does occur, a considerable degree of nonspecific adhesion is possible, for cells adhere very well to a wide range of nonliving substrates of widely differing nature such as various glasses, cellulose acetates, polyacrylates and vinyl esters, polytetrafluorethylene, polyethylenes and polypropylenes, fluorocarbons, stainless steels, gold, platinum, and so on. It is possible that the mechanism of adhesion to such nonliving substrates is different from that for other cells, although Curtis (1967) has noted the many similarities between the two situations. It is also clear that cells adhere to others of a different type both in experimental situations and in the intact animal. Armstrong (1971) reports that there appears to be no difference in the electron microscope appearance of contacts between homologous and heterologous cells.

Until recently, the only evidence for the occurrence of specific adhesion has been the segregated positions of cells in the aggregates arising from a mixed cell suspension. Since many other mechanisms could account for the finding of either sorted-out aggregates or aggregates of one cell type alone, such evidence does not provide an unequivocal answer to the question of the occurrence of specific adhesion. Roth and Weston (1967) and Roth (1968) introduced a test for specific adhesion in which the relative proportions of homologous and heterologous cells that could be collected to the surface of a preexisting aggregate, were measured. These experiments showed that considerably more homologous cells adhered than heterologous ones. Curtis (1970) noted that the experiment was open to other interpretations. Curtis (1969) introduced a quantitative test for the occurrence of specific adhesion based on the use of aggregation kinetics. In several papers (summarized by Curtis and van de Vyver, 1971), he reported that no evidence for specific adhesion of cells could be found. However, Curtis and van de Vyver described a system found in the sponge *Ephydatia fluviatilis* in which the cells produce factors that specifically alter the quantitative strength of cell adhesiveness. Cell adhesion itself remains nonspecific in this system, but the level of adhesiveness shown by a cell is specifically controlled. This feature of the system appears to explain many earlier results, such as those describing the existence of specific aggregation factors hypothesized as being specific cementing agents for cells (see Humphreys, 1963; Moscona, 1968; Antley and Fox, 1970). Clearly, the question of the occurrence of specific adhesion is not yet resolved, but no compelling evidence in favor of the concept appears to have been obtained so far. However, Crandall and Brock (1968) discovered that two mating types of the yeast *Hansenula wingeii* agglutinate in a specific manner and showed that these cell wall reactions require the presence of specific mannan proteins. This appears to be a definite example of a specific adhesion.

I. Mechanisms of Adhesion

There are basically three possible mechanisms for the adhesion of surfaces.

1. Adhesion by sensitization (i.e., bridging or cements). Here the surfaces are bridged by molecules that are bi- or polyfuntional, so that one end of the molecule binds to one surface and another part to the second surface. The agglutination produced by antibodies appears to be a reaction of this type. It is hard to list typical features of this adhesion but the following are often found. (a) A gap is present be-

tween the particles, whose width corresponds to the mean distance between functional groups on the bridging molecule. (b) The strength of the adhesion is determined by the relative proportions of bridging agent and sites on the surfaces that can bind the agent. In the presence of sufficient bridging agent to combine with more than half the binding sites, adhesions weaken. (c) Adhesions are extremely strong if the density of bonds between the surfaces is anything but very low. (d) The bonding is very sensitive to specific changes in local chemical conditions, and it is possible to ensure considerable specificity of adhesion.

2. Adhesion by molecular contact. At first sight such a mechanism might not seem to differ appreciably from that described above. If the bridging agent is considered to be part of the surface, then it might be argued that adhesion by bridging is no more than a type of molecular contact between surfaces. However, the term molecular contact has come to include, from its use by colloid chemists, the concept that two surfaces, whose properties are very different from that of the surrounding medium, make so close an approach that they form or nearly form one homogeneous structure. To illustrate this idea in a biological context, two cell surfaces would form a molecular contact if the lipid layers were to approach to within 5 Å. Typical features of such contacts are: (a) a very small gap between cell surfaces; (b) considerable energy of adhesion arising from close range van der Waals forces, hydrophobic bonding, and various other bonding interactions between the surfaces; (c) irreversibility of the adhesions.

3. Adhesion in the secondary minimum. This type of adhesion is characterized by a gap varying between, for example, 60 Å upward to 300 Å between the surfaces. The gap is filled with the medium that surrounds the particles. The biologist unfamiliar with colloid chemistry may at first be surprised that such adhesions could exist. The reason for their existence is that in certain situations adhesion is controlled by an interplay of electrostatic forces of repulsion and a force of attraction arising from the London–van der Waals dispersion force. This is termed the Derjaguin–Landau–Verwey–Overbeek theory (DLVO theory). Since the two forces decline with distance between surfaces in differing ways, it is possible to have interaction energy curves of the type shown in Figure 6. It can be seen from this figure that adhesion can occur (provided the electrostatic repulsive energy is not too great), both when the surfaces are separated by no gap (molecular contact or adhesion in the primary minimum), and when a gap occurs between the surfaces (adhesion in the secondary minimum of the potential energy diagram). It is appreciated that the London force between assemblies of atoms and larger particles has a long-range character. This force arises from the electronic and molecular vibrations, respectively, of individual atoms and molecules. Recent papers by Napper (1967) and by Ninham and Parsegian (1970) describe this force in detail. Visser (1968) also discusses the force from a theoretical point of view and reviews experimental evidence for its action.

Such adhesions have the following characteristics. (a) A gap is found between the particles even at their closest approach. The gap is filled with the surrounding medium. The dimensions of the gap are sensitive to any treatment that affects the electrostatic forces of repulsion between the surfaces. Thus divalent cations aid adhesion since they reduce repulsive forces. (b) The adhesions are not specific. (c) The adhesions are easily reversible, that is, redispersable. (d) The energy of adhesion is low.

If the electrostatic forces are reduced sufficiently, Brownian motion energy will bring the particles into the primary minimum region of attraction and the secondary mini-

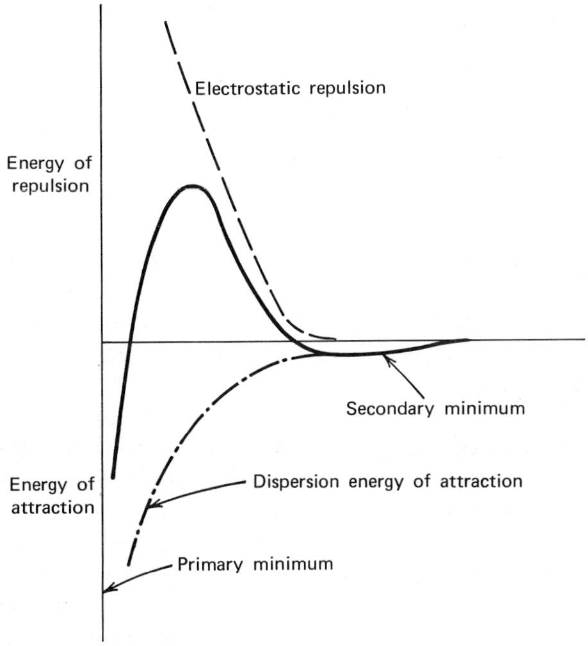

Figure 6. Diagram showing the way in which the electrostatic energy of repulsion between two surfaces and the London force of attraction can sum (continuous line) to give two positions (separations of the surfaces of the particles) at which adhesions form, the primary and the secondary minima. These two positions are separated by a potential energy barrier which may prevent a secondary minimum adhesion from turning into a primary minimum adhesion. Abscissa, distance between surfaces of the particles.

mum adhesion will be replaced with a primary minimum one. The main parameter that determines the London force of attraction is the difference between the dielectric constants of the surfaces and those of the medium. The parameters that determine the electrostatic forces of repulsion are the static dielectric constant of the medium, the surface potential of the surfaces, and the concentration and valency of counterions in the medium. Increase in surface potential, decrease in counterion valency, or decrease in concentration all act to increase the electrostatic force of repulsion and the range over which it acts.

The next step is to consider each theory in terms of the biological evidence. Electron microscopy suggests that there are two main types of adhesion, close contacts in which the cell surfaces (lipid layers) come so close that the surfaces are effectively in molecular contact, and adhesions in which a gap of between 60 and 300 Å is found between the lipid parts of the plasmalemmas. The zonula occludens appears to agree well with a type of adhesion in which molecular contact takes place, both from the dimensions of the adhesion and from the fact that such cell adhesions appear to be irreversible and of great strength. The zonula and macula adherens types of adhesion are much weaker, but unfortunately this feature does not clearly distinguish between the bridging and the secondary minimum explanations although the first explanation can be used only in cases in which the density of bridging molecules is very low. The dimensions of the gap found in such adhesions make it attractive to consider the possibility that a secondary minimum mechanism is in operation. The close relationship between counterion concentration and gap thickness found for a cell–glass adhesion by Curtis (1964) strongly indicates that a secondary minimum system operates in such an adhesion. However, the question whether the electron microscope appearance of the zonula adherens supports one theory of adhesion or another will be resolved only when the presence of cross-linking molecules in the gap can be proved or disproved. Tracer molecule experiments show that the gap must be of low viscosity, which implies that the density of cross-bridging molecules is either very low or zero.

Measurement of the energy of adhesion from reaggregation kinetics (Curtis, 1969), suggests that two types of adhesion can be formed. First are those with energies of the order of 10^{-5}–10^{-4} J m^{-2} which may perhaps correspond to primary minimum adhesions. The second type are weaker by a factor of 10^4–10^5 and may be secondary minimum type adhesions.

When we consider the data from experiments on the effects of various reagents on cell adhesion, we are faced with the problem that in nearly all cases no investigation using electron microscopy has been made of the type of adhesions being formed, and few investigations have been made of the type of adhesions being broken in tissue dispersion experiments. However, the few high-resolution electron micrographs that have been made of aggregates suggest that the cells are in contact by zonulae adherentes or by maculae adherentes. For this reason we assume that the biochemical work on cell adhesion refers to these two types of adhesion.

The bridging (cementing) theory of cell adhesion appears to be supported by the following lines of evidence. (1) the effect of trypsin and other enzymes in separating cells, and (2) the apparent metabolic dependence of reaggregation on such processes as glutamine incorporation or protein synthesis. If specific adhesion occurs, this would strongly suggest that cements are formed between cells, because the other two theories of adhesion do not predict specificity. However, we have seen that the evidence for a metabolic dependence is unsatisfactory. The effects of removal of cations on the dispersion of tissues and their presence on the aggregation of cells can be accounted for on the basis of the cementing theory if it is assumed that the cement is in some way sensitive to the presence of divalent cations. The "calcium bridge" theory (according to which the actual bridging agent is the calcium atom) is, however, most improbable and is not supported by experimental evidence (see Curtis, 1967). The bridging theory explains the existence of a gap between plasmalemmae in some adhesions, although it does not account for the dimensions of the gap particularly well nor does it explain the effects of various reagents on this.

It seems improbable that the zonulae or maculae adherentes types of adhesion can be explained as examples of adhesion by molecular contact, because this theory would predict that the adhesions would be irreversible, extremely strong, and without a gap between plasmalemmas.

The idea that these adhesions represent secondary minimum adhesions receives support from the following evidence: (1) the lack of evidence of a metabolic involvement in adhesion (except where repair processes have been stimulated); (2) the effects of the nature of the ionic environment on cell adhesion; (3) the measured strength of such adhesions; (4) the dimensions and nature of the gap; (5) their reversibility; and (6) the apparent lack of specificity in cell adhesion. The effects of low ionic strength or lack of divalent cations in reducing cell adhesiveness strongly suggests that an electrostatic repulsive force plays a role in cell adhesion. Unfortunately, the few attempts to test this by measuring cell surface charge in relation to changes in the external medium have produced equivocal results. Armstrong and Jones (1968) have suggested that the fact that different divalent cations affect adhesiveness to different degrees disproves the involvement of an electrostatic repulsive force in adhesion. In fact, such a finding provides strong support for the concept since the double-layer theory (see Overbeek, 1952) states that it is normal to expect such differences within a class of ions of constant valency. However, although the secondary minimum concept is the only one to implicate specifically a role for electrostatic forces of repulsion, evidence for their involvement does not necessarily prove this concept. It is possible to conceive of a role for electrostatic forces in the other theories of cell adhesion. One difficulty in testing the hypothesis that electrostatic forces play a role in cell adhesion arises from the fact that differences in adhesiveness between cells, whether natural or chemically produced, may result from changes in other forces such as the London force. For example, an agent such as lysolecithin might reduce the electrostatic forces, which would favor adhesion, but the simultaneous changes in lipid structure might reduce the forces of attraction to a greater extent, resulting in an apparently anomalous effect if the result were to be considered only from the electrostatic point of view.

A further difficulty in all studies of cell adhesion arises from the fact that some biochemical feature of a cell that appears to be associated with cell adhesion may have only an incidental and not a causal connection. For example, Beug et al. (1970) demonstrated that a glycoprotein on the cell surface is associated with the ability of such cells to form adhesions. It might be concluded that this fact argues against a secondary minimum interpretation, but if the association is merely incidental, the finding has little bearing on the mechanism of cell adhesion.

The chief objection to the application of the secondary minimum theory to cell adhesions of the macula or zonula adherens type is the fact that trypsin and certain other proteolytic enzymes produce cell dispersal. Although it is not impossible to conceive of some mechanism by which the trypsinization either increases the electrostatic forces or reduces the London force as a result of changes in cell surface structure, there is currently no evidence for such a system. At present, it is rather hard to cite any biochemical work that clearly establishes the mechanism of adhesion used in the formation of zonula occludens, gap junctions, or septate desmosomes. The reason for this is that these types of adhesion are relatively rare and normally involve only a small region of a cell's contact surface. The fact that such adhesions are unredispersable (irreversible) and very strong argues in favor of the idea that these adhesions are of the molecular contact type. If this is true, they would be expected to form between two surfaces having low surface potentials.

Clearly, work on cell adhesion has nearly reached that fascinating point at which it should be possible to make a clear decision between the correctness of the various theories.

J. Electrical Properties: Electrical Coupling of Cells (Low-Resistance Pathways)

Loewenstein (1966) and Furshpan and Potter (1968) discovered that very many cell types have contacts which are of very low electrical resistance. Rapid permeation of ions and small molecules can take place between such cells. It appears that the gap junction is probably the structure that provides the pathway between cells (Hudspeth and Revel, 1971). Loewenstein (1966) showed that the formation of this coupling requires the presence of divalent cations. Presumably, a cell adhesion must be established which is then converted into a gap junction. It seems probable that the phenomenon of metabolic cooperation between genetically dissimilar cells is an instance of functional coupling between cells. Metabolic cooperation was discovered by Subak–Sharpe et al. (1969). At present, it is unclear whether or not such coupling between cells plays a role of any great importance in the life of cell populations either *in vivo* or *in vitro*, but it is conceivable that it might be of importance in inductive and regulative interactions of cells during differentiation. Although it is clear that coupling can take place between cells that have been entirely separate at some time in their life cycle, it is possible that some examples of coupling represent the incomplete separation of daughter cells after division, leaving an intercellular bridge of cytoplasm as a pathway (Fawcett, 1961).

3. SECONDARY FUNCTIONS OF CELL SURFACE

A. Phagocytosis and Pinocytosis

Many types of cell, perhaps indeed all, possess the ability to form indentations in the plasmalemma which encircle either drops of the external medium (pinocytosis) or solid particles (phagocytosis). The plasmalemmal pocket is then pinched off from the cell surface and passes into the cytoplasm. These two related processes may be referred to jointly as endocytosis. Clearly, endocytosis is a complex process which involves the triggering of the inpocketing process. The adhesion of the particle to the cell surface is a prerequisite of phagocytosis. A variety of factors affects the inpocketing. For example, the viscosity and elasticity of the plasmalemma determine the extent to which the plasma membrane can be bent; as pocketing is completed, an adhesion must be formed between either side of the pocket; this is followed by breaking of the membrane so that the pocket is released as a vacuole. These processes are shown in detail in Figure 7. The source of the process that causes the membrane to inpocket is unclear, although it is not unreasonable to attribute it to a contractile process operated perhaps by microfilaments just below the plasmalemma. A process rather similar to endocytosis takes place in the release of virus from cells for those viruses that bear a coat of plasmalemma; in these cases a pocket of plasmalemma surrounds the virus particle and is then detached from the surface. The ability of the surface to break off small pockets that fuse to form spheres is also shown by cells during antibody attack (Goldberg and Green, 1960). It should be noted that endocytosis and other processes bear some resemblance to cell fusion (see Section 3.E) in that a plasmalemma breaks and establishes a new connection and continuity with a nearby piece of plasmalemma. Secretion may well take place as a reverse form of pinocytosis.

A considerable amount of work has been carried out on the nature of the factors that stimulate phago- or pinocytosis in a cell. Pinocytosis has been reviewed by Chapman-Andresen (1965), while Karnovsky (1962) has summarized recent work on phagocytosis. Unfortunately, no clear picture has emerged of the mechanisms behind either process. Probably, this is because a complex of cell surface properties and some subsurface properties are involved. For example, phagocytosis requires the adhesion of the particle to the surface, possibly by a zonula occludens type of adhesion (see Curtis, 1967). However, it is probable that factors that might stimulate adhesion act to reduce membrane mobility so that one is faced with the possibility of a factor

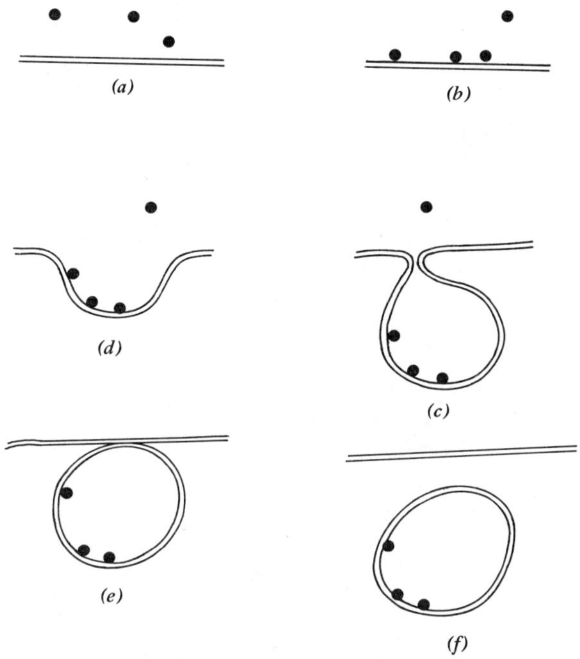

Figure 7. Schematic diagram of the stages of pino- or phagocytosis. The cell surface is shown in section. Particles to be phagocytosed or molecules to be pinocytosed (dots) in the vicinity of the surface (*a*) form an adhesion (*b*). The surface then starts to infold (*c*). At (*d*) the inpocketing has reached the point where the two sides of the pocket form a close adhesion at the neck. By (*e*) the inpocketing has shown scission of the cell surface and the formation of membrane fusions, both to complete the vacuole and to restore the cell surface; the vacuole is still in adhesion with the inner side of the plasmalemma. Finally, this adhesion is lost and the vacuole moves off into the cytoplasm (*f*) where it may be digested.

acting in two opposite directions on phagocytosis. Although pinocytosis does not necessarily require the adhesion of materials to the cell surface, it seems probable from the nature of those substances that are pinocytosed most easily that an electrostatic binding with the surface takes place.

B. Cell Shape

Clearly, surface properties might play an important role in the maintenance of cell shape if the internal mechanical properties of the cytoplasm were too weak to have effect. Early workers thought that the surface tension of the plasmalemma might determine cell shape, but it now seems that the surface tension is too low to play any important role. Internal osmotic pressure, and the presence and action of microfilaments and microtubules (see Wessells et al., 1971) are probably much more influential in the determination of cell shape than are other factors. However, changes in shape may result in considerable increases or decreases in cell surface area. Such changes can in part probably be met by alterations in cell surface packing, but larger changes require either the synthesis of new cell surface or its resorption, or the alternative mechanism of storing assembled cell surface within the cell, for example, in the endoplasmic reticulum and releasing or resorbing surface into this store. Bennett (1963) has suggested that this mechanism takes place. Recent biochemical work argues that cell surface synthesis can be extremely rapid and this may provide a means of preparing the cell for shape changes. At present, however, it is unclear whether surface is synthesized *in situ* or not, or whether the synthesis or insertion of new membrane is localized or generally spread over the whole of the plasmalemma. It is clear, however, that contact reactions involving the cell surface can control cell shape (see Curtis, 1961) , but it is not known whether the changes result from adhesive interactions of the cell or more subtle mechanisms.

C. Cell Lysis

Cell lysis results from the formation of a hole or holes in the plasmalemma. This may be caused simply by excess hydrostatic pressure inside the cell, which stretches the plasmalemma until it breaks. However, the most interesting forms of lysis result from those agents that alter plasmalemmal properties so that the normal membrane becomes unstable. As a result, the membrane breaks into pieces or develops holes in it. In some forms of lysis, the membrane rounds up into small spheroids leaving large areas of naked cytoplasm exposed to the medium (Goldberg and Green, 1960). A more general appearance is the development of irregular rents, tears, and holes which may be visible only at the level of the electron microscope. There appear to be three main mechanisms by which membrane breakdown can take place. First, direct substitution of components of the membrane can occur, resulting in a change in intramembrane forces of cohesion, so that the membrane either expands so rapidly that it tears itself, or that it becomes so weak mechanically that random thermal or mechanical motions break it. It seems probable that lysolecithin acts in this manner (Dingle and Lucy, 1962). Second, substances may remove membrane components without replacing them; the consequences are very similar. Third, substances may dissolve in the plasmalemma, adding to and altering its structure.

Lipid solvents and surface-active substances (detergents) probably dissolve in the membrane and thereby alter its structure. The various proteins that cause membrane breakdown are enzymes that lyse various components of the plasmalemma and other nonenzymic proteins that probably bind to the surface (see Becker, 1960). The most remarkable of the proteins that cause lysis are those involved in the complement series [see Kabat (1968) for a detailed account]. If complement (found in serum) is present, an immune reaction with a cell surface component leads to lysis. The mechanism by which this operates is still not understood.

Lysis and cell surface damage may also arise indirectly because lysolecithin is constantly being formed from lecithin in the plasmalemma (Fischer et al., 1967). Simultaneously, lysolecithin is reacylated to lecithin by acyl transferases. If the process of reacylation is affected, lysolecithin may accumulate with the consequence that lysis begins when the lysolecithin level has reached a certain value.

D. Cell Movement and Cell Behavior

Many aspects of cell movement are discussed in Chapter 16. However, the normal movement of cells on solid surfaces shown, for example, by amebas or by fibroblasts, blasts, is not discussed there, and for this reason it is treated here. Since such cell movement is not primarily a cell surface function we cannot devote much space to consideration of its mechanism. Cells moving in this way are characterized by the fact that they continually protrude pseudopods which appear to form adhesions with the substrate near their tips. A contraction then takes place, pulling the cell toward the attached end of the pseudopod. Pseudopods can show a wide variation of form; for example, fibroblasts form a broad, thin lamellopod which continually shows ruffling activity (Abercrombie et al., 1970). This ruffling activity slowly moves the leading edge of the fibroblast forward in a series of protrusions and attachments. The basic contractile mechanism is probably located in the microfilaments just below the cell surface (Goldman and Follett, 1970). Adhesion clearly plays an important role in this mechanism, and it is probable that the viscoelastic properties of the surface may determine such matters as the size and rate of formation of protrusions. It is possible that a variety of other mechanisms of movement is used by some cells (see review by Wolpert and Gingell, 1968), but there is little experimental evidence to support such systems.

The main classes of cell behavior that have been described are: (1) contact inhibition of movement (Abercrombie and Heaysman, 1954); (2) chemotaxis; (3) chemokinesis; (4) contact guidance (P. Weiss, 1941); and (5) contact promotion (see Curtis, 1967); peripolesis. All these types of behavior have been discovered and mainly analyzed by studies on tissue culture. Nevertheless, there is fairly good reason (see Curtis, 1967; Trinkaus, 1969) to believe that such processes take place inside the tissues of organisms. There they would be expected to play an important part in determining the placing of cells in their correct positions in the adult. Malignant invasion of tumor cells may result from a breakdown in this normal behavior (Abercrombie, 1961).

Contact inhibition of movement occurs when two locomoting cells make contact. The contact results in the cessation of movement of one or both cells for a period of time, pseudopodal activity apparently being inhibited. The phenomenon is a statistical one, and the frequency with which this inhibition takes place on the formation of a contact is a property of the two cells involved. It is also affected by the nature of the medium around the cells and by the total population density of cells, which may represent the effect of the increasing number of contacts that involve three or more cells as the density rises. If free space is available, the cells will move away from their contact because new pseudopods can form in any part of the cell not in contact with another cell. As a result, contact inhibition both prevents the overlap of cells, leading to monolayering, and promotes the outward spread of a cell sheet from regions of high population density. It is clear, however, that in many cells continued movement results from the presence of contact with another cell at the hind end of the first cell (see Shaffer, 1964; Abercrombie and Gitlin, 1965). The mechanism of contact inhibition of movement is still not understood, although it may involve an adhesive reaction of the cells and appears to require very close proximity of two cells. The phenomenon occurs both between heterologous and homologous cells.

Chemotaxis requires the orientation of a cell in relation to a chemical gradient; cell movement then takes the cell either up-gradient (positive chemotaxis) or down-gradient (negative chemotaxis). Unfortunately, it is unclear at present whether or not such a system involves the cell surface except to the extent that the chemical involved presumably can enter the cell by permeation. The best understood and most amazing examples of chemotaxis occur in the cellular slime molds (*Acrasiales*) in which the chemotactic agent appears to be cyclic AMP (Konijn et al., 1968). Chemokinesis is a similar phenomenon in which the concentration of the active agent or its gradient affects either the rate of turning or the rate of movement of the cells (see description by Curtis, 1967).

Contact guidance is the oriented movement of cells on a substrate which shows some type of orientation. It was discovered by Weiss (1941). Dunn (in preparation) suggests that contact guidance is in effect nothing more than the operation of contact inhibition of movement under a special set of circumstances. The phenomenon of contact promotion or peripolesis is the converse of contact inhibition. Unfortunately, the phenomenon has never been investigated quantitatively. Cells showing contact promotion of movement would be expected to have a tendency to form aggregates in which cell movement is active.

A considerable number of other biological phenomena may operate either solely or partially through the cell surface. For example, the inhibition of mitosis found in densely packed cell cultures appears to depend on contact between the cells, but the nature of this is as yet undetermined (Abercrombie, 1970). It should also be remembered that some substances can be either synthesized or assembled at the cell surface, for example, collagen.

E. Cell Fusion

Cell fusion has been known for many years in the formation of a zygote by fusion of gametes. Colwin and Colwin (1963) showed that plasmalemmal fusion occurs in this process. Skeletal muscle forms by the fusion of myoblasts, and presumably the fusion of the plasmalemmas of the cells is an essential feature of the process. However, Barski et al., (1961) were the first workers to obtain the experimental fusion of cells. Early work required the use of inactivated Sendai virus to obtain cell fusions, but recently Lucy (1970) has shown that lysolecithin treatment induces fusion. Lucy has suggested that a phase change in the lipids may take place after lysolecithin treatment such that the bilayer structure of the plasmalemma is converted to a micellar structure. If this takes place at a cell contact of the zonula occludens type, the first result will be a gap junction (see Curtis, 1972). If the gap junction then breaks down, cell fusion will result. Experimentally induced cell fusions have of course considerable promise as a research tool for the investigation of gene action in cells. It is worth noting that endocytosis, secretion, virus release, and even the final stages of cell division require the fusion of membranes of a single cell. Cell fusion clearly must involve cell adhesion, but little is known about this.

REFERENCES

Abercrombie, M. (1961). Behaviour of normal and malignant connective tissue cells *in vitro*. *Can. Cancer Conf.*, **4,** 101.

Abercrombie, M. (1970). Contact inhibition in tissue culture. *In vitro*, **6,** 128.

Abercrombie, M., and G. Gitlin (1965). The locomotory behaviour of small groups of fibroblasts. *Proc. Roy. Soc.* (*London*), **B162,** 289.

Abercrombie, M., and J. E. M. Heaysman (1954). Observations on the social behaviour of cells in tissue culture. II. "Monolayering" of fibroblasts. *Exptl. Cell Res.*, **13,** 276.

Abercrombie, M., J. E. M. Heaysman, and S. M. Pegrum (1970). The locomotion of fibroblasts in culture. *Exptl. Cell Res.*, **60,** 437.

Antley, R. M., and A. S. Fox (1970). The relationship between RNA and protein synthesis and the aggregation of *Drosophila* embryonic cells. *Develop. Biol.*, **22,** 282.

Armstrong, P. B. (1970). A fine structural study of adhesive cell junctions in heterotypic cell aggregates. *J. Cell Biol.*, **47,** 197.

Armstrong, P. B., and D. P. Jones (1968). On the role of metal cations in cellular adhesion: cation specificity. *J. Exptl. Zool.*, **167,** 275.

Barski, G., S. Sorieul, and F. Cornefert (1961). "Hybrid" type cells in combined cultures of two different mammalian cell strains. *J. Natl. Cancer Inst.*, **26,** 1269.

Becker, F. S. (1960). Studies on the hemolytic properties of protamine. *J. Gen. Physiol.*, **44,** 433.

Bennett, H. S. (1963). Morphological aspects of extracellular polysaccharides. *J. Histochem. Cytochem.*, **11,** 14.

Beug, H., G. Gerisch, S. Kempff, V. Riedel, and G. Cremer (1970). Specific inhibition of cell contact formation in *Dictyostelium* by univalent antibodies. *Exptl. Cell Res.*, **63,** 147.

Born, G. V. R., and M. Hume (1967). Effects of the numbers and sizes of platelet aggregates on the optical density of plasma. *Nature*, **215,** 1027.

Bretscher, M. S. (1971). Major human erythrocyte glycoprotein spans the membrane. *Nature*, **231,** 229.

Brightman, M. W. (1965). The distribution within the brain of ferritin injected into cerebrospinal fluid compartments. I. Ependymal distribution. *J. Cell Biol.*, **26,** 99.

Brightman, M. W., and S. L. Palay (1963). The fine structure of ependyma in the brain of the rat. *J. Cell Biol.*, **19,** 415.

Brooks, D. E., J. S. Millar, G. V. F. Seaman, and P. S. Vassar (1967). Some physicochemical factors relevant to cellular interactions. *J. Cell. Physiol.*, **69,** 155.

Burger, M. M., and A. R. Goldberg (1967). Identification of a tumor specific determinant on neoplastic cell surfaces. *Proc. Natl. Acad. Sci., U.S.*, **57,** 359.

Chapman-Andresen, C. (1965). The induction of pinocytosis in amoebae. *Arch. Biol. (Liege)*, **76,** 189.

Collins, M. (1966). Electrokinetic properties of dissociated chick embryo cells. II. Calcium ion binding by neural retinal cells. *J. Exptl. Zool.*, **163,** 39.

Colwin, A. L., and L. Colwin (1963). Role of the gamete membranes in fertilization in *Saccoglossus kowalevski* (Enteropneusta). I. The acrosomal region and its changes in the early stages of fertilization. *J. Cell Biol.*, **19,** 477.

Cook, G. M. W. (1968). Glycoproteins in membranes. *Biol. Rev.*, **43,** 363.

Cornell, R. (1969). Cell-substrate adhesion during cell culture. *Exptl. Cell Res.*, **58,** 289.

Crandall, M. A., and T. D. Brock (1968). Molecular aspects of specific cell contact. *Science*, **161,** 473.

Curtis, A. S. G. (1961). Control of some cell-contact reactions in tissue culture. *J. Natl. Cancer Inst.*, **26,** 253.

Curtis, A. S. G. (1963). The effect of pH and temperature on cell re-aggregation. *Nature*, **200,** 1235.

Curtis, A. S. G. (1964). The adhesion of cells to glass. A study by interference reflection microscopy. *J. Cell Biol.*, **19,** 199.

Curtis, A. S. G. (1967). *The Cell Surface: Its Molecular Role in Morphogenesis.* Pp. x + 405. Logos, London.

Curtis, A. S. G. (1969). The measurement of cell adhesiveness by an absolute method. *J. Embryol. Exptl. Morphol.*, **22,** 305.

Curtis, A. S. G. (1970). On the occurrence of specific adhesion between cells. *J. Embryol. Exptl. Morphol.*, **23,** 253.

Curtis, A. S. G. (1972). Intra- and inter-membrane interactions of the cell surface. *Subcellular Biochem.*, in press.

Curtis, A. S. G., and M. F. Greaves (1965). The inhibition of cell aggregation by a pure serum protein. *J. Embryol. Exptl. Morphol.*, **13,** 309.

Curtis, A. S. G., and G. Van De Vyver (1971). The control of cell adhesion in a morphogenetic system. *J. Embryol. Exptl. Morphol.*, **26,** 295.

Dan, K. (1947). Electrokinetic studies of marine ova. V, VI, and VII. *Biol. Bull.*, **93,** 259.

Dingle, J. T., and J. A. Lucy (1962). Studies on the mode of action of excess of vitamin A. *Biochem. J.*, **84,** 611.

Edwards, J. G., and J. A. Campbell (1971). The aggregation of trypsinized BHK21 cells. *J. Cell Sci.*, **8,** 53.

Elton, R. A., and C. A. Tickle (1971). The analysis of spatial distributions in mixed cell populations: a statistical method for detecting sorting out. *J. Embryol. Exptl. Morphol.*, **26,** 135.

Farquhar, M. G., and G. Palade (1963). Junctional complexes in various epithelia. *J. Cell Biol.*, **17,** 375.

Fawcett, D. (1961). Intercellular bridges. *Exp. Cell Res., Suppl.*, **8,** 174.

Fischer, H., E. Ferber, I. Haupt, A. Kohlschutter, M. Modolell, P. G. Munder, and R. Sonak (1967). Lysophosphatides and cell membranes. *Protides Biol. Fluids*, **15,** 175.

Frye, L. D., and M. Edidin (1970). The rapid intermixing of cell surface antigens after formation of mouse-human heterokaryons. *J. Cell Sci.*, **7,** 319.

Furshpan, E. J., and D. D. Potter (1968). Low-resistance junctions between cells in embryos and tissue culture. *Curr. Top. Develop. Biol.*, **3,** 95.

Galtsoff, P. S. (1925). Regeneration after dissociation. (An experimental study on sponges). I. Behaviour of dissociated cells of *Microciona prolifera* under normal and altered conditions. *J. Exptl. Zool.*, **42,** 183.

Garvin, J. E. (1968). Effects of divalent cations on adhesiveness of rat polymorphonuclear neutrophils *in vitro*. *J. Cell Physiol.*, **72,** 197.

Gerisch, G. (1960). Zellfunktionen und Zellfunktionswechsel in der Entwicklung von *Dictyostelium discoideum*. I. *Arch. Entwicklungsmech.*, **152,** 632

Giudice, G. (1965). The mechanism of aggregation of embryonic sea urchin cells: a biochemical approach. *Develop. Biol.*, **12,** 233.

Glaeser, R. M., J. E. Richmond, and P. W. Todd (1968). Histotypic self-organization by trypsin-dissociated and EDTA-dissociated chick embryo cells. *Exptl. Cell Res.*, **52,** 71.

Goel, S. C., and A. Jurand (1971). The structures at the epithelio-connective tissue junction: a comparison of light and electron microscopic observations. *Proc. Roy. Soc. Edinburgh*, **B71,** 1.

Goldberg, B., and H. Green (1960). The cytotoxic action of immune gamma globulin and complement on Krebs ascites tumor cells. I. Ultrastructural studies. *J. Exptl. Med.*, **109,** 505.

Goldman, R. D., and E. A. C. Follett (1970). Birefringent filamentous organelle in BHK-21 cells and its possible role in cell spreading and motility *Science*, **169,** 286.

Goodenough, D. A., and J.-P. Revel (1970). A fine structural analysis of intercellular junctions in the mouse liver. *J. Cell Biol.*, **45,** 272.

Hornby, J. E. (1971). Ph.D. thesis. University of London.

Hudspeth, A. J., and J. P. Revel (1971). Coexistence of gap and septate junctions in an invertebrate epithelium. *J. Cell Biol.*, **50,** 92.

Humphreys, T. (1963). Chemical dissolution and *in vitro* reconstruction of sponge cell adhesion. I. Isolation and functional demonstration of the components involved. *Develop. Biol.*, **8,** 27.

Jones, B. M. (1965). Inhibitory effect of *p*-benzoquinone on the aggregation behaviour of embryo-chick fibroblast cells. *Nature*, **205,** 1280.

Jones, B. M. (1966). A unifying hypothesis of cell adhesion. *Nature*, **212,** 362.

Kabat, E. A. (1968). *Structural Concepts in Immunology and Immunochemistry*. Holt, New York.

Karnovsky, M. L. (1962). Metabolic basis of phagocytic activity. *Physiol. Rev.*, **42,** 143.

Konijn, T. M., D. S. Barkley, Y. Y. Chang, and J. T. Bonner (1968). "Cyclic AMP": a naturally occurring acrasin in the cellular slime molds. *Am. Naturalist*, **102,** 225.

Loewenstein, W. R. (1966). Permeability of membrane junctions. *Ann. N.Y. Acad. Sci.*, **137,** 441.

Lucy, J. A. (1970). The fusion of biological membranes. *Nature*, **227,** 815.

Martinez-Palomo, A. (1970). The surface coats of animal cells. *Intern. Rev. Cytol.*, **29,** 29.

Mitchison, J. M., and M. M. Swann (1955). The mechanical properties of the cell surface. III. *J. Exptl. Biol.*, **32,** 734.

Moscona, A. (1961). Rotation mediated histogenetic aggregation of dissociated cells. *Exptl. Cell Res.*, **22,** 455.

Moscona, A. A. (1968). Cell aggregation properties of specific cell ligands and their role in the formation of multicellular systems. *Develop. Biol.*, **18,** 250.

Moscona, M. H., and A. A. Moscona (1966). Inhibition of cell aggregation *in vitro* by puromycin. *Exptl. Cell Res.*, **41,** 703.

Napper, D. H. (1967). Modern theories of colloid stability. *Sci. Prog.* (*Oxford*), **55,** 91.

Ninham, B. W., and V. A. Parsegian (1970). Van der Waals forces. Special characteristics in lipid-water systems and a general method of calculation based on the Lifshiftz theory. *Biophys. J.*, **7,** 646.

Oppenheimer, S. B., M. Edidin, C. W. Orr, and S. Roseman (1969). An l-glutamine requirement for intercellular adhesion. *Proc. Natl. Acad. Sci., U.S.*, **63,** 1395.

Overbeek, J. Th. G. (1952). In *Colloid Science* (H. R. Kruyt, ed.), pp. 115–244. Elsevier, Amsterdam.

Pardoe, G. I., and G. Uhlenbruck (1971). Genetics and immunochemistry of blood group antigens. *Med. Lab. Technol.*, **28,** 1.

Phillips, H. M., and M. S. Steinberg (1969). Equilibrium measurements of embryonic chick cell adhesiveness. I. Shape equilibrium in centrifugal fields. *Proc. Natl. Acad. Sci., U.S.*, **64,** 121.

Rand, R. P., and A. C. Burton (1964). Mechanical properties of the red cell membrane. I. Membrane stiffness and intracellular pressure. *Biophys. J.*, **4,** 115.

Roseman, S. (1970). The synthesis of complex carbohydrates by multi-glycosyltransferase systems and their potential function in intercellular adhesion. *Chem. Phys. Lipids.*, **5,** 270.

Roth, S. (1968). Studies on intercellular adhesive selectivity. *Develop. Biol.*, **18,** 602.

Roth, S., and J. A. Weston (1967). The measurement of intercellular adhesion. *Proc. Natl. Acad. Sci., U.S.*, **58,** 974.

Rous, P., and F. S. Jones (1916). A method for obtaining suspensions of living cells from the fixed tissues, and for the plating out of individual cells. *J. Exptl. Med.*, **23,** 549.

Schiefferdecker, P. (1886). Methode zur Isolirung von Epithelzellen. *Z. Mikroskop.*, **3,** 483.

Sedar, A. W., and J. G. Forte (1964). Effects of calcium depletion on the junctional complex between oxyntic cells of gastric glands. *J. Cell Biol.*, **22,** 173.

Shaffer, B. M. (1964). In *Primitive Motile Systems in Cell Biology*, p. 387 (R. D. Allen and N. Kamiya, eds.). Princeton Univ. Press, Princeton.

Subak-Sharpe, H., R. R. Burk, and J. D. Pitts (1969). Metabolic co-operation between biochemically marked mammalian cells in tissue culture. *J. Cell Sci.*, **4,** 353.

Sweeley, C. C. (1970). *Chemistry and Metabolism of Sphingolipids* (C. C. Sweeley, ed.). North Holland, Amsterdam.

Townes, P. L., and J. Holtfreter (1955). Directed movements and selective adhesion of embryonic amphibian cells. *J. Exptl. Zool.*, **128,** 53.

Trinkaus, J. P. (1969). *Cells into Organs. The Forces That Shape the Embryo.* Prentice-Hall, Englewood Cliffs, New Jersey.

Visser, J. (1968). Adhesion of colloidal particles. *Rept. Prog. Appl. Chem.*, **53,** 714.

Weiss, L. (1961). The measurement of cell adhesion. *Exptl. Cell Res., Suppl.*, **8,** 141.

Weiss, P. (1941). Nerve patterns: The mechanics of nerve growth. *Growth, Suppl.*, **5,** 163.

Wessells, N. K., B. S. Spooner, J. F. Ash, M. O. Bradley, M. A. Luduena, E. L. Taylor, J. T. Wrenn, and K. M. Yamada (1971). Microfilaments in cellular and developmental processes. *Science*, **171,** 135.

Wilkins, D. J., R. H. Ottewill, and A. D. Bangham (1962). On the flocculation of sheep leucocytes. I and II. *J. Theoret. Biol.*, **2,** 165.

Wilson, E. V. (1907). On some phenomena of coalescence and regeneration in sponges. *J. Exptl. Zool.*, **5,** 245.

Winzler, R. J. (1970). Carbohydrates in cell surfaces. *Intern. Rev. Cytol.*, **29,** 77.

Wolpert, L., and D. Gingell (1968). Cell surface membrane and amoeboid movement. *Symp. Soc. Exptl. Biol.*, **22,** 169.

15

MECHANISMS OF MUSCLE CONTRACTION

C. C. Ashley

1. GENERAL INTRODUCTION

It seemed natural for biologists in previous centuries to wonder if the process of contraction was a property of every living cell, both animal and plant. We now know that not every animal cell is capable of contraction, and certainly the mechanism of movement in higher plants does not appear to be similar to that in animal cells. It is of interest, however, that movement at the microscopic level, such as is involved in the process of cell division and associated with the movement of intracellular components, may have features common to both plants and animals (Miller, 1968). This chapter deals with the way in which gross movement is brought about in animal cells, and in this connection the specific mechanisms associated with ciliary activity are not mentioned since they are discussed elsewhere (see Chapter 16). Considerable reference is made to the activity of vertebrate skeletal muscle as this has been studied in the greatest detail and much is known about its molecular architecture.

2. SKELETAL MUSCLE

A. Organization and Structure of the Contractile Proteins

1. Introduction. The nineteenth century microscopist William Bowman, better known perhaps for his work on kidney structure, was among the first to describe the fine structure of muscle (Bowman, 1840). He observed that muscles were composed of a series of smaller units or fibers, each containing many nuclei. The fibers were apparently single cells enclosed within a delicate outer membrane or sarcolemma. Single fibers from vertebrate muscles are generally in the range 10–100 μ in diameter and slightly eliptical in cross section (1 μ = 10^{-4} cm). Bowman also observed that the fibers could be subdivided into finer fibrils (*myofibrils*) when the outer membrane was removed or destroyed and that these myofibrils, which are usually 1–2 μ in diameter, were imbedded in an amorphous matrix. He described the myofibrils, of which there may be many thousands per fiber, as having transverse or cross-striations, and detected changes in the striation pattern during contraction. It was not until a few years later that the suggestion was made that the myofibrils might be divided into still finer components of submicroscopic dimensions, the myofilaments (Brücke, 1858; Kölliker, 1888).

These morphological features are illustrated diagrammatically in the cross section of a large muscle fiber, where only a few myofibrils are shown for simplicity (Figure 1). This fiber from an invertebrate is much larger than an average vertebrate fiber and can be 1000–2000 μ in diameter and several centimeters in length. Nevertheless, apart from certain modifications concerned with excitation (Section 2.C.3), it closely resembles a vertebrate fiber in structure.

2. Structure of the Myofibril. When myofibrils were examined in more detail with the light microscope, the cross-striations were observed to be regions of high and low refractive index. The regions of high refractive index were birefringent or *anisotropic* when examined with plane-polarized light and had an optical axis parallel to the long axis of the myofibril. This region constitutes what is now termed the A band. The zone of lower refractive index was isotropic (i.e., having virtually the same optical properties in all directions) and forms what is now termed the I band. The terminology frequently used in the past, that is, "light" and "dark," when referring to these regions has led to confusion. The precise appearance of the cross striations in living muscle depends on the position of focus of the microscope (A. F. Huxley, 1957), and in the electron microscope upon the staining properties of the section. These observations made a century ago with the light microscope still describe the essential features of muscle structure as it is known today. With the advent of the electron microscope, the fine structure of the myofibril has been investigated in much more detail, confirming the idea that the myofibrils contain still finer components, the myofilaments.

The two proteins that make up the main proportions by weight of the myofibrils are actin and myosin. The fact that myosin can be selectively extracted from the myofibril by suitable salt solutions (Hasselbach, 1953), and that after extraction the characteristic refractive index and anisotropy of the A band is absent, implies that myosin is normally located in this region (Hanson and Huxley, 1955). Similar biochemical evidence suggests that actin is located within the I band (Hanson and Huxley, 1953, 1955). In the I band electron microscope observations indicate that actin is in the form of thin filaments, while in the A band both thick and thin filaments can be seen. The

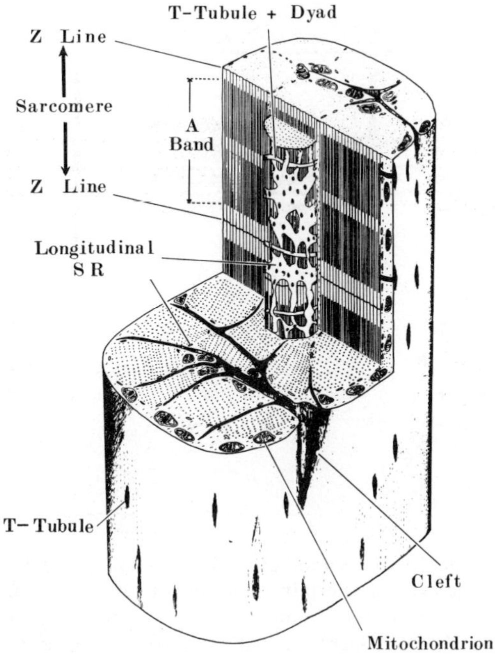

Figure 1. Diagrammatic structure of a large invertebrate muscle fiber. The myofibrils are subdivided into individual sarcomeres. The sarcomeres are bounded by Z lines and contain myofilaments of actin and myosin. (From Ashley, 1971.)

thick filaments have tapered ends (Fig. 2*a*), extend the full length of the A band and contain myosin. The myofilaments of actin and myosin are contained within the repeating unit of the myofibril, the sarcomere, which is bounded by dense Z lines. Thus the periodic distribution of actin and myosin, producing zones of different refractive index, gives rise to the characteristic cross-striated appearance of voluntary muscle (Hanson and Huxley, 1953, 1955).

A comprehensive picture of the arrangement of the actin and myosin filaments within the sarcomere has emerged from more detailed electron microscope and x-ray diffraction studies. In longitudinal electron microscope sections, the myosin or thick filaments have a series of projections along their length except for regions at the center of the sarcomere. These projections or cross-bridges extend at right angles from the long axis of the filament, at least in fixed material, and appear to come into contact with the actin (Fig. 2*a*). The thinner actin filaments are attached to the Z line at one end and terminate in the center of the sarcomere at the H zone. Consequently, the actin and myosin filaments overlap in certain regions of the sarcomere. The I band is defined as extending from the edge of the A band of one sarcomere to the beginning of an adjacent A band (Fig. 2*a*). The lengths of the myofilaments in the electron microscope vary considerably because of shrinkage of the specimen during fixation. However, observations on living muscle fibers combined with electron microscope observations on carefully fixed material indicate that in most vertebrate muscles the actin filaments are close to 2.1 μ long through the Z line and approximately 80 Å thick, while the myosin filaments are 1.6 μ by 120 Å (H. E. Huxley, 1963; Page, 1964). The heads of the cross-bridges of the myosin project an additional 70–100 Å from the surface of

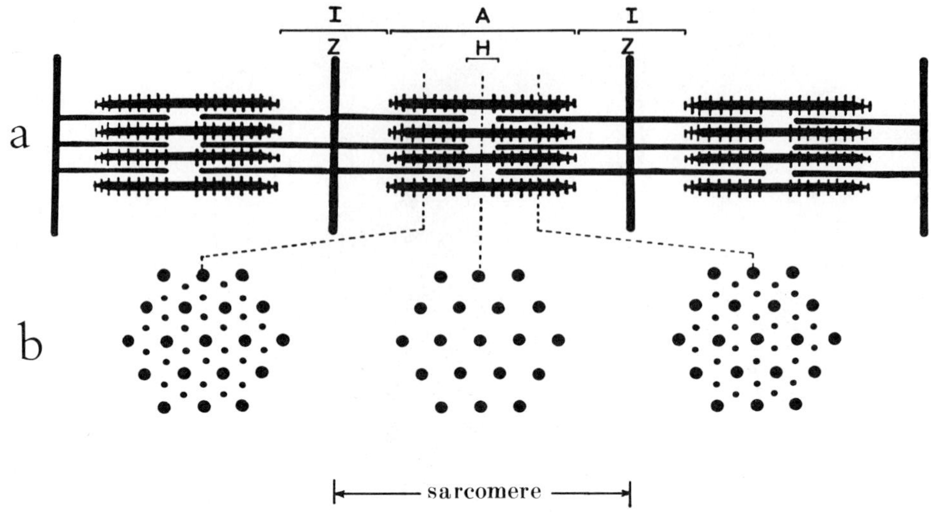

Figure 2. Diagrammatic structure of a sarcomere in longitudinal (*a*) and transverse (*b*) section showing the overlapping arrangement of the thin actin and thick myosin filaments. The cross-bridges project from the backbone of the myosin filaments. In transverse section through the overlap region, a double hexagonal array of myofilaments is seen. Myosin–myosin bridges in the M region are not illustrated. (From H. E. Huxley, 1969.)

the thick filaments. The remaining distance between the end of the head of the cross-bridge and the surface of the actin filament is variable. It depends upon the degree of overlap between the filaments, hence on the sarcomere length. For example, it can vary by at least 40 Å for sarcomere lengths of 2.0 μ–2.8 μ (H. E. Huxley, 1969). The variable spacing between the actin and the myosin filaments is a crucial factor in the sliding filament hypothesis for contraction which is discussed in more detail in Section 2.A.5.

Detailed examination of the sarcomere in cross section reveals a complicated array of myofilaments (H. E. Huxley, 1953b), of which the precise organization depends upon the position of the section. In vertebrate muscle a cross section through the overlap region (the A-I zone) reveals a *double* hexagonal array of actin and myosin, therefore each myosin is surrounded by six actin filaments (Fig. 2*b*). The actin filaments occupy the trigonal position of the lattice so that each is shared by three myosins. If the section is taken in the H zone, only the hexagonal array of myosins is observed, while at the I band only the thin actin filaments can be seen (Hanson and Huxley, 1955; H. E. Huxley, 1960). Although the hexagonal array of myosin filaments is generally observed in both vertebrate and invertebrate skeletal muscle, the actin array can vary. In some invertebrates each myosin can be surrounded by up to 9 to 12 actins (Peachey, 1968), even if the muscle is fixed at rest lengths where there is only a single overlap zone (see Section 2.A.5).

In parallel with the observations made with the electron microscope, early x-ray diffraction studies were also extremely informative. Both living and dried muscle gave sharp x-ray reflections, indicating that the muscle proteins were contained in ordered structures within the muscle cell. In particular, meridional reflections indicative of axial periodicities along the myofibril (H. E. Huxley, 1953a, 1960) remained constant both during stretch and during contraction. In addition, sharp equatorial reflections suggested a hexagonal array of myofilaments and transverse symmetry within the fiber. Thus the x-ray evidence indicated that the muscle proteins were organized in a complicated filamentous array within the fiber. The constancy of the meridional reflections suggested that the actin and perhaps myosin filaments did not alter their lengths during contraction, and supported the idea of a sliding rather than a coiling of the myofilaments (H. E. Huxley, 1960).

3. Structure of Myosin and the Thick Filament. By a combination of x-ray diffraction, electron microscopy, and antibody labeling, much is now known about the organization of the myosin molecules within the thick filament. Each filament consists of a series of longitudinally arranged molecules 1500 Å long and approximately 20 Å in breadth (1 Å = 10^{-8} cm). At one end the molecule has a thickened region which is approximately 200 Å long and 40 Å wide (H. E. Huxley, 1963; Rice et al., 1966). The molecular weight of myosin has been the subject of much investigation as the molecule readily undergoes transformation into a series of aggregates. Currently accepted figures for vertebrate myosin are close to 500,000 (Perry, 1967). The precise number of myosin molecules within a single myosin filament is, however, uncertain. The number of cross-bridges per filament has been estimated from electron microscope observations as 200 to 220 (H. E. Huxley, 1957) and, assuming one cross-bridge per myosin of molecular weight 500,000, the molecular weight of the filament should be 105×10^6. Recent estimates of the molecular weight of the thick filament are closer to 160×10^6, and the discrepancy between the values is almost large enough to suggest two myosin molecules per cross-bridge (H. E. Huxley, 1963). In insect flight muscle there is evidence for more than one myosin molecule per cross-bridge (Chaplain and Tregear, 1966; Reedy, 1967) (see Section 3.B.2), but it is fair to say that in vertebrate muscle the situation is still uncertain (Squire, 1971).

Investigation into the substructure of the myosin molecule has been possible by using procedures in which a myosin solution (Section 2.B.1) is subjected to brief digestion with proteolytic enzymes such as trypsin or a derivative (Mueller and Perry, 1962; Lowey et al., 1966). The characteristic biochemical function of myosin, essential to its activity in muscle, is its ability to split high-energy phosphate compounds such as adenosine triphosphate (ATP) (Section 2.B.1). Brief tryptic digestion produced a solution of lower viscosity which consisted of two components each having a lower molecular weight than myosin, while the overall ATPase activity was unimpaired (A. G. Szent-Györgyi, 1960). A longer incubation period with trypsin resulted in a "myosin" solution which lacked this characteristic activity.

The components resulting from the brief tryptic digestion are termed *meromyosins* and can be distinguished from each other by differential centrifugation. The faster sedimenting, heavier component retains the ATPase activity and also myosin's ability to combine with actin. This is heavy meromyosin or HMM (Mihályi and Szent-Györgyi, 1953; A. G. Szent-Györgyi, 1953). The slower-sedimenting, lighter component is devoid of ATPase activity and is termed light meromyosin or LMM. Analysis of the meromyosins and comparison with the known amino acid composition of intact myosin indicates that one HMM and one LMM are produced per myosin molecule fragmented (Kominz et al., 1954). The fragments show different solubilities in salt solutions and organic solvents, and this can be used as a convenient way of preparing larger quantities for more detailed investigations (A. G. Szent-Györgyi, 1960).

Examination of the LMM fragment indicates that it is composed of two identical polypeptide chains coiled around each other (Lowey and Cohen, 1962) and has a molecular weight of about 150,000. The length of this rod-shaped fragment is approximately 900 Å and it represents the main structural part of the myosin molecule. The HMM fragment is more complex and consists of a globular "head" region and a small rod or "tail" region. The HMM fragment can be split into the head and tail regions by additional enzymic digestion (Mueller and Perry, 1962). The head region or S_1 sub-fragment is a globular protein with a molecular weight of about 120,000 while the tail, sub-fragment S_2, has a lower molecular weight of about 60,000 (Lowey et al., 1966, 1967). The S_2 sub-fragment, as with LMM, is composed of two polypeptide chains coiled around each other. Stoichiometrically, one HMM yields *two* S_1 sub-fragments and *one* S_2 sub-fragment, the characteristic ATPase being retained only by the S_1. Essentially the complete myosin molecule is a dimeric structure consisting of the identical singlet units (m), each with a head (S_1) and tail (HMm plus LMm). This has been illustrated schematically in Figure 3. More recently, lower molecular weight polypeptide chains—the "light chains—have been detected, particularly associated with the S_1 sub-unit where they may control the ATPase activity (Dreizen et al., 1967; Weeds and Lowey, 1971).

Within each of the polypeptide chains of the LMM and S_2 fragments, the amino acids are arranged in a configuration which is for the most part α helix (Lowey and Cohen, 1962). This certainly increases the structural stability of these molecules. The regions where the polypeptide chain is not as well organized seem to be the positions at which the molecule is more susceptible to enzymic attack. It is, however, precisely these regions that are vital in producing cross-bridge flexibility, allowing it to adapt to a changing filament lattice during shortening (Section 2.A.5).

The way in which the cross-bridges are arranged along the myosin filament has been deduced from an analysis of x-ray diffraction diagrams of living muscle (H. E. Huxley and Brown, 1967) (Fig. 4*a*). The heads of the HMM (Fig. 6), the cross-bridges, project from the backbone of the filament along helical paths at intervals of 143 Å. Each pro-

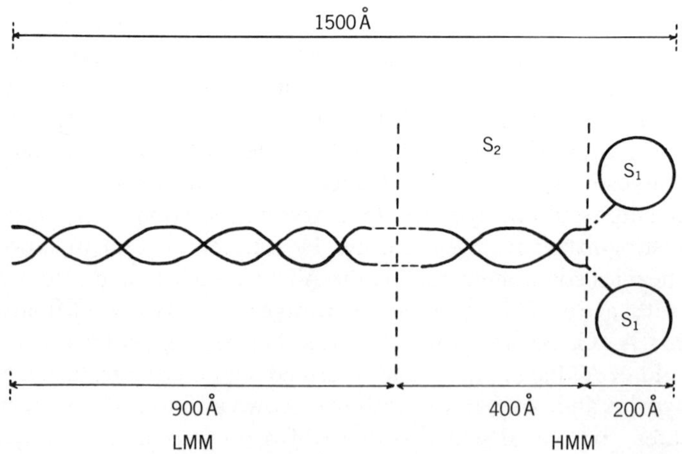

Figure 3. Diagrammatic structure of the *dimeric* myosin molecule. The dashed lines represent the probable positions of enzymic attack. The dimenions of the sub-fragments are only approximate. The *singlet* myosin molecule (m), consists of the single polypeptide chains LMm, HMm–S_2 plus an S_1. (Modified from Lowey et al., 1969).

jection has an identical one on the other side of the filament, so that there are three pairs of cross-bridges for every half-turn of the helix (H. E. Huxley, 1969).

Perhaps of equal interest is the packing of the myosin molecules within the thick filament to produce this arrangement of cross-bridge pairs. Various models for the structure of the myosin filament have been proposed based on antibody labeling experiments (Pepe, 1967a, b, 1971). The doubt about the precise number of myosin molecules per thick filament makes any model rather uncertain. It is possible to explain some of the known facts about the structure of the thick filament if it is assumed that the myosin molecules aggregate in a simple tail-to-tail fashion. Electron microscope observations of small "synthetic" myosin filaments suggest this type of aggregation and explain the absence of cross-bridges in the center of the filament (Fig. 4*b*) (H. E. Huxley, 1963, 1964a). In the complete thick filament, the arrangement must be more complex in order to explain the *myosin–myosin* bridges in the M zone (H. E. Huxley, 1965). These structures produce the increased density of this region seen in the electron microscope and presumably help to stabilize the A band (Knappeis and Carlsen, 1968).

An aggregation of myosin molecules from the center of the filament outwards (Fig. 4*b*) certainly explains the reversal of polarity on passing from one side of the

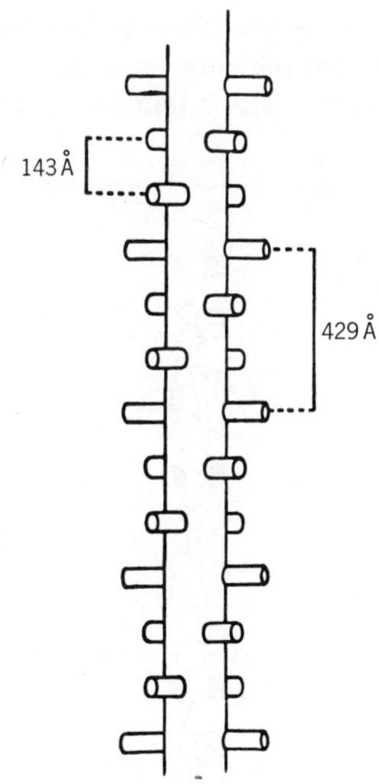

(*a*)

Figure 4*a*. Organization of the cross-bridges along the length of the thick filament. Two bridges occur at each level and three pairs every 429 Å. (From H. E. Huxley and Brown, 1967.)

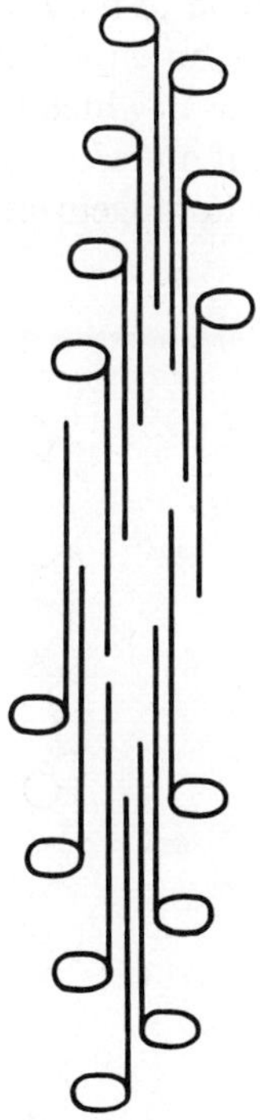

(*b*)

Figure 4*b*. Possible organization of myosin molecules in a 'synthetic' myosin filament. Here the cross-bridges are not paired and the myosins are drawn as singlets. Note the reversal of the polarity of the globular-S_1 on either side of of the bare central shaft. (Modified from H. E. Huxley, 1963, 1964a.)

M zone to the next (H. E. Huxley, 1963; Pepe, 1967a). This is matched by a reversal of polarity of the actin filaments on either side of the Z line (Section 2.A.4). The importance of this reversal is that the thin filaments can move, via the activity of the cross-bridges, in *opposite* directions on either side of the M zone, so that the sarcomere can shorten from both ends.

4. **Structure of F-Actin and the Thin Filament.** The structure of the thin actin filament is relatively simple compared with the complexities of myosin. It is composed of two strands of F-actin wound around each other (Hanson and Lowy, 1963) in the configuration of a right-handed helix (Depue and Rice, 1965). Each F-actin strand is a chain of basic G-actin subunits, and under the electron microscope the pitch of the helix is 350 Å, providing an integral number of 13 G-actin subunits per turn (Hanson and Lowy, 1964). However, more recent x-ray diffraction observations on living muscle indicate that the pitch of the helix is somewhat greater, in the range 360–370 Å, and provides for a nonintegral number of subunits per turn (H. E. Huxley and Brown, 1967; Hanson, 1968) (Fig. 5*a*). The G-actin monomers appear roughly spherical under the electron microscope with a diameter of 55 Å and a molecular weight of about 55,000. The nature of the linkages between each G-actin unit is unknown, but it does not involve the direct participitation of adenosine diphosphate (ADP) which is normally bound to the F-actin strands (Section 2.B.1). The change in polarity of the thin filaments on either side of the Z line, referred to in Section 2.A.3, was observed in preparations in which the actin was still attached to the Z line. This can be achieved, for example, if myosin is selectively extracted and then the remaining muscle structure disrupted by homogenization. If this preparation is treated with a solution containing

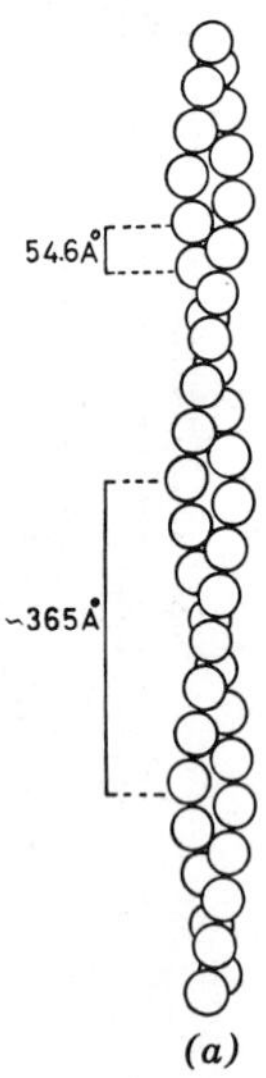

(*a*)

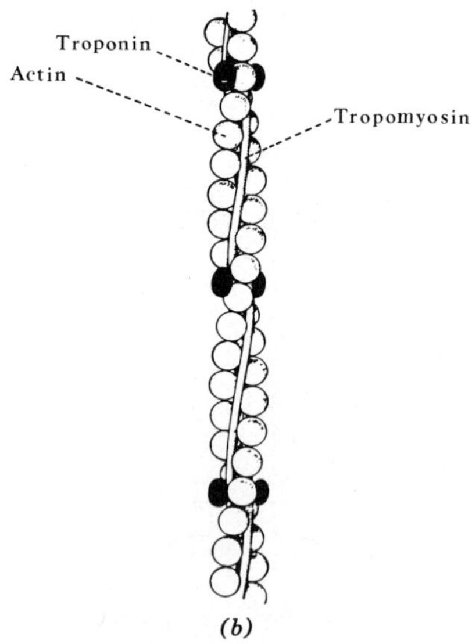

(*b*)

Figure 5*a*. Structure of the thin actin filament. Two strands of F-actin are wound around each other in a helical configuration. The pitch of the filament (ca. 365 Å) provides for a nonintegral number of G-actin subunits per turn. (From H. E. Huxley and Brown, 1967.)

Figure 5*b*. Structure of thin filament showing a possible arrangement of molecules of tropomyosin B and troponin. (From Ebashi et al., 1969.)

HMM or S_1 subunit, a characteristic arrowhead or chevron formation is seen along the thin filaments (Section 2.B.2). The chevrons point in opposite directions on either side of the Z line (H. E. Huxley, 1963; Moore et al., 1970).

The structure of the Z line has also been the subject of speculation in recent years and various models have been proposed to account for the well-known zigzag appearance in longitudinal sections (Knappeis and Carlsen, 1962; Reedy, 1964). It seems certain that the Z line contains F-actin strands which are formed by the unwinding of the thin filament as it approaches this position. Also, antibody labeling experiments show that it contains the protein tropomyosin B (Kelly, 1967).

This protein which makes up approximately 4% of the weight of the myofibrillar protein, at least from rabbit muscle was first described by Bailey (1948). It can be located both in the Z line and in the I band by using a fluorescent antibody labeling technique (Pepe, 1966). The tropomyosin-B molecule resembles LMM in many respects but is somewhat smaller, approximately 400 Å long. It readily forms aggregates as does myosin, so that it may form a continuous filament by tail-to-tail or side-to-side aggregation within the grooves of the thin filament. Certainly, it appears to be periodically distributed along the length of the thin filament (Ebashi, 1968; Ebashi et al., 1969), and biochemical evidence supports this finding since tropomyosin B combines with F-actin rather than myosin (Ebashi and Endo, 1968).

In addition to tropomyosin B, there is in the I band at least one more protein of importance, namely, troponin. This is a globular molecule with a molecular weight of 50,000, it is composed of several subunits and constitutes approximately 3% of the myofibrillar proteins (Ebashi et al., 1969). Antibody staining experiments indicate that troponin molecules are distributed along the thin filament at intervals of 400 Å (Ohtsuki et al., 1967; Ebashi, 1968). The tentative structure of the actin filament with attached troponin and tropomyosin B is shown in Figure 5*b* in which there are two molecules of each protein per 400 Å. The part that both tropomyosin B and troponin are thought to play in the regulation of events of contraction is discussed in Section 2.B.2.

The other protein found in abundance in certain muscles is tropomyosin A or *paramyosin.* It is especially prevalent in invertebrate smooth muscle cells in which it may be involved in the characteristic physiological properties of these muscles, that is, their ability to maintain tension for prolonged periods of time, the so-called "catch" mechanism (Section 4).

5. **Concept of the Sliding Filament Theory of Contraction.** Observations made on living muscle fibers at rest and during contraction, as well as the x-ray diffraction analysis already discussed, were instrumental in the formulation of a sliding filament hypothesis for contraction (Huxley and Hanson, 1954; Huxley and Niedergerke, 1954; A. F. Huxley, 1957).

The observations made on living muscle fibers through the use of interference and polarizing light microscopes permitted the lengths of the A and I bands to be determined with considerable accuracy during phases of contraction. It was observed that at any sarcomere length the size of the A band remained constant (Huxley and Niedergerke, 1954), while there was a considerable change in the length of the I band. At extreme degrees of shortening, the I band was almost obliterated, with its edges coming into close contact with the Z lines at either end of the sarcomere. When the fiber was stretched passively, the A band again remained constant in length but the I band increased. In addition, a lighter region, the H zone, appeared at the center of the A band and increased in size with progressive stretching of the muscle. When the muscle shortened excessively, dense "contraction bands" appeared on either side of the Z discs, as well as at the center of the sarcomere in the region of the H zone.

The reasons for these findings are apparent if reference is made to Figures 2 and 8, and the assumption is made that neither the myosin *nor* the actin filaments change their length appreciably during contraction. The clear H zone is the area between the ends of the actin filaments from adjacent Z lines, and it is made larger and easily visible by stretching. The contraction bands seen at short sarcomere lengths represent the "crumpling" of the ends of the myosins at the Z line or the double overlapping in the H zone of actin filaments from adjacent Z lines (see also Fig. 15*b*). An increased number of actin filaments surrounding each myosin are observed in electron microscope sections of the H zone at very small sarcomere lengths. One important point is that the distance from the edge of an H zone in one sarcomere to the nearest edge of the adjacent H zone should remain constant independent of the state of the sarcomere. This was confirmed in the original experiments with the interference microscope. In addition, it was proposed that a set of finer filaments, the S filaments, crossed the H zone linking the ends of the actin filaments (Hanson and Huxley, 1955; A. F. Huxley, 1957). In more recent years the existence of these finer filaments has been questioned, as they have not been unequivocally demonstrated in electron microscope sections (H. E. Huxley, 1968a). If present, however, they would add to the elastic component within the fiber as well as help to maintain the organization of the sarcomere when stretched to the point where the filaments no longer overlap (A. F. Huxley, 1957).

It was proposed that shortening and the development of force by the muscle were produced by the interaction of these overlapping sets of myofilaments. The greater the overlap, the greater the number of sites where interaction could occur (Huxley and Hanson, 1954). It seemed more than likely that the cross-bridges, the HMM myosin heads, played some vital role in this interaction (see Davies, 1963). As already mentioned, the spacing between the ends of the cross-bridges and the surface of the actin is not constant but varies with sarcomere length. As the sarcomere does not change its *volume* during shortening, the lattice spacing must change with sarcomere length, the filaments moving further apart during contraction (H. E. Huxley, 1953a; Elliott et al., 1963). It is known that the force produced by a muscle decreases in linear fashion as the sarcomere distance is increased from rest length (l_0) to the point where the filaments no longer overlap (Gordon et al., 1966a, b; see also Section 2.D.3). It seems likely that each cross-bridge develops a given amount of tension whatever the sarcomere length (H. E. Huxley, 1969; see Hill, 1970). The cross-bridge must adapt in some way to overcome the change in lattice spacing, hence the distance to the adjacent actin filaments. This problem has been clarified by the knowledge of the detailed structure of the myosin molecule (Lowey et al., 1966, 1967), in particular the possibility of a flexible joint between the LMM and HMM, as well as between S_1 and S_2 subfragments. A way in which the cross-bridges might act has recently been outlined (H. E. Huxley, 1968b, 1969, 1971) and is illustrated pictorially in Figure 6. At long sarcomere spacings the cross-bridges need only be partially extended from the surface of the myosin filament (Fig. 6*b*). In shorter sarcomeres (Fig. 6*a*), the lattice spacing increases and it would be necessary for the cross-bridge to extend further from the surface of the myosin during a contraction to make contact with the actin (Fig. 6*c*). Here the "flexible" coupling between the myosin subunits is of importance.

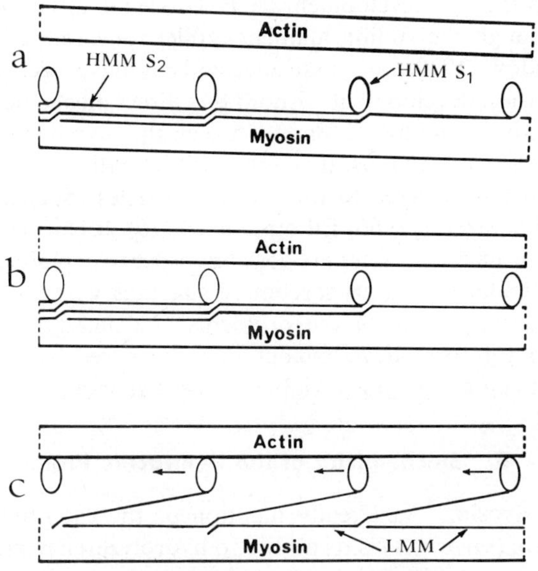

Figure 6. Diagrammatic structure of the position of the cross-bridges at two different interfilament spacings. In (*a*) 250 Å which corresponds to a sarcomere length of ca. 2.0 μ and (*b*) 200 Å which corresponds to a sarcomere length of ca. 3.1 μ in resting frog sartorius muscle. In contraction or during rigor (*c*) the cross-bridges would attach to the actin by bending at the flexible joints. Myosins drawn as singlets for convenience (Modified from H. E. Huxley, 1969).

More recent low-angle x-ray diffraction experiments indicate that a change in the x-ray diagram can occur in resting muscle compared with that from a muscle in a rigor contraction (see Section 2.B.4). In rigor, produced by the lack of the high-energy phosphate compound ATP, there is a disappearance of certain meridional reflections, which indicates a change in the helical arrangement of the cross-bridges along the thick filament (Huxley and Brown, 1967). There is also a marked change in the *intensity* of certain equatorial reflections so that up to 30% of the original mass of the myosin molecule has been estimated to move from the vicinity of the myosin filament to the actin in rigor (H. E. Huxley, 1967, 1968b). As the major part of the mass of the myosin molecule is the HMM fragment and since physiological evidence indicates that rigor muscles are inextensible, this is strong evidence that the cross-bridges are attached to the actin filaments in rigor. Similar results have been found for insect flight muscle (Section 3.B.2). Finally, x-ray diffraction experiments from *actively* contracting muscle have provided first-hand evidence that the meridional reflections, which are associated with the cross-bridges, show marked changes in intensity compared with resting muscle (Elliott et al., 1967; Huxley and Brown, 1967). Thus rigor produces an extreme state in which all the cross-bridges become attached to the actin filaments, while in actively contracting vertebrate muscle the evidence indicates an *asynchronous* cyclical activity of the bridges (Huxley and Brown, 1967).

In summary, there is strong experimental evidence to support the idea that muscle shortens by a sliding of the myofilaments and that the myosin cross-bridges move to the vicinity of the actin during contraction. It is also likely that the cross-bridges provide the vital mechanical link between myofilaments which is directly responsible for shortening and tension development (A.F. Huxley and Simmons, 1971; Podolsky et al., 1969). Although the sliding filament model for contraction is now generally accepted (see Sandow, 1970), alternate mechanisms have been proposed to explain shortening and tension development without this direct mechanical link. These invoke electrostatic attractive or repulsive forces between the myofilaments, which are modified by the binding of calcium ions. It is not possible to discuss these at any length and the interested reader is referred to the relevant articles (Spencer and Worthington, 1960; Ingels and Thompson, 1966; Elliott et al., 1970). It is worth noting that most of these electrostatic mechanisms were proposed in order to explain contraction and force development, particularly at short sarcomere spacings where the cross-bridges were considered inoperative. Many of the problems associated with the original cross-bridge mechanism for force development have been resolved by the concept of a flexible link which can adapt to a variable filament spacing.

B. Biochemistry of the Contractile Proteins

1. Actin and Myosin. As already mentioned, the essential feature of the biochemistry of myosin (*myosin A*) is its ability to hydrolyze enzymically the high-energy phosphate compound ATP. This observation, made approximately 30 years ago (Engelhardt and Ljubimowa, 1939), implied that the muscle cell could control the supply of energy for contraction by controlling the activity of myosin ATPase. When the reaction was investigated in more detail, it was found that only the terminal (or γ) phosphate group was involved (Eq. 1) and this step resulted in the release of approximately 10 kcal/mole of ATP hydrolyzed.

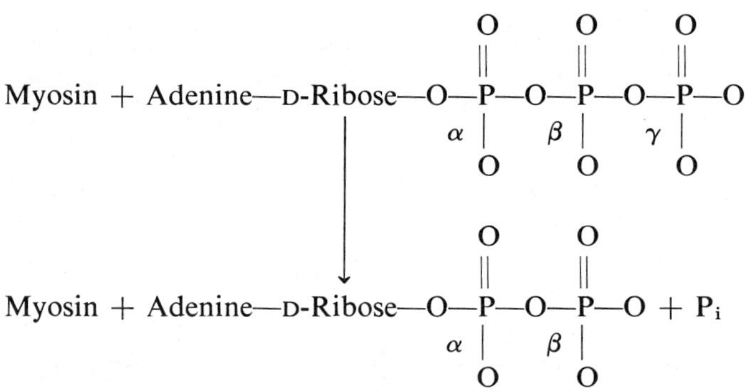

(1)

Although ATP is the physiological substrate for the reaction, myosin is capable of splitting other nucleoside triphosphates such as uridine triphosphate (UTP) or guanosine triphosphate (GTP), but at different rates (Hasselbach, 1957).

The reaction is strongly influenced by the ionic environment, particularly by the presence of the divalent ions calcium and magnesium as well as by actin. Magnesium ions have virtually no activating effect (Trayer and Perry, 1966), and the ATPase activity of myosin remains at a low level. There is evidence that myosin ATPase is actually inhibited by tightly bound magnesium (Offer, 1964; Keily and Martonosi, 1968). In the presence of calcium ions, the myosin ATPase is activated and ATP rapidly hydrolyzed. The precise position of the sites on the myosin molecule where calcium, magnesium, and ATP exert their controlling action is not completely known. There are, however, two ATP binding sites per HMM molecule (Eisenberg and

Moos, 1968; Schliselfeld and Bárány, 1968), one on each of the S_1 subunits (Eisenberg et al., 1968; Lowey and Luck, 1969; Lowey et al., 1969). It is this subunit that constitutes the major part of the myosin cross-bridge seen in the electron microscope (Section 2.A.3) and which retains the ability to combine with actin.

Myosin at physiological ionic strengths ($I = 0.1$–0.2) forms a "gel" which consists of an interlacing network of myosin filaments each containing a variable number of molecules. At higher ionic strengths ($I > 0.2$), myosin forms a colloidal solution, although the precise ionic strength at which this occurs depends upon the pH (A. G. Szent-Györgyi, 1960).

Actin, as monomeric G-actin, can be transformed into polymeric F-actin by increasing the salt concentration in the G-actin solution. The polymerization reaction can be represented as:

$$\text{(G-Actin—ATP)} \xrightarrow[I = 0.1-0.2]{} \text{(F-Actin—ADP)} + P_i \tag{2}$$

The ATP attached to the G-actin is readily exchangeable with ATP in the supporting medium and is not essential for the polymerization reaction (Kasai et al., 1965). In the hydrolysis (Eq. 2), inorganic phosphate (P_i) is released and ADP remains bound to F-actin. This ADP can be removed without disruption of the structure of the actin strand (Bárány et al., 1961), but it is not readily exchangeable with ATP in the medium (A. G. Szent-Györgyi and Prior, 1966; Laki, 1971). It is significant that the bound ADP becomes exchangeable in the presence of myosin and calcium when the actomyosin complex is superprecipitated (see Section 2.B.2). A conformational change in the actin filament brought about by the binding of calcium ions to the troponin molecules may well initiate this increased exchange (Moos et al., 1966; A. G. Szent-Györgyi, 1968).

The biochemical evidence indicates that myosin has the ability to split ATP enzymically but that this reaction is under the control of.the divalent cations calcium and magnesium as well as the molecules actin and troponin (Sections 2.B.2 and 3).

2. The Behavior of Actomyosin Systems. At high ionic strengths ($I > 0.2$), when F-actin and myosin are mixed, the actomyosin solution has a viscosity considerably greater than the sum of the individual components. This suggests that a higher-molecular-weight component has been produced, and electron microscope observations confirm that myosin molecules are now attached to the actin strands. The arrangement of myosin along the F-actin produces a characteristic arrowhead or chevron appearance (see Section 2.A.4). This is more clearly demonstrated if HMM is used instead. Perhaps there is less steric hindrance in the attachment of the smaller HMM molecule. Electron microscope examination of the acto-HMM complex reveals that most, if not all, of the G-actin subunits have an attached HMM (H. E. Huxley, 1963, 1964a).

When ATP is added to an actomyosin solution in the presence of magnesium ions, a decrease in viscosity and light scattering occurs, and the ATPase activity remains at a low level. This is the *plasticizing* effect of ATP, and the actomyosin solution now contains separated myosin and F-actin (Portzehl et al., 1950; Ikemoto et al., 1966). This effect may be attributed to the activity of bound Mg–ATP, since the nucleotide forms one-to-one complexes with divalent ions at physiological pH values.

When Mg–ATP is added to actomyosin in the presence of calcium, there is little decrease in viscosity, and a vigorous hydrolysis of ATP occurs. The activity of this actomyosin ATPase is many times greater than that of myosin alone at similar calcium concentrations. Thus ATP can be considered to have a dual role in the actomyosin system. In the absence of calcium ions, ATP and magnesium promote the separation of F-actin and myosin, and this is associated with a low ATPase activity. In the presence of calcium, the components are not permanently separated, and ATP acts as the substrate for the ATPase. In an actomyosin *solution*, however, there is no analog for the process of contraction.

At physiological ionic strengths (I = 0.1–0.2), the situation is more complex and resembles more closely the interaction between actin and myosin in intact muscle. In this range of salt concentrations, actomyosin is no longer soluble and forms a gel which in suitable ionic conditions can shrink to a small fraction of its volume. This process, termed *syneresis* (or superprepitation), clearly resembles the process of contraction in intact muscle. The precise ionic requirements for syneresis depend upon the nature of the F actin used to produce the gel. If *natural* actomyosin is used, the addition of low concentrations of Mg-ATP induces neither an increased ATPase activity nor a syneresis (Levy and Ryan, 1967). Natural actomyosin either can be produced from impure F actin (containing *native* tropomyosin, see later) plus myosin or can be extracted in solution directly from muscle at high ionic strength. The extract is then allowed to gel by reducing the salt concentration. At higher concentrations, Mg-ATP produces a distinct *clearing phase* in which the light scattering properties of the gel are decreased (Levy and Ryan, 1967). Flow-birefringence studies indicate that actomyosin is now dissociated (Maruyama and Gergely, 1962a, b). When calcium ions are present initially, the clearing phase is much reduced (Noda and Maruyama, 1960; Levy and Ryan, 1967) and it rapidly gives way to syneresis and a rapid ATPase activity (Levy and Fleischer, 1965; Kominz, 1971). The amount of calcium required to produce syneresis in natural gels is extremely small, in the micromolar range. It is therefore difficult to investigate the properties of the ATPase reaction without the use of either stabilized calcium solutions (see Section 2.B.3) or reagents and glassware that have been treated to remove small traces of calcium.

When HMM is used instead of myosin, the resulting acto-HMM complex is soluble at physiological ionic strengths. Viscosity measurements can be made on this solution when there is an activated ATPase, since there is now no syneresis. The viscosity of the acto-HMM solution in the presence of calcium ions is low and more characteristic of a dissociated complex (Leadbeater and Perry, 1963). This appeared to contradict the idea that F-actin and myosin must be associated in some way to produce the ATP hydrolysis typical of actomyosin. More recent evidence indicates that an appreciable amount of acto-HMM is formed even in these low-viscosity solutions. It is found that the ATPase activity continued to increase as the concentration of HMM in the solution is increased and, unlike *actomyosin*, maximum ATPase activity of acto-HMM is not reached until the HMM concentration is many times that of actin (Eisenberg and Moos, 1968; A. G. Szent-Györgyi, 1968). The association between HMM and the G actin must only be transitory, and in solution, where diffusion from the active site can occur, an excess of HMM is required to ensure maximum ATP hydrolysis. The transient and presumably cyclical interaction between actin and HMM to produce ATPase and syneresis must resemble the cyclical attachment of the cross-bridges to actin in active muscle (Lynn and Taylor, 1971).

If an actomyosin gel is prepared from myosin and highly purified F-actin, it is no longer sensitive to calcium ions. These *desensitized* complexes hydrolyze ATP at a high rate in the presence of Mg–ATP alone, even when chelating agents are present to remove traces of calcium. These gels show almost maximal syneresis under these conditions (Weber and Winicur, 1961). It seems that ATP and magnesium ions are no longer able to inhibit the ATPase and syneresis reactions, and in turn calcium can no longer control this inhibition. The essential factor that restores calcium control to the gel is a protein complex termed *native tropomyosin* (Ebashi, 1963). This can be extracted intact from muscle and when added to desensitized gels restores the controlling action of micromolar amounts of calcium. It resembles tropomyosin B (Bailey, 1948) in many of its properties, except that pure tropomyosin B is physiologically inactive. An additional component, troponin, is present in native tropomyosin, and this binds calcium ions selectively. The protein has a molecular weight of 50,000 and with tropomyosin B is located on the thin filament (Section 2.A.4 and Fig. 5*b*). The affinity of troponin for one or two calcium ions is high, in the range 10^4–10^6 M^{-1} (Ebashi and Endo, 1968; Fuchs and Briggs, 1968; Ebashi et al., 1969).

In summary, although natural actomyosin gels differ markedly from myofibrils as far as the arrangement of the myofilaments is concerned, their behavior at the biochemical level is very similar (Section 2.B.3). In the absence of calcium and the presence of bound Mg–ATP, there is little interaction between the proteins. The ATPase activity is low, and this is equivalent to the relaxed state. When calcium ions are bound to troponin and possibly other sites, the inhibitory effect on the interaction is removed. A transitory complex between HMM and actin is formed which results in a vigorous ATP splitting and syneresis.

3. Behavior of the Myofibrils. A preparation of myofibrils containing an orderly array of myofilaments as well as essential factors such as native tropomyosin can be prepared by homogenization of fresh or glycerol-extracted muscle.

The biochemical behavior of myofibrils resembles that of *natural* actomyosin gels. In the presence of Mg–ATP and suitable chelating agents, ATPase activity is low and no contraction is observed. When calcium is added to the preparation, ATP splitting increases and this is accompanied by contraction (Weber and Herz, 1963; Weber et al., 1963, 1964). The complexing agent EGTA [ethylene glycol-bis(β-aminoethyl ether)-*N,N′*-tetraacetate], which chelates calcium ions preferentially to magnesium, provides a convenient method to investigate the details of calcium activation. Solutions containing low calcium concentrations in the range 0.01–10 μM (pH 6.8–7.1) can be prepared using calcium and excess EGTA in known proportions. The precise ionized calcium concentration in the solution is governed by the following relationship

$$K_{\text{Ca-EGTA}} = [\text{Ca-EGTA}]/[\text{Ca}][\text{EGTA}] \quad (3)$$

Thus if calcium and EGTA are mixed in a ratio of 1:3, the ionized calcium concentration in the mixture at pH 7.1 is

$$[\text{Ca}]_{\text{free}} = [\text{Ca-EGTA}]/K_{\text{Ca-EGTA}}[\text{EGTA}] = 1/7.6 \times 10^6 \times 2 = 6.6 \times 10^{-8}\text{M} \quad (4)$$

where $K_{\text{Ca-EGTA}}$ is the apparent binding constant of the chelating agent. Calcium "buffer" solutions resist changes in free calcium concentration and are reasonably tolerant to dilution in a manner similar to that of hydrogen ion buffers (see Weber and Winicur, 1961; Portzehl et al., 1964).

A typical response curve of myofibrillar ATPase activity as a function of calcium concentration is shown in Figure 7*a*. At calcium concentrations of about 0.1 μM, the

ATPase activity is low but it increases to close to 100% at 2–3 μM. The interesting point about the curve is its well-defined S-shape, so that the ATPase activity rises slowly initially and then more rapidly over a narrow range of calcium concentrations, eventually reaching saturation (Weber et al., 1964). The relationship between ATPase activity and calcium is apparently not a simple one, particularly when the amount of calcium bound to the myofibrils over these ranges is also considered. It is likely that more than one calcium ion and probably two are involved in the activation of the ATPase (Weber and Herz, 1963; Ebashi and Endo, 1968; Bendall, 1969; Ashley and Moisescu, 1972).

More recently, the relationship between myofibrillar tension and calcium concentration has been investigated by using the Natori preparation (Natori, 1954). In this technique part of the surface membrane of a single muscle fiber is removed, exposing the contractile apparatus which then can be directly activated by the application of a calcium-buffer solution. The relationship between calcium concentration and tension follows an S-shaped curve similar to that of myofibrillar ATPase (Hellam and Podolsky, 1969) (see Fig. 7*b*). Tension rises rapidly over a small range of calcium concentrations, and full activation occurs at a concentration of 2–3 μM. The exact relationship between tension and ATPase activity is still uncertain. Nevertheless, the fact that *in vitro* both are controlled by micromolar amounts of calcium suggests that in intact muscle there must be components capable of regulating the internal calcium concentration within the range 0.1–3 μM (see Section 2.C.4 and 5).

4. **Glycerol-Extracted Fiber Systems.** The outer membrane of a muscle fiber can be rendered permeable to solutes by treatment for several weeks with a 50% glycerol solution at low temperatures. This procedure leaves the myofibrils intact and able to function normally but removes most of the soluble components of the fiber. Biochemically, the system behaves very similarly to a natural actomyosin gel or a myofibrillar preparation, in that it is activated in the presence of Mg–ATP by small amounts of calcium. The contraction can be reversed by the presence of a calcium-chelating agent (Watanabe and Sleator, 1957), and this is accompanied by a decline in the ATPase activity (Bendall, 1958). The maximum force that can be obtained from a fully activated preparation is at least as great as that from an intact muscle stimulated tetanically (Weber, 1952). If at any stage ATP is removed from the bathing solutions, the muscle becomes inextensible and enters a rigor state characterized by permanent links between the myofilaments (Section 2.A.5). When ATP is restored, in the absence of calcium, the fibers again become extensible, indicating that there is now little interaction between the actin and myosin. Many of the early experiments using glycerol-extracted muscle preparations led to ambiguity since contraction could be achieved simply by adding ATP (A. Szent-Györgyi, 1951). Probably sufficient calcium was present contaminating the ATP and other solutions to produce this response. Glycerinated preparations have been used in more recent years to investigate contraction in insect flight muscle (Jewell and Rüegg, 1966) and are discussed in Section 3.C.

C. Excitation of the Contractile System

1. **Introduction.** It was Ringer who first demonstrated that salines that lacked calcium could not maintain the contraction of an isolated frog heart (Ringer, 1883). When calcium ions were restored, the relaxed heart began to contract, but the precise way in which calcium affected the preparation was unknown. Later experiments

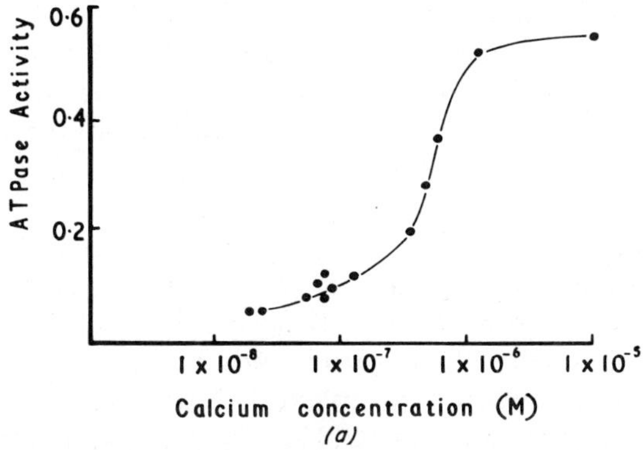

Figure 7*a*. Myofibrillar ATPase activity versus calcium concentration. Myofibrils from rabbit muscle. (From Weber et al., 1964.)

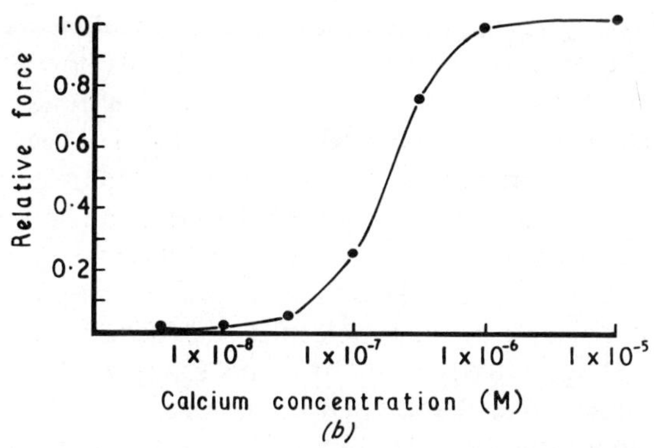

Figure 7*b*. Myofibrillar tension versus calcium concentration. Myofibrils from frog muscle. (From Hellam and Podolsky, 1969.)

showed that a lack of calcium did not affect the spontaneous electrical activity of the heart (Mines, 1913) when the mechanical events had ceased. We now know that this ion is involved both in the electrical as well as in the mechanical events of the cardiac cycle (see Sonnenblick and Stam, 1969). Nevertheless, Ringer's observations made almost a century ago were the first evidence for calcium's important role in muscular contraction.

In more recent years calcium has been implicated in the activity of skeletal as well as cardiac muscle. The direct injection of this ion into single muscle fibers caused a pronounced contraction (Kamada and Kinosita, 1943; Heilbrunn and Wiercinski, 1947), while other physiologically occurring ions were ineffective. As a result of these and other experiments, Heilbrunn proposed that calcium played an important intermediary part in the activation of muscle. He suggested that as a result of the electrical events on the cell membrane calcium was released from the outer regions of the cell and then diffused inward, activating the myofibrils (Heilbrunn et al., 1934; Heilbrunn, 1940). However, the activation time for a frog muscle was much faster than could be accounted for by a diffusion of calcium from the outer membrane to the center of the fiber (Hill, 1948, 1949a). Either some process other than simple diffusion was involved, (Sandow, 1952) or the calcium required for activation was stored at sites within the muscle close to the contractile proteins. How could such sites be affected by the electrical triggering process on the outer membrane?

2. Evidence from the Electron Microscope. Investigations revealed that the myofibrils were not just imbedded in an amorphous matrix as suggested by Bowman (Section 2.A.1) but were surrounded by a fine network of tubules which ran longitudinally (Bennett and Porter, 1953). This network resembled the endoplasmic reticulum seen in noncontractile cells. In muscle the reticulum was "smooth" and did not react with reagents able to detect RNA. In muscle cells this network of tubules was termed sarcoplasmic reticulum (SR), but its physiological function was not obvious.

Later, it became evident that the SR was composed of at least two parts. The first was the longitudinal SR surrounding each myofibril. Making contact with this was the second part, a series of transversely running vesicles (Porter and Palade, 1957) which improved fixatives revealed were delicate tubules extending to the outer membrane (see Fig. 1) (Sabatini et al., 1963). The lumina of these transverse tubules were open to the saline, since large protein molecules such as ferritin (110 Å) or fluorescent dyes (Endo, 1964; H. E. Huxley, 1964b) could easily penetrate from the external medium. None of these molecules entered the lumina of the longitudinal SR. It is of interest that elements of the SR were first described by the light microscopists of the nineteenth century, but much of this work was neglected until confirmed and extended by electron microscope observations (see Veratti, 1902; Porter, 1961; Smith, 1961).

The network of tubules connecting the inside of the muscle with the outer membrane seemed ideal for the conduction of a wave of excitation inward. This idea has support from the experiments to be described in the following section.

3. Local Activation Experiments. Normally, frog fast muscle is activated by an all-or-none action potential, an actively propagated response, on the outer cell membrane of the individual muscle fibers. The ionic basis for the action potential is similar to that in the membrane of the giant axon from the squid *Loligo* (Hodgkin, 1958, 1964). In nerve as in muscle, small passive electrotonic potentials can be initiated at low-intensity stimulation, and these are insufficient to trigger a propagated response. When such stimuli are applied to localized spots on the surface membrane of a frog muscle fiber opposite a Z line, they result in a local contraction of the myofibrils lying beneath the surface. This contraction is always symmetrical and involves the complete A band, hence two sarcomeres (Fig. 8*a*). Increasing the intensity of the pulse produced contraction in sarcomeres up to 10 μ below the surface membrane, yet those only 1–2 μ horizontally from the pipet electrode were not activated. When the pipet was applied to the surface membrane, but at a position opposite an A band, no local contraction was elicited (Huxley and Taylor, 1958).

When single crab muscle fibers were used instead of frog, a local contraction was obtained only when the pipet was placed on the surface membrane opposite the junction between the A and I bands (Fig. 8*b*). Stimulation opposite the Z line or opposite the A band was ineffective (Huxley and Taylor, 1958). The contraction obtained when the pipet was opposite the junction between the A and I bands differed not only in

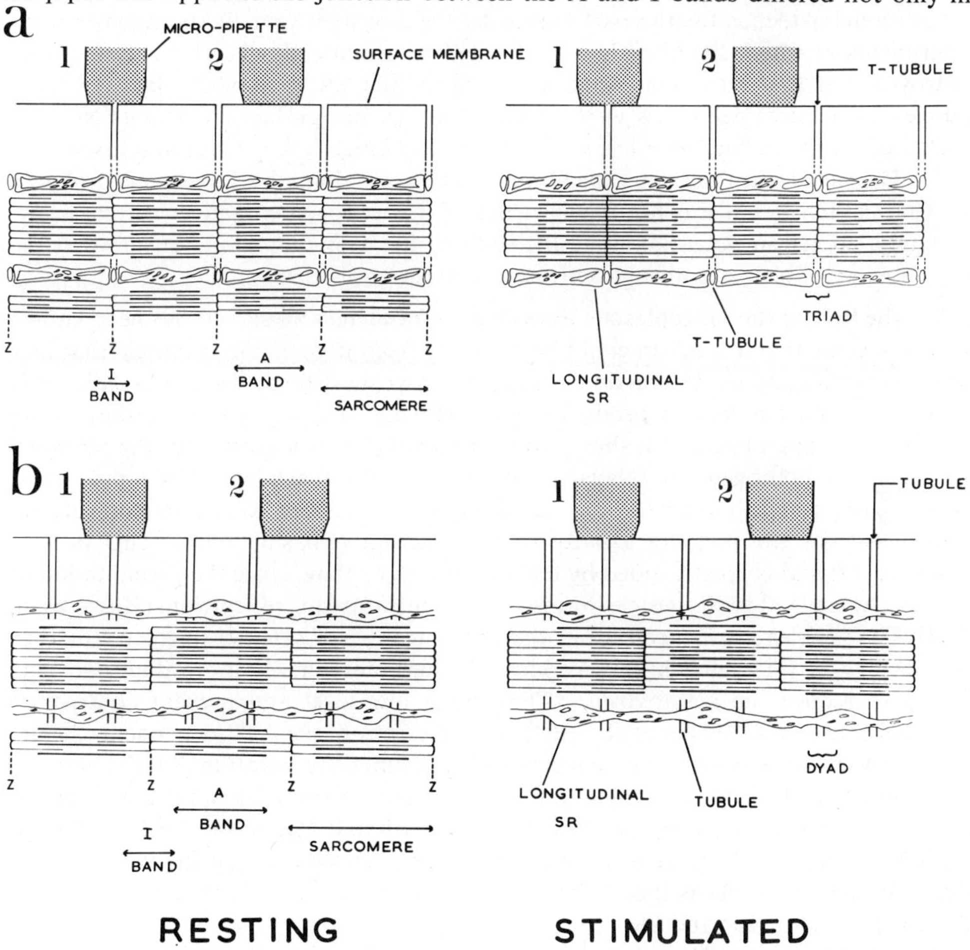

Figure 8. Local activation experiments with frog (*a*) and crab muscle (*b*). In (*a*1) the micropipet (ca. 2-μ tip) was applied to the surface membrane opposite a Z line and stimulation resulted in a symmetrical contraction. Stimulation at position (*a*2) opposite an A band was ineffective. In (*b*1) stimulation opposite an A–1 junction initiated a contraction in one sarcomere, but opposite a Z line (*b*2) was ineffective.

position from that obtained in frog, but it also involved one-half of an I band, hence only one sarcomere (Fig. 8*b*). There appeared to be a structure associated with the Z line in frog and with the A–I junction in crab that conducted the potential change inward. The diagrammatic organization of the SR in both crab and frog shown in Figure 8 is based on light and electron microscope observations. Here the T or activating tubules open onto the surface membrane at the position suggested by the local activation experiments. The reason for the symmetry of contraction in frog is also clear, since the T tubules make contact at the level of the Z line with the dilatations (*lateral cysternae*) of the longitudinal SR from two adjacent sarcomeres, forming a "triad." In crab muscle the SR organization is simpler and the T tubules or their equivalent make contact with the longitudinal system from only one sarcomere at any one time, forming a "dyad" (Figs. 1 and 8*b*). It should be noted that the position of these contact regions varies considerably from species to species (Huxley and Straub, 1958; Smith, 1966).

Additional evidence for the part played by the T system in activation comes from experiments in which the fibers were subjected to osmotic shock. The T system was destroyed, but the outer membrane and longitudinal SR remained intact. In these muscles mechanical responses were absent although normal action potentials could be elicited on the surface membrane (Eisenberg and Gage, 1967; Gage and Eisenberg, 1969). The signal from the outer membrane that passes along the T system is probably an electrical pulse, but it is not certain whether it is propagated actively or passively. Certainly, an action potential along the walls of the T system would ensure that the innermost myofibrils are always activated (Adrian et al., 1969).

4. **The Role of the Sarcoplasmic Reticulum as a Calcium Store.** It has been known for many years that a fraction could be obtained from homogenized muscle that prevented syneresis and the interaction of actin and myosin (Marsh, 1951; Bendall, 1953; Portzehl, 1957), but how it produced this "relaxing" activity was uncertain. More recently, the fraction has been shown to contain vesicles produced from the SR when the muscle is homogenized (Nagai et al., 1960; Hasselbach and Makinose, 1961, 1962; Ebashi and Lipmann, 1962). The ability of the vesicles to accumulate calcium is the basis for the relaxing activity of the fraction (Ebashi, 1961). This process requires ATP and is greatly aided by the presence of oxalate, since the precipitation of calcium oxalate within the vesicle lowers the concentration of free calcium (Hasselbach, 1964). The vesicles *in vitro* can lower the concentration of calcium in the medium to $1\text{–}5 \times 10^{-8}$ *M* (Hasselbach and Makinose, 1963), and this is more than adequate *in vivo* to reduce the actomyosin ATPase to a low level and prevent contraction (Portzehl et al., 1964; Weber et al., 1964). The efficiency of the ATP-dependent calcium pump on the vesicle membrane increases with calcium concentration. At a concentration below 5×10^{-8} *M*, one calcium ion is transported per ATP split, but at higher concentrations the ratio increases to 2:1 (Hasselbach and Makinose, 1963;see Weber, 1966; Weber et al., 1966). It is not certain what replaces oxalate in the intact muscle although one possibility is the phosphate anion, but it is unlikely that calcium phosphate is precipitated in the SR.

Direct confirmation that calcium can be stored within the SR of intact muscle has been possible in recent years. Experiments in which muscles have exchanged part of their internal calcium for ^{45}Ca, or that have been glycerinated and made permeable to calcium-precipitating agents, show dense accumulations of calcium within the lumina of the longitudinal SR. These deposits are mainly in the lateral cysternae of the triads in resting frog muscle (Costantin et al., 1965; Pease et al., 1965). During contraction and subsequent relaxation, a translocation of the calcium can be detected (Winegrad, 1965a, b; Connolly et al., 1971).

The precise way in which calcium is released from these sites by a signal from the T system is uncertain. It is possible that the membranes of the longitudinal SR can be directly depolarized by the signal from the T system and that this is sufficient to release the stored calcium which lies only 1–2 μ from the myofibrillar binding sites. Alternatively, there may be an intermediate chemical step, perhaps the release of a small "trigger" amount of calcium which then acts to release the main calcium stores (see Endo et al., 1970; Ford and Podolsky, 1970).

The mechanisms involved in the reaccumulation of calcium are also not fully understood. In addition to the ATP-dependent calcium pump across the membranes of the SR, there is a rapid ATP-dependent binding of calcium to the surface membranes of the vesicles (Ohnishi and Ebashi, 1964), which may not be related to the pump mechanism (Carvalho, 1968). This surface-bound fraction might provide calcium for rapid release during a twitch contraction, while during a tetanus calcium would be mobilized from within the SR.

5. **Direct and Indirect Estimates of Calcium Changes during Contraction.** It was not possible to produce direct evidence for the calcium release theory of Heilbrunn until the advent of suitable isotopes of calcium. In experiments using ^{45}Ca, muscles were allowed to exchange part of their internal calcium for the isotope. The muscle was then transferred from the saline containing ^{45}Ca to an inactive one, and the efflux of the isotope followed over a period of time. The efflux soon attained a low value characteristic of the resting fiber but increased appreciably on contraction (Woodward, 1949; Bianchi and Shanes, 1959; Shanes and Bianchi, 1960). The result is readily interpreted if ^{45}Ca was accumulated by the elements of the SR during the initial "loading" period. The subsequent contraction mobilized calcium from the SR, and some of this was lost from the sarcoplasm across the outer membrane. In experiments in which the influx of calcium was investigated, it increased during contraction to about an equal extent as the efflux (Bianchi and Shanes, 1959). There is therefore a turnover of the fiber calcium during contraction but apparently no *net* change.

More recently, direct evidence for calcium release and reaccumulation during contraction has been obtained from the use of calcium-sensitive indicators (Jöbsis and O'Connor, 1966; Ashley and Ridgway, 1968, 1970a, b). In particular, a technique using the bioluminescent protein aequorin, which is sensitive to calcium ions, has provided unequivocal evidence for these changes within isolated muscle fibers. The extent of the calcium changes that lead to contraction can also be estimated for the first time (Ashley, 1970a, b, 1971). The protein, which is obtained from the jellyfish *Aequorea* (see Shimomura et al., 1962, 1963), can be introduced into large muscle cells by microinjection. The large fibers from the barnacle *Balanus nubilus* and the crab *Maia squinado* are often 2–3 cm in length and 1–2 mm wide and are well suited to these procedures. If these large fibers are stimulated after injection with aequorin, they emit a bluish light which is visible to the dark-adapted eye (Ashley, 1971). This light emission represents the interaction of aequorin molecules within the fiber sarcoplasm with calcium released from the SR by the stimulus. The time course of the calcium release and reaccumulation (the *calcium transient*) is shown in Figure 9 (trace 2) and

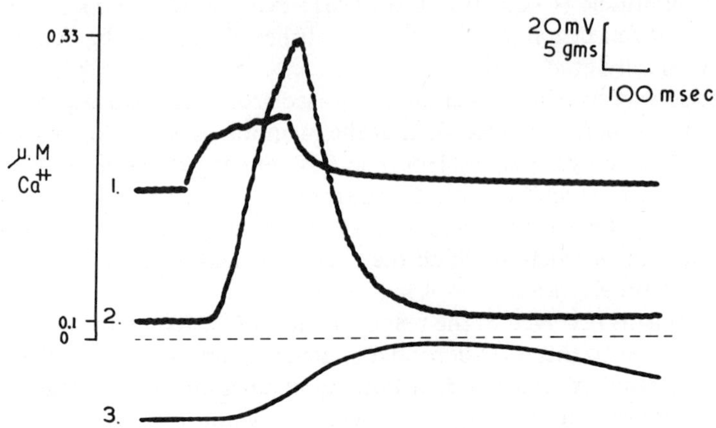

Figure 9. Sequence of events in E–C coupling in a single barnacle muscle fiber. Trace 1, the passive electrical response on the outer membrane; trace 2, the change in calcium inside the fiber (*calcium transient*) as monitored by aequorin; and trace 3, the tension response. (From Ashley, 1971.)

can be compared to that of the membrane response and tension. The calcium in the sarcoplasm rises rapidly during the stimulus and precedes the onset of detectable tension. The peak of the calcium transient occurs at the maximum rate of rise of tension. This finding indicates that calcium controls the *rate* of tension development in these fibers. The calcium transient is virtually complete when the fiber develops peak tension. It is likely that at this time calcium is still attached to the myofibrils, including troponin, so that subsequent relaxation must occur at resting calcium concentrations.

The light emission from the fibers in the resting state indicates that the calcium concentration in the sarcoplasm is in the range $0.7–1.0 \times 10^{-7}$ *M* (Ashley, 1970a, 1971). This figure agrees with the values arrived at by other methods, such as the injection of stabilized calcium solutions (Portzehl et al., 1964; Ashley, 1967). The light emission from the aequorin molecules within the fiber increases as the square of the calcium concentration, so that the size of the calcium transient is easily estimated. When a fiber is maximally activated, for example, by raised potassium salines, the light emission reaches a peak value equivalent to $1–2 \times 10^{-6}$ *M* (Ashley, 1970a). At lower intensity stimulation the change in calcium concentration is smaller; a typical calcium transient is illustrated in Figure 9. One further point related to a kinetic analysis of the transient is that two calcium ions are required per tension site for activation (Ashley and Moisescu, 1972). These experiments do not predict the precise positions at which the calcium ions are bound, but at least one of these sites must be on the troponin–tropomyosin B complex (Section 2.B.2).

D. Mechanical Events during Contraction

In a typical vertebrate skeletal muscle fiber, activation and contraction occur as a result of a single propagated action potential along the surface membrane (Section 2.C.3), and this results in a discrete mechanical response or twitch from the muscle. The *twitch* can be either a shortening against a constant load (*isotonic*) or, in the absence of external shortening, the development of tension when the muscle is restrained (*isometric*).

1. The Twitch Response. The time course of an isometric twitch is shown in Figure 10*b* and lasts about 1 sec in frog at 0°C. The onset of the rise of the tension response is preceded by a transient fall, *latency relaxation*, the size of which is only a small fraction of peak tension and is not visible in Figure 10*b* (Sandow, 1947, 1965). Its relationship in time is between the action potential and the rising phase of tension, that is, within the *latent period*. This suggests that it is related to the intermediate events of excitation–contraction (E–C) coupling such as the release of calcium into the fiber sarcoplasm (Section 2.C.5 and E.4).

The time course of the isotonic twitch is complicated since the precise mechanical response depends upon the afterload applied to the muscle. Shortening occurs only when the muscle has developed sufficient tension to lift the afterload. The velocity of shortening, as might be expected, depends upon the value of this load and is faster at lighter loads. The shortening response is completed sooner, even at light loads, than the prolonged tension response of the isometric twitch (Jewell and Wilkie, 1960) (see Figs. 10 and 12*a*). This discrepancy between the time course of isotonic and isometric twitches implies that at a time when shortening is complete the contractile elements may still be "active" and able to bear tension. Shortening is thus a poor guide as to the state of the contractile machinery. The disparity between isotonic and isometric twitches, as well as other factors, led to the suggestion that muscle can be considered to consist of an *active* contractile element (CE) together with *passive* elastic and viscous elements which are arranged in series and parallel with the CE. The CE, which resides within the sarcomere, is activated by calcium release and can develop tension and shorten. The activity of the CE is transmitted to the recording apparatus via components such as the tendons, elements of the SR, and others which act as elastic elements and behave in a passive fashion. The time course of the activity of the CE has been termed the *active state* of the muscle (Gasser and Hill, 1924; Hill, 1938) and can be

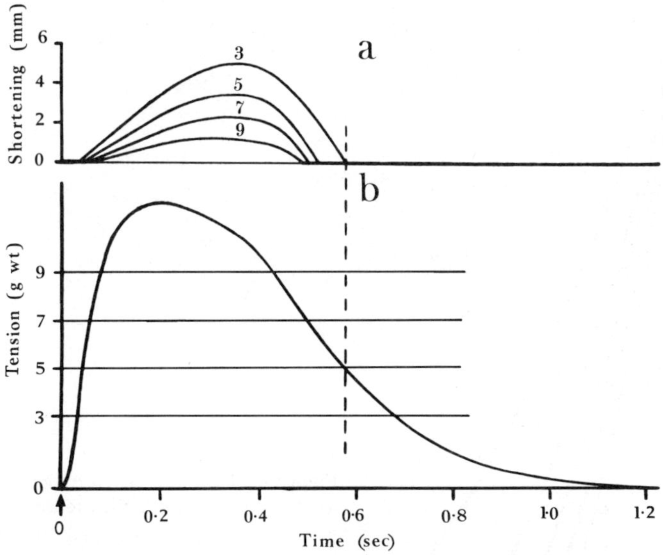

Figure 10. Comparison of the time course of an isotonic (*a*) and isometric (*b*) twitches in frog muscle stimulated at the arrows by a single shock. In (*a*) the muscle contracts against afterloads of 3, 5, 7, and 9 g. Horizontal lines in (*b*) correspond to the afterloads in (*a*). (From Jewell and Wilkie, 1960.)

defined as the tension that can be tolerated by the muscle without lengthening. The methods used to distinguish the activity of the CE free from the damping produced by internal viscosity and elasticity are discussed in Section 2.D.2. The separate existence of the passive elastic components can be demonstrated only when the muscle is active, since in the resting state the muscle is readily extensible and both the CE and elastic components offer little resistance to stretch. When the muscle is stimulated either with a single shock or a series of shocks, the CE develops tension and becomes inextensible, while the elastic components are unaffected. If during stimulation the muscle is prevented from shortening by a mechanical "stop," and after a fixed period of time the "stop" is removed, the muscle shortens but in two distinct phases (Fig. 11). There is an initial rapid phase of shortening attributable to the elastic components which were stretched initially during the preceding isometric phase (Wilkie, 1956a, b). The extent of this phase depends upon the initial load and the muscle tension at the time of the release. This is followed by a slower phase of shortening attributable to the activity of the CE. Analysis of the rapid shortening phase in these experiments indicates that the elastic components behave in a purely *passive* manner during contraction (see Jewell and Wilkie, 1958).

It is possible to record both tension and the shortening response simultaneously in a twitch (see Jewell and Wilkie, 1960). Tension rises rapidly (Fig. 12*b*) until it reaches the value of the afterload when shortening occurs. During the time course of the shortening response, tension remains constant and starts to decay only when shortening is complete (Fig. 12*a* and *b*). Since the velocity of shortening depends upon the value of the load, experimental record curves can be constructed relating the maximum velocity of shortening to the value of the afterload (force). The force–velocity curves have a characteristic shape (Fig. 12*b*) and can be obtained from intact muscle or isolated

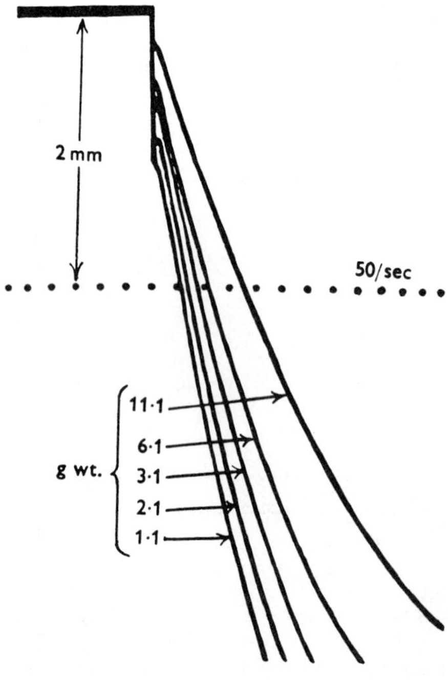

Figure 11. Isotonic release experiment with various loads in frog muscle. Shortening is downward and in all cases release occurred 100 msec after stimulation. The initial rapid phase attributable to the elastic elements is followed by a slower phase of shortening attributable to the contractile elements. (From Wilkie, 1956b.)

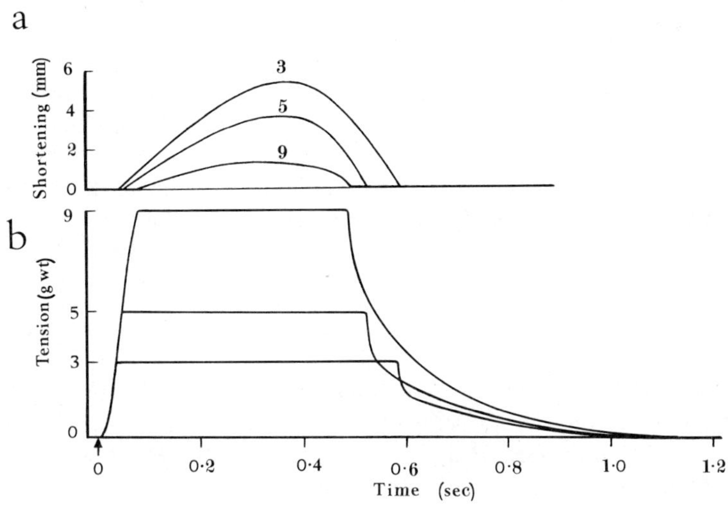

(a)

Figure 12*a*. Simultaneous measurements of shortening (*a*) and tension (*b*) following a single stimulus pulse applied at (↑). The twitches are against afterloads of 3, 5, and 9 g. (From Jewell and Wilkie, 1960.)

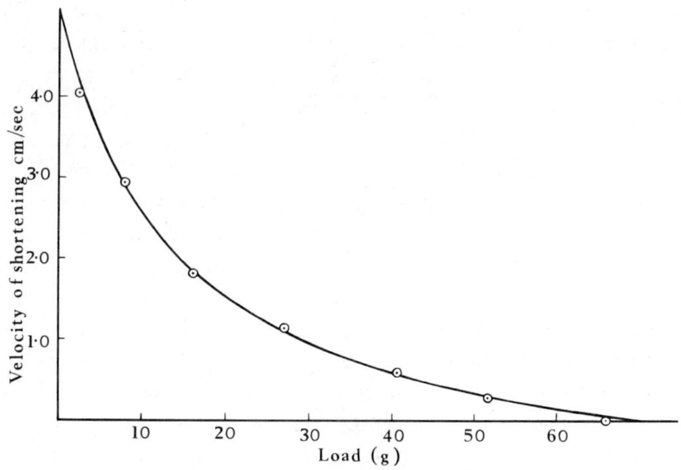

(b)

Figure 12*b*. Force–velocity curve for frog muscle. (From Hill, 1938.)

bundles of myofibrils (see Podolsky and Teichholz, 1970; Julian, 1971). They are described mathematically in several ways, although perhaps the best known is the hyperbolic form first suggested by Hill (1938). An equally successful exponential relationship has been described by Auber (1956).

The hyperbolic form of Hill can be expressed as:

$$Pv + va = (P_0 - P)b \tag{5}$$

thus

$$(P + a)v = (P_0 - P)b \tag{6}$$

where v is the velocity of shortening, P_0 a constant, is the maximum tetanic tension, P is the load on the muscle, and a and b are also constants. In its simple form the Hill equation tried to relate the rate of the work done (Pv) plus that of the heat output (va) during contraction to the difference between the maximum and actual load ($P_0 - P$) (Eq. 5). When the afterload is zero and virtually no external work is performed, the velocity of shortening is close to v_{max} and the relationship simplifies to:

$$v_{max} = \frac{P_0 b}{a} \tag{7}$$

The constants a and b can be assessed from independent experiments.

At one stage it was thought possible to assess the constant a from thermal measurements during shortening (see Hill, 1938, 1949b, 1953b). More recent observations indicate that the heat output during shortening is more complicated than early experiments indicated and that it depends upon the value of the afterload (Hill, 1964a). The Hill equation can be modified to take into account the new value of the constant a (now termed α; see Section 2.E.3). The agreement of the Hill equation with experimental observations however is still only satisfactory over a certain range of shortenings (Hill, 1964b).

2. Tetanic Response and the Time Course of "Active State." When a single stimulus is repeated at sufficiently short intervals of time (10–20/sec in frog at 0°C), twitch responses fuse to produce a tetanus. The magnitude of a tetanus is greater either isotonically or isometrically than a twitch. An isometric tetanus produces tensions in the range 1–2 kg/cm^2. This increased mechanical response is a result of prolonged active state. The time course is long enough so that the elastic components are fully stretched and external tension is maximal. While in a twitch active state is short, the elastic elements are not fully stretched and tension is smaller (Hill, 1953c).

Although in frog muscle its time course varies, the *intensity* of active state is the same in a twitch as in a tetanus (Hill, 1949a). The time course of active state in either case can be assessed fairly readily since it is equal to the tension of the muscle when there is no change in length (see Section 2.D.1). In active muscle the only time when there is no change in length is at *peak* tension. At this time the internal elastic components are fully stretched and no longer "mask" the activity of the CE. Experimentally, the muscle is stimulated but prevented from shortening by a mechanical "stop." During this period the muscle contracts isometrically and develops tension, but this is not recorded. At intervals of time after stimulation the muscle is released and allowed to shorten to a new length. Following this *quick-release* the elastic components shorten rapidly and tension falls to zero. At this new length the CE redevelops tension, and this is recorded as a modified twitch (Hill, 1953a). The results of such a procedure are shown in Figure 13. A line can be drawn through the peaks of the tension responses and indicates the time course of the decline in active state (Ritchie, 1954a, b).

The onset of active state is extremely rapid in frog muscle and the quick-release method is more suited to studying the decay rather than the rising phase of this process. A *quick-stretching* procedure is often employed to study the rising phase of

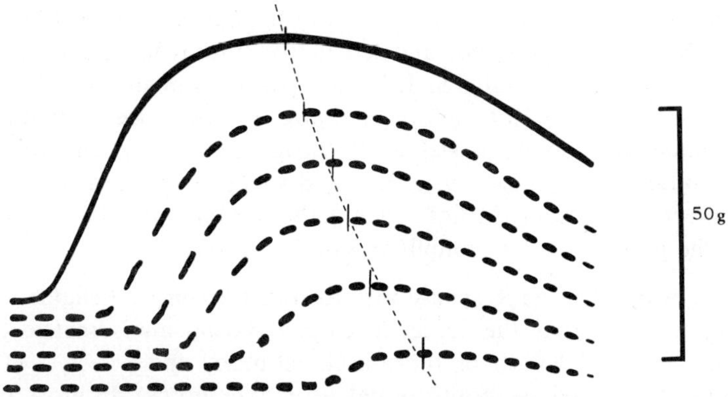

Figure 13. Quick-release experiment in frog muscle. The stimulus starts the trace and the intensity is modulated at 50 cps. The top (solid) trace is the normal isometric twitch. The subsequent traces are produced by a quick release (ca. 3 mm) of the muscle approximately 10, 40, 60, 100, and 175 msec after the stimulus. The dashed line plots the time course of the decay of active state. (Modified from Ritchie, 1954b.)

active state (Hill, 1949a, 1950) in conjunction with a slowly contracting muscle in which, as might be expected, active state rises more slowly. The underlying idea behind these experiments is that the full activity of the CE is not seen until the elastic components have first been stretched. If these components could be extended *artificially* during the latent period, tension should rise much faster and the maximum values should indicate directly the activity of the CE. Tension rises extremely rapidly, and stretching just at the end of the latent period in frog (Fig. 14*a*) or in toad (Fig. 14*b*)

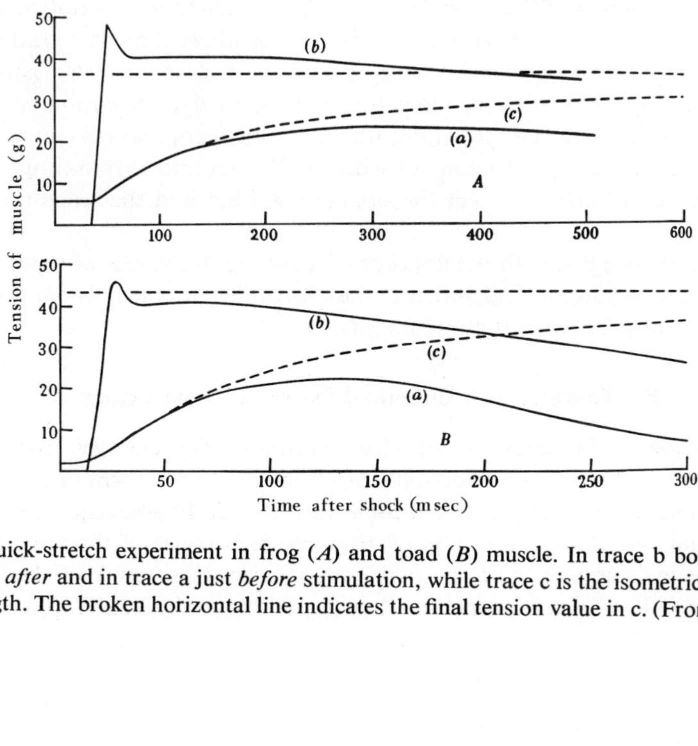

Figure 14. Quick-stretch experiment in frog (*A*) and toad (*B*) muscle. In trace b both muscles are stretched just *after* and in trace a just *before* stimulation, while trace c is the isometric tension at the stretched length. The broken horizontal line indicates the final tension value in c. (From Hill, 1949a.)

muscle led to a tension that was equal or occasionally just in excess of the maximum tetanic tension; however, this soon decayed to the normal value (Hill, 1950).

These and other experiments indicate that active state reaches a maximum value extremely rapidly after the stimulus and, in frog at least, remains at full intensity for 40 msec at 0°C or 10 msec at 10°C before declining. Temperature, certain anions such as nitrate, bromide, and iodide, as well as the drugs caffeine and adrenalin, affect its time course (Goffart and Ritchie, 1952; Hill and Macpherson, 1954). It seems likely that anions affect active state by influencing the action potential and thereby act indirectly via the process of E–C coupling (Sandow, 1965).

3. Variation of the Tension Response with Resting Sarcomere Length. It has been known for many years that the force developed during an isometric contraction depends upon the initial length of the muscle (Ramsey and Street, 1940) and that maximum force is obtained at about normal resting length (l_0). In more recent years this finding has been investigated in more detail, with particular reference to the sliding filament theory. In these experiments, performed with single muscle fibers, the sarcomere lengths, hence the degree of overlap between the myofilaments, was observed directly and maintained at a constant value during contraction (Gordon et al., 1966a, b; see also Podolsky, 1964). Virtually any change in muscle length, either greater or smaller than l_0, resulted in a smaller given tension response; the results are summarized in Figure 15*a* and *b*. Maximum tension could be obtained over a small range of sarcomere spacings from 2.0 to 2.25 μ (B to C). The effect of stretching the fiber, thereby increasing the sarcomere spacing, was to produce a continuous decline in the given tension response until at 3.65 μ the tension was zero. This decline in tension can be readily explained by assuming that the force depends upon the degree of overlap between the actin and myosin filaments. The values for the lengths of the myosin can be taken as 1.6 μ and that for actin through the Z line as 2.1 μ (Section 2.A.2). Thus at 3.65 μ there is no longer overlap between the myofilaments, hence no tension (A and a). At sarcomere lengths shorter than l_0, the decline in the given response shows a number of "corners" (C and D) at which there is a noticeable change in the decline of tension (Fig. 15*a*). The first is at 2.0 μ where there is a gradual overlap between the actins from opposite Z lines, c and d—a *double overlap* zone (Section 2.A.5). This must interfere with the efficiency of cross-bridge formation between actin and myosin. Also, at 1.67 μ (D) there is another marked change which can be attributed to myosin filaments coming into contact with the Z lines and "crumpling." At 1.05 μ the ends of the actin filaments meet the opposite Z lines and the tension response is zero *E* and *f*.

These experiments agree with the concept of a sliding of the sets of myofilaments as the basis for the contraction and indicate that force development is dependent upon the extent of overlap between these filaments.

E. Thermal and Chemical Events of Contraction

1. Introduction. The chemical and mechanical energy changes associated with contraction discussed in previous sections appear finally in the form of heat and work. The work output from a contracting muscle can be readily assessed, but the accompanying thermal events are more difficult to measure because of their small size. The pioneer in the measurements of thermal events in muscle was Hill. He developed sensitive thermopiles and galvanometers that responded to the small rapid temperature

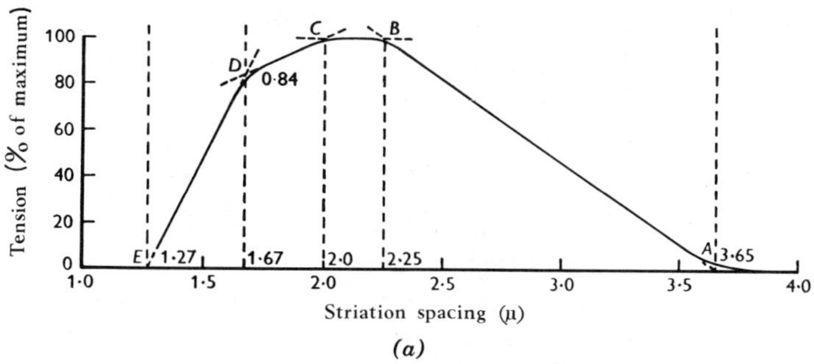

Figure 15*a*. Schematic diagram showing the variation in tension with sarcomere length in a single frog muscle fiber as a result of stimulation. (From Gordon et al., 1966b.)

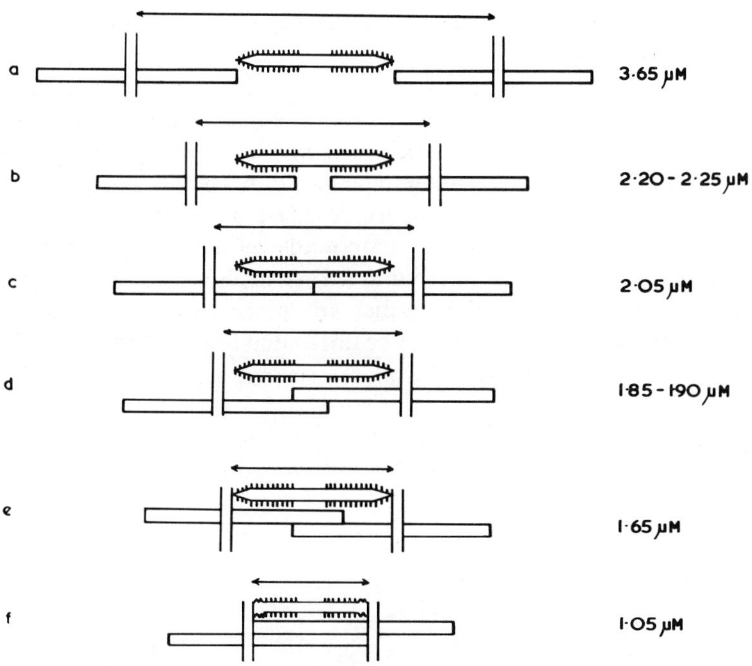

Figure 15*b*. Diagrammatic arrangement of the overlap between actin and myosin filaments at various sarcomere lengths. (Modified from Gordon et al., 1966b.)

changes associated with contraction (ca. 1×10^{-3} °C) upon which heat measurements are based.

The thermal events can be conveniently divided into several separate but not always clearly defined phases: first, heat output associated with the early stages of contraction before a detectable mechanical response is observed, the *heat of activation;* second, a *heat of shortening* associated with mechanical events, that is, isotonically when a muscle shortens and lifts a load, or isometrically when the series elastic components are stretched; this mechanical change is also associated with a steady rate of heat output when the shortening is complete (a *maintenance heat*); and finally during the time that the muscle is restored to its original mechanical state, a *heat of relaxation*.

The thermal events are evidently a sensitive guide as to the underlying processes of contraction. However, their interpretation biochemically or in terms of current information about muscle function is sometimes more difficult.

2. Thermal Events Associated with Isometric Contractions. The thermal events associated with an isometric contraction are portrayed in Figure 16. Here the trace shows the *rate* of heat output for a single isometric twitch from a toad muscle at 0°C. Heat evolution is most rapid soon after the stimulus, reaching a maximum rate of approximately 19 mcal/(g)(sec) (off scale) during the latent period (A). It then falls to a low plateau value (at B) of about 1 mcal/(g)(sec) which is maintained until the muscle starts to relax. When relaxation begins, the *rate* of heat evolution increases (at C) to about 2 mcal/(g)(sec) and then falls to zero during the final phase of relaxation (Hill, 1938, 1949c).

The early rapid phase of heat output, the activation heat (A), coincides with the time that intermediary processes in E–C coupling are occurring. The recent biochemical finding (Section 2.E.4) that the amount of early ATP breakdown is insufficient to account for this large heat output indicates that the heat of activation is likely related to processes such as the release of calcium from the SR into the fiber sarcoplasm (Kushmerick et al., 1969). The actual maintenance of tension is associated with a much lower rate of heat output, the maintenance heat (B). This must represent in part the breakdown of high-energy phosphate compounds by actomyosin and in part ATP usage by systems such as the calcium pump across the SR membrane. Finally, the increased rate of heat output seen during the early phase of relaxation (C), the heat of relaxation, is mechanical in origin. It can be attributed to the shortening of the internal elastic elements that were stretched during the rising phase of tension (Hill, 1938).

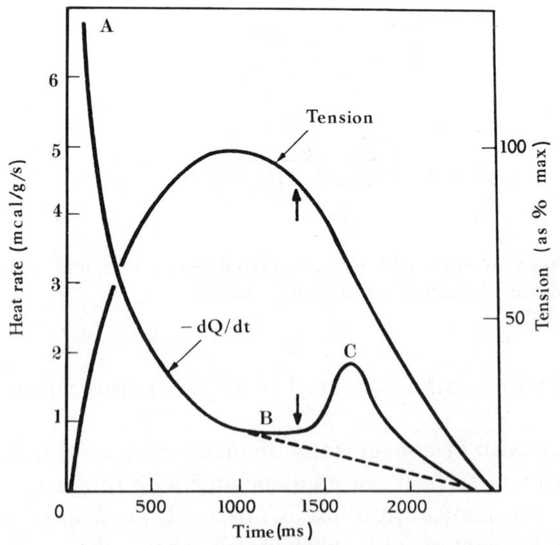

Figure 16. The *rate* of heat output ($-dQ/dt$ during an isometric twitch of a toad muscle at 0°C. Maximum rate (A), plateau value (B), and an increased rate (C) as a result of relaxation at arrows. (From Bendall, 1969.)

The heat output during an isometric tetanus shows the same main phases as in an isometric twitch. There is initially a large activation heat which declines to a low rate characteristic of the maintenance heat. The plateau (B) of the maintenance heat is, however, more prolonged during a tetanus, presumably because of an extended active state (Section 2.D.2; Hill and Woledge, 1962; Woledge, 1968). A heat of relaxation is also observed when tension declines and the internal elastic elements are allowed to shorten.

3. Thermal Events Associated with Isotonic Contractions. The amount of *external* shortening that occurs during an isometric contraction is small providing the recording apparatus is noncompliant. Estimates of the amount of *internal* shortening of the muscle during activity attributable to the elastic components indicates that it is no more than 3% of the rest length, even at maximum tetanic tensions (P_0) (Hill, 1953a,b). In an *isotonic* contraction much greater changes in length are of course possible, and this presents some difficulties for accurate heat measurements. The reason is that the part of the muscle that is initially over the thermopile is displaced during shortening by a region of the muscle at a slightly higher temperature and artifacts are produced. The difficulty can be overcome if "dummy" thermocouples are included in the thermopile on either side of those used for recording. Thus all muscle likely to move into the recording region is maintained initially at the same temperature (Hill, 1938).

The thermal events associated with an isotonic contraction are best demonstrated if the muscle is initially allowed to contract isometrically so that a steady *maintenance* heat is established. The muscle is then allowed to shorten a predetermined amount. The shortening is accompanied by an increased rate of heat production (Fig. 17)

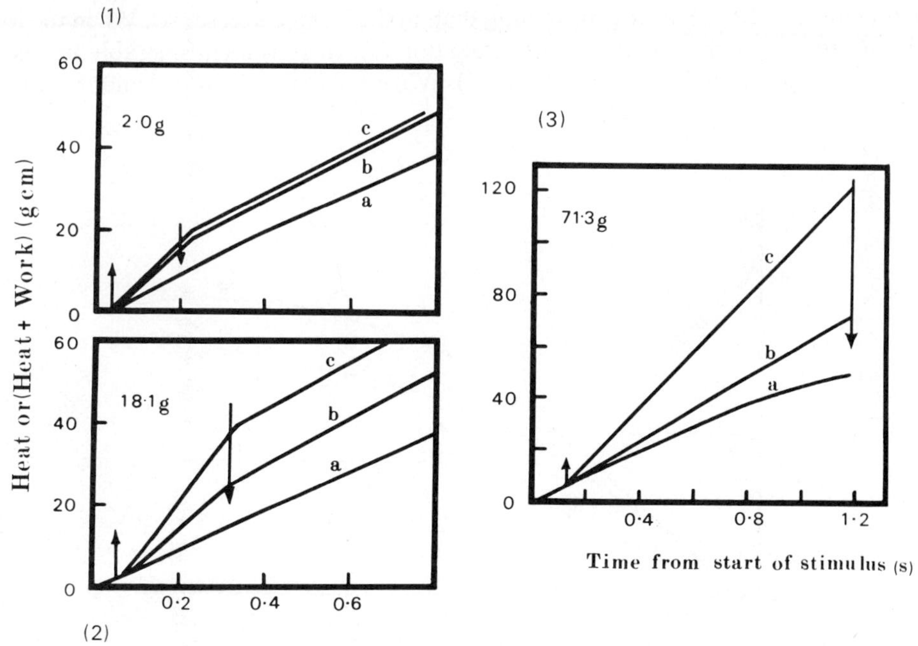

Figure 17. Heat output during isotonic shortening against various afterloads. (1) 2.0, (2) 18.1, and (3) 71.3 g. a, Isometric heat output at same length; b, heat output during shortening; c, heat plus work output during shortening. The vertical arrows indicate the onset and cessation of shortening which was 7mm in (1)–(3). (Modified from Hill, 1964a.)

which lasts as long as the shortening persists and then falls to the steady maintenance rate (Hill, 1938, 1964a). As with the isometric twitch, there is an early rapid rate of heat evolution, the heat of activation that precedes the maintenance heat. In his original paper, Hill (1938) came to the conclusion that the heat of shortening was a simple linear function of the amount of shortening, independent of the load lifted, so that the heat of shortening was equal to the quantity ax, where a was a constant and x was the distance shortened. More recent experiments indicate that there were considerable errors in the early heat measurements that led to these conclusions (Hill, 1964a). It is now found that when a muscle shortens, the heat of shortening depends not only upon the distance but also upon the load (Fig. 17, traces 1, 2, and 3). More heat is produced in a heavily loaded contraction, in which the velocity of contraction is small, than in a lightly loaded contraction (Hill, 1964a). Therefore heat experiments cannot readily be used to provide a value for the constant a appearing in the characteristic Hill equation (Eq. 6; Section 2.D.1). The heat of shortening is now expressed in a more complicated way, as a function both of the load and the degree of shortening:

$$\text{Heat of shortening} = \alpha x \tag{8}$$

$$= (0.16P_0 + 0.18P)x \tag{9}$$

where P_0 is maximum isometric tension, x is the distance shortened, P is the load, and the constant α replaces a in the original relationship.

The time course for the heat output during a complete isotonic contraction in which the load is allowed to perform work during relaxation, or alternatively in which the load is clamped and the muscle is not permitted to extend, is indicated in Figure 18. As expected, there is initially a higher rate of heat output from the isotonically contracting muscles (B, C) during shortening than in the isometric case (A). When the load falls and reextends the muscle during relaxation (c), there is a considerably increased rate of heat output, a heat of relaxation (C). When the load is clamped and the muscle

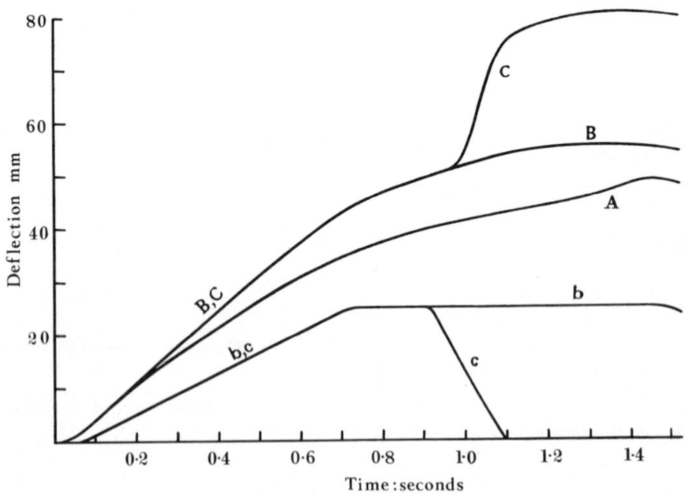

Figure 18. Heat output during a complete twitch. During an isometric (A), isotonic with a clamped load (B) and isotonic with load performing work during relaxation (C). The traces b and c are the respective mechanical responses. (From Hill, 1938.)

not allowed to reextend (b), a much smaller heat of relaxation is observed (B). A small heat of relaxation is also seen in the isometric contraction toward the end of trace A.

The heat of relaxation in the loaded isotonic case is equivalent to the work done by the load in falling. The small "humps" seen in the clamped load and in the isometric experiments represent the reextension of the internal elastic elements stretched by the initial onset of tension (Section 2.D.2 and E.2). When the total heat output was measured for a complete, lightly loaded isotonic contraction, and either the load clamped after shortening occurred or the thermal equivalent of the work done calculated, the total heat output differed very little from that observed during an isometric contraction (Hill, 1949c; Carlson et al., 1963; Mommaerts, 1969). It was essentially a fixed amount of heat (mainly heat of activation) plus the external work done. There was no apparent "extra" heat in the isotonic contraction attributable to the heat of shortening. Either the extra heat associated with shortening was absorbed during the relaxation phase, or the heat output during an isometric contraction was distributed differently in time than that from an isotonic contraction. If this latter idea is correct, the increased heat output during a lightly loaded contraction should decline more rapidly than in an isometric or heavily loaded contraction (Carlson et al., 1963). Recent experiments give credence to this idea, in that there is a positive feedback from tension in an isometric contraction which prolongs activity and results in an extended phase of heat production compared to the isotonic case (Hill, 1964c, d). Hill observed that as the heats of shortening and maintenance are not easily distinguished perhaps what is termed heat of shortening is in fact an increased heat of maintenance during shortening.

In summary, the concept that the total energy ouptut during a contraction is a "fixed amount of heat (activation) plus the work done" (see Fenn, 1923, 1924) now seems to be too narrow a definition because it is applicable only to the *complete* contraction–relaxation cycle and not to any *instant* in time. A more general statement is required, and in the light of the more recent work it can be expressed as (Hill, 1964d; Mommaerts, 1969; Woledge, 1971):

$$\text{Energy output during contraction} = A + W + ax + f(P, t) \tag{10}$$

where A is the activation heat, W is the work done, ax the "old" heat of shortening and, finally, a quantity $f(P, t)$ which is a function of both tension and time and which includes a contribution from the heat of maintenance.

Isotonically, the equation might be expanded and the terms $ax + f(P, t)$ replaced by a maintenance heat, plus the expression for the "new" heat of shortening (αx) (Eq. 9). While isometrically the work term W and the heat of shortening would be small, leaving only heat of activation plus the variable heat depending on tension and time (P, t).

4. **Chemical Changes Associated with Contraction.** In Section 2.B.2 the properties of actomyosin ATPase were described, and it was evident that both ATPase activity and tension production were controlled by calcium ions. It was noted that the energy required for contraction was provided by the ATP molecule in isolated myofibrils, but it has been difficult to detect the breakdown of ATP within the living muscle cell. The reasons for this are that the predicted ATP breakdown per twitch, if it is assumed to be providing all the energy, is only a small fraction (~0.5 μmole/g) of the total muscle ATP (approximately 4–5 μmole/g). In addition, the ATP concentration is

carefully controlled in the normal muscle by a rapid rephosphorylation reaction. This step, which is illustrated below (Eq. 11, step 1), utilizes a subsidiary phosphate compound (*phosphagen*) as an energy store. In vertebrates the phosphagen is generally creatine phosphate (CrP), while in most invertebrates it is arginine phosphate (ArP).

$$\text{CrP} + \text{ADP} \underset{}{\overset{\text{Phosphotransferase}\ 1}{\rightleftharpoons}} \underset{\text{creatine}}{\overset{}{\text{ATP} \atop +}} \xrightarrow{\text{Actomyosin ATPase}\ 2} \text{ADP} + \text{P}_i \qquad (11)$$

The changes in CrP or creatine produced as a result of the rephosphorylation reaction can be used as an indirect index of the changes in ATP concentration during contraction. When used to detect changes associated with a long series of twitches or tetani, the method is only successful if the muscles are first metabolically poisoned. Iodoacetic acid (IAA) is frequently used (Lundsgaard, 1930), as it prevents the formation of metabolically derived ATP needed for the rephosphorylation of creatine. In short time course experiments, such as associated with a single twitch, these side reactions are unimportant providing the muscle reactions are halted by rapid freezing soon after the contraction cycle.

In the case of an isotonic twitch, it is found that the breakdown of CrP is linearly related to the work done (Cain et al., 1962; Infante et al., 1965) (see Fig. 19). In an isometric contraction the breakdown of CrP is related both to the duration of the stimulus and to the rest length of the muscle, hence to the developed tension (Infante et al., 1964; see Section 2.D.3).

In more recent years it has been possible to inhibit the rapid transphosphorylation reaction (Eq. 11, step 1) by an inhibitor of the enzyme creatine phosphotransferase (Cain et al., 1962). In these experiments either the disappearance of ATP or the appearance of inorganic phosphate (P_i) can be used as a direct guide to the hydrolysis

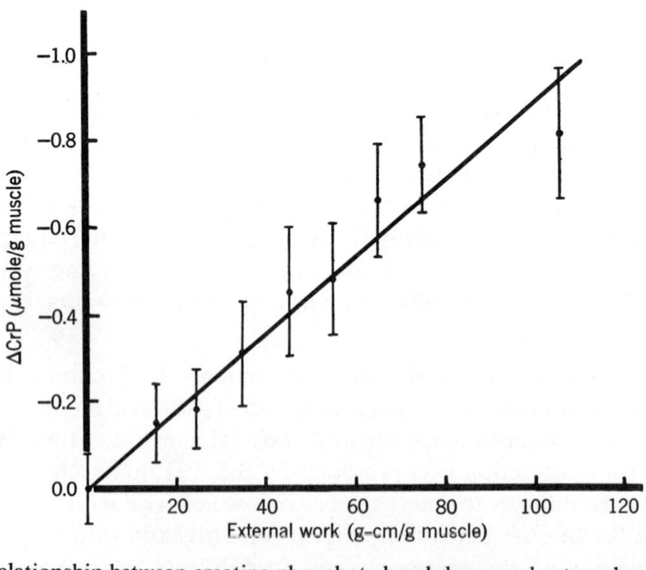

Figure 19. The relationship between creatine phosphate breakdown and external work in frog muscle at 0°C. (From Cain et al., 1962.)

of ATP by employing the specific inhibitor dinitrofluorobenzene (DNFB). These experiments provide critical data for an assessment of the relationship between ATP breakdown and the thermal events of contraction.

ATP usage during an isotonic contraction was roughly proportional to the work done as observed when the change in CrP was measured (Kushmerick and Davies, 1969). The amount of ATP splitting during an unloaded isotonic contraction was much smaller, 0.06 ± 0.009 μmole ATP/g (Kushmerick et al., 1969), and this value has been corrected for the small amount of work done against inertial and frictional forces. In an isometric contraction the amount of ATP hydrolyzed is similar to the change in CrP content of IAA-poisoned muscles and is related to the duration of the tetanic stimulation and to the tension up to stimulus durations of at least 2–3 sec (Infante et al., 1964; Kushmerick and Davies, 1969). In addition, ATP breakdown occurs almost entirely within the rising phase of isometric tension with very little occurring during the relaxation phase of the contraction cycle (Mommaerts and Wallner, 1967).

The amount of ATP broken down during a contraction reflects the activity of many different systems within the fiber and not just the activity of the force-generating apparatus. When little or no external work is performed, hence the tension is negligible, the heat of activation or, more importantly, the heat of shortening might play an important part in ATP consumption. A value of 1.2 ± 0.05 mcal/g (Gibbs et al., 1966) for the heat of activation is available, and if the $-\Delta F$ for the hydrolysis of ATP is approximately 10 kcal/mole, the predicted ATP breakdown if this is the sole basis for the heat of activation is 0.12 μmole/g. The heat of activation accounts for *twice* as much ATP as was actually hydrolyzed under these conditions. Similarly, the amount of ATP required to be consumed to account for the heat of shortening can be calculated (see Eq. 9; Section 2.E.3), and this again is higher than the experimental value (Kushmerick et al., 1969). Consequently, neither the heat of activation nor the heat of shortening can be entirely attributed to the breakdown of ATP molecules. It is of interest to speculate as to the basis of these thermal events. Probably, the heat of activation is related to ion movements associated with the action potential and, more importantly, with the release of calcium from the SR (Kushmerick et al., 1969). At present, it is not possible to explain the heat of shortening so readily, but it is likely related to the activity of the sliding of the myofilaments during contraction. The lack of ATP breakdown during isometric relaxation confirms that the heat of relaxation is mainly mechanical in origin and reflects reextension of the internal elastic elements.

What therefore is the cause of the ATP breakdown in an unloaded isotonic twitch (approximately 0.06 μmole/g)? Calculations suggest that a major part of this ATP may be used for the calcium pump across the membrane of the SR (Section 2.C.4). If it is assumed that during full activation every troponin molecule has one calcium, approximately 0.1 μmole Ca/g of muscle would need to be released from the SR. A similar amount of ATP would be required to restore this amount of calcium to the SR sometime during the contraction cycle (Kushmerick et al., 1969).

3. INSECT FLIGHT MUSCLE

A. Introduction

Certain orders of insects have specialized muscle used for powered flight which, although cross-striated, can be distinguished physiologically from vertebrate skeletal muscle. Histologically, flight muscle contains myofibrils which are a little larger (ca. 5 μ in diameter), unbranched, and more easily separated from each other than myofibrils from vertebrate muscle. Small sarcomeres are also a distinguishing feature of the tissue (Tiegs, 1955; Edwards, 1960). The muscle is termed *fibrillar*, and the individual fibrils (myofibrils) are separated from each other by numerous mitochondria. In general, insects possessing flight muscle capable of contracting at high frequencies is found to be fibrillar. These contractions can be in exceptional cases as high as 2000–3000/sec (Sotavalta, 1953), although the contraction frequency is more usually in the range 100–200/sec. Early electrophysiological observations on this muscle revealed that the number of nerve impulses received by the muscle was much lower than the number of contractions (Pringle, 1949). The absence of the usual one-to-one correlation of electrical impulse with mechanical response suggested that flight muscle differed in its physiological properties from vertebrate skeletal muscle. This *asynchronous* muscle has been found in the flight muscle of most flies, bees, and wasps, as well as in the sound-producing muscles of certain cicadas (Boettiger, 1960; Pringle, 1967a).

Not all flight muscles behave in this manner. In dragonflies and moths there is a normal correlation between the nerve impulse and the contraction. This *synchronous* muscle appears to have normal physiological properties and histologically can be either fibrillar or afibrillar in structure. The term afibrillar is confusing as it implies the absence of myofibrils. In fact, the two types of muscle differ quantitatively rather than qualitatively in their histological structure. The myofibrils in afibrillar muscle are usually smaller, the sarcomeres longer, and the SR more extensive (Edwards et al., 1956; Smith, 1966).

B. Organization and Structure of the Sarcomere

1. Appearance of the Sarcomere. The flight muscle sarcomere when examined with the electron microscope reveals an interdigitating array of thick (myosin) and thin (actin) filaments. The I band of the sarcomere is much reduced compared with vertebrate skeletal muscle, so that at rest there is virtually maximum overlap of the myofilaments. The amount of shortening that can occur is quite small, of the order of 1–3% of the resting length, depending on the source of the muscle (Boettiger, 1960). In certain flight muscles an additional structure links the ends of each myosin filament to the Z line (Auber and Couteaux, 1963) (Fig. 20).

In a cross section through the sarcomere in the A–I zone there is the familiar double hexagonal array of actin and myosin (Fig. 21). The actin filaments are not, however, in the trigonal lattice position and are relatively more numerous, each actin being shared by two myosins compared to vertebrate muscle in which the ratio is 1:3 (Worthington, 1961). The arrangement of actin in flight muscle, in which the actin and myosin are no longer "staggered," permits a thin section containing alternating filaments to be obtained (see Fig. 21). This "myac" layer can provide information on the distribution of cross-bridges along the thick filament (Reedy, 1967, 1968). Additionally, the orientation of the cross-bridges both at rest and in the rigor state (Section 2.A.5) can be observed.

2. Actin, Myosin, and the Cross-Bridges. As with vertebrate muscle, a combination of electron microscope and x-ray diffraction studies has provided a wealth of information concerning the fine structure of the myofilaments. The thick filament is similar to vertebrate myosin in that the cross-bridges are opposed at each level and that

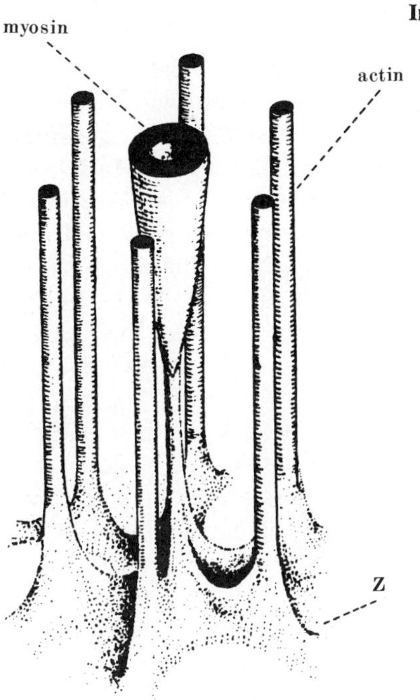

Figure 20. Structure that may link the myosin filaments with the Z line in insect flight muscle. (From Auber and Couteaux, 1963.)

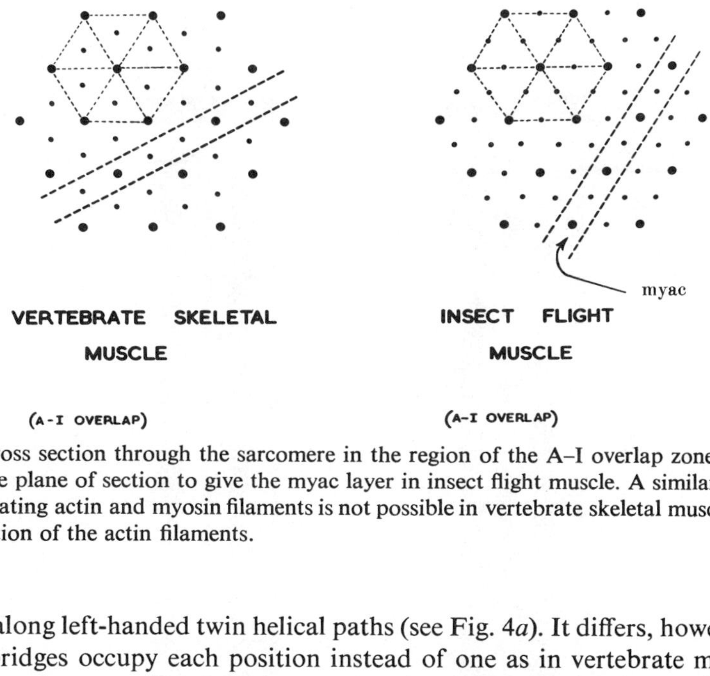

Figure 21. Cross section through the sarcomere in the region of the A–I overlap zone. The dashed line shows the plane of section to give the myac layer in insect flight muscle. A similar section containing alternating actin and myosin filaments is not possible in vertebrate skeletal muscle because of trigonal position of the actin filaments.

they occur along left-handed twin helical paths (see Fig. 4*a*). It differs, however, in that two cross-bridges occupy each position instead of one as in vertebrate muscle. As a consequence, there are four bridges at every level. The cross-bridge origins repeat every 146 Å as in the vertebrate case (Reedy, 1967, 1968). The actin filament consists of two strands of F-actin wound around each other in a right-handed helical configuration, but in flight muscle the helix has a pitch of 388 Å. The actin and myosin filaments are of equal pitch and remain in phase throughout the overlap zone, but the screw sense of the two filaments is opposed, in that the actin is right-handed while myosin is left-handed.

Observation of the myac layer with the electron microscope indicates a marked change in the orientation of the cross-bridges in resting as compared with rigor-contracted muscle (Fig. 22) (Reedy et al., 1966). In the resting state the cross-bridges project at 90° from the backbone of the myosin filament (Fig. 22*b*), in which every cross-bridge is a double structure. In rigor contraction, produced in a glycerinated preparation by lack of ATP, the cross-bridges assume chevron or arrowhead configurations and are linked to the nearby actin filaments (Fig. 22*a*). The cross-bridges are now at an angle of 45° to the myosin backbone, with the ends of the bridges attached approximately 150 Å closer to the center of the sarcomere than their origins on the myosin. The increased resistance to stretch of rigor muscle suggests that the cross-bridges form permanent links between the sets of myofilaments. The distance between the points of attachment of the cross-bridge along the actin filament is 388 Å, and is identical to the pitch of the actin filament. Observations of the myac layer in cross section reveal a "flared-X" configuration of the four cross-bridges at every level. Current estimates, however, of the amount of myosin per filament (Chaplain and Tregear, 1966) suggest six myosin molecules per four cross-bridges (Reedy, 1967).

Rigor muscle also provides informative x-ray data since the chevron formation of the bridges produces characteristic reflections which are not altered by stretching the muscle up to 5% (Reedy, 1967, 1968). This indicates that the bridges are capable of bearing tension without bending when in this configuration and implies that they

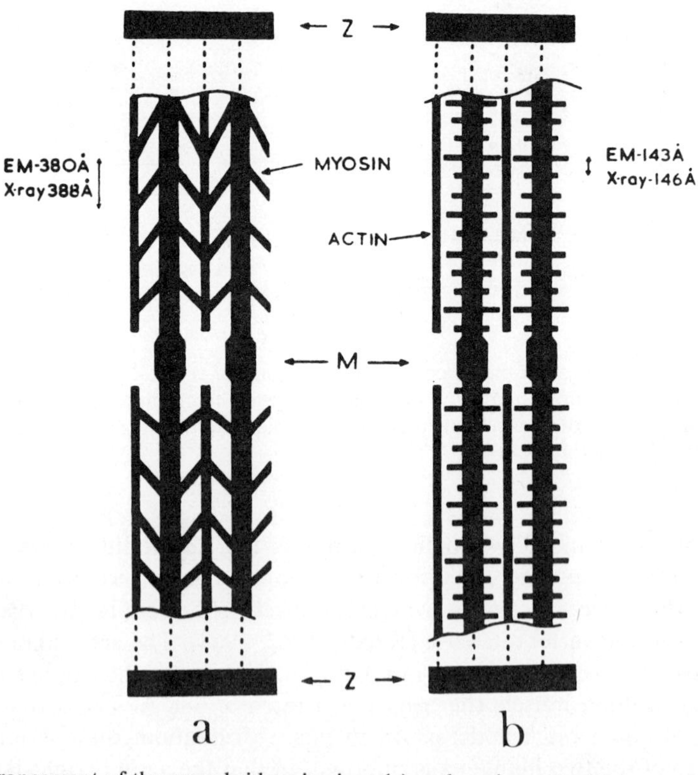

Figure 22. Arrangement of the cross-bridges in rigor (*a*) and resting state (*b*) from an analysis of the myac layer in glycerinated insect flight muscle. (From Reedy et al., 1966.)

function in active muscle by pushing rather than by pulling the actin filaments toward the center of the sarcomere (Reedy, 1967). In active muscle, the regularity of the cross-bridge structure, together with the fact that both sets of filaments have identical pitches, suggests that the cross-bridges may operate synchronously, while in vertebrate skeletal muscle the cross-bridges appear to act asynchronously (Huxley and Brown, 1967). The movement of the heads of the cross-bridges towards the thin filament has been detected in flight muscle during activity (Miller and Tregear, 1970).

C. Biochemical and Mechanical Observations

The initial electrophysiological observations on intact insects employed wire electrodes implanted into the flight muscle and indicated a lack of correlation between the electrical and mechanical events. When the exoskeleton of these animals was further disrupted by dissection, the oscillatory activity of the muscle was destroyed. In certain species, in particular the bee *Bombus* (Boettiger, 1957a, b) and lamellicorn beetles (*Oryctes*) (Machin and Pringle, 1959, 1960), as well as in the sound-producing muscles of some cicadas (Pringle, 1954), the *in situ* preparation is sufficiently well preserved so that more detailed physiological investigations can be undertaken. The large muscles from the lamellicorn beetle are particularly useful having a low frequency of wing oscillation which is no greater than 40–50/sec (Darwin and Pringle, 1959). The mechanical requirements for investigating these contractions are consequently less stringent than at the higher frequencies.

Resting flight muscle is inextensible compared with vertebrate skeletal muscle, and this is likely related to the structures linking the ends of the myosin filaments to the Z line (Pringle, 1967a) (Section 3.B.1). Active tension in the muscle can be developed isometrically, in which case the external recording system must have a very low compliance, of the order of 1/100 that of the muscle (Machin and Pringle, 1959), and here tension develops without oscillation. The maximum value for P_0 is 0.25 kg/cm^2 and is much less than that observed in vertebrate muscle (1–2 kg/cm^2). When the muscle is allowed to undergo shortening and the load has inertia, the muscle undergoes cyclical changes in length, that is, it oscillates. These changes are most readily illustrated by a length–tension (L–T) diagram, in which both parameters are portrayed simultaneously on the X and Y axes of an oscilloscope (Fig. 23*a*). Direct observation indicates that the direction of the "loop" is counterclockwise, so that if the lengthening and shortening sections are analyzed separately, tension is greater during shortening than during lengthening. In this freely oscillating state, work is done by the muscle and the area of the loop is a measure of this work (Boettiger, 1957b; Pringle, 1967a, b). In vertebrate skeletal muscle similar loops can be obtained only by imposing an oscillatory cycle upon the muscle, and in this case the direction of the loop is clockwise. Thus tension is always greater during lengthening than during shortening (see Section 2.D.2), and work is absorbed (Rack, 1966).

More recently, flight muscles from tropical water bugs (*Lethocerus*) have been described which have exceptionally large muscle fibers, often 70 μ in diameter and 1–2 cm in length. These muscles are well suited to glycerol extraction procedures (see Section 2.B.4). After extraction the fiber bundles can be attached to force transducers (Jewell and Ruegg, 1966; Ruegg, 1967), and under suitable mechanical and biochemical conditions can be made to undergo oscillatory contractions. If the muscles are prevented from shortening, tension and ATPase activity develop biphasically with increasing calcium concentration. The ATPase activity lags behind the tension over the range 1×10^{-8} to 2×10^{-7} *M* calcium (Chaplain, 1967). At higher calcium concentrations the fibers are in a *high-tension* state in which both tension and ATPase start to rise more rapidly, and this can be reversed by increasing the ATP concentration in the bathing medium. It has been suggested that the high-tension state is related to the build-up of ADP at tension-generating sites (Chaplain, 1967; Pringle, 1967a), and this biphasic relationship is not seen in freshly isolated myofibrils. The absolute changes in

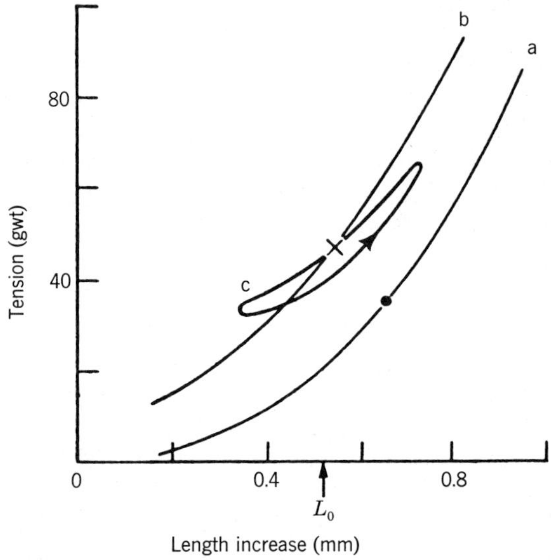

Figure 23*a*. A typical length–tension (L–T) diagram obtained during the oscillatory contraction of insect flight muscle. Line a, unstimulated; line b, stimulated under isometric conditions and loop (c), oscillation with an inertial load; (●) is the starting point when unstimulated and (×) when the muscle is stimulated isometrically before the inertial load is applied. (From Machin and Pringle, 1959.)

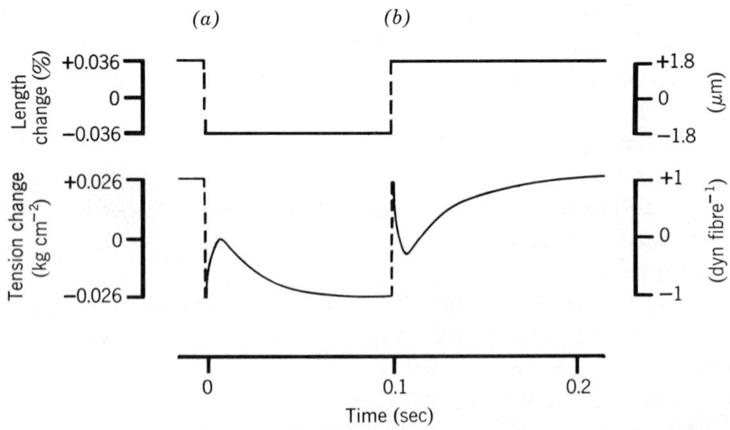

Figure 23*b*. Tension changes following quick release (*a*) and quick-stretch (*b*) applied to glycerinated insect flight muscle. Calcium concentration, 1.1×10^{-7}M. (From Jewell and Rüegg, 1966).

ATPase activity with calcium concentration differ in flight muscle compared with myofibrils from vertebrate or invertebrate skeletal muscle or even nonoscillatory flight muscle (Maruyama et al., 1968). The sensitivity of flight muscle myofibrils to calcium, however, appears to be under the control of the tropomyosin B–troponin complex (Maruyama et al., 1968; see also Sections 2.A.4 and B.2).

A glycerol-extracted muscle can be made to oscillate if it is allowed to shorten and the load has some inertia. The biochemical requirements for oscillation are a concentration of calcium greater than 2×10^{-8} M and the presence of ATP and magnesium ions. When the calcium in the solution is increased, the amplitude but not the frequency of the contractions increases. Contractions in the range of 4–45 cps were obtained, depending upon the characteristics of the load (Jewell and Rüegg, 1966). The fact that a glycerol-extracted preparation can oscillate implies that this phenomenon is a function of the contractile apparatus and not just of the E–C coupling system. The P_0 for active muscle at high amplitude oscillations approaches the maximum tension of vertebrate muscle (Pringle and Tregear, 1969). In mechanical terms oscillation must involve length and tension changes that are out of phase with each other, and this phenomenon can be demonstrated by quick-stretch or quick-release experiments (see Section 2.D.2). After a quick release (Fig. 23*b*), there is a rapid fall in tension, then some redevelopment, but this is rapidly followed by a prolonged fall despite the isometric conditions. In a quick *stretch*, tension rises rapidly (Fig. 23G), declines somewhat, and then rises more markedly. The delayed increase in tension is associated with a delayed increase in ATPase activity (Rüegg and Tregear, 1966; Rüegg, 1967). Similar delayed changes in tension after a quick stretch can be demonstrated with an *in situ* muscle preparation (Boettiger, 1957b). In a vertebrate skeletal muscle, a quick stretch produces no *delayed increase* in tension (Fig. 14; Section 2.D.2).

In conclusion, the activity of insect flight muscle, unlike vertebrate skeletal muscle, is not directly controlled by rapid changes in internal calcium concentration. Providing that in intact muscle the calcium concentration is kept at a value greater than threshold by nerve activity, the muscle continues to oscillate. The mechanisms that bring about delayed changes in tension are not fully understood, but they directly involve the contractile proteins. One possibility is that the structures linking the thick filaments to the Z line transmit tension changes which alter the configuration of the myosin filament and permit an increased interaction with actin (Pringle, 1967b; see also Julian, 1969; Thorson and White, 1969).

4. SOME GENERAL REMARKS

It has not been possible in the space available in this chapter to discuss the mechanisms of contraction of either cardiac or smooth muscle. The evidence from both the light and electron microscope indicates that cardiac muscle is cross-striated and that the individual sarcomeres contain an interdigitating array of actin and myosin filaments. As with skeletal muscle, the electrical events are coupled in a one-to-one fashion with the mechanical events and calcium ions play an essential intermediary role mediated by the troponin–tropomyosin B complex (Sonnenblick and Stam, 1969). Despite these similarities, the details of the E–C coupling process and certain mechanical properties of cardiac muscle distinguish it from skeletal muscle (Sonnenblick, 1965; Brady, 1967; Tanz et al., 1967).

In vertebrate smooth muscle the electron microscope has only recently provided evidence for two distinct sets of myofilaments (Lowy and Small, 1970; Rice et al., 1970), in which the myosin in cross section can be in the form of a flat ribbon rather than a rod. The biochemistry and detailed physiology of vertebrate smooth muscle differ from both that of cardiac and skeletal muscle (Bülbring et al., 1970). The smooth muscles from certain invertebrates have been investigated in some detail because of their ability to maintain tension for prolonged periods of time with the expenditure of little extra energy (Johnson et al., 1959); the so-called *catch mechanism*. These muscles have actin and myosin filaments which are more readily identified than in vertebrate smooth muscle and contain the additional protein paramyosin or tropomyosin A (Lowy et al., 1964). The precise relationship between paramyosin and the catch mechanism is still a matter of speculation (Rüegg, 1964, 1968; Twarog, 1967a, b).

ACKNOWLEDGMENTS

Heinemann Books Ltd., London and the editors of the following journals are gratefully acknowledged for permission to reproduce published work: *Endeavour*, *J. Microscopie*, *J. Molecular Biology*, *J. Physiology*, *Nature*, *Q. Reviews of Biophysics*, *Science*, and *Proceedings of the Royal Society*. I am also grateful to the Medical Research Council for a grant.

REFERENCES

Adrian, R. H., L. L. Costantin, and L. D. Peachey (1969). Radial spread of contraction in frog muscle fibres. *J. Physiol. (London)*, **204,** 231.

Ashley, C. C. (1967). The role of cell calcium in the contraction of single cannulated muscle fibres. *Am. Zoologist*, **7,** 647.

Ashley, C. C. (1970a). An estimate of calcium concentration changes during the contraction of single muscle fibres. *J. Physiol. (London)*, **210,** 133P.

Ashley, C. C. (1970b). In *Membranes and Ion Transport* (E. E. Bittar, ed.), Vol. 2, pp. 1–31. Wiley, New York.

Ashley, C. C. (1971). Calcium and the activation of skeletal muscle. *Endeavour*, **30,** 18.

Ashley, C. C., and E. B. Ridgway (1968). Simultaneous recording of membrane potential, calcium transient and tension in single muscle fibres. *Nature*, **219,** 1168.

Ashley, C. C., and E. B. Ridgway (1970a). On the relationships between membrane potential, calcium transient and tension in single barnacle muscle fibres. *J. Physiol. (London)*, **209,** 105.

Ashley, C. C., and E. B. Ridgway (1970b). In *Calcium and Cellular Function* (A. Cuthbert, ed.), pp. 42–53. Macmillan, New York.

Ashley, C. C., and D. G. Moisescu (1972). A model for the action of calcium in muscle. *Nature, New Biology*, (in Press).

Auber, J., and R. Couteaux (1963). Ultrastructure de la strie Z dans des muscles de Diptères. *J. Microscopie*, **2,** 309.

Auber, X. (1956). Le mécanisme contractile *in vivo:* Aspects mécaniques et thermiques. *J. Physiol. (Paris)*, **48,** 105.

Bailey, K. (1948). Tropomyosin: A new asymmetric protein component of the muscle fibril. *Biochem. J.*, **43,** 271.

Bárány, M., B. Nagy, F. Finkelman, and A. Chrambach (1961). Studies on the removal of the bound nucleotide of actin. *J. Biol. Chem.*, **236,** 2917.

Bendall, J. R. (1953). Further observations on a factor (the "Marsh" factor) effecting relaxation of ATP-shortened muscle-fiber models and the effect of Ca and Mg ions upon it. *J. Physiol.* (*London*) **121,** 232.

Bendall, J. R. (1958). Relaxation of glycerol-treated muscle fibres by ethylene diamine tetraacetate. *Arch. Biochem. Biophys.*, **73,** 283.

Bendall, J. R. (1969). *Muscles, Molecules and Movement.* Heinemann, London.

Bennett, H. S., and K. R. Porter (1953). An electron microscope study of sectioned breast muscle of the domestic fowl. *Am. J. Anat.*, **93,** 61.

Bianchi, C. P., and A. M. Shanes (1959). Calcium influx in skeletal muscle at rest, during activity and during potassium contracture. *J. Gen. Physiol.*, **42,** 803.

Boettiger, E. G. (1957a). In *Physiological Triggers* (T. H. Bullock, ed.), pp. 103–116. American Physiological Society, Washington, D.C.

Boettiger, E. G. (1957b). In *Recent Advances in Invertebrate Physiology* (B. T. Sheer, ed.), pp. 117–142. Univ. of Oregon Press, Eugene.

Boettiger, E. G. (1960). Insect flight muscles and their basic physiology. *Ann. Rev. Entomol.*, **5,** 1.

Bowman, W. (1840). On the minute structure and movements of voluntary muscle. *Phil. Trans. Roy. Soc.* (*London*), **B130,** 457.

Brady, A. J. (1967). Length-tension relations in cardiac muscle. *Am. Zoologist*, **7,** 603.

Brücke, E. (1858). Untersuchungen über den Bau der Muskelfasern mit Hülfe des polarisirten Lichtes. *Denkschr. Akad. Wiss. Wien. Kl.*, **15,** 69. Cited in Bennett, H. S. (1955). *Am. J. Phys. Med.*, **34,** 46.

Bülbring, E., A. F. Brading, A. W. Jones, and T. Tomita (eds.) (1970). *Smooth Muscle.* Arnold, London.

Cain, D. F., A. A. Infante, and R. E. Davies (1962). Chemistry of muscle contraction. Adenosine triphosphate and phosphorylcreatine as energy supplies for single contractions of working muscle. *Nature*, **196,** 214.

Carlson, F. D., D. J. Hardy, and D. R. Wilkie (1963). Total energy production and phosphocreatine hydrolysis in the isotonic twitch. *J. Gen. Physiol.*, **46,** 851.

Carvalho, A. P. (1968). Calcium-binding properties of sarcoplasmic reticulum as influenced by ATP, caffeine, quinine and local anesthetics. *J. Gen. Physiol.*, **52,** 622.

Chaplain, R. A. (1967). Insect actomyosin ATPase—A regulatory protein. *Arch. Biochem. Biophys.*, **121,** 154.

Chaplain, R. A., and R. T. Tregear (1966). The mass of myosin per cross-bridge in insect fibrillar flight muscle. *J. Mol. Biol.*, **21,** 275.

Connolly, R., W. Gough, and S. Winegrad (1971). Characteristics of the isometric twitch of skeletal muscle immediately after a tetanus. A study of the influence of the distribution of calcium within the sarcoplasmic reticulum on a twitch. *J. Gen. Physiol.*, **57,** 697.

Costantin, L. L., C. Franzini-Armstrong, and R. J. Podolsky (1965). Localization of calcium-accumulating structures in striated muscle fibers. *Science*, **147,** 158.

Darwin, F. W., and J. W. S. Pringle (1959). The physiology of insect fibrillar muscle. I. Anatomy and innervation of the basalar muscle of lamellicorn beetles. *Proc. Roy. Soc.* (*London*), **B151,** 194.

Davies, R. E. (1963). A molecular theory of muscle contraction. *Nature*, **199,** 1068.

Depue, R. H., and R. V. Rice (1965). F-Actin is a right-handed helix. *J. Mol. Biol.*, **12,** 302.

Dreizen, P., L. C. Gershman, P. P. Trotta, and A. Stracher (1967). Myosin. Subunits and their interactions. *J. Gen. Physiol.*, **50,** 85s.

Ebashi, S. (1961). Relaxation and the removal of calcium by muscle particulate fraction. *J. Biochem.* (*Tokyo*), **50,** 77.

Ebashi, S. (1963). Third component participating in the superprecipitation of natural actomyosin. *Nature*, **200,** 1010.

Ebashi, S. (1968). In *Symposia Biologica Hungarica* (E. Ernst and F. B. Straub, eds.), Vol. 8, pp. 77–87. Akadémiai Kiado, Budapest.

Ebashi, S., and M. Endo (1968). Calcium ions and muscle contraction. *Progr. Biophys. Mol. Biol.*, **18,** 123.

Ebashi, S., and F. Lipmann (1962). ATP-linked concentration of calcium ions in a particulate fraction of rabbit muscle. *J. Cell Biol.*, **14,** 389.

Ebashi, S., M. Endo, and I. Ohtsuki (1969). Control of muscle contraction. *Quart. Rev. Biophys.*, **2,** 351.

Edwards, G. A. (1960). Insect micromorphology. *Ann. Rev. Entomol.*, **5,** 17.

Edwards, G. A., H. Ruska, P. de Souza Santos, and A. Vellejo-Freire (1956). Comparative cytophysiology of striated muscle with special reference to the role of the endoplasmic reticulum. *J. Biophys. Biochem. Cytol., Suppl.*, **2,** 143.

Eisenberg, R. S., and P. W. Gage (1967). Frog skeletal muscle fibres: Changes in electrical properties after disruption of transverse tubular system. *Science*, **158,** 1700.

Eisenberg, E., and C. Moos (1968). The adenosine triphosphatase activity of acto-heavy meromyosin. A kinetic analysis of actin activation. *Biochemistry*, **7,** 1486.

Eisenberg, E., C. R. Zobel, and C. Moos (1968). Subfragment 1 of myosin: Adenosine triphosphatase activation by actin. *Biochemistry*, **7,** 3186.

Elliott, G. F., J. Lowy, and C. R. Worthington (1963). X-ray diffraction studies of resting, living muscle. *J. Mol. Biol.*, **6,** 295.

Elliott, G. F., J. Lowy, and B. M. Millman (1967). Low angle X-ray diffraction studies of living striated muscle during contraction. *J. Mol. Biol.*, **25,** 31.

Elliott, G. F., E. Rome, and M. Spencer (1970). A type of contraction hypothesis applicable to all muscles. *Nature*, **226,** 417.

Endo, M. (1964). Entry of a dye into the sarcotubular system of muscle. *Nature*, **202,** 1115.

Endo, M., M. Tanaka, and Y. Ogawa (1970). Calcium induced release of calcium from the sarcoplasmic reticulum of skinned skeletal muscle fibers. *Nature*, **228,** 34.

Engelhardt, W. A., and M. N. Ljubimowa (1939). Myosine and adenosine triphosphatase, *Nature*, **144,** 668.

Fenn, W. O. (1923). A quantitative comparison between the energy liberated and the work performed by the isolated sartorius muscle of the frog. *J. Physiol. (London)*, **58,** 175.

Fenn, W. O. (1924). The relation between the work performed and the energy liberated in muscular contraction. *J. Physiol. (London)*, **58,** 373.

Ford, L. E., and R. J. Podolsky (1970). Regenerative calcium release within muscle cells. *Science*, **167,** 58.

Fuchs, F., and N. F. Briggs (1968). The site of calcium binding in relation to the activation of myofibrillar contraction. *J. Gen. Physiol.*, **51,** 655.

Gage, P. W., and R. S. Eisenberg (1969). Action potentials, afterpotentials, and excitation-contraction coupling in frog sartorius fibers without transverse tubules. *J. Gen. Physiol.*, **53,** 298.

Gasser, H. S., and A. V. Hill (1924). The dynamics of muscular contraction. *Proc. Roy. Soc. (London)*, **B96,** 398.

Gibbs, C. L., N. V. Ricchiuti, and W. F. H. M. Mommaerts (1966). Activation heat in frog sartorius muscle. *J. Gen. Physiol.*, **49,** 517.

Goffart, M., and J. M. Ritchie (1952). The effect of adrenaline on the contraction of mammalian skeletal muscle. *J. Physiol. (London)*, **116,** 357.

Gordon, A. M., A. F. Huxley, and F. J. Julian (1966a). Tension development in highly stretched vertebrate muscle fibres. *J. Physiol. (London)*, **184,** 143.

Gordon, A. M., A. F. Huxley, and F. J. Julian (1966b). The variation in isometric tension with sarcomere length in vertebrate muscle fibres. *J. Physiol. (London)*, **184,** 170.

Hanson, J. (1968). Recent X-ray diffraction studies of muscle. *Quart. Rev. Biophys.*, **1,** 177.

Hanson, J., and H. E. Huxley (1953). The structural basis of the cross-striations in muscle. *Nature*, **172,** 530.

Hanson, J., and H. E. Huxley (1955). The structural basis of contraction in striated muscle. *Symp. Soc. Exptl. Biol.*, **9,** 228.

Hanson, J., and J. Lowy (1963). The structure of F-actin and of actin filaments isolated from muscle. *J. Mol. Biol.*, **6,** 46.

Hanson, J., and J. Lowy (1964). The structure of actin filaments and the origin of the axial periodicity in the I-substance of vertebrate striated muscle. *Proc. Roy. Soc.* (*London*), **B160,** 449.

Hasselbach, W. (1953. Electronmikroskopische Untersuchungen an Muskelfibrillen bei totaler and partieller Extraction des L-Myosins. *Z. Naturforsch.*, **8b**, 449.

Hasselbach, W. (1957). Die Nukleosidetriphosphatase-Aktivität von L-Myosin und Aktomyosin in Abhängigkeit von den ionalen Bedingungen. *Biochim. Biophys. Acta*, **25,** 365.

Hasselbach, W. (1964). Relaxing factor and the relaxation of muscle. *Prog. Biophys. Mol. Biol.*, **14,** 167.

Hasselbach, W., and M. Makinose (1961). Die Calciumpumpe der "Erschlaffungsgrana" der Muskels und ihre Abhängigkeit von der ATP-Spaltung. *Biochem. Z.*, **333,** 518.

Hasselbach, W., and M. Makinose (1962). ATP and active transport. *Biochem. Biophys. Res. Commun.*, **7,** 132.

Hasselbach, W., and M. Makinose (1963). Über den Mechanismus des Calciumtransportes durch die Membranen des sarkoplasmatischen Reticulums. *Biochem. Z.*, **339,** 94.

Heilbrunn, L. V. (1940). The action of calcium in muscle protoplasm. *Physiol. Zool.*, **13,** 88.

Heilbrunn, L. V., and F. J. Wiercinski (1947). The action of various cations on muscle protoplasm. *J. Cellular Comp. Physiol.*, **29,** 15.

Heilbrunn, L. V., D. Mazia, and H. B. Steinbach (1934). Free and bound calcium in stimulation and injury. *Anat. Record*, **60,** 32.

Hellam, D. C., and R. J. Podolsky (1969). Force measurements in skinned muscle fibers. *J. Physiol.* (*London*), **200,** 807.

Hill, A. V. (1938). Heat of shortening and dynamic constants of muscle. *Proc. Roy. Soc.* (*London*), **B126,** 136.

Hill, A. V. (1948). On the time required for diffusion and its relation to processes in muscle. *Proc. Roy. Soc.* (*London*), **B135,** 446.

Hill, A. V. (1949a). The abrupt transition from rest to activity in muscle. *Proc. Roy. Soc.* (*London*), **B136,** 399.

Hill, A. V. (1949b). Work and heat in a muscle twitch. *Proc. Roy. Soc.* (*London*), **B136,** 220.

Hill, A. V. (1949c). The energetics of relaxation in a muscle twitch, *Proc. Roy. Soc.* (*London*), **B136,** 211.

Hill, A. V. (1950). The development of the active state of muscle during the latent period. *Proc. Roy. Soc.* (*London*), **B137,** 320.

Hill, A. V. (1953a). The mechanics of active muscle. *Proc. Roy. Soc.* (*London*), **B141,** 104.

Hill, A. V. (1953b). The "instantaneous" elasticity of active muscle. *Proc. Roy. Soc.* (*London*), **B141,** 161.

Hill, A. V. (1953c). The "plateau" of full activity during a muscle twitch. *Proc. Roy. Soc.* (*London*), **B141,** 498.

Hill, A. V. (1964a). The effect of load on the heat of shortening of muscle. *Proc. Roy. Soc.* (*London*), **B159,** 297.

Hill, A. V. (1964b). The efficiency of mechanical power development during muscular shortening and its relation to load. *Proc. Roy. Soc.* (*London*), **B159,** 319.

Hill, A. V. (1964c). The effect of tension in prolonging the active state in a twitch. *Proc. Roy. Soc.* (*London*), **B159,** 589.

Hill, A. V. (1964d). The variation of total heat production in a twitch with velocity of shortening. *Proc. Roy. Soc.* (*London*), **B159,** 596.

Hill, A. V., and L. Macpherson (1954). The effect of nitrate, iodide and bromide on the duration of the active state in skeletal muscle. *Proc. Roy. Soc.* (*London*), **B143,** 81.

Hill, A. V., and R. C. Woledge (1962). An examination of absolute values in myothermic measurements. *J. Physiol.* (*London*), **162,** 311.

Hill, T. L. (1970). Sliding filament model of muscular contraction. V. Isometric force and interfilament spacing. *J. Theoret. Biol.*, **29,** 395.

Hodgkin, A. L. (1958). Ionic movements and electrical activity in giant nerve fibres. *Proc. Roy. Soc.* (*London*), **B148,** 1.

Hodgkin, A. L. (1964). *The Conduction of the Nervous Impulse*. Liverpool Univ. Press, Liverpool.

Huxley, A. F. (1957). Muscle structure and theories of contraction. *Prog. Biophys. Biophys. Chem.*, **7**, 255.

Huxley, A. F., and R. Niedergerke (1954). Structural changes in muscle during contraction. Interference microscopy of living muscle fibres. *Nature*, **173**, 971.

Huxley, A. F., and R. M. Simmons (1971). Proposed mechanism of force generation in striated muscle. *Nature*, **233**, 533.

Huxley, A. F., and R. W. Straub (1958). Local activation and interfibrillar structures in striated muscle. *J. Physiol. (London)*, **143**, 40P.

Huxley, A. F., and R. E. Taylor (1958). Local activation of striated muscle fibres. *J. Physiol. (London)*, **144**, 426.

Huxley, H. E. (1953a). X-ray analysis and the problem of muscle. *Proc. Roy. Soc. (London)*, **B141**, 59.

Huxley, H. E. (1953b). Electron microscope studies of the organisation of the filaments in striated muscle. *Biochim. Biophys. Acta*, **12**, 387.

Huxley, H. E. (1957). The double array of filaments in cross-striated muscle. *J. Biophys. Biochem. Cytol.*, **3**, 631.

Huxley, H. E. (1960). In *The Cell* (J. Brachet and A. E. Mirsky, eds.), Vol. 4, pp. 365–481. Academic, New York.

Huxley, H. E. (1963). Electron microscope studies on the structure of natural and synthetic protein filaments from striated muscle. *J. Mol. Biol.*, **7**, 281.

Huxley, H. E. (1964a). Structural arrangements and the contraction mechanism in striated muscle. *Proc. Roy. Soc. (London)*, **B160**, 442.

Huxley, H. E. (1964b). Evidence for the continuity between the central elements of the "triads" and extracellular space in frog sartorius muscle. *Nature*, **202**, 1067.

Huxley, H. E. (1965). In *Muscle* (W. M. Paul, E. E. Daniel, C. M. Kay, and G. Monckton, eds.), pp. 3–28. Pergamon, Oxford.

Huxley, H. E. (1967). Recent X-ray diffraction and electron microscope studies of striated muscle. *J. Gen. Physiol.*, **50**, 71.

Huxley, H. E. (1968a). In *Symposia Biologica Hungarica* (E. Ernst and F. B. Straub, eds.), Vol. 8, pp. 54–55. Akadémiai Kiado, Budapest.

Huxley, H. E. (1968b). Structural difference between resting and rigor muscle: Evidence from intensity changes in the low-angle equatorial X-ray diagram. *J. Mol. Biol.*, **37**, 507.

Huxley, H. E. (1969). The mechanism of muscular contraction. *Science*, **164**, 1356.

Huxley, H. E. (1971). The structural basis of muscular contraction. *Proc. Roy. Soc. (London)*, **B178**, 131.

Huxley, H. E., and W. Brown (1967). The low-angle X-ray diagram of vertebrate striated muscle and its behaviour during contraction and rigor. *J. Mol. Biol.*, **30**, 383.

Huxley, H. E., and J. Hanson (1954). Changes in the cross-striations of muscle during contraction and stretch and their structural interpretation. *Nature*, **173**, 973.

Infante, A. A., D. Klaupiks, and R. E. Davies (1964). Length, tension and metabolism during short isometric contractions of frog sartorius muscles. *Biochim. Biophys. Acta*, **88**, 215.

Infante, A. A., D. Klaupiks, and R. E. Davies (1965). Phosphorylcreatine consumption during single-working contractions of isolated muscle. *Biochim. Biophys. Acta*, **94**, 504.

Ingels, N. P., and N. P. Thompson (1966). An electrokinematic theory of muscle contraction. *Nature*, **211**, 1032.

Ikemoto, N., S. Kitagawa, and J. Gergely (1966). Electron microscopic investigation of the interaction of actin and myosin. *Biochem. Z.*, **345**, 410.

Jewell, B. R., and J. C. Rüegg (1966). Oscillatory contraction of insect fibrillar muscle after glycerol extraction. *Proc. Roy. Soc. (London)*, **B164**, 428.

Jewell, B. R., and D. R. Wilkie (1958). An analysis of the mechanical components in frog's striated muscle. *J. Physiol. (London)*, **143**, 515.

Jewell, B. R., and D. R. Wilkie (1960). The mechanical properties of relaxing muscle. *J. Physiol. (London)*, **152**, 30.

Jöbsis, F. F., and M. J. O'Connor (1966). Calcium release and reabsorption in the sartorius muscle of the toad. *Biochem. Biophys. Res. Commun.*, **25,** 246.

Johnson, W. H., J. S. Kahn, and A. G. Szent-Györgyi (1959). Paramyosin and contraction of "catch muscles." *Science*, **130,** 160.

Julian, F. J. (1969). Activation in a skeletal muscle contraction model with a modification for insect fibrillar muscle. *Biophys. J.*, **9,** 547.

Julian, F. J. (1971). The effect of calcium on the force-velocity relation of briefly glycerinated frog muscle fibres. *J. Physiol.* (*London*), **318,** 117.

Kamada, T., and H. Kinosita (1943). Disturbances initiated from naked surface of muscle protoplasm. *Japan. J. Zool.*, **10,** 469.

Kasai, M., E. Nakano, F. Oosawa (1965). Polymerization of actin free from nucleotides and divalent cations. *Biochim. Biophys. Acta*, **94,** 494.

Keily, B., and A. Martonosi (1968). Kinetics and substrate binding of myosin adenosine triphosphate. *J. Biol. Chem.*, **243,** 2273.

Kelly, D. E. (1967). Models of muscle Z-band fine structure based on a looping filament configuration. *J. Cell Biol.*, **34,** 827.

Knappeis, G. G., and F. Carlsen (1962). The ultrastructure of the Z disc in skeletal muscle. *J. Cell Biol.*, **13,** 323.

Knappeis, G. G., and F. Carlsen (1968). The ultrastructure of the M line in skeletal muscle. *J. Cell Biol.*, **38,** 202.

Kölliker, A. (1888). Zur Kenntnis der quergestreiften Muskelfasern. *Z. Wiss. Zool.*, **47,** 689. Cited in Bennett, H. S. (1955). *Am. J. Phys. Med.*, **34,** 46.

Kominz, D. R. (1971). In *Contractile Proteins and Muscle* (K. Laki, ed.), pp. 253–261. Dekker, New York.

Kominz, D. R., A. Hough, P. Symonds, and K. Laki (1954). The amino-acid composition of actin, myosin, tropomyosin and the meromyosins. *Archiv. Biochem. Biophys.*, **50,** 148.

Kushmerick, M. J., and R. E. Davies (1969). The chemical energetics of muscle contraction. II. The chemistry, efficiency and power of maximally working sartorius muscles. *Proc. Roy. Soc.* (*London*), **B174,** 315.

Kushmerick, M. J., R. E. Larson, and R. E. Davies (1969). The chemical energetics of muscle contraction. I. Activation heat, heat of shortening and ATP utilization for activation-relaxation processes. *Proc. Roy. Soc.* (*London*), **B174,** 293.

Laki, K. (1971). In *Contractile Proteins and Muscle* (K. Laki, ed.), pp. 97–133. Dekker, New York.

Leadbeater, L., and S. V. Perry (1963). The effect of actin on the magnesium-activated adenosine triphosphatase of heavy meromyosin. *Biochem. J.*, **87,** 233.

Levy, H. M., and M. Fleischer (1965). Studies on the superprecipitation of actomyosin suspensions as measured by the change in turbidity. *Biochim. Biophys. Acta*, **100,** 479.

Levy, H. M., and E. M. Ryan (1967). The contractile and control sites of natural actomyosin. *J. Gen. Physiol.*, **50,** 2421.

Lowey, S., and C. Cohen (1962). Studies on the structure of myosin. *J. Mol. Biol.*, **4,** 293.

Lowey, S., and S. M. Luck (1969). Equilibrium binding of adenosine diphosphate to myosin. *Biochemistry*, **8,** 3195.

Lowey, S., L. Goldstein, and S. Luck (1966). Isolation and characterization of a helical sub-unit from heavy meromyosin. *Biochem. Z.*, **345,** 248.

Lowey, S., L. Goldstein, C. Cohen, and S. Luck (1967). Proteolytic degradation of myosin and the meromyosins by a water-insoluble polyanionic derivative of trypsin. *J. Mol. Biol.*, **23,** 287.

Lowey, S., H. S. Slayter, A. G. Weeds, and H. Baker (1969). Substructure of the myosin molecule. I. Subfragments of myosin by enzymatic degradation. *J. Mol. Biol.*, **42,** 1.

Lowy, J., and J. V. Small (1970). The organization of myosin and actin in vertebrate smooth muscle. *Nature*, **227,** 46.

Lowy, J., B. M. Millman, and J. Hanson (1964). Structure and function in smooth tonic muscles of lamellibranch molluscs. *Proc. Roy. Soc.* (*London*), **B160,** 525.

Lundsgaard, E. (1930). Untersuchungen über Muskelkontraktionen ohne Milchsaürebildung. *Biochem. Z.*, **217,** 162.

Lymn, R. W., and E. W. Taylor (1971). Mechanism of adenosine triphosphate hydrolysis by actomyosin. *Biochemistry*, **10,** 4617.

Machin, K. E., and J. W. S. Pringle (1959). The physiology of insect fibrillar muscles. II. Mechanical properties of a beetle flight muscle. *Proc. Roy. Soc.* (*London*), **B151,** 204.

Machin, K. E., and J. W. S. Pringle (1960). The physiology of insect fibrillar muscle. III. The effect of sinusoidal changes of length on a beetle flight muscle. *Proc. Roy Soc.* (*London*), **B152,** 311.

Marsh, B. B. (1951). A factor modifying muscle fiber synaeresis. *Nature*, **167,** 1065.

Maruyama, K., and J. Gergely (1962a). Factors influencing clearing and superprecipitation: Adenosinetriphosphatase and birefringence of flow studies. *J. Biol. Chem.*, **237,** 1100.

Maruyama, K., and J. Gergely (1962b). Dissociation of actomyosin during the clear phase. *J. Biol. Chem.*, **237,** 1095.

Maruyama, K., J. W. S. Pringle, and R. T. Tregear (1968). The calcium sensitivity of ATPase activity of myofibrils and actomyosins from insect flight and leg muscles. *Proc. Roy. Soc.* (*London*), **B169,** 229.

Mihályi, E., and A. G. Szent-Györgyi (1953). Trypsin digestion of muscle proteins. I. Ultracentrifugal analysis of the process. *J. Biol. Chem.*, **201,** 189.

Miller, A., and R. T. Tregear (1970). Evidence concerning crossbridge attachment during muscle contraction. *Nature*. **226,** 1060.

Miller, P. L. (ed.) (1968). *Aspects of Cell Motility*. Symposium of the Society for Experimental Biology, Vol. 22. Cambridge Univ. Press, Cambridge.

Mines, G. R. (1913). On functional analysis by the action of electrolytes. *J. Physiol.* (*London*), **46,** 188.

Mommaerts, W. F. H. M. (1969). Energetics of muscular contraction. *Physiol. Rev.*, **49,** 427.

Mommaerts, W. F. H. M., and A. Wallner (1967). The breakdown of adenosine triphosphate in the contraction cycle of the frog sartorius muscle. *J. Physiol.* (*London*), **193,** 343.

Moore, P. B., H. E. Huxley, and D. J. DeRosier (1970). Three-dimensional reconstruction of F-actin, thin filaments and decorated thin filaments. *J. Mol. Biol.*, **50,** 279.

Moos, C., J. E. Estes, and E. Eisenberg (1966). Exchange of F-actin bound nucleotide in the presence and absence of myosin. *Biochem. Biophys. Res. Commun.*, **23,** 347.

Mueller, H., and S. V. Perry (1962). The degradation of heavy meromyosin by trypsin. *Biochem. J.*, **85,** 431.

Nagai, T., M. Makinose, and W. Hasselbach (1960). Der physiologische Erschlaffungsfaktor und die Muskelgrana. *Biochim. Biophys. Acta*, **43,** 223.

Natori, R. (1954). The property and contraction process of isolated myofibrils. *Jikeikai Med. J.*, **1,** 119.

Noda, H., and K. Maruyama (1960). The effect on myosin B of KCl concentration and pH. *Biochim. Biophys. Acta*, **41,** 393.

Offer, G. W. (1964). The antagonistic action of magnesium ions and ethylenediaminetetraacetate on myosin A ATPase (potassium activated). *Biochim. Biophys. Acta*, **89,** 566.

Ohnishi, T., and S. Ebashi (1964). The velocity of calcium binding of isolated sarcoplasmic reticulum. *J. Biochem.* (*Tokyo*), **55,** 599.

Ohtsuki, I., T. Masaki, Y. Nonomura, and S. Ebashi (1967). Periodic distribution of troponin along the thin filament. *J. Biochem.* (*Tokyo*), **61,** 817.

Page, S. (1964). Filament lengths in resting and excited muscles. *Proc. Roy. Soc.* (*London*), **B160,** 460.

Peachey, L. D. (1968). Muscle. *Ann. Rev. Physiol.*, **30,** 401.

Pease, D. C., D. J. Jenden, and J. N. Howell (1965). Calcium uptake in glycerol-extracted rabbit psoas muscle fibers. II. Electron microscopic localization of uptake sites. *J. Cellular Comp. Physiol.*, **65,** 141.

Pepe, F. (1966). Some aspects of the structural organization of the myofibril as revealed by anti-body staining methods. *J. Cell Biol.*, **28,** 505.

Pepe, F. (1967a). The myosin filament. I. Structural organisation from anti-body staining observed in electron microscopy. *J. Mol. Biol.*, **27,** 203.

Pepe, F. (1967b). The myosin filament. II. Interaction between myosin and actin filaments observed using antibody staining in fluorescent and electron microscopy. *J. Mol. Biol.*, **27**, 227.

Pepe, F. (1971). Structure of the myosin filament of striated muscle. *Prog. Biophys. Mol. Biol.*, **22**, 75.

Perry, S. V. (1967). The structure and interactions of myosin. *Prog. Biophys. Mol. Biol.*, **17**, 325.

Podolsky, R. J. (1964). The maximum sarcomere length for contraction of isolated myofibrils. *J. Physiol. (London)*, **170**, 110.

Podolsky, R. J., A. C. Nolan, and S. A. Zaveler (1969). Cross bridge properties derived from muscle isotonic velocity transients. *Proc. Natl. Acad. Sci.*, **64**, 504.

Podolsky, R. J., and L. E. Teichholz (1970). The relation between calcium and contraction kinetics in skinned muscle fibres. *J. Physiol. (London)*, **211**, 19.

Porter, K. R. (1961). The sarcoplasmic reticulum. Its recent history and present status. *J. Biophys. Biochem. Cytol., Suppl.*, **10**, 219.

Porter, K. R., and G. E. Palade (1957). Studies on the endoplasmic reticulum. III. Its form and distribution in striated muscle cells. *J. Biophys. Biochem. Cytol.*, **3**, 269.

Portzehl, H. (1957). Bewirkt das System Phosphokreatin-Phosphokinase die Erschlaffung des lebenden Muskels? *Biochim. Biophys. Acta*, **24**, 474.

Portzehl, H., G. Schramm, and H. H. Weber (1950). Aktomyosin und seine Komponenten. I. Mitt. *Z. Naturforsch.*, **5b**, 61.

Portzehl, H., P. C. Caldwell, and J. C. Rüegg (1964). The dependence of contraction and relaxation of muscle fibers from the crab *Maia squinado* on the internal concentration of free calcium ions. *Biochim. Biophys. Acta*, **79**, 581.

Pringle, J. W. S. (1949). The excitation and contraction of the flight muscles of insects. *J. Physiol. (London)*, **108**, 226.

Pringle, J. W. S. (1954). The mechanism of the myogenic rhythm of certain insect striated muscles. *J. Physiol. (London)*, **124**, 269.

Pringle, J. W. S. (1967a). The contractile mechanisms of insect fibrillar muscle. *Prog. Biophys. Mol. Biol.*, **17**, 1.

Pringle, J. W. S. (1967b). Evidence from insect fibrillar muscle about the elementary contractile process. *J. Gen. Physiol.*, **50**, 139.

Pringle, J. W. S., and R. T. Tregear (1969). Mechanical properties of insect fibrillar muscle at large amplitudes of oscillation. *Proc. Roy. Soc. (London)*, **174**, 33.

Rack, P. M. H. (1966). The behaviour of a mammalian muscle during sinusoidal stretching. *J. Physiol. (London)*, **183**, 1.

Ramsey, R. W., and S. F. Street (1940). The isometric length-tension diagram of isolated skeletal muscle fibers of the frog. *J. Cellular Comp. Physiol.*, **15**, 11.

Reedy, M. K. (1964). Discussion meeting. *Proc. Roy. Soc. (London)*, **B160**, 458.

Reedy, M. K. (1967). Cross-bridges and periods in insect flight muscle. *Am. Zoologist*, **7**, 465.

Reedy, M. K. (1968). Ultrastructure of insect flight muscle. I. Screw-sense and structural grouping in the rigor cross-bridge lattice. *J. Mol. Biol.*, **31**, 155.

Reedy, M. K., K. C. Holmes, and R. T. Tregear (1966). Induced changes in orientation of the cross-bridges of glycerinated insect flight muscle. *Nature*, **207**, 1276.

Rice, R. V., A. C. Brady, R. H. Depue, and R. E. Kelly (1966). Morphology of individual macromolecules and their ordered aggregates by electron microscopy. *Biochem. Z.*, **345**, 370.

Rice, R. V., J. A. Moses, G. M. McManus, A. C. Brady, and L. M. Blasik (1970). The organisation of contractile filaments in a mammalian smooth muscle. *J. Cell Biol.*, **47**, 183.

Ringer, S. (1883). A further contribution regarding the influence of the different constituents of the blood on the contraction of the heart. *J. Physiol. (London)*, **4**, 29.

Ritchie, J. M. (1954a). The duration of the plateau of full activity in frog muscle. *J. Physiol. (London)*, **124**, 605.

Ritchie, J. M. (1954b). The effect of nitrate on the active state of muscle. *J. Physiol. (London)*, **126**, 155.

Rüegg, J. C. (1964). Tropomyosin-paramyosin system and "prolonged contraction" in a molluscan smooth muscle. *Proc. Roy. Soc. (London)*, **B160**, 536.

Rüegg, J. C. (1967). ATP-driven oscillation of glycerol-extracted insect fibrillar muscle: Mechanochemical coupling. *Am. Zoologist*, **7,** 457.

Rüegg, J. C. (1968). Contractile mechanisms in smooth muscle. *Symp. Soc. Exptl. Biol.*, **22,** 45.

Rüegg, J. C., and R. T. Tregear (1966). Mechanical factors affecting the ATPase activity of glycerol-extracted insect fibrillar flight muscle. *Proc. Roy. Soc.* (*London*), **B165,** 497.

Sabatini, D. D., K. Bensch, and R. J. Barrnett (1963). Cytochemistry and electron microscopy. The preservation of cellular ultrastructure and enzymatic activity by aldehyde fixation. *J. Cell Biol.*, **17,** 19.

Sandow, A. (1947). Latency relaxation and a theory of muscular mechano-chemical coupling. *Ann. N.Y. Acad. Science*, **47,** 895.

Sandow, A. (1952). Excitation contraction coupling in muscular response. *Yale J. Biol. Med.*, **25,** 176.

Sandow, A. (1965). Excitation-contraction coupling in skeletal muscle. *Pharmacol. Rev.*, **17,** 265.

Sandow, A. (1970). Skeletal muscle. *Ann. Rev. Physiol.*, **32,** 87.

Schliselfeld, L. H., and M. Bárány (1968). The binding of adenosine triphosphate to myosin. *Biochemistry*, **7,** 3206.

Shanes, A. M., and C. P. Bianchi (1960). Radio-calcium release by stimulated and potassium-treated sartorius muscles of the frog. *J. Gen. Physiol.*, **43,** 481.

Shimomura, O., F. H. Johnson, and Y. Saiga (1962). Extraction, purification and properties of aequorin, a bioluminescent protein from the luminous hydromedusan, *Aequorea. J. Cellular Comp. Physiol.*, **59,** 223.

Shimomura, O., F. H. Johnson, and Y. Saiga (1963). Further data on the bioluminescent protein, aequorin. *J. Cellular Comp. Physiol.*, **62,** 1.

Sleigh, M. A. (1971). In *Cell Biology in Medicine* (E. E. Bittar, ed.), pp. 525. Wiley, New York.

Smith, D. S. (1961). Reticular organizations within the striated muscle cell. *J. Biophys. Biochem. Cytol., Suppl.*, **10,** 61.

Smith, D. S. (1966). The organisation and function of the sarcoplasmic reticulum and T-system of muscle cells. *Prog. Biophys. Mol. Biol.*, **16,** 107.

Sonnenblick, E. H. (1965). Current problems in cardiac muscle physiology. *Federation Proc.*, **24,** 1396.

Sonnenblick, E. H., and A. C. Stam (1969). Cardiac muscle: Activation and contraction. *Ann. Rev. Physiol.*, **31,** 647.

Sotavalta, O. (1953). Recordings of high wing-stroke and thoracic vibration frequency in some midges. *Biol. Bull.*, **104,** 439.

Spencer, M., and C. R. Worthington (1960). A hypothesis of contraction in striated muscle. *Nature*, **187,** 388.

Squire, J. M. (1971). General model for the structure of all myosin-containing filaments. *Nature*, **233,** 457.

Szent-Györgyi, A. (1951). *Chemistry of Muscular Contraction*. Academic, New York.

Szent-Györgyi, A. G. (1953). Meromyosins, the subunits of myosin. *Arch. Biochem. Biophys.*, **42,** 305.

Szent-Györgyi, A. G. (1960). In *Structure and Function of Muscle* (G. H. Bourne, ed.), Vol. 2, pp. 1–54. Academic, New York.

Szent-Györgyi, A. G. (1968). The role of actin-myosin interaction in contraction. *Symp. Soc. Exptl. Biol.*, **22,** 17.

Szent-Györgyi, A. G., and G. Prior (1966). Exchange of adenosine diphosphate bound to actin in superprecipitated actomyosin and contracted myofibrils. *J. Mol. Biol.*, **15,** 515.

Tanz, R. D., F. Kavaler, and J. Roberts (eds.) (1967). *Factors Influencing Myocardial Contractility*. Academic, New York.

Thorson, J., and D. C. S. White (1969). Distributed representations for actin-myosin interaction in the oscillatory contraction of muscle. *Biophys. J.*, **9,** 360.

Tiegs, O. W. (1955). The flight muscle of insects—Their anatomy and histology with some observations on the structure of striated muscle in general. *Phil. Trans. Roy. Soc.* (*London*), **B238,** 221.

Trayer, I. P., and S. V. Perry (1966). The myosin of developing skeletal muscle. *Biochem. Z.*, **345,** 87.

Twarog, B. M. (1967a). The regulation of catch in molluscan muscle. *J. Gen. Physiol.*, **50,** 157.

Twarog, B. M. (1967b). Factors influencing contraction and catch in *Mytilus* smooth muscle. *J. Physiol.* (*London*), **192,** 847.

Veratti, E. (1902). English translation: Investigations on the fine structure of striated muscle fiber. (1961). *J. Biophys. Biochem. Cytol.*, *Suppl.*, **10,** 3.

Watanabe, S., and W. Sleator (1957). EDTA relaxation of glycerol-treated muscle fibers and the effects of magnesium, calcium and manganese ions. *Arch. Biochem. Biophys.*, **68,** 81.

Weber, A. (1966). Energized calcium transport and relaxing factors. *Curr. Top. Bioenerg.*, **1,** 203.

Weber, A., and R. Herz (1963). The binding of calcium to actomyosin systems in relation to their biological activity. *J. Biol. Chem.*, **238,** 599.

Weber, A., and S. Winicur (1961). The role of calcium in the superprecipitation of actomyosin. *J. Biol. Chem.*, **236,** 3198.

Weber, A., R. Herz, and I. Reiss (1963). On the mechanism of the relaxing effect of fragmented sarcoplasmic reticulum. *J. Gen. Physiol.*, **46,** 679.

Weber, A., R. Herz, and I. Reiss (1964). The regulation of myofibrillar activity by calcium. *Proc. Roy. Soc.* (*London*), **B160,** 489.

Weber, A., R. Herz, and I. Reiss (1966). Study of kinetics of calcium transport by isolated fragmented sarcoplasmic reticulum. *Biochem. Z.*, **345,** 329.

Weber, H. H. (1952). Is the contracting muscle in a new elastic equilibrium? *Proc. Roy. Soc.* (*London*), **B139,** 512.

Weeds, A. G., and S. Lowey (1971). Substructure of the myosin molecule. *J. Mol. Biol.*, **61,** 701.

Wilkie, D. R. (1956a). The mechanical properties of muscle. *Brit. Med. Bull.*, **12,** 177.

Wilkie, D. R. (1956b). Measurement of the series elastic component at various times during a single twitch. *J. Physiol.* (*London*), **134,** 527.

Winegrad, S. (1965a). Autoradiographic studies of intracellular calcium in frog skeletal muscle. *J. Gen. Physiol.*, **48,** 455.

Winegrad, S. (1965b). The location of muscle calcium with respect to the myofibrils. *J. Gen. Physiol.*, **48,** 997.

Woledge, R. C. (1968). The energetics of tortoise muscle. *J. Physiol.* (*London*), **197,** 685.

Woledge, R. C. (1971). Heat and chemical change in muscle. *Prog. Biophys. Mol. Biol.*, **22,** 37.

Woodward, A. A. (1949). The release of radioactive ^{45}Ca from muscle during stimulation. *Biol. Bull.*, **97,** 264.

Worthington, C. R. (1961). X-ray diffraction studies on the large-scale molecular structure of insect muscle. *J. Mol. Biol.*, **3,** 618.

16

CELL MOTILITY

Michael A. Sleigh

1. INTRODUCTION

Motility is one of the fundamental attributes of living things. Movements of cells and movements within cells take many characteristic forms in addition to those associated with the contraction of muscle, which has been described in Chapter 15. The commonly recognized forms of movement fall into two classes, one in which there is a change in shape of fibrous structures visible in the light microscope, and a second in which the ground substance of the cell, together with its particulate inclusions, shows streaming movements that are much more extensive than Brownian motion, yet cannot be seen to involve a change in shape of fibrous structures. Both classes may be further subdivided into two sections; streaming of cytoplasm may occur over the surfaces of structures that are more-or-less permanent, or cytoplasmic flow may be associated with changes in shape of a more fluid cell structure; fibrous structures may change in shape without change in length of the fiber system, or the fiber system may shorten or lengthen during the change in shape.

Movement within cells also occurs at the level of changes in membrane configuration, for example, in association with the formation of vacuoles in phagocytosis and pinocytosis, as well as in the formation of vacuoles around Golgi complexes and changes in the nuclear envelope before and after nuclear division. Mechanisms involved here may include growth and probably molecular rearrangements in the extension or retraction of existing membranes; fibrous elements, comparable to those shown in Figure 1, may be present in the membrane or attached to it and may alter the configuration of membranous structures. Such movements are not well known and receive little further discussion. Movements that are not visible with the light microscope but are only inferred from other evidence are not described here; information on the movement of substances through membranes is found in Chapter 3.

Cell cytoplasm is a fluid containing fibrous structures, cell organelles, and particles up to the size of membranous vacuoles. This fluid may have a low viscosity and flow easily, when it is said to be in the sol state; or it may become stiffer and more jellylike with presumably a change in chemical organization and water content, and it is then said to be in the gel state. Fibrous structures are commonly found in regions having the characteristics of a gel, but in many cases the elementary fibrils of a fibrous system may still be surrounded by a fluid allowing more-or-less free movement of the fibrils relative to each other.

Fibrous structures formed by the polymerization of protein molecules are normally found in cytoplasm either in the form of solid microfilaments (commonly 2–10 nm in diameter), or as cylindrical fibrils (commonly approximately 20 nm in diameter); the latter are called microtubules because in transverse section they are seen to have a dense wall surrounding a less dense center. Microtubules and many intracellular microfilaments appear to be made up of globular protein molecules. In microfilaments the globular molecules are polymerized in linear arrays (Fig. 1*a*), several of which may be twisted together to form a thicker filament—thus the actin filament of striated muscle is commonly made up of two chains of globular actin molecules (Fig. 1*b*). In the microtubule 12 or 13 longitudinal rows of globular protein molecules form the wall of the cylindrical fibril (Fig. 1*c*). The structural properties of these two fibrous elements are clearly different; the microfilament can resist extension but has little resistance to longitudinal compression or lateral distortion, while the microtubule has greater resistance to longitudinal compression and lateral distortion as well as resistance to extension. Microtubular fibrils may be very long indeed, extending to 1 mm or more in length, and certainly have "skeletal" functions in some cases at least.

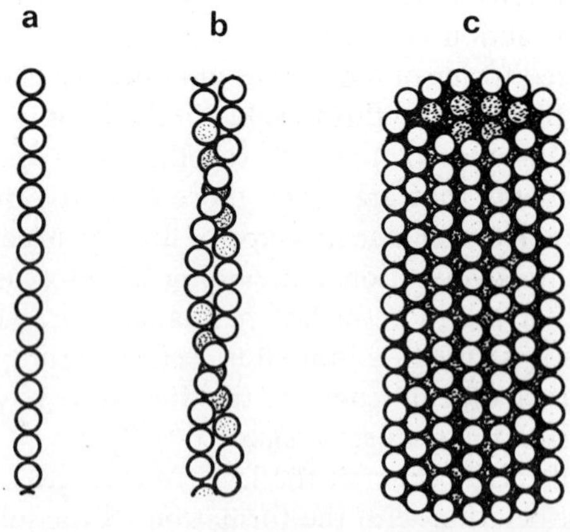

Figure 1. The arrangement of globular protein molecules in (*a*) a single microfilament, (*b*) the two-stranded filament of muscle actin (Davies, 1963), and (*c*) the microtubule (Ringo, 1967).

In this account of cell motility, a range of examples covering various forms of movement within cells is described. Wherever possible, the description of observed movements is correlated with ultrastructural evidence of the presence of fibrous structures and with any available information on the biochemical characteristics and mechanism of the movement.

2. CYTOPLASMIC STREAMING

A. Movements of Cytoplasm without Cell Locomotion

Although electron micrographs give the impression that cells contain a wealth of permanent structures, the cytoplasm of most cells is involved in more-or-less active continuous movement. In many motile cells it is not easy to distinguish cytoplasmic movements produced in the cytoplasm by changes in cell shape from movements of cytoplasm that have their origin within the cytoplasm itself. In plant cells with rigid walls and in other cells with firm pellicles, the active cyclosis of cytoplasm and the streaming or rotation of cytoplasmic organelles are undoubtedly intrinsic. Cytoplasmic streaming in a variety of animal cells has rather similar characteristics and may be used for specific purposes, such as food intake and distribution in some protozoan cells, or the dispersal and concentration of pigment in certain melanophore cells. Rebhun (1967) has reviewed evidence concerning these movements and suggests that they are caused by the generation of forces between fiber structures and other cytoplasmic components.

1. Cytoplasmic Flow Associated with Bundles of Microfilaments. The most intensively studied example of cyclosis in plant cells is that in *Nitella* and related members of the algal family Characeae. In the large cylindrical cells of the internodes and rhizoids, the layer of cytoplasm that lines the wall and surrounds the vacuole (Fig. 2) is subdivided into an outer gel ectoplasmic layer which contains the chloroplasts in the internode cell, and an inner sol endoplasmic layer which streams actively around the cells at speeds of up to 50 or 60 μ/sec. If parts of the ectoplasmic layer are damaged by mild centrifugation, the movement ceases in the damaged regions of the cell but not in the intact parts, so that cytoplasm may accumulate in the damaged regions; re-formation of the ectoplasmic layer is accompanied by the reappearance of stream-

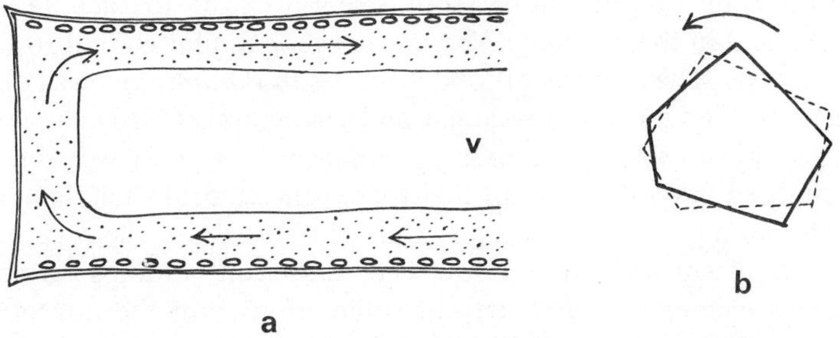

Figure 2. (*a*) Diagram of a portion of an internode cell of *Nitella;* the side walls of the cell are lined with fixed chloroplasts over which the cytoplasm streams around the central vacuole (v). (*b*) A rotating polygonal loop of fibrils as seen in cytoplasm obtained from a cell of *Nitella*.

ing (Hayashi, 1964). It is concluded that the motive force for streaming of cytoplasm is generated at the boundary between ectoplasm and endoplasm, and electron micrographs obtained by Nagai and Rebhun (1966) show bundles of microfilaments in this position. These bundles, which were many microns long and contained 50 to 100 microfilaments each about 5 nm in diameter, were oriented with their long axes parallel to the direction of the cytoplasmic streaming.

Cytoplasm removed from these cells is found to contain fibrils formed into polygonal loops which rotate in the cytoplasmic drops (Kuroda, 1964) (Fig. 2*b*). In centrifuged cells of *Nitella*, Kamitsubo (1966a, b) observed the formation of these polygons from linear fixed fibrils in the position of those described by Nagai and Rebhun. He also found that the loops were of two types—naked fibrils that rotated freely at a speed of about 50 μ/sec (15°C) and fibrils coated with cytoplasmic granules that propagated bending waves around the loop at about 10 μ/sec (15°C). Kamitsubo reported that granules could be seen moving along the fixed fibrils in the direction of cytoplasmic streaming, and that cytoplasmic streaming was rapid along the fibrils and slow or absent between the fibrils. There is substantial evidence that the movement of cytoplasm results from forces generated between free cytoplasmic components and fixed microfilamentous fibrils. Normally, most chloroplasts in these cells are fixed in the ectoplasmic zone, but detached chloroplasts and nuclei rotate vigorously and are probably associated with fragments of the fibrous system.

Streaming movement of cytoplasmic particles is also a common feature of the cells of higher plants. In cytoplasm of cells of the oat coleoptile, O'Brien and Thimann (1966) found long fibrous bundles, approximately 0.1–0.2 μ thick, composed of microfilaments 5–7 nm in diameter. The orientation of these bundles corresponded with the main flow of cytoplasm around the cell. Streaming movements were also vigorous in strands of cytoplasm that crossed the vacuole of these cells, and these strands were seen in electron micrographs to contain well-developed bundles of longitudinal filaments. Transvacuolar strands of cells of *Nicotiana* and *Euphorbia*, studied by Mahlberg (1964), were commonly less than 1 μ in diameter and contained granular particles that flowed at speeds of up to 19 μ/sec (24–25°C). The rate of streaming was not dependent upon the diameter of the strand, and sometimes streams of particles were seen to flow in opposite directions in the same strand, in a manner comparable to the motion of granules in pseudopods of Foraminifera described in Section 2.A.2. In many plant cells large organelles including chloroplasts are carried around the cell by cytoplasmic streaming; this can easily be demonstrated in well-illuminated leaves of *Elodea*.

Comparable movements have been described in animal cells, particularly in protozoa. The motion of granules along strands of cytoplasm in vacuolated regions of radiolarian cells and in the flagellate *Noctiluca* is very similar to that in transvacuolar strands of plant cells. The cyclosis of food vacuoles in *Paramecium* was observed more than a century ago, and a recent investigation by Koenuma (1963) has shown that the endoplasmic cytoplasm streams over the ectoplasmic layer in very much the same manner as described in *Nitella*, except that there is no central vacuole in this cell. The cytoplasmic streaming is fastest, at approximately 5 μ/sec, in the region of the interface between ectoplasm and endoplasm. The ectoplasm contains microtubular and microfilamentous elements in a variety of orientations, but the microtubular fibrils are mostly too superficial in the pellicle to be near the interface with the endoplasm.

2. Cytoplasmic Flow Associated with Microtubular Fibrils. Streaming movements of plant cytoplasm are less commonly associated with the presence of microtubular fibrils, but in the coenocytic alga *Caulerpa*, Sabnis and Jacobs (1967) found large bundles of microtubules in cytoplasm in which active streaming movements at 3–5 μ/sec were observed. Similar bundles of microtubules were found by MacGregor and Stebbings (1970) in the syncytial trophic tubes of the ovary of the insect *Notonecta*, in which approximately 30,000 microtubules occur within a tube 15 μ in diameter; food materials and ribosomal RNA travel down these tubes to the developing oocyte.

The feeding activities of some ciliate protozoa involve an intake of food material through a tube of fibers. Thus the tentacle of such suctorian ciliates as *Acineta* and *Tokophrya* contains a cylinder of microtubular fibrils in a characteristic array, which divides the tentacle into two concentric tubes (Fig. 3); particles may be seen to pass outward in the outer tube, while food vacuoles taken into the body at the tentacle tip pass inward in the central tube (Rudzinska, 1965; Bardele and Grell, 1967). It seems reasonable to assume that the fibrous cylinder is involved in the active uptake of food, although it probably also serves as the skeletal support of the extended tentacle. A more complex cylindrical structure formed largely from microtubular elements was described in the cytopharyngeal basket of *Nassula* by Tucker (1968) (Figs. 4 and 5). In this case algal filaments were drawn in through the basket at speeds of 7–20 μ/sec, but their active motion ceased once the filament lost contact with the walls of the basket. Tucker suggested that cytoplasmic streaming through the basket might draw the filament along, but it was not possible to confirm this because particles could not be seen near the refractile basket structures.

Color change in teleost fish is based upon cytoplasmic streaming. The melanophore cells of *Fundulus* are extended into many slender processes within which the melanin pigment granules migrate outward to darken the skin of the fish, or migrate toward the cell center to concentrate the pigment and lighten the skin (Fig. 6). The channels in which the melanin granules migrate along the cell processes were found by Bikle et al. (1966) to be surrounded by zones of cytoplasm containing numerous longitudinally oriented microtubules. These microtubules were present in the walls of the channels in cell processes when the pigment was concentrated as well as when it was dispersed.

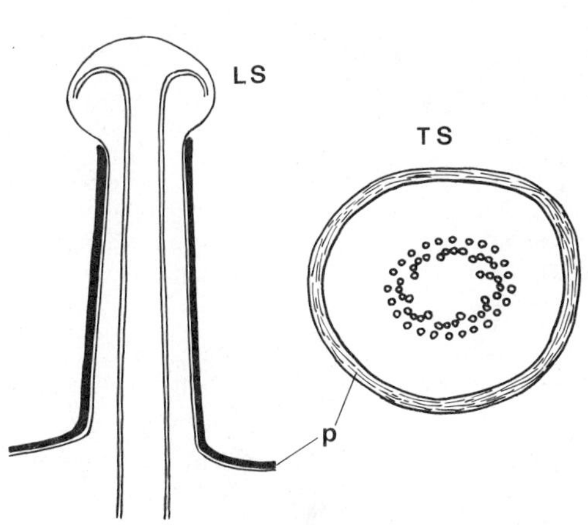

Figure 3. Diagrams of longitudinal (LS) and transverse (TS) sections of a suctorian tentacle. The ciliates upon which these organisms prey adhere to the head of the tentacle, and cytoplasm from the prey is carried in through the cylinder of microtubular fibrils. There is a pellicle (p) around most of the body.

Figure 4. Electron micrograph of a cross section through the pharyngeal basket of *Nassula*. The microtubular bundles form the walls of a cylinder approximately 60 μ long and 15 μ wide, held together at the level shown here by an annulus of dense material linking the bundles. Algal filaments are drawn through the center of the cylinder. The scale represents 1μ. (From Tucker, 1968.)

The microtubules may provide support to maintain the shape of the cell processes, and they could also be associated with provision of the motive force for pigment migration. Kinosita (1963) reported that the speed of migration of pigment granules in melanophores of *Oryzias* ranged between 10 and 40 μ/min and was very uniform.

The axopod of the heliozoan ameba *Actinosphaerium* (=*Echinosphaerium*) provides an example of a structure in which cytoplasm streams around the outside of an array of microtubules rather than within a cylinder of these fibrils. Along the center of the axopod is a highly organized system of longitudinal microtubules (see Section

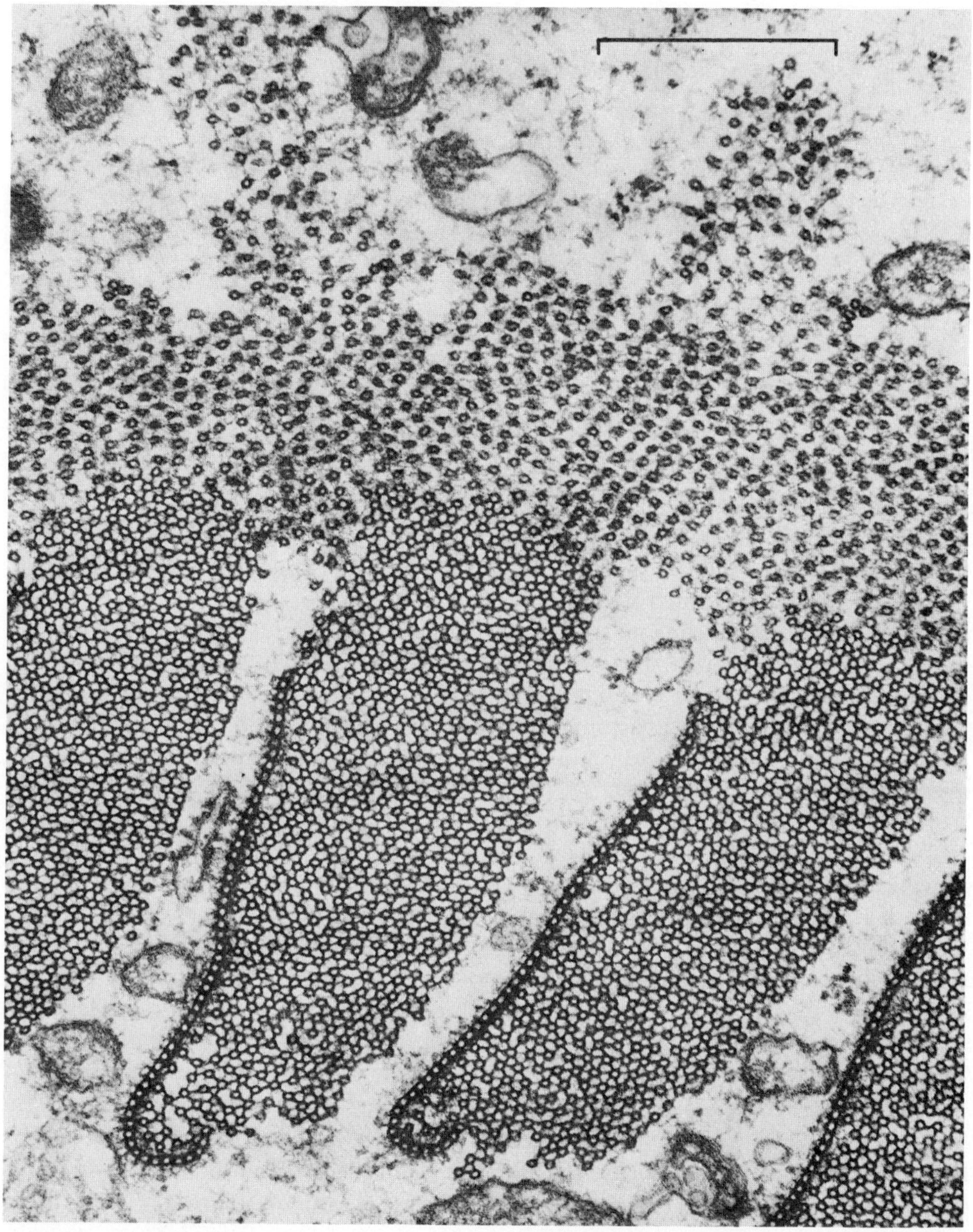

Figure 5. Electron micrograph of part of a cross section through the pharyngeal basket of *Nassula* at a level different from that shown in Fig. 4. At least four different patterns of microtubule packing are seen in different parts of the micrograph; the different regions presumably have different functions. The inside of the basket is at the bottom of the micrograph. The scale represents 0.5 μ. (From Tucker, 1968.)

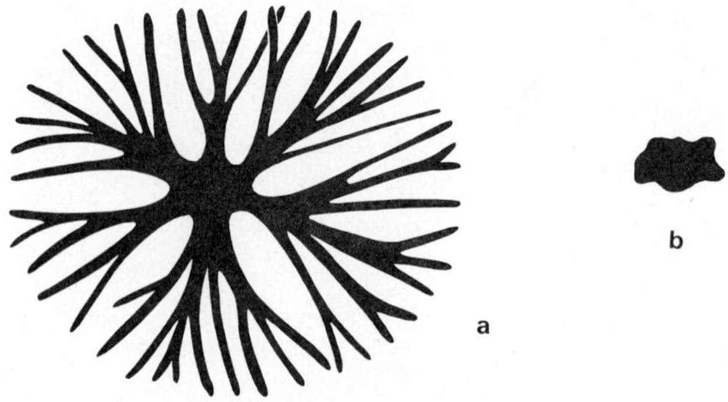

Figure 6. Diagram showing the pigment distribution of a melanophore cell in expanded (*a*) and contracted (*b*) states.

3.A.1 and Figs. 7 and 11), and this is surrounded by a layer of cytoplasm containing mitochondria and other small organelles. Streams of cytoplasm move longitudinally up and down the axopod at speeds of up to 20 μ/sec, with the most rapid movement immediately around the microtubular axis (Jahn and Bovee, 1969), although particles at a considerable distance from the fibrous axis can also be seen to move.

The slender pseudopods of foraminiferans show comparable streaming of particles. These pseudopods extend from the shell of the protozoan as a reticulum of fine threads which branch and anastomose (Fig. 10*c*). The finest branches of the network are only about 1 μ in diameter, and examination shows that particles of the size of mitochondria and perhaps bacterial food inclusions stream in both directions along the very thin strands at speeds of up to 10 or 20 μ/sec. Where the threads anastomose to form larger branches of the network, it is possible to see several streams of particles moving in either direction, so that it is assumed that the single-thread units remain more-or-less intact. In addition to the movements of granules along the pseudopods, there is also continuous extension and retraction of the individual pseudopods so that the shape and interconnections of the network are changing all the time. While changes in shape of pseudopods may involve cytoplasmic streaming to transport materials, the streaming movements are still continuous in stationary pseudopods. These pseudopods have proved extremely difficult to fix for electron microscopy, but micrographs obtained by McGee-Russell and Allen (1969), by using a stabilization procedure prior to fixation, showed that within the cell membrane of the pseudopod there were numerous longitudinal microtubular fibrils, often arranged in one or more bundles, as well as some microfilaments. According to the hypothesis of Jahn and Rinaldi (1959), each fine pseudopod is constructed of a thin strand of cytoplasm looped back on itself (Fig. 8*a*), running as a conveyor belt with its attached particles and generating motive force along its length where the two limbs of the strand, moving in opposite directions, lie in close contact. Allen (1964) has noted that the movement is not as regular as Jahn and Rinaldi suggest. Perhaps it might be more realistic to suggest that the cytoplasm with its particulate contents flows along the surfaces of the bundles of microtubular fibrils (Fig. 8*b*), which could have the dual skeletal functions of maintaining the shape of the pseudopod and resistance to forces exerted in the propulsion of cytoplasm. If this is true, the functions of microtubular fibrils in the foraminiferan pseudo-

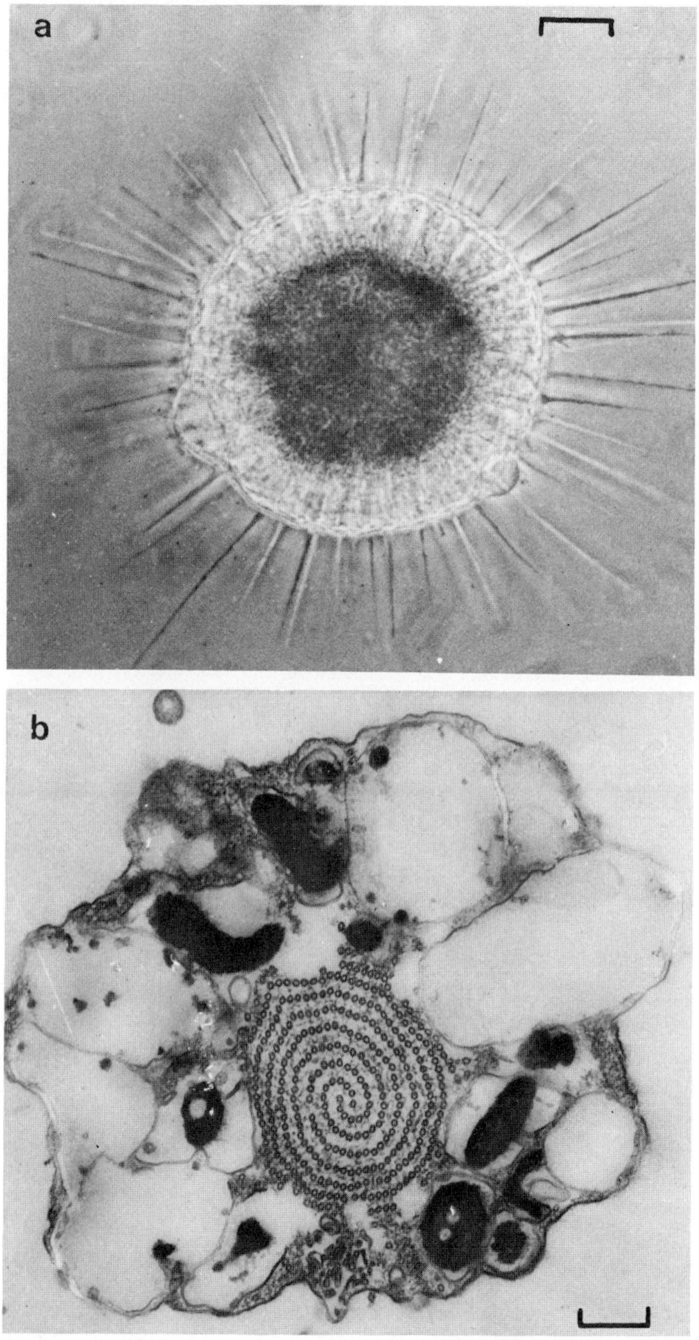

Figure 7. The heliozoan *Actinosphaerium*. (*a*) Micrograph of the live organism. The scale represents 100 μ. (*b*) Electron micrograph of a cross section through an axopod of *Actinosphaerium* showing the central microtubules and peripheral cytoplasm. The scale represents 0.2 μ. (Micrograph by A. C. Macdonald.)

a

b

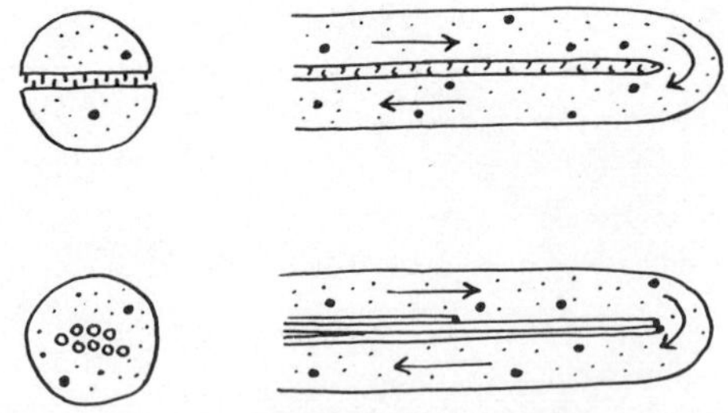

Figure 8. Diagrams illustrating hypotheses of the mechanism of cytoplasmic streaming in foraminiferans discussed in the text. (*a*) Transverse and longitudinal sections of a pseudopod according to the "conveyor belt" theory in which shearing forces are assumed to be exerted between the two limbs of the strand in the position of the short, hooked lines. (After Jahn and Rinaldi, 1959.) (*b*) Streaming of cytoplasm over the microtubular fibrils of a pseudopod.

pod have features similar to those described for tentacles of suctoria, processes of melanophore cells, and axopods of heliozoa (as far as they have been discussed).

Neurons in tissue culture display features similar to the foraminiferan pseudopodial system in the extension and retraction of fine branches and in the motility of particles in streams within the pseudopodial branches (Nakai, 1964). Electron micrographs of neurons reveal the presence of longitudinal microtubules and microfilaments, and in the region of the origin of the axon from the cell body there are loose bundles of microtubules with cross-links between adjacent fibrils (Palay et al., 1968); Bertolini et al. (1970) found regular arrays of 9 or 10 microfilaments around neuronal microtubules. The streaming of cytoplasm in the neurons, and therefore presumably around the microtubules, can be seen in the movement of particles. It has been reported, for example, that granules in hypothalamic neurons may travel 2 mm in a minute or less, that is, at a speed of at least 30 μ/sec (Jasinski et al., 1966). These streaming movements of cytoplasm in neurons are important because of the great distances over which metabolites and other materials must be transported within nerve cells (Schmitt, 1968).

There seems to be sufficient evidence from these examples to show that microtubular fibrils may be associated with cytoplasmic streaming in situations in which they also play a skeletal role. The form of molecular activity that generates the forces for streaming is not known, and it remains a possibility that the microtubules provide the more obvious structural elements of these systems while motive forces are supplied by small numbers of microfilaments that may be easily overlooked in the presence of the larger microtubules.

B. Movements of Cytoplasm Resulting in Locomotion

This form of cell motility is broadly described as "ameboid movement," and is probably most familiar in the locomotion of *Amoeba proteus* and related protozoa. In these forms there is an ectoplasmic layer surrounding more fluid endoplasm, and changes in shape of the ectoplasmic envelope are associated with the flow of endoplasm from one part of the body to another. Locomotion of this general type, but often with important differences in the details of movement, is found in various cells having an ameboid body form, including many embryonic cells and cells in tissue

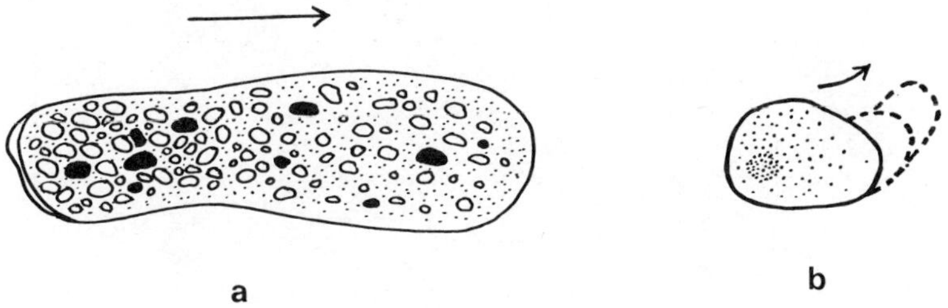

Figure 9. Diagrams showing two amebas in motion. (*a*) The body form of *Pelomyxa palustris* (length approximately 2000 μ). (*b*) The movement of an "eruptive" ameba showing the outlines of two spurts in the forward movement (length approximately 20 μ).

culture as well as such independent cells as macrophages. The flowing of cytoplasm in the plasmodial slime molds has similar characteristics and is described in Section 2.B.2. In spite of the wide range in the form of these movements, suggestions have been made that a common biochemical mechanism is involved; it is conceivable that the same components may be used in different ways to propel the cytoplasm.

1. Ameboid Movement. The large herbivorous ameba *Pelomyxa palustris* has a simple saclike body form (Fig. 9*a*) (Griffin, 1964). Numerous inclusions are present in the endoplasmic region, while the ectoplasm is more homogeneous. As the ameba moves, the endoplasm flows forward through a tube of ectoplasm, and when the endoplasm comes to the surface at the anterior end it gelates and transforms to ectoplasm, extending the tube forward. At the rear end of the body, ectoplasmic materials are solated, pass into the endoplasm, and move forward in the endoplasmic flow. These transformations and the flow of endoplasm result in movement of the ameba because of occasional points of attachment of the ectoplasmic tube to the substratum. The hyaline cap, which is an accumulation of water between the ectoplasmic surface and the cell membrane, is formed at the rear end of this ameba and is believed to originate from the syneresis of water during contraction of the cytoplasm. It is therefore suggested that in this ameba a posterior contraction, perhaps involving ectoplasmic proteins immediately prior to or in the act of solation, provides a force which is transmitted hydraulically through the endoplasm, driving it forward through the ectoplasmic sleeve.

In several smaller amebas a similar contraction-hydraulic mechanism is probably also an adequate explanation for the movement, although in some of these cases the movement is no longer a smooth-flowing one but a series of sudden eruptions in each of which a bulge of cytoplasm bursts out to form a new pseudopod or extend the existing one (Fig. 9*b*). One of these smaller amebas, *Acanthamoeba*, has been the subject of biochemical investigations on fibrous proteins. Weihing and Korn (1969) extracted a protein with a molecular weight of about 55,000, which was shown by analysis to have an amino acid composition similar to that of the actin in muscle (see Section 4). *Acanthamoeba* actin can be isolated in the G form as globular molecules and is transformed to the fibrous F-form in 0.1 *M* potassium chloride. In negatively stained preparations the F-actin from *Acanthamoeba* appears as filaments about 6 nm in diameter and up to 3 μ long and has a double-helical substructure. When a suspension of these actin filaments is mixed with purified myosin from rabbit skeletal muscle, a great increase in viscosity occurs which disappears on addition of

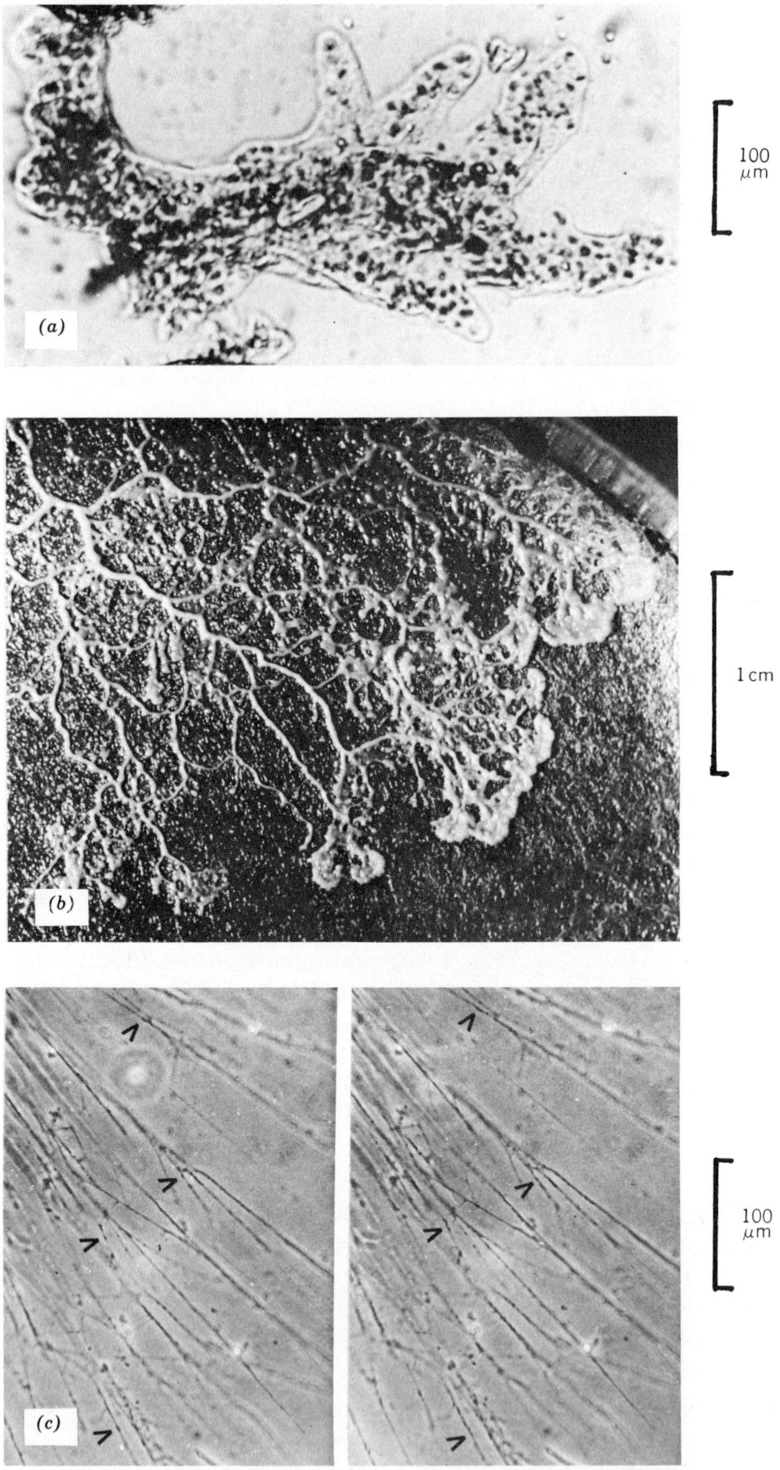

Figure 10. (*a*) Micrograph of a specimen of *A. proteus* in motion showing several active pseudopods. (*b*) A portion of the plasmodium of the myxomycete *Badhaemia*. (*c*) The reticulopodial system of a miliolid foraminiferan. Two frames from a cinefilm with an interval between them of about 1 sec. Particles on various branches of the network have moved during this time; some of these are indicated by the arrows.

ATP. The attachment of heavy meromyosin particles from rabbit muscle to *Acanthamoeba* actin filaments in an arrowhead array can be demonstrated in electron micrographs, and the formation of the arrowhead pattern is blocked by Mg-ATP or Mg-pyrophosphate (Pollard et al., 1970). These amebas therefore contain proteins of the actin type with properties very similar to those of muscle actin; it seems reasonable to suggest that filaments of an actin protein are involved in the movement of *Acanthamoeba*, although the protein could conceivably perform some other function.

The large carnivorous amebas *A. proteus* and *Chaos carolinensis* have several pseudopods which show some degree of independence in movement (Fig. 10*a*). Each pseudopod consists of an ectoplasmic tube within which there is a forward flow of endoplasm at speeds of up to 30 μ/sec; this endoplasm transforms to ectoplasm near the advancing tip of the pseudopod. The transformation of ectoplasm to endoplasm takes place mainly at a wrinkled uroid region at the rear of the organism, although with the frequent extension and retraction of pseudopods in these forms some solation of ectoplasm is likely to take place in any region of the body that is retracting. The finding of Goldacre and Lorch (1950) that forward movement of *Amoeba* is accelerated by an injection of ATP in the tail region suggests that a tail-contraction mechanism may be involved. However, several observations, for example, that movements of endoplasm normally begin at the pseudopod tip, that endoplasmic flow at the tip of a pseudopod may continue at the same time as withdrawal of endoplasm from the rear of the pseudopod, that stimuli applied to the tip of the pseudopod may modify only the movement in the tip region, and that hyaline caps are found at the tips of advancing pseudopods, suggest that the tip is closely involved in the movement; indeed, it seems probable that the site of application of the motive force is near the tip of the pseudopod in these forms. Allen (1968) has put forward further evidence in favor of such frontal contractions in which forces generated during the endoplasm–ectoplasm transition are presumed to result in the endoplasm being drawn forward through the ectoplasmic tube. The streaming of ameboid cytoplasm in the absence of tail contraction was strikingly demonstrated by Allen et al. (1960), who studied the movement of "skinned" cytoplasm of *C. carolinensis* in narrow quartz tubes.

More recently, attention has turned to the study of isolated fractions of *A. proteus*. This work was initiated by Wolpert et al. (1964), who concluded that the cell membrane played a passive role in locomotion, so that the origin of forces for locomotion should be sought in the cytoplasm. Cytoplasmic extracts prepared by cooling, followed by centrifugation to remove nuclei, food vacuoles, crystals, and most of the surface membranes, showed no traces of active streaming, but on addition of ATP movements of cytoplasm began. These movements took the form of cytoplasmic streams flowing over considerable distances, sometimes as fast as 80 μ/sec, or in some cases a large area of the extracted cytoplasm appeared to gel into a sheet which contracted at up to 160 μ/sec to form a small mass, squeezing out more fluid materials. The extracted cytoplasm showed the following four properties when treated with ATP: gelation, cytoplasmic streaming, contraction with the ability to transmit tension, and syneresis. The gel regions of an active preparation contained fibrils.

In an extension of this study, Pollard and Ito (1970) found that extracted cytoplasm maintained at 0°C was nonmotile and contained filaments 16 nm thick and occasionally some smaller filaments about 7 nm thick. On warming to 22°C movement began, and the viscosity increased markedly with the appearance in the cytoplasm of numerous 7-nm filaments which later aggregated into bundles visible in the light microscope as birefringent fibers. The movement was accelerated by the addition of ATP, and the number of filaments was also increased. In the absence of 16-nm filaments, the 7-nm filaments still formed and aggregated, but the cytoplasm did not move, even after the addition of ATP. The thin filaments were of indefinite length and were probably about 5 nm thick, aggregating to form 7-nm filaments; these filaments were cold-labile, and complete polymerization to form filaments, presumably from precursors in the cytoplasm, required heat and ATP. Aggregation of these filaments into larger fibers was stimulated by the presence of such calcium chelators as EGTA or EDTA. These thin filaments were similar to muscle actin in size and shape, and specifically bound purified heavy meromyosin to form arrowhead arrays on the filaments; such arrays could be dissociated with Mg-ATP. The thicker (16-nm) filaments were about 0.5 μ long; they were stable between 0 and 22°C and were present in motile and nonmotile amebas as well as in extracted cytoplasm. These filaments were similar to myosin in muscle, having a filamentous substructure and apparently lateral projections. It was concluded that the 16-nm filaments are always present, and that the 5- to 7-nm filaments form on gelation of the cytoplasm and interact with the thicker filaments to cause movement. Any gel region may therefore be capable of contraction in the presence of ATP, and either frontal contraction or rear contraction, or a mixture of both, may control the flow of cytoplasm in the ameba. The existence of regions of different consistency in the endoplasmic zone of *Amoeba* or *Chaos*, resulting in localized streamlets of more fluid cytoplasm flowing between stiffer stationary regions, raises the possibility that these amebas may have a spongy texture, with contractile activity spread through the more gelated parts of the endoplasm as well as the ectoplasm.

The movement of mammalian fibroblasts in culture appears to be basically similar to that of some amebas. In detail it may be rather different, however, because Ambrose and Forrester (1968) have described undulations of the cell surface that pass backward from the advancing margin of the cell. It is not clear how these waves may cause locomotion; one might assume them to be analogous, on a quite different scale, to the waves passing over the foot of a snail as it crawls, although in fact the waves on these larger animals move forward over the surface of the body and cause forward locomotion in a manner explained by Lissmann (1945). However, Wolpert and Gingell (1968) concluded that the cell surface does not provide the force for ameboid movements, although it may be deformed by cytoplasmic forces. A significant difference between the movement of these cells and those of the larger amebas is that in tissue cells pseudopodial extension and attachment to the substrate is followd by retraction, drawing the body of the cell toward the attachment of the pseudopod in a manner similar to that described for *Difflugia* (see Section 3.A.2). The presence of actinlike filaments with the ability to form arrowhead patterns with heavy meromyosin has been reported in leukocytes (Senda et al., 1969), and in fibroblasts and a variety of other cells (Ishikawa et al., 1969).

2. Streaming in Myxomycete Plasmodia. In the plasmodial slime molds such as *Physarum*, there is a network of tubular channels (Fig. 10*b*) having some superficial similarities to the foraminiferan pseudopodial reticulum but with differences in size and structure and probably in mechanism of movement. Each strand of the plasmodium of *Physarum* is comparable to a section of a pseudopod of *Chaos;* it may be 1 mm or more in diameter with an ectoplasmic gel layer around the endoplasm which may flow at speeds of up to 1350 μ/sec (Kamiya, 1959). The advancing fan-shaped fringe of the plasmodium has a somewhat more complex organization of the same components as the tubular network, and the whole syncytial network is enclosed by a single membrane. The flow of endoplasm in any tubular channel periodically reverses, and the direction and speed of flow was shown by Kamiya and Kuroda (1958) to depend on the difference in pressure between the ends of that length of plasmodial tube. Within the network of the plasmodium, there are some regions that contract, generating pressure, and whose ectoplasmic walls may become thinner as the ectoplasm solates, while in other expanding regions the reverse transformation may take place and the ectoplasmic layer of the wall thickens.

There is evidence for a cyclical oscillation of contraction and relaxation of elements in the ectoplasm of the plasmodial tubes. Kamiya (1968) found that normally these elements appeared to constrict the tube, but when the tube was subjected to longitudinal tension the contractile elements gradually reorientated themselves in a longitudinal direction and continued their periodic contractions. These contractions also occurred when glycerine-extracted preparations of both intact plasmodia and strands under longitudinal tension were placed in appropriate solutions containing ATP. The walls of the tubes contained fibrillar material which could be seen in the electron microscope as bundles of numerous filaments, each about 7 nm in diameter. In ATP-treated glycerinated preparations, the fiber bundles appeared to be more compact and more closely aggregated than in those not treated with ATP.

Fibrillar material has been extracted from plasmodia of *Physarum* by Hatano and Oosawa (1966). They found that the protein component of the filaments was an actin, which they called plasmodial actin, whose amino acid composition was similar to that of actin from skeletal muscle [and similar to that of *Acanthamoeba* actin isolated by Weihing and Korn (1969); see Section 4], and whose properties are also similar to those of muscle actin, for example, in the ability to bind ATP. Filaments extracted from *Physarum* by Nachmais et al. (1970) were up to 1.2 μ long and approximately 5 nm thick. As extracted, they carried "whiskers" of attached material which was released upon treatment with Mg-ATP to leave unbranched beaded filaments. In the presence of the heavy meromyosin component of rabbit muscle myosin, the ATP-treated filaments became covered with a continuous array of projections to form the characteristic arrowhead pattern, displaying a structural polarity in the interaction between actin filaments and myosin. The myosin component in *Physarum* probably exists as single molecules which appear to have an elongate shape similar to that of the muscle myosin molecule; the whiskers on extracted actin filaments were probably molecules of plasmodial myosin.

The correlation of fibril formation with the production of motive force in *Physarum* seems to be confirmed (Kamiya, 1968). It is less clear how the fibrils actually cause endoplasmic movements. One possibility is that the actin filaments slide on one another because of forces generated through bridging myosin molecules, and that filament bundles contract and exert pressure on the endoplasm. Alternatively, it is possible that the matrix material around the filaments slides along the bundles of fibrils to generate the endoplasmic flow.

3. MOVEMENT INVOLVING CHANGE IN SHAPE OF FIBER SYSTEMS

Many cells contain organelles that are visible in the light microscope and have a fibrous substructure which can be shown by electron microscopy to be based on microfilaments and microtubules. The fibrous nature of most of these organelles can also be inferred from the characteristic birefringence observed in them in the polarizing microscope. Some of these fibrous structures have skeletal functions and appear to be incapable of change in shape or production of movement. Other fibrous structures change in shape to deform the organelles and parts of the cell in which they occur; such deformations may cause local movements or locomotion of the cell. In some cases deformation takes the form of bending or coiling of the organelle in such a manner that the length of the fibrous structure does not change visibly. In other cases the deformation of the organelle involves a visible change in length of the fibrous structure, with the result that different parts of the cell connected by the fibers can be seen in the light microscope to move closer together or further apart. With further knowledge it may be possible to state that some movements of the latter type result from the sliding of fibrils of constant length, as in the case of the contraction of myofibrils of striated muscle, but it seems certain that some of these fiber systems shorten in other ways.

A. Fiber Systems That Cause Movement by Change in Length

Since the form and probably the mechanism of change in length of fibers composed of microtubules differs from that of fibers composed of microfilaments, movements caused by organelles with these different substructures are discussed under separate subheadings. It appears that microtubules normally extend for the full length of an organelle, while microfilamentous structures are thought to contain short, overlapping filaments.

1. Systems in Which the Main Elements are Microtubular. Large heliozoan amebas of the family Actinophryidae (e.g., *Actinophrys*, *Actinosphaerium*) have numerous stiff pseudopods (axopods) which may be several hundred microns in length and which are supported by axial rods shown by Kitching (1964) to be constructed of longitudinal microtubules arranged in two interlocking spiral sheets (Figs. 7 and 11). The central microtubules extend the full length of the axopod, but peripheral microtubules tend to be shorter since the bundle of fibrils narrows distally. The streaming movements of the cytoplasm around these microtubule bundles has already been described. Locomotion of these amebas is achieved by changes in length of axopods which attach to the substratum. The body glides and rolls as the axopods that make contact with and attach to the substratum in front of the organism retract at speeds of approximately 20 μ/min, and attached axopods at the rear elongate at approximately 7 μ/min (Watters, 1968). The locomotion of the organisms is therefore slow; Watters observed speeds of between 5 and 100 μ/min.

The axopods of these heliozoans are labile structures; they were found to be retracted at low temperatures (Tilney and Porter, 1967), under high hydrostatic pressures (Kitching, 1957; Tilney et al., 1966), in colchicine (Tilney, 1968), and in the presence of cupric and nickel ions (Roth and Shigenaka, 1970). On return of the organisms to normal conditions, the axopods were re-formed. An axopod never re-formed in the absence of the axial bundle of fibrils, and changes in length of the axopods were found to reflect changes in length of the fibril bundle. Changes in length of the fibril bundle did not involve any change in diameter or substructure of the microtubular elements; it is believed that the microtubular fibrils were disassembled and the globular protein molecules transported away

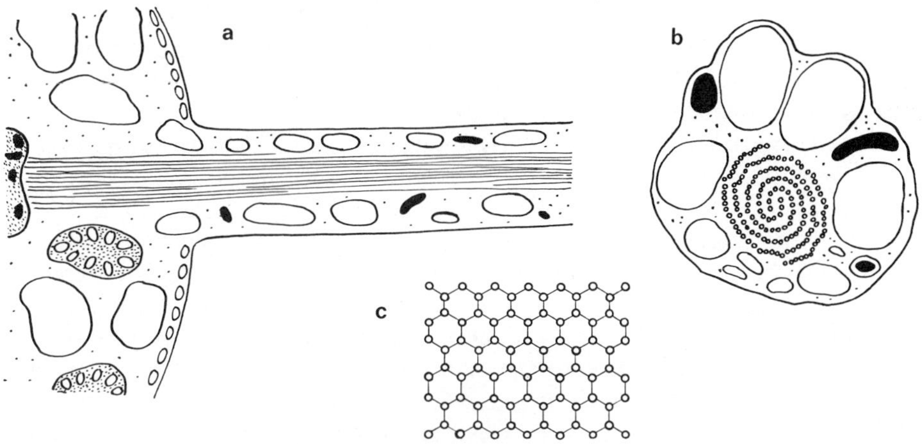

Figure 11. Diagrams showing axopod structure in Heliozoa and Radiolaria. The longitudinal section of the basal part of an axopod of *Actinosphaerium* (*a*) shows the bundle of microtubular fibrils extending along the center of the axopod from the surface of one of the nuclei. In (*b*) the arrangement of the microtubular fibrils within the bundle is seen in a cross section of an axopod of *Actinosphaerium* (cf. Fig. 7*b*). (*c*) The arrangement of microtubular fibrils in the fiber bundles of axopodia of *Sticholonche* as shown in micrographs by Hollande et al. (1967).

by cytoplasmic streaming to the cell body, where large amounts of particulate and fine fibrillar material accumulated following experimentally induced retraction of the axopods. This material disappeared again on extension of the axopod fibrils, which is assumed to involve reconstruction of the microtubular fibrils from the globular subunits. It is interesting that retraction in these experiments tends to occur at rates of approximately 10–30 μ/min, while extension tends to be slower at approximately 5–10 μ/min.

The change in length of axopods during locomotion of *Actinophrys* and *Actinosphaerium* is also believed to result from a change in the length of the axial bundle of microtubular fibrils. It is assumed therefore that these axopods retract by disassembly (depolymerization) of the microtubules into component protein molecules and elongate by reconstruction (polymerization) of the microtubules from molecular subunits. It is not clear whether the addition or removal of subunits takes place proximally, distally, or at both ends of the microtubular fibrils.

By contrast, the axopods of simpler heliozoans, for example, members of the Centrohelida such as *Raphidiophrys* and *Acanthocystis* (Bardele, 1970), and of radiolarians, for example, *Sticholonche* (Hollande et al., 1967), have a bundle of axial fibrils in which the microtubules are arranged in an open hexagonal array (Fig. 11*c*). The microtubular array of the extended axopods of *Sticholonche* was found to be replaced in retracted axopods by a fibrillar or lamellar bundle folded in a zigzag pattern. It would be interesting to know more details of the form of contraction and the structure of the components. The centrohelid axopods were observed to retract within seconds, but structural details of retracted axopods are not available to determine whether folding or depolymerization of microtubules was involved.

The microtubular fibrils that form the mitotic spindle at cell division share some properties with those in the axopod of *Actinosphaerium*. Pease (1946) found that the mitotic spindle disappeared under high hydrostatic pressures, and Inoué (1952a, b) showed that both low temperature and colchicine reduced the birefringence of the mitotic spindle. For these and other reasons, it seems appropriate to consider the mitotic movements of chromosomes in this section, although a variety of mechanisms has been suggested to account for these movements.

Evidence concerning the movements of chromosomes during mitosis has been carefully analyzed by Forer (1969). Before the breakdown of the nuclear membrane, a spindle of fibers is formed between the asters that surround the centrioles of animal cells, or in the "clear zone" of plant cells. During the metaphase of mitosis, the chromosomes become arranged among the fibers near the equator of the spindle, and in the subsequent anaphase stage, the divided chromosomes move toward the poles of the spindle at speeds of up to 5 μ/min. During the anaphase movement it can be seen that there are in fact two types of spindle fibers—continuous fibers approximately 0.2 μ in diameter which extend from one pole of the spindle to the other, and chromosomal fibers 0.2–1.5 μ in diameter which connect the kinetochores of individual chromosomes to the nearer pole (Fig. 12). Both types of fibers contain numerous microtubules, as well as other filamentous and vesicular material. Careful microdissection and microirradiation experiments have shown that intact chromosomal fibers are necessary for the anaphase movements, but there is uncertainty about the origin of the force that moves the chromosomes. Bajer (1968) suggested, "as a useful working hypothesis," that the chromosomes are drawn toward the poles when the chromosomal fibers slide along the continuous fibers where they intermingle in the pole region of the spindle. Several earlier investigators have described streaming movements of cytoplasm around the spindle fibers and have suggested that the chromosomes might be carried in this flow toward the poles (see Jahn and Bovee, 1969).

There are many interesting variants of the process of mitosis. In the division of the micronucleus of the ciliate *Nassula*, described by Tucker (1967), the nuclear membrane persists and the spindle is entirely intranuclear (Fig. 13). The metaphase micronucleus has a diameter of approximately 6 μ and contains chromatin material near the equator of a nuclear spindle composed of microtubules. Anaphase lasts approximately 55 min (at 22°C), and during this phase the micronucleus elongates until it is a dumbbell-

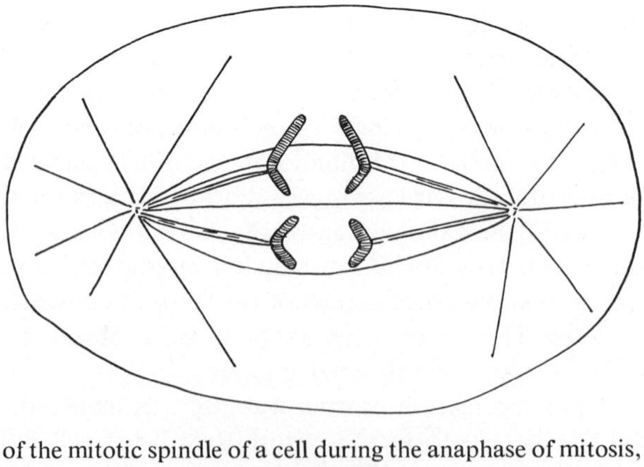

Figure 12. Diagram of the mitotic spindle of a cell during the anaphase of mitosis, showing continuous fibers extending between the poles of the spindle and shorter chromosomal fibers connecting the chromosomes to the poles of the spindle.

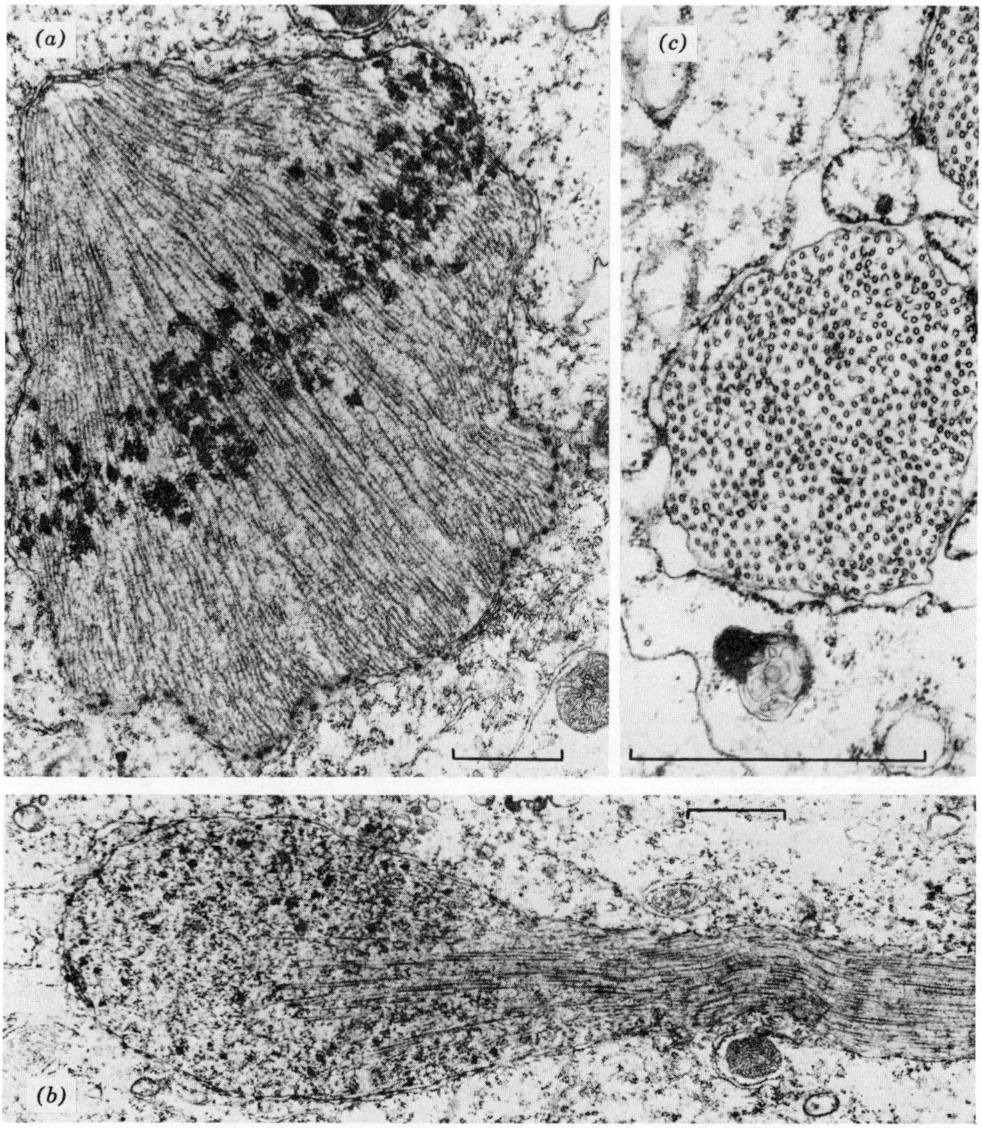

Figure 13. Electron micrographs of micronuclei of *Nassula* at different stages during nuclear division. (*a*) Longitudinal section through a nucleus at metaphase. (*b*) Longitudinal section through one end of the separation spindle in late anaphase. (*c*) Transverse section through the separation spindle. The scales each represent 1 μ. (From Tucker, 1967.)

shaped structure approximately 100 μ long. Elongation of the continuous fibers of the spindle carries the two masses of chromatin apart at a speed of approximately 2 μ/min. This long "separation spindle" is approximately 1 μ in diameter and consists of a compact bundle of microtubules; its thickness and the number of microtubules within it do not change during elongation, so that it must be assumed that the microtubules themselves are elongating. A similar separation spindle has been reported in the ciliate *Blepharisma* by Jenkins (1967), and the spindle of *Trypanosoma* appears to push the two halves of the nucleus apart (Vickerman and Preston, 1970). In the flagellate *Barbulanympha*, Cleveland (1954) suggested that movement apart of the chromosomes might result from pulling by chromosomal fibers and pushing by continuous fibers.

Many such observations suggest that during mitosis the microtubular fibers of the spindle change in length by the addition or removal of subunits in a manner comparable to the extension or retraction of microtubules in the axopod of *Actinosphaerium*. Inoué and Sato (1967) summarize a wide range of evidence in support of the hypothesis that the microtubules of the spindle represent a labile association of molecules. A pool of protein subunits is available in the cell which can be added to microtubules (presumably at the ends) to extend them, or such subunits can be returned to the pool on shortening of the microtubules. The forces required to move chromosomes are small—approximately 10^{-8} dyne (Forer, 1969). Inoué and Sato believe that the elongation of microtubules by polymerization can provide the necessary forces to move chromosomes. It is interesting that the retraction of spindle fibers in low concentrations of colchicine or by slow cooling is sufficient to move chromosomes and, similarly, that the extension of spindle fibers after cooling and other treatments also moves the chromosomes. Another possibility, as Inoué and Sato note, is that the forces for movement are generated by a separate mechanism, but that the extension and shortening of the spindle fibers are necessary for the other mechanism to function.

Other data are available on the rate of extension of microtubular structures. For example, the complex rods that form the main part of the pharyngeal basket of *Nassula* elongate during growth at approximately 0.4 μ/min, which Tucker (1970) notes represents the addition of about 20 subunits per second to each of the microtubules of each rod. The resorption of the rods of old baskets takes place at approximately 0.7 μ/min, and the elongation of microtubules is most rapid within 1 or 2 hours after the old basket has disappeared. Again, it seems likely that the disassembly contributes to a pool of subunits and that the polymerization draws subunits from the pool

2. Systems in Which the Main Elements are Microfilamentous. Exceedingly rapid contractions occur in the fiber systems of some ciliate protozoa. The most familiar of these is probably that which results in the coiling of the stalk of *Vorticella* and some related peritrichs (Fig. 14*a* and *b*). The body of the ciliate is borne at the end of a stalk several hundred microns long. The stalk is seen in transverse section to have an outer region which is extracellular and is composed of a thin cylindrical sheath surrounding a zone of fibrous matrix material, which is reinforced with longitudinal fibrils at one side, and an inner cytoplasmic region surrounded by a membrane and containing a fibrous strand (the spasmoneme) at the side of the cylinder opposite the reinforced part of the outer region (Fig. 15). The spasmoneme contains longitudinally oriented microfilaments 2–5 nm in diameter arranged in loose bundles and evenly distributed tubules approximately 50 nm in diameter (Favard and Carasso, 1965; Amos, 1970). Numerous mitochondria are present in the stalk cytoplasm. The wall thickening and the spasmoneme spiral opposite one another along the length of the stalk.

Stimulation of the ciliate causes retraction of the body toward the attached end of the stalk by a contraction of the spasmoneme, which causes coiling of the stalk. At the same time fibers within the body of the ciliate also contract. A reduction in height of

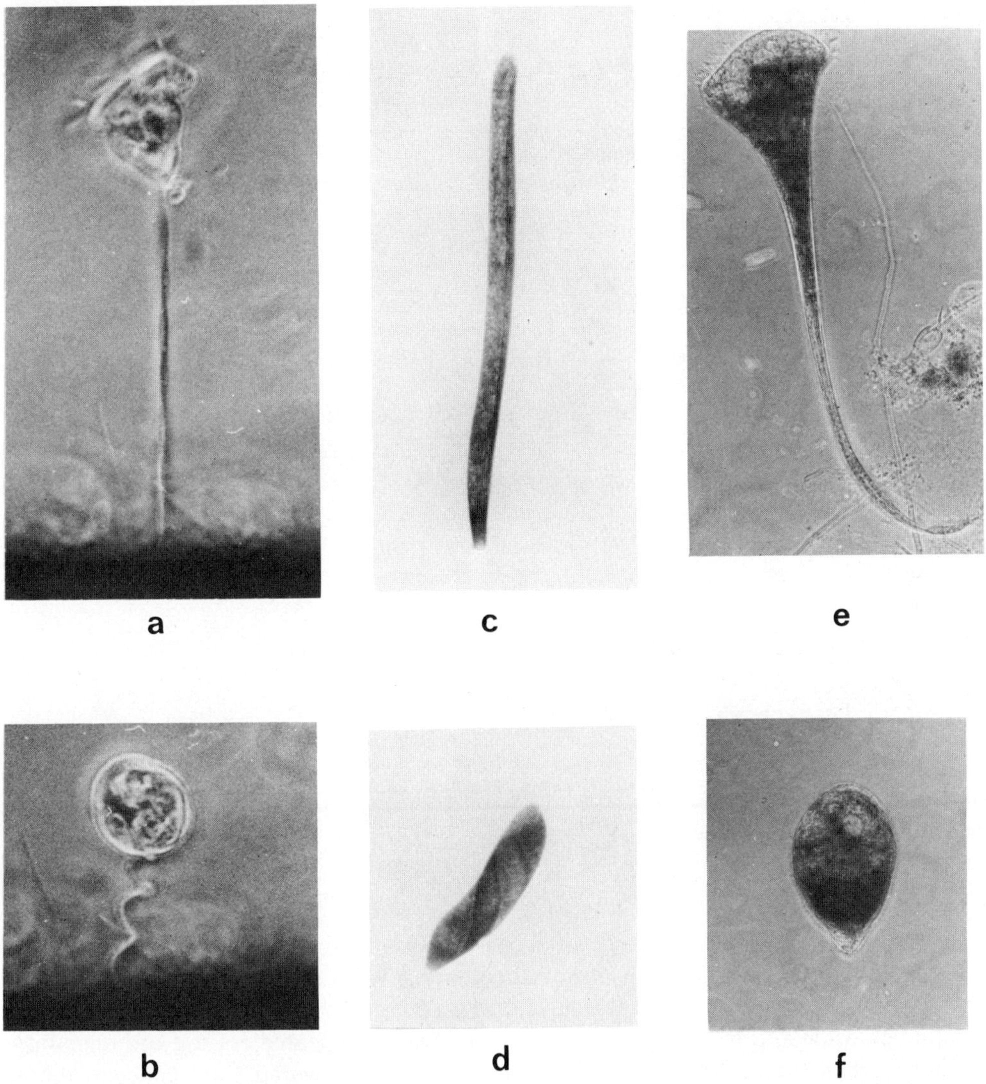

Figure 14. Micrographs of contractile ciliate protozoa. *Vorticella* has a long stalk (*a*) which coils on contraction (*b*). The cylindrical *Spirostomum* (*c*) twists as it shortens (*d*). *Stentor* is trumpet-shaped (*e*), and contraction of myonemes makes the body almost spherical (*f*).

the stalk to 10–15% of its extended length is achieved in approximately 5 msec by a shortening of the spasmoneme to 30–40% of its original length—a rate of shortening of approximately 20–50 mm/sec (Sugi, 1961; Amos, 1970; Jones et al., 1970b). The thickened region of the wall of the stalk lies at the outside of the coil, and the inner thinner part is thrown into folds. A sharp reduction in the birefringence of the spasmoneme occurs during contraction, but the source of this change is uncertain for the filaments appear longitudinal in contracted spasmonemes (Amos, 1970). Relaxation of the stalk takes 1–2 sec, and may involve active extension of the spasmoneme as well as elastic extension of the sheath.

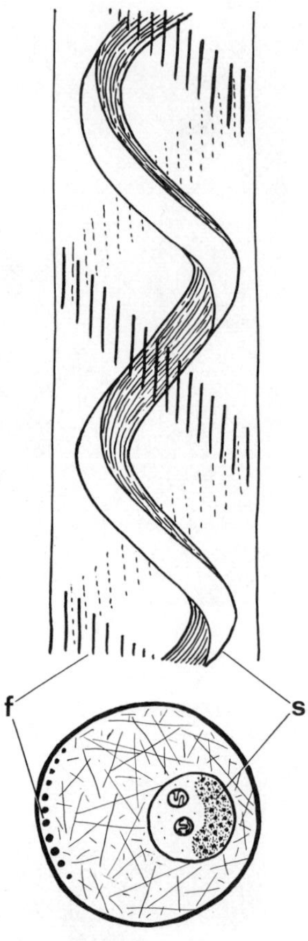

Figure 15. Diagrams of the stalk of *Vorticella* showing reinforcing fibrils (f) and the spasmoneme (s) as they appear in a portion of the stalk and in a transverse section of the stalk.

Additional interest is centered on the contraction of the peritrich spasmoneme because the fibrillar contraction does not require ATP. Levine (1956) first observed that glycerine-extracted preparations of *Vorticella* stalks could be made to coil in calcium chloride and to uncoil again in 4 m*M* EDTA. Repeated coiling and uncoiling occurred when the stalks were subjected to raised and lowered Ca^{2+} concentrations by immersion alternately in calcium chloride and EDTA, but the addition of ATP had no effect on the coiling or uncoiling. The calcium-induced contractions of glycerinated preparations were found by Hoffmann-Berling (1958) to continue in 2 millimolar salyrgan, which completely inhibits the ATPase system of muscle, and Amos (1971) obtained other evidence indicating that ATP is not directly involved in the contraction of the spasmoneme. It is concluded that in life the spasmoneme contracts when the concentration of free calcium ions around the filaments rises to a threshold level, and Hoffmann-Berling suggested that this may involve a structural change in the polymeric protein molecules of the filaments, resulting from the neutralization of charges by Ca^{2+}. Since the birefringence falls on contraction, one might suspect that the filaments fold rather than slide to shorten the spasmoneme. Carasso and Favard (1966) demonstrated that calcium was present in the tubule system that ramifies between the filaments, and it is thought that these tubules may actively accumulate Ca^{2+} from the spasmoneme cytoplasm and may release it upon stimulation, thereby causing contraction. Amos calculated that the work done in contraction of the spasmoneme could be derived from the energy stored when Ca^{2+} is accumulated against a concentration gradient. Hence ATP from the mitochondria in the stalk cytoplasm may be used in ion pumping and thereby serve as an indirect source of energy for contraction.

The giant ciliates *Spirostomum* and *Stentor* show rapid body contractions in response to mechanical or other excitation (Fig. 14*c–f*); the contraction is followed by a slower extension. The extended cylindrical body of a *Spirostomum* 1.5 mm long shortened to an ellipsoid 0.65 mm long in about 4 msec with a maximum speed of approximately 200 mm/sec, and extended again in about 10 sec with a maximum speed of approximately 0.2 mm/sec at 20°C (Jones et al., 1966, 1970a; Cragg and Sleigh, 1970). *Stentor* has a trumpet-shaped body which contracts to an almost spherical ball; at 20°C an extended cell of *Stentor coeruleus* contracted in 50 msec from a length of 1.4 mm to a length of 0.5 mm with a maximum speed of shortening of 63 mm/sec, and extended again over a period of 80 sec with a maximum rate of extension of 17 μ/sec (Cragg and Sleigh, 1970).

Two pellicular fiber systems are present in each ciliate—M fibers composed of microfilaments, and Km fibers composed of microtubules; they have been studied in *Spirostomum* by Finley et al. (1964) and Ettienne (1970), and in *Stentor* by Randall and Jackson (1958), Bannister and Tatchell (1968), and Huang (1970). Cilia occur along the body in longitudinal rows called kineties, and from the basal bodies of cilia a row of microtubular fibrils runs obliquely backward to join a stack of rows of microtubules which originate from the more anterior basal bodies of the same kinety (Fig. 16*a*). In a contracted *Stentor* there are many rows in each stack, while in an extended *Stentor* there are only a few rows in each stack, so it is assumed that when the ciliate contracts, and the cilia move closer together, the rows of microtubules slide over one another similarly to the sections of a fireman's ladder (Fig. 16*b* and *c*). Links connect the microtubules within the rows and between the rows of a stack.

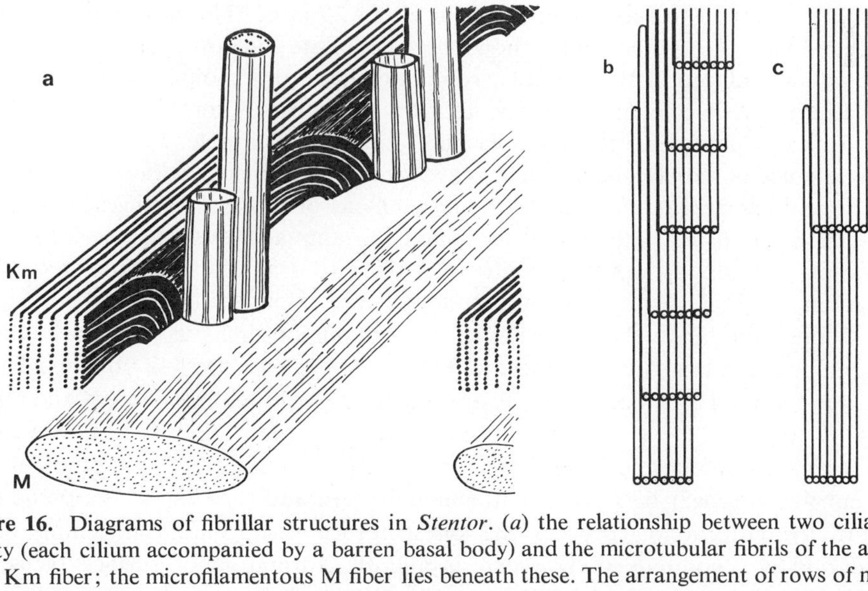

Figure 16. Diagrams of fibrillar structures in *Stentor*. (*a*) the relationship between two cilia of a kinety (each cilium accompanied by a barren basal body) and the microtubular fibrils of the associated Km fiber; the microfilamentous M fiber lies beneath these. The arrangement of rows of microtubules in a Km stack from contracted (*b*) and extended (*c*) animals.

The M fibers lie beneath these microtubular Km fibers and form a network of bundles composed of fine filaments 3–10 nm in diameter. These bundles are mostly organized into longitudinal tracts in the stalk region of *Stentor;* in *Spirostomum* the network is more diffuse, but the filaments are largely oriented along the long axis of the body. The M fibers appear to be comparable to the spasmoneme of *Vorticella;* they are associated with membranous vesicles and mitochondria and can be seen from the following discussion to share other features.

Bannister and Tatchell found by light microscopy that when *Stentor* contracts both fiber systems shorten, thicken, and remain straight. Following contraction the M fibers lengthen before the Km fibers and are thrown into lateral convolutions that disappear only as the body lengthens—the Km fibers remain straight throughout. The M fibers are thought to be the contractile elements, responsible for the rapid shortening of the body, and capable of some intrinsic extension because of the way in which they are thrown into folds. The rows of microtubules that constitute the Km fibers are thought to slide over one another to accommodate the changes in length of the kinety on contraction and extension; active sliding of these rows of microtubules over one another (without change in length of the fibrils themselves) could be responsible for the extension of the ciliate and could also provide the mechanism for the slow extensions and contractions that occur in this mobile-bodied ciliate. It is not clear whether or not the Km fiber system could operate in a comparable manner in *Spirostomum*, in which it is disposed spirally and little or no sliding of the fibrils can take place as the spiral shortens and widens when the M fibers contract.

The excitation of contraction appears to require a threshold level of free calcium ions which can be released experimentally in a variety of ways (Sleigh, 1970). When a *Spirostomum* was stimulated electrically following microinjection of the bioluminescent protein aequorin (see Chapter 15), light emission commenced and increased with continued electrical stimulation until the ciliate contracted (Ettienne, 1970). The level of light emission of aequorin was related to the amount of free Ca^{2+} present. Following the end of electrical stimulation, the light emission, and therefore the free Ca^{2+}, returned to the starting level in about 25 msec. The membranous vesicles associated with the M fibers were shown to accumulate calcium oxalate crystals when the ciliate was placed in 10 millimolar sodium oxalate; it is thought that these vesicles normally act as a calcium store and as sites of reabsorption of calcium from the cytoplasm comparable to the sarcoplasmic reticulum of muscle.

Some amebas secrete shells which may become very heavy because of continued secretion of calcareous material or the attachment of mineral particles. In *Difflugia* the shell is encrusted with sand grains, and the animal moves in a series of jerks, extending pseudopods and then contracting them to drag the shell along (Wohlman and Allen, 1968). Long, cylindrical pseudopods are extended from the shell (Fig. 17) in a manner similar to the extension of pseudopods of *Amoeba* or *Chaos*, although the endoplasm is less granular and the flowing movements are less obvious. The extended pseudopod attaches to the substratum at one or more "sole" areas and rapidly shortens to draw the shell toward the attachment point. Before attachment the pseudopod is more or less isotropic, but at attachment several longitudinal birefringent fibers appear between the attachment point and the base of the pseudopod. During the contraction process the birefringence disappears; shortening of the pseudopod is accompanied by syneresis and the formation of surface blebs. The birefringent fibers are about 1 μ in diameter and can be seen in the Nomarski inter-

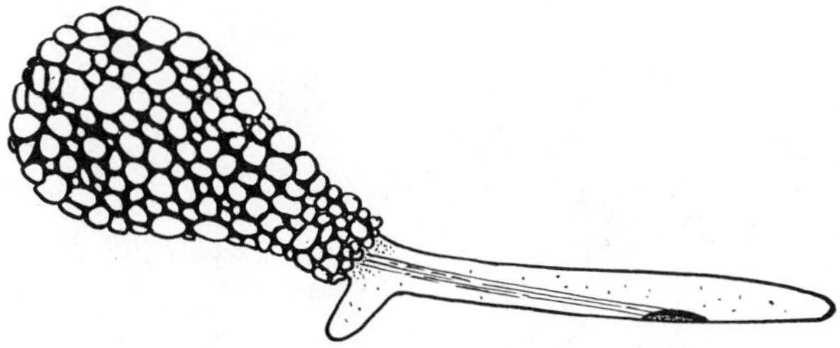

Figure 17. Diagram of *Difflugia* illustrating the shape of pseudopods and the position of the fibrous bundle that forms between the main body and an attachment point on the pseudopod.

ference microscope as well as in polarized light. The electron microscope reveals 1-μ-thick bundles of microfilaments 5–8 nm in diameter in attached pseudopods, and also some randomly oriented microfilaments. It is thought that the birefringence appears when filaments that were previously randomly arranged become aligned because of flowing movements and the development of tension, and the loss of birefringence during contraction could result from some form of folding phenomenon as suggested for *Vorticella*.

Another form of contraction that appears to involve microfilaments is seen in the retraction of the larval tail at metamorphosis of the ascidian *Amaroucium* (Cloney, 1966). In this case the epidermal cells covering the tail shorten over a period of about 6 min from a total length of 1.1 mm to about 60 μ, during which time the individual cells shorten from a length of about 26 μ to only 1 or 2 μ. Shortening of the cell is accompanied by extensive folding of the surface membrane. During the shortening the birefringence of an anisotropic zone near the cell surface becomes more pronounced. Electron microscopy of these cells reveals abundant microfilaments 5–7 nm in diameter near the cell surface, and the increase in birefringence suggests that these microfilaments become better aligned and perhaps more highly concentrated by sliding together during the shortening of the cells.

In the recent literature there is a profusion of other examples of contractile cells or contractile regions of cells that contain microfilaments in situations that suggest a role in their contractile mechanism. Thus the cleavage furrow of the dividing newt's egg is found over a broad band of microfilaments, each 8–10 nm in diameter, which are oriented along the line of the furrow (Selman and Perry, 1970). The band extends around the circumference of the cell and shortens as the cleavage furrow dips in from the surface to divide the cell. Similar observations have been made on the eggs of coelenterates (Schroeder, 1968; Szollosi, 1968) and the axolotl (Bluemink, 1970). It had been demonstrated earlier that contraction of cortical cytoplasm beneath the furrow was involved in cleavage and that the microfilaments seem to be responsible for the contraction. Similar contractile rings of 4- to 7-nm filaments are found at the cleavage furrows of the ciliate *Nassula* (Fig. 18*b* and *c*) (Tucker, 1971) and of mammalian (HeLa) cells in culture (Schroeder, 1970). Cell cleavage in fibroblast and HeLa cells is prevented by treatment with cytochalasin B, although nuclear fissions continue (Carter, 1967; Schroeder, 1970). Schroeder has also found that the filaments of the contractile ring bind heavy meromyosin, as do various forms of actin filaments (Section 2.B.1). In the morphogenetic movements of amphibian embryos, contraction is associated with the presence of arrays of microfilaments around the blastopore

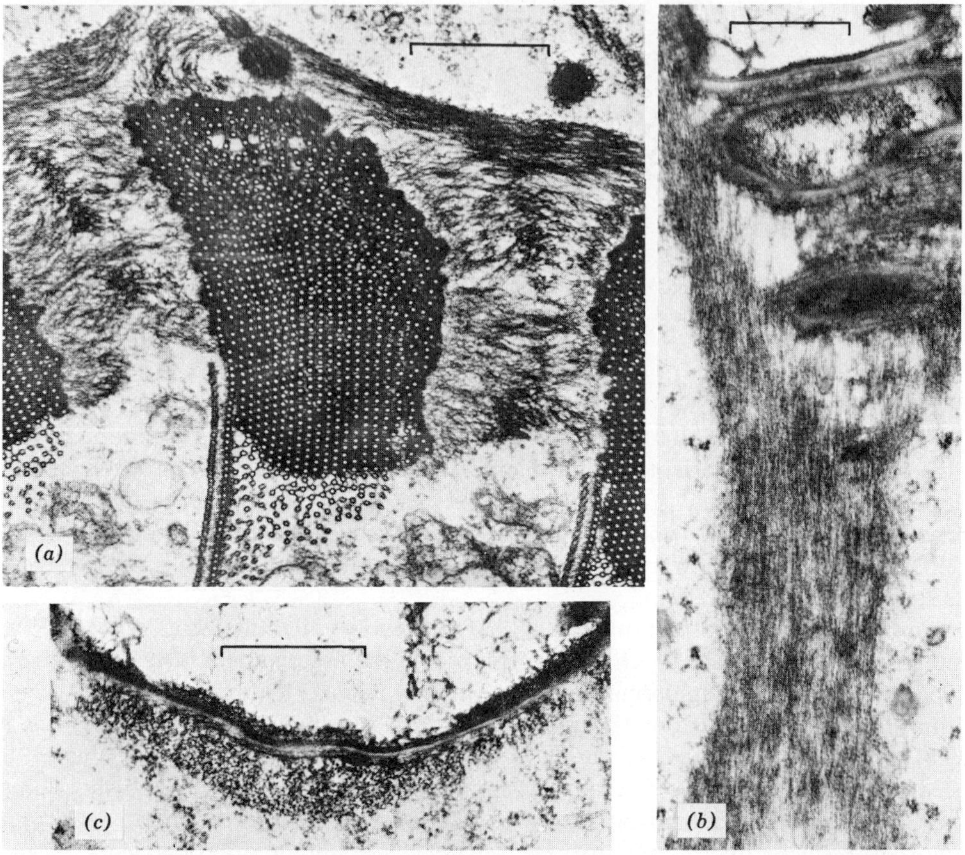

Figure 18. Microfilamentous contractile structures. (*a*) Longitudinal section of a fibrillar structure that contracts repeatedly each time the organism feeds. The band of fibrous material seen here connects the tops of the fibrillar bundles that form the walls of the cylindrical pharyngeal basket of *Nassula;* narrowing of the top of the cylinder appears to result from contraction of the fibers. (From Tucker, 1968.) (*b*) and (*c*) show, respectively, longitudinal and transverse sections of parts of the contractile ring of *Nassula* during fission of the ciliate; this fibrillar structure contracts only once. Each scale represents 0.5 μ. (From Tucker, 1971.)

(P. C. Baker, 1965) and in the inrolling neural tube (P. F. Baker and Schroeder, 1967). Contraction of the blood clot in vertebrates is associated with microfilamentous structures, both in the thrombocytes of fish (Shepro et al., 1969) and in mammals, in which cytochalasin B has an inhibitory effect on clot retraction (Shepro et al., 1970). While these structures contract only once, a thick band of filaments is apparently involved in the repetitive contractions that constrict the upper end of the pharyngeal basket of *Nassula* during feeding (Fig. 18*a*) (Tucker, 1968). In all these cases a fairly rapid change in shape is associated with structures in which microfilaments are oriented with their long axes in the direction of contraction.

In a rather different category are the gliding movements of blue-green algae. A filament of *Oscillatoria* glides over the substrate at speeds of about 2–3 μ/sec. The cell wall is not seen to be altered in shape but, according to Halfen and Castenholz (1970), there is a shear zone between the cell wall and the surrounding sheath. The superficial part of the cell wall includes a 16-nm layer containing a parallel array of filaments 5–8 nm thick oriented in a right-handed spiral and continuous across cell boundaries. The algal filament shows a right-handed rotation in both forward and reversed gliding which seems to be related to the microfilaments, but the mechanism is obscure. The gliding movements of diatoms (commonly at approximately 10 μ/sec) seem to depend on a quite different mechanism involving attachment and traction by means of a secreted substance (Harper and Harper, 1967).

B. Fiber Systems that Change in Shape without Change in Length

It is likely that in some of the contractile structures discussed in the previous section the individual fibrous elements slide along one another during contraction and do not themselves change in length although the fibrous array as a whole is shortened; however, in none of these cases is it certain that the length of the fibrils remains constant. Other motile structures contain continuous longitudinal fibrils and show shape changes in which neither the length of the structure nor the length of the fibrils changes appreciably. These structures are based on arrays of microtubules and appear to depend upon the stiffness of the component microtubular elements for maintenance of length. The most familiar organelles of this type are cilia and flagella, but there are several other forms of microtubular structures that can be considered under this heading.

Cilia and flagella are projections from cells that are used in swimming movements or in the propulsion of fluids by animals and plants of almost all groups. Undulatory waves are propagated along flagella, and a single flagellum is adequate to propel small cells such as spermatozoa; small numbers of flagella are also functionally effective in propulsion, and some protozoa carry very large numbers of flagella. Cilia are normally found in larger numbers because their unilateral lashing movements are better suited to the swimming of larger organisms and the propulsion of fluids over surfaces. A great variety of beating patterns is shown by these organelles, different examples providing many intermediates between the typical flagellum and the typical cilium (Sleigh, 1968) (Fig. 19).

The movement of water by flagella is achieved in a variety of ways (Holwill, 1966a). More-or-less symmetrical, planar undulations or helical undulations may be propagated along flagella from the base to the tip or, more rarely, from the tip to the base (Fig. 20). The water propelled by these propagated waves normally flows in the direction of the waves, but if the flagellum carries stiff lateral projections, the direction of flow of the water is reversed (Holwill and Sleigh, 1967). Flagella commonly beat at 10–40 cycles/sec, and the waves are propagated along the flagella at between 100 and 1000 μ/sec.

Each cycle of the beat of a typical cilium consists of an active phase in which the extended organelle swings quickly through a large arc, and a recovery phase during which a wave of bending travels up the cilium from the base and returns the cilium more slowly through a smaller volume of water to the starting position (Fig. 21). This beat often takes place in one plane, but in several cases it has been reported that the cilium swings to one side during the recovery stroke. A simple cilium 12 μ long may beat 30 times per second with an angular velocity of 12°/msec and a tip velocity of 2.5 mm/sec in the effective stroke, while the rate of propagation of the bend in the recovery stroke may be about 350 μ/sec. In a large compound cilium 600 μ long beating 20 times per

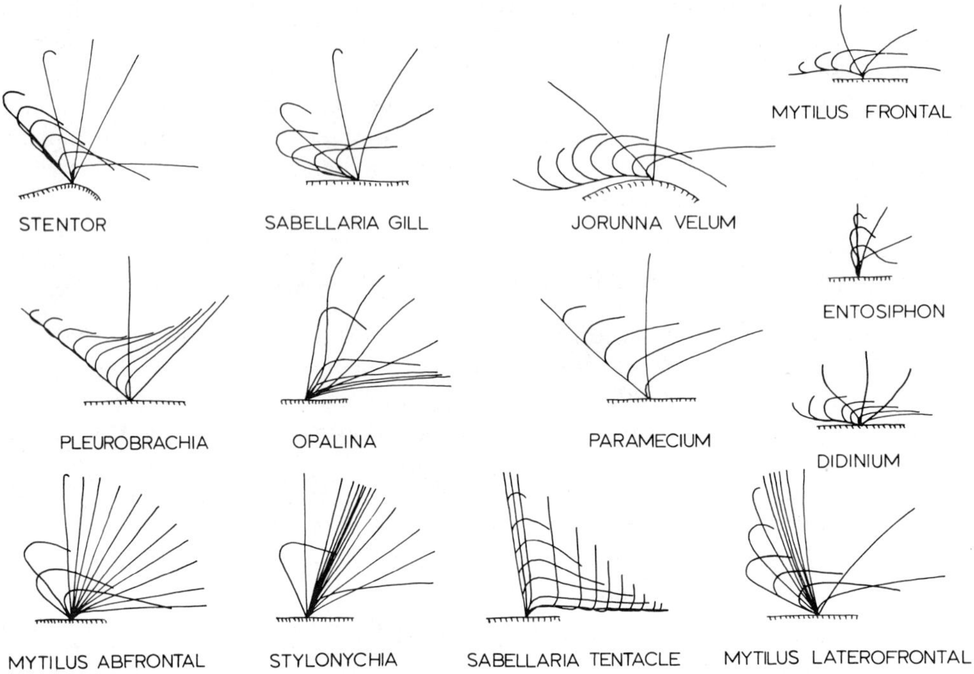

Figure. 19. Patterns of ciliary beating. Where information was available the profiles in each sequence were drawn at equal time intervals to show up the slower and faster parts of the beat; detailed information on many of the cilia can be found in Sleigh (1968), from which this figure is taken.

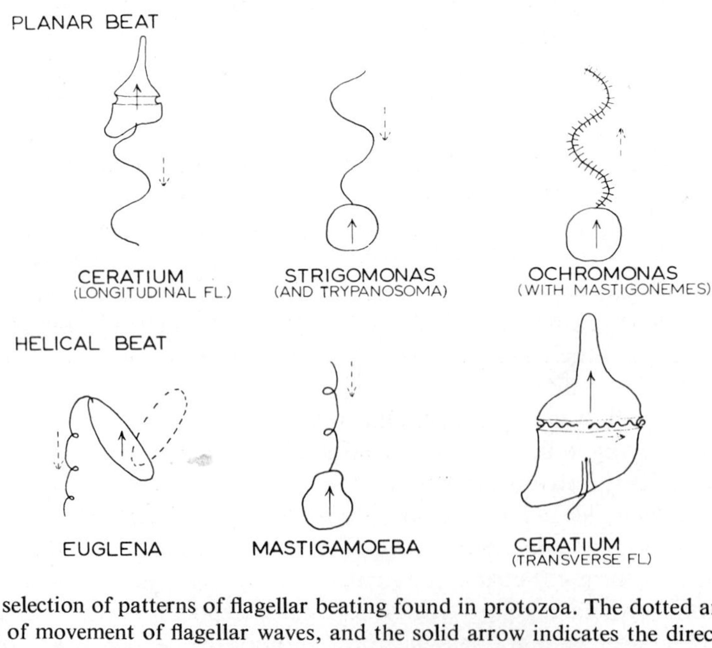

Figure 20. A selection of patterns of flagellar beating found in protozoa. The dotted arrow indicates the direction of movement of flagellar waves, and the solid arrow indicates the direction of movement of the organism. [Based on information from Jahn et al. (1963) for *Ceratium*, from Holwill (1964, 1966b) for *Strigomonas* and *Euglena*, and from Bovee (1964) for *Mastigamoeba*.]

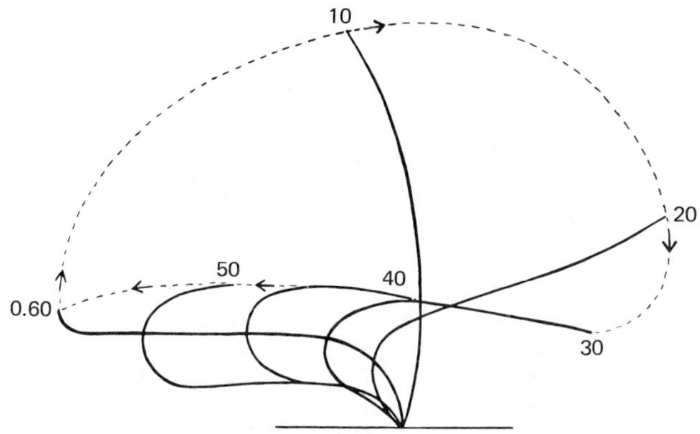

Figure 21. Six stages in the cycle of beat of a cilium from *Sabellaria*. The numbers indicate the time in milliseconds from the beginning of the effective stroke. (From Sleigh, 1971.)

second, the angular velocity in the effective stroke may be 10°/msec, and the bend may be propagated at 17 mm/sec in the recovery stroke (Sleigh, 1971).

The lengths of cilia and flagella vary widely, ranging from 1 or 2 μ to several millimeters, although both types of organelle are most commonly approximately 20 μ in length. Longer cilia occur where a number of ciliary units move together in a bundle, forming a compound structure, but some single flagellar sperm tails are several millimeters long. Unmodified organelles of both types are approximately 0.25 μ in diameter and contain the same basic arrangement of longitudinal fibrils enclosed within a membrane continuous with the surface membrane of the cell. The fibrils in a cilium form a 9 + 2 pattern that is best seen in transverse section (Fig. 22). The two separate central microtubules are normally found only in the part of the cilium that projects from the cell surface, while the nine double microtubules that form the peripheral cylinder extend within the cell, where each doublet becomes a triplet and is cross-linked to the other triplets of the cylinder in the basal body. Within the ciliary shaft each peripheral doublet is composed of two subfibrils; the A subfibril appears to be a complete microtubule and carries pairs of lateral arms projecting toward the adjacent doublet and strands projecting radially toward the central fibrils, while the B subfibril is an incomplete microtubule attached to the A subfibril. The pairs of lateral arms are borne at intervals of about 17 nm along the A subfibril and are composed of the ATPase protein dynein (Gibbons, 1965). The protein units (tubulins) of the microtubular structures forming the central fibrils, A subfibrils, and B subfibrils are all somewhat different chemically and have different amino acid compositions (see Section 4); for example, Stephens (1970) found that it was possible to separate tubulins of A and B subfibrils because the B protein dissolved in a low-ionic-strength medium within 2 min at 40°C, leaving intact A microtubules.

The movement of a cilium involves bending of the cylinder of fibrils. In earlier accounts of the movement, it was assumed that the fibrils at the inside of the bend shortened, and/or those at the outside of the bend extended, but Satir (1965, 1968) obtained persuasive evidence in support of the conclusion that the peripheral fibril doublets slide upon one another when the cilium bends. The fibrils are anchored firmly in the basal body and, as the cilium bends to one side or the other, the peripheral

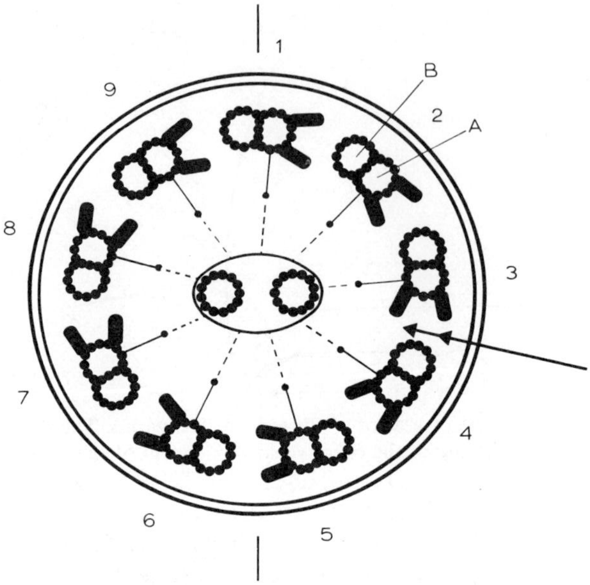

Figure 22. Details of ciliary structure as seen in a transverse section, indicating the A and B subfibrils of doubled number 2, and the presumed position of the formation of links between the dynein arms on doublet 3 and the B subfibril of doublet 4 (double arrowhead). (From Sleigh, 1971.) (See also Fig. 29*a*.)

fibrils nearest the cell body protrude furthest at the ciliary tip, while the fibrils furthest from the cell surface, which therefore must pass around the outside of the bend in the cilium, terminate lowest on the shaft (Fig. 23). The peripheral fibrils probably do not change in length when a cilium bends. Flagellar undulations could also take place without change in length of the peripheral fibrils (Brokaw, 1970). The bending of these organelles may therefore be caused by active sliding of the peripheral doublets relative to one another. In the active stroke of a ciliary beat, the doublets would have to slide against one another throughout the full length of the cilium, but in the recovery stroke sliding would be largely confined to the curved part of the cilium, and the zone of sliding would progress up the cilium as it unrolls (Fig. 24). To achieve the angular velocity observed in the effective stroke of a cilium, one fibril would have to slide with respect to its neighbor at between 1 and 20 μ/sec, and a similar rate of sliding would be required in the recovery stroke. It seems likely that active sliding results from the formation of links between peripheral fibrils through the lateral arms and that energy for bending is derived from the breakdown of ATP by activity of the dynein ATPase; these links could therefore have a function analogous to that of the bridges formed by the heads of myosin molecules between myosin filaments and actin filaments of striated muscle. The ATPase of the dynein arms is activated by both Mg^{2+} and Ca^{2+}. It has been estimated that the energy required for ciliary movement could be provided if each dynein arm broke down one molecule of ATP in each cycle of beat. The demands for ATP along the length of the axoneme could be satisfied by diffusion (Brokaw, 1966; Nevo and Rikmenspoel, 1970) and, in addition, there appears to be a mechanism for the regeneration of ATP using a phosphagen, since Watts and Bannister

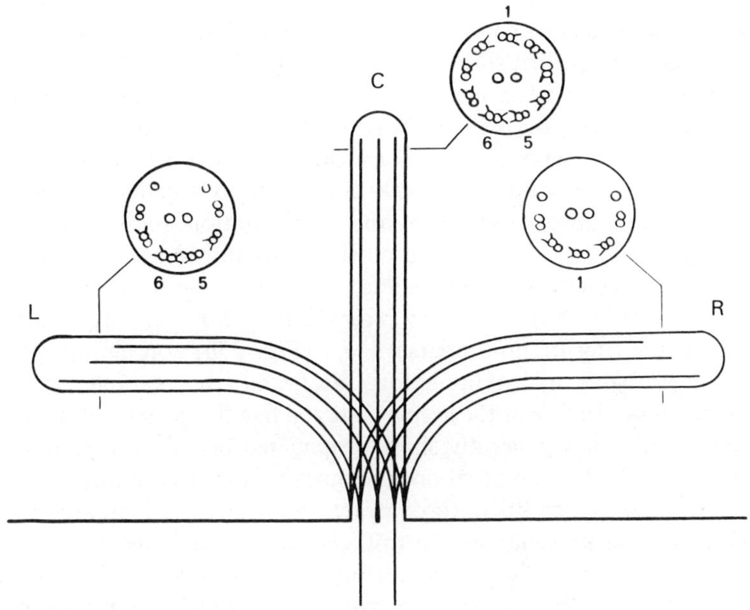

Figure 23. Idealized diagram illustrating the conclusions of Satir (1965) concerning the sliding of ciliary fibrils. At the center, cilium C is straight, the fibrils terminate at the same level, and in a transverse section near the tip all peripheral doublets appear the same. At the left cilium L is bent toward peripheral doublets 5 and 6 (identified by the bridge links between the two doublets), which project furthest and are entirely at the level of the transverse section, while fibril 1 terminates first and is missing from the transverse section. At the right cilium R is bent toward doublet 1, which projects furthest and is entire in the transverse section, while doublets 5 and 6 are missing. (From Sleigh, 1971.)

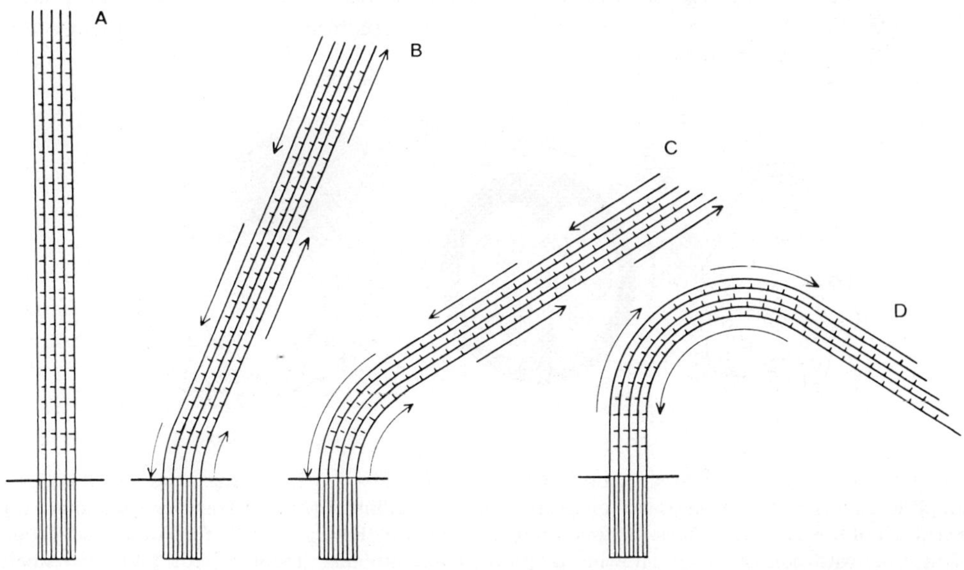

Figure 24. The presumed positions and directions of sliding of ciliary fibrils when straight (*A*), during the effective stroke (*B* and *C*), and during the recovery stroke (*D*). (From Sleigh, 1971.)

(1970) found arginine kinase in the cilia of *Tetrahymena*. New structural information on the radial links in flagellar axonemes justifies speculation concerning their displacement or distortion during fibril sliding (Kiefer, 1970; Warner, 1970).

Flagellar sperm tails are often modified in quite complex ways. Very often mitochondria or mitochondrial derivatives form part of the tail, running longitudinally alongside the 9 + 2 axoneme or spiraling around it; these are assumed to serve as energy sources and probably do not directly contribute to flagellar motility. The basic 9 + 2 axoneme is sometimes modified by the augmentation, reduction, or loss of the central structures, but the nine doublet fibrils normally remain intact. Additional longitudinal fibrous elements outside the axoneme include an outer ring of nine single microtubular fibrils (Fig. 25*a*) found, for example, in sperm of many insects (Phillips, 1970), and an outer ring of nine coarse fibers (Fig. 25*b*) containing homogeneous material that is probably microfilamentous, as found, for example, in mammalian sperm (Fawcett, 1968). In few of these spermatozoa has the pattern of movement been studied in any detail. They generally show propagated helical waves, and in the best known example, the bull spermatozoon, the motion is clearly influenced by the contribution of the outer coarse fibers to contraction and by the increased resistance to bending caused by the presence of the mitochondrial spiral and the fibrous sheath (Gray, 1958; Rikmenspoel, 1965). The curious sperm tail of the fungus gnat *Sciara* described by Phillips (1966) contains an axoneme in which there may be 70 to 90 microtubular doublets, each accompanied by a single microtubule (Fig. 25*c*); the doublets arise from a giant centriole. Detailed information about the motion of this and many other sperm could be most interesting and might give valuable information on mechanisms of motility.

The activity of cilia and flagella is under the control of the organism or cell that bears them; the organelles may be inhibited or activated, and nerves or hormones may be involved in this control (Sleigh, 1966; Kinosita and Murakami, 1967). Cilia normally occur in groups, and several flagella sometimes occur together; in both cases the movement of the organelles normally shows coordination of direction, rate, and phase. The organelles of a group tend to beat in the same direction at the same frequency and

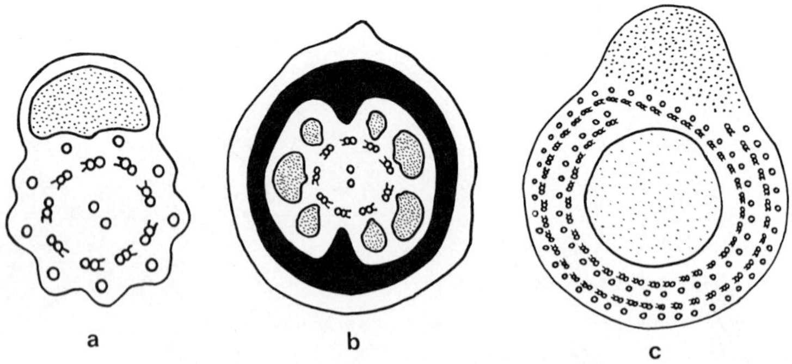

Figure 25. Diagrams of fibrillar structures in some spermatozoa. (*a*) Transverse section of the sperm tail of the caddis fly *Neuronia*. (Based on micrographs by Phillips, 1970.) (*b*) Transverse section of the sperm tail of the guinea pig *Cavia*. (Based on micrographs by Fawcett, 1968.) (*c*) Transverse section of the spermatozoon of *Sciara* showing a spiral of microtubular doublets with associated single microtubules; at this level in the body, the spiral encloses nuclear material and mitochondrial material lies alongside. (Based on micrographs by Phillips, 1966.)

show synchronous beating, metachronous beating, or both synchrony and metachrony within the group of organelles. The coordination of rhythm appears to be most commonly mediated by hydrodynamic interaction between moving organelles, although there is evidence in one or two cases that the basal cytoplasm is concerned in metachronal coordination (Kinosita and Murakami, 1967; Sleigh, 1969); changes in direction of beating of protozoan cilia are correlated with changes in the membrane potential of the cell (Naitoh and Eckert, 1969).

The flagella responsible for the movement of many bacteria are of simpler construction than those in eukaryote organisms. Evidence regarding the structure of bacterial flagella and hypotheses concerning their mode of movement have been reviewed by Lowy and Spencer (1968). Bacterial flagella are cylindrical filaments which range in diameter between approximately 12 and 20 nm, according to species and strain. They have the form of a left-handed helix, the diameter of the helix being approximately 0.5 μ and the pitch usually being between 1.5 and 2.5 μ. There is some variation in the ultrastructure of the filaments, since in some species the flagella are composed entirely of the globular protein flagellin (see Section 4), while in flagella of other species additional components are thought to be present. The flagella of *Salmonella* appear to contain only flagellin which has a molecular weight of about 40,000; isolated flagellin molecules can be made to reaggregate to form filaments that are indistinguishable in the electron microscope from native flagella. Flagellin molecules are 4–5 nm in diameter and are arranged to form the walls of a cylinder in a manner comparable to the arrangement of tubulin molecules in microtubules, but with only about 8 rows of molecules in flagella of *Salmonella typhimurium* and about 10 rows in flagella of *Pseudomonas fluorescens*. These naked flagellar filaments lie outside the membrane of the bacterial cell, to which the basal end of the helix is anchored.

A single bacterial flagellum is of course too small to be seen in the light microscope, but the flagella may aggregate in bundles whose motion can be studied by light microscopy. These helical bundles appear to rotate and, since there is no structural evidence that the flagellum is rotated at its attachment to the cell, Lowy and Spencer suggest that the evidence favors hypotheses based upon shape changes in rows of molecules aligned at a small angle to the long axis of the flagellum. The apparent rotation of the flagellum could result from intramolecular changes (contractions) within these somewhat oblique rows, such that complete rows contract in sequence; alternatively, sequential changes in the orientation of nonspherical molecules within these spiral rows could have the same effect. In either case the sequential changes must be excited from the basal end and, presumably, the energy required for the movement must be imparted to the flagellum at its base.

An organelle with some superficial similarity to the eukaryote flagellum, but with a different structure and a very different form of movement, is the haptonema found on members of one of the groups of algae. A haptonema moves by coiling and uncoiling (Fig. 26*a* and *b*). Leadbeater (1971) found that retraction by coiling is rapid, lasting only approximately 10 msec, and that extension by uncoiling is slow, requiring 2–3 sec. Incidentally, some flagella have also been seen to coil in a comparable manner. The shaft of a haptonema contains about seven microtubular fibrils arranged approximately in a ring and surrounded by three membranes; the outermost of these membranes is continuous with the cell membrane, while the two inner membranes enclose a vesicular cavity and appear to represent an extension of the endoplasmic reticulum

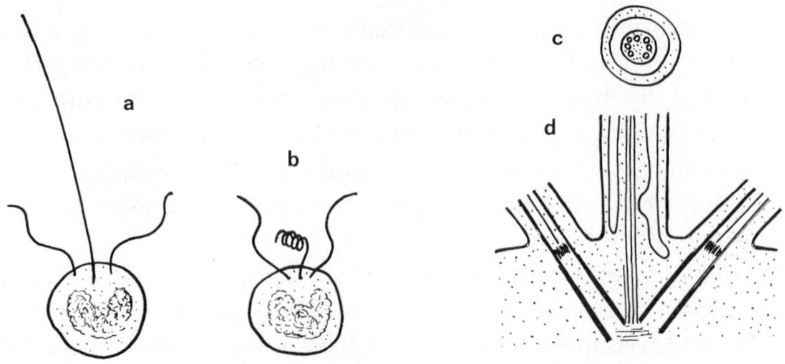

Figure 26. Diagrams of the form and structure of a haptonema. The haptonema in extended (*a*) and contracted (*b*) positions on a flagellate of the Haptophyceae. The structure of a haptonema seen in transverse section (*c*) and in a longitudinal section of the base (*d*) where the relationship between fibrils of haptonema and flagella can be seen. [Based on Manton (1968) and other papers.]

of the cell (Manton, 1968). The fibrils of the haptonema terminate at the base alongside the basal bodies of the flagella; presumably, the coiling of the haptonema could be caused by sliding forces generated between these microtubules. The haptonema appears to be adhesive—it may serve a function in attachment or possibly in the capture of food, since many of these algae are phagotrophs as well as autotrophs.

In the flagellates *Oxymonas*, *Saccinobaculus*, and *Notila*, there is a giant bundle of longitudinal microtubular fibrils which forms an axostyle down the center of the body (Fig. 27) (Grimstone and Cleveland, 1965). The axostyle may perform a variety of movements, but most frequently it undulates sinusoidally, with waves passing backward along the fibril bundle from the anterior end at speeds of the order of 100 μ/sec (McIntosh et al., 1970). These movements may also cause the body to undulate, and they provide the main means of locomotion of the cell, the flagella being less effective. The ribbon-shaped axostyle of a large flagellate may be several hundred microns long and is seen in transverse section as a stack of about 30 rows of microtubules, with

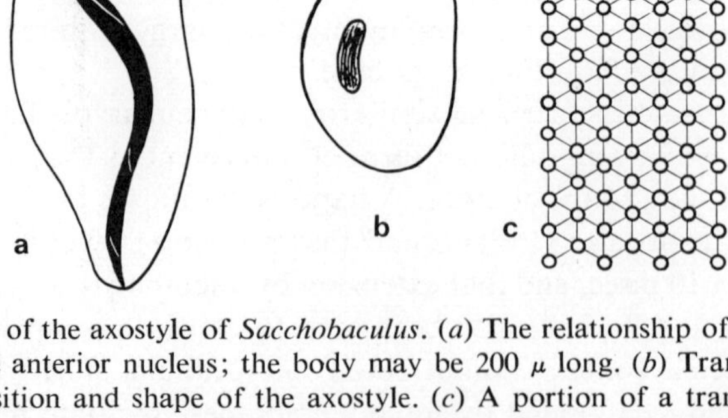

Figure 27. Diagrams of the axostyle of *Sacchobaculus*. (*a*) The relationship of the axostyle with the four flagella and the anterior nucleus; the body may be 200 μ long. (*b*) Transverse section of the cell showing the position and shape of the axostyle. (*c*) A portion of a transverse section of the axostyle showing the rows of microtubules; bending takes place in the direction indicated by the arrows, at right angles to the rows. (Based primarily on information in Grimstone and Cleveland, 1965.)

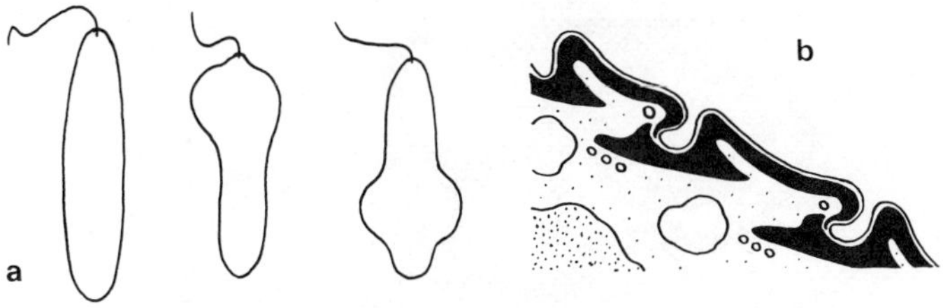

Figure 28. Diagrams of the movement and structure of *Euglena*. (*a*) Stages in the progression of a peristaltic wave of euglenoid movement along the body. (*b*) Portion of a transverse section showing interlocking pellicular strips and associated microtubular fibrils.

about 150 fibrils in each row; generally, the microtubules are more-or-less hexagonally packed within the bundle, and cross-links between fibrils occur between rows and within rows. It is not known whether the microtubules slide or contract during bending of the fibril bundles, but sliding would provide an adequate explanation of the bending.

There are other examples of locomotion involving changes in body shape in which the role of fibrous systems is less obvious. Undulatory movements of the body may be used in swimming or movement through soft substrate, while "peristaltic" movement may be used for progression through mud. The latter form of movement is shown by *Euglena* (Fig. 28) and some of its relatives, in which the pellicle is reinforced by thickened strips underlaid by microtubular fibrils; the part played by these various structures in movement is not understood. Undulatory movements of the body of *Trypanosoma* are associated with the waves of bending that travel from tip to base along the flagellum which is attached laterally; it is not clear whether or not the layer of longitudinal microtubules that lies immediately under the cell membrane (Fig. 29*a*) has any function in movement other than maintaining body shape. A similar array of pellicular microtubules is found in the cigar-shaped bodies of some sporozoa, including many sporozoites and the trophozoites of *Selenidium* (Fig. 29*b*), whose movement involves undulations of the body.

In a wide range of spermatozoa, there are microtubular arrays not associated with centrioles and not part of a 9 + 2 fibrillar axoneme. In some cases these microtubules are transient and are present only during the elongation of the spermatid, and in particular during the elongation of the nucleus, for example, in the dragonfly (Kessel, 1966). Persistent microtubules occur in the flatworm *Dugesia*, in which the elongate head contains 30 to 40 longitudinal microtubules arranged peripherally (Fig. 30*a*) and both the head and the two flagella undulate (Silveira and Porter, 1964). In another flatworm, *Plagiostomum*, the flagella are missing but the microtubular system is present around the nucleus in the long head (Christensen, 1961). The sperm of coccid bugs are shed in syncytial bundles containing many sperm, each consisting of an elongate structure approximately 300 μ long and 0.1–0.3 μ thick (Robison, 1966). Each sperm contains 40 or more longitudinal microtubules forming a scroll-like sheath seen in transverse section as a spiral formed of 1½, 2½, or up to 7 or 8 turns (Fig. 30*b*). On release from the bundle, the separated sperm swim with undulations that are propagated along the body from anterior to posterior ends. Even more unusual motile structures in arthropod sperm are the fine "cellular processes" of tick sperm (Rothschild, 1961; Reger, 1962) and the contractile bands of the sperm of

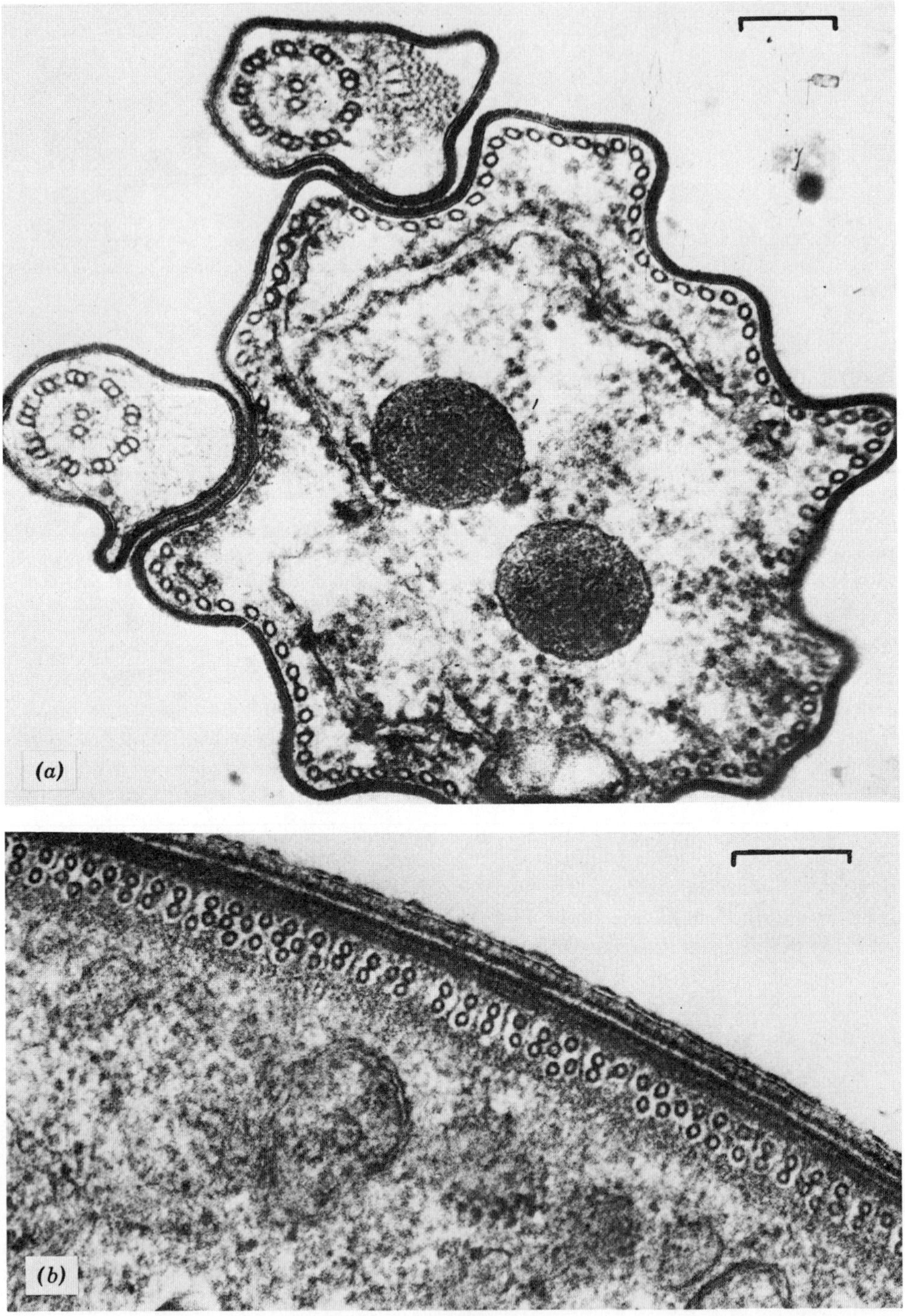

Figure 29. (*a*) Electron micrograph of a transverse section through *Trypanosoma* (fixed during division, after replication of the flagellum) showing the zone of microtubules lying immediately under the surface membrane of the body. (From Vickerman, 1969.) (*b*) Electron micrograph of part of a transverse section of the protozoon *Selenidium* showing the zone of microtubules beneath the cell membranes. Each scale represents 0.2 μ. (Micrograph by A. G. H. Dorey.)

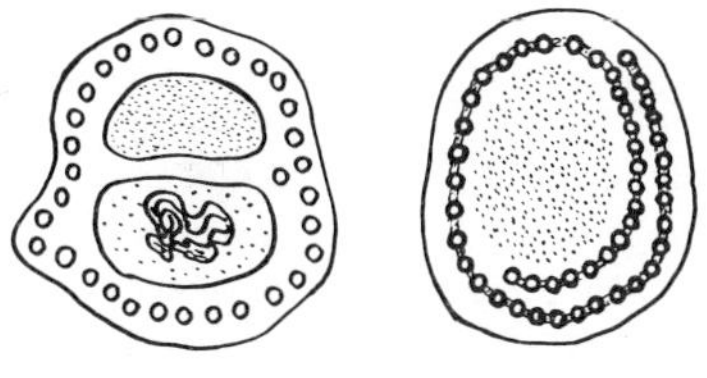

Figure 30. Diagrams of structure of spermatozoa. (*a*) Cross section through the head of a sperm of *Dugesia* in which a row of microtubules underlies the membrane and surrounds nuclear and mitochondrial structures. (Based on micrographs by Silveira, 1970.) (*b*) Cross section through a spermatozoon of the bug *Parlatoria* showing a spiral of microtubules surrounding core material. (Based on micrographs by Robison, 1966.)

ostracods (Gupta, 1968); in both cases fine tubular fibrils approximately 10 nm in diameter are found in the motile structures, but their role in motility is not clear.

4. CONCLUDING REMARKS

In the vast majority of the examples described, fibrous structures clearly form a principal part of the machinery of movement. These fibrous structures have been subdivided into two categories according to whether the elements are microtubular or microfilamentous, but both are constructed from globular protein molecules with a molecular weight in the region of 40,000 to 50,000, and it has been questioned whether the proteins involved belong to the same or different classes. The amino acid composition of several of these proteins has now been elucidated, and some of them are shown in Figure 31 together with the composition of other proteins for comparison. It seems clear that there is a class of actinlike proteins that polymerize as microfilaments which form arrowhead complexes with heavy meromyosin and have amino acid compositions that differ about as much from one another as do the actins from muscles of different animal species. The tubulins form another class of proteins, whose pattern of amino acid composition is quite similar to the actins but different from that of flagellin, myosin, or collagen, and which form into microtubular fibrils. These tubulin fibrils are sensitive to low temperature, high pressure, colchicine, and some metal ions, while the functioning of microfilaments is disrupted by cytochalasin B (Schroeder, 1970; Yamada et al., 1970). The composition of flagellin is quite different from that of any of the other proteins illustrated. Four characteristic groups of proteins involved in motility, actins, tubulins, flagellins, and myosins can thus be recognized. In addition, there are known to be other muscle proteins, and it is likely that dynein and perhaps the protein forming the contractile strands in ciliate myonemes may belong to other categories of movement-producing proteins.

Another set of relationships between the various examples discussed concerns the rates of movement summarized in Table 1. Here there appear to be three or four main categories. Where cytoplasm, particles, or fibers move over the surface of a fiber, irrespective of whether it is microfilamentous or microtubular, the speed or movement lies between approximately 1 and 50 μ/sec, most commonly approximately 10 μ/sec. Another form of movement is associated with the growth or disassembly of microtubular fibrils, and in this case the speed of movement is slower, between 0.03 and 0.4 μ/sec. One form of much faster motion, the cytoplasmic streaming of plasmodial slime molds, could be caused indirectly by pressure exerted on the cytoplasm when a

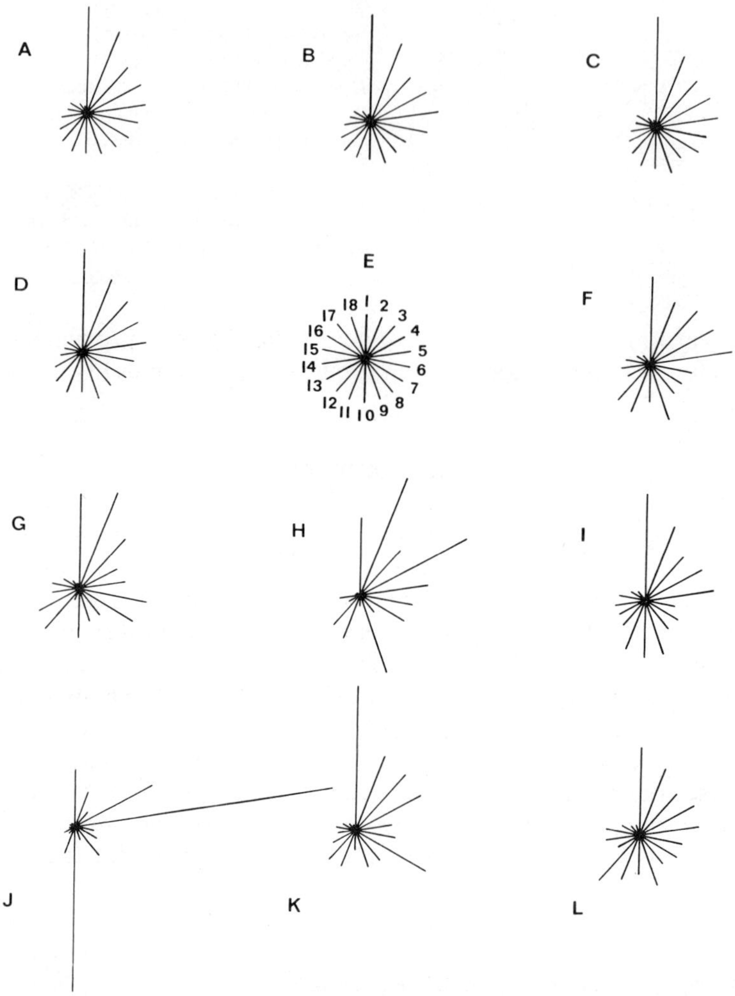

Figure 31. A comparison of the amino acid compositions of some proteins. Each line represents an amino acid according to the angular positions indicated in (*E*), and the length of each line represents the relative proportion of that amino acid in the protein based on the number of moles of that amino acid in 1000 moles of amino acid recovered from the protein. (*A*) Tubulin from A subfibrils of sperm tails of *Strongylocentrotus droebachiensis* (Stephens, 1970). (*B*) Tubulin from B subfibrils of sperm tails of *Strongylocentrotus droebachiensis* (Stephens, 1970). (*C*) Tubulin from central fibrils of sperm tails of *Strongylocentrotus purpuratus* (Shelanski and Taylor, 1968). (*D*) Tubulin from microtubules of mouse neuroblastoma cells (Olmsted et al., 1971). (*E*) *1*, Glutamic acid; *2*, aspartic acid; *3*, leucine; *4*, alanine; *5*, glycine; *6*, valine; *7*, lysine [(includes *N*-methyllysine in (*H*) and hydroxylysine in (*J*)]; *8*, arginine; *9*, threonine; *10*, proline [includes hydroxyproline in (*J*)]; *11*, serine; *12*, isoleucine; *13*, phenylalanine; *14*, tyrosine; *15*, methionine; *16*, histidine [includes 3-methylhistidine in (*F*) and (*L*)]; *17*, one-half cystine; *18*, tryptophan. (*F*) Actin from *Acanthamoeba* (Weihing and Korn, 1969). (*G*) Protein from mitotic spindles of *Strongylocentrotus droebachiensis* (Stephens, 1967). (*H*) Flagellin from *Salmonella* (McDonough, 1965). (*I*) Actin from *Physarum* (Hatano and Oosawa, 1966). (*J*) Collagen from human tendon (Eastoe, 1967). (*K*) Myosin from rabbit striated muscle (Kominz et al., 1954). (*L*) Actin from rabbit striated muscle (Johnson and Perry, 1968).

Table 1. Rates of Movement in Various Cells

Form of movement[a]	Speed (μ/sec)	Reference
Movements in which the displacement of structures with respect to individual fibers can be measured:		
Sliding of actin filaments along myosin filaments, striated muscle (R)	7	Wolpert (1965)
Cytoplasmic streaming in *Chara* (mf) (C)	60	Hayashi (1964)
Cytoplasmic streaming in *Nicotiana* (mf?) (C)	5–10	Mahlberg (1964)
Cytoplasmic streaming in *Paramecium* (mf?) (C)	5	Koenuma (1963)
Sliding of ciliary peripheral doublets along one another (mt) (R)	1–20	Sleigh (1971)
Food intake through pharyngeal basket of *Nassula* (mt) (C)	7–20	Tucker (1968)
Cytoplasmic streaming in *Caulerpa* (mt) (C)	3–5	Sabnis and Jacobs (1967)
Cytoplasmic streaming in *Actinosphaerium* axopod (mt) (C)	20	Jahn and Bovee (1969)
Cytoplasmic streaming in nerve cells (mt? or perhaps mf) (C)	30	Jasinski et al. (1966)
Cytoplasmic streaming in Foraminifera (mt?) (C)	8–15	Jahn and Rinaldi (1959)
Pigment migration in fish melanophores (mt?) (R)	0.2–0.7	Kinosita (1963)
Pseudopod extension and retraction of *Actinosphaerium* (mt) (R)	0.1–0.4	Watters (1968)
Mitotic movement of chromosomes (mt extension and retraction) (S)	0.1	Forer (1969)
Mitotic movement of chromosomes in *Nassula* (mt extension) (S)	0.03	Tucker (1967)
Movements in which the displacement measured probably results from the activity of a whole fibrous system and does not directly indicate the movement relative to individual fibers:		
Cytoplasmic flow of *Chaos* (mf system) (C)	30	Allen and Roslansky (1959)
Cytoplasmic flow of *Physarum* (mf system) (C)	1,350	Kamiya (1959)
Stalk shortening in *Vorticella* (mf system) (R)	20–50,000	Amos (1970)
Body shortening of *Stentor* (mf system) (R)	63,000	Cragg and Sleigh (1970)
Body shortening of *Spirostomum* (mf system) (R)	200,000	Jones et al. (1966)
Shortening of ectoderm cells of *Amaroucium* (mf system) (S)	0.07	Cloney (1966)
Contractile ring of first cleavage of sea urchin egg (mf system) (S)	1	Wolpert (1965)
Body elongation of *Stentor* (mt system) (R)	17	Cragg and Sleigh (1970)

[a] mf, microfilaments; mt, microtubules; R, movements that may be repeated many times; C, movements that may be continued intermittently or continuously for a long period; S, movements that are performed only once by that structure.

fibrillar system in the wall of the tubular plasmodium contracts by an interaction of sliding fibrils; the rate of movement of the fibrils could lie within the range commonly found for other movements involving sliding fibrils. In the rapid contraction of the filament systems of the *Vorticella* spasmoneme and the *Spirostomum* myoneme, it has not yet been possible to determine the behavior of the filaments; if the motion involves the folding of long filaments, a very rapid movement may be needed, but if it results from the sliding of many short, overlapping filaments, the units of the system could move at a speed within the usual range.

At the present stage of our researches into cell motility, there is clear evidence for two fundamentally different forms of motility—that involving extension or retraction of fibrils by growth or disassembly, and that involving the active progression of cytoplasmic components along fixed fibrils. The latter appears in many forms, including muscle, cytoplasmic streaming, and probably cilia, and can be subdivided according to the form of fibrils into microtubular systems in which the fibrils are required to resist compression as well as extension and microfilamentous systems in which the fibrils resist only extension. Other possible mechanisms of movement may also exist, for example, the contraction of ciliate myonemes by the folding of filaments, and the deformation of helical tubular fibrils of bacterial flagella by molecular deformations or rearrangements.

REFERENCES

Allen, R. D. (1964). In *Primitive Motile Systems in Cell Biology* (R. D. Allen and N. Kamiya, eds.), pp. 407–431. Academic, New York.

Allen, R. D. (1968). Differences of a fundamental nature among several types of amoeboid movement. *Symp. Soc. Exptl. Biol.*, **22,** 151.

Allen, R. D., and J. D. Roslansky (1959). The consistency of amoeba cytoplasm and its bearing on the mechanism of amoeboid movement. I. An analysis of endoplasmic velocity profiles of *Chaos chaos* (L.). *J. Biophys. Biochem. Cytol.*, **6,** 437.

Allen, R. D., J. W. Cooledge, and P. J. Hall (1960). Streaming in cytoplasm dissociated from the giant amoeba *Chaos chaos*. *Nature*, **187,** 896.

Ambrose, E. J., and J. A. Forrester (1968). Electrical phenomena associated with cell movements. *Symp. Soc. Exptl. Biol.*, **22,** 237.

Amos, W. B. (1970). Aspects of contraction in the peritrich stalk. Ph.D. Thesis, University of Cambridge.

Amos, W. B. (1971). Reversible mechanochemical cycle in the contraction of *Vorticella*. *Nature*, **229,** 127.

Bajer, A. (1968). Chromosome movement and fine structure of the mitotic spindle. *Symp. Soc. Exptl. Biol.*, **22,** 285.

Baker, P. C. (1965). Fine structure and morphogenetic movements in the gastrula of the treefrog, *Hyla regilla*. *J. Cell Biol.*, **24,** 95.

Baker, P. F., and T. E. Schroeder (1967). Cytoplasmic filaments and morphogenetic movement in the amphibian neural tube. *Develop. Biol.*, **15,** 432.

Bannister, L. H., and E. C. Tatchell (1968). Contractility and the fiber systems of *Stentor coeruleus*. *J. Cell Sci.*, **3,** 295.

Bardele, C. F. (1970). Comparative ultrastructural studies on Centrohelida. *J. Protozool., Suppl.*, **17,** 10.

Bardele, C. F., and K. G. Grell (1967). Elektronenmikroskopische Beobachtungen zur Nahrungsaufnahme bei dem Suktor *Acineta tuberosa* Ehrenberg. *Z. Zellforsch. Mikroskop. Anat.*, **80,** 108.

Bertolini, B., G. Monaco, and A. Rossi (1970). Ultrastructure of a regular arrangement of microtubules and neurofilaments. *J. Ultrastruct. Res.*, **33,** 173.

Bikle, D., L. G. Tilney, and K. R. Porter (1966). Microtubules and pigment migration in the melanophores of *Fundulus heteroclitus* L. *Protoplasma*, **61,** 322.

Bluemink, J. G. (1970). The first cleavage of the amphibian egg. An electron microscope study of the onset of cytokinesis in egg of *Ambystoma mexicanum*. *J. Ultrastruct. Res.*, **32,** 142.

Bovee, E. C. (1964). In *Primitive Motile Systems in Cell Biology* (R. D. Allen and N. Kamiya, eds.), pp. 189–219. Academic, New York.

Brokaw, C. J. (1966). Mechanics and energetics of cilia. *Am. Rev. Respirat. Diseases*, **93,** No. 3, Part 2, 32.

Brokaw, C. J. (1970). Bending moments in free-swimming flagella. *J. Exptl. Biol.*, **53,** 445.

Carasso, N., and P. Favard (1966). Mise en évidence du calcium dans les myonèmes pédonculaires de ciliés péritriches. *J. Microscopie*, **5,** 759.

Carter, S. B. (1967). Effects of cytochalasins on mammalian cells. *Nature*, **213,** 261.

Christensen, K. (1961). Fine structure of an unusual spermatozoon in the flatworm *Plagiostomum*. *Biol. Bull.*, **121,** 416.

Cleveland, L. R. (1954). Hormone-induced sexual cycles of flagellates. XII. Meiosis in *Barbulanympha* following fertilization, autogamy and endomitosis. *J. Morphol.*, **95,** 557.

Cloney, R. A. (1966). Cytoplasmic filaments and cell movements: Epidermal cells during ascidian metamorphosis. *J. Ultrastruct. Res.*, **14,** 300.

Cragg, P. A., and M. A. Sleigh (1970). Unpublished observations.

Davies, R. E. (1963). A molecular theory of muscle contraction. Calcium-dependent contractions with hydrogen bond formation plus ATP-dependent extension of part of the myosin-actin cross-bridges. *Nature*, **199,** 1068.

Dorey, A. G. H. (1970). A study of some species of *Selenidium* (Sporozoa, Archigregarinida) from polycheates of the family Terebellidae. Ph.D. Thesis, University of Bristol.

Eastoe, J. E. (1967). In *Treatise on Collagen* (G. N. Ramachandran, ed.), Vol. 1, pp. 1–72. Academic, New York.

Ettienne, E. M. (1970). Control of contractility in *Spirostomum* by dissociated calcium ions. *J. Gen. Physiol.*, **56,** 168.

Favard, P., and N. Carasso (1965). Mise en évidence d'un reticulum endoplasmique dans le spasmonème de ciliés péritriches. *J. Microscopie*, **4,** 567.

Fawcett, D. W. (1968). Topographical relationship between the plane of the central pair of flagellar fibrils and the transverse axis of the head in guinea-pig spermatozoa. *J. Cell Sci.*, **3,** 187.

Finley, H. E., C. A. Brown, and W. A. Daniel (1964). Electron microscopy of the ectoplasm and infraciliature of *Spirostomum ambiguum*. *J. Protozool.*, **11,** 264.

Forer, A. (1968). In *Handbook of Molecular Cytology* (A. Lima-de-Faria, ed.), pp. 553–601. North Holland, Amsterdam.

Gibbons, I. R. (1965). Chemical dissection of cilia. *Arch. Biol. (Liège)*, **76,** 317.

Goldacre, R. J., and I. J. Lorch (1950). Folding and unfolding of protein molecules in relation to cytoplasmic streaming, amoeboid movement and osmotic work. *Nature*, **166,** 497.

Gray, J. (1958). The movement of the spermatozoa of the bull. *J. Exptl. Biol.*, **35,** 96.

Griffin, J. L. (1964). In *Primitive Motile Systems in Biology* (R. D. Allen and N. Kamiya, eds.), pp. 303–321. Academic, New York.

Grimstone, A. V., and L. R. Cleveland (1965). The fine structure and function of the contractile axostyles of certain flagellates. *J. Cell Biol.*, **24,** 387.

Gupta, B. L. (1968). Aspects of motility in the non-flagellate spermatozoa of freshwater ostracods. *Symp. Soc. Exptl. Biol.*, **22,** 117.

Halfen, L. N., and R. W. Castenholz (1970). Gliding in a blue-green alga: A possible mechanism. *Nature*, **225,** 1163.

Harper, M. A., and J. F. Harper (1967). Measurements of diatom adhesion and their relationship with movement. *Brit. Phycol. Bull.* **3,** 195.

Hatano, S., and F. Oosawa (1966). Isolation and characterization of plasmodial actin. *Biochim. Biophys. Acta*, **127,** 488.

Hayashi, T. (1964). In *Primitive Motile Systems in Cell Biology* (R. D. Allen and N. Kamiya, eds.), pp. 19–28, Academic, New York.

Hoffmann-Berling, H. (1958). Der Mechanismus eines neuen, von der Muskelkontraktion verschiedenen Kontractionszyklus. *Biochim. Biophys. Acta*, **27,** 247.

Hollande, A., J. Cachon, M. Cachon-Enjumet, and J. Valentin (1967). Infrastructure des axopodes et organisation generale de *Sticholonche zanclea* (Hertwig) (Radiolaire, Sticholonchidae). *Protistologica*, **3,** 155.

Holwill, M. E. J. (1964). The motion of *Strigomonas oncopelti*. *J. Exptl. Biol.*, **42,** 125.

Holwill, M. E. J. (1966a). Physical aspects of flagellar movement. *Physiol. Rev.*, **46,** 696.

Holwill, M. E. J. (1966b). The motion of *Euglena viridis:* The role of flagella. *J. Exptl. Biol.*, **44,** 579.

Holwill, M. E. J., and M. A. Sleigh (1967). Propulsion by hispid flagella. *J. Exptl. Biol.*, **47,** 267.

Huang, B. (1970). Ultrastructure of the cortical fiber systems in *Stentor coeruleus* relaxed in EGTA. *J. Cell Biol.*, **47,** 92a.

Inoué, S. (1952a). The effect of colchicine on the microscopic and submicroscopic structure of the mitotic spindle. *Exptl. Cell Res., Suppl.*, **2,** 305.

Inoué, S. (1952b). Effects of temperature on the birefringence of the mitotic spindle. *Biol. Bull.*, **103,** 316.

Inoué, S., and H. Sato (1967). Cell motility by labile association of molecules. The nature of mitotic spindle fibers and their role in chromosome movement. *J. Gen. Physiol.*, **50,** No. 6, Part 2, 259.

Ishikawa, H., R. Bischoff, and H. Holtzer (1969). Formation of arrowhead complexes with heavy meromyosin in a variety of cell types. *J. Cell Biol.*, **43,** 312.

Jahn, T. L., and E. C. Bovee (1969). Protoplasmic movements within cells. *Physiol. Rev.*, **49,** 793.

Jahn, T. L., and R. A. Rinaldi (1959). Protoplasmic movement in the foraminiferan, *Allogromia laticollaris*, and a theory of its mechanism. *Biol. Bull.*, **117,** 100.

Jahn, T. L., W. M. Harmon, and M. Landman (1963). Mechanisms of locomotion in flagellates. I. *Ceratium*. *J. Protozool.*, **10,** 358.

Jarosch, R. (1962). In *Physiology and Biochemistry of Algae* (R. A. Lewin, ed.), pp. 573–581. Academic, New York.

Jasinski, A., A. Gorbman, and T. J. Hara (1966). Rate of movement and redistribution of stainable neurosecretory granules in hypothalamic neurons. *Science*, **154,** 776.

Jenkins, R. A. (1967). Fine structure of division in ciliate protozoa. I. Micronuclear mitosis in *Blepharisma*. *J. Cell Biol.*, **34,** 463.

Johnson, P., and S. V. Perry (1968). Chemical studies on the cysteine and terminal peptides in tryptic digests of actin. *Biochem. J.*, **110,** 207.

Jones, A. R., T. L. Jahn, and J. R. Fonseca (1966). Contraction of protoplasm. I. Cinematographic analysis of the anodally stimulated contraction of *Spirostomum ambiguum*. *J. Cell. Physiol.*, **68,** 127.

Jones, A. R., T. L. Jahn, and J. R. Fonseca (1970a). Contraction of protoplasm. III. Cinematographic analysis of the contraction of some heterotrichs. *J. Cell Physiol.*, **75,** 1.

Jones, A. R., T. L. Jahn, and J. R. Fonseca (1970b). Contraction of protoplasm. IV. Cinematographic analysis of the contraction of some peritrichs. *J. Cell Physiol.*, **75,** 9.

Kamitsubo, E. (1966a). Motile protoplasmic fibrils in cells of Characeae. I. Movement of fibrillar loops. *Proc. Japan Acad.*, **42,** 507.

Kamitsubo, E. (1966b). Motile protoplasmic fibrils in cells of Characeae. II. Linear fibrillar structure and its bearing on protoplasmic streaming. *Proc. Japan Acad.*, **42,** 640.

Kamiya, N. (1959). *Protoplasmic Streaming*. Protoplasmatologia XIII, No. 3a, p. 1. Springer, Vienna.

Kamiya, N. (1968). The mechanism of cytoplasmic movement in a myxomycete plasmodium. *Symp. Soc. Exptl. Biol.*, **22,** 199.

Kamiya, N., and K. Kuroda (1958). Studies on the velocity distribution of the protoplasmic streaming in the myxomycete plasmodium. *Protoplasma*, **48,** 1.

Kessel, R. G. (1966). The association between microtubules and nuclei during spermiogenesis in the dragonfly. *J. Ultrastruct. Res.*, **16,** 293.

Kiefer, B. I. (1970). Development, organization, and degeneration of the *Drosophila* sperm flagellum. *J. Cell Sci.*, **6,** 177.

Kinosita, H. (1963). Electrophoretic theory of pigment migration within fish melanophore. *Ann. N.Y. Acad. Sci.*, **100,** 992.

Kinosita, H., and A. Murakami (1967). Control of ciliary motion. *Physiol. Rev.*, **47,** 53.

Kitching, J. A. (1957). Effects of high hydrostatic pressures on *Actinophrys sol* (Heliozoa). *J. Exptl. Biol.*, **34,** 511.

Kitching, J. A. (1964). In *Primitive Motile Systems in Cell Biology* (R. D. Allen and N. Kamiya, eds.), pp. 445–456. Academic, New York.

Koenuma, A. (1963). The velocity distribution of the cyclosis in *Paramecium caudatum*. *Annotationes Zool. Japon.*, **36,** 66.

Kominz, D. R., A. Hough, P. Symonds, and K. Laki (1954). The amino acid composition of actin, myosin, tropomyosin and the meromyosins. *Arch. Biochem. Biophys.*, **50,** 148.

Kuroda, K. (1964). In *Primitive Motile Systems in Cell Biology* (R. D. Allen and N. Kamiya, eds.), pp. 31–40. Academic, New York.

Leadbeater, B. (1971). Observations by means of cine photography on the behaviour of the haptonema in planktonic flagellates of the class Haptophyceae. *J. Marine Biol. Assoc., U.K.*, **51,** 207.

Levine, L. (1956). Contractility of glycerinated vorticellae. *Biol. Bull.*, **111,** 319.

Lissman, H. W. (1945). The mechanism of locomotion in gastropod molluscs. II. Kinetics. *J. Exptl. Biol.*, **22,** 37.

Lowy, J., and M. Spencer (1968). Structure and function of bacterial flagella. *Symp. Soc. Exptl. Biol.*, **22,** 215.

McDonough, M. W. (1965). Amino acid composition of flagellin of antigenically distinct *Salmonella* flagellar proteins. *J. Mol. Biol.*, **12,** 342.

McGee-Russell, S. M., and R. D. Allen (1969). Combined observations by light and electron microscopy of the motile cytoplasm of *Allogromia*, a marine foraminiferan. *J. Cell Biol.*, **43,** 88a.

MacGregor, H. C., and H. Stebbings (1970). A massive system of microtubules associated with cytoplasmic movement in telotrophic ovaries. *J. Cell Sci.*, **6,** 431.

McIntosh, J. R., S. Cleland, and E. S. Ogata (1970). Motion and structure of the axostyle in *Saccinobaculus*. *J. Cell Biol.*, **47,** 134a.

Mahlberg, P. G. (1964). In *Primitive Motile Systems in Cell Biology* (R. D. Allen and N. Kamiya, eds.), pp. 43–67. Academic, New York.

Manton, I. (1968). Further observations on the microanatomy of the haptonema in *Chrysochromulina chiton* and *Prymnesium parvum*. *Protoplasma*, **66,** 35.

Nachmais, V. T., H. E. Huxley, and D. Kessler (1970). Electron microscope observations on actomyosin and actin preparations from *Physarum polycephalum*, and on their interaction with heavy meromyosin subfragment I from muscle myosin. *J. Mol. Biol.*, **50,** 83.

Nagai, R., and L. I. Rebhun (1966). Cytoplasmic microfilaments in streaming *Nitella* cells. *J. Ultrastruct. Res.*, **14,** 571.

Naitoh, Y., and R. Eckert (1969). Ciliary orientation: Controlled by cell membrane or by intracellular fibrils? *Science*, **166,** 1633.

Nakai, J. (1964). In *Primitive Motile Systems in Cell Biology* (R. D. Allen and N. Kamiya, eds.), pp. 377–385. Academic, New York.

Nevo, A. C., and R. Rikmenspoel (1970). Diffusion of ATP in sperm flagella. *J. Theoret. Biol.*, **26,** 11.

O'Brien, T. R., and K. V. Thimann (1966). Intracellular fibers in oat coleoptile cells and their possible significance in cytoplasmic streaming. *Proc. Natl. Acad. Sci., U.S.*, **56,** 888.

Olmstead, J. B., K. Carlson, R. Klebe, F. Ruddle, and J. Rosenbaum (1970). Isolation of microtubule protein from cultured mouse neuroblastoma cells. *Proc. Natl. Acad. Sci., U.S.*, **65,** 129.

Palay, S. L., C. Sotelo, A. Peters, and P. M. Orkand (1968). The axon hillock and the initial segment. *J. Cell Biol.*, **38,** 193.

Pease, D. C. (1946). High hydrostatic pressure effects upon the spindle figure and chromosome movement. II. Experiments on the meiotic divisions of *Tradescantia* pollen mother cells. *Biol. Bull.*, **91,** 145.

Phillips, D. M. (1966). The fine structure of *Sciara coprophila* sperm. *J. Cell Biol.*, **30,** 499.

Phillips, D. M. (1970). Insect sperm: Their structure and morphogenesis. *J. Cell Biol.*, **44,** 243.

Pollard, T. D., and S. Ito (1970). Cytoplasmic filaments of *Amoeba proteus*. I. The role of filaments in consistency change and movement. *J. Cell Biol.*, **46,** 267.

Pollard, T. D., E. Shelton, R. R. Weihing, and E. D. Korn (1970). Ultrastructural characterization of F-actin isolated from *Acanthamoeba castellanii* and identification of cytoplasmic filaments as F-actin by reaction with rabbit heavy meromyosin. *J. Mol. Biol.*, **50,** 91.

Randall, J. T., and S. F. Jackson (1958). Fine structure and function in *Stentor polymorphus*. *J. Biophys. Biochem. Cytol.*, **4,** 807.

Rebhun, L. I. (1967). Structural aspects of saltatory particle movement. *J. Gen. Physiol.*, **50,** No. 6, Part 2, 223.

Reger, J. F. (1962). A fine-structure study on spermiogenesis in the tick *Amblyomma dissimili*, with special reference to the development of motile processes. *J. Ultrastruct. Res.*, **7,** 550.

Rikmenspoel, R. (1965). The tail movement of bull spermatozoa. Observations and model calculations. *Biophys. J.*, **5,** 365.

Ringo, D. L. (1967). The arrangement of subunits in flagellar fibres. *J. Ultrastruct. Res.*, **17,** 266.

Robison, W. G. (1966). Microtubules in relation to the motility of a sperm syncytium in an armoured scale insect. *J. Cell Biol.*, **29,** 251.

Roth, L. E., and Shigenaka, Y. (1970). Microtubules in the heliozoan axopodium. II. Rapid degradation by cupric and nickelous ions. *J. Ultrastruct. Res.*, **31,** 356.

Rothschild, N. M. V. (1961). Structure and movements of tick spermatozoa. *Quart. J. Microscop. Sci.*, **102,** 239.

Rudzinska, M. A. (1965). The fine structure and function of the tentacle in *Tokophrya infusionum*. *J. Cell Biol.*, **25,** 459.

Sabnis, D. D., and W. P. Jacobs (1967). Cytoplasmic streaming and microtubules in the coenocytic marine alga *Caulerpa prolifera*. *J. Cell Sci.*, **2,** 465.

Satir, P. (1965). Studies on cilia. II. Examination of the distal region of the ciliary shaft and the role of the filaments in motility. *J. Cell Biol.*, **26,** 805.

Satir, P. (1968). Studies on cilia. III. Further studies on the cilium tip and a "sliding filament" model of ciliary motility. *J. Cell Biol.*, **39,** 77.

Schmitt, F. O. (1968). Fibrous proteins—Neuronal organelles. *Proc. Natl. Acad. Sci., U.S.*, **60,** 1092.

Schroeder, T. E. (1968). Cytokinesis: Filaments in the cleavage furrow. *Exptl. Cell Res.*, **53,** 272.

Schroeder, T. E. (1970). The contractile ring. I. Fine structure of dividing mammalian (HeLa) cells and the effects of cytochalasin B. *Z. Zellforsch. Mikroskop. Anat.*, **109,** 431.

Selman, G. G., and M. M. Perry (1970). Ultrastructural changes in the surface layers of the newt's egg in relation to the mechanism of its cleavage. *J. Cell Sci.*, **6,** 207.

Senda, N., N. Shibata, N. Tatsumi, K. Kondo, and K. Hamada (1969). A contractile protein from leucocytes. Its extraction and some of its properties. *Biochim. Biophys. Acta*, **181,** 191.

Shelanski, M. L., and E. W. Taylor (1968). Properties of the protein subunit of central-pair and outer-doublet microtubules of sea urchin flagella, *J. Cell Biol.*, **38,** 304.

Shepro, D., F. A. Belamarich, F. B. Merk, and F. C. Chao (1969). Changes in thrombocyte ultrastructure during clot retraction. *J. Cell Sci.*, **4,** 763.

Shepro, D., F. A. Belamarich, L. Robblee, and F. C. Chao (1970). Antimotility effect of cytochalasin B observed in mammalian clot retraction. *J. Cell Biol.*, **47,** 544.

Silveira, M. (1970). Characterization of an unusual nucleus by electron microscopy. *J. Submicroscop. Cytol.*, **2,** 13.

Silveira, M., and K. R. Porter (1964). The spermatozoids of flatworms and their microtubular systems. *Protoplasma*, **59,** 240.

Sleigh, M. A. (1966). The coordination and control of cilia. *Symp. Soc. Exptl. Biol.*, **20,** 11.

Sleigh, M. A. (1968). Patterns of ciliary beating. *Symp. Soc. Exptl. Biol.*, **22,** 131.

Sleigh, M. A. (1970). Some factors affecting the excitation of contraction in *Spirostomum. Acta Protozool.*, **7,** 335.

Sleigh, M. A. (1971). Cilia. *Endeavour*, **30,** 11.

Stephens, R. E. (1967). The mitotic apparatus. Physical chemical characterization of the 22S protein component and its subunits. *J. Cell Biol.*, **32,** 255.

Stephens, R. E. (1970). Thermal fractionation of outer fiber doublet microtubules into A- and B-subfiber components: A and B tubulin. *J. Mol. Biol.*, **47,** 353.

Sugi, H. (1961). Volume change during contraction in the stalk muscle of *Carchesium. J. Fac. Sci. Univ. Tokyo, Sect. IV*, **9,** 155.

Szollosi, D. (1970). Cortical cytoplasmic filaments of cleaving eggs: A structural element corresponding to the contractile ring. *J. Cell Biol.*, **44,** 192.

Tilney, L. G. (1968). Studies on the microtubules in Heliozoa. IV. The effect of colchicine on the formation and maintenance of axopodia and the redevelopment of pattern in *Actinosphaerium nucleofilum* (Barrett). *J. Cell Sci.*, **3,** 549.

Tilney, L. G., and K. R. Porter (1967). Studies on microtubules in Heliozoa. II. The effect of low temperature on these structures in the formation and maintenance of the axopodia. *J. Cell Biol.*, **34,** 327.

Tilney, L. G., Y. Hiramoto, and D. Marsland (1966). Studies on the microtubules in Heliozoa. III. A pressure analysis of the role of these structures in the formation and maintenance of the axopodia of *Actinosphaerium nucleofilum* (Barrett). *J. Cell Biol.*, **29,** 77.

Tucker, J. B. (1967). Changes in nuclear structure during binary fission in the ciliate *Nassula. J. Cell Sci.*, **2,** 481.

Tucker, J. B. (1968). Fine structure and function of the cytopharyngeal basket in the ciliate *Nassula. J. Cell Sci.*, **3,** 493.

Tucker, J. B. (1970). Morphogenesis of a large microtubular organelle and its association with basal bodies in the ciliate *Nassula. J. Cell Sci.*, **6,** 385.

Tucker, J. B. (1971). Microtubules and a contractile ring of microfilaments associated with a cleavage furrow. *J. Cell Sci.*, **8,** 557.

Vickerman, K. (1969). On the surface coat and flagellar adhesion in trypanosomes. *J. Cell Sci.*, **5,** 163.

Vickerman, K., and T. M. Preston (1970). Spindle microtubules in the dividing nuclei of trypanosomes. *J. Cell Sci.*, **6,** 365.

Warner, F. D. (1970). New observations on flagellar fine structure. The relationship between matrix structure and the microtubule component of the axoneme. *J. Cell Biol.*, **47,** 159.

Watters, C. (1968). Studies on the motility of the Heliozoa. I. The locomotion of *Actinosphaerium eichhorni* and *Actinophrys* sp. *J. Cell Sci.*, **3,** 231.

Watts, D. C., and L. H. Bannister (1970). Location of arginine kinase in the cilia of *Tetrahymena pyriformis. Nature*, **226,** 450.

Weihing, R. R., and E. D. Korn (1969). Ameba actin: The presence of 3-methylhistidine. *Biochem. Biophys. Res. Commun.*, **35,** 906.

Wohlman, A., and R. D. Allen (1968). Structural organization associated with pseudopod extension and contraction during cell locomotion in *Difflugia. J. Cell Sci.*, **3,** 105.

Wolpert, L. (1965). Cytoplasmic streaming and amoeboid movement. *Symp. Soc. Gen. Microbiol.*, **15,** 270.

Wolpert, L., and D. Gingell (1968). Cell surface membrane and amoeboid movement. *Symp. Soc. Exptl. Biol.*, **22,** 169.

Wolpert, L., C. M. Thompson, and C. H. O'Neill (1964). In *Primitive Motile Systems in Cell Biology* (R. D. Allen and N. Kamiya, eds.), pp. 143–168. Academic, New York.

Yamada, K. M., B. S. Spooner, and N. K. Wessells (1970). Axon growth: Roles of microfilaments and microtubules. *Proc. Natl. Acad. Sci., U.S.*, **66,** 1206.

17

CELL AGGREGATION

A. A. Moscona

1. INTRODUCTION

The organization of cells in metazoan systems into tissues and organs depends on the ability of individual cells to be linked in specific ways and ordered into characteristic patterns. These multicellular patterns arise in the course of embryonic development and differentiation; the mechanisms that direct the assembly of cells into tissues are dependent upon the genomic, phenomic, and environmental factors that control embryonic development.

One of the basic aspects of embryonic development is that throughout its progression cells and cell groups leave their place of origin and move, individually or in groups, to new sites; by migrating through tissues, between tissue layers or between cells, or by way of the bloodstream they arrive in specific, sometimes quite distant areas. There the cells reassemble and reassociate with each other and with local cells and form new structural entities within which differentiation, histogenesis, and morphogenesis take place.

The overall processes involved in the assembly of cells into tissue are too complex for study by simple morphological or biochemical dissection. In order to unravel the laws that govern assembly and association of cells into tissues and organs, morphological and biochemical dissection must proceed hand in hand with synthesis, that is, with attempts to construct, under controllable, experimental conditions, complex systems from simple units. This involves compounding living cells into multicellular systems and synthesizing tissues from cellular units outside the living body.

While this may have seemed an unrealistic expectation not too long ago, it became a feasible proposition with the advent of the necessary experimental methods; today, construction of tissues *in vitro*, by recombination and aggregation of cells in suspension, has become an important tool in studies on cell dynamics, cell sociology, and the properties of the cell surface in relation to mechanisms of cell assembly, differentiation, cell communication, and multicellular organization. This chapter briefly surveys some of these problems, describes selected experimental results, and emphasizes future prospects and issues; more detailed information is provided by reviews and numerous articles on various aspects of this subject.

2. DISSOCIATION AND REASSOCIATION OF CELLS IN THE EMBRYO

Before describing experimental studies *in vitro*, and in order to frame the nature and importance of the problems discussed here, we discuss briefly some of the more outstanding examples of normal dispersal, recombination, and aggregation of cells within the organism during embryonic development.

An excellent example of this is provided by the embryonic history of the primordial germ cells (PGC), that is, the precursor cells of the gametes. This history has been studied in some detail in the chick embryo in which the PGC arise outside the future gonad, outside the embryo proper, and are first seen as relatively large cells in the extraembryonic endoderm in an area called the "germ cell crescent." Surgical removal of this crescent in the early embryo results in a sterile but otherwise normal organism. In the early embryo the PGC dissociate themselves from the neighboring endodermal cells, enter adjacent capillaries, and pass into the blood stream. Although they circulate freely throughout the vascular system, they converge and assemble in a vein nearest the site of the future gonad, attach to the inner wall of this vein, exit, migrate into the gonad site, and settle there. They associate with the somatic cells of the gonad and proceed to develop into the definitive germ cells. This sequence of events was confirmed by experiments (Simon, 1960) utilizing pairs of joint (parabiosed) early chick embryos, one of which had had its PGC surgically removed before parabiosis. PGC from the intact embryo passed through the joint circulation into the deficient embryo also and populated its gonad as well. This also occurred when the intact partner in the parabiosis was a duck embryo and was joined with a deficient chick embryo; the duck PGC entered the joint circulation and settled selectively not only in the duck gonad but also in the gonad of the chick parabiont.

These results illustrate the kind of changes in affinity, or contact selectivity, that embryonic cells undergo in the course of differentiation: in this case loss of affinity by the PGC for their original neighboring cells in the endoderm and the acquisition of new affinity resulting in their ability to survey and recognize the site of the future gonad, converge, and reassemble there. An important aspect of the selectivity of the mechanisms responsible for the preferential lodging of PGC is indicated by the fact that duck PGC circulating in the chick embryo blood "sensed" the homologous "homing" site in spite of the generic difference between these two organisms. This implies that the nature of the signals responsible for the selective site recognition by these cells reflects the functional and phenotypic homology of the cells across the taxonomic difference. We return to this subject in a later discussion.

There are many other examples of assembly and reaggregation of cells in the embryo at "preselected" sites. Among them are: the emigration of groups of cells from the anterior neural crest and their reassembly at specific sites in the skull where they form various cartilages of the chondrocranium (Hörstadius, 1950); the coordinated mass movements of cells that form the sternum (Fell, 1939; Chen, 1952); and the migrations of prospective heart cells to the future site of the heart (DeHaan, 1965). The morphogenetic movements and aggregations of slime mold cells (Bonner, 1952; Sussman and Sussman, 1956; Gerisch, 1968) belong essentially in the same class of phenomena. Thus in treating cell aggregation as a topic in experimental biology, one must not lose sight of the fact that it represents a counterpart of normal developmental processes in which cell dissociation, aggregation, and selective, histogenetic assembly of cells are of fundamental importance.

3. THE NATURE OF THE CELL SURFACE

The discussion of these matters naturally focuses on the cell surface, since it is obviously the site of the mechanisms through which cells survey their environment, cellular and acellular, effect mutual recognition, and make transient or permanent contacts. A detailed morphological and biochemical description of the cell surface is given in Chapter 14; therefore only certain relevant comments are made here.

The surface of embryonic cells is not a constant entity; on the contrary, it undergoes various changes in structure, composition, and function in the course of development and differentiation. These changes are reflected in changes in the selective adhesivity of cells, their movements and migrations, differential uptake of metabolites, intercellular communication, and so on. The embryonic cell surface is usually thought of as a three-dimensional mosaic of cooperatively interrelated entities, which changes with phenotypic diversification and specialization of the cells. Some of these changes are dictated directly by gene action; others occur in response to micro- and macromolecular factors in the external environment of the cells, including contact with other cells, or other inducers, and with immunogens, antibodies, or viruses. Perturbations in the cell surface might in turn feed back signals into the internal regulatory circuitry of the cell and elicit internally new cycles of events; it is this reciprocal interplay between the outside and the inside of the cell, with the cell surface playing a central role in this communication, that is largely responsible for driving the cell along the path of differentiation and morphogenesis.

Among the properties of the embryonic cell surface that are directly relevant to cell assembly and cell interactions in development is the mutual adhesiveness of cells. The mechanisms of cell adhesion can be subdivided into two main categories: the entities at the cell surface which, by interacting between adjacent cells, function directly in linking cells to each other and are therefore termed *cell ligands;* and the genetic and epigenetic regulatory mechanisms which control the biosynthesis, assembly turnover, and other properties of these ligands.

The outstanding characteristic of the mechanism of developmental adhesiveness of embryonic cells is *selectivity*. By this we mean the capacity of cells to recognize and identify other cells, to distinguish like from unlike, or complementarity from non-complementarity and, accordingly, to adhere and interact histogenetically, or not to adhere. The capacity of cells for selective recognition or affinity arises in the course of embryonic development and differentiation. As the various cell lines and cell types arise in the embryo, the cell surfaces become differentially specified and encoded with identity markers and with selective propensities for mutual adhesiveness.

It should be noted that cell recognition involves not only "self-recognition" between cells with identical phenotypes but also affinities between cells of different but functionally matching or complementary types. Examples of the latter are the contacts established between motor or sensory nerve fibers and muscles or sensory organs (Hamburger, 1962) resulting in specific neuromuscular and neurosensory junctions. In this class also are the associations of PGC with the somatic cells of the gonad, the preferential localization of lymphocytes in the spleen, and the numerous specific associations between mesenchymal and epithelial cells in the embryo which result in inductive epithelial-mesenchymal interactions. The contacts and connections established between different nerve cells in the brain, or between neural retina cells and cells in the optic centers in the brain, represent further examples of recognition between dissimilar but functionally matching or complementary types of cells. These are just a few examples from the wide spectrum of selective cell recognition phenomena, which reflect the properties of specific constituents on cell surfaces involved in linking cells with each other; differences in qualitative, quantitative, topographical, and temporal patterns of these constituents are reflected in differences in complementarities and affinities of cells.

The most compelling experimental evidence for the role of cell surface specification in selective adhesiveness and histogenetic association of cells is provided by work on embryonic cell aggregation *in vitro*; in these studies embryonic tissue cells are dissociated and dispersed into suspensions and are then made to assemble into tissuelike systems under conditions suitable for analyzing the mechanisms involved in cell adhesion, recognition, multicellular organization, and histogenesis.

4. PREPARATION OF CELL SUSPENSIONS

Procedures for the dissociation of embryonic tissues into suspensions of live, single cells are based on treatments that modify or degrade the bonds between or within intercellular ligands, or the coupling of ligands to the cell surface, so as to permit the separation of the cells and their release as discrete units. The structural and chemical complexity of intercellular materials, their tissue-dependent differences, and changes with differentiation make it difficult to design a general method of cell dissociation that could be effectively applicable to all tissue. Since the nature of intercellular attachments varies not only with tissue and cell type but also in different areas of the same cell's surface, ideally, individually tailored dissociation methods should be devised for each kind of tissue or cell population to obtain highest cell yield and viability. In practice, however, most of the methods for cell dissociation are based on standard procedures involving enzymic and/or chemical treatment of the tissue in combination with mechanical disruption. These procedures have been reviewed often (for example, Moscona, 1962b; Moscona et al., 1965; Weiss, 1967; Bhargava, 1968) and are considered here only briefly.

That intercellular bonds are weakened by removal of *divalent cations* was observed by Herbst (1900) who demonstrated that sea urchin embryos exposed to calcium-free seawater could be separated into free cells (see also Giudice, 1965). More recently, calcium and magnesium depletion was used successfully for the preparation of single-cell suspension of marine sponges (Humphreys, 1963; Moscona, 1963a, 1968).

Embryonic or adult vertebrate tissues cannot be completely dissociated into single cells solely by treatment with solutions free of divalent cations. Treatment with Versene (ethylenediaminetetraacetate) or other chelators of divalent cations accompanied by mechanical agitation facilitates the dissociation of early embryonic tissues (Zwilling, 1954; Feldman, 1955). However, caution is essential in the use of chelators because depletion of cations from cells may have persistent detrimental consequences resulting in cell death or in abnormal changes in cell functions (Kalant and Young 1957; Moscona and Moscona, 1967).

Although divalent cations play a role in the stability of intercellular contacts, particularly in early embryonic stages, it is not known how these cations function in this respect, or why their depletion becomes less effective as a means of cell dissociation in later embryos and adult tissues. The notions about direct coupling of cells through "calcium bridges" are no longer tenable. There is evidence that in some systems calcium may be essential for the stability or structure of the organic molecules that constitute the intercellular ligands (Moscona, 1968). Conceivably, calcium may also play a role in the activity of enzymes involved in the energetics of the maintenance of the overall shape of cells, or in the biosynthesis of ligands; therefore its depletion might cause the cells to round off. Thus depletion of divalent cations can effect cell separation directly only when the stability of the cell-linking mechanisms depends critically on a certain level of this cation. With few exceptions (such as marine sponges), this seems to be the case only in the early stages of embryonic development; later, the tissues become increasingly resistant to cell dissociation by cation depletion (Easty and Mutolo, 1960; Laws and Stickland, 1961). Presumably, this is because of qualitative or quantitative changes in the cell-linking mechanisms, such as the formation of various desmosomal and other types of junctional complexes and the encasement of cell groups in fibrous (collagenous, elastic, and so on) reinforcement networks.

From these considerations it is apparent that the significance of studies on cell dissociation and preparation of cell suspensions is not only pragmatic, but that their essential goal is elucidation of mechanisms involved in the linking of cells into tissues. This applies also to the use of enzymes for cell dissociation. The effectiveness of enzymic treatments depends of course on the availability of appropriate substrates in intercellular bonds, on the relative nontoxicity of the treatment, and on the reversibility of its adverse effects, if any.

Proteolytic enzymes (trypsin, chymotrypsin, pronase) have proved to be by far the most generally effective agents for the preparation of cell suspensions from intact tissues. The susceptibility of intercellular bonds to trypsin suggests that proteins attacked by this enzyme are important in cell binding. The trypsinization procedure as originally described (Moscona, 1952; Dulbecco and Vogt, 1954) utilized crude preparations of trypsin; however, it has since been amply demonstrated that pure trypsin is equally effective (Moscona, 1961), and that inhibitors of trypsin abolish the ability of this enzyme to dissociate cells (Moscona, 1963b). Pure trypsin is not rapidly lethal to uninjured cells if carefully used for the limited time necessary for cell dissociation. Viability studies (Hanks and Wallace, 1958; Rosenbluth and Pappas, 1958), developmental assays (Moscona, 1965; Moscona and Moscona, 1967), and electron microscope examinations (Harris and Leone, 1966; Sheffield and Moscona, 1970) indicated that the majority of trypsin-dissociated cells do not show irreversible morphological or physiological damage. Careless handling of the cells while in trypsin, or prior to their exposure to the enzyme, can result in mechanical or chemical injury, cell death, and release of DNA which forms massive gels and hampers cell separation.

Several factors should be remembered in considering the use of trypsin for cell dissociation. Different cell types (or cell populations) differ in their sensitivity to dissociation by trypsin; this applies also to differences in cell age and functional state of the cells. Considerable variations in amounts and nature of impurities exist among various commercial preparations of trypsin, and some of these are toxic to cells. The susceptibility of cells to dissociation with trypsin is affected by various experimental conditions such as micro- and macromolecular composition of culture medium, temperature, physical handling of the tissue, pH, ratio of enzyme to amount of tissue, and so on.

Some cells are not readily separable with trypsin; for example, sponge cells (Moscona, 1968) can be dissociated with the wide-range protease pronase, but not with trypsin; presumably, the cell-linking mechanisms in these organisms contain proteins that are either insensitive to tryptic degradation, or are masked and inaccessible to this enzyme. In various late embryonic and adult vertebrate tissues, fibrous networks and matrices encasing cells and groups of cells hinder the complete release of cells by trypsinization; in such cases incubation in collagenase or elastase, before or following tryptic treatment, was found helpful in releasing cells from the tissue.

Although carbohydrate-containing proteins are present in the cell surface, the use of specific carbohydrate-degrading enzymes in the preparation of cell suspensions has not been systematically explored. Tests with sialidases and hyaluronidases have yielded negative results; this matter awaits further study.

5. AGGREGATION OF EMBRYONIC VERTEBRATE CELLS *IN VITRO*

Suspensions of trypsin-dissociated embryonic chick or mouse tissue cells, maintained under suitable culture conditions, tend to aggregate into multicellular masses and reconstruct tissuelike patterns capable of histotypic differentiation (Moscona, 1952, 1960). Such histogenetic aggregation of embryonic cells can be studied by the following procedures.

A. Self-Aggregation

When a freshly prepared suspension of embryonic tissue cells dispersed in a suitable culture medium is allowed to settle down in a stationary culture plate, the cells settle out, form clusters, and aggregate (Moscona, 1952; Moscona and Moscona, 1952; Trinkaus and Lentz, 1964). The initial clustering is attributable to random collisions between cells, to active cell movements, and to contractions of protoplasmic filaments between cells; subsequently, fusion of clusters and of cells with clusters contributes to the increase in the mass of the emergent aggregates. It is evident that differences in size and cellular composition of aggregates are determined not only by differences in cell adhesiveness but also by differences in cell mobilities which in turn are influenced by composition of the culture medium, characteristics of the medium–substratum interface, and other difficult-to-control, incidental experimental factors. Thus self-aggregation of cells, as it is practiced at present, lacks the prerequisites of a precisely reproducible and rigorous experimental system. For more detailed analysis experimental conditions are preferred in which the processes of cell aggregation are not complicated by secondary factors, and which are more readily amenable to control and measurement.

B. Cell Aggregation by Rotation

In this procedure [rotation-mediated cell aggregation (Moscona, 1960, 1961a)], suspensions of dissociated cells in a culture medium are placed in Erlenmeyer flasks which are swirled gently on a gyratory shaker at an optimal temperature. The spinning flow of the liquid in the flasks concentrates the cells and cell clusters in the center of the vortex and causes a high frequency of cell collisions. If the cells are mutually adhesive, their aggregation is enhanced by these conditions. The rate of cell aggregation by rotation, and the size and number of aggregates depend on the properties of the cells, on the composition of the medium, and on the balance between the two sets of forces acting on the cells and generated by the swirling liquid: those forces causing concentration of the cells in the vortex and promoting their aggregation—if the cells are mutually adhesive; and the shearing forces that tend to separate and disperse the cells. The balance between these forces depends on the shape and size of the culture flask, the amount of culture medium, and the diameter and speed of rotation. In testing the same kind of cells at high and low rotation speeds, it is found that at higher rotation speeds (when shearing is greater) only those cells aggregate that upon collision attach firmly enough to resist separation by the shearing forces exerted on them; at low rotation speeds the same cells aggregate faster and form larger aggregates. Conversely, different populations of more and less adhesive cells exposed to the same speed of rotation show significant differences in aggregation; the more adhesive cells rapidly form large aggregates comprising practically all the cells in the suspension; the less adhesive cells produce only small aggregates which include only those cells in the population that are more adhesive than the rest; nonadhesive cells remain in suspension. By appropriate handling of this procedure, a mixed population of cells can be separated into more and less adhesive cells by differential aggregation at graded speeds of rotation.

The various environmental and physical parameters that affect cell aggregation by rotation can be adequately standardized and controlled; the same is true of the cellular parameters (cellular composition of the suspension, developmental and functional states of the cells, mode of preparation of the cell suspension). The results obtained by this aggregation procedure are uniformly reproducible and diagnostically characteristic of the set of conditions being tested. By introducing single, defined, environmental or cellular variables into this system, their effects on specific aspects of cell aggregation and histogenesis in cell aggregates can be analyzed.

Not unexpectedly, cells dissociated from different tissues of the same embryo show striking differences in aggregability when tested under identical conditions; these differences result in characteristically distinct *aggregation patterns* defined by rates of cell aggregation, number, size, shape, and internal structure of aggregates (Moscona, 1961a). The tissue-specific aggregation patterns reflect differences in the adhesivity of cells dissociated from different tissues and exposed to the given test conditions.

In addition to tissue-specific differences in cell aggregation, embryonic cells exhibit age-dependent (differentiation-dependent) differences in aggregability (Moscona, 1962a; Garber, 1967; Garber et al., 1968); trypsin-dissociated cells from older embryos aggregate less effectively than cells from the same tissue of a younger embryo. Evidently, incompletely differentiated cells can regenerate surface components lost by trypsinization and essential for reaggregation, while cells that have become fully committed to specialized biosynthesis are less capable or incapable of effectively regenerating these constituents in a manner obligatory for histogenetic reassociation.

The procedure of cell aggregation by rotation can reveal differences in aggregability of cells dissociated by different procedures (Moscona and Moscona, 1967), or resulting from precultivation of cells in monolayer (Kuroda, 1963), or from neoplastic transformation (Pejsachowicz et al., 1964; Pessac and Alliot, 1970). This method can be applied to studies on interactions of neoplastic cells and combinations of normal and neoplastic cells (Moscona, 1957a, b; Schleich, 1967).

6. DEVELOPMENT AND DIFFERENTIATION IN CELL AGGREGATES

The kind of cell aggregation discussed here is referred to as "histogenetic" when recombined embryonic cells reform tissuelike patterns of association conducive to further development and expression of their genetic potential for differentiation. Such developmental progression distinguishes histogenetic cell aggregation from various cell clumping, agglutination, and flocculation phenomena in which cells do not interact to form developing tissue.

The formation of cell aggregates from cell suspensions in rotation cultures can be subdivided, for convenience of discussion, into three phases; needless to say, these phases are not clearly demarcated and overlap, and the time factors involved may vary, depending on the composition of the cell population and experimental conditions.

1. Initial random adhesion of freshly dissociated cells into loose clusters or *primary aggregates* constitutes one phase. If different cell types are present in the suspension, their initial associations are usually indiscriminate, that is, their adhesiveness at this stage is nonspecific. The cells in primary aggregates are in motion relative to each other, which results in their progressive assortment into more stable and more specific associations and groupings. Primary aggregates fuse with each other and continue to incorporate cells from the suspension.

2. The stage of *secondary aggregates* is reached as the cells become arranged into specific groupings and tissuelike patterns.

3. In *tertiary aggregates* histodifferentiation continues and, given suitable maintenance conditions, histological development and differentiation may proceed beyond the stage at which the cells were dissociated from their tissue of origin (Moscona and Garber, 1968; Seeds and Vatter, 1971).

The outstanding feature of development in cell aggregates is the progressive sorting out of cells, resulting in their grouping according to type, homology, or functional complementarity, and in their histological organization. The possibility of sorting out of cells was first suggested by the classic experiments of Holtfreter (1939, 1944) on selective affinities of amphibian blastula and gastrula cells. That the same principle also holds for tissue cells from higher vertebrate embryos was demonstrated conclusively by studying the development of aggregates of trypsin-dissociated embryonic chick cells; in these aggregates cells from different tissues were combined (for review, see Moscona, 1965). Different cells could be identified by their morphologies and characteristic products, and their disposition within the composite aggregates could be followed. Thus it was found that in composite aggregates of cartilage and muscle-forming cells (Moscona and Moscona, 1952), or of cartilage and kidney-forming cells (Moscona, 1956), the cells from each tissue sorted out preferentially into separate areas and re-formed their characteristic tissues. Their sorting out was further confirmed by labeling the nuclei of cells from one of the tissues with isotope and following changes in their location during the development of the composite aggregates by radioautography (Trinkaus, 1961).

A variety of such studies has demonstrated conclusively that not only do cells of different histotypes sort out in composite aggregates into tissue-specific groupings, but that these groupings end up in predictable regions of the aggregate (peripheral, central, or intermediate) depending on the cellular combination (Moscona, 1956, 1957a, b, 1961a, 1962a, b; Trinkaus, 1961). Thus in combinations of cartilage cells with liver, kidney, or melanoma cells, the cartilage cells were always found in the center of the aggregate, the other cells forming the external zone; however, in co-aggregates of liver with melanoma cells, the melanoma cells converged internally. The existence of such "regionalizations" in composite, heterotypic, cell aggregates was later corroborated with several different cell combinations, and a scheme of sorting out "hierarchies" of cells was suggested (Steinberg, 1963).

However, sorting out of cells takes place not only when cells from different tissues are coaggregated but also in the reconstruction of a single tissue from a cell suspension. This has been examined electron microscopically in the aggregation of embryonic chick neural retina cells (Sheffield and Moscona, 1969, 1970). The neural retina consists of several morphologically distinguishable kinds of cells, with distinct polarities, arranged in layers and joined by various intercellular junctions and by materials between cells. Dissociation with trypsin degrades the ligands responsible for this multicellular organization and converts the tissue into a cell suspension, but the dispersed cells retain their original identities. Although in the early aggregates the cells assemble at first without a recognizable order and are randomly interspersed, within hours they *sort out* and become organized into patterns characteristic of neural tissue. In addition, prospective photoreceptor cells re-form a layer linked by junctions of the outer limiting membrane, and the bipolar and ganglion cells re-form their layers; eventually, the aggregates contain typical neuroretinal configurations which continue to differentiate, synaptic connections are formed, and an advanced state of retinal development is attained.

Thus in the course of histogenetic reaggregation, the dissociated cells undergo a transition from nonspecific stickiness, ensuing from their dissociation with trypsin and resulting in their initial random, nonspecific adhesions, to selective adhesiveness which reflects a return of *specification* of cell surfaces, is conducive to cell recognition, and results in the sorting out of cells into histotypic patterns. It further appears that the dissociated retina cells "remember" their original topographical dispositions within the tissue and in aggregates again become arranged in the appropriate layers. A similar sequence of cell sorting out was described in detail for aggregates of embryonic skin cells (Garber and Moscona, 1964).

It is noteworthy that under certain conditions sorting out of cells in heterotypic aggregates may be impaired, resulting in the absence of the expected histogenetic products. Thus the development of integumental differentiation in aggregates of embryonic chick skin cells is "suppressed" if the skin cells are commingled and coaggregated with cells from other embryonic tissues. In such combinations the regrouping of the cells required for the formation of feathers is prevented (Moscona and Moscona, 1965). The severity of such suppressions and the developmental characteristics of the resulting structures depend on the kind of cells combined with the skin cells. For discussion of the significance of such suppressive cell interactions in the control of normal embryonic morphogenesis and of other aspects of this problem, see Moscona and Moscona (1965), Garber and Moscona (1967), Moscona and Garber (1968), and Shimada (1968).

7. COAGGREGATION OF CELLS FROM DIFFERENT SPECIES

Some of the most significant and intriguing information about the role of cell surface specification in sorting out of cells and tissue organization comes from coaggregating embryonic cells from different species, such as mouse and chick. The cytological differences between their cells provide permanent, "built-in," markers making it possible to distinguish these cells individually even when they are commingled and interspersed within a composite aggregate. Such bispecific aggregates can be compounded from homologous cells (for example, mouse and chick cartilage cells) or heterologous cells (for example, mouse cartilage cells and chick kidney cells). One of the major questions answered by these studies was whether homologous cells from different species can recognize their histotypic similarity and form joint tissues, or whether species differentials take precedence in cell recognition.

In bispecific aggregates of *heterologous* cells, cells from each tissue were found to sort out into separate groupings and to reconstruct the different tissues, each consisting only of cells from the corresponding species. This was a predictable outcome, since heterologous cells even from the same embryo sort out in composite aggregates into histotypic groupings. However, it showed that in such bispecific combinations there was no unexpected impairment of, or interference with, the mechanisms of cell recognition and sorting out. This clarification was important in view of the results with bispecific combinations of homologous cells; in these, the cells did not sort out sharply according to species; on the contrary, they tended to form bispecific tissues which consisted of mosaics of cells and cell groups from the two species. Such "chimeric," bispecific chick-mouse tissues were synthesized *in vitro* from kidney cells, cartilage cells, skin cells (Moscona, 1957a, b, 1961a, 1962a, b 1964; Garber and Moscona, 1964, 1967; Garber et al., 1968; Moscona and Garber, 1968). Recently, it was shown that cells from comparable regions of mouse and chick brain can recognize their homologies and form bispecific mosaic neural tissue (Garber and Moscona, 1972).

The principle of preferential recognition between homologous embryonic cells, not only within the same species but also between different species (mouse and chick), was confirmed by Roth (1968) using a modified cell aggregation system. His studies also confirmed the loss of surface specification by cells after dissociation with trypsin and the return of specification in the course of reaggregation.

Transspecies recognition between homologous cells in aggregates and their ability to interact developmentally was clearest when the cells in the combination were properly matched with respect to their stage of differentiation (or developmental age), so that their phenotypic homology was maximal. Cells in discrepant stages of differentiation, even though from homologous tissues, tended to sort out separately, and this tendency was greater with increased discrepancy between their developmental stages.

Taken together, these findings showed that (1) specification of embryonic cell surfaces, as expressed in their preferential adhesiveness, recognition, and sorting out preferences, is closely related to the phenotypic state of the cells; and (2) the mechanism of this recognition can function even across considerable taxonomic distance, suggesting that homologous phenotypes from mouse and chick embryos share identity markers which enable them to interact developmentally in spite of the generic difference. It appears that in the evolution of these organisms certain features of the mechanism responsible for histotypic recognition, adhesion, and developmental organization of cells were conserved. How these similarities are related to the species differences that distinguish these cells, to histocompatibility differences, and to other distinctions that mark the cell surface remain to be explained.

8. MECHANISMS OF CELL RECOGNITION AND SPECIFIC CELL ADHESION

From the information already mentioned, it is apparent that selective cell adhesion and cell recognition represent essentially two facets of the same phenomenon and may therefore have a common underlying mechanism. Accepting this possibility, we proposed the working hypothesis that selective cell adhesion is mediated by specific macromolecular entities, localized at the cell surface and between cells, which function as *intercellular ligands*. These ligands are part of the glycoprotein "coat" of the cell surface (the so-called "extracellular material"), and they link adjacent cells by interactions of complementary sites. Strong and persistent interactions result in the cells remaining linked; weak or transient interactions result in separation or continued mobility of the cells. The hypothesis suggests that the properties, amounts, and distribution patterns of ligands on the cell surface change in the course of cell differentiation; cells of different phenotypes display on their surfaces qualitatively and quantitatively different ligand populations which *specify* the cell surface in terms of its adhesive and cognitive properties. The characteristics of ligands are controlled, as are other cell features, by the cell's genetic and regulatory mechanisms, in correspondence with the cell's differentiation; in addition, ligands may also be affected by external factors, such as contact with other cells, or with factors in the extracellular milieu; such effects, even when limited to a small part of the cell surface, could modify the overall properties of the cell surface, considering the cooperative nature of the cell membrane; these changes might in turn be signaled into the cell and trigger further modifications in cell function, mobility, rate of replication, or specific biosynthesis.

The immediate usefulness of this tentative hypothesis was that it reduced the complex problems of cell adhesion and recognition to the testable question concerning the existence on the cell surface of entities with the properties postulated for intercellular ligands. If such cell-linking ligands exist, it should be possible to isolate from cells materials with characteristic ligandlike activities; their removal or depletion from the cell surface should reduce the mutual adhesiveness of the cells; their regeneration should restore specific adhesiveness to the cells; their addition to depleted cells should enhance specific cell reaggregation.

Assuming that trypsin dissociates tissue cells by degrading ligand proteins, one would expect that, in order to be able to reaggregate, trypsin-dissociated cells must regenerate lost ligands. If this regeneration requires synthesis, then inhibition of metabolic and biosynthetic processes in dissociated cells by low temperature or by metabolic inhibitors should prevent their normal histogenetic reaggregation. The following evidence supports this assumption.

Reaggregation of trypsin-dissociated embryonic cells is temperature dependent; below 15°C it is almost completely inhibited (reversibly) even though the cells are brought into contact by rotation in flask cultures (Moscona, 1961b, 1962a; Moscona and Moscona, 1966). Inhibition of protein synthesis in trypsin-dissociated embryonic cells by translational inhibitors (puromycin, cycloheximide) stops (reversibly) their reaggregation in the initial stage of nonspecific adhesiveness (Moscona and Moscona, 1963, 1966). Inhibition of RNA synthesis in trypsin-dissociated cells stops reaggregation after approximately 4 hr (Moscona and Moscona, 1963). These and similar results (Glaeser et al., 1968; Antley and Fox, 1970) demonstrate the requirement for continued protein and RNA synthesis in histogenetic cell reaggregation.

Some of this biosynthesis is directed toward regeneration of glycoproteins at the cell surface removed by trypsinization and presumably required for reaggregation (Glaeser et al., 1968; Sheffield and Moscona, 1970). Particularly interesting in this context are recent studies with suspensions of teratoma cells which require exogenous glutamine for the synthesis of carbohydrate-containing surface materials; when these cells are deprived of glutamine or inhibited with 6-diazo-5-oxynorleucine, their reaggregation is prevented (Oppenheimer et al., 1969).

Since biosynthesis is essential for normal cell reaggregation, it is to be expected that genetic controls are involved. The requirement for RNA synthesis has already been mentioned. Further evidence that histogenetic cell associations and sorting out are subject to gene-controlled processes was obtained in studies on aggregation of brain cells from mouse embryos of the "reeler" mutant. This mutation results in an alignment defect of certain cells in the cerebrum and cerebellum, presumably because they fail to sort out correctly from the other cells and do not take up their normal positions (DeLong, 1970). A similar misalignment of these cells also occurs in aggregates of dissociated brain cells from the mutant embryos; in contrast, in aggregates of normal brain cells there is correct sorting out, alignment, and positioning of the cells (DeLong and Sidman, 1970). One might thus speculate that the mutation results in a change in the nature of ligands or in their temporal sequence of appearance on the affected cells, thereby causing a modification of the recognition properties normally exercised by these cells.

9. ISOLATION FROM LIVE CELLS OF MATERIALS WITH LIGAND ACTIVITY

More direct evidence for the ligand hypothesis was obtained by isolating from live cells materials which when added to homologous cells in suspensions enhanced their aggregation in a specific way.

A. Embryonic Retina and Brain Cells

Cells in suspension prepared from neural retinas of 10-day-old chick embryos were "washed" in a physiological salt solution either by swirling them rapidly on a shaker, or after monolayering the cells in petri dishes (for details, see Moscona, 1962a; Lilien and Moscona, 1967; Lilien, 1968, 1969). The supernatant solution was used as a medium for cell aggregation; it was found to contain a factor which strikingly enhanced the aggregation of retina cells, causing them to form markedly larger aggregates than those formed in the controls, that is, in the absence of the factor. The ligand hypothesis requires the factor to be specific for the kind of cells from which it was obtained; and in fact the retina cell aggregation factor was found to enhance only the aggregation of neural retina cells and had no such effect on cells from other tissues even from the same embryo. Similar results were obtained by Daday and Creaser (1970), who used a somewhat different procedure for the preparation of the factor.

A cell aggregation-enhancing factor was also prepared from cerebrum cells of mouse embryos (Garber and Moscona, 1969, 1972). The biological specificity of this factor could be more narrowly defined and proved to be even more impressive than the specificity of the retina factor; the cerebrum factor enhanced only the aggregation of embryonic cerebrum cells and had no such effect even on cells from other regions of the mouse brain. Moreover, this factor enhanced the aggregation of cerebrum cells from the chick embryo (but of no other chick tissue cells). In these respects this factor's selectivity paralleled the preferential affinities of embryonic mouse cerebrum cells, as displayed in their sorting out in composite aggregates; when commingled with cells from other tissues, including cells from other brain areas, the cerebrum cells tend to sort out into separate groupings; however, when they are commingled with chick embryo cerebrum cells, the homologous cells from both species remain interspersed and cooperate in the formation of a bispecific neural mosaic (Garber and Moscona, 1972).

The biochemical and biological characterization of these specific cell aggregation-enhancing factors is now in progress. The evidence available indicates that they may be glycoproteins, that they attach to receptor sites on the cell membrane and cross-link cells, thereby enhancing cell attachment and cell aggregation, and that they also function as intercellular ligands.

B. Sponge Cells

If the ligand hypothesis reflects generally valid principles not limited to embryonic cells, its basic assumptions should also apply to other systems in which cell affinities and selective sorting out of cells are demonstrable and can be effectively analyzed by the procedures of cell aggregation *in vitro*. Cells of marine sponges have been a classic object of studies on cell aggregation since the pioneering work of Wilson (1907; see also Moscona, 1962b). Sponge tissues can be easily dissociated into cell suspensions without enzymic treatment, by mechanical dispersion, that is, by pressing through a fine sieve; this excludes side effects that might occur in using trypsin, as in the preparation of vertebrate cell suspensions. Sponge cells can be readily obtained in sufficient quantities for biochemical work, and their aggregation can be effectively studied by the procedures of rotation in flasks. Furthermore, sponge cells contain bright pigments and different sponge species are differently colored, so that in mixtures of cells from different species the cells are identifiable by their intrinsic colors.

Most of the work described here was done with cells from the red sponge *Microciona prolifera* (Humphreys, 1963; Moscona, 1963a, 1967, 1968).

Mechanically dispersed *Microciona* cells aggregate by rotation in natural or artificial seawater at 25°C (the "normal" temperature for this organism), rapidly adhere, and form aggregates; if such aggregates are maintained in seawater under suitable conditions, they develop into miniature sponges. However, such mechanically dispersed *Microciona* cells, also adhere into large masses when swirled in flasks at 4°C (naturally at this low temperature no further development of these cell masses occurs, because of inhibition of cell movements and metabolic processes required for morphogenesis). This is in contrast to trypsin-dissociated embryonic vertebrate cells which, as explained earlier, do not aggregate at suboptimal temperatures beyond the initial clumping stage because of ineffective regeneration of ligands. However, mechanically dispersed sponge cells retained sufficient functional ligands on their surfaces to be able to readhere even at a suboptimal temperature. Assuming this to be the case, one would expect that (1) a more complete removal of ligands from the sponge cells should reduce their capacity to readhere and bring their behavior more in line with that of trypsin-dissociated cells, and (2) the addition of ligand material to depleted cells should increase their mutual adhesiveness and enhance their aggregation.

These predictions were confirmed by work in our laboratory (Humphreys, 1963; Moscona, 1963a, 1968) and by others (Gasic and Galanti, 1966; Loewenstein, 1967). By washing mechanically dispersed sponge cells in cold calcium- and magnesium-free seawater, a material with ligand activity was isolated from them. The "washed" cells were resuspended in normal seawater, to restore the lost cations, and tested for capacity to aggregate. At 4°C such washed cells formed only minute clusters, much smaller than the large masses formed by unwashed cells. From the "wash water" a macromolecular factor was recovered which, when added to sponge cell suspension (after restoration of divalent cations) strikingly enhanced their aggregation at 4°C as well as at 25°C. A similar factor was prepared from cells of another sponge, *Haliclona occulata*.

Selectivity of effect is one of the basic assumptions of the ligand hypothesis; the testing of this question was of special interest in the case of the sponge factors because of the particular sorting-out properties of sponge cells. It has long been known that when suspensions of freshly dispersed *Microciona* and *Haliclona* cells are commingled the cells of each species sort out and form separate aggregates (see Moscona, 1968). The effectiveness of species-specific sorting out of sponge cells in bispecific cell aggregates probably differs for different species and different combinations of species; however, with *Microciona* and *Haliclona* the efficiency is very high, and this made possible a crucial test of the specificity of the ligand preparations. For if the function of the isolated ligand materials reflect with fidelity cell affinities, then *Microciona* ligand preparation should enhance only the aggregation of *Microciona* cells but not of *Haliclona* cells, and vice versa. This proved to be the case; within the limits of the available tests, each preparation was found to be highly specific for the cells of the species from which it was obtained. This close correspondence between the effects of these ligand preparations and the normal overall propensities and sorting-out-preferences of their donor cells suggested that the preparations contained constituents which were functionally related to the actual mechanisms involved in the preferential adhesiveness, recognition, and histogenetic aggregation of these cells (Humphreys,

1963; Moscona, 1963a). More recently, similarly species-specific aggregation factors were isolated from several other species of sponges (McClay, 1971).

The factor responsible for the characteristic ligand activity of the preparation is sedimentable by high-speed centrifugation into pellets of mucoidal consistency. It is not destroyed by RNAase, DNAse, neuraminidase, collagenase, lysozyme, or hyaluronidase. Following additional purification the factor was examined by electron microscopy and found to consist of unit particles 20–25 Å in diameter. Initial biochemical analysis of this material (Margoliash et al., 1965) showed it to consist predominantly of glycoprotein (or glycoproteins) and revealed that the preparations from the two species differed quantitatively in their sugar composition. While it is reasonable to assume that the particles represent the ligand activity of the factor, direct proof of this is not yet available; nor is it known whether the ligand activity resides in individual particles or in some polymeric combinations.

It is of special interest that the *Microciona* factor retains its characteristic ligand activity after treatment with formaldehyde and that it also can react with formaldehyde-killed cells. These and related results (Moscona, 1968) demonstrate that the active constituents in the factor function as true intercellular ligands, that is, they enhance cell aggregation not by stimulating intracellular metabolic processes but directly, by attaching to the cell surface and "cross-linking" the cells into aggregates. These conclusions were supported by evidence from a variety of serological studies (using antisera prepared against purified sponge factor) indicating that the factor is localized on the cell surface of cells (Moscona, 1968).

Another means of comparing the embryonic and the sponge systems in this context would be to use enzyme-dissociated sponge cells. Unlike embryonic cell ligands, which are degraded by trypsin, sponge ligand preparations are quite resistant to trypsin; however, they are readily inactivated by pronase, a wide-spectrum protease. This suggested that, if the pronase-sensitive constituents in the ligand preparation were also responsible for holding cells together in the sponge tissue, pronase should dissociate sponge cells. This prediction was experimentally confirmed (Moscona, 1968). Pronase dissociates sponge cells by degrading their intercellular ligands. In accordance with the theoretical predictions from this finding, pronase-dissociated sponge cells, having been depleted of surface ligands, did not aggregate in the cold, in contrast to mechanically dispersed sponge cells; at normal temperatures aggregation started after a lag period, presumably with the onset of regeneration of surface materials. Additional support for this is the finding that sponge cells freshly dissociated with pronase are not agglutinated by the specific antiserum against the homologous ligand preparation but become agglutinable with time, as their aggregability returns (Moscona, 1968). Of further significance is the fact that pronase-dissociated sponge cells respond to the aggregation-promoting effect of the ligand preparation only after a lag period; this indicates that regeneration of cell surface constituents removed by pronase (receptors for ligands, ligand precursors, or ligands) is essential for the binding of exogenously supplied ligands to the cell surface. A similar situation was described for trypsin-dissociated embryonic retina cells (Lilien, 1968, 1969).

10. CONCLUDING REMARKS

The experimental evidence discussed briefly in this chapter, taken with a variety of related findings in other systems, supports the notion that interactions between macromolecular entities, localized at the cell surface and between cells and referred to as intercellular ligands, are an essential aspect of the mechanism of histogenetic linking and association of metazoan cells. The strongest evidence for this is the isolation from cells of preparations with activities of the postulated ligands. Such preparations have been obtained from such diverse cells as mouse and chick embryonic cells and marine sponge cells. While there are important differences in the detailed properties of these disparate systems, the fundamental similarities and consistencies are impressive, suggesting a generality of the underlying principles. The relevant reservations expressed when the hypothesis concerning the role of extracellular materials in cell specification and aggregation was first considered (Moscona, 1960, 1962a) have been largely answered; the way is now clear for detailed chemical characterization of intercellular ligands and examination of their exact mode of function. This may not prove to be as simple a task as it might appear, for reasons discussed below. Equally interesting should be information about control of ligand biosynthesis in different cells, and the temporal and spatial regulation of ligands on the cell surface by genetic and epigenetic factors. Further work should clarify the relationship of the intercellular ligands to other constituents and structural specialization at the cell surface, such as diverse intercellular junctions (desmosomes, tight junctions, and so on) and different receptors present in the cell periphery (Moscona, 1971).

Among the most intriguing questions for future exploration is the role of cell ligands in cell interactions dependent on selective cell adhesion. There is increasing evidence that specific contact may affect communication between cells (Loewenstein, 1967), that contact-regulating macromolecules at the cell surface may be involved in the phenomena of "contact inhibition" of movement and cell replication (Inbar and Sachs, 1969; Burger and Noonan, 1970; Moscona, 1971), and that in inductive interactions between embryonic cells, contact-mediating entities on the cell surface may play a crucial role (Grobstein, 1962). Changes in cell adhesiveness are clearly involved in neoplastic transformation, and detailed knowledge about the properties of surface materials on embryonic, adult, and malignant cells would be of obvious significance to the understanding of neoplasia (Moscona, 1971), as well as normal differentiation.

In view of the possibility that intercellar ligands may be of glycoprotein nature, it has been suggested that protein–carbohydrate interactions may be an essential feature of ligand function and specificity (Moscona, 1962a, 1963a, 1968). Carbohydrates are involved in a variety of specific interactions in living systems; one outstanding example of this is the blood group substances in which differences in the terminal sugar provide an "identity code" for the cells (Morgan, 1960). In this connection one recalls the role of carbohydrate-containing entities in specific interactions between sperm and egg, viruses and cells, antigens and antibodies, and in the "homing" of lymphocytes into the spleen (Gessner and Ginsburg, 1964). It is tempting to speculate that the carbohydrate moieties of intercellular ligands might serve in cell recognition, and that interactions with complementary protein moieties on adjacent cells would result in specific cell adhesion.

That such interactions might be controlled by specific enzymes is suggested by recent evidence for the existence of glycosyl transferases in the cell surface reported by Roth et al. (1971). These investigators suggest that the protein moiety of the ligand mechanism is an enzyme and that enzyme–substrate (protein–carbohydrate) interactions represent the ligand mechanism discussed above. This is an attractive notion; however, present evidence is insufficient to exclude the possibility that glycosyl transferases control the synthesis and turnover of the carbohydrate moiety of the ligands, rather than acting directly as structural elements of the cell-binding mechanism.

Molecular discrimination based on phenotypically determined chemical differences in ligand entities on the cell surface may not suffice to account for the details of cell recognition in multicellular systems. In order to explain cell surface specification, the complexity of cellular organization, and the sorting out of cells that occurs even within a single organ (for example, in the brain), one must postulate, in addition to chemical differences in ligands, another order of diversity; a diversity that could provide a high degree of specificity and modifiability within the fairly limited number of chemical permutations of the kind that can be envisaged for ligands. Such a diversity could be generated by unique topographical distributions of ligands on the surfaces of different cells at different times in development. Thus recognition and selective adhesion of cells would depend not only on the presence of stereochemically complementary ligand entities on adjacent cell surfaces but also on their unique four-dimensional arrangement in the cell surface; it can be easily imagined how such cell surface mosaics might result in a cell recognition code of great diversity stemming, as the genetic code, from a relatively small primary alphabet.

The existence of unique topographical distributions of specific markers on the cell surface is supported by the findings of Boyse and Old (1969) and Boyse et al. (1968) that certain specific cell surface antigens are not uniformly distributed but are confined to discrete regions of the surface, and that the proportion of the occupied surface is characteristic for a particular antigen and cell type. That specification of cell surfaces can change with development or with cell function is supported by the evidence of changes in the affinities of isotypic cells in the course of their differentiation (Moscona and Garber, 1968), and by the appearance of neuronal specificities at specific times in development (Jacobson, 1968). The findings that various receptors are situated at different levels of the cell surface (Inbar and Sachs, 1969) and that their availability changes in the course of development (Moscona, 1971; Kleinschuster and Moscona, 1972) and in neoplastic cell transformation (Burger and Goldberg, 1967; Inbar and Sachs, 1969) all are consistent with the four-dimensional conception of the cell surface advanced above.

Thus elucidation of topographical and temporal organization and differentiation of ligand entities on the cell surface, together with analysis of their chemical composition and structure, in correlation with the biological-developmental properties of cells, are some of the immediate research aims in this area. Answers to these problems hold a key to detailed understanding of mechanisms of cell specification and of cell interactions in development; progress in this area would undoubtedly be important also for studies on cell and tissue transplantation, immunological surveillance, and certain aspects of neoplasia. While these are admittedly far-reaching expectations, they are encouraged by the current accomplishments of studies on cell aggregation and experimental tissue synthesis.

REFERENCES

Antley, R. M., and A. S. Fox (1970). The relationship between RNA and protein synthesis and the aggregation of *Drosophila* embryonic cells. *Develop. Biol.*, **22,** 282.

Bhargava, P. M. (1968). Preparation, properties and uses of single cell suspensions from animal tissues. *Sci. Culture*, **34,** 105.

Bonner, J. T. (1952). The pattern of differentiation in amoeboid slime molds. *Am. Naturalist*, **86,** 79.

Boyse, E. A., and L. J. Old (1969). Some aspects of normal and abnormal cell surface genetics. *Ann. Rev. Genet.*, **3,** 269.

Boyse, E. A., L. J. Old, and E. Stockert (1968). An approach to the mapping of antigens on the cell surface. *Proc. Natl. Acad. Sci., U.S.*, **60,** 886.

Burger, M. M., and A. R. Goldberg (1967). Identification of a tumor-specific determinant on neoplastic cell surfaces. *Proc. Natl. Acad. Sci., U.S.*, **57,** 359.

Burger, M. M., and K. D. Noonan (1970). Restoration of normal growth by covering of agglutinin sites on tumor cell surface. *Nature*, **228,** 512.

Chen, J. M. (1952). Studies on the morphogenesis of the mouse sternum. I. Normal embryonic development. II. Experiments on the origin of the sternum and its capacity for self-differentiation *in vitro. J. Anat.*, **86,** 373.

Daday, H., and E. H. Creaser (1970). Isolation of a protein responsible for aggregation of avian embryonic cells. *Nature*, **226,** 970.

DeHaan, R. L. (1965). Morphogenesis of the vertebrate heart. In *Organogenesis* (R. L. DeHaan and H. Ursprung, eds.), p. 377. Holt, New York.

DeLong, G. R. (1970). Histogenesis of fetal mouse isocortex and hippocampus in reaggregating cell cultures. *Develop. Biol.*, **22,** 563.

DeLong, G. R., and R. L. Sidman (1970). Alignment defect of reaggregating cells in cultures of developing brains of reeler mutant mice. *Develop. Biol.*, **22,** 584.

Dulbecco, R., and M. Vogt (1954). Plaque formation and isolation of pure lines with poliomyelitis viruses. *J. Exptl. Med.*, **99,** 167.

Easty, G. C., and V. Mutolo (1960). The nature of the intercellular material of adult mammalian tissues. *Exptl. Cell Res.*, **21,** 374.

Feldman, M. (1955). Dissociation and reaggregation of embryonic cells of *Triturus alpestris. J. Embryol. Exptl. Morphol.*, **3,** 251.

Fell, H. B. (1939). The origin and developmental mechanics of the avian sternum. *Phil. Trans. Roy. Soc. London*, **B229,** 407.

Garber, B. (1967). Aggregation *in vivo* of dissociated cells. II. Role of developmental age in tissue reconstruction. *J. Exptl. Zool.*, **164,** 339.

Garber, B., and A. A. Moscona (1964). Aggregation *in vivo* of dissociated cells. I. Reconstruction of skin in the chorioallantoic membrane from suspensions of embryonic chick and mouse skin cells. *J. Exptl. Zool.*, **55,** 179.

Garber, B., and A. A. Moscona (1967). Suppression of feather morphogenesis in co-aggregates of skin cells from embryos of different ages. *J. Exptl. Zool.*, **164,** 351.

Garber, B. B., and A. A. Moscona (1969). Enhancement of aggregation of embryonic brain cells by extracellular materials from cultures of brain cells. (9th Annual Meeting, American Society for Cell Biology, Detroit, Michigan, 1969, abstr. No. 96.) *J. Cell Biol.*, **43,** 41a.

Garber, B. B., and A. A. Moscona (1972). Reconstruction of brain tissue from cell suspensions. I. Aggregation patterns of cells dissociated from different regions of the developing brain. II. Specific enhancement of aggregation of embryonic cerebral cells by supernatant from homologous cell cultures. *Develop., Biol.*, **27,** 217.

Garber, B., E. Kollar, and A. A. Moscona (1968). Aggregation *in vivo* of dissociated cells. III. Effect of state of differentiation of cells on feather development in hybrid aggregates of embryonic mouse and chick cells. *J. Exptl. Zool.*, **168,** 455.

Gasic, G. J., and N. L. Galanti (1966). Proteins and disulfide groups in the aggregation of dissociated cells of sea sponges. *Science*, **151,** 203.

Gerisch, G. (1968). Cell aggregation and differentiation in *Dictyostelium. Curr. Top. Develop. Biol.*, **3,** 157.

Gessner, B. M., and V. Ginsburg (1964). Effect of glycosidases on the fate of transfused lymphocytes. *Proc. Natl. Acad. Sci., U.S.*, **52,** 750.

Giudice, G. (1965). The mechanism of aggregation of embryonic sea urchin cells: A biochemical approach. *Develop. Biol.*, **12,** 233.

Glaeser, R. M., J. E. Richmond, and P. W. Todd (1968). Histotypic self-organization by trypsin-dissociated and EDTA-dissociated chick embryo cells. *Exptl. Cell Res.*, **52,** 71.

Grobstein, C. (1962). Interactive processes in cytodifferentiation. *J. Cellular Comp. Physiol., Suppl. 1*, **60,** 35.

Hamburger, V. (1962). Specificity in neurogenesis. *J. Cellular Comp. Physiol., Suppl. 1*, **60,** 81.

Hanks, J. H., and J. H. Wallace (1958). Determination of cell viability. *Proc. Soc. Exptl. Biol.*, **98,** 188.

Harris, C. C., and C. A. Leone (1966). Some effects of EDTA and tetraphenylboron on the ultrastructure of mitochondria in mouse liver cells. *J. Cell Biol.*, **28,** 405.

Herbst, C. (1900). Über das Auseinandergehen von Furchungs und Gewebezellen in Kalkfreiem Medium. *Arch. Entwicklungsmech. Organ.*, **9,** 424.

Hörstadius, S. (1950). *The Neural Crest.* Oxford Univ. Press, London.

Holtfreter, J. (1939). Gewebeaffinität ein Mittel der Embryonalen Formbildung. *Arch. Exptl. Zellforsch.*, **23,** 169.

Holtfeter, J. (1944). Experimental studies on the development of the pronephros. *Rev. Can. Biol.*, **3,** 220.

Humphreys, T. (1963). Chemical dissolution and *in vitro* reconstruction of sponge cell adhesions. I. Isolation and functional demonstration of the components involved. *Develop. Biol.*, **8,** 27.

Inbar, M., and L. Sachs (1969). Interaction of the carbohydrate-binding protein conconavalin A with normal and transformed cells. *Proc. Natl. Acad. Sci., U.S.*, **63,** 1418.

Jacobson, M. (1968). Development of neuronal specificity in retinal ganglion cells in *Xenopus*. *Develop. Biol.*, **17,** 202.

Kalant, H., and F. G. Young (1957). Metabolic behaviour of isolated liver and kidney cells. *Nature*, **179,** 816.

Kleinschuster, S. J., and A. A. Moscona (1972). Interactions of embryonic and fetal neural retina cells with carbohydrate-binding phytoagglutinins: cell surface changes with differentiation. *Exptl. Cell* Res., **70,** 397.

Kuroda, Y. (1963). Changes in aggregation and differentiation of cartilage cells grown in monolayer cultures. *Exptl. Cell Res.*, **30,** 446.

Laws, J. O., and L. H. Stickland (1961). The adhesion of liver cells. *Exptl. Cell. Res.*, **24,** 240.

Lilien, J. E. (1968). Specific enhancement of cell aggregation *in vitro*. *Develop. Biol.*, **17,** 657.

Lilien, J. E. (1969). Toward a molecular explanation for specific cell adhesion. *Curr. Top. Develop. Biol.*, **4,** 169.

Lilien, J. E., and A. A. Moscona (1967). Cell aggregation: Its enhancement by a supernatant from cultures of homologous cells. *Science*, **157,** 70.

Loewenstein, W. R. (1967). On the genesis of cellular communication. *Develop. Biol.*, **15,** 503.

Margoliash, E., T. R. Schenck, M. P. Hargie, S. Borokas, W. R. Richter, G. H. Barlow, and A. A. Moscona (1965). Characterization of specific cell aggregating materials from sponge cells. *Biochem. Biophys. Res. Commun.*, **20,** 383.

McClay, D. R. (1971). An autoradiographic analysis of the species specificity during sponge cell reaggregation. *Biol. Bull.*, **141,** 319.

Morgan, W. T. J. (1960). A contribution to human biochemical genetics: The chemical basis of blood-group specificity. *Proc. Roy. Soc.* (*London*), **B151,** 308.

Moscona, A. (1952). Cell suspensions from organ rudiments of chick embryos. *Exptl. Cell Res.*, **3,** 535.

Moscona, A. (1956). Development of heterotypic combinations of dissociated embryonic chick cells. *Proc. Soc. Exptl. Biol.*, **92,** 410.

Moscona, A. (1957a). The development *in vitro* of chimeric aggregates of dissociated embryonic chick and mouse cells. *Proc. Natl. Acad. Sci., U.S.*, **43,** 184.

Moscona, A. (1957b). Formation of lentoids by dissociated retinal cells of the chick embryo. *Science*, **125,** 598.

Moscona, A. A. (1960). Patterns and mechanisms of tissue reconstruction from dissociated cells. In *Developing Cell Systems and Their Control* (D. Rudnick, ed.), p. 45. Ronald, New York.

Moscona, A. (1961a). Rotation-mediated histogenetic aggregation of dissociated cells: A quantifiable approach to cell interactions *in vitro*. *Exptl. Cell Res.*, **22,** 455.

Moscona, A. (1961b). Effect of temperature on adhesion to glass and histogenetic cohesion of dissociated cells. *Nature*, **190,** 408.

Moscona, A. A. (1962a). Analysis of cell recombinations in experimental synthesis of tissues *in vitro*. *J. Cellular Comp. Physiol., Suppl. 1*, **60,** 65.

Moscona, A. A. (1962b). Cellular interactions in experimental histogenesis. *Intern. Rev. Exptl. Pathol.*, **1,** 371.

Moscona, A. A. (1963a). Studies on cell aggregation: Demonstration of materials with selective cell-binding activity. *Proc. Natl. Acad. Sci., U.S.*, **49,** 742.

Moscona, A. A. (1963b). Inhibition by trypsin inhibitors of dissociation of embryonic tissue by trypsin. *Nature*, **199,** 379.

Moscona, A. A. (1964). Studies on stability of phenotypic traits in embryonic integumental tissues and cells. In *The Epidermis* (W. Montagna, ed.), p. 83. Academic, New York.

Moscona, A. A. (1965). Recombination of dissociated cells and the development of cell aggregates. In *Cells and Tissues in Culture* (E. N. Willmer, ed.), p. 489. Academic, New York.

Moscona, A. A. (1967). Aggregation of sponge cells: Cell-linking macromolecules and their role in the formation of multicellular systems. *In Vitro*, **3,** 13. (21st Annual Meeting, Tissue Culture Association, Philadelphia, 1967.)

Moscona, A. A. (1968). Cell aggregation: Properties of specific cell ligands and their role in the formation of multicellular systems. *Develop. Biol.*, **18,** 250.

Moscona, A. A. (1971). Embryonic and neoplastic cell surfaces: Availability of receptors for concanavalin A and wheat germ agglutinin. *Science*, **171,** 905.

Moscona, A. A., and B. B. Garber (1968). Reconstruction of skin from single cells and integumental differentiation in cell aggregates. In *Epithelial–Mesenchymal Interactions* (R. W. Fleischmajer and R. E. Billingham, eds.), p. 230. Williams and Wilkins, Baltimore.

Moscona, A., and H. Moscona (1952). The dissociation and aggregation of cells from organ rudiments of the early chick embryo. *J. Anat.*, **86,** 287.

Moscona, A., O. A. Trowell, and E. N. Willmer (1965). In *Cells and Tissues in Culture* (E. N. Willmer, ed.), p. 19, Academic, New York.

Moscona, A. A., and M. H. Moscona (1966). Aggregation of embryonic cells in a serum-free medium and its inhibition at suboptimal temperatures. *Exptl. Cell Res.*, **41,** 697.

Moscona, A. A., and M. H. Moscona (1967). Comparison of aggregation of embryonic cells dissociated with trypsin or Versene. *Exptl. Cell. Res.*, **45,** 239.

Moscona, M. H., and A. A. Moscona (1963). Inhibition of adhesiveness and aggregation of dissociated cells by inhibitors of protein and RNA synthesis. *Science*, **142,** 1070.

Moscona, M. H., and A. A. Moscona (1965). Control of differentiation in aggregates of embryonic skin cells: Suppression of feather morphogenesis by cells from other tissues. *Develop. Biol.*, **11,** 402.

Moscona, M. H., and A. A. Moscona (1966). Inhibition of cell aggregation *in vitro* by puromycin. *Exptl. Cell Res.*, **41,** 703.

Oppenheimer, S. B., M. Edidin, C. W. Orr, and S. Roseman (1969). An L-glutamine requirement for intercellular adhesion. *Proc. Natl. Acad. Sci., U.S.*, **63,** 1395.

Pejsachowicz, B., B. Halpern, H. Febvre, and G. Barski (1964). *Compt. Rend.*, **259,** 4891.

Pessac, B., and F. Alliot (1970). Aggregation of mouse malignant tissue culture cells at +6°. *Exptl. Cell Res.*, **60,** 305.

Rosenbluth, J., and G. D. Pappas (1958). Measurement of osmotic fragility of fibroblasts. *Proc. Soc. Exptl. Biol.*, **98,** 38.

Roth, S. (1968). Studies on intercellular adhesive selectivity. *Develop. Biol.*, **18,** 602.

Roth, S., E. S. McGuire, and S. Roseman (1971). Evidence for cell-surface glycosyltransferases: Their potential role in cellular recognition. *J. Cell Biol.*, **51,** 536.

Schleich, A. (1967). Studies on aggregation of human ascites tumor cells. *European J. Cancer*, **3,** 243.

Seeds, N. W., and A. E. Vatter (1971). Synaptogenesis in reaggregating brain cell cultures. *Proc. Natl. Acad. Sci., U.S.*, **68,** 3219.

Sheffield, J. B., and A. A. Moscona (1969). Early stages in the reaggregation of embryonic chick neural retina cells. *Exptl. Cell Res.*, **57,** 462.

Sheffield, J. B., and A. A. Moscona, (1970). Electron microscopic analysis of aggregation of embryonic cells: The structure and differentiation of aggregates of neural retina cells. *Develop. Biol.*, **23,** 36.

Shimada, Y. (1968). Suppression of myogenesis by heterotypic and heterospecific cells in monolayer cultures. *Exptl. Cell Res.*, **51,** 564.

Simon, D. (1960). Contribution á l'étude de la circulation et du transport des gonocytes primaires dans les blastodermes d'oiseau cultivés *in vitro. Arch. Anat. Microscop. Morphol. Exp.*, **1,** 93.

Steinberg, M. S. (1963). Reconstruction of tissues by dissociated cells. *Science*, **141,** 401.

Sussman, M., and R. R. Sussman (1956). Cellular interactions during the development of the cellular slime molds. In *Cellular Mechanisms in Differentiation and Growth* (D. Rudnick, ed.), p. 125. Princeton Univ. Press, Princeton.

Trinkaus, J. P. (1961). Affinity relationships in heterotypic cell aggregates. In *La Culture Organotypique*, *Colloq. Intern. Centre Natl. Rech. Sci.*, 101, p. 209. Paris.

Trinkaus, J. P., and J. P. Lentz (1964). Direct observations on type-specific segregation in mixed cell aggregates. *Develop. Biol.*, **9,** 115.

Weiss, L. (1967). *The Cell Periphery, Metastasis and Other Contact Phenomena.* North-Holland, Amsterdam.

Wilson, H. V. (1907). On some phenomena of coalescence and regeneration in sponges. *J. Exptl. Zool.*, **5,** 245.

Zwilling, E. (1954). Dissociation of chick embryo cells by means of a chelating compound. *Science*, **120,** 219.

PART 5
CELL INJURY

18

CELL INJURY

J. Chayen and Lucille Bitensky

1. INTRODUCTION

In recent years cellular biology has emerged as a major unifying subject in biology and medicine. Yet in fact it has a long history which has been well summarized by Sharp (1934) and by Guilliermond (1941). Cells were first described by Hooke (1665), who studied the structure of sections of cork. Moldenhawer (1812) separated cells from tissue by maceration and thus showed that cells had walls of their own and did not arise as cavities within what we now would call the cell wall material. Certain features of the cell theory had been taught in Paris since 1796, mainly by Lamarck (1809). Yet it appears to have been Dutrochet (1824) who first emphasized the primary importance of cells. Robert Brown (1831) first recognized the nucleus as an important cellular structure.

The Cell Theory, which is often taken as the starting point of cellular biology, was initiated by Schleiden (1838) and finally formulated by his pupil Schwann (1839). The importance of cells, and of cellular biology, in pathology was clearly stated by Virchow (1858). The cell theory was finally overtaken by "the protoplasm doctrine" which emphasized the primary importance of units of protoplasm rather than of the cell walls. According to Sharp (1934), "It is not too much to say that the identification of protoplasm as the material substratum of life processes was one of the most significant events of the nineteenth century." The doctrine was given popular expression by T. H. Huxley (1868).

Great advances were made in fundamental cellular biology throughout the last century by botanists, zoologists, and histologists, the cardinal importance of cellular biology being fully recognized by the great biologists of the nineteenth century. Not least among these cellular biologists was Ehrlich, now chiefly remembered for his development of the use of histological dyes and his pioneer work on chemotherapy. Consequently, by the early years of the present century, there was a vast wealth of knowledge about cells, embodied in Wilson's great work on this subject. To this knowledge was soon added the understanding of chromosome behavior, which in basic form was recorded in the second edition of Wilson's *The Cell in Development and Inheritance* (1925). This work is still unsurpassed as a manual of general cellular biology and has a place in the modern biological library.

Even this brief bibliography makes one wonder why there is a general impression that cellular biology is a new subject. The reason is that by about 1925 cellular biologists had asked, and answered, most of the questions they knew how to ask and for which they had the tools and techniques for providing answers. For this is one of the essentials of scientific method. First, one must have sufficient knowledge to ask the pertinent question. Then, one needs the technology with which to answer that question. Science requires both speculation and investigation; speculation without investigation may demonstrate the wit or the polemic skill of the dielectic, but it is not science any more than is detailed investigation without cerebration. This view has been well summarized by two dicta attributed to the physicist Lord Kelvin and to the geneticist Bateson, respectively, namely, "All science is measurement," countered by, "But not all measurement is science." This double requirement is demonstrable in the work of the cellular biologists of the first two or three decades of the present century. During the first years the Lewises in particular developed the technique of growing cells in single layers in proliferative tissue culture (Lewis and Lewis, 1924). They, and Strangeways (1924), developed these methods so that they could study the behavior of individual growing cells. But at that stage what questions could they ask? They knew nothing of the importance of nucleic acids in the life cycle of cells, nor did they have the cytochemical and biophysical tools which were essential for answering basic questions. Their work came to its real fruition (e.g., Brachet, 1950; Brachet and Mirsky, 1961) only after the organic and biochemists had given us the chemical knowledge to know which questions to ask of these proliferating cells, and when Caspersson (1950) and others had developed the apparatus that made it possible to measure the amounts of significant chemical substances which occurred in single specified cells. This is the reason why cellular biology appears to be a very recent subject. It is only because cellular biologists had to wait for biochemistry and biophysics to catch up with them and to supply the knowledge and technology which was essential before cellular biology could advance again. While waiting for these advances, cellular biology concerned itself with cytogenetics, and this waiting period produced the basic understanding of chromosome structure and of chromosome mechanics (e.g., Belar, 1928; 1929; Darlington, 1937, 1946).

The cellular biologists of the last century were well aware of the possible errors and of the limitations of the results they obtained. This is seen in their detailed discussions of the nature of protoplasm and of the probable artifacts in their preparations (Flemming, 1882). Similarly, there were debates as to whether cross-striations in muscle in proliferative culture were coagulation type—or another type—of artifact. But with the advent of modern cellular biology, there is an even greater need for rigorous skepticism. This follows from the fact that cellular biology nowadays requires the focusing of biophysical and biochemical methods and concepts on the living cell. All the disciplines of chemistry, physics, general biology (physiology, morphology, and genetics), and even of pathology and pharmacology, impinge on the study of the cell. Each of these has its own concepts, criteria, and limitations, well known to its own practitioners but not necessarily obvious to the cellular biologist. For this reason it is not enough to state that Dr. X has shown that a particular phenomenon occurs. It is essential to know in which cells, at which stage of their life, and under precisely which conditions Dr. X observed these phenomena. We will try to clarify this statement with three examples.

The first occurred recently in work we were doing on the response of distal small blood vessels in cardiac muscle following their temporary occlusion. We were told that catecholamines (such as adrenalin) played no part in the subsequent dilation of these blood vessels. On enquiry it transpired that this statement really meant that a particular substance, which normally blocked β-receptor sites, did not affect the response. But it is likely under these particular conditions that this substance has an effect of its own; its biological function is not restricted to blocking only β-receptor sites. To a pharmacologist the initial statement was correct but only in the general context of organ pharmacology. To the cellular biologist that particular experiment was inconclusive, that is, no one had tested the effect of catecholamines directly and the only evidence was indirect and susceptible of alternative interpretation.

The second example concerns the use of "specific inhibitors." To a biochemist who specializes in studying the synthesis of protein by isolated fractions of cells, actinomycin D has a specific effect in that it blocks transcription of the code from nuclear DNA to RNA (e.g., see Sibatani, 1966); in this way it halts protein synthesis which depends on the transcribed RNA. But to a cellular biochemist, actinomycin D may have many toxic effects on cellular metabolism. Hence to the first type of biochemist, any effect of actinomycin D is described as an effect on protein synthesis. If he is dealing solely with isolated DNA and ribosomes, it may well be correct. However, any effect which actinomycin D may produce on the biochemistry of whole cells must not be assumed to be due solely to this cause (as discussed by Whisson, 1967).

The third example is closer to Heisenberg's famous uncertainty principle as used in atomic nuclear physics. This principle applies very frequently when living systems are studied. In its general statement the principle states that anything an observer does in order to observe a system may itself alter the system. An example of this occurred in early studies in which ultraviolet microscopy was used to investigate the state and quantity of nucleic acids in cell nuclei. Walker and Davies (1950) showed that the amount of ultraviolet absorption, and hence the apparent concentration of DNA, varied according to the dose of ultraviolet light used in studying the cells. In the older form of ultraviolet microscope, cells were subjected to continuous ultraviolet irradiation while they were investigated, their image being observed on a fluorescent screen which converted the invisible ultraviolet image into a visible image. During this investigation the optical density of the nuclei altered very appreciably.

Consequently, we introduce this chapter on the cellular biology of cell injury with a plea for precise thinking and for scientific skepticism. The plethora of publications makes for slovenly reporting because it often is difficult to read all the publications in as great detail as the subject merits. Consequently it really is essential, when generalizing a finding, to be a little critical of that finding and to know the limitations within which that finding is wholly correct. In reviewing the huge subject of cell injury, we are well aware that we cannot do it full justice and that our reporting will be rather sketchy. All we can hope to do is to indicate what we believe are the highlights and interesting lines of thought and study in this subject. We hope to interest the reader sufficiently for him to read reports of the original research on the subject for himself—with due skepticism.

2. STRUCTURAL CHANGES IN CELL INJURY

A. Introduction

There are relatively few structural responses which cells can make in response to injury. Similarly, there are only few changes in structure that can occur when cells are stimulated to unusual activity. For example, it has been known for many years that cells that are metabolically very active may have many-lobed nuclei (Sharp, 1934) and that their mitochondria may differ from those of less active cells. In some cells the Golgi apparatus is said to migrate to the part of the cytoplasm from which secretion takes place; in others its structure alters according to the secretory activity of the cell. (This subject has been reviewed by Bourne, 1951; also see Beams and Kessel, 1968.) However, these are all nonspecific responses of the cell. All they indicate is that there is some unusual degree of metabolic activity. Equally, it has long been known that one of the earliest responses of cells to injury is the "cloudy swelling" of pathology (Cameron, 1952, pp. 305–312; see Section 2.D). But this too is a very nonspecific response of cells to a wide range of deleterious conditions. The same is true of nuclear changes; if cellular injury affects the nucleus, it can cause only one of a few changes, namely, loss of regular outline or loss of optical homogeneity leading finally to clumping of the chromatin and pyknosis. The nucleus may swell or contract. It must be realized, however, that this lack of specificity has proved to be of great advantage to the pathologist; it means that one of only a few possible structural changes is likely to become observable if injury of any kind has occurred to the cells being studied.

The range of observable structural responses a cell makes has been greatly extended by electron microscopy although, even with this tool, it is often difficult or impossible to guess at the functional significance of such changes as are observed. It should be appreciated that the electron microscope is only one of a number of biophysical tools that can give useful information concerning changes in cellular structure and texture. Indeed, the cytologists' armamentarium has been only poorly exploited in recent years.

Within the limitations imposed by structural change (as contrasted with functional changes), much can be learned if the available equipment is properly used. Hence it may be helpful to consider what methods are available, how they may be used, and what are their limitations.

B. Methods for Studying Structural Changes

1. Conventional Light Microscopy. The normal transmitted-light microscope allows structure to be distinguished provided: (1) that the structure is sufficiently large and discrete from other structures so that it can be resolved; and (2) that there is sufficient contrast between the structure and the surrounding matter, or the illuminated field, in which it lies.

The ability to resolve a structure is mainly controlled by the resolving power of the objective used. Magnification, whether it be of the objective or of the eyepiece, is secondary to this primary factor because once a structure has been resolved by the objective it must be magnified sufficiently to be seen (resolved) by the eye of the observer. But if it has not been resolved by the objective, no amount of magnification will make it visible.

The resolution of the objective (*d*) is the smallest distance that can occur between two objects and still allow them to be recognized as separate objects. It is dependent on the wavelength (λ) of the light used and on the numerical aperture (N.A.) of the objective:

$$d = \frac{0.6\lambda}{\text{N.A.}}$$

So if we use a good oil immersion objective having an N.A. of 1.3 and we use green light (i.e., $\lambda = 550$ nm), then

$$d = \frac{0.6 \times 550}{1.3} = 254 \text{ nm, or } 2540 \text{ Å, or } 0.25\ \mu$$

It should be noted that this value of approximately 0.25 μ can be obtained only if the microscope is set up and used correctly. If the objective is a good "air-dry" objective of N.A. 0.85, the microscopist can resolve only approximately 400 nm, that is, he can detect two objects as two discrete objects only if they are at least 0.4 μ apart. If they are closer together he would see only one structure.

Unfortunately, it is not readily feasible to increase the N.A. of objectives appreciably above 1.3. Instead one can reduce the wavelength by using ultraviolet light. This imposes some practical difficulties, but these are readily surmounted. When photographs are taken with an objective of N.A. 1.3 but with ultraviolet light of wavelength 280 nm, the resolution becomes 160 nm (1600 Å or 0.16 μ). This is still far from the resolution of the electron microscope which, as generally used, is approximately 50 Å (i.e., 1/10 the thickness of the tissue section; see 2.B.6), but the advantage of ultraviolet microscopy is that it can be used for living cells (e.g., see Ludford and Smiles, 1954).

The actual size of the object controls whether it is seen as a structure or whether only its presence can be detected. This question of "detection" is often confused with "resolution."

Suppose we have an object isolated from neighboring objects by a space of 0.25 μ. The microscopist, using green light, can detect it, that is, the oil immersion objective

(N.A. 1.3) resolves it from the surrounding objects as a distinct object. He can see it as a structure only if it itself is 0.25 μ wide, in which case his objective resolves each edge separately. However, a structure can be *detected* even when it cannot be fully resolved.

In this discussion it should be noted that the magnification of the objective is relatively unimportant. It is the N.A. that determines what the objective will resolve.

The other major determining factor is the contrast between the object to be resolved and the background in which it lies. If you have an optically homogeneous object lying in an optically homogeneous material, you will not resolve any detail with a normal conventional light microscope however high is the N.A. of the objective, or however short the wavelength of the light, unless there is contrast between the object and its background.

Generally, in normal microscopy contrast is obtained from one of two effects: either by optical properties in the specimen such as diffraction and "scatter," or by making the object absorb light differentially from its background. The first, namely, diffraction, is enhanced by closing the condenser diaphragm or racking down the condenser; neither procedure is recommended except as a rough aid in examining the tissue or cell. If you can see a structure because it diffracts light, it means that you are not truly seeing the structure but only its shadow. Generally, an object produces a diffraction image because it has a different refractive index from its surroundings; if this is so, phase-contrast microscopy (as discussed in Section 2) should be used. On the other hand if the object is scattering light, you should benefit from using dark-ground microscopy (Section 2.B.3).

In general microscopic practice contrast is obtained by staining the specimen so that different parts absorb light differently. The parts can then be seen because of the contrast between them and their surroundings. For example, you stain a cell with a mixture of dyes such as hematoxylin and eosin or methyl green and pyronin. With the latter mixture the nucleus stains blue-green, that is, it absorbs red light. The cytoplasm stains red, that is, it transmits red light and absorbs other wavelengths. Consequently, the nucleus can now be seen because it absorbs red light, whereas the cytoplasm transmits red light. The chromosomes on the mitotic spindle stain blue; they are separated by pink-colored spindle matter and the space between them is greater than 0.25 μ. Hence the microscope can resolve them (it could before they were stained), but now they have sufficient contrast from their background for them to be readily seen. The chromatids within a chromosome may be separated by less than 0.25 μ, the space between the blue chromatids being pink. The fact that there are two chromatids can be appreciated because of the contrast in color (absorption) and possibly because of diffraction effects, even if complete resolution is impossible. Consequently, you may guess that there are two chromatids lying together, but you should not place too much reliance on what you think you see unless they are resolvable, that is, they are 0.25 μ apart. There is a long tendentious literature on this subject in which different workers have claimed different structures for chromatids which in fact could not have been resolved but only detected with the microscopes which these workers used.

Another way of providing contrast in a cell is to use the natural absorption of ultraviolet light shown by different parts of the cell. Instead of staining the chromosomes, you can photograph them with light of approximately 260-nm wavelength. The nucleic acids strongly absorb light of this wavelength, so that they appear naturally "stained." The great advantages of this method are that resolution is increased (by reducing wavelength) and contrast enhanced (by the natural absorption of the nucleic acids), and yet the cells are kept alive. In a sense, this is a physical "vital" dye.

For other structures so-called vital dyes can be used, which penetrate, or are actively taken up, into living cells and stain either the whole cell differentially, or certain cellular components specifically. Janus green has been much used as a vital dye for staining mitochondria. The way in which this dye acts has been fully studied by Lazarow and Cooperstein (1953a, b), but it is probable that even at low concentrations (e.g., 1/100,000) it is ultimately toxic to cells. Other investigators have used fluorescent dyes (see Bräutigam and Grabner, 1949). Strugger (1938; Höfler, 1949) used rhodamine B and rhodamine 6G to great advantage in plant cells especially because these dyes are taken up more strongly by lipid-containing structures and so make them visible. Allison and Young (1969) have used acridine orange and a number of other dyes as vital dyes to make lysosomes visible. This is a pretty use of vital dyes because lysosomes concentrate these dyes when they are used at sufficiently low concentration (1 in 10^6 w/v) and so protect the cytoplasm from any lethal effect they might have. Neutral red has also been much used to stain the Golgi apparatus (e.g., see Bourne, 1951; Beams and Kessel, 1968), as well as, more latterly, the lysosomes (see Allison and Young, 1969).

It seems likely that the vacuoles of the Golgi apparatus, which concentrated neutral red, were only one component of the whole Golgi complex (Baker, 1944). It is possible that this component is that part of the apparatus in which the newly formed lysosomes are found (Novikoff, 1963; also see Cohn and Fedorko, 1969). But with all these dyes care must be taken if they are to be truly vital dyes and not act by killing the cells and then staining the cytoplasm or the nucleus. Even ultraviolet absorption microscopy can result in the death of cells from irradiation damage unless used with considerable care.

2. Phase-Contrast Microscopy and Interferometry. Instead of using vital dyes, or the natural absorption of light of specific wavelengths, it is possible to use natural differences in refractive index to give contrast to components of living (and of fixed) cells. This concept of refractive index needs a little explanation.

The velocity of light in a vacuum (C) is equal to 3×10^{10} cm/sec. It is approximately the same in air, as far as the microscopist is concerned. Light travels more slowly through transparent matter because it is retarded by the matter. Its velocity is then C/n, where n is the refractive index of the matter. Hence the definition of refractive index of a material is the velocity of light in vacuum divided by the velocity of light through that material. Consequently, material of higher refractive index will retard light more than will other material of lower refractive index. How does this concept help us in cellular biology? The refractive index of water is approximately 1.334. Each gram of protein dissolved in 100 ml of water increases the refractive index of the water by a constant amount known as the specific refractive increment and equal to 0.0018 (approximately). Thus a 20% solution of protein in water has a refractive index of $1.334 + 20(0.0018)$, or 1.3376. It is a fortunate chance that the specific refractive increment of nucleoprotein and of lipoprotein are almost the same as that of pure protein. Different regions of cells have different concentrations of these proteins, and consequently they have different refractive indices. In fact, when you stop down the aperture of the condenser, or rack it down out of focus, you accentuate the effects produced by the different refractive indices of the various parts of the cell. Phase-contrast microscopy achieves this without misusing the microscope. Interference microscopy is a quantitative form of phase-contrast microscopy in that it allows the refractive index of each part of the cell to be measured. In actual practice it is usually used to measure the dry mass of any part of the cell because you measure how much the light has been retarded while passing through that particular region. But it can also be used qualitatively as variable "interference-contrast" microscopy. Phase-contrast microscopy has been fully described by Barer (1956, 1968) and interference microscopy by Chayen (1967); for detailed accounts of both types of microscopy, see Ross (1967) and Chayen and Denby (1968). The point to be stressed here is that contrast, which is needed if structures are to be seen, is achieved by using the natural differences in the refractive indexes of the various cellular components. Moreover, both forms of microscopy, namely, phase-contrast and interference, can readily be used on living cells with minimum disturbance to these cells.

A great wealth of excellent cellular biology has come from studies in which living cells were photographed at regular intervals by time-lapse cinephotography working through a phase-contrast microscope. Many fine films have been made which demonstrate movements of chromosomes and of cellular organelles during normal and abnormal mitosis or cellular growth. Some of this work is recorded in a volume by Hughes (1952) which describes mitosis and in the studies on mitosis and the effect of mitotic poisons in plant cells by Bajer (1956, 1964, 1965). The study of mitochondria in particular is well represented by the outstanding films and publications of Chèvremont (1956), Gey (1956), Frederic (1958), and Pomerat (1961).

All these methods allow structural changes to be studied in living cells by direct observation of the structures. We will now turn to certain methods that give precise information concerning structural changes but which are rather more indirect. They are not less valuable for being indirect.

3. Dark-Ground Illumination. This form of microscopy was much used to demonstrate the presence of organelles in living cells, where the discontinuity between the organelle and the cytoplasmic "cell sap" caused light to be scattered. Mitochondria and nuclear membranes were especially clear (Strangeways and Canti, 1927). It was especially valuable in the studies of Horning (1951) on the distribution of intracellular minerals. In this work Horning microincinerated sections so that all the carbon and nitrogen was volatilized, leaving only the mineral components which showed clearly by their scattering under dark-ground illumination.

Coagulated protoplasm also scatters light, whereas undenatured protoplasm, excluding its organelles, is transparent. This fact has been used in many studies (see Ludford, 1951) to disclose, for example, how acidic and alkaline conditions can cause reversible coagulation of nucleoplasm.

4. Polarized Light Microscopy. Although the detailed theory of this form of microscopy is very complex (Hartshorne and Stuart, 1970), its basic understanding and practice are quite simple. Ordinary light vibrates in all directions, so that if A shines a torch on B the rays travel in a straight line from A to B and vibrate in all planes perpendicular to AB (Fig. 1).

In fact, the light path AB is a series of waves which vibrate in all planes perpendicular to AB. It is possible to absorb all the vibrations except those in one plane. This produces plane-polarized light and is effected either by a Nicol prism, made from a specially cut calcite crystal, or by passing ordinary light through a Polaroid plate. Such a plate is made of long-oriented molecules of polyvinyl alcohol (PVA)

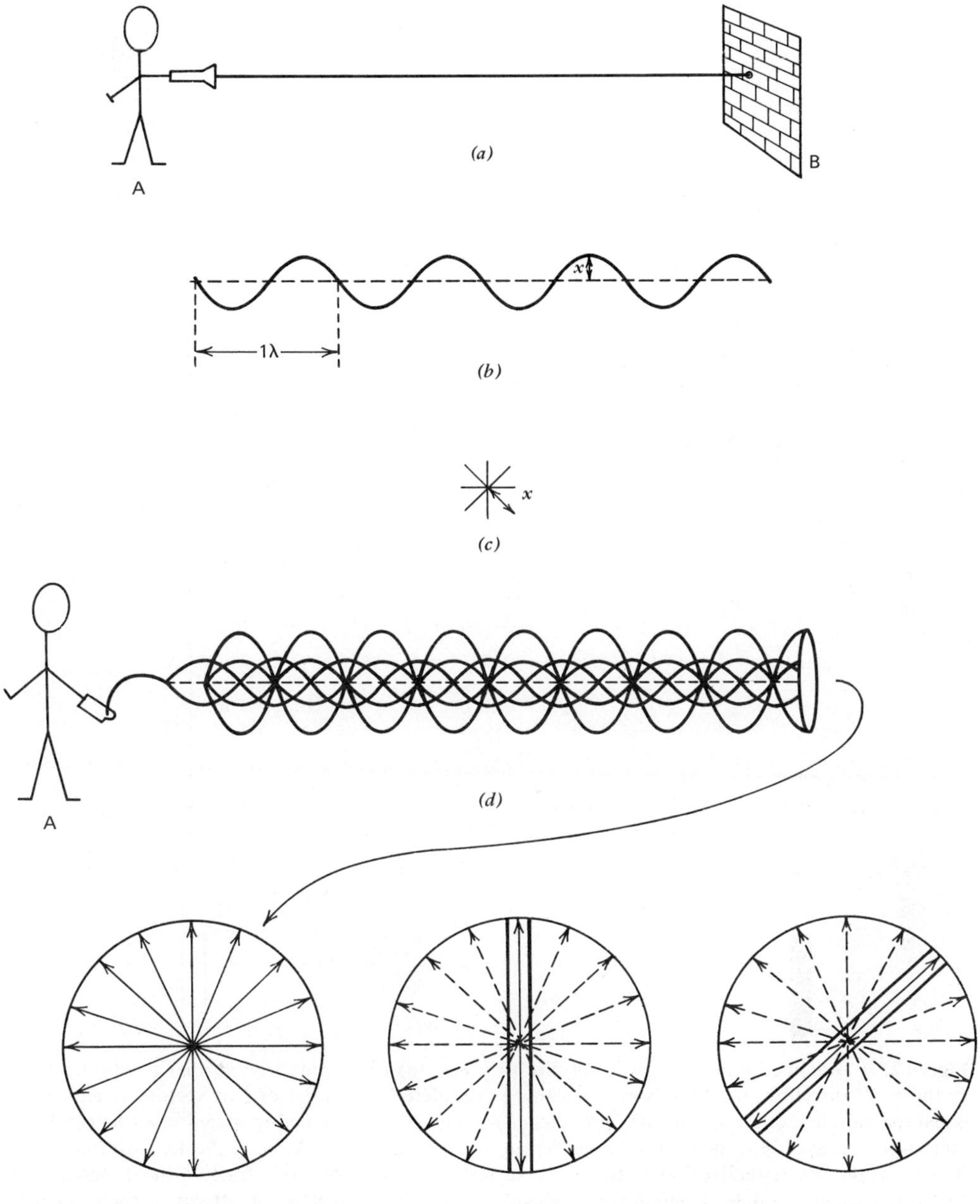

Figure 1. (*a*) A shines a torch on the wall B. The beam of light consists of waves (*b*) of amplitude a x and wavewavelength λ; the color of the light is due to the wavelength (for green light λ is approximately 550 nm; for red light approximately 650 nm; and for yellow light approximately 450 nm). When viewed from A the waves of light vibrate in all directions with their amplitude α perpendicular to the line of the light beam (AB) as indicated in (*c*). The simplest analogy is that of a man swinging a rope and making it oscillate in all directions (*d*). When viewed from B the oscillations are seen rising and falling in all directions, as indicated in (*e*). A slit placed at B will restrict all vibrations except those either directly in the plane of the slit (f) and (*g*) or those which are almost in the same plane, that is, those which have a vector in this plane (see Fig. 2). A second slit, placed at right angles to the first, will completely dampen the oscillation of the rope.

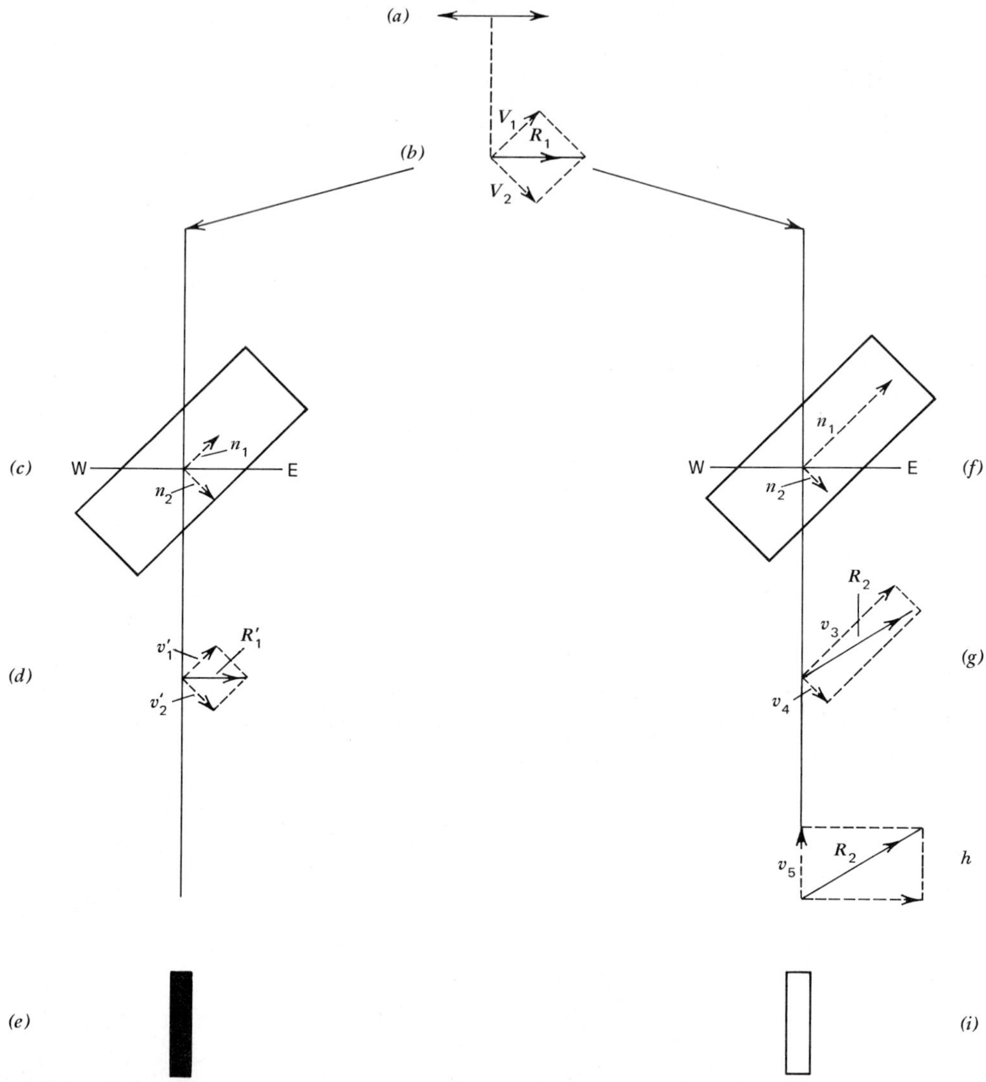

Figure 2. A simplified scheme illustrating birefringence. (*a*) Light polarized so that it vibrates only in the E–W direction. (*b*) This vibration can be considered to consist of two vectors v_1 and v_2 of equal magnitude, each at 45° to the E–W direction. Since they are equal in magnitude, the resultant vibration R_1 is exactly E–W. (*c*) First consider an isotropic structure placed at the 45° position. Since it is isotropic, its refractive index along the structure is equal to the refractive index across the structure (i.e., $n_1 = n_2$). (*d*) Consequently, when light, as in (*b*), traverses this structure, there is equal reduction in the velocity of each vector. Hence v_1 is still equal to v_2 even though they may be slightly less than v_1 and v_2. However, being equal in magnitude, the resultant vibration R_1 is still exactly E–W. Since no force can have a vector at right angles to itself, no vibration or vector can pass an analyzer mounted in the N–S position so the microscope field remains black (*e*). (*e*) Now consider an anisotropic object in the same field, as (*b*). Since the object is anisotropic, the refractive index along the structure (n_1) is different from that across the structure (n_2); this means that the velocity of light differs according to whether it vibrates along or across the structure. (*f*) Consequently, the two vectors will now be dissimilar, and being retarded relative to the other. In this diagram v_2 has been retarded, so that its magnitude after passing through the structure (v_4) is drawn shorter than v_3. The resultant of these dissimilar vectors (R_2) is no longer E–W; its direction is determined by the parallelogram of forces diagram. (*g*) The resultant light (R_2) can be considered to be composed of two

suitably stained so as to absorb practically all the vibrations except those that vibrate along one particular plane. The simplest analogy is to imagine a man swinging and vibrating a rope in all directions. Then place a slit around the rope so that he can vibrate the rope only in one direction, for example, up and down. In this analogy the slit takes the place of the polarizer. If another slit is placed at right angles to the first, none of these waves will pass through it. This second slit represents the part played by the analyzer which is a second plate of Polaroid material which can be set so that the two Polaroid plates (polars) are "crossed," namely the orientated molecules of the analyzer are aligned at right angles to those of the polarizer. Under these conditions the microscope field seen through the ocular is black. If a birefringent object is introduced into this field and orientated correctly with reference to the two polars it will have various degrees of brightness, depending on its thickness and its degree of birefringence. A birefringent object is one which can rotate the angle of vibration of the polarized light so that it becomes possible for some components of its vibration to pass the analyzer set in the crossed position.

To understand even superficially how this comes about, we have to use the Triangle Law of Vectors, which may be better known as the Triangle or Parallelogram Law of Forces. Suppose the light is polarized so that, in passing through the condenser into the specimen and through the objective, it is vibrating East and West in the microscope field. It will have vectors at 45° on either side of E–W, and because these vectors are equal the resultant will be a net vibration exactly East and West.

A birefringent, or anisotropic, object is one with two dissimilar refractive indices. This means that the speed of light along two axes in the object are unequal. An isotropic object has only one refractive index, that is, its refractive index is the same in all planes. Consequently, when light traverses an isotropic object, the velocity of its vibration is altered equally in all directions, so that any change in V_2 (Fig. 2) is also found in V_1, and the resultant plane of polarization is unaltered. Consequently, it still remains dark when viewed with crossed polars. On the other hand, when light passes through an anisotropic object, the speed of light along one plane is accelerated relative to the other. This becomes most marked if one optic axis of the specimen is placed at the 45° position, that is, NW–SE.

The vibration of the resultant light has been bent towards the N–S position and it now has a component in this direction; consequently, the object appears bright. The greater the difference between the velocity of light along and across the object (i.e., the greater the birefringence of the object), the greater will be the alteration in the resultant direction of vibration of the light; hence the N–S component will be greater and the object will appear brighter.

Birefrigence is very important in cellular biology. Even at a time when there was doubt as to whether the spindle fibers really existed or whether they were only fixation artifacts (see Schrader, 1944), polarization microscopy was able to show that the spindle contained oriented long molecules aligned along what were apparently the spindle "fibers" (Swann, 1951a, b; also Inoué and Dan, 1951). Quantitative measurement by Swann demonstrated how the orientation of these fibrous proteins, and those of the

---

vectors, one N–S and the other E–W. The magnitude of these vectors in determined by drawing the parallelogram that fits R_2. (*h*) R_2 has a vector in the N–S direction. This vector is transmitted by the analyzer set to the N–S position, and the microscope field appears illuminated. It must be appreciated that this is an oversimplified scheme and should be used only as a first approximation to the ideas of polarizing microscopy.

asters, varied with the movement of the chromosomes during mitosis. Even today electron microscopy is not very effective at showing the precise structure of these molecular structures which now appear to include microtubules (Mazia, 1961); however, their presence can be demonstrated, if not directly visualized, by polarized light microscopy.

In voluntary and cardiac muscle, the birefringent or anisotropic protein molecules (and groups of molecules or "micelles") of myosin are oriented parallel to the long axis of the muscle fibril. Consequently, that part of the muscle which contains the myosin is bright between crossed polars when the muscle is placed with its fibrils at the 45° (NW–SE or NE–SW) position. These bright regions, or cross-striations, were called the A bands because they were anisotropic bands. The interband regions, which contain mainly actin, were called I bands because they were isotropic. It now appears likely that contraction of muscle may involve changes in the orientation of the birefringent myosin molecules (Harriman and Stocks, 1970). This type of change in orientation can be measured precisely with a polarized light microscope fitted with a suitable compensator for measuring the amount of birefringence (see Chayen and Denby, 1968) even though the molecules or micelles that cause the birefringence cannot be resolved by light microscopy. Electron microscopy, on the other hand, can resolve the micelles of myosin in the muscle fibrils (H. E. Huxley, 1961), but the measurement of changes of orientation would be very difficult, if not impossible. Polarized light microscopy detected birefringent cellulose in the young cell walls of plants, and careful measurement showed the way in which these molecules were oriented in such walls (Preston, 1951). This indirect evidence–indirect because the cellulose molecules or micelles were not actually resolved—was later fully confirmed by electron microscopy. But whereas the latter generally required fixed, dehydrated, and stained thin sections, in which changes could have been induced by the processing, polarized light microscopy allows you to deduce quantitatively the presence of such molecules even in living cells.

5. X-Ray Microscopy. It has been emphasized that resolution through a microscope is increased if the wavelength of the light is reduced. The wavelength of x rays can be rendered very short by using a high voltage. Unfortunately, matter refracts x rays so weakly that it has not been possible to construct refracting objectives for x-ray microscopes, and such reflecting lenses as have been made have given resolutions less than that achieved by the normal light microscope (Cosslett, 1965). Some attempts have been made to make x-ray shadow projection microscopes without lenses, and there is no reason why these could not be constructed to give a resolution of, for example, 0.06 μ. The difficulty is that the image produced by this type of microscopy may have considerable resolution but little magnification. The image is photographed and the negative is magnified by photographing it through a normal light microscope, so that the resolution obtainable through the latter again becomes the limiting resolving power. (For fuller discussion of x-ray microscopy in cellular biology, see Chayen and Denby, 1968.)

The special advantages of x-ray microscopy lie not in improved resolution, although this may yet be achieved, but in its use for measuring the dry mass of parts of a section and for mineral analysis. Its use for "weighing" regions of a section has been superceded by interference microscopy which can be used on living cells, whereas x-ray microscopy requires dried specimens. Comparisons of the two types of mass measurement have been made, notably by Sandritter and Müller (1959) and by Müller et al. (1959a, b).

But there are considerable potentialities in measuring specific elements, such as phosphorus or calcium, by means of specific x-ray absorption microscopy. These methods have been discussed by Engström (1956, 1959, 1962).

6. Electron Microscopy. Electrons have wavelike properties, and the wavelength (in Ångstrom units) associated with them is equal to $12/\sqrt{v}$, where v is the accelerating voltage. Thus if we use 90 kV (90,000 V), the wavelength λ will be 0.04 Å. The theoretical resolution is derived from the same equation as for the light microscope, so that it is equal to $0.6\lambda/\text{N.A.}$ The drawback in electron optics is that electron lenses have such severe aberrations that they work reasonably well only at an extremely low N.A., namely, approximately 0.01 to 0.001 (this should be compared with the N.A. of 1.3 readily attainable with the light microscope). With a wavelength of 0.04 Å and an N.A. of 0.001, the limit of resolution is 24 Å (or 0.24×10^{-2} μ, that is, 10^2 times better than can be achieved with the light microscope).

As with the light microscope, the ability to actually resolve structure depends on the resolution of the microscope and on the amount of contrast in the object to be resolved. To increase contrast it is conventional to fix the specimen, or to "stain" it after fixation, with a heavy metal such as osmium. Such metals scatter electrons and so give the appearance of absorption.

Electrons are scattered by air, so the specimen must be viewed in a vacuum. They are scattered and slowed up by matter, and these slow electrons do not focus at the correct image plane because the focal length of the electron lenses depends on the speed of the electrons. Consequently, the out-of-focus electrons blur the image and deprive it of the contrast necessary for full actual resolution. To overcome this problem the specimen must be very thin. The actual thickness varies with the speed of the electrons, and hence with the voltage used in accelerating them onto the specimen. With the usual electron microscopes now operating, namely at up to 100,000 V, a resolution of 24 Å probably requires that the section and its support should not be thicker than 240 Å (1 Ångstrom unit $= 10^{-1}$ nm $= 0.0001$ μ). With the newer higher-voltage electron microscopes, the specimen can be thicker and still give 24-Å resolution.

But it is evident that the tissue must be sectioned into very thin slices, and this requires special techniques of embedding and sectioning. It must be fixed in special fixatives to allow for these procedures to be done and to allow the tissue to have enough contrast to show detail. Thus, on the one hand, the electron microscope offers wonderful improvement in resolving power and in total magnification (200,000× magnification is now fairly normal). On the other hand, the procedures of fixation and embedding remove a considerable proportion of the tissue components. It is not unusual to lose between 50 and 100% of the lipid components (Cope and Williams, 1967), and it must be realized that the artifacts caused by the loss of this material are also magnified 200,000 times. Moreover you rarely see the specimen but rather the contrast imposed by osmication or similar electron-staining procedures. Although this is not much more serious than looking at stained specimens with the light microscope, it can lead to considerable errors of interpretation. One of the most common concerns the description of pores in a membrane. These may be true pores or holes, or they may only look like holes if they consist of tissue lying encircled by osmium-darkened lips. Only precise microdensitometry can distinguish between these possibilities, and unfortunately it is rarely used in electron microscopy.

Because the electron microscope uses lenses of such very low N.A., there is a remarkably large depth of focus in the image and this can lead to difficulties in interpretation. Structures that appear to be contiguous can in fact be at different levels in the section. However, all these problems are well discussed by the leading electron microscopists, and for reference to this form of microscopy the reader is referred to such works as those of Sjöstrand (1967), Kay (1965), and Pease (1964).

C. The Nature of the Noninjured Cell

There is no such thing as "the typical cell," but characteristic features of cells can be understood by reference to two types of cell which have been much studied in life, namely, the fibroblast grown in proliferative tissue culture and the staminal hair of *Tradescantia*.

When either of these cells is mounted in a physiologically acceptable medium and viewed through a normal light microscope with the condenser used at almost full aperture, remarkably little is seen. However good the resolving power of the optics used, there is too little contrast in the cells: the detail has been resolved optically but cannot in fact be seen. If the aperture of the condenser is diminished to give stopped-down illumination, structure becomes visible because of diffraction effects. In the rather thick, refractile plant cells, such illumination is often adequate, but phase-contrast illumination is essential for the thin (flat) fibroblasts grown from animal tissues.

1. Nucleoli. By phase-contrast microscopy the fibroblast can be seen as a flat, irregularly shaped grey object, often with processes. The most obvious feature is the nucleus which is a large, round, or oval plate or sphere which seems to be empty and of a lighter color than the rest of the cell except that it contains one or two dense, round bodies, the nucleoli. In many cells the nucleoli are re-formed at telophase by the accumulation of material around a specific site, the nucleolar "organizer," on one chromosome of each haploid set (McClintock, 1934). Consequently, it often happens that there are two nucleoli in each restitution nucleus, one from each nucleolar organizer of each haploid set. Should these lie close to one another, or come together during interphase, they may fuse to form one large nucleolus (Chayen et al., 1953). This is significant because the nucleoli move around inside the nucleus as if the nucleoplasm were entirely fluid. It may not be easy to see this happen while you watch the living cell, but it can be seen very readily when cinephotographs of the cell are made, for example, one every 10 sec (time-lapse cinephotography), and the film projected at some speed in order to speed up the rather slow living movements. Interest in the cellular biology and chemistry of nucleoli has grown recently and is discussed more fully in Section 2.D.3. (also see Vincent, 1955).

2. Heterochromatin. Occasionally, other bodies will be seen inside the nucleus, some of which may seem almost indistinguishable from nucleoli. In suitably stained preparations these will be seen to be aggregates of heterochromatin. This material requires a little explanation; from the beginning of prophase to the end of metaphase, chromatin material, namely, deoxyribonucleoprotein, condenses from the diffuse state in which it occurred during interphase and appears as discrete chromosomes. After telophase it disperses—at least as far as the light microscope can see—and becomes diffuse. In forming the structures which we recognize as chromosomes, the chromatin becomes spiralized into tight spirals; maximum spiralization is seen in the small, dense metaphase chromosomes which are formed by this tight spiralization of the long prophase chromosomes. After anaphase the chromosomes despiralize, but the heterochromatin does not lose this tightly spiralized and condensed form even during interphase. Full details of these phenomena were described by Darlington (1937) and by White (1948); superb photographs can be found in Darlington and La Cour (1969). This heterochromatin, namely, that chromatin material which behaves differently from the rest of the deoxyribonucleoprotein, has been said to have different properties from the rest of the chromatin (euchromatin). It may actually despiralize during metaphase and thus have a condensation cycle precisely opposite from that of the euchromatin. Such heterochromatin has been considered to be a site of metabolically turned-over DNA (La Cour, et al., 1956). Mather (1949) has suggested that heterochromatin is composed mainly of polygenes, while euchromatin contains predominantly the Mendelian or oligogenes. Serra (1949) has powerful evidence that heterochromatin close to the site of a Mendelian gene can suppress the activity of that gene. When he removed the heterochromatin by breaking the chromosome at that point by x-radiation, the gene showed its full and normal activity again in contrast to its lack of expression when the heterochromatin was close to it. The Y chromosome is said to be mainly heterochromatic (Schultz, 1941); the Barr bodies, that is, the double-X chromosomes, also behave heteropycnotically and this property is used in determining the sex of the animal (Barr, 1966). Caspersson and co-workers (Caspersson et al., 1969a, b) showed that certain fluorochromes can be used to demonstrate heterochromatic regions. Zech (1969) showed that quinacrine mustard is particularly valuable for staining the Y chromosome of human blood cultures. La Cour and Chayen (1958) considered that the phospholipid component of the heterochromatic segments was very different from that found associated with euchromatin, and Chayen et al. (1957) suggested that this condensed and relatively available lipid matter might be the site at which lipophilic carcinogenic agents, such as benzopyrene, could enter the nucleus.

Again, the idea that the heterochromatin could be the site of entry of some carcinogenic agents becomes more realistic if we watch the movement of the cell, preferably by time-lapse cinephotomicrography. The nucleus is not still but rotates and moves in the cytoplasm, with the nucleoli and heterochromatic bodies rotating and moving inside the nucleus. Frederic (1958), in particular, showed in living cells that the nucleoli often migrate from the inside of the nucleus up to the nuclear membrane where they may remain, apparently lightly attached, for an appreciable time. In his film mitochondria can be seen wriggling through the cytoplasm and becoming lightly but definitely attached to the outside of the nuclear membrane often at the same point as the nucleolus is attached inside the membrane. There appears to be transfer of material, and then the mitochondrion seems to tear itself away from the nuclear membrane.

3. Chromatin. If you continue watching the nucleus for so long that the oxygen tension in the medium becomes reduced, you frequently see a marked change in its chromatin. Instead of the nucleus appearing as an apparently empty, light structure, it becomes clumped with a granular meshwork which, with suitable staining, can be shown to be chromatin. This is readily seen in the cells of *Tradescantia* staminal hairs and also occurs in fibroblasts. This change is said to be reversible (Lewis, 1923) and is one way in which the state of the deoxyribonucleoprotein can be altered; increases in ionic strength also precipitate the nucleoprotein which can be dispersed again by sucrose or similar molecules (Ris and Mirsky, 1949). Irradiation with ultraviolet light also causes condensation of the chromatin (as discussed above).

The electron microscope has not been very helpful in elucidating the structure of the chromosomes during the dispersed state of the chromatin, that is, during interphase (see Wolff, 1969). More surprisingly, it has yielded very little information concerning the structure of mitotic or meiotic chromosomes (but see Kaufmann and Deepesh, 1956). This is probably because the methods of fixation have been developed mainly for cytoplasmic detail and these procedures are not ideal for nuclear material. Some valuable results on the electron microscopy of nucleoli and of chromosome material associated with them have been obtained (e.g., see La Cour, 1966; La Cour and Wells, 1967). Electron microscopy has been valuable in disclosing the structure of the nuclear membranes (see Mirsky and Osawa, 1961) which often seem to consist of two structures, one acting as a support to give rigidity and the other as an apparently porous envelope. It has been questioned, however, whether or not these are true pores (Kautz and De Marsh, 1955). It is of interest that microdissection has shown that some nuclei do have a skin which can be deformed by microdissection needles, whereas others, mainly plant cells, do not seem to have such a physical membrane (Wada, 1950). Brown (1951) failed to find a nuclear membrane in certain plant cells even with polarized light. It is possible that the interface between a viscous nucleus, containing nucleoprotein, and a less viscous cytoplasm could behave as if it were a physical membrane without being a skin or a true membrane. Moreover, it is not unlikely that a surface structure could be formed from such a system by a number of stimuli including slow fixation.

A more definite problem concerning the nuclear surface is the fact that electron microscopy frequently shows a cytoplasmic membrane around the nucleus, this membrane being derived from the endoplasmic reticulum (ER) (Fig. 3). This implies that the nucleus is situated in a pouch of the ER. This pouch, being of the cavity of the ER, is connected to the exterior of the cell. This picture, derived from fixed sections, is apparently at extreme variance with the cinefilms of the nucleus moving in the cytoplasm and mitochondria attaching and detaching themselves. How can they penetrate across the walls of this "pouch"? No doubt this problem will be resolved, but the reader should be aware of the fact that we still know far too little of what might be called cellular physiology or *functional* cellular biology.

4. Mitochondria and Lysosomes. We should return to the living cell seen under phase contrast illumination. Apart from the nucleus, the most striking objects are bodies lying in the cytoplasm. In fibroblasts these bodies are mainly long and very thin (approximately 5 by 0.5 μ), although rodlike and granular forms also can be seen. These are the mitochondria. It was realized very early that the granular, rodlike filamentous structures were various forms of a single type of organelle, and the name mitochondrion (*mitos*, a thread, and *chondrios*, a granule) was coined for this organelle. In plants they have been studied fully by Guilliermond (1941) and his school, who showed that the granular form (chondrosomes) was typical of young, growing cells, while the filamentous form (chondroconts) occurred mainly in differentiated cells. In fibroblasts, particularly when they are studied by phase-contrast cinemicrography, the filamentous forms are seen to be in continuous wriggling motion, changing shape, branching, re-forming threads, and moving up to and away from the nucleus. One of the first signs of cellular damage, which can be induced even by excessive

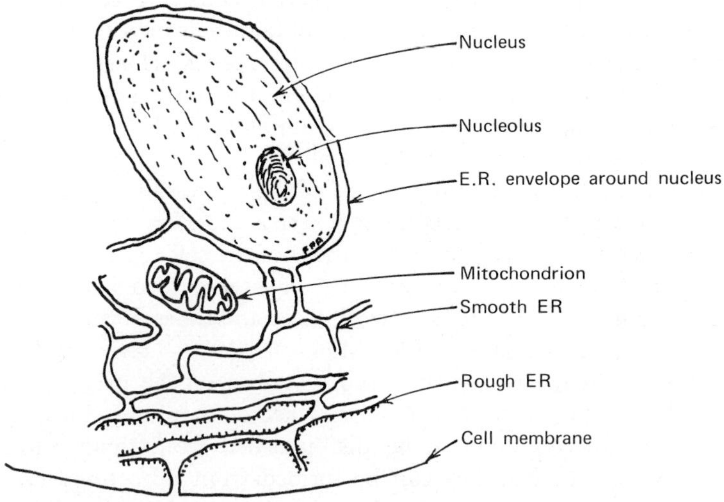

Figure 3. Diagrammatic representation of the structure of part of a cell drawn from three different electron micrographs. It purports to show the endoplasmic reticulum (ER) envelope around the nucleus and how this nuclear space apparently communicates with the exterior of the cell. Obviously you cannot follow the maze from the outside into the cell and around the nucleus in any one electron micrograph. It is also not absolutely certain that the nuclear envelope does communicate with the external environment of the cell. This diagram also emphasizes how the mitochondrion is apparently enmeshed in the membranes of the ER.

radiation with the normal light used for their examination, is that these threads disintegrate into small rods and granules but this change can often be reversed. It is believed that the wriggling motion involves the mitochondrial membrane ATPase (adenosine triphosphatase).

It is noteworthy that most of the electron micrographs of mitochondria, which have been separated by homogenization and differential centrifugation, show round granular structures. This may be a result of cellular injury induced by homogenization procedures which could cause the filamentous forms to break up into shorter forms. The electron microscope shows that these granular forms, which are also the characteristic forms seen in ultrathin sections, contain separate compartments with deep cristae mitochondriales increasing the surface area of the inner membrane. The structure and function of mitochondria are dealt with in Chapter 4.

Among the other cytoplasmic bodies seen in the living fibroblasts are small glowing bodies approximately 0.5 μ in diameter. These were formerly called lipochondria because their refractile nature suggested that they contained high concentrations of lipids. It is possible that lipid granules or lipid-rich granules do occur as separate structures, but the present tendency is to equate these lipochondria with lysosomes (Lane, 1968). This also follows from the elegant and careful work of Munro et al. (1964) and of Allison (1968a; Allison and Mallucci, 1965; Allison and Young, 1969) on living cells studied by phase-contrast microscopy and by cytochemistry. Lysosomes are of immense significance in cellular biology; their structure and function have been excellently reviewed in three volumes by Dingle and Fell (1969) (see also Chapter 7).

5. Genetics of Cytoplasmic Organelles. There is considerable evidence that mitochondria are self-replicating organelles which contain their own genetic determinants. They therefore possess cytoplasmic genes or "plastogenes" (Darlington and Mather, 1949). They are known to contain a form of DNA different from that of the chromosomes (Borst and Kroon, 1969), although the amount of DNA seems to be too little to account for the genetic determination of all the diverse proteins (i.e., that of the enzymes as well as of the structural proteins of the membranes and of the hydrogen transport chain) that mitochondria contain. There is therefore much scope for genetic interactions between the nuclear genes and those of the self-replicating mitochondria. Plant cells contain plastids, the best known being the chloroplasts. These too have their own genes, and the interactions between these and those of the nucleus have been extensively studied (e.g., Rhoades, 1946; Michaelis, 1959; Davies and Sleigh, 1970). There are many excellent electron microscope studies on the origins of chloroplasts dealing with whether they can arise *de novo*, whether they must form from "plastid primordia" which are too small to be distinguished from small mitochondria by light microscopy, or whether they can be formed from mitochondria. The current view seems to be that they can form only from plastid primordia which can be distinguished from mitochondria only by electron microscopy (see Hodge et al., 1956; von Wettstein, 1959). Chloroplasts too contain DNA.

In contrast to the self-replicating nature of chloroplasts and mitochondria, lysosomal enzymes seem to be produced by the ER and packaged into primary lysosomes by the Golgi apparatus (Novikoff, 1963; also see Cohn and Fedorko, 1969). The complex system comprising the Golgi apparatus, the ER, and the lysosomes is often referred to as the GERL system.

6. The Golgi Apparatus. In some cells clusters of fatty droplets, seen in the living cells, may indicate the location of what has become known as the Golgi apparatus (Baker, 1957; Lacy and Challice, 1957). In the past this "apparatus" has been the topic of many long and bitter controversies. It was first observed by Golgi as a meshwork in the Purkinje cells of the brain of the barn owl (Golgi, 1898) after special fixation and staining. Later workers found various meshworks or complex systems, which could be coated with silver or osmium under specified conditions, in most animal cells but never in plant cells in which only dictyosomes were found (Sharp, 1934; Guilliermond, 1941). With the widespread use of electron microscopy of ultrathin sections of tissue fixed in osmium tetroxide, and with other heavy metals, large and apparently fatty (osmiophilic) clusters of membranes were found in most animal and even in plant cells (Dalton, 1953; Beams and Kessel, 1968), and these were considered to represent the true nature of the Golgi complex (Baker, 1957; Lacy and Challice, 1957). There is no doubt that the "apparatus" argued by such major light microscopists as Baker (1944), Gatenby (1931), and Bourne (1951) was an artifact in the truest sense of the word, namely, something made by the technique itself. But although the staining methods may have enlarged and modified the fundamental cellular system, they could do so only where there was some organelle, or system to act on. This system, as visualized by the artifact generating staining methods, showed interesting changes with changes in the cellular activity. For example, the polarity of the "apparatus" is affected by endocrine secretion (see Foster, 1942) and alters in different functional areas of the kidney tubules (McManus, 1944). What little evidence seems certain indicates that the "Golgi apparatus" contains a high concentration of lipid matter, sometimes of an unusual nature (Baker, 1957). Such lipids are notoriously difficult to fix adequately (Baker, 1958, 1965), so that there is no certainty that the Golgi apparatus disclosed by present methods of electron microscopy is any more truly representative of this organelle in life than were the artifacts of the light microscopists. However, the recent development of freeze-etching techniques for electron microscopy, in which fractured surfaces of cells can be investigated without fixation, already suggests that the present picture of the Golgi apparatus may be a fair representation of the structure as it occurs in life. This subject is fully reviewed in Chapter 6.

7. The Cytosol. The cell obviously seems to contain two major structural elements. The nucleus stands out as an apparently separate entity floating in the cytoplasm. The term "cytoplasm" is often taken to refer to two entities: by "cytoplasm" some writers mean all the rest of the cell's protoplasm apart from the nucleus; others use this term for the clear fluid in which the various cytoplasmic organelles are deployed (see Guilliermond, 1941). To avoid confusion we will refer to this fluid, namely, the fluid component of the cytoplasm, as the cytosol. And indeed, it is a strange fluid. It is thixotropic, that is, it is a colloidal fluid which readily and reversibly solidifies (or gels) and liquifies (becomes a sol again). It gels very readily with slight changes in externally applied acidity or pressure (Cameron, 1952, p. 226) and goes into the sol condition under high hydrostatic pressure (Brown and Marsland, 1936). In plants the cytoplasm can be seen to be in constant, flowing motion; this protoplasmic streaming has been studied in detail by Kamiya (1959).

So we have the cell composed of nucleus and cytoplasm, bounded by a cellular membrane. The cytoplasm consists of a clear fluid, the cytosol, in which float the mitochondria, lysosomes, Golgi apparatus, and other structures (inclusion bodies, centrosphere). In the light microscope, even under phase-contrast or dark-ground illumination, the cytosol appears to be structureless. But it was found, first by the use of acidic dyes and later by ultraviolet absorption microscopy, that certain regions of the cytosol could be characterized by the presence of high concentrations of ribose nucleic acid RNA. Such regions were known as the "ergastoplasm." The elegant work of Brachet (1950) showed that this RNA could be centrifuged down to the centrifugal pole of some cells, leaving a clear cytosol. In his outstanding pioneer work, Claude (1946a, b) demonstrated that cells could be homogenized and the nuclei and large organelles, such as mitochondria, separated by relatively low-speed differential centrifugation. This procedure left what seemed to be the clear cytosol, but high-speed centrifugation then sedimented particles of apparently small size which contained the bulk of the RNA of the cytoplasm. Claude (1943) referred to such particles as "microsomes"; they varied in size from 0.3 μ down to approximately 500 Å (see de Robertis et al., 1948, p. 85). The larger of these microsomes were resolvable by normal light and dark-ground microscopy and probably corresponded to the earlier cytologists' "microsome" (Baker, 1945; de Robertis et al., 1948, pp. 47 and 85; Brachet, 1950, p. 239). From such "microsomes" Littlefield et al. (1955) separated ribosomes which apparently contained the bulk of the RNA of the "microsomal fraction." Chantrenne (1947) showed that after mitochondria had been centrifuged out of an homogenate the supernatant fluid yielded a complete spectrum of particles, depending on the conditions of ultracentrifugation. The term microsome is still used variously by different workers, although it has tended to be used more for the smallest particles that lie below the resolution of the light microscope.

It is of some interest that when the microsomal fraction from plant cells was examined by electron microscopy without prior fixation it was found to be composed of small spheres or discs between 200 and 400 Å in diameter (Brown et al., 1952; Hackett, 1955). On the other hand, the long series of outstanding studies by Palade and his co-workers (Palade, 1956; Palade and Siekevitz, 1956) [also see Hodge et al. (1957) for similar studies with plant cells] showed that when the microsomal fraction was suitably fixed, sectioned, and studied by electron microscopy it was composed of various membranes of vesicles and cysternae as was the ER in sections of fixed cells. They postulated that when cells were homogenized the ER broke up and became rounded off into the structures which biochemists called "microsomes." It will be recalled that Palade (1956) and Porter (1961) showed by electron microscopy that the cytoplasm of all types of cells, after suitable fixation, embedding, and ultrathin sectioning, was composed of an ER. This ER was defined by Palade (1956) as "a continuous network of membrane-bound cavities permeating the entire cytoplasm from the cell membrane to the nucleus." There are usually two types of ER in each cell: a rough ER in which the membranes are liberally studded with ribosomes, and a smooth ER which has no ribosomes. Other ribosomes, either singly or in clusters or polyribosomes, may also occur in the general cytosol unattached to membranes. Assumably, when the cell is homogenized, all these components and their fragments contribute to the spectrum of particles that can be sedimented under different conditions of ultracentrifugation.

There seems little doubt that the ribosomes are the main site of protein synthesis in the cell and that the sequence of amino acids assembled on them is determined by messenger RNA (mRNA) which is transcribed from the genetic DNA which ordains which proteins can be made (see Williamson, 1969). It is claimed (Redman et al., 1966) that the ribosomes which are attached to the rough ER produce protein for export from the cell, whereas the unattached ribosomes produce protein for the cell itself.

Thus, from electron microscopists, the final picture we have of the cytosol is a peculiarly static, formalized, and structured mass of tubes, vesicles, and cysternae. They communicate with the outside environment of the cell and enclose the nucleus in what amounts to an extension of the outside environment deep inside the cell (Fig. 3). Regions corresponding to the ergastoplasm have ribosomes liberally studded on the cytoplasmic side of the reticulum and with other ribosomes lying free in the cytoplasm. It is difficult to relate this image to what one sees in the living cell unless the ER itself is in a fluid and constantly changing flux, with the cytoplasmic organelles and the nucleus breaking the reticulum, or with the reticulum constantly dissolving and re-forming, as the organelles move through it.

In essence, this picture of the cytoplasm is the generally accepted version even though it is difficult to relate it to events in living cells. In the main, it has proved a valuable model and may turn out to represent precisely the fine structure of the cytoplasm. On the other hand it has been built on studies made with a small range of fixatives. It will be recalled that about the end of the last century there was great controversy over the nature of the cytoplasm as seen in the light microscope. It was found that the cytosol had very peculiar physicochemical properties, which meant that it responded to certain fixatives in strange, even bizarre, ways. Consequently, although the present image of the cytosol, and of the cytoplasmic organelles, is probably correct, it is not unbecoming to retain a trace of skepticism at least until the picture established by electron microscopy as it is at present can explain the motion of the cytoplasmic organelles through the cytosol and the peculiar physicochemistry of the cytosol.

8. The Centriole and the Spindle. In most animal cells, but only rarely in plant cells, a small body which lies close to the interphase nucleus seems to play a major part in the formation of the mitotic spindle. This body has been called the centrosome; it consists of a small granule, the centriole, which lies in a mass known as the centrosphere (see Sharp, 1934). It has been shown by light microscopy and confirmed by electron microscopy that the centriole is usually double even during interphase, having divided during the later stages of the previous mitosis (Sharp, 1934, p. 153). In particularly favorable cells centrioles can be observed in the living cell (Hughes, 1952, p. 108). Electron microscopy has disclosed that the centriole consists of nine tubules arranged in a cylinder approximately 1500 Å in diameter. Each tubule is approximately 200 by 5000 Å. Frequently they occur paired, with the sister centrioles at right angles, not parallel, to one another; this orientation is believed to determine the axes of cell division. This subject, and that of mitosis and the spindle generally, have been fully reviewed by Mazia (1961), whose outstanding work in isolating spindles contributed considerably to knowledge of this cellular structure.

During the early stages of prophase, each centrosome begins to orientate cytoplasmic proteins to form the spindle. At the end of prophase, when the nuclear membrane disappears, there is considerable evidence that nuclear material becomes incorporated into the body of the incipient spindle. It has been suggested (Chayen et al., 1953; Ludford and Smiles, 1954) that the nuclear component of the spindle is derived from hydrated protein from "the middle zone" of the nucleus.

In animal cells the spindle is of the amphiastral type and the centrioles orientate spindle proteins which lie between them but also cause orientation all around them, sending out rays known as "asters." In plants the spindle is generally of the anastral type in that there are no obvious centrioles. The centromeres (or kinetochores) of chromosomes also have the ability to orientate cytoplasm into spindle filaments. In animal cells or, more precisely, in amphiastral spindles, both centrioles and centromeres seem to contribute to the formation of the spindle. In anastral spindles the whole spindle formation seems to be done by the kinetochores.

For many years there were great arguments as to whether or not the spindle really existed in life and particularly whether or not the spindle fibers were fixation artifacts. These arguments were reviewed by Schrader (1944). Two lines of investigation have now given us clear evidence concerning the nature of the spindle. The first was Mazia's (1961) isolation of spindles separate from the rest of the cell and in sufficient concentration to allow fairly full biochemical analysis of them. The second began with a careful polarized light microscopic analysis of living sea urchin eggs undergoing mitosis (Swann, 1951a, b; also see Hughes, 1952); this has culminated in the beautifully detailed polarized light microscopy of Inoué and his collaborators (Inoué and Dan, 1951; Inoué, 1953, 1959; Inoué and Bajer, 1961). It now seems very likely that the spindle contains proteins which are highly orientated as if in threads or fibers, and that the orientation of these threads changes as the chromosomes move at anaphase. The exact dynamics have been investigated and, although it may be true that we are not sure of exactly how the traction fibers of the spindle operate, it seems fairly certain that there are traction fibers (i.e., thick fibers) attached to the kinetochores of the chromatids (Cornman, 1944); [also see Östergren et al. (1960), who also discuss possible pumping movements in the spindle]. At present, electron microscopy has contributed little to our knowledge of the spindle except that microtubules may also be involved; their function is not yet clear.

When the chromosomes have moved apart at anaphase and during telophase, fascinating changes occur in them and in the spindle. There are some sound reports of material which is similar to nucleoprotein (by its staining and its ultraviolet-absorbing properties) being released from the anaphase chromosomes and forming around the equatorial plate of the spindle (Jacobson and Webb, 1952a; Davies, 1952). It is questionable whether this material is simply ribonucleoprotein or whether it contains pteridinelike material. It is noteworthy in this context that one important function of folic acid is to provide a factor which is essential if anaphase chromosomes are to separate (Jacobson and Webb, 1952b). One function of "dummy" folic acidlike substances, or folic acid antagonists, is to inhibit anaphase movement, and they have been used for this purpose to slow the mitotic rate of malignant cells (Jacobson, 1964). But it must be added that folic acid is also an essential cofactor for the synthesis of DNA.

In most animal cells the actual cell division occurs by invagination of the cell wall. In living cells studied by phase-contrast time-lapse cinephotomicrography, the cell membrane begins to "bubble" and even apparently to "boil" during metaphase and anaphase; this demonstrates at least a change in the state of the cell membrane. These changes are fully described by Hughes (1952). In plant cells the more common method of cell division is by the formation of a cell plate, at the equatorial plate, apparently composed of pectic material and seemingly the forerunner of the middle lamella. Particularly in plant cells, the spindle then grows laterally to form what looks like a huge barrel, the phragmoplast, and the new cell wall grows from the center of the equatorial plate outward, being extended as the spindle grows laterally (Sharp, 1934). The chromosomes in such a spindle may finally be separated by a contraction of the phragmoplast which squeezes the chromosomes still further apart. This is the *Stemmkorper* theory of Bělař (as discussed by Hughes, 1952).

As with so much of cellular biology, one can present a descriptive generalized picture of what seems to happen. In fact, we still know incredibly little of what forces are involved in cell division even though the whole process goes on in the microscope under our gaze. We discuss very briefly some aspects of this process in Section 2.C.10.

9. The Cell Membrane. A great deal of interest in recent years has been focused on the nature of the surface of the protoplasm of the animal and plant cell. In particular, many workers have investigated how cell membranes adhere to certain cell membranes but are apparently unaffected, or even repelled by, other cell surfaces (see Chapter 14). Other research has concentrated on the immunological properties of cell surfaces. However, here we will refer to another aspect of cell membranes which is of vital importance in cellular biology, namely, "drinking movements" or pinocytosis, first described by Lewis (1931; also see Holter, 1959, 1961a). With speeded-up time-lapse cinephotomicrography and with phase-content illumination, it can be seen that the surface of cells in tissue culture is not a static line or membrane. It undulates and produces miniature pseudopodia which engulf small droplets of the surrounding medium which then pass into the general cytoplasm. These drinking movements of the cell membrane can be stimulated by certain substances and inhibited by others. They represent a way by which small and large molecules can enter the cell without having to diffuse through the membrane. It is difficult as yet to evaluate the significance of pinocytosis in cellular physiology.

In general, the uptake of nutrients and of drugs, and the effect of anesthetics and narcotics, more or less follows rules based on the protein and lipid nature of cellular membranes (Davson and Danielli, 1943). With new evidence from electron microscopy and from cellular pharmacology, there is a definite need to modify the earlier concepts of the double bimolecular leaflet as the sole structural and functional element of thc membrane. If pinocytosis can be stimulated sufficiently, the membrane ceases to be a static barrier. This process of pinocytosis has been particularly well studied in amebae (Chapman-Andresen, 1962, 1967; Holter, 1965a, b). Insulin (at as low a concentration as 0.004 μg/ml) and certain amino acids and polypeptides, including serum albumin, have been shown to stimulate pinocytosis, whereas metabolic inhibitors such as iodoacetate, sodium fluoride, dinitrophenol (DNP), and antimycin A inhibit this process (Holter, 1961b; Chapman-Andresen, 1962, 1967).

Pinocytosis has been described most directly in animal cells, and generally in those that occur as simple (single) cells whether naturally, like amebae, or artificially, as in fibroblasts grown in monolayers. Micropinocytosis has been described by electron microscopists in the cells of many animal tissues. There is some evidence that it also occurs in plant cells in solid tissues, although it is difficult to understand how the cell membrane can show drinking movements when it is so intimately bound up with the cell wall.

10. The Process of Mitosis. We will now have to consider what is known of mitosis. Many excellent books and reviews have been written on this subject (e.g., Hughes, 1952; Mazia, 1961). However, some brief account must be given here so as to explain those forms of cellular injury that have their origin in disturbances of the mitotic cycle.

As should be done, wherever possible in cellular biology, we will consider what happens in a cell which we can watch, on a warm stage of the microscope, as it progresses from one restitution stage to the next. Most work of this type has been done on fibroblasts grown in proliferative tissue culture (Hughes, 1952; Walker and Yates, 1952a, b). For plant cells much of this type of study has been made either on living staminal hairs, which at normal ambient temperature undergo mitosis in single-cell strands (Barber, 1939; Fig. 4), or on living endosperm (Bajer, 1956; 1964, 1965).

After telophase the restitution nucleus forms and the centriole divides, remaining as a double structure in the centrosphere. Nucleolar material accumulates around the nucleolar organizer, often appearing to come partly from the chromosomes themselves (McClintock, 1934; also see Zirkle, 1928; Sharp, 1934, p. 119). The nuclear membrane reforms around the daughter nuclei, and cell division, namely cytokinesis, takes place as discussed above. At this stage each nucleus contains the diploid number of chromatids. Ultraviolet light absorption microscopy of the living cells shows that they contain the $2c$ amount of DNA, that is, twice the amount present in sperm or in the tetrads of the last pollen mother cell division, (Walker and Yates, 1952a, b; Chayen, 1960), and this is confirmed by measurement of the Feulgen stain (for DNA) done on cells fixed at this stage (Swift, 1953; Deeley et al., 1957). During interphase there is usually first a stable period as regards the DNA content; the chromatids have disappeared from view and the DNA is dispersed fairly homogeneously. Autoradiographic studies (Howard and Pelc, 1951, 1953) and ultraviolet absorption measurements (Walker and Yates, 1952a, b) show that there is a distinct period of quiescence (or a gap; i.e., G_1) followed by synthesis (S) and then by quiescence (or a second gap, G_2) as far as the DNA content of these nuclei is concerned (Fig. 5). Interference microscopy, however, indicates that protein synthesis may

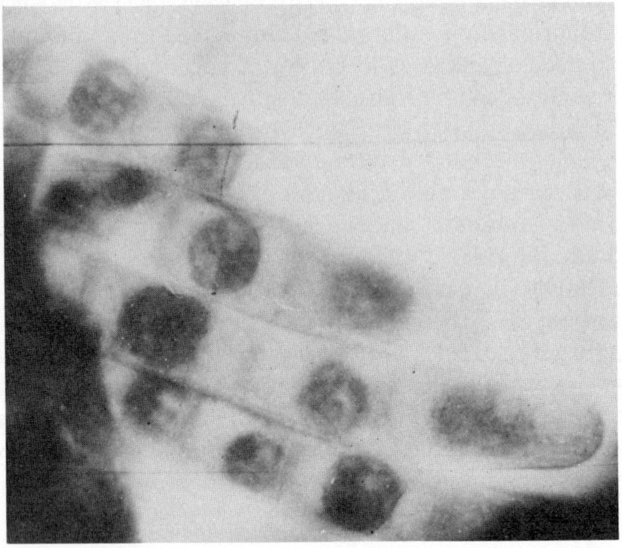

Figure 4. Living cells of *Tradescantia* staminal hairs photographed at 265 nm. The nucleic acids absorb strongly at this wavelength and consequently the nuclei are prominent. One cell is undergoing anaphase.

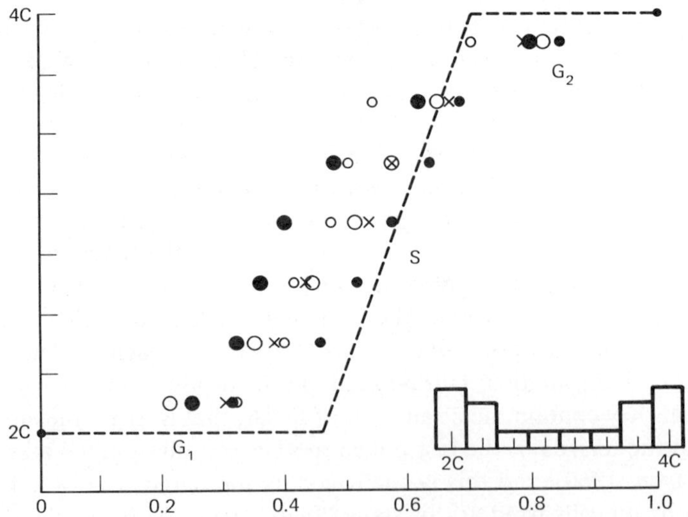

Figure 5. DNA synthesis. The data from five experiments on nuclei from a particular segment of the bean root are plotted. Ordinate: average nuclear DNA content (in arbitrary microdensitometry units); abscissa: fractions of the total interphase time (T). The dotted line shows the synthesis curve derived by Howard and Pelc (1953) for the same region of the bean root, studied by the use of ^{32}P. G_1: 0–0.47T (about 6–12 hr); S: 0.475–0.71T (6–8 hr); G_2: 0.71–1.0T (about 8 hr). In the small schematic histogram the abscissa corresponds to the ordinate of the main figure. (From Deeley et al., 1957.)

occur at a different rate from that of DNA during interphase (Richards, 1960; Prescott, 1961; but see Sandritter et al., 1960). Consequently, before there are obvious signs of prophase, there has been a doubling of the mass of the nucleus, and this includes an increase in DNA from the 2*c* to the 4*c* amount (1*c* being the amount found in the haploid sperm). Probably, the first structural evidence that prophase may begin is an enlargement of the clear nucleoplasm in which the nucleoli are situated (Ludford and Smiles, 1954; also see Chayen et al., 1953). This protein material of what has been called "the middle zone" of the nucleus (Chayen, 1952), or the perinucleolar zone, is probably the nuclear component of the spindle. Its increase in size has been regarded as being due to its hydration and this, at least as much as the increased mass, accounts for the fact that mitotically active interphase nuclei and prophase nuclei have greater volume than have the equivalent resting nuclei. The presence of enlarged nuclei in what is normally a nonmitotic population of cells, either in tissue or exfoliated into body fluids, has been used as an indication of malignancy (Spriggs, 1970), that is, it indicates the possibility of mitotically active nuclei in a region of tissue that should contain predominantly resting nuclei.

When living cells are observed, the nucleus is seen to rotate during prophase (Hughes, 1952). By this time the centrosomes have separated and have begun to form cytoplasmic proteins into orientated spindle material. Each chromosome is seen to consist of two chromatids coiled around each other in relic spirals (Darlington, 1937). Thus there are 2*n* chromosomes (1*n* being the chromosome number of the sperm), each of which is double, so giving 4*n* chromatids. Equally, there is the 4*c* amount of DNA.

By the end of prophase the chromosomes have begun to coil into denser chromosome masses; the nuclear membrane has disappeared and the spindle is being fully formed with both cytoplasmic and nuclear components.

The next stage, often called precongress, involves the apparent attachment of the kinetochores (centromeres) of the chromosomes to the traction fibers of the spindle. Lima de Faria (1958, 1959) has made a detailed study in living and in fixed cells of the kinetochore and how it is oriented with respect to these fibers. It seems that at this stage the kinetochore is double, as is the chromosome. But the chromosome consists of two separate chromatids held together by the double, but still fused, kinetochore. Each chromosome then undergoes movements which look as if it is being pulled first to one spindle pole and then to the other, until finally each comes to rest apparently at the position of equilibrium; this is normally at the equatorial plate of the spindle, midway from each pole (see Östergren, 1950; also Östergren, 1951 concerning meiotic chromosomes). At this stage the chromosomes are "congressed" at the equatorial plate, and they are fully spiralized to achieve maximum diminution in length to give the typical normal metaphase condition. Anaphase would seem to be initiated by cleavage of the skin which holds the double kinetochores together, and the tension or pull on the chromatids is now unidirectional (following the views of Lima-de-Faria and of Östergren). As has been indicated, pteridines may play some part in the active movement of the chromatids at anaphase. Whatever the mechanism, one sees a sudden pull apart of homologous chromatids from the metaphase plate and the typical anaphase movement of these to the opposite poles of the spindle. At metaphase there were 2*n* chromosomes, each double (i.e., 4*n* chromatids), and the metaphase plate contained the 4*c* amount of DNA. At anaphase each spindle pole has 2*n* chromosomes, but each is a single chromatid (i.e., $2n$ chromatids) and each daughter nucleus has the $2c$ amount of DNA.

Some idea of the time taken for different types of cell to pass through each of the major stages of mitosis may be helpful. According to Hughes (1952), who has tabulated the results from many workers, they are as follows.

In the growing root cells of *Vicia faba* grown at 19°C: prophase, 90 min; metaphase, 31 min; anaphase, 34 min; telophase, 34 min; total time, 3–4 hr. Interphase in these cells may take about 21 hr (personal observation). In chick cells grown in culture, the times are: prophase, 19–25 min; metaphase, 4–7 min; anaphase, 3.5–6 min; telophase, 7.5–14 min. Howard and Pelc (1953), working with roots of *V. faba*, suggested that the time taken from prophase to the end of restitution was 4 hr; G_1 was 12 hr; synthesis of DNA occurred during the next 6 hr and was followed by a second "gap" (G_2) of 8 hr. If the roots are grown under such conditions that mitosis occurs randomly throughout the meristematic region and throughout the day, then the probability of seeing a metaphase plate is proportional to the fraction of the whole mitotic period (from one prophase to the next) taken up by metaphase. Consequently it is possible in fixed preparations to see an unusually large number of metaphases either if more cells are made to divide or if the period occupied by metaphase is abnormally long relative to the whole mitotic cycle; or if the cells become entirely stopped at metaphase so that they cannot enter anaphase but other cells are not precluded from altering from interphase through prophase and reaching the metaphase state. Various forms of cell injury can cause each of these alterations.

During mitotic interphase there is reduplication of the DNA and of the chromatids. But if the cell differentiates, or ceases to divide, this need not happen. The resting nucleus can remain with the $2c$ amount of DNA. However, in many dicotyledonous plants the meristematic cells (relatively embryonic cells) of the root increase their DNA content from the $2c$ to the $4c$ and even to the $8c$ value when they differentiate (Deeley et al., 1957). These cells, when forced to divide again by the action of natural or artificial hormones, go into prophase with each chromosome containing four chromatids instead of two. These double chromosomes are known as diplochromosomes (d'Amato, 1950, 1952a, b). In the liver of the rat, the "resting" nuclei of the hepatic cells may have $2c$, $4c$, or $8c$ amounts of DNA (Naora, 1957).

The oxidation–reduction state of the cytoplasm, particularly its —SH component, may be of critical importance for mitosis (see Brachet, 1950; Chayen, 1953). Synthetic processes require oxygen, but mitosis itself, once begun, can continue even under anaerobic conditions. Apparently, the cell draws on previously synthesized high-energy compounds (Bullough and Johnson, 1951; Swann, 1957; Bullough and Laurence, 1960). Frequently, adverse conditions may affect prophase and cause the nucleus to revert to an interphaselike condition. Cellular injury to proliferating cells can affect all the steps we have considered so that, from a structural abnormality alone, it is often difficult to tell precisely which component, of the many processes needed for mitosis, has been affected. Some examples of injury at specific sites, or to specific mechanisms, in the cell are given in Section 2.D.2.

The functional aspects of meiosis have been much less studied than mitosis. The special problems of the mechanics of chromosome behavior during normal and abnormal meiosis have been considered in some detail (see Darlington, 1936, 1937). Abnormal synthesis for meiosis, or abnomal chromosomal behavior, can lead to abnormal spermatozoa and infertility (Leuchtenberger et al., 1955, 1956; Leuchtenberger, 1960).

11. The Cell as a Dynamic Complex. We conclude this survey of the noninjured cell, as seen by cellular biologists, with the impression of a dynamic unity with its parts in constant motion and change. To the biochemist the cell comprises various fractions which can be analyzed in chemical detail if a large number are smashed and the parts separated and concentrated by differential centrifugation. He then has a nuclear fraction, a mitochondrial fraction, a lysosomal fraction, a microsomal fraction (from which he can separate components of the ER), and a "supernatant" fraction, namely, dispersed matter which cannot readily be centrifuged to give a solid pellet. These fractions are invaluable for determining the typical chemistry of an average organelle component of an average cell. They provide a means of defining the chemicals and the enzymes that may be present. If the cell were solely the sum of its parts, these techniques would analyze "the cell." But inspection of the living cell shows it to be a dynamic system; the filamentous mitochondria are continuously changing their shape, fragmenting, rejoining to form long threads, branching, forming globules along or at the end of the filament, and moving through the cell up to the nucleus and away again. The nucleoli revolve inside the nucleus and move to and from the nuclear membrane, as if drawn thence by itinerant mitochondria. The protoplasm, in plant cells at least, is in constant flow; the cell membrane of animal cells varies in the extent to which it undulates in "drinking movements." The chemical analysis of homogenized cells, and even the ultrastructural analysis of fixed cells, can tell little of the dynamic life of the cell, that is, of cellular biology. The chemical analysis of homogenates has been likened to the behavior of a man who blows up a house with a large grenade in order to see what is inside it. From analyzing the pieces he can perhaps determine what materials went to its construction and what furniture it had, but he can only surmise as to the life that was lived in that house. The electron microscopist has a totally static view vaguely reminiscent of a Victorian family photograph, with the components frozen into formal rigidity. Thus electron microscopy and conventional biochemistry must be seen as two essential tools of cellular biology which have meaning only as partial and static answers to the complex dynamic whole that is a cell. We have already seen that time-lapse cinephotomicrography is another essential tool for understanding the dynamic cell. In Section IV we consider the form of microchemistry that can be used to give information on the biochemistry of a cell (not of the average of a vast number of different cells) and, more important, on the interactions that go to make the cell a single entity, a whole that is greater than the sum of its parts. But first we will consider some of the structural changes that occur in cells that undergo injury.

D. The Nature of the Injured Cell: Structural Changes

1. Nuclear Changes. (*a*) *Nuclear Pathology.* The following terms are often used to describe major nuclear changes (see Leuchtenberger, 1950; Cameron, 1952).

(i) *Pyknosis:* the appearance of small shrunken and densely staining nuclei; the chromatin is compact and homogeneous.

(ii) *Karyorrhexis:* the chromatin splits into fragments which either fill the nucleus or lie dispersed in the cell.

(iii) *Karyolysis:* the chromatin becomes markedly reduced so that in the extreme situation the nuclei have no chromatin at all, or seem to be entirely lost.

In diagnostic pathology these appearances are used to diagnose irreversible cell damage which leads to cell death (i.e., necrosis). Cameron (1952, p. 401) described a number of detailed studies made on tissue which had been rendered ischemic (e.g., the kidney deprived of its blood supply by tying off the renal artery) in which all these changes were observed.

These are the last, irreversible changes and show necrosis. In general pathology they are seen typically in cells that have been deprived of their oxygen supply, for example, by a blood clot (thrombus or embolus) blocking one of the arterial vessels supplying these cells. If a cell is killed suddenly, it shows none of these signs; in fact, it is a "fixed cell"—just as if it had been coagulated by chemical fixation. If it dies slowly, or if the dead cell remains in the body, these nuclear changes may occur (see Boyd, 1947, p. 49). Similar nuclear changes are found when cells are killed by bacterial toxins or by various physical and chemical agents. Consequently, these changes are of value to the diagnostic histopathologist; when he observes them, he knows that they represent either slow cell death or autolysis, hence he is made to look for the causative factor. However, they tell us little about the alterations in the biology of the cells that caused these changes except the fact that these cells are now dead. At present, there seems to be little information as to how the nuclear material becomes broken up and dispersed (in karyorrhexis), or how the chromatin disappears (in karyolysis). The latter might be expected to be caused by release of lysosomal enzymes inside the dead cell. Such release of lysosomal enzymes into the cell, with resulting autolytic changes, were reported by de Duve (1959) in ligated liver and by Dianzani (1963) in the later stages of necrosis induced in liver by carbon tetrachloride. Similar results were obtained with various hepatotoxins by Slater et al. (1963). Changes in the DNA–protein complex of nuclei in the late stages of pyknosis, particularly with loss of the protein component and depolymerization of both components of nucleohistone, have been reported by Leuchtenberger (1950; also see Vendrely et al., 1958). However, it must be emphasized that very little is known about the mechanism of any of these changes and what governs which of these changes will be shown by any one nucleus in a cluster of dead cells.

In pathology emphasis is placed on the irreversible nuclear changes of cell death. But nuclei also show more subtle changes. They show coagulation and gelation, which can be reversible, when exposed to acidic pH (pH 4.6 in the studies of Lewis, 1923; 0.05% acetic acid in those of van Herwerden, 1924, 1925); to various salts, bases, some anesthetics, and saponins; and to radiation with x and γ rays (Cameron, 1952, p. 258). They shrink after ultraviolet irradiation; this effect has many features of pyknosis in living cells (Walker and Davies, 1950). They swell when they occur in an area where the mitotic index is elevated (Chayen et al., 1953).

We now consider nuclear changes from the point of view of the cellular biologist studying how cells respond to injurious conditions. He may see the following.

(i) INCREASED NUCLEAR VOLUME. This may be due to increased mass of the nuclei, that is, to increase in DNA content and in protein content as occurs when cells prepare for division. These increases do not necessarily mean that the cells will undergo mitosis. They may imply that replication of the chromosomes has occurred so that, if all other conditions are favorable, the cells will divide. Or it may mean that these cells will remain as tetraploids or polyploids in the resting condition, as occurs in the liver of rats (Swift, 1953; Naora, 1957), or in the differentiating cells of the roots of the broad bean *V. faba* (Deeley et al., 1957). But other factors may cause the amount of DNA per nucleus to increase. For example, Leuchtenberger and Leuchtenberger (1960; also Leuchtenberger et al., 1959) reported cells with 14 times the normal amount of DNA in cells in human skin infected with a virus (the common wart or *Verruca vulgaris*); HeLa tissue culture cells had up to 10 times their normal amount of DNA when infected with adenoviruses. The apparent protein content of nuclei was found to be markedly elevated as a result of surgical stress and of a number of other pathological processes.

A major factor that seems to influence nuclear volume is the degree of hydration of the "middle zone" protein. The point of particular interest is that the swelling does not occur only in those nuclei which are preparing for prophase. This phenomenon was first analyzed by Chayen et al. (1953) in plant cells, and confirmed in animal cells by Ludford and Smiles (1954). It seems likely that some of the enlargement of nuclei in malignant cancer cells is attributable to this change in the "middle zone" protein, and hence probably to the degree of hydration of these nuclei.

(ii) DECREASED NUCLEAR VOLUME. This could be caused by dehydration of the "middle zone" protein or by gelation of the nucleoplasm (as discussed above). It could result from loss of DNA or of chromosomal protein. It has been suggested that an early effect of radiation may be loss of DNA (Walker and Davies, 1950; Leuchenberger and Leuchtenberger, 1960); DNA can also be lost if the previous mitosis was abnormal (see Section 2.D.4.c). Nuclear protein is not as constant as is nuclear DNA, and the amount of protein per unit DNA and even the type of protein can vary greatly during the life of the cell and between different cell types of the same animal (for review, see Mirsky and Osawa, 1961).

(iii) NUCLEAR OUTLINE. There is a great amount of older literature which suggests that the shape of the nucleus is an indicator of the state of the cell. In certain cells the nucleus produces outgrowths and processes when the cell is actively secreting. In others the shape of the nucleus reflects the polarity of the cells, or the compression to which the cells are exposed. For example if you have a mass of dividing cells in which all the spindles are oriented in one direction, the mass will grow in one dimension and this polarity, together with the compression induced by the neighboring cells, can produce a typical cellular and nuclear shape (Fig. 6). The shape of cells and nuclei in the various layers of the skin seems to reflect these polarity-compression phenomena (Fig. 7).

The skin is an especially interesting example because there is normally a chemical polarity with a gradient from a more reductive to a more oxidative state as the cells move to the surface (Chayen et al., 1971b). The outer keratinizing layers have a high proportion of their sulfur-containing amino acids in the —S—S (oxidized) form, whereas the basal cells have a high concentration in the reduced (—SH) form. In psoriasis this chemical polarity does not occur, and the outer layers have an abnormally high proportion of their sulfur-containing amino acids in the reduced (—SH) state. These outer cells therefore never produce true keratin. Their nuclei remain round and are not elongate and flat, as in normal keratinizing epithelium. Obviously, this chemical or redox gradient (see Section 3.B.3) does not explain the disease; but it must be taken into account as one factor when explaining the cellular changes that occur.

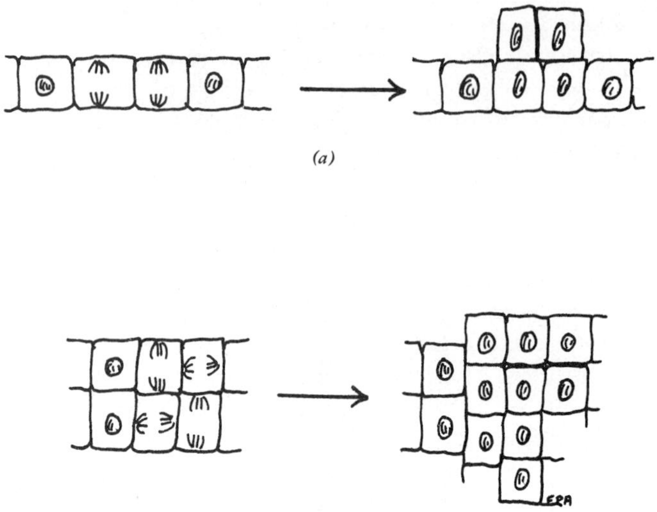

(a)

(b)

Figure 6. Diagram showing how the polarity of the spindles in a growing tissue could affect the shape of the cells and of the newly formed tissue. (*a*) Spindles oriented at right angles to the line of basal cells; this produces orderly growth from the basal layer with some orientation upward, as in the basal layer of the skin in Fig. 7. This phenomenon is often seen in human carcinoma *in situ* of the human cervix. (*b*) When the orientation of the spindles is haphazard, a round growth is formed with equal pressure from all sides. This is frequently found in secondary growths of cancers before they begin to infiltrate along regions of least reistance.

The outline of nuclei may change with the oxygen tension of the cell. Frequently, when cells such as fibroblasts are grown in monolayer tissue culture and are deprived of oxygen, the nuclear outline becomes crenated or angular. Coagulation of the nucleoplasm may produce a similar effect.

(iv) CHANGES IN CHROMATIN. As discussed above, the chromatin may be dispersed homogeneously throughout the nucleus, or it may be condensed. This distribution of chromatin is altered drastically by chemical fixation, so that it must be studied in unfixed, uncoagulated cells. For all that, under certain conditions the chromatin of defined cells may stain much more densely than it normally does even after the same fixative. Often this is taken to imply that there is more DNA in such nuclei. In some cases this has proved to be correct by cytophotometry of such nuclei after they were stained by the Feulgen reaction, or by an equally quantitative cytochemical method for DNA such as Kurnick's (1950) methyl green method or the gallocyanine-chromalum method of Sandritter et al. (1963). Such deeply staining nuclei were thus proved to be either polyploid or to contain excessive amounts of DNA for other reasons (see Leuchtenberger and Leuchtenberger, 1960). But frequently, densely staining nuclei, even when stained by quantitative cytochemical methods for DNA, were found to contain normal amounts of DNA. The reason they appeared to be unusually darkly stained could be due to one of two possibilities.

The first possibility concerns the nature of the processes by which staining is achieved. All methods depend on freeing the DNA, at least in part, from the protein to which it is normally bound. In the Feulgen reaction, if the DNA-protein binding is

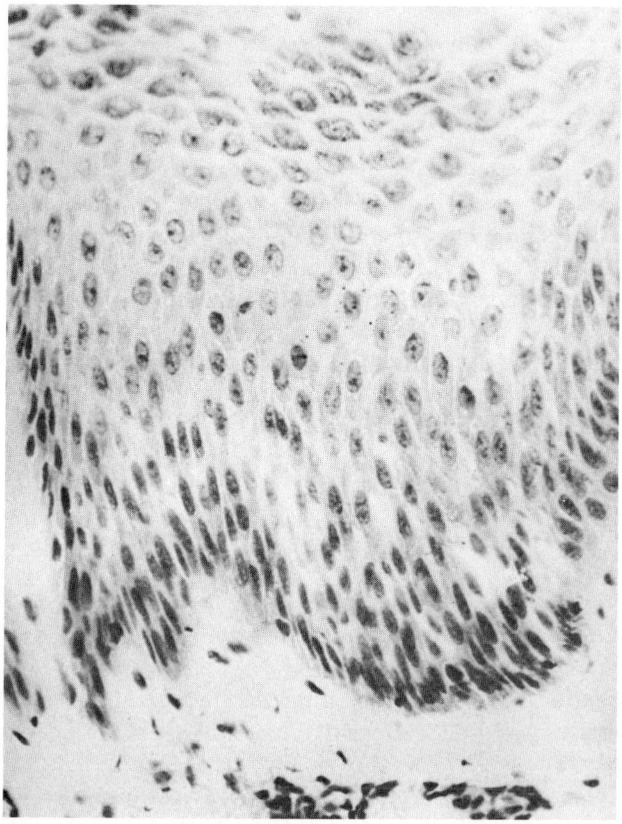

Figure 7. Photograph of a section through mildly psoriatic human skin, stained with hematoxylin and eosin. The cells and nuclei of the basal layer of the epidermis are elongated toward the surface of the skin (as in Fig. 6*a*), whereas those of the outer layers are flattened at right angles to the surface; these cells should normally form the keratin layer which in nonpsoriatic skin is even more flattened. The elongation of the basal nuclei is seen better in psoriatic epidermis than in normal skin because there is an increased mitotic rate which accentuates the effect shown in Fig. 6*a*. ×550.

unusually strong, an unusually long acid hydrolysis is required to achieve optimal response of the DNA to the Feulgen reagent. If the binding is unusually weak, the "normal" time of hydrolysis may produce exceptional amounts of reactive DNA. The other cytochemical methods depend even more directly on the number of DNA–phosphate groups, free from linkage with protein, that are available for staining. Hence changes in the denseness of the nuclear stain can be brought about, even with a constant amount of DNA per nucleus, if the protein moiety is altered. Very occasionally, we have found that reduced amounts of Feulgen stain in nuclei were due to abnormally poorly polymerized DNA in these nuclei; poorly polymerized DNA is degraded readily by the acid hydrolysis used in the Feulgen reaction, so that some of the DNA has been removed during the reaction.

The second possibility concerns an optical artifact which all cellular biologists should guard against. When you see a darkly stained nucleus you are actually seeing a high degree of color per unit area. Suppose one nucleus has an area of 10 μ^2 (radius 1.8 μ) and contains 10 pg of DNA. The denseness of stain will be equivalent to 1 pg

DNA/μ^2. But suppose the nucleus expands to 20 μ^2 and still retains 10 pg of DNA. The radius has increased only to 2.5 μ. The denseness of stain is now equivalent to 0.5 pg DNA/μ^2 so the nucleus appears half as strongly stained as it was formerly. Yet it still contains the same actual amount of DNA.

This "concentration artifact" gave rise to a well-known error. Before cytophotometry was established, cytologists looked to see how well mitotic nuclei stained. They noted that interphase nuclei stained very weakly, prophase chromosomes more strongly, and that the staining was most intense at metaphase. It seemed as if the substance responsible for the staining, namely, DNA, came onto the chromosomes only during prophase and reached maximum concentration at metaphase. When in later years it became possible to measure the amount of staining precisely by cytophotometry, it was found that the interphase nuclei, just before prophase, contained as much DNA—or as much stainable matter and indeed as much stain—as did the interphase chromosomes. The latter had more stain per unit area because the same amount of stain was now concentrated into the very tightly coiled and compacted metaphase chromosomes.

(v) CONCLUSION. These then are the most common structural changes which nuclei may show. Their structure changes in only relatively few ways even though it is possible that a vast number of cellular dysfunctions may induce these changes. But at least, when one of these structural changes is observed, it indicates strongly that the cell is suffering some injury. To disclose the exact nature of the injury requires other types of investigation, possibly concerning the metabolic status of the cell; such investigations are discussed in Sections 4 and 5. However, at this juncture one feature must be stressed. Many workers who diagnose disease use cellular criteria in their diagnoses. They are not cellular biologists, but they use whatever criteria may be of help in their very difficult and responsible task. Their lines of reasoning may be expressed in the following way. The cells of a particular organ, or at a particular location in the body, normally show certain characteristics. Cancerous cells, when found at the same location, show characteristic differences from the normal cells at this site. If, when there is no invasive malignancy, they detect cells that show the same type of characteristics as do the malignant cells, these unknown cells—having appearances similar to those of cancer cells—must be regarded as malignant or at least as premalignant. Now this statement is undoubtedly an oversimplification of an extremely complex method of diagnosis, but it is a near approximation to the cellular biological rationale of their work. This is not an appropriate place to discuss the problems of "the premalignant cell"; here we intend only to stress the possible dangers of arguing by this type of "circumstantial analogy" and to emphasize how important it can be to understand the cellular biology which may cause the circumstantial analogy. As we have said, nuclear change of any particular type can be induced by several cellular metabolic changes; it is not specific to any one type of metabolic dysfunction. Consequently, the nuclear change "characteristic" of malignancy is characteristic only if we compare the nuclei of normal, healthy cells with those of cancer cells. We may well find that other cellular metabolic disorders produce nuclear changes similar to those found in the cancer cells. At least we must be cautious in describing any nuclear-abnormality found in one type of cancer cell as representing the characteristic, or primary, nuclear change of malignancy; it could be secondary, or it could be a generalized response of nuclei to the altered conditions in that tissue.

(*b*) *Micronuclei and Other Effects of Mitotic Aberration.* Under some abnormal conditions cells which normally contain one nucleus contain more nuclei, all in the common cytoplasm. Sometimes these nuclei are dissimilar in size, so that near the main nucleus a smaller sphere of chromatin is found; this is a micronucleus. In many cases which have been examined in detail, these micronuclei are formed from fragments of chromosomes, or from chromosomes whose kinetochores failed to attach to the spindle. There seem to be turbulences around the spindle (see Hughes, 1952, p. 125) [Östergren et al. (1960) postulate pumping movements which eject unattached particles from the region occupied by the spindle]; these turbulences remove such chromosomes or their fragments. When at telophase the remaining chromosomes re-form daughter nuclei, the nonspindle chromosome material also coalesces to produce micronuclei. If the chromosome fragments include a nucleolar organizer region, they can reconstitute a nucleolus inside the micronucleus (Fig. 8).

Chromosome fragments are frequently caused by X-irradiation which can break off regions of chromosomes. Sometimes the broken ends fuse together. If such fusion occurs between two broken chromosomes, each of which contains a kinetochore, it is possible that at anaphase one kinetochore will be moved to one pole of the spindle while the other will be attracted to the other. If this happens, an "anaphase bridge" will be seen made up of the stretched chromosome material between the two kinetochores (Fig. 9). When this "bridge" snaps, the chromatin that breaks off may either form a micronucleus, or it may form a rather odd extension of one of the daughter nuclei. Anaphase bridges could also be formed if the phospholipid covering of the chromosomes (La Cour et al., 1958) becomes sticky and causes adhesion of neighboring chromosomes.

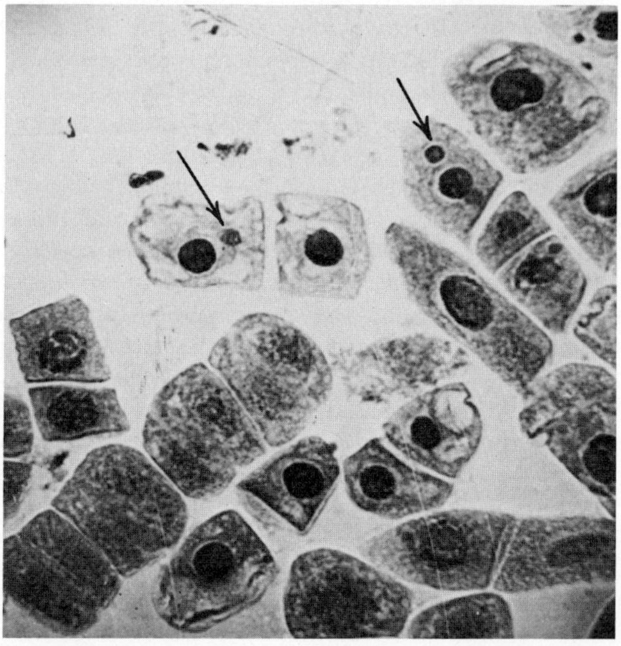

Figure 8. Micronuclei in plant cells (bean root meristem) treated with cobalt. Acetic carmine squash. Approximately ×500.

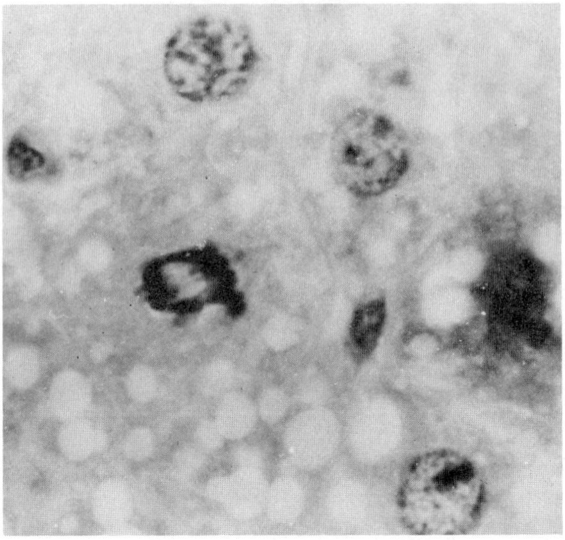

Figure 9. Anaphase bridge in a section of normal regenerating rat liver stained with hematoxylin and eosin. Frequently, mitotic abnormalities such as this and others are found when animal and plant tissues are induced to divide again after having been in the "resting" condition for long periods. Approximately ×1200.

Although X-irradiation is the most studied method of producing chromosomal abnormalities, it is of more interest in general cell biology that lysosomal enzymes can cause breaks (Allison, 1968a, 1969); moreover, certain substances can break chromosomes at specific sites [e.g., maleic hydrazide caused breakage at the heterochromatic points of plant chromosomes in the studies of McLeish (1953)]. Of even greater interest was the demonstration that the normal reproduction of chromosomes depended on the metabolic sufficiency of the cytoplasm. This was shown in the elegant studies of Rutishauser and La Cour (1956). They produced a hybrid between two species. The chromosomes of one, in the cytoplasm of the other, were fully normal, but the chromosomes of the second, developing in the nuclei in the cytoplasm of the first, invariably produced a large number of broken chromosomes at mitosis. Apparently, the cytoplasm of the first species was either unable to completely synthesize the material of the chromosomes of the second species or it provided an unfavorable microenvironment for them. Among the abnormalities found were: micronuclei caused by chromosomes that remained outside the spindle, acentric fragments, anaphase bridges; and giant nuclei caused partly by endomitotic divisions. It is likely that this type of effect could occur in many ova even when the paternal and maternal chromosomes are of the same species, either if the paternal chromosomes require a higher degree of synthetic activity than can be produced by the maternal cytoplasm, or if the metabolism of the ovum is diminished.

Multipolar spindles can produce abnormal nuclei. Normal spindles possess two poles, and the daughter chromatids go to either pole. Cell division (cytokinesis) occurs between the poles and results in two daughter cells each with a nucleus with identical chromosome composition. Occasionally, however, a cell will go into mitosis with three, or even more, spindle poles; this effect can be produced experimentally by subjecting cells to low concentrations of narcotic drugs (Östergren, 1950b, 1951b). It is of interest that Allison and Nunn (1968) showed that certain anesthetics reversibly depolymerize microtubules such as occur in the spindle. The daughter chromosomes seem to be allotted in an arbitrary fashion to each of the poles of such multipolar spindles. Cytokinesis usually produces only two cells, so at least one of these daughter cells will contain two nuclei. Occasionally, the two nuclei in one cell become almost confluent and give bizarre shaped nuclei. Other bizarre nuclei can be produced by abnormally short spindles when cytokinesis fails to separate the daughter nuclei. Double nuclei, which resemble cottage loaves, have been reported to occur in astrocytes in the brain of rats subjected to portacaval shunt which short-circuits the detoxicating systems of the liver. Cavanagh and Kyu (1969) suggested that these nuclei are due to spindle abnormalities as a result of which the chromosomes did not separate correctly at anaphase but merely reformed as two confluent nuclei. They reported that ammonium ions and other substances have a similar effect on the spindle.

(*c*) *Nucleolar Changes.* Quite early in the modern phase of cell biology, it was noted that nucleoli increased in size in cells that were much involved in protein synthesis (see Caspersson, 1950), and there was much speculation concerning the function of nucleolar RNA in this process. This subject has been reviewed by Mirsky and Osawa (1961), who emphasize that the rapport between nucleus and cytoplasm, and variations in their interaction, always influences the state of the nucleolus. It seems true that nucleoli frequently become enlarged and strongly stainable by basic dyes (suggesting increased concentration of RNA) in synthesizing and proliferating cells. With the present, and possibly undue, emphasis on protein synthesis by the cytoplasm, the significance of these nucleolar changes is not clear. Some of these questions have been reviewed by Mitchison (1966). Electron microscopical studies (particularly by La Cour and Wells, 1967; also Birnstiel et al., 1963; Marinozzi and Bernhard, 1963) indicate that the nucleolus may contain ribosomes associated with loops of DNA (in plants, at least). It is possible that these ribosomes are involved in protein synthesis as are their counterparts in the cytoplasm. Equally, there is a suggestion that the nucleolus is the site at which ribosomes destined for the cytoplasm are made (Mitchison, 1966).

Various antibiotics, which may have potential antitumor activity by virtue of their blocking protein synthesis, have been shown to have a marked effect on the structure of nucleoli (Simard and Bernhard, 1966). These include daunomycin, actinomycins, and mitomycins (Rees, 1968a).

(*d*) *Nuclear Chemical Changes.* In this section we have been concerned with the structure of the nucleus and with changes induced in this structure by injurious conditions or agents. But structure cannot be divorced from chemistry. Indeed, we see "structure" either because of its physicochemical properties, such as refraction, or because of its chemical nature, as when we stain the nucleus. Hence it is proper to enquire into the structural chemistry of the nucleus and how this may alter in cell injury, taking "injury" in its widest sense.

(i) GENERAL COMPOSITION. Chemical analyses have been made of nuclei obtained from a large number of plant and animal tissues by homogenization and differential centrifugation. Simple analysis has shown that the main constituents are DNA and protein with relatively small amounts of RNA and of free phospholipid. As is now well known, the DNA is considered to be the genetic component (e.g., Sinsheimer, 1957; but also see Chayen, 1958; Gahan and Chayen, 1965). In most diploid cells

($2n$ number of chromatids), there is twice the amount of DNA (the $2c$ value) found in the sperm in animals (Leuchtenberger and Leuchtenberger, 1960) or in the tetrads at the end of the second meiotic division of pollen mother cells in plants (Chayen, 1960). As discussed above, during interphase there is a distinct period when the DNA doubles so that at prophase, when there is the $4n$ number of chromatids ($2n$ chromosomes, present as double chromatids), there is the $4c$ amount of DNA (see Walker and Yates, 1952a, b; Deeley et al., 1957). This concept of DNA as the stable substance of the gene is undergoing some modification because it now transpires that some fraction of the nuclear DNA may fulfil a metabolic function (Pelc, 1959, and in the press; Lima-de-Faria, 1962; Roels, 1966; also see Section D.4.c). Hence it now seems likely that the DNA content of any given nucleus may be compounded of the basic gene material plus copies of those genes which are required for the present metabolic needs of the cell. It is still not certain where the synthesis of DNA occurs. Chèvremont et al. (1960a, b) showed that when the normal function of mitochondria in chick fibroblasts grown *in vitro* was halted, much of the new DNA required to increase the nuclear content from the old $2c$ value to the new $4c$ value was found in the cytoplasm, apparently in the mitochondria. Gahan and Chayen (1965) reported similar findings in plant roots. If these results are correct, it follows that one facet of cell injury could result from a disturbance of DNA synthesis and of the passage of the DNA from the cytoplasm to the nucleus. But irrespective of the site of its synthesis, many valuable drugs act by disturbing DNA synthesis either by acting as "dummy" molecules for its synthesis (e.g., the cytotoxic drug 6-mercaptopurine) or as inhibitors of the synthetic pathway [e.g., methotrexate and aminopterin (Holland, 1961)].

The proteins of nuclei from animal cells have been examined in minute detail (reviewed by Chayen, 1961; Mirsky and Osawa, 1961). They are of two main classes: the histones which are basic proteins, and the other nonbasic nuclear proteins. The former are basic because they contain unusually high proportions of basic amino acids, usually arginine but also lysine. It is possible that in chromatin these basic groups are linked to the acidic phosphate groups of the DNA. It is said that certain antibiotics may cause dissociation of these acidic and basic moieties in nucleoproteins (Rees, 1968b), and that some carcinogenic agents, notably nitrosamines, may act by methylating the DNA and so disturbing the nucleoprotein (Magee and Swann, 1969).

There is considerable evidence that the proportion of histone to nonhistone nuclear protein increases when cells cease to function very actively and decreases when the cells become metabolically very functional again (Mirsky and Osawa, 1961). In spermatozoa there is very little or even no nonhistone protein, and the histonelike protein which is present may be a modified and simplified histone or protamine. In spermatozoa of some animals, the protamine may be so simple that it consists of a polymer of arginine with few other amino acids. This type of evidence has led to the concept that histone blankets the genic DNA. Paul (1967; Paul et al., 1971), in a series of truly elegant studies, has indicated that the ability of chromosomal DNA to direct the synthesis of specific proteins depends on how free the DNA is from the histone. Following these suggestions it would seem that the gene becomes active only when the histone, which blankets the specific genic DNA, becomes prised free from that DNA which can then form messenger RNA, as it were in its own image, to be sent into the cytoplasm to direct the ribosomal synthesis of protein. This process of suppression of genic activity by histone or by other proteins, and of its derepression when the histone separates from the DNA, could be reflected in changes in the staining of the chromatin because basic dyes can stain DNA only when the DNA–phosphate moieties are not linked tight'y to protein.

Nuclei contain only small amounts of free phospholipids, namely, that phospholipid matter which can be extracted by lipid solvents. It has been shown, however, that the "protein" component of plant and an'mal nuclei contains very significant amounts of "bound phospholipids" (Wang et al., 1953; Chayen et al., 1959a, b; Chayen, 1961). These lipids can be isolated from the protein only after severe extraction procedures which often require the digestion of the protein. They can be of great importance in cell injury because of the physical properties they give to the chromatin. If they remain tightly linked to the protein, the lipid–protein complex can be hydrophilic, but if they become more loosely associated the complex can become markedly lipophilic. The significance of these changes is considered later. Another phospholipid–protein complex is said to act as a sheath around the chromosomes (La Cour and Chayen, 1958; La Cour et al., 1958).

(ii) CHANGES IN PHOSPHOLIPIDS. Work mainly by Lloyd Thomas and his associates (Bruemmer and Thomas, 1957; Smith et al., 1957) and by La Cour and Chayen (La Cour et al., 1958; Chayen et al., 1959b, c) suggests that there are two types of phospholipids in resting and dividing chromosomes. Superficially covering mitotic chromosomes there seems to be a phospholipid–protein complex which acts as a sheath around each chromosome or chromatid. This whole lipid–protein complex is unusual in that it is readily extracted by 10% formaldehyde in water. During interphase this complex remains around the heterochromatin; it can be stained by most methods for phospholipids, particularly by the acid hematein method (Chayen, 1968a). The chemistry of this complex has been studied after extraction from nuclei from sheep lungs (Altman et al., 1964). There is some evidence that in human bronchial epithelium its physical nature becomes altered in various conditions in such a way that it changes from a lipophobic to a lipophilic lipid–protein complex. When it is lipophilic, it can concentrate such carcinogenic agents as 3,4-benzopyrene, and it has been suggested that this could represent a first step in carcinogenesis (Chayen et al., 1957; 1959c).

The second type of nuclear phospholipids are the so-called "masked" phospholipids. Thomas (Bruemmer and Thomas, 1957; Smith et al., 1957) showed that liver nuclei remained completely identifiable as nuclei even after all the DNA–protein and RNA–protein had been extracted from them. Their structure appeared to depend on lipid–protein in which the lipids were so tightly bound to the protein that they could not be removed by normal fat-solvent procedures. This tight binding, apparently by chemical bonding, has been investigated in some detail (Chayen et al., 1959b, c; 1961; Chayen, 1968a). In normal chromatin the phospholipids are so intimately bound to protein that it is difficult to demonstrate them by cytochemical methods, but many forms of cellular injury are reflected by a rapid release of this bonding, so "unmasking" the phospholipids. Various carcinogenic agents, and even dietary changes, can cause this unmasking (Chayen, 1968a).

(iii) CHANGES IN DNA CONTENT. It used to be a basic tenet of faith that the amount of DNA per chromatid was constant for any given species. This was shown to be untenable when Deeley et al. (1957) found that mitotic figures and nuclei from the roots of *Trillium grandiflorum* could contain widely varied amounts of DNA (Fig. 10). This was a good test case because the chromosomes of this plant are very large,

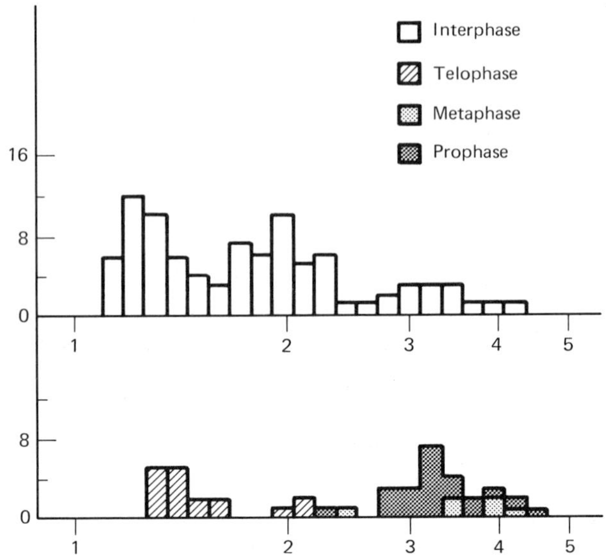

Figure 10. Histograms of the Feulgen stain measured in interphase (upper histogram) and mitotic (lower histogram) nuclei from the same root meristem of *Trillium grandiflorum* treated at the same time. Ordinates: number of cells; abscissas: cellular DNA content in log arbitrary units. It is seen that prophase values vary from just over 2 to 4.5; metaphase values of 2.5, 3.5, and 4 were recorded even though these cells were definitely neither polyploid nor aneuploid. (From Deeley et al., 1957.)

relatively few in number, and readily recognizable. Consequently, the variations in DNA values were unlikely to be due to loss or gain of chromosomes or of parts of chromosomes. In a second study, La Cour et al. (1956) reinvestigated the effect of growing plants at suboptimal and at supraoptimal temperatures. It had been shown previously that cold treatment resulted in the heterochromatic regions of chromosomes becoming completely unstained during mitosis. La Cour et al. (1956) found that, provided the plant roots continued to undergo mitosis at suboptimal temperatures, they did so with only approximally 80% of their normal DNA content as measured by Feulgen photometry. Ultraviolet absorption microscopy confirmed that the heterochromatic regions lacked DNA (Fig. 11). Two features seemed to be of particular importance. First, the loss of DNA was found only as long as the cells could continue to undergo mitosis; when the suboptimal temperature halted mitotic activity, the DNA value per nucleus was normal. Second, a loss of 10–20% of nuclear DNA, depending on the species, occurred irrespective of whether the heterochromatin occurred in blocks (as in Fig. 11) or was diffuse. La Cour et al. (1956) therefore suggested that the DNA which was lost was metabolized, that is, there was a turnover of some 20% of the DNA of each nucleus; the breakdown could continue, in mitotically active cells, even at suboptimal temperatures which retarded or halted the resynthesis of this DNA. Support for these ideas came from the biochemical work of Sampson and Davies (1966a, b; Sampson et al., 1963) on plant roots. Pelc (in press) has reviewed the evidence for metabolic DNA in animals, and it is very likely that his concept of damage to DNA will become highly significant in understanding cell injury and aging (Pelc, 1970). When a particular strain of rats fed on a specified diet was sub-

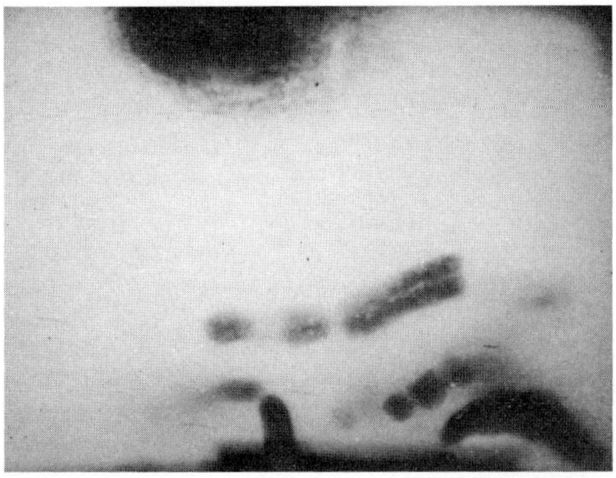

Figure 11. Ultraviolet light absorption photograph of a mitotic chromosome of the plant *Trillium grandiflorum* grown at suboptimal temperature. The DNA absorbs light of the selected wavelength (265 nm) and appears black. The heterochromatin is seen as nonabsorbing blocks in the chromosome apparently having lost its charge of DNA because of the cold treatment. The high resolution and contrast achievable by ultraviolet light absorption microscopy is evident in this photograph with the twin chromatids clearly visible.

jected to cold stress, Viola-Magni (1966) found as much as 32% of the DNA of the nuclei of the adrenal medulla was lost from the nuclei but was re-formed when the stress was removed (also see Pelc and Viola-Magni, 1969). The whole subject of metabolic DNA, and of changes in the amount of DNA per nucleus induced by cellular activity in response to hormones, has been much studied by the Belgian group under Fautrez and Roels in Ghent. According to this group, the DNA content of a nucleus is not strictly constant; rather it is an equilibrium value around which significant variations occur (Roels, 1966).

In cancer cells there may be much variation in the DNA content per nucleus. This has been extensively studied (e.g., Atkin and Richards, 1956; Richards et al., 1956; Sandritter et al., 1958, 1966; Sandritter and Kleinhans, 1964). Some cancers contain the same amounts of DNA per nucleus as do the noncancerous dividing cells, for example, a normal 2*c* amount for that animal rising to the 4*c* value before prophase. Others are frankly polyploid. Yet others have cells that may vary in their DNA content and show a wide spread of values. At one time it was thought that cancers with such aneuploid and trisomic cells might represent a more malignant form. It now seems more likely that the aneuploid and trisomic cells are "injured" cells which contain broken chromosomes, or have lost chromosomes, or have gained extra chromosomes through spindle abnormalities. It seems likely that these abnormal cells cease to divide after a few divisions and it is only the "stem line" of diploid or of fully polyploid or stable aneuploid cells that maintains the active cancer (Makino, 1957; Sandritter et al., 1966).

2. Mitochondrial Changes Induced by Cell Injury

(*a*) *Conventional Histopathology.* Cellular injury has stimulated much research and speculation among histopathologists. The mass of information produced has been well surveyed by Cameron (1952: p. 304 et seq.). It has been noted that cells can suffer injury and show altered structure and function and that this injury may yet be reversible or may lead to cell death. Most of the work has been done by methods in conventional use in histopathology. By such techniques two major types of cell injury have been distinguished, namely, degenerations and infiltrations, the former apparently resulting from breakdown of normal cytological structure and the latter resulting from osmotic or ionic imbalances which result in protein or water inclusion droplets in the cytoplasm.

The mildest form of cell degeneration is "cloudy swelling" in which the cell becomes filled with droplets or granules which many think are swollen and fragmented mitochondria, and possibly fragments of the Golgi apparatus. "Hydropic degeneration" is considered to be closely related to cloudy swelling but involves increase in the cytoplasmic water content or "intracellular edema." Certain cells, particularly those of the liver, respond to cellular injury by showing what is called "fatty change" or "fatty metamorphosis."

(*b*) *The Effect of Substances on Mitochondrial Shape and Movement.* As Cameron (1952, p. 221) stated: "It can safely be said that mitochondria are very sensitive indicators of injury." This is particularly clearly seen when single fibroblast cells grown in tissue culture are studied. If the medium in which they are grown becomes acid (pH 4.4), the mitochondria become varicose and may even swell to form vesicles; they recover if the medium is quickly replaced by one of normal pH. If the medium is made alkaline, the mitochondria change into thick rods and granules. Slight cellular damage, even excess light used to study the cells, can change the shape and size of the mitochondria so that filamentous mitochondria break up into rods and these break up further to form granules. Similar changes are seen when many cellular poisons are added to the culture medium. Acids, or potassium cyanide (which cuts off the oxygen supply to the mitochondria by inhibiting mitochondrial cytochrome oxidase), stop the movement of mitochondria. These responses of mitochondria to damaging conditions have been reviewed by Novikoff (1961a).

In a series of elegant studies on fibroblasts in tissue culture, Frederic and Chèvremont (1952; Frederic, 1958) showed that normally mitochondria move around the cell. They frequently touch against the nuclear membrane to which they seem to become temporarily stuck. Frequently, the nucleolus moves toward this point of attachment. The mitochondria later pull away from the nuclear membrane and resume their movement throughout the cytoplasm. Any one filamentous mitochondrion may shorten or elongate, produce side arms, or partially break into shorter filaments and re-form again. Certain poisons stop this movement; some cause the mitochondria to fragment into small rods or granules. When this change was induced by adding acid deoxyribonuclease to the cells, Chèvremont et al. (1960a, b) found that DNA accumulated on these mitochondrial granules. Moreover, they found almost the full 2*c* amount of DNA, required for the next cell division, actually in these granules, in the cytoplasm and not in the nucleus which remained with its telophase 2*c* amount of DNA. From this evidence it seems possible that normally mitochondria make at least some of the DNA required for the next nuclear division and pass it on to the nucleus when they touch onto it during their normal perambulation about the cell. It must be assumed at present that these mitochondrial movements are mediated by mitochondrial ATPase activity (or by a particular contractile protein as discussed by Parsons, 1965), so that it is likely that any substances that inhibit this enzyme, or possibly even the flow of energy through the mitochondrial hydrogen transport chain, may affect mitochondrial mobility and the supply of mitochondrial products to the nucleus.

(*c*) *Electron Microscope Evidence.* Light microscopy demonstrates that alteration in the structure of mitochondria is an early sign of cell damage (as discussed above). It was therefore hoped that electron microscopy would reveal ultrastructural changes in cell damage that would advance our knowledge of the specific sites of attack of different types of damaging agents. Unhappily, this does not yet seem to have come about. Shafiq and Gorycki (1968), in reviewing this subject, reported that comparatively little is known concerning changes undergone by mitochondria in diseased conditions. A commonly found condition is the swollen mitochondrion, which has been reported in many muscle diseases, in which the swelling occurs only in the matrix of the mitochondrion. After poisoning with aflatoxin (Theron, 1965), or after administration of thyroxine, the swelling also occurs inside the cristae (see Shafiq and Gorycki, 1968). Zaimis et al. (1969) found that thyroxine produced marked mitochondrial pleomorphism in cardiac tissue. Some were fused together in long rows; others were exceptionally long and wide, with unusually long cristae. Typically, the mitochondria from thyroxine-treated tissue had considerable enfolding of their walls. Similar changes in mitochondria of injured cells have been reported (see Trump and Ericsson, 1965). Another typical abnormality, found by many workers who have studied mitochondria in damaged or diseased cells, is the presence of dense paracrystalline matter either in the matrix or inside some of the cristae (Shafiq and Gorycki, 1968; Trump and Ginn, 1969). Sometimes this may be caused by deposition of calcium; but in dying cells it is common to find characteristic dense aggregations which usually appear to begin close to a crista (see Trump and Ericsson, 1965).

Trump and Ericsson (1965), who have reviewed the subject of ultrastructural consequences of cell injury, considered that the pathogenesis of structural changes induced in mitochondria by cell injury was still extremely speculative. The question of calcium in mitochondria has been reviewed by Lehninger (1970). Ericsson (1969) has discussed the electron microscopy of cloudy swelling induced in kidney tubule cells by ischemic anoxia. He found that the mitochondria were enlarged and rounded-up but became normal again after 2–4 hr of anoxia. He doubted whether mitochondrial changes were the sole factor in cloudy swelling as observed by light microscopy. In extreme cases, as in livers of rats which had been injected with very large doses of cortisone for 5 days and then starved for 1 day, the swollen mitochondria may lose their cristae entirely. In spite of that, they seem to function normally as assessed biochemically (see Novikoff, 1961a).

3. The Golgi Apparatus in Cell Damage. There has been so much controversy as to the nature of this cellular organelle or system of organelles, and so much doubt cast as to the possibility that what is seen is an artifact, that it is difficult to rely indubitably on reports on changes induced in it by cell damage.

In living fibroblasts the Golgi apparatus cannot be distinguished except possibly by the appearance of a reticulate area when cells degenerate. Generally, it seems to lie close to the centriole (Cameron, 1952, p. 105, p. 70 et seq.), although it may be seen as a stained meshwork surrounding the whole nucleus in certain cells after they have been fixed and treated by special methods. During mitosis it is said to fragment, to lie around the spindle, and finally to become apportioned to each daughter cell. There have been many reports of its significance in cell secretion (see Cameron, 1952, p. 71). Gillman (1934, 1935) claimed that it changed its activity in accord with the different secretory phases of thyroid cells. Some authors have suggested that the Golgi apparatus moves from one side of a cell to the other in such a way that it always resides closest to the secretory activity of the cell. Gatenby (1931) believed that the Golgi apparatus was the most sensitive of all organelles in responding to damage or deleterious conditions. In degenerating cells it breaks up into granules from which fat droplets arise, according to Ludford (1927; 1951). Similar changes have been recorded in cells of the renal tubules after experimentally induced damage and in nerve cells which have been treated with poisons or irradiation (Cameron, 1952, p. 222).

Electron microscopists have not reported appreciable changes in the Golgi apparatus of cells subjected to injury, other than some increased prominence of the complex or some dilatation of the cisternae and vesicles (Trump and Ericsson, 1965).

4. Lysosomal Changes in Cell Damage. So far we have tried to restrict our discussions to those changes that can be seen in the *structure* of cellular organelles when the cell responds to damaging conditions. This has been possible when considering chromosomes, nuclei, mitochondria, or Golgi apparatus, all of which were described as structural components of cells and were studied as such for many years before the scope of cytology became so greatly expanded by cytochemistry. In contrast, lysosomes are defined as subcellular organelles, about 0.25 to 0.5 μ in diameter, which contain acid hydrolases in a latent form. Consequently, their definition depends on two functional attributes: the presence of acid hydrolases and, more significantly, the sequestration of these enzymes from the rest of the cell within a semipermeable membrane (i.e., they act as osmometers according to Beaufay et al., 1964) which is impermeable to these enzymes. Their structural attribute by light microscopy, namely, that they are granules, is not characteristic. It has been necessary therefore to use some cytochemical or functional attribute in most studies on structural changes in these organelles.

In electron microscopy it had been suggested (see Novikoff, 1961b) that some or possibly all of the "dense bodies" found in sections fixed with osmium tetroxide and embedded in methacrylate were probably lysosomes. However at present it seems likely that dense bodies and lysosomes are not certainly and invariably the same structures, so that many electron microscopists prefer to call "lysosomes" only those bodies (dense or otherwise) that can be stained cytochemically for acid phosphatase or for other lysosomal enzymes (Daems et al., 1969). The choice of acid phosphatase stems from two facts. First, it is one of the most characteristic acid hydrolases of lysosomes and, second, it can be demonstrated cytochemically by a reaction involving the deposition of lead. The value of lead is that it is very electron-dense so that it is a good "electron stain." Indeed, this cytochemical test has led electron microscopists away from the concept of the lysosome as a single structure to the lysosome as a vacuolar apparatus comprised of many types of subcellular structure.

The present concept of the lysosomal vacuolar apparatus is summarized elsewhere in this volume (Chapter 7). The whole subject has been very fully reviewed in a three-volume work edited by Dingle and Fell (1969).

(*a*) *The Bitensky Fragility Test.* Investigation of the activity of lysosomes and of lysosomal enzymes requires some subtlety. By their nature, although they contain powerful lytic enzymes, these enzymes are latent because, although in an active form, they are isolated from the rest of the cell by a relatively impervious membrane. This means not only that the enzymes cannot attack the cytoplasm but that substrate cannot readily enter the lysosome to be acted on by the intralysosomal enzymes. Consequently, if we have intact, normal lysosomes, and we test lysosomal activity against a substrate that does not readily penetrate membranes (such as β-glycerophosphate), we should find no activity. However, the lysosomal enzymes all have optimal activity on the acidic side of neutrality, that is, they are acidic hydrolases. Hence the reactions are done at an acidic pH.

At such pH values, especially in acetate buffer, the membrane becomes progressively more permeable with prolonged time of incubation so that, slowly, the substrate can enter the lysosome and be acted on by the intralysosomal enzymes. Bitensky (1962; 1963a, b) used this effect in her fragility test for lysosomal acid phosphatase. In this test you incubate at the acid pH (pH 5.0) for progressively increasing times. You observe the least time required to give a just distinct lysosomal response. This time is compounded of the time required at this pH to make the lysosomal membrane permeable to the β-glycerophosphate and of the amount of acid phosphatase activity hitherto latent in the lysosomes. The same conditions are used on other tissues that may have been subjected to damaging agents or conditions. The lysosomes in these cells may react at half the incubation time needed for normal cells, so showing an increased fragility of their lysosomal membranes in that they need less modification in the acidic buffer solutions to disclose the same activity. The kinetics of this test and its extension have been discussed by Chayen (1968b).

In conventional biochemistry lysosomes are isolated from tissue by homogenization and differential centrifugation. These processes themselves may result in some damage to the lysosomes. The biochemist then measures how much of the lysosomal enzyme activity is "free," that is, that can be detected in the supernatant fluid of the lysosomal fraction. He then disrupts the lysosomal membranes and measures the total activity of the selected enzyme. This gives him a ratio between free and latent activity or, in common parlance, the "free-to-bound" ratio. With the advent of quantitative cytochemistry of lysosomes (Bitensky et al., in press; Chayen et al., 1971a), a similar free-to-bound ratio can be obtained, particularly for lysosomal naphthylamidase activity. In this reaction the substrate, leucine-2-naphthylamide, can penetrate intact membranes, albeit rather slowly, so that the "free" activity is a measure of how much substrate can penetrate into the lysosomes in the intact cell, or into the lysosomes in section, and this is compared with the maximal activity obtained when the lysosomal membranes are most permeable but before there is significant loss of enzyme out of the lysosomes. The use of these methods is discussed in Sections 5.A and 5.B.

(*b*) *Vital Staining*. Before considering what evidence can be obtained from electron microscopy, it should be noted that lysosomes can be seen by normal light microscopy in living cells. As already discussed, they are found among the lipochondria, namely small refractile granules which are very apparent by phase-contrast microscopy and which have been "stained" by their acid phosphatase activity (Munro et al., 1964; Lane, 1968). In a series of elegant studies, Allison (e.g., Allison and Young, 1969) showed that these bodies accumulate acridine dyes, particularly acridine orange, added to the culture medium of the living tissue-cultured cells. Moreover, Dingle and Barrett (1969) showed that when lysosomes are isolated from cells by homogenization and differential centrifugation the acridine orange is concentrated characteristically in the lysosomal fraction. In cells in tissue culture which are being damaged by viruses, Allison and Mallucci (1965) have observed changes in the particles that stain with euchrysine. Similarly, Allison (1969) noted that many carcinogenic agents are taken up into lysosomes, and he showed that if lysosomal deoxyribonuclease were to be liberated into the cytoplasm it could cause chromosomal breakage and hence mutation. This might produce a somatic mutation to the cancerous condition, but there is no definite evidence as yet that lysosomal deoxyribonuclease is liberated into the cytoplasm or that it can cause the number of chromosome breakages and gene deletions that might be considered necessary to produce one which would induce that cell to become a cancer cell.

Allison's work (see Allison, 1969) on silicosis is possibly more clear-cut. He showed that macrophages in tissue culture take up silica particles into their lysosomes. Because of interaction between silica and the lysosomal membrane (Allison, 1968a), the lysosomes rupture and release the silica particles into the cytoplasm. It is reasonable to expect that, on rupture, the lysosomal enzymes also are ejected into the cytoplasm and could be the cause of the death of the cell. If this is correct, it remains one of the very few examples in which there is strong evidence for lysosomal enzymes damaging their own cell. Another possible example is that described by Slater and Riley (1966; also see Allison et al., 1966). In contrast, there are now many examples that show how lysosomal enzymes of one cell damage neighboring cells or intercellular material. One of these is reviewed in Section 5.A. Even so, it is not certain how the lysosomal enzymes are released from the first cell and enter the second; whether they diffuse, or whether they move as small packages from one cell to the other (Dingle, 1968).

The functional role of lysosomes in cell damage has been reviewed by Weissmann (1967), Allison (1968b, c), and by Chayen and Bitensky (1971) with particular reference to their function in inflammation and rheumatoid arthritis. Their involvement in storage diseases and possibly in lipofuscin is discussed in Section D.6.a.

(*c*) *Electron Microscopy of Lysosomes in Cell Damage.* In contrast to the amount of information concerning lysosomes in cell damage obtained by light microscopy and cytochemistry, there is surprisingly little information obtained from electron microscopy. This is not to belittle the observations of changed form, from primary to secondary lysosomes or to autophagic vacuoles, for example, which have been reported (see Novikoff, 1961b; Ghadially and Roy, 1969). These, however, are normal functional changes. But whereas light microscopy and cytochemistry can readily show changes in the fragility or function of lysosomal membranes in various conditions where cells are subjected to damaging conditions (e.g., Bitensky, 1963a; Allison and Mallucci, 1965; Dumonde et al., 1965; Slater and Riley, 1966; also see Section 5), yet the lysosomal membrane in the electron microscope seems remarkably constant. Assumably the fixation and processing required for electron microscopy reduces all lysosomal membrane structure to a common form. In general, at present, electron microscopy can demonstrate changes in the number, or intracellular distribution, of lysosomes but not definite alteration in function. For example, Abraham et al. (1967; also Grasso et al., 1968) found unusual numbers of lysosomes in damaged heart muscle; Barland et al. (1964) found increased numbers of lysosomes in rheumatoid synovial cells. Hendy et al. (1969) reported that chloroquine induced abnormal lysosomes which contained myeloid forms in heart muscle; the effect was reversible.

5. Changes in the Endoplasmic Reticulum in Cell Damage. This subject has been extensively reviewed by McLean et al. (1965) and by Smuckler and Arcasoy (1969). The following account draws heavily on these able articles. Most of the information comes from studies on rodent, generally rat, liver. As Smuckler and Arcasoy point out, much of it ignores the heterogeneity of the liver and this may account for some of the discrepancies that make this whole subject somewhat unclear.

The ER, as seen in hepatocytes, is made up of lipoprotein membranes (approximately 50–70 Å wide) which take the form of vesicles, tubules, and stacked, flattened cisternae. The membranes appear to be continuous with the envelope that surrounds the nuclear membrane, and there is some doubt as to whether they communicate with the plasmalemma (cell membrane). The two parts of the ER are distinguished by the fact that one part, the rough ER, has ribosomes (about 150 Å spheres) studded along its surface, often in groups variously spaced according to the age and physiological state. This part of the ER appears to be involved in protein synthesis. The second part of the ER is similar in appearance but lacks ribosomes. This is the smooth ER and apparently contains the drug-metabolizing enzymes and the NADPH oxidases, including the cytochrome P_{450} system.

The effect of many toxins on the liver has been studied in considerable biochemical and cytomorphological detail. These toxins include carbon tetrachloride, thioacetamide, dimethylnitrosamine, and ethionine. All those which have been studied by such combined investigations have shown that the response of the ER is the earliest indication of cell injury as observed in the electron microscope. It may be noted, however, that biochemical and cytochemical changes can be detected appreciably before these structural changes appear (see Section 5). With all the toxins the same general structural changes are seen; vacuoles are formed; there is loss of ribosomal granules from the rough ER; the normal parallel arrays of the ER become deformed and more haphazard; and there is an unusual approximation of the ER to the mitochondria. In general, these changes, particularly the loss of ribosomes from the rough ER, are associated with failure of protein synthesis *in vivo* as well as when examined in liver slices. However, there are significant exceptions to this generalization. All the evidence suggests that changes in the ER and in protein synthesis occur very readily but that they are equally readily reversible; they do not imply that the cells will die (McLean et al., 1965).

The sequence of events is of some interest. When carbon tetrachloride is the hepatotoxin, changes in the ER are seen 30 min after its administration. One hour after it has been given, fat begins to accumulate as triglyceride; 2 hr after its administration, there is inhibition of protein synthesis (i.e., 1½ hr after the changes in the ER). Mitochondrial function can be detected, biochemically, to be disturbed 10 hr after administration of carbon tetrachloride, and 2 hr later hepatocyte necrosis is apparent. The last two effects led McLean et al. (1965) to suggest that cell death may be more related to mitochondrial function than to any other cellular structure.

When dimethylnitrosamine is administered to induce liver cell injury, it must first be demethylated by the ER, which requires NADPH and molecular oxygen (i.e., the cytochrome P_{450} system) for this process. The demethylation produces the actual toxic substance which appears to be diazomethane; its toxic effect seems to be due to its powerful methylating action. This is a particular example of "lethal synthesis" (see Section 3.B.4). Changes in the ER occur 2 hr after poisoning with this substance; loss of glycogen and the formation of fat is seen 6 hr after administration, and necrosis 12–18 hr after poisoning.

Phenobarbital, injected daily into rats, produces remarkable changes in the ER which begin within the first day. There is a great increase in the smooth ER with its tubules and cisternae distributed randomly. There is also some increase in the rough ER. Associated with these changes there is enhanced incorporation of amino acids and also an increase in the NADPH-linked electron transport. There is no change in the NAD-linked system which indicates some selectivity in the 200–500% increase in membranes and in the other enzymes "induced" by this drug. When the drug is withdrawn, there is a rapid decrease in the "induced" enzymes but no concomitant change in the membrane structures (Smuckler and Arcasoy, 1969).

The response of the ER to toxins and to drugs is another example of our argument concerning the limited structural changes available to a cell in responding to injury. There are only few ways in which the ER responds structurally to injurious agents even though it may have very many different biochemical or functional responses. Because of this, it is not possible to say that a particular structural response indicates that a particular cellular chemical function has been activated. For example, phosphorus poisoning produced the same changes in the ER as did the other hepatotoxins mentioned above, but there was no *in vitro* failure of ribosomal incorporation of amino acids [McLean et al. (1965) reporting the results of Smuckler and Barker], although there was an *in vivo* failure of incorporation. Other evidence suggests that failure of incorporation of amino acids for one protein can occur at the same time that incorporation for another is unaffected. Hence the link between disturbance of the rough ER and loss of "protein synthesis" is not necessarily a complete phenomenon.

The whole question is far from clear. McLean et al. (1965) suggested that the structure and biochemistry of a cell resembles a net which if pulled at any point becomes distorted in all its links. At present, interest centers on that link which is represented by protein synthesis. However, as they stress, protein synthesis (or its lack) may be insignificant in the production of cellular necrosis. In their analogy, it too will be distorted, like a distant link in the net, by some major cellular change which will actually cause the death of the cell. This analogy is valuable because it stresses the integrated nature of the cell.

6. Inclusion Bodies in Cell Injury. In cells that are responding to, or undergoing damage from, injurious agents or conditions, there frequently appear abnormal structures which are never otherwise found in normal cells. We have already seen how "cloudy swelling" produces abnormal structures, but these are apparently distortions of mitochondria or other cellular organelles. There is considerable suggestive evidence that "wear and tear" pigments and "lipofuscin" may also be distorted forms of cellular organelles, as are the abnormal inclusion bodies found in some storage diseases. But other inclusion bodies may be completely new structures foreign to the normal cell.

(*a*) *Lysosomal Involvement.* The clearest demonstration that lysosomal defects can produce intracellular inclusion bodies comes from the study of the so-called storage diseases. Certain human beings have a genetically controlled lack of a particular enzyme. When this enzyme is a glucosidase (acid α-1,4-glucosidase), vast accumulations of glycogen are found as inclusion bodies in the liver; this disease, namely, the type-II glycogen storage disease, is lethal. It has been shown (Hers and van Hoof, 1969) that the inclusion bodies are vastly swollen lysosomes. It appears that glycogen taken up into lysosomes cannot be hydrolyzed because of the absence of this enzyme and that the undegraded glycogen continues to accumulate inside the lysosomes. In Gaucher's and in Tay-Sachs diseases, there is an abnormal accumulation of cerebrosides and of gangliosides, respectively, and these apparently also are caused by enzyme defects. Both diseases are characterized by the presence of foamy cells swollen with intracellular inclusion bodies, probably of lysosomal origin, of the nondegraded material. Several other metabolic diseases are known in which lysosomes have not been implicated. In xanthinuria, patients lack xanthine oxidase and the muscle of such patients may have cellular inclusions composed of hypoxanthine and xanthine (Chalmers et al., 1969). Crystals of various urates have been found in the muscle of patients suffering from gout, in which there is unbalanced uratc metabolism (Watts et al., 1971). In cystinosis the inclusion bodies are composed of cystine (Frazier et al., 1965; Spear et al., 1971). The protein inclusion bodies in cells in the distal convoluted tubules found in nephrosis, or in other conditions in which proteinuria occurs, are apparently due to excessive resorption of protein (Ericsson, 1969). The damage in this case is not in these cells but in those of the glomerulus which permit excessive amounts of protein to leak into the glomerular filtrate (Cameron, 1952, p. 332; Lannigan et al., 1969).

A subject that has long perplexed pathologists is the pigment or family of pigments, found particularly in cells of the heart and liver. These pigments almost always occur in these tissues taken from old animals or humans, so that they are often referred to as "wear-and-tear" pigments. They are also often found at postmortem even in younger patients who have suffered from a chronic debilitating disease. Lipofuscin granules isolated from human myocardium contained a brightly fluorescent phospholipid which seemed to be an oxidized cephalin. They contained acid phosphatase, catheptic and esterase activity. Novikoff (1961b), who has well reviewed this topic, concluded that lipofuscin granules from many organs may be considered to be altered lysosomes in which oxidized cephalin and other insoluble or indigestible residues have accumulated. Many pathologists believe that there is a gradual change from one type of pigment to another, lipofuscin being the last stage of this sequence (see Pearse, 1960).

(*b*) *Other Inclusion Bodies.* The best known inclusion bodies are those produced by viral infection of cells. When cells become infected by a virus, which may be of ribonucleoprotein or of deoxyribonucleoprotein, the metabolism of the cells becomes altered so as to produce the virus. In some very elegant studies, Fraenkel-Conrat has taken the nucleic acid from one strain of virus and added it to the protein of a second strain; he infected cells and found that the new virus made by these cells had both the nucleic acid and the protein of the first strain of virus, not of the second. It therefore seems that the "code" is carried by the nucleic acid which need not be DNA. (This work has been reviewed by Thomas, 1960.) The infected cells also produce inclusion bodies, which may be found either in the nucleus or in the cytoplasm depending on the virus, and which often have a morphology characteristic of that virus. The structures can vary in size from a few microns to 20μ in diameter; they may be either amorphous, or crystalline as they are in the case of tobacco mosaic virus. Some are typical of well-known human diseases. These may even be eosinophilic, as are the Guarnieri bodies of variola and of vaccinia, although more often viral inclusion bodies are basophilic (see Cameron, 1952, p. 284; Bernhard, 1964).

Inclusion bodies have been found in a wide range of uninfected cells which have been subjected to damaging conditions and even in otherwise apparently healthy tissues (Cameron, 1952, p. 286). It is therefore possible that they represent a nonspecific response of the cell to deleterious conditions which is different from the way the cell responds to direct trauma or to bacterial infection.

3. THE BIOCHEMISTRY OF CELL INJURY

There are two completely different facets of the biochemistry of cell injury, each of which is predominantly the domain of different types of investigators. The first facet is seen when there is no doubt that there is cell injury because there is obvious structural change. As far as the practising medical pathologist is concerned, the structural changes, including the form of necrosis, if any, allows him to categorize the disease or injurious process, so that he can give it a name which will inform his clinical colleague. There is little applied or medical need for biochemical studies in this type of lesion. To the academic biochemical pathologist, however, it is of great interest to understand the biochemical faults that produced these morphological changes. The second facet is seen in those conditions in which structural change at the histological or cytological level is less evident or apparently absent but in which there is a distinct physiological disturbance. In such examples it becomes of considerable importance to determine the biochemical dysfunction or the chemical lesion. This is the province of the chemical pathologist or clinical biochemist.

A. Clinical Biochemistry

Unlike his academic counterpart, the chemical pathologist cannot take large samples from all the main organs and test their detailed biochemical functions. He has to rely on the abnormal substances, or proportions of substances, present in the blood, urine, or other fluids of the living patient. In the simplest case, if cells of a particular tissue die, at least some of their contents will be detectable in the blood. For example, cell damage in the myocardium or the liver results in leakage of enzymes from these cells. Transaminases have been found to be especially useful markers of this altered cellular permeability, and clinical biochemists, finding a raised serum transaminase activity, can decide precisely whether it arises from damaged cells in the heart muscle or in the liver. The study of isoenzymes, and the different pattern of isoenzymes derived from the main organs, has further helped the chemical pathologist to decide where the main cellular lesion lies. Overactive glands can be detected by measuring the endocrine imbalance in the body fluids. Unbalanced metabolism can be demonstrated by the metabolites voided from the disturbed cells; the extreme examples of this are the inborn (genetic) errors of metabolism such as phenylketonuria, cysteinuria, and xanthinuria (see Levene and Luft, 1964; Stanbury et al., 1966).

Chemical pathology therefore provides much information concerning the biochemistry of cell injury even though it is indirect evidence. Its great importance and its many contributions lie outside the scope of this chapter.

B. Basic Biochemistry

The biochemist who is interested in the chemistry of cell injury often approaches the problem from the first facet described above. He first produces a situation that leads to cell injury, to cell alteration, or to cell death. He then studies the chemical changes that precede the structural lesion he has induced. For example, he may feed animals with a carcinogenic agent and after a particular time of feeding he may obtain tumors. He then studies the target tissue at times of feeding prior to the appearance of the tumors. Or he may give the animals a substance that is toxic to the liver, or the kidneys. He will find that cell death occurs at a specified time after the ingestion of a particular dose of toxin. He will then study the biochemistry of the organ at times between the administration of the toxin up to, and possibly including, the time at which cell death is significantly present.

There is a vast literature concerned with this type of study which is more properly the domain of toxicology and experimental pathology; it is too voluminous to be reviewed here (see, for example, Peters, 1963; Farber and Magee, 1966). The results of this work, however, and the concepts derived from it, are of great relevance to the cellular biologist. Consequently, a brief appraisal of some of the most significant concepts is given.

1. Nuclear Chemistry. There is a considerable body of work, mainly stemming from Magee (Magee and Farber, 1962; Magee and Hultin, 1962), which suggests that the critical effect of certain types of carcinogenic agents may be in causing methylation of the nuclear DNA. The methylated DNA might account for a mutation that favors a cancerous state for the affected cell (but see Magee and Swann, 1969). Magee (1964, 1966) considered it possible that many substances can cause cell injury by becoming converted by the cells into alkylating agents. This is an example of lethal synthesis (discussed in Section 3.B.4).

Other workers have been concerned with the molecular biology of the nucleoproteins of the chromosomes. Clifford and Rees (1967) suggested that the carcinogenic agent aflatoxin B_1 interacts with DNA in the nucleus and prevents transcription of the DNA by the enzyme RNA polymerase. Actinomycin D binds to DNA and behaves in a similar way. It appears to become wedged in the groove of the DNA helix and prevents the RNA polymerase from transcribing that section of the DNA which has interacted with the actinomycin (Reich and Goldberg, 1964; Rees, 1966; Rees, 1968a, b). Also of great interest are those studies in which it has been shown that when DNA is transcribed the RNA so produced can be made to undergo molecular hybridization with the DNA from which it has been made. This hybridization, and apparently this transcription, can occur only at those sites in the chromosomes where the DNA has been at least partially freed from its associated protein (see Paul, 1967; Paul et al., 1971). This work has led to the idea that localized freeing of chromosomal DNA allows specific genes at these sites to become active. This would explain the phenomenon of "puffs" which appear at certain distinct sites on many polytene chromosomes of Diptera (Beerman, 1952, 1962) as possibly representing localized sites of gene activation. It is conceivable that such gene activation may be involved in carcinogenesis and in the activity of cells which have been subjected to reversible damage or to certain drugs. Some steroid hormones are also said to affect the DNA–protein association of chromosomes and so may induce gene activation (see Mueller, 1970). The effect of steroid hormones on cellular biochemistry has been reviewed by Grant (1969).

These results and concepts are novel and exciting. It is by no means certain that any one mechanism definitely operates, at least as the causative factor, in any given case of cell injury or carcinogenesis. However, they open interesting possibilities which are currently being investigated in many laboratories.

2. The Chemistry of the Smooth Endoplasmic Reticulum. One of the dramatic advances in cellular chemistry in recent years has been the demonstration that certain drugs, notably barbiturates, are detoxicated in the smooth ER in the liver cells by an enzyme system which involves the cytoplasmic cytochrome P_{450}. When rats are treated for several days with barbiturates, or with some other substances such as the insecticide DDT (see McLean and McLean, 1966), they are detoxicated by a cytoplasmic enzyme system of the type shown in Figure 12. But very rapidly there is an increased rate of detoxication which is caused, apparently, by enzyme induction.

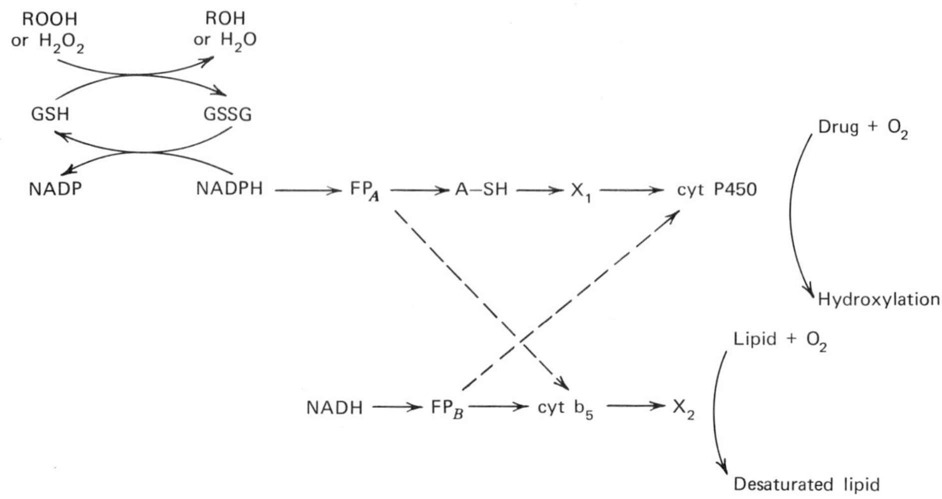

Figure 12. A schematic representation of hydrogen-transport in the cytoplasm. Hydrogen from NADH is transported via cytochrome b_5 to an unknown substance X_2; it is used in lipid desaturation. Hydrogen from NADH can also pass from the flavoprotein (FP_B) of this hydrogen-transport system to cytochrome P_{450}. The hydrogen-transport system which is most worked on at present is that from NADPH. It can be used to reduce oxidized glutathione (GSSG) which, in the reduced form (GSH) reduces ROOH by the action of glutathione peroxidase. ROOH can be lipid, or fatty acid, or unsaturated hydroperoxides; ROH is the product of this reduction. Equally, the hydrogen from NADPH can be transported through the microsomal hydrogen-transport system to cytochrome P_{450} which plays a major part in the hydroxylation of many drugs. This hydrogen-transport system is not well characterized and seems to include a flavoprotein (FP_A), a sulfhydryl-containing substance (A—SH), and one or more undefined compounds (X_1). (After Slater, 1968; Pinto and Bartley, 1969.)

This implies that the presence, or perhaps the detoxication of such substances apparently induces the production of more enzymes of this system (see Remmer, 1965; Slater, 1968). Concomitantly there is a striking increase in the amount of smooth ER in the liver cells that show this "enzyme induction" (see Remmer and Merker, 1963; Orrenius et al., 1965; Jones and Fawcett, 1966; Smuckler and Arcasoy, 1967). It is of some interest that similar effects appear to occur in the liver of humans subjected to barbiturates or to DDT (Thompson et al., 1969). Some caution must be exercised in considering these results. This is particularly important because complete proof of the induction of enzyme synthesis would require net synthesis of cytochrome P_{450} and of its related enzymes. Hence it can still be debated whether or not the increased enzyme activity is entirely attributable to increased synthesis of this enzyme or enzyme system and not to changed activity of the cytochrome P_{450} system (e.g., a change in its physicochemical properties such as an alteration in the Michaelis constant of one or more of its components).

The whole subject of detoxication by this enzyme complex of the smooth ER and of enzyme induction is well worth fuller study. Reference should be made to such detailed works as Conney (1967) and Parke (1968).

The recent studies by Brodie and his associates give a good indication of the importance of this detoxication system in understanding the response of the liver to certain drugs and even to some hepatotoxins (Brodie et al. 1971a, b). They showed that halogenated aromatic hydrocarbons cause centrilobular necrosis in the liver because they are detoxicated by the NADPH-dependent microsomal system (see Fig. 12). Induction of new membranes of the smooth ER occurs mainly in the centrilobular regions of the liver, and induction of such membranes enhances the toxicity of these compounds. The detoxication of these substances is a good example of lethal synthesis (discussed in Section3.B.4) because it produces epoxides and other active metabolites; the epoxides conjugate with reduced glutathione. On the other hand allyl alcohol is not detoxicated by this system but is converted to the hepatotoxic aldehyde, namely, acrolein. As shown by Rees and Tarlow (1967), this occurs mainly in the periportal region which has a high concentration of the alcohol dehydrogenase which is responsible for this conversion. Consequently, allyl alcohol causes predominantly periportal necrosis.

3. **Redox Balance.** A major concept that has been brought back to the forefront relatively recently is the possibility that one of the most sensitive cellular systems involved very early in some types of cellular injury is the redox balance. This concept requires brief explanation.

In general, any group or substance that can be reversibly oxidized and reduced undergoes oxidation or reduction at a particular electrode potential. This is understandable simply if we consider a solution of ferric and ferrous ions. If we insert two suitable electrodes, as used in polarography, and apply an increasing potential difference across them, no current will flow until a particular electrode potential difference is obtained. At this electrode potential ferric ions will be reduced to ferrous ions and current will flow. The electrode potential required to obtain flow of current is the electrode potential of the change $Fe^{3+} \rightleftharpoons Fe^{2+}$. There is a similar specific electrode potential for the change NADP $\rightleftharpoons$ NADPH; this is about -300 mV. A system such as NAD $\rightleftharpoons$ NADH (also about -300 mV) can reduce a second reduction–oxidation system of a lower (less negative) electrode potential if the difference between electrode potentials of the two systems is not too great. This is the basis of electron, or hydrogen, transport in the mitochondrion where hydrogen (or electrons) moves from redox systems of more negative to more positive electrode potential until it reacts with atmospheric oxygen (e.g., see Lehninger, 1964).

We see, then, that any cell contains a large number of redox systems. Many are segregated from the rest of the cell into discrete multiredox carrier systems; of these, the electron (or hydrogen) transport system of mitochondria is probably the most clearly defined (see Lehninger, 1964). In such complexes each redox system is affected mainly by the state of the redox system on either side, that is, the more negative and the more positive side. But in the cytoplasm generally (i.e., in the cytosol) there is a possibility that the redox systems equilibrate with each other to give an overall approximate redox equilibrium for the cell (see Hill, 1956). One of the most significant of these redox systems is the NADP $\rightleftharpoons$ NADPH system. Another is the cytoplasmic and the mitochondrial NAD $\rightleftharpoons$ NADH balance (Krebs and Gascoyne, 1968). It has been difficult to measure these ratios (e.g., NAD/NADH) accurately by conventional biochemistry because of oxidation or reduction of the system during extraction and measurement and because the coenzyme is readily destroyed during these processes. However, it is possible to establish these ratios indirectly by measuring the concentration of substrates (Williamson et al., 1967; Williamson et al., 1968; Ballard, 1971). It has been shown that many liver toxins produce a rapid loss of either NAD, NADP, or ATP and inhibit their resynthesis (see Slater, 1968). Clarke and Pinder (1969) showed that the NADPH/NADP ratio fell sharply in the liver of rats 1 hr after they had been injected with the liver cell toxin ethionine.

It is reasonable to expect that at least some of the redox systems of the cytosol will ultimately equilibrate with the —SH/—S—S— redox balance first for glutathione and finally for the tissue proteins (Chayen, 1953). There is much evidence (see Chayen, 1953) that the —SH/—S—S— equilibrium of glutathione is in equilibrium with the ascorbate–dehydroascorbate system (i.e., the reduced and oxidized form of ascorbic acid) and that changes in the redox balance of cells, particularly in the —SH/—S—S— ratio, are of great significance in the onset of mitosis (Brachet, 1950, Chapter 5). Chayen et al. (1969a) showed that the cellular redox balance, as assessed from the —SH/—S—S— ratio, could directly affect the functional state of lysosomal membranes, and Butcher and Chayen (1971) attempted to link imbalance of this ratio with an altered NADPH/NADP ratio. It should also be apparent that inside a cell the effects of pH may be relatively limited because there may be little free water. In cells at least much of the water is likely to be bound in colloidal form, or in a coacervate colloidal system as suggested by Booij and Bungeberg de Jong (1956). Moreover, electron microscopy shows the cell to consist of a multitude of membranes and other surfaces; Lehninger (1966) calculated that from 40 to 90% of the mass of cells consists of membranes. Under these conditions hydrogen ion concentration is likely to be significant more in relation to the redox balance which can control the amount of hydrogen available at the active site of an enzyme or the reduced or oxidized state of hydrogen (electron) carriers (for a discussion of pH at surfaces, see Small, 1954; intracellular pH is fully and clearly discussed by Bittar, 1964, 1967).

4. The Concept of Lethal Synthesis. The concept of "the biochemical lesion" (Gavrilescu and Peters, 1931) has been of fundamental importance in modern cellular biology. Peters (1969) followed this line of argument: if you find a pathological change in the living animal or plant which runs parallel to biochemical change studied *in vitro*, for example, in homogenates or in sections, there is a good chance that this represents the biochemical lesion that underlies the pathology. Obviously, it is advisable to then study the biochemical change in life, as is done in clinical biochemistry. Moreover, as Peters emphasized, it is helpful if the lesions are reversible so that inhibition of the biochemical lesion results in the return to normal function of the tissues. His classic studies on convulsions in pigeons demonstrate his thesis admirably. In this work (see Peters, 1963), he showed that biochemical lesions of the pyruvate oxidase system were found in the brain tissue in convulsive birds. Thiamine deficiency also produced convulsions which could be cured in a matter of minutes when thiamine was injected into the subarachnoid space of the brain. The relationship between thiamine and the pyruvate oxidase system, which included the Krebs' cycle and related systems, now seems fairly clear (Peters, 1969), as does the evidence that thiamine deficiency, for example, induced by a thiamine antagonist, precedes both enzyme change and convulsions.

Thiamine deficiency is a good example of a lesion induced by a deficiency of an essential substance. Similar cases are those now grouped together as "inborn errors of metabolism" many of which involve the genetically controlled lack of an enzyme. There are also many examples of the reverse effect, namely, cases in which a substance attacks an enzyme system or essential active groups of a cell. For example, tervalent arsenical compounds attack —SH groups in certain enzymes; this effect can be reversed by the addition of substances that have vicinal —SH groups (see Peters, 1969).

A third class of biochemical lesion is where a nontoxic agent becomes metabolized *in vivo* to become toxic, so producing a biochemical lesion. This synthesis of a toxic substance by the tissues was called "lethal synthesis" by Peters in his Royal Society Croonian Lecture of 1952 (see Peters, 1963; Slater, 1968). For example, fluoroacetate, which is relatively nontoxic to isolated enzyme systems, becomes converted by tissues into fluorocitrate which inhibits aconitase (which is the enzyme of the Krebs' cycle that converts citrate to isocitrate). A newer extension of this concept, which may have very wide-reaching significance, is the possibility that free radicals can be formed from many substances when they come into a suitable intracellular environment. This phenomenon is clearly explained by Slater (1969). It will be recalled that X-irradiation owes much of its cellular damaging effects to its ability to induce the formation of free radicals. Slater emphasizes that free radicals, of similar importance in cell injury, can be formed without any form of irradiation; their formation from otherwise bland agents can be greatly influenced by the redox state of the cell or system.

We have already considered how lethal synthesis may be responsible for the cell injury induced by nitrosamines (see above).

5. Effect of Changes in Ionic Balance. There have been many theories that have involved flow (loss or gain or change in localization) of ions as the major factor in cellular injury. Heilbrunn (1943, 1956) elaborated the calcium release theory of stimulation and anesthesia (Heilbrunn, 1943, p. 535) to explain many aspects of cellular physiology. Judah et al. (1964) put forward a tentative hypothesis for the initial stages of cell damage; alteration to the cell membrane causes mobilization of calcium in the membranes. This directly allows increased flow of sodium into the cell and a reduction in cellular ATP as a result of increased demands on the sodium pump mechanism. Consequently, the cell becomes progressively less able to restore its energy supply, and hence more calcium can migrate in from the external medium. The basic idea is that it is an excess of calcium which kills cells, possibly through its effect on mitochondria (Lehninger, 1964, 1970). But there is little certain evidence that excessive concentrations of calcium are present when liver cells are killed by a number of different hepatotoxic agents (Judah, 1969).

Changes in ionic balance can have very significant effects on intracellular pH, as discussed clearly by Bittar (1964, 1967).

6. Cell Membranes and Pinocytosis. Pinocytosis (or endocytosis) seems to be a common property of animal and possibly of plant cells. Many new words, mainly of complex Greek origins, have been coined to express the various types of pinocytotic activity found in different cells. As Holter (1961a) wrote: ". . . the revival of the Greek language has been quite considerable in an attempt to describe more or less specialized instances of phenomena that seem to be quite adequately covered by the two oldest and most widespread terms, namely phagocytosis and pinocytosis." His view is that both represent different aspects of the same basic phenomenon and differ mainly as regards the nature of the ingested matter—whether solid or water.

The ingested vacuoles, particularly when they contain solid matter (by phagocytosis) have been seen to fuse with opaque granules which, in the ameba *Chaos chaos* (Chapman-Andresen, 1967; Chapman-Andresen and Nilsson, 1967), in fibroblasts (Munro et al., 1964), and in polymorphs and macrophages (Cohn and Fedorko, 1969) have been shown to be lysosomes. So there is probably a direct relationship between pinocytotic (phagocytotic) activity of the surface membrane of the cell and the lysosomes (e.g., Dingle, 1968). The modification of these processes by drugs has been considered by de Duve (1968).

It will be appreciated that for many years cell physiologists have been concerned with the function of cell membranes and with the physicochemical properties of substances which can or cannot enter cells (e.g., see Davson and Danielli, 1952). Pinocytosis appears to offer a way for substances to enter the cytoplasm without penetrating through an intact cell membrane. It is likely that in most cells this is a relatively little-used route, but it can be stimulated by various agents. For example, insulin has been shown to induce pinocytosis in cells of adipose tissue and in HeLa cancer cells (see Holter, 1961a). Chapman-Andresen (1962) showed that inorganic salts, amino acids, and some proteins could stimulate pinocytosis in amebas and drew attention to the part played by the mucopolysaccharides of the outer surface of the plasmalemma (cell membrane); the first phase of induced pinocytosis appeared to be an interaction between the mucopolysaccharide and the inducing agent. It is possible therefore that cell injury can involve change in pinocytotic activity leading to increased or diminished penetration of materials and ions from the external environment of cells into their cytoplasm. It is also likely that protective drugs could be made to concentrate inside injured cells which have relatively impermeable cell membranes by stimulating pinocytosis in such cells. In relation to cell injury, Chapman-Andresen (1967) showed that metabolic inhibitors, such as potassium cyanide (1 mM), dinitrophenol (0.04 mM), or antimycin A (10 μg/ml), inhibited pinocytosis in amebas. The reported effects of polyanions and of polycations on the membranes and pinocytotic activity of various cells has been reviewed by Allison (1968b), who also considered the effects of narcotics and of other surface-active substances on cell membranes. His concepts of membrane alteration by such substances merit careful consideration by workers interested in this subject.

4. MULTIPHASE CHEMISTRY OF CELL INJURY

A. Heterogeneous versus Homogeneous Chemistry

Before the advent of homogenization and differential centrifugation (e.g., Claude, 1941; 1946a, b), most biochemistry was "homogeneous biochemistry." By this is meant that biochemistry predominantly dealt with single substances in solution, that is, in a single homogeneous phase. Once it became common to isolate cellular organelles, it became obvious that many enzymes acted in concert with others and that biological chemistry had to include the study of multienzyme complexes which acted rather differently, and often more efficiently, than did the same enzymes when they were randomly dispersed. These ideas were well discussed by Mahler (1953), Green (1958), Lehninger (1964), and Roodyn (1968). It was seen, moreover, that enzymes behaved differently when attached to their natural cell structure than when they were isolated from it, and that lipids to which they are attached in the cell may greatly alter their reactivity. Siekevitz (1962) listed many examples to illustrate this phenomenon. In relation to an electron micrograph of pancreas, in which the cytoplasm is a maze of membranes, he wrote ". . . gone is the idea of chemical reactions in a one-phase, soluble environment, for where is such an environment in the cell?" This concept was extended when it became possible to attach isolated and purified enzymes to solid structures, or matrices, the physical nature of which could be determined experimentally. It was then found that the activity could be varied greatly according to the density of charge around the enzyme on the solid surface or matrix (Hornby et al., 1966; Filippusson and Hornby, 1970).

Quarles and Dawson (1969) reported that the pH requirements of a phospholipase varied considerably depending on the nature of the surface on which it acted, and quoted other examples. They pointed out that the pH in the surface phase (pH_S) adjacent to a negatively charged interface differs from the pH present in the bulk phase (pH_B) by the expression

$$pH_S = pH_B + (\epsilon\psi/2.3kT)$$

where T is the absolute temperature; k is Boltzmann's constant; ψ is the potential in the surface phase (see Davies and Rideal, 1961), and ϵ is the electronic charge in the surface phase. The evidence in their own example indicated that some other factor, such as a conformational change in the enzyme molecule, played a major part in the altered pH optima that they observed. Indeed, it is likely that changes in the conformation of enzymes, induced by agents which either alter the structure or the physicochemistry of cells, may play a significant part in altered enzyme activities observed in injured cells.

Redox changes (as discussed above) could have very significant effects on the charges at the enzyme sites. In this type of study, the enzyme, or other active group, is situated on or in a solid structure, be it an isolated mitochondrion or an experimental matrix, and the substrate is added in the fluid (bulk) phase. This, then, is heterogeneous phase biochemistry. The results from such studies have led Roodyn (1968) to wonder whether there is any point in persevering with methods of homogeneous phase biochemistry in attempting to understand the integrated biochemistry of structured organelles such as mitochondria. The concepts of structure and function, and of the many phases present in the cell, were well summarized by Bittar (1964): "In any case, it is well recognized that a cellular enzyme is a part of structure and that structure and function are inseparable at the molecular level. Since the cell is a multiphase system, and the bulk of the available knowledge is based on enzyme activities in the test tube, the next line of attack is obviously one of charting enzyme and pH relationships in terms of regional distribution in the cell."

Another aspect of the difference between homogeneous and heterogeneous phase biochemistry is seen in the comparative studies of enzyme systems in homogenates and in tissue slices (Campbell and Stone, 1957; Brew and Campbell, 1967; Hems et al., 1968). There is an old and well-established form of biochemistry in which enzyme activities and metabolic processes are studied in slices of tissue or of organ. The slices are about 200 μ thick, and there may be difficulties due to penetration of substrate into such slices and to solubilization of the slice in the biochemical reaction medium. This tissue slice biochemistry is heterogeneous phase biochemistry, and it frequently produces results that differ significantly from those obtained with "purified" systems isolated by differential centrifugation after homogenization. Possibly the most nearly ideal form of heterogeneous phase biochemistry is that form which uses whole organs which are perfused with substrates and measures the metabolites in the outflowing perfusate. This type of study has been greatly advanced by sensitive methods which use radioactive tracer techniques (Chain, 1965).

1. Elution Methods. Considerable advances in heterogeneous phase biochemistry were made mainly by Glick and co-workers, who used unfixed sections up to 20 μ thick and reacted them with histochemical procedures for the measurement of activity of certain enzymes (e.g., Jardetsky and Glick, 1956; Glick, 1962, 1963). These successful advances were extensions of the micromethods, first developed at the Carlsberg Laboratories by Lindestrøm-Lang and Holter, to the microbiochemistry of tissue sections. The range of these methods, and of the related techniques of Lowry (1963, 1964), is now a major feature of heterogeneous phase microbiochemistry.

The work begun by Defendi and Pearson (1955) and by Jardetsky and Glick (1956) was continued by workers associated with the present authors. Jones et al., (1963) showed that it was possible to adapt histochemistry so that tissue sections approximately 10 μ thick could be used to study the biochemistry of certain dehydrogenase enzymes. These workers demonstrated dehydrogenase activity in the usual histochemical manner (as reviewed by Seligman, 1963; also see Novikoff, 1963), namely, by the amount of a tetrazolium salt which, on accepting hydrogen from the redox systems associated with the dehydrogenase, became precipitated as a colored formazan. They eluted the formazan from the sections and measured it spectrophotometrically. The amount of formazan so produced was related to the nitrogen content of the sections measured by a micro-Kjeldahl procedure. Altman, in a series of studies, has developed this approach into a precise measure of the amount of hydrogen generated by the dehydrogenase. He has described procedures which measure quantitatively the same activity in single sections of rat liver (3 $\times$ 5 mm in area and 10 μ thick) as can be measured by conventional biochemistry (Altman, 1969a; 1971). He has applied these techniques to a number of enzymes, and it is likely that the range of enzyme activities measurable in such sections by these techniques can be considerably extended (also see Altman, 1969b, c).

B. Multiphase Chemistry

A further development in tissue section biochemistry is the study of the activity of one multienzyme complex in the presence of the activity of other enzyme systems in the same cell. This work arose from the following observations: when 10 μ sections of undenatured tissue are incubated in a biochemical reaction medium for measuring the activity of one enzyme, much of the undenatured protoplasm becomes dispersed into the medium. This disruption of the tissue can be stopped if a sufficient concentration of a colloidal stabilizer such as polyvinyl alcohol (PVA) is included in the reaction medium (Altman and Chayen, 1965). However, even though the colloidal stabilizer protects the structural proteins and nucleic acids of the tissue, the so-called soluble enzymes are lost from the cells and can be detected in the reaction medium above the sections, unless the cells are performing oxidative work (Altman and Chayen, 1966). It seemed, therefore, that energy produced by any one system in a cell could influence the molecular integrity (Chayen, 1968b) of the whole cell. This concept has been developed by Butcher in a series of studies. The basic philosophy of these investigations was as follows: suppose you consider two oxidative pathways A and B in a cell. Conventional biochemistry isolates A from B and measures each separately. The question is whether the oxidative activity of the cell for these systems is the sum of its parts, namely, A + B, or whether it may be greater than A + B when each is measured separately. System A may be pentose shunt dehydrogenation, for example, the oxidation of glucose 6-phosphate and the production of NADPH; this oxidation is performed by the enzyme glucose-6-phosphate dehydrogenase which is either "soluble" in the cytosol, or weakly linked to the ER. System B may be the oxidation of glutamate and the passage of hydrogen (or electrons) from the coenzyme of this enzyme, glutamate dehydrogenase, along the mitochondrial hydrogen transport system. Butcher (Butcher and Chayen, 1968) has shown that there is, indeed, interaction between oxidation in the ER and oxidative activity in the mitochondria, that is, that in cells there is an integration between the different phases of the cytoplasm so that the effect of B in the presence of A is greater than would be obtained from the sum of the parts (i.e., greater than A + B) measured separately. This treatment of the cell as a multifold heterogeneous phase system constitutes the basic concept of multiphase biochemistry, namely where the tissue remains intact as a solid matrix made up of different systems each with its own phase boundary but each capable of interacting as an integrated system (see Chayen and Bitensky, 1968). If this concept is correct, then it will be obvious that some forms of cell injury may begin as a breakdown in these subcellular interactions which could not be detected if A and B were measured separately.

1. The Need to Relate Activity to Histology. There is another important feature of the biochemistry of tissue that can be discovered only with considerable effort by conventional homogenate or tissue slice biochemistry. This concerns the distribution of activity in the various cells or cell types that constitute the tissue sample; should the proportion of cell types vary in the tissue during the course of the experimental procedures, any alteration in chemical activity may reflect only this histological change and not an actual chemical change in the original cells of the tissue. This general statement can be clarified by considering a particular example.

When rats are fed with a suitable diet slightly deficient in protein, to which is added the liver carcinogenic agent dimethylaminoazobenzene (DMAB), it is said that liver cancers develop after about 120 days. Much work has been done on this experimentally induced carcinogenesis, and all workers agree that the succinate dehydrogenase activity per unit mass or volume of liver is halved after about 90 days of feeding, that is, well before cancerous growths become apparent (e.g., Fiala and Fiala, 1959; Jones, 1963). This gives rise to the concept that liver cell cancer is induced by a loss of mitochondria and the effect this can have on the metabolism of the liver cells (Elson, 1958). When the cellular activity of succinate dehydrogenase was examined (Jones et al., 1961; Jones, 1965), it was readily shown that the activity of this enzyme in the liver hepatocytes was not diminished; nor was it reduced in the secondary growths of malignant liver cells found in other organs such as the lung. The reduction in succinate dehydrogenase activity per unit mass of liver after 90 days of feeding with DMAB was due to the fact that there had been a very great proliferation of cells of the portal tract. Consequently, a unit mass of "liver" now contained a far higher proportion of these cells (bile duct and connective tissue cells), which, even in normal liver, have a low level of succinate dehydrogenase activity. So it became evident that the decreased activity of this enzyme per unit mass of liver was a "tissue dilution artifact" (Chayen et al., 1961) resulting from the dilution of the normal hepatocyte activity by the relative inactivity of the portal tract cells.

It follows therefore that it is advisable to be able to relate chemical activity to the histology of the section and to be able to determine which cells in a tissue slice or section are the truly active ones and which are the less active cells. In studies on lysosomes (e.g., in the spleen), it may be that macrophages have very labile lysosomes and the rest of the tissue (e.g., lymphocytes) has more stable lysosomes. Experimentally induced changes in the proportion of macrophages to lymphocytes might result in more of the lysosomes being labile. This result might be interpreted as implying that the experimental conditions labilized lysosomes, whereas in fact they had merely increased the number of macrophages per unit mass of spleen. This kind of effect emphasizes the need to be able to analyze the chemistry of each cell in a section of tissue and not solely to measure the overall activity of the section or slice or homogenate.

2. Microchemical Pathology. We have already emphasized that the clinical biochemist has had to develop ingenious methods for analyzing blood and urine to determine which organ is behaving abnormally. It is akin to analyzing the nutritional status and metabolic normality of a household by examining the garbage, the drains, and the sewers. The difficulty which necessitates these indirect methods is the fact that it is rarely possible to obtain a sufficiently large sample of the suspectedly abnormal organ from a living patient in order to prove the metabolic defect in the target organ. The methods of multiphase biochemistry, with their emphasis on the measurement of the activity of each cell within a section, now make it possible to make up to 20 enzyme assays on a single needle or drill biopsy.

C. The Techniques of Multiphase Chemistry

We now must consider, albeit briefly, how one can investigate the biochemistry of an intact tissue section, for example, from a needle biopsy; relate the overall activity to that of each of the cells in the section; and measure accurately the activity of each cell. This has entailed the derivation of a whole series of methods, as discussed in some detail by Chayen and Bitensky (1968) and by Chayen et al. (1970).

1. To Cut Sections. In order to cut sections, the tissue must be hardened. Conventionally, this is achieved by embedding it in a supporting wax and for this it usually must be dehydrated. The process of dehydration and embedding would be disastrous to the tissue structure unless the specimen had been fixed and so denatured that it could withstand these processes. But fixation grossly alters the chemical structure of protoplasm (e.g., see Wolman, 1955), and, in altering chemical groups, including those at the active sites of enzymes, it usually causes more or less inhibition of enzyme activity (e.g., see Appendix 1 in Chayen et al., 1969b). Consequently, this procedure cannot be adopted when one is concerned to measure the biochemistry of tissue. The alternative is to harden the tissue by cooling it. This usually involves freezing the tissue water, and this produces ice artifact (see Chayen and Bitensky, 1968).

It has been found possible, however, to cool small samples of tissue, which can be as large as $5 \times 5 \times 3$ mm, by precipitate immersion in a particular grade of hexane at $-70°C$. Under these conditions the tissue is supercooled (Chayen et al., 1969b) so that no ice forms inside the tissue, neither in the extracellular spaces nor inside the cells. (For a discussion of tissue supercooling, see Lynch et al., 1965.) The theory is that a relatively inert medium is required at sufficiently low temperature, which makes intimate contact with the tissue (so that there can be no insulating gas or air between the tissue and the medium). The medium must have a high thermal conductivity to remove heat rapidly away from the tissue.

It was also shown that such supercooled tissue can be sectioned, for example, at 10 μ, in a suitable cryostat, namely on a microtome housed in a cabinet whose ambient temperature is approximately −25°C, provided that the metallic knife is cooled to approximately −70°C. Under these conditions heat produced during the cutting is apparently conducted through the knife rather than through the tissue. The section is then flash-dried onto a glass slide; this involves bringing the slide, from laboratory temperature, up to the section so that there is a small gap of about 2 mm between the section and the slide. The temperature gradient across this gap is about 90°C, and the supercooled water boils off the section and condenses onto the knife, ejecting the section on to the slide. Consequently, the supercooled water cannot freeze and produce ice artifact in the tissue (see Chayen and Bitensky, 1968).

2. To Incubate Sections. When unfixed or undenatured 10 to 20 μ sections are immersed in an aqueous buffer of near neutrality containing the substrate for biochemical assay of enzyme activity, as much as 50% of the nitrogenous matter of the tissue dissolves or disperses into the buffer. Enzymes, especially "soluble" enzymes, are lost from the section into the aqueous medium within a few minutes. This is a well-known phenomenon; because of it many workers have considered it impossible to study enzyme activity inside unfixed tissue sections (e.g., Farber et al., 1956; Sternberg et al., 1956; Novikoff, 1963). This problem was overcome when Altman and Chayen (1965, 1966) demonstrated that the addition of a sufficient concentration of a chemically inert colloid stabilizer, PVA, could retain inside the section all the nitrogenous matter and all the "soluble" enzymes studied up to the present. This result has been confirmed by many workers. The effect is not confined to PVA [see Stuart et al., (1969) for a polypeptide; Stuart and Simpson (1970) using Ficoll].

3. To Relate Activity to Histology. Thus it is possible to cut sections of undenatured tissue and to incubate them in a biochemical reaction medium for testing enzyme activity, or reactive groups, without losing any appreciable part of the tissue. In conventional biochemistry the end products of the reaction are allowed to accumulate in the reaction medium, and their concentration in this solution is measured. But although this gives the overall activity of the tissue section or slice, it does not allow the investigator to determine the relative activities of each cell or cell type. To achieve this result it is necessary to precipitate the end-product of the biochemical reaction in the cell that has produced the reaction. To take a simple example, in conventional biochemistry you can incubate a mince of tissue (e.g., muscle) under anaerobic conditions with sugar-substrates, at the required pH, and with methylene blue. The last-mentioned is a hydrogen-acceptor which takes the place of atmospheric oxygen and is reduced to the colorless form. The amount of decolorized methylene blue is measured colorimetrically. It should be noted that in this case the indicator, or hydrogen-acceptor (electron-acceptor), is in solution so that there is no means of knowing which cells are more, and which are less, active. In multiphase biochemistry you begin with a section of the tissue which is a good histological preparation in that you can distinguish each cell (particularly under phase-contrast illumination). You incubate it at the required pH in the presence of a sugar-substrate with PVA to protect the section. In place of methylene blue you include a tetrazolium salt which is readily soluble, which is a hydrogen-acceptor, and which becomes precipitated as a colored formazan when it is reduced. Thus the two systems are comparable. But you can see by microscopic inspection, which cells have the most activity in that they are more intensely colored. So it is a simple task to relate activity to the intimate histology of the section.

It will be obvious that multiphase biochemistry depends on the methods of histochemistry for obtaining a colored precipitate in the active cells. Most methods have had to be revised to ensure that the amount of activity is directly proportional to the amount of colored end-product precipitated, and it has been necessary to check that the precipitation is sufficiently rapid to allow the end product to be precipitated in the cell which produced that activity, that is, that diffusion from one cell to another cannot occur. These questions have been discussed fully by Chayen et al. (1969b), who list a wide range of active chemical groups and of enzymes which can be studied by these techniques.

4. To Measure Activity per Cell. We have already seen that the amount of formazan precipitated by a section of liver tested for glucose-6-phosphate dehydrogenase activity and eluted from the section gives the same activity for that enzyme as can be obtained by more conventional biochemistry (Altman, 1969). The limit of analysis by the elution technique is approximately 1×10^{-6} *g* of formazan per square centimeter of a 10 μ section. Yet if we take the area of a liver cell to be 600 μ^2 (i.e., 30×20 μ), we would have to increase the sensitivity of analysis by about 10^6 times to make a comparable measurement of the amount of formazan per cell in a liver section. This increase in sensitivity is achieved by relatively simple biophysical methods of measurement. The main procedure used at present involves the use of the scanning and integrating microdensitometer first designed by Deeley (1955).

In this instrument (Fig. 13) the section is placed on the stage of a conventional microscope and inspected by normal light optics. The wavelength of the light can be varied by means of a simple monochromator. The selected cell is placed at the center

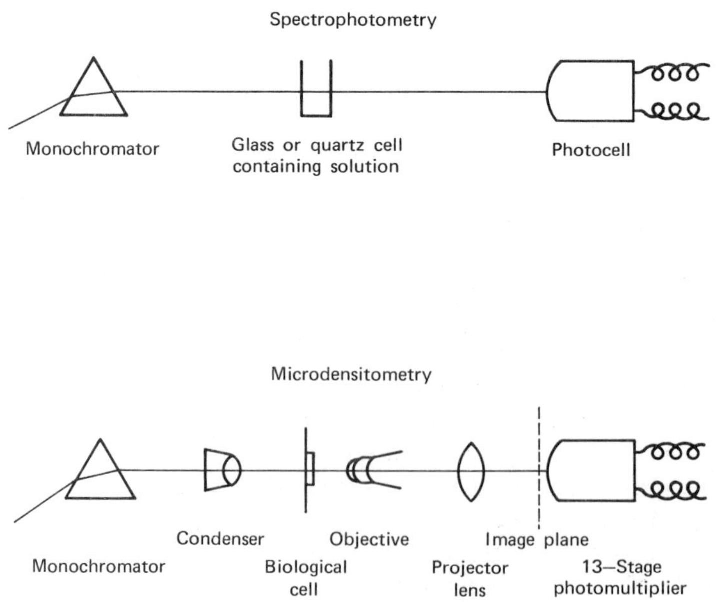

Figure 13. A diagrammatic comparison of spectrophotometry with microdensitometry. (From Chayen et al., 1970.)

of the field, and an aperture at the level of the eyepiece is inserted to block out all the field except the selected cell. This image is thrown on to the surface of a 13-stage photomultiplier via a scanning device. The scanner must be sufficiently small to ensure that, even though the end-product may be precipitated inhomogeneously in the cell, each point of the scan is optically homogeneous (for discussion of errors in measurement caused by optical inhomogeneity see Gomori, 1952; Chayen and Denby, 1968). Sensitivity is increased both by the magnification of the microscope system and by the multiple-stage photomultiplier. By suitable calibration it is possible to relate the arbitrary units of such measurements to extinction per unit area or per cell. By the use of such microdensitometry, it has been possible to measure 1×10^{-12} g of DNA (stained by the Feulgen nucleal reaction) with an instrumental accuracy or reproducibility of $\pm 3\%$ (see Chayen and Denby, 1968).

To measure uncolored end-products, it is possible to use microscopic interferometry (see Chayen, 1967; Chayen and Denby, 1968). This is a microscopic device which measures dry mass of part or of all of a cell. Barter et al. (1955) measured alkaline phosphatase activity in selected parts of a section which split phosphate from glycerophosphate at an alkaline pH. The phosphate was precipitated by calcium in the medium, and the mass of precipitated calcium phosphate was measured by microscopic interferometry.

An even more valuable use of microscopic interferometry is its application for measuring the water content of cells (see Ross, 1967; Chayen et al., 1970).

D. Effects of Hormones and Drugs on Cells

There are many examples to show that hormones or drugs, which have a decisive effect when given to the whole animal in life, may have much less or even no detectable effect on isolated chemical systems. For example, Weissmann (1969) could find no alteration in lysosomal membrane stability when aspirin was added to isolated lysosomes, although Chayen et al. (1970) showed that aspirin did stabilize lysosomal membranes if added to human skin maintained *in vitro*. Similarly, although hydrocortisone had an only just detectable effect on isolated lysosomes when added at a concentration of 10^{-3}–10^{-4} M (in the absence of dimethyl sulfoxide), it has a definitely stabilizing action even at 10^{-6} M when administered *in vivo* (Weissman, 1969). There is therefore considerable advantage in testing the effects of hormones and drugs by adding them to intact tissue and testing their subsequent effects either by conventional or by multiphase biochemistry. The special advantage of the latter is that it can be used to make many measurements of activities in the very small samples of tissue needed for such tests.

Many studies on the effect of hormones or drugs have been made on intact cells grown in proliferative tissue culture. Although such investigations have yielded much valuable information, the objection to them is that they represent the response of tissue culture cells rather than that of tissue cells. For example, when chick-embryo heart is grown in proliferative culture, the cells that grow out in a beautiful monolayer contain an appreciable proportion of fibroblasts. Their response to hormones may not reflect that of the heart, the function of which depends largely on the muscle component rather than on the admittedly appreciable content of connective tissue cells. Other studies have been made on cells separated from selected tissue and maintained in mass culture in bottles or roller tubes. Again it could be argued that cells in a tissue react with their neighbors and with their stroma and so behave unlike cells which are separated in this way (see Fell, 1969).

For all these reasons we prefer to maintain biopsies of tissue in nonproliferative maintenance culture, following the methods developed by Trowell (1959; also see in Willmer, 1965). A very interesting alternative is the procedure described by Wolff and Wolff (see in Thomas, 1970). Our experience of the Trowell technique is that biopsies of adult human tissues can be maintained *in vitro* for at least 24 hr in such a state that they resemble, both structurally and chemically, the tissue it has been excised from in the body. During the period of maintenance culture, hormones or drugs at specified concentrations can be added to the medium over which the tissue is maintained and their effects can be measured by the methods of multiphase microbiochemistry described above. For plants, of course, the whole process is simplified by

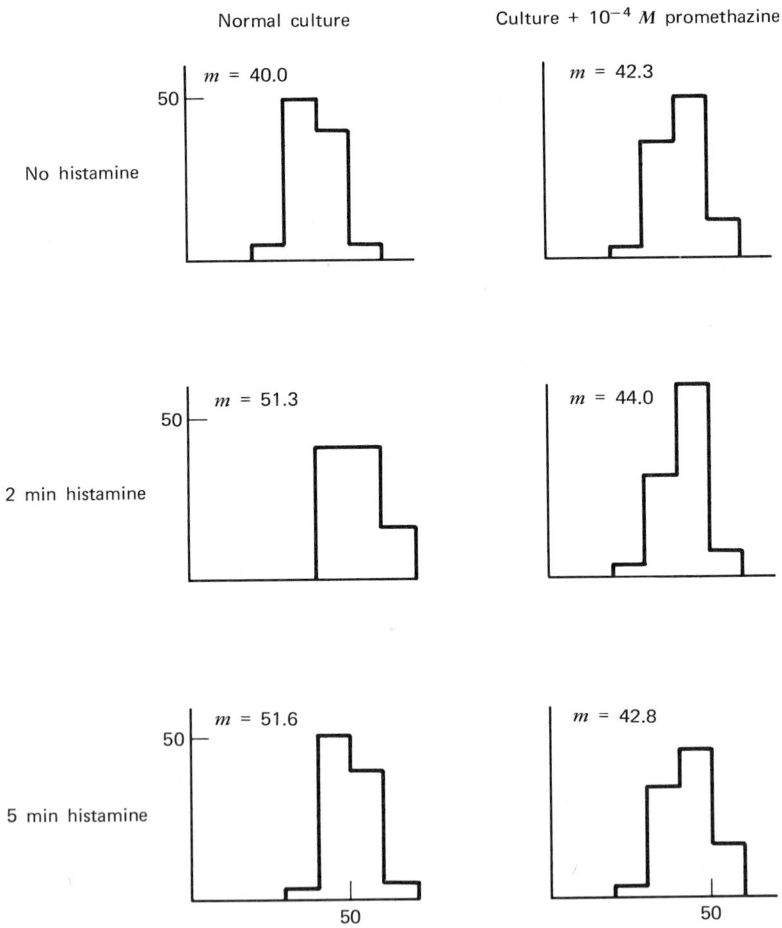

Figure 14. Histograms of the monoamine oxidase activity in rat liver mainttained in nonproliferative culture for 5 hr in the presence and absence of promethazine (10^{-4} *M*) and then subjected to histamine (10^{-4} *M*) for 2 and 5 min. Ordinates: number of readings which show that activity marked as abscissas. In the absence of promethazine, histamine increases the apparent activity of this mitochondrial enzyme assumably, by making the mitochondrial membrane more permeable to the substrate. This effect is nullified by the presence of promethazine.

many excellent techniques for maintaining parts of plants and allowing them to grow as an organized tissue, or even to produce complete plants *in vitro* (e.g., as reviewed by Street, 1965).

Some simple examples of the value of the Trowell technique may emphasize the value of this approach to what may be called "cellular pharmacology," and to the study both of the response of tissue to injury and to therapy:

1. Rat liver can be maintained in Trowell maintenance culture. The act of cutting out the pieces results in obvious cell damage,with large watery vacuoles in the cytoplasm and loss of glycogen. Provided that insulin is added to the culture medium, the cells re-form their normal quota of glycogen and the vacuoles disappear after about 3 hr. After 5 hr of maintenance culture, when the cultured tissue appears completely normal, a sample is removed and the medium is replaced by an exactly comparable medium which also contains 10^{-4} *M* histamine. After 2 min another sample is removed. When the sections are tested for the mitochondrial monoamine oxidase activity, it is found that histamine has increased the apparent activity of this enzyme, apparently by altering the permeability of the mitochondrial membranes. If the surface-protective antihistamine phenergan (promethazine) has been included in the original culture medium, it appears to protect the mitochondrial membranes against damage by histamine so that in such tissue histamine does not have this effect (Fig. 14).

2. In human skin maintained in Trowell maintenance culture, histamine (10^{-4} *M* acting for 2–5 min) opens lysosomal membranes so that the "free" activity of lysosomal naphthylamidase is enhanced (i.e., latent activity is converted into "free" activity because histamine alters the lysosomal membranes and renders them permeable to substrate). Aspirin (10^{-5} *M*) in the culture medium protects the lysosomal membranes against 10^{-4} *M* histamine acting for 2 min but is not able to protect fully against this substance acting for 5 min. Other antiinflammatory agents may protect against this concentration of histamine acting even over 5 min. It becomes possible therefore to measure the effect of antiinflammatory agents objectively, by their ability to retard the effect of a known concentration of histamine acting on human tissue *in vitro*. The parameters for such grading of antiinflammatory drugs are (a) their protective effect on lysosomes, (b) their protective effect on mitochondria, and (c) their ability to enable the tissue to withstand the tissue edema induced by histamine (Chayen et al., 1970).

5. THE CELLULAR BIOLOGICAL APPROACH TO THE PROBLEM OF CELL INJURY

In this section we consider a few examples to illustrate how cellular biological analysis can extend our understanding of cellular injury and pathology. We have already referred to various examples that illustrate how cellular biology, at the simplest level, has led to understanding certain disease or toxic changes. The most obvious examples were those that involved lysosomal changes, and of these the type-II glycogen storage disease is probably the simplest example.

We now consider some rather more complex forms of cellular injury.

A. Rheumatoid Arthritis

This is a condition that affects many joints in humans, although the effect may be most severe in any one joint. Its effects are seen as inflammation and swelling of the joint, with erosion of the cartilage of that joint. There is a great increase in the lining tissue of the joint, particularly of its loose connective tissue component, which may fill the joint cavity. At the points at which the articular cartilage is eroded, the loose connective tissue becomes stimulated to form granulation tissue (capillaries and fibroblasts) which grows over the eroded surface.

If a biopsy is taken from the synovial tissue, there is usually little difficulty in deciding whether there is overt rheumatoid arthritis. The synovial lining cells usually are increased in number from about two cells thick to six to eight cells deep. The underlying tissue contains some, or often masses of, chronic infiltrating cells such as plasma cells, lymphocytes, and macrophages. Thus the diagnosis of rheumatoid arthritis that has been made clinically can be confirmed, in many cases at least, with some degree of certainty. Consequently, this is a situation in which the cellular injury has been given a name and the clinician knows with what he has to deal. But we still know next to nothing about the abnormal function of the synovial tissue. Admittedly, the presence of concentrations of plasma cells indicates that immunological reactions may be involved, but it is not certain whether these are primary responses or are secondary to some other cellular changes, such as the presence of degradation products of collagen, or of lysosomal enzymes (Weissmann, 1966).

Electron microscopy (Ghadially and Roy, 1969) confirms these changes without adding much to our understanding of the functional changes involved. Experimental pathologists have shown that conditions akin to rheumatoid arthritis can be induced in some animals by producing an immunological response in joints (Dumonde and Glynn, 1962), or by injecting inflammatory substances such as caragheenin or turpentine (reviewed by Gardner, 1960) into joints, or by injecting lysosomal contents into joints (Weissmann et al., 1969). Yet it is still not known what starts the rheumatoid state. Some workers believe it may be initiated by infective agents (as reviewed by Duthie, 1969).

The erosion of cartilage begins from the point at which synovial lining cells lie against the cartilage (Ball, 1968). It therefore seemed likely that these cells might be the source of the erosive influence rather than cells in the synovial fluid which would attack the whole surface of the cartilage equally, or cells inside the cartilage (the chondrocytes) which would cause erosion from inside outward. It has been amply demonstrated that lysosomal enzymes can erode cartilage (e.g., Weissmann and Spilberg, 1968; Dingle, 1969). It was therefore obviously necessary to investigate whether or not the lysosomes of the synovial lining cells were abnormal in rheumatoid joints. This could not readily be done by homogenate biochemistry because a homogenate of rheumatoid synovial tissue contains lysosomes from a large number of cell types not found in nonrheumatoid synovia. Hence any difference between lysosomes isolated from nonrheumatoid and from rheumatoid synovial tissues could be due to differences in cell types rather than to changes in the lysosomes of any one cell type. Moreover, the synovial lining cells, when present in a given specimen (they are often lost from any one specimen) constitute only a very small fraction of the sample, so that changes in the function of the lysosomes in these cells may well be masked by lack of change in the lysosomes of the rest of the tissue.

As indicated above, multiphase biochemistry is ideally suited to this type of problem. It precipitates the end product in the cells that possess the activity being analyzed, and microdensitometry allows measurement of the amount of activity in each cell. Thus activity can be precisely related to histology. Such investigations showed that the lysosomes of rheumatoid synovial lining cells were totally abnormal in that their membranes were virtually functionless (Chayen et al., 1971a). It was shown, moreover, that there was an altered redox balance in rheumatoid synovial lining cells (Chayen et al., 1969a), the balance being displaced toward an excess of hydrogen in the rheumatoid cells. This effect was confirmed by the pH characteristics of these tissues when maintained in nonproliferative culture (Poulter et al., 1970). Addition of hydrogen acceptors to the culture medium corrected the redox imbalance and also changed the lysosomal membranes into functional membranes (Chayen et al., 1969a).

We see therefore that this type of investigation has extended knowledge of the functional cellular disturbance in this condition. It has shown that the likely cause of erosion of cartilage is the attack on the cartilage by lysosomal enzymes emanating from the abnormal lysosomes of the synovial lining cells. Furthermore, the reason these lysosomal membranes are not functioning as relatively impermeable (or semipermeable) barriers might be the change in the redox balance of the cytoplasm of the synovial lining cells which converts —S—S— bonds to the noncross-linked —SH state. All these abnormalities can be corrected by the use of hydrogen acceptors which correct the redox imbalance.

Clearly, the next phase of these investigations is to find the cause of the redox imbalance. We have already indicated above that the cytoplasmic redox balance can be controlled by the relative proportions of oxidized and reduced coenzyme (NADP). Butcher and Chayen (1971) have now shown that while the rate of oxidation of NADPH in rheumatoid and nonrheumatoid synovial lining cells is approximately the same, the rate of reduction of NADP is far greater in the rheumatoid than in the nonrheumatoid synovial lining cells. This could be the basis for the abnormal redox state of the rheumatoid cells. The role of lysosomes in rheumatoid disease and the value of cellular biological investigations for such disease processes has been reviewed by Chayen and Bitensky (1971).

In conclusion, we would argue that a disease such as rheumatoid arthritis is a functional disturbance, irrespective of the one or the many stimuli which initiate this disturbance. As far as we can see, only cellular biology can unravel the intricate details of this type of functional disturbance or "cellular injury."

B. Acute Inflammation

The study on rheumatoid arthritis is a somewhat unusual example of cell injury in that the obvious injury is to the cartilaginous matrix nearby the synovial lining cells which are altered in function but do not show other signs of being injured themselves. In contrast, in acute inflammation cells may be directly injured by inflammatory substances.

When mouse skin is painted with croton oil, or its active ingredient crotonaldehyde, an immediate inflammation develops. If a 2% solution of the aldehyde in oil is used, the skin becomes red in about 5–15 min. Inflammation is defined macroscopically as: redness, swelling, heat, pain, and usually impairment of function (see Boyd, 1947). The first three result from a chain of events taking place in the blood vessels and capillaries of the affected region which leads to an increased blood flow (redness and heat) and leakage out of the blood vessels of fluid and inflammatory cells (swelling). Until actual necrosis takes place, the injury to the cells of the skin is less obvious than the dramatic events of the vascular response. However, the cells are injured, as can be shown by cellular chemical studies.

When sections of the skin are examined, the following effects are seen: the mitochondrial and lysosomal membranes of the epidermis and of dermal cells become very permeable and may even rupture, with loss of enzymes from these organelles. These effects occur as early as do the vascular changes and hence are unlikely to be a secondary response to these changes. There is also a great increase in the water content of the dermis. Similar effects are found in mouse skin following subcutaneous injection of histamine (Diengdoh, 1965). Moreover, the same cellular changes are found in skin which has been maintained in nonproliferative culture and which has then been subjected to 10^{-4} *M* histamine for 2 min. Under these conditions no vascular effect can occur because the sample of skin *in vitro* has been separated from its vasculature. Consequently, cellular chemistry has been able to disclose cell injury induced by an inflammatory agent which was not apparent on histological examination. Moreover, it has shown that the same effects occur when the inflammation was induced in the intact animal as in the isolated biopsy maintained *in vitro*.

The advantages of this type of study become evident when assessing the effect of antiinflammatory drugs. We are concerned to know how these drugs affect inflammation in the tissues of man. There is ample evidence showing that the responses of the tissues of one animal may be very different from those of another animal. Hence it is advisable to test antiinflammatory drugs, or indeed any other drug, on human tissues, and it is fortunate that such tissues can readily be maintained *in vitro*. Moreover, it is helpful to have parameters of inflammation which can be measured, so that quantitative, objective assessment can be made of the potency of antiinflammatory agents in combating these inflammatory changes.

In practice many samples are taken from one strip of human skin and maintained in separate culture vessels for 24 hr. To some are added various concentrations of antiinflammatory agents. After 24 hr the medium is replaced by the same medium to which 10^{-4} *M* histamine has been added, the new medium being flowed in over the tissue. After various times (usually 2 and 5 min), samples are removed, chilled, sectioned, and tested for (1) freely available lysosomal enzyme activity, which is usually acid phosphatase or lysosomal naphthylamidase; (2) freely available mitochondrial activity (usually monoamine oxidase is used for this test); and (3) tissue water in the dermis, which is measured by interferometry. So, for example, inflammation increases the naphthylamidase and the monoamine oxidase activities because it changes the relatively impermeable lysosomal and mitochondrial membranes, making them more permeable to substrate. This is measured in the samples that have not had antiinflammatory agents added to the culture medium. The ability of the drugs to protect these membranes against the damage induced by histamine is shown by how little effect histamine has had on these enzyme activities in the samples which have been maintained in the presence of these drugs (see Fig. 15). Similarly, histamine increases the water content of the dermis, and the effect of the drugs in combating the increased water content can be measured (see Chayen et al., 1970).

It is conceivable that such studies can be extended by first maintaining the skin *in vitro*, then causing cellular injury by adding histamine or other inflammatory agents (such as bradykinin), and finally adding antiinflammatory drugs and observing the extent to which they restore the state of the cells to their original normality. Another extension of such studies is to add suspected injurious agents, such as certain food additives, or other drugs, to tissue maintained *in vitro* and to determine at what

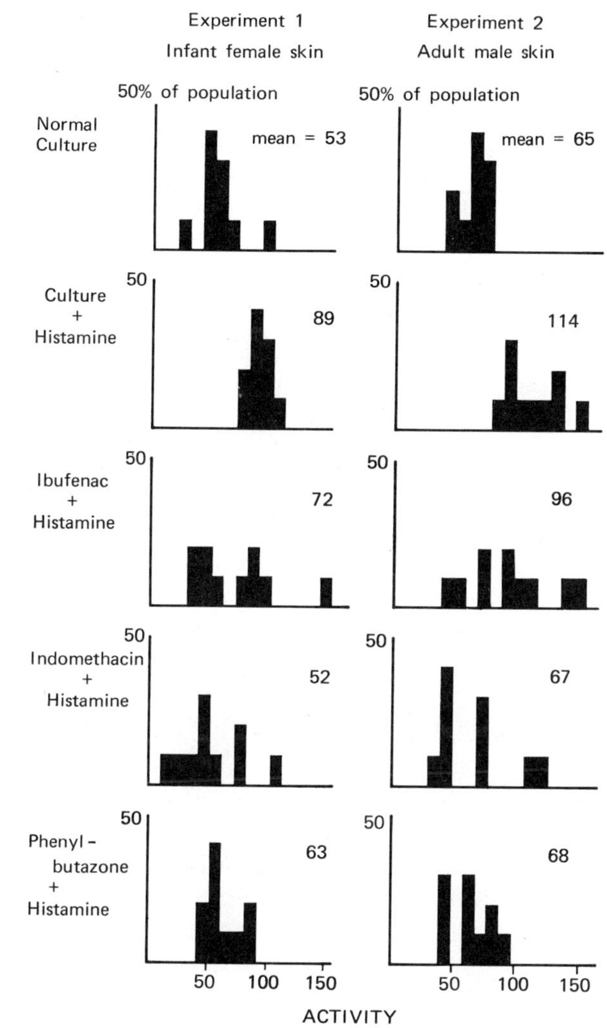

Figure 15. Histograms showing how 10^{-4} *M* histamine affects the "proteolytic" activity in various fields of the superficial dermis (number of readings) in two different samples of human skin cultured normally and in the presence of three antiinflammatory agents. Activity is recorded as optical density (relative absorption).

concentration these substances may alter the cellular chemistry beyond that of normal physiological variation.

We have considered these two examples in some detail to show how cellular biology enlarges our knowledge of functional injury or alteration to cells and to demonstrate the cellular chemists' attitude to problems of cell injury. Cellular biological concern for cellular injury begins with the effect on the whole organism; obvious damage to one cell type in a particular tissue could be due, in fact, to altered function in a totally different organ. This is particularly well seen in those cases in which cell damage is induced in an organ because of altered function of the adrenal gland, either of the cortex (steroid production) or of the medulla (catecholamine production). In the next step of the investigation, the cellular biologist must determine any histological changes that may occur, such as we have seen in the example concerning rheumatoid arthritis. If these changes are present, they can lead him to the target cells that are most or critically affected. But frequently, there may be no histological abnormality and he must find the affected cells either by their altered cytological structure, studied by all forms of light and electron microscopy, or by their altered cellular chemistry. Finally, he must follow the detailed changes in cellular chemistry to such an extent that allows him to understand the altered function of the cells, and so of the tissues, and ultimately of the organism.

C. Other Examples

We now briefly review some studies that illustrate some aspects of these concepts of cellular biology.

1. Myocardial Dysfunction. After prolonged open heart surgery, a proportion of cases showed a morbidity (Kloster et al., 1965) which on detailed examination, appeared to be due to poorly functioning left ventricular muscle. Drill biopsies from the left ventricle of patients undergoing this type of surgery were examined by cellular chemical procedures to test whether this was indeed the critical target tissue and, if so, to discover the functional lesion. In cases which showed such morbidity, it was found that there was a breakdown in energy transport from the muscle mitochondria to the muscle fibers and a dissociation of the phospholipids from cellular lipoprotein complexes (Niles et al., 1964; Niles et al., 1966; Chayen et al., 1966). Rather similar changes have been observed in experimentally induced myocardial dysfunction that lacked histological change (Niles and Barnhouse, 1967; Niles et al., 1968; Zaimis et al., 1969; Kakari, 1970).

2. Liver Cell Injury. It is usually relatively easy for clinicians and clinical pathologists to distinguish between jaundice caused by obstruction of the bile duct and that resulting from liver cell damage. The differentiation depends on characteristic changes which occur in the serum and urine of the patient as well as on clinical features. However, in a small proportion of cases, this distinction is difficult to make. It is important to distinguish between these two types of jaundice because the treatment of each is different. It has been shown by Bitensky (1967) that when cellular chemical tests are applied to sections from needle biopsies of such livers it is possible to demonstrate liver cell injury, for example, by the abnormal intracellular distribution of alkaline phosphatase and of bile salts.

3. Traumatic and Hemorrhagic Shock. In a long series of outstanding studies, Fine had shown that when death occurs from this form of shock the probable cause is bacterial endotoxins liberated in and absorbed by the gut. These bacterial endotoxins are normally detoxicated before they can damage smooth muscle of blood vessels and other organs (e.g., see Fine, 1958a, b; Fine et al., 1959). These investigations had already indicated that the spleen was a major site of such detoxication. The problem was to understand how the endotoxin was detoxicated and why it ceased to be when the animal (or human) went into irreversible shock. One piece of evidence was obtained by cellular chemistry. This showed that the lysosomes of the macrophages of the spleen became activated (and so very permeable) during the reversible phase of shock but became ruptured under conditions which led to irreversible shock (Bitensky et al., 1963). As long as the reticuloendothelial system could be protected, the state of shock could be reversed (Palmerio and Fine, 1969).

D. Conclusion

The various examples which have been discussed have been selected to emphasize certain aspects of the cellular biology of cell injury.

1. Injury to the cells of one organ may be induced by changes in the cells of a totally different organ. The cellular biologist should never forget that he is, primarily, a biologist. He should therefore study the organism as a whole even though his final analysis will be of cellular changes.
2. Originally, cellular injury was assessed predominantly by structural changes. The most obvious sign of cell damage is necrosis. But the cytologist who studied the organelles inside living cells was able to detect early structural signs of cell damage. Some of these were reversible; others were the inevitable prelude to cell death. The electron microscope has greatly extended the possibility of sensing cell damage by alterations in fine structure.
3. Cell injury is composed of a chain of events which could begin with genetic alterations, be they at the level of chromosome aberrations, or changes at the molecular level of the DNA, or its association with chromosomal protein. This chain of events could extend from specific alterations in the molecular organization of subcellular membranes or of organelle membranes down to alteration in the metabolic chemistry of the cell. The new developments of multiphase cellular chemistry have extended the possibility of detecting small functional changes in cells due to injurious agents. Because these methods are both specific and quantitative they allow the investigator to determine what variations occur in the metabolic activity relative to the normal physiological variation in the life of the cell. Consequently, deviations beyond the normal range must be viewed with suspicion as likely to lead to frank cell injury and possibly to cell death.
4. It must be appreciated that the cell is an integrated, multiphase, multicomponent system in which the various components are geared to each other. Hence it is likely that any dramatic alteration in any one factor, such as a great change in cyclic AMP activity, can occur only in conjunction with changes in other cellular systems. It is probably only one of many functional features contributing to the response of the cell to injury.

REFERENCES

Abraham, R., M. Morris, and J. Smith (1967). Histochemistry of lysosomes in rat heart muscle. *J. Histochem. Cytochem.*, **15**, 596.

Allison, A. C. (1968a). In *The Biological Basis of Medicine* (E. E. Bittar and N. Bittar, eds.), Vol. 1, pp. 209–242. Academic, New York.

Allison, A. C. (1968b). Some effects of pharmacologically active compounds on membranes. *Brit. Med. Bull.*, **24**, 135.

Allison, A. C. (1968c). In *Scientific Basis of Medicine*, pp. 18–30. Athlone, London.

Allison, A. C. (1969). In *Lysosomes in Biology and Pathology* (J. T. Dingle and H. B. Fell, eds.), Vol. 2, pp. 178–206. North Holland, Amsterdam.

Allison, A. C., and L. Mallucci (1965). Histochemical studies of lysosomes and lysosomal enzymes in virus-infected cell cultures. *J. Exptl. Med.*, **121,** 463.

Allison, A. C., and J. F. Nunn (1968). Effect of general anaesthetics on microtubules: A possible mechanism of anaesthesia. *Lancet*, **ii,** 1326.

Allison, A. C., and M. R. Young (1969). In *Lysosomes in Biology and Pathology* (J. T. Dingle and H. B. Fell, eds.), Vol. 2, pp. 600–628. North Holland, Amsterdam.

Allison, A. C., I. A. Magnus, and M. R. Young (1966). Role of lysosomes and of cell membranes in photosensitization. *Nature*, **209,** 874.

Altman, F. P. (1969a). A comparison of dehydrogenase activities in tissue homogenates and tissue sections. *Biochem. J.*, **114,** 13P.

Altman, F. P. (1969b). The quantitative elution of nitro-blue formazan from tissue sections. *Histochemie*, **17,** 319.

Altman, F. P. (1969c). The use of eight different tetrazolium salts for a quantitative study of pentose shunt dehydrogenation. *Histochemie*, **19,** 363.

Altman, F. P. (1971). *An Introduction to the Use of Tetrazolium Salts in Quantitative Enzyme Cytochemistry*. Koch-Light Laboratories, Colnbrook, Bucks, England.

Altman, F. P., and J. Chayen (1965). Retention of nitrogenous material in unfixed sections during incubation for histochemical demonstration of enzymes. *Nature*, **207,** 1205.

Altman, F. P., and J. Chayen (1966). The significance of a functioning hydrogen-transport system for the retention of "soluble" dehydrogenases in unfixed sections. *J. Roy. Microscop. Soc.*, **85,** 175.

Altman, F. P., S. E. Hampson, and R. Chayen (1964). Phospholipids in formalin extracts of nuclei isolated from sheep lungs. *Nature*, **202,** 1215.

d'Amato, F. (1950). Differenziazione istologica per endopoliploidia nelle radici di alcune monocotiledoni. *Caryologia*, **3,** 11.

d'Amato, F. (1952a). New evidence on endopolyploidy in differentiated plant tissues. *Caryologia*, **4,** 121.

d'Amato, F. (1952b). Polyploidy in the differentiation and function of tissues and cells in plants. *Caryologia*, **4,** 311.

Atkin, N. B., and B. M. Richards (1956). Deoxyribonucleic acid in human tumours as measured by microspectrophotometry of Feulgen stain: A comparison of tumours arising at different sites. *Brit. J. Cancer*, **10,** 769.

Bajer, A. (1956). Ciné-micrographic studies on mitosis in endosperm. II. Chromosome, cytoplasmic and Brownian movements. *Chromosoma*, **7,** 558.

Bajer, A. (1964). Ciné-micrographic studies on dicentric chromosomes. *Chromosoma*, **15,** 630.

Bajer, A. (1965). Ciné-micrographic analysis of cell plate formation in endosperm. *Exptl. Cell Res.*, **37,** 376.

Baker, J. R. (1944). The structure and chemical composition of the Golgi element. *Quart. J. Microscop. Sci.*, **85,** 1.

Baker, J. R. (1945). *Cytological Technique*, 3rd edition. Methuen, London.

Baker, J. R. (1957). The Golgi controversy. *Symp. Soc. Exptl. Biol.*, **10,** 1.

Baker, J. R. (1958). Fixation in cytochemistry and electron-microscopy. *J. Histochem. Cytochem.*, **6,** 303.

Baker, J. R. (1965). The fine structure produced in cells by fixatives. *J. Roy. Microscop. Soc.*, **84,** 115.

Ball, J. (1968). In *Rheumatic Diseases* (J. J. R. Duthie and W. R. M. Alexander, eds.), pp. 124–126. Edinburgh Univ. Press, Edinburgh.

Ballard, F. J. (1971). Regulation of gluconeogenesis during exposure of young rats to hypoxic conditions. *Biochem. J.*, **121,** 169.

Barber, H. N. (1939). The rate of movement of chromosomes on the spindle. *Chromosoma*, **1,** 33.

Barer, R. (1956). In *Physical Techniques in Biological Research* (G. Oster and A. W. Pollister, eds.), Vol. 3, pp. 30–90. Academic, New York.

Barer, R. (1968). *Lecture Notes on the Use of the Microscope*, 3rd edition. Blackwells, Oxford.

Barland, P., A. B. Novikoff, and D. Hamerman (1964). Fine structure and cytochemistry of the rheumatoid synovial membrane, with special reference to lysosomes. *Am. J. Pathol.*, **44,** 853.

Barr, M. L. (1966). The significance of sex chromatin. *Intern. Rev. Cytol.*, **19,** 35.

Barter, R., J. F. Danielli, and H. G. Davies (1955). A quantitative cytochemical method for estimating alkaline phosphatase activity. *Proc. Roy. Soc.* (London), **B144,** 412.

Beams, H. W., and R. G. Kessel (1968). The Golgi apparatus: Structure and function. *Intern. Rev. Cytol.*, **23,** 209.

Beaufay, H., P. Jacques, P. Baudhuin, O. Z. Zellinger, J. Berthet, and C. de Duve (1964). Tissue fractionation studies. 18. Resolution of mitochondrial fractions from rat liver into three distinct populations of cytoplasmic particles by means of density equilibration in various gradients. *Biochem. J.*, **92,** 184.

Beerman, W. (1952). Chromomerenkonstanz und spezifische Modifikationen der Chromosomenstruktur in der Entwicklung und Organdifferenzierung von *Chironomus tentans. Chromosoma*, **5,** 139.

Beermann, W. (1962). *Riesenchromosomen.* Protoplasmatologia, VI/D. Springer, Vienna.

Bělař, K. (1928). In *Methodik der Wissenschaftlichen Biologie* (T. Péterfi, ed.), pp. 638–735, Springer. Berlin.

Bělař, K. (1929). Beiträge zur Kausalanalyse der Mitose. *Arch. Entwicklungsmech. Organ.*, **118,** 359.

Bernhard, W. (1964). In *Cellular Injury* (A. V. S. de Reuck and J. Knight, eds.), pp. 209–247. Churchill, London.

Birnstiel, M. L., M. I. H. Chipchase, and B. B. Hyde (1963). The nucleolus, a source of ribosomes. *Biochim. Biophys. Acta*, **76,** 454.

Bitensky, L. (1962). The demonstration of lysosomes by the controlled temperature freezing-sectioning method. *Quart. J. Microscop. Sci.*, **103,** 205.

Bitensky, L. (1963a). In *Lysosomes* (A. V. S. de Reuck and M. P. Cameron, eds.), Churchill, London.

Bitensky, L. (1963b). Cytotoxic action of antibodies. *Brit. Med. Bull.*, **19,** 241.

Bitensky, L. (1967). In *The Liver* (A. E. Read, ed.), pp. 61–69. Butterworths, London.

Bitensky, L., J. Chayen, G. J. Cunningham, and J. Fine (1963). Behaviour of lysosomes in haemorrhagic shock. *Nature*, **199,** 493.

Bitensky, L., R. G. Butcher, and J. Chayen (1972). In *Lysosomes in Biology and Pathology* (J. T. Dingle, ed.), Vol. 3. North Holland, Amsterdam, in press.

Bittar, E. E. (1964). *Cell pH.* Butterworths, London.

Bittar, E. E. (1967). In *In Vivo Techniques in Histology* (G. H. Bourne, ed.), pp. 1–39. Williams and Wilkins, Baltimore.

Booij, H. L., and H. G. Bungeberg de Jong (1956). *Biocolloids and Their Interactions with Special Reference to Coacervates and Their Related Systems.* Protoplasmatologia. Springer, Vienna.

Borst, P., and A. M. Kroon (1969). Mitochondrial DNA: Physicochemical properties, replication and genetic function. *Intern. Rev. Cytol.*, **26,** 108.

Bourne, G. H. (1951). In *Cytology and Cell Physiology* (G. H. Bourne, ed.), 2nd edition, pp. 232–286. Oxford Univ. Press, London.

Boyd, W. (1947). *Textbook of Pathology*, 5th edition. Kimpton, London.

Brachet, J. (1950). *Chemical Embryology* (L. Barth, translator). Wiley, New York.

Brachet, J. (1952). The role of the nucleus and cytoplasm in synthesis and morphogenesis. *Symp. Soc. Exptl. Biol.*, **6,** 173.

Brachet, J., and A. E. Mirsky (eds.) (1961). *The Cell.* Academic, New York.

Braütigam, F., and A. Grabner (eds.) (1949). *Beiträge zur Fluoreszenzmikroskopie.* Fromme, Vienna.

Brew, K., and P. N. Campbell (1967). Studies on the biosynthesis of protein by lactating guinea-pig mammary gland; characteristics of the synthesis of α-lactalbumin and total protein by slices and cell-free systems. *Biochem. J.*, **102,** 265.

Brodie, B. B., A. K. Cho, G. Krishna, and W. D. Reid (1971a). Drug metabolism in man: Past, present and future. *Proc. N.Y. Acad. Sci.*, in press.

Brodie, B. B., W. D. Reid, A. K. Cho, G. Sipes, G. Krishna, and J. R. Gillette (1971b). Possible mechanism of liver necrosis caused by aromatic compounds. *Proc. Natl. Acad. Sci., U.S.*, **68,** 160.

Brown, D. E. S., and D. A. Marsland (1936a). The viscosity of amoeba at high hydrostatic pressure. *J. Cellular Comp. Physiol.*, **8,** 159.

Brown, D. E. S., and D. A. Marsland (1936b). Amoeboid movement at high hydrostatic pressure. *J. Cellular Comp. Physiol.*, **8,** 167.

Brown, G. L., S. F. Jackson, and J. Chayen (1952). Cytoplasmic particles in bean root cells. *Nature*, **171,** 1113.

Brown, R. (1831). Quoted in Sharp (1934).

Brown, R. (1951). Isolation of nuclei from plant cells. *Nature*, **168,** 941.

Bruemmer, N. C., and L. E. Thomas (1957). Cellular lipoproteins II. A "ghost-cell" preparation. *Exptl. Cell Res.*, **13,** 103.

Bullough, W. S., and M. Johnson (1951). The energy relations of mitotic activity in adult mouse epidermis. *Proc. Roy. Soc. (London)*, **B138,** 562.

Bullough, W. S., and E. B. Laurence (1960). The control of mitotic activity in mouse skin. *Exptl. Cell Res.*, **21,** 394.

Butcher, R. G., and J. Chayen (1968). The value of intact tissue sections for studying metabolic interactions between the cytoplasm and mitochondria. *Exptl. Cell Res.*, **49,** 656.

Butcher, R. G., and J. Chayen (1971). Unbalanced production of NADPH in rheumatoid synovial lining cells. *Biochem. J.*, **124,** 19P.

Cameron, G. R. (1952). *Pathology of the Cell.* Oliver and Boyd, Edinburgh.

Campbell, P. N., and N. E. Stone (1957). The synthesis of serum albumin and tissue proteins in slices of rat liver and liver tumour. *Biochem. J.*, **66,** 19.

Caspersson, T. (1950). *Cell Growth and Cell Function.* Norton, New York.

Caspersson, T., L. Zech, E. J. Modest, G. E. Foley, U. Wagh, and E. Simonsson (1969a). Chemical differentiation with fluorescent alkylating agents in *Vicia faba* metaphase chromosomes. *Exptl. Cell Res.*, **58,** 128.

Caspersson, T., L. Zech, E. J. Modest, G. E. Foley, U. Wagh, and E. Simonsson (1969b). DNA-binding fluorochromes for the study of the organization of the metaphase nucleus. *Exptl. Cell Res.*, **58,** 141.

Cavanagh, J. B., and M. H. Kyu (1969). Colchicine-like effect on astrocytes after portacaval shunt in rats. *Lancet*, **ii,** 620.

Chain, E. B. (1965). Landmarks and perspectives in biochemical research. *Brit. Med. J.*, **i,** 209, 273.

Chalmers, R. A., R. W. E. Watts, L. Bitensky, and J. Chayen (1969). Microscopic studies on crystals in skeletal muscle from two cases of xanthinuria. *J. Pathol.*, **99,** 45.

Chantrenne, H. (1947). Hétérogénéité des granules cytoplasmiques du foie de souris. *Biochim. Biophys. Acta*, **1,** 437.

Chapman-Andresen, C. (1962). Studies on pinocytosis in amoebae. *Compt. Rend. Trav. Lab. Carlsberg*, **33,** 75.

Chapman-Andresen, C. (1967). Studies on endocytosis in amoebae. The distribution of pinocytically ingested dyes in relation to food vacuoles in *Chaos chaos.* I. Light microscopic observations. *Compt. Rend. Trav. Lab. Carlsberg*, **36,** 161.

Chapman-Andresen, C., and J. R. Nilsson (1967). Studies on endocytosis in amoebae. The distribution of pinocytically ingested dyes in relation to food vacuoles in *Chaos chaos.* II. Electron microscopic observations using Alcian blue. *Compt. Rend. Trav. Lab. Carlsberg*, **36,** 189.

Chayen, J. (1952). Structure of root meristem cells of *Vicia faba. Symp. Soc. Exptl. Biol.*, **6,** 290.

Chayen, J. (1953). Ascorbic acid and its intracellular localization, with special reference to plants. *Intern. Rev. Cytol.*, **2,** 78.

Chayen, J. (1958). The quantitative cytochemistry of DNA and its significance in cell physiology and heredity. *Exptl. Cell Res., Suppl.*, **6,** 115.

Chayen, J. (1960). The localization of deoxyribose nucleic acid in cells of the root meristem of *Vicia faba. Exptl. Cell Res.*, **20,** 150.

Chayen, J. (1961). In *Biochemists' Handbook* (C. Long, ed.), pp. 746–762. Spon, London.

Chayen, J. (1967). In *In vivo Techniques in Histology* (G. H. Bourne, ed.), pp. 40–67. Williams and Wilkins, Baltimore.

Chayen, J. (1968a). "The histochemistry of phospholipids and its significance in the interpretation of the structure of cells." In *Cell Structure and Its Interpretation* (S. M. McGee-Russell and K. F. A. Ross, eds.), Arnold, London.

Chayen, J. (1968b). In *Cell Structure and Its Interpretation* (S. M. McGee-Russell and K. F. A. Ross, eds.), pp. 265–274. Arnold, London.

Chayen, J., and L. Bitensky (1968). In *The Biological Basis of Medicine* (E. E. and N. Bittar, eds.), Vol. 1, pp. 337–368. Academic, London.

Chayen, J., and L. Bitensky (1971). Lysosomes and inflammation with particular reference to rheumatoid disease. *Ann. Rheumatic Diseases*, **30,** 522.

Chayen, J., and E. F. Denby (1968). *Biophysical Technique as Applied to Cell Biology*. Methuen, London.

Chayen, J., H. G. Davies, and U. J. Miles (1953). Observations on some plant interphase nuclei. *Proc. Roy. Soc. (London)*, **B141,** 190.

Chayen, J., L. F. La Cour, and P. B. Gahan (1957). Uptake of benzpyrene by a chromosomal phospholipid. *Nature*, **180,** 652.

Chayen, J., G. J. Cunningham, and D. P. Winstanley (1959a). Chemical changes in human bronchial epithelium and their relation to bronchial cancer. *Nature*, **183,** 1743.

Chayen, J., P. B. Gahan, and L. F. La Cour (1959b). The nature of a chromosomal phospholipid. *Quart. J. Microscop. Sci.*, **100,** 279.

Chayen, J., P. B. Gahan, and L. F. La Cour (1959c). The masked lipids of nuclei. *Quart. J. Microscop. Sci.*, **100,** 325.

Chayen, J., R. Chayen, G. J. Cunningham, and L. Bitensky (1961). Bound lipids and their significance in cell biology. *Pathol. Biol.*, **9,** 925.

Chayen, J., G. R. N. Jones, L. Bitensky, and G. J. Cunningham (1961). Histological variation as a source of biochemical error. *Biochem. J.*, **79,** 34P.

Chayen, J., F. P. Altman, L. Bitensky, M. V. Braimbridge, T. Kadas, and P. J. Wells (1966). A study of the changes in hydrogen transport in an isolated rat heart preparation. *J. Roy. Microscop. Soc.*, **86,** 151.

Chayen, J., L. Bitensky, R. G. Butcher, and L. W. Poulter (1969a). Redox control of lysosomes in human synovia. *Nature*, **222,** 281.

Chayen, J., L. Bitensky, R. G. Butcher, and L. W. Poulter (1969b). *A Guide to Practical Histochemistry*. Oliver and Boyd, Edinburgh.

Chayen, J., L. Bitensky, R. G. Butcher, L. W. Poulter, and G. S. Ubhi (1970). Methods for the direct measurement of anti-inflammatory action on human tissue maintained *in vitro*. *Brit. J. Dermatol., Suppl. 6*, **82,** 62.

Chayen, J., L. Bitensky, R. G. Butcher, and B. Cashman (1971a). Evidence for altered lysosomal membranes in synovial lining cells from human rheumatoid joints. *Beitr. Pathol.*, **142,** 137.

Chayen, J., L. Bitensky, and G. S. Ubhi (1971b). Quantitative cytochemical studies on psoriasis and the effect of certain drugs. *J. Clin. Trials*, **8,** 35.

Chèvremont, M. (1956). *Notions de Cytologie et Histologie*. Desver, Liège.

Chèvremont, M., E. Baeckeland, and S. Chèvremont-Comhaire (1960a). Contribution cytochimique et histoautoradiographique à l'étude du métabolisme et de la synthèse des ADN dans des cellules animales cultivées *in vitro*. *Biochem. Pharmacol.*, **4,** 67.

Chèvremont, M., E. Baeckeland, and S. Chèvremont-Comhaire (1960b). Contribution à l'étude du métabolisme et de la synthèse cytoplasmique d'acides déoxyribonucléiques en culture de tissus. *Bull. Acad. Roy. Med. Belg.*, **15,** 349.

Clark, J. B., and S. Pinder (1969). Control of the steady-state concentrations of the nicotinamide nucleotides in rat liver. *Biochem. J.*, **114,** 321.

Claude, A. (1941). Particulate components of the cytoplasm. *Cold Spring Harbour Symp. Quant. Biol.*, **9,** 263.

Claude, A. (1943). Distribution of nucleic acids in the cell and the morphological constitution of cytoplasm. *Biol. Symp.*, **10,** 111.

Claude, A. (1946a). Fractionation of mammalian liver cells by differential centrifugation. I. Problems, methods and preparation of extract. *J. Exptl. Med.*, **84,** 51.

Claude, A. (1946b). Fractionation of mammalian liver cells by differential centrifugation. II. Experimental procedures and results. *J. Exptl. Med.*, **84,** 61.

Clifford, J. I., and K. R. Rees (1967). The action of aflatoxin B on the rat liver. *Biochem. J.*, **102,** 65.

Cohn, Z. A., and M. E. Fedorko (1969). In *Lysosomes in Biology and Pathology* (J. T. Dingle and H. B. Fell, eds.), Vol. 1, pp. 43–63. North Holland, Amsterdam.

Conney, A. H. (1967). Pharmacological implications of microsomal enzyme induction. *Pharmacol. Rev.*, **19,** 317.

Cope, G. H., and M. A. Williams (1967). Quantitative studies on neutral lipid preservation in electron microscopy. *J. Roy. Microscop. Soc.*, **88,** 259.

Cornman, I. (1944). A summary of evidence in favor of the traction fiber in mitosis. *Am. Naturalist*, **78,** 410.

Cosslett, V. E. (1965). *Reports on Progress in Physics*. Vol. 28. Institute of Physics and Physical Society, London.

Daems, W. Th., E. Wisse, and P. Brederoo (1969). In *Lysosomes in Biology and Pathology* (J. T. Dingle and H. B. Fell, eds.), Vol. 1, pp. 64–114. North Holland, Amsterdam.

Dalton, A. J. (1953). Electron microscopy of tissue sections. *Intern. Rev. Cytol.*, **2,** 403.

Darlington, C. D. (1936). The external mechanics of the chromosomes. *Proc. Roy. Soc.* (*London*), **B121,** 264.

Darlington, C. D. (1937). *Recent Advances in Cytology*, 2nd edition. Churchill, London.

Darlington, C. D. (1946). *The Evolution of Genetic Systems*. Cambridge Univ. Press, Cambridge.

Darlington, C. D., and L. F. La Cour (1969). *The Handling of Chromosomes*, 5th edition. Allen and Unwin, London.

Darlington, C. D., and K. Mather (1949). *The Elements of Genetics*. Allen and Unwin, London.

Davies, H. G. (1952). The ultra-violet absorption of living chick fibroblasts during mitosis. *Exptl. Cell Res.*, **3,** 453.

Davies, J. T., and E. K. Rideal (1961). *Interfacial Phenomena*, p. 94. Academic, New York.

Davies, D. D., and M. A. Sleigh (eds.) (1970). *Control of Organelle Development*. Symposium of the Society for Experimental Biology, Vol. 24, Cambridge Univ. Press, Cambridge.

Davson, H., and J. F. Danielli (1952). *The Permeability of Natural Membranes*, 2nd edition. University Press, Cambridge.

Deeley, E. M. (1955). An integrating microdensitometer for biological cells. *J. Sci. Instr.*, **32,** 263.

Deeley, E. M., H. G. Davies, and J. Chayen (1957). The DNA content of cells in the root of *Vicia faba*. *Exptl. Cell Res.*, **12,** 582.

Defendi, V., and B. Pearson (1955). Quantitative estimation of succinic dehydrogenase activity in a single microscopic tissue section. *J. Histochem. Cytochem.*, **3,** 61.

Dianzani, M. U. (1963). In *Lysosomes* (A. V. S. de Reuck and M. P. Cameron, eds.), pp. 335–351. Churchill, London.

Diengdoh, J. V. (1965). Histochemical studies on monoamine oxidase in skin. *J. Roy. Microscop. Soc.*, **85,** 103.

Dingle, J. T. (1968). Vacuoles, vesicles and lysosomes. *Brit. Med. Bull.*, **24,** 141.

Dingle, J. T. (1969). In *Lysosomes in Biology and Pathology* (J. T. Dingle and H. B. Fell, eds.), Vol. 2, pp. 421–436. North Holland, Amsterdam.

Dingle, J. T., and A. J. Barrett (1969). In *Lysosomes in Biology and Pathology* (J. T. Dingle and H. B. Fell, eds.), Vol. 2, pp. 555–566. North Holland, Amsterdam.

Dingle, J. T., and H. B. Fell (eds.) (1969). *Lysosomes in Biology and Pathology*. North Holland, Amsterdam.

Dumonde, D. C., and L. E. Glynn (1962). The production of arthritis in rabbits by an immunological reaction to fibrin. *Brit. J. Exptl. Pathol.*, **43,** 37.

Dumonde, D. C., L. Bitensky, G. J. Cunningham, and J. Chayen (1965). The effects of antibodies on cells. I. Biochemical and histochemical effects of antibodies and complement on ascites tumour cells. *Immunology*, **8,** 25.

Duthie, J. J. R. (1969). In *Textbook of the Rheumatic Diseases* (W. S. C. Copeman, ed.), pp. 259–322. Livingstone, Edinburgh.

Dutrochet, R. J. H. (1824). *Recherches Anatomiques et Physiologiques sur la Structure intime des Animaux et des Végétaux et sur leur Motilité*. Paris. Cited in Sharp (1934).

Duve, C. de (1959). In *Subcellular Particles* (T. Hayashi, ed.), pp. 128–159. Ronald, New York.

Duve, de C. (1968). In *The Interaction of Drugs and Subcellular Components in Animal Cells* (P. N. Campbell, ed.), pp. 155–169. Churchill, London.

Elson, L. A. (1958). Dynamic aspects of chemical carcinogenesis. *Brit. Med. Bull.*, **14,** 161.

Engström, A. (1956). In *Physical Techniques in Biological Research* (G. Oster and A. W. Pollister, eds.), Vol. 3, pp. 489–544. Academic, New York.

Engström, A. (1959). In *Analytical Cytology* (R. C. Mellors, ed.), 2nd edition, pp. 343–379. McGraw Hill, New York.

Engström, A. (1962). *X-ray Microanalysis in Biology and Medicine*. Elsevier, Amsterdam.

Ericsson, J. L. E. (1969). Electron microscopy of renal tubule degeneration. *Methods Achievement Exptl. Pathol.*, **4,** 171.

Farber, E., and P. N. Magee (eds.) (1966). *Biochemical Pathology Lab. Invest.*, **15,** 1.

Farber, E., W. H. Sternberg, and C. E. Dunlap (1956). Histochemical localization of specific oxidative enzymes. III. Evaluation studies of tetrazolium staining methods for diphosphopyridine nucleotide diaphorase, triphosphopyridine nucleotide diaphorase and the succindehydrogenase system. *J. Histochem. Cytochem.*, **4,** 284.

Fell, H. B. (1969). Role of biological membranes in some skeletal reactions. *Ann. Rheumatic Diseases*, **28,** 213.

Fiala, S., and A. E. Fiala (1959). On the correlation between metabolic and structural changes during carcinogenesis in rat liver. *Brit. J. Cancer*, **13,** 136.

Filippusson, H., and W. E. Hornby (1970). The preparation and properties of yeast β-fructofuranosidase chemically attached to polystyrene. *Biochem. J.*, **120,** 215.

Fine, J. (1958a). The role of the intestine in traumatic shock. *Am. J. Gastroenterol.*, **29,** 596.

Fine, J. (1958b). *The Liver in Traumatic Shock*. American Institute of Biological Sciences Publ. No. 4, p. 585.

Fine, J., E. D. Frank, H. A. Ravin, S. H. Rutenberg, and F. B. Schweinburg (1959). The bacterial factor in traumatic shock. *New Engl. J. Med.*, **260,** 214.

Flemming, W. (1882). *Zellsubstanz, Kern und Zelltheilung*, Leipzig. Cited in Sharp (1934).

Foster, G. L. (1942). The effect of implanting tablets of synthetic oestrogens on the histology and cytology of the anterior pituitary of immature rats. *J. Endocrinol.*, **3,** 79.

Frazier, P. D., P. S. Liteman, and J. E. Seegmiller (1965). Identification of L-cystine crystals in bone marrow by x-ray diffraction. *J. Lab. Clin. Med.*, **65,** 108.

Frederic, J. (1958). Recherches cytologiques sur le chondriome normal ou soumis à l'expérimentation dans des cellules vivantes cultivées *in vitro*. *Arch. Biol.* (*Liège*), **69,** 167.

Frederic, J., and M. Chèvremont (1952). Recherches sur les chondriosomes de cellules vivantes par la microscopie et la microcinematographie en contraste de phase. *Arch. Biol.* (*Liège*), **63,** 109.

Gahan, P. B., and J. Chayen (1965). Cytoplasmic deoxyribonucleic acid. *Intern. Rev. Cytol.*, **18,** 223.

Gardner, D. L. (1960). The experimental production of arthritis, a review. *Ann. Rheumatic Diseases*, **19,** 297.

Gatenby, J. B. (1931a). The prozymogen granules ("vacuome") of R. R. Bensley in the *Pseudotriton* pancreas, and the modern neutral-red cytology. *Am. J. Anat.*, **48,** 421.

Gatenby, J. B. (1931b). *J. Exptl. Zool.*, **58,** 259; *J. Roy. Microscop. Soc.*, **51,** 221. Cited in Cameron (1952).

Gavrilescu, N., and R. A. Peters (1931). *Biochem. J.*, **25,** 1397. Cited in Peters (1969).

Gey, G. (1954/55). Some aspects of the constitution and behavior of normal and malignant cells maintained in continuous culture. *Harvey Lectures*, **50,** 154.

Ghadially, F. N., and S. Roy (1969). *Ultrastructure of Synovial Joints in Health and Disease*. Butterworths, London.

Gillman, J. (1934). *Anat. Record*, **60,** 209. Cited in Cameron (1952).

Gillman, J. (1935). *S. African J. Med. Sci.*, **1,** 97. Cited in Cameron (1952).

Glick, D. (1962–1963). *Quantitative Chemical Techniques of Histo- and Cytochemistry*. Vols. I and II. Wiley, New York.

Golgi, C. (1898). Sur la structure des cellules nerveuses. *Arch. Ital Biol.*, **30,** 60. Cited in Sharp (1934).

Gomori, G. (1952). *Microscopic Histochemistry*. Univ. of Chicago Press.

Grant, J. K. (1969). Action of steroid hormones at cellular and molecular levels. *Essays Biochem.*, **5,** 1.

Grasso, P., A. Muir, L. Golberg, and E. Batstone (1968). Studies on brown FK. IV. Cytopathic effects of brown FK on cardiac and skeletal muscle in the rat. *Food Cosmetic Toxicol.*, **6,** 13.

Green, D. E. (1958). Organization in relation to enzyme function. *Symp. Soc. Exptl. Biol.*, **10,** 30.

Guilliermond, A. (1941). *The Cytoplasm of the Plant Cell* (L. R. Atkinson, translator), Chronica Botanica, Waltham, Massachusetts.

Hackett, D. P. (1955). Recent studies on plant mitochondria. *Intern. Rev. Cytol.*, **4,** 143.

Harriman, D. G. F., and P. L. Stocks (1970). Changes in birefringence of human skeletal muscle during contraction and relaxation. *Beitr. Pathol. Anat. Allgem. Pathol.*, **141,** 187.

Hartshorne, N. H., and A. Stuart (1970). *Crystals and the Polarizing Microscope*, 4th edition. Arnold, London.

Heilbrunn, L. V. (1943). *An Outline of General Physiology*, 2nd edition. Saunders, Philadelphia.

Heilbrunn, L. V. (1956). *The Dynamics of Living Protoplasm*, Academic, New York.

Hems, R., M. Stubbs, and H. A. Krebs (1968). Restricted permeability of rat liver for glutamate and succinate. *Biochem. J.*, **107,** 807.

Hendy, R. J., R. Abraham, and P. Grasso (1969). The effect of chloroquine on rat heart lysosomes. *J. Ultrastruct. Res.*, **29,** 485.

Hers, H. G., and F. van Hoof (1969). In *Lysosomes in Biology and Pathology* (J. T. Dingle and H. B. Fell, eds.), Vol. 2, pp. 19–40. North Holland, Amsterdam.

Herwerden, M. A. van (1924). *Biol. Zbl.*, **44,** 579. Cited in Cameron (1952).

Herwerden, M. A. van (1925). *Arch. Exptl. Zellforsch.*, **1,** 145. Cited in Cameron (1952).

Hill, R. (1956). In *Modern Methods of Plant Analysis* (K. Paech and M. V. Tracey, eds.), pp. 393–415. Springer, Berlin.

Hodge, A. J., J. D. McLean, and F. V. Mercer (1956). A possible mechanism for the morphogenesis of lamellar systems in plant cells. *J. Biophys. Biochem. Cytol.*, **2,** 597.

Hodge, A. J., E. M. Martin, and R. K. Morton (1957). The structure of some cytoplasmic components of plant cells in relation to the biochemical properties of isolated particles. *J. Biophys. Biochem. Cytol.*, **3,** 61.

Höfler, K. (1949). In *Beiträge zur Fluoreszenzmikroskopie* (F. Bräutigam and A. Grabner, eds.), pp. 46–70. Fromme, Vienna.

Holland, J. F. (1961). Symposium: Antimetabolites. VIII. Folic acid antagonists. *Clin. Pharmacol. Therap.*, **2,** 374.

Holter, H. (1959). Pinocytosis. *Intern. Rev. Cytol.*, **8,** 481.

Holter, H. (1961a). In *Biological Structure and Function* (T. W. Goodwin and O. Lindberg, eds.), Vol. 1, pp. 157–168.

Holter, H. (1961b). In *Biological Approaches to Cancer Chemotherapy* (R. J. C. Harris, ed.), pp. 77–88. Academic, New York.

Holter, H. (1965a). In *Intracellular Membranous Structure* (S. Seno and E. V. Cowdry, eds.), pp. 451–465. Japan Society for Cell Biology, Okayama.

Holter, H. (1965b). Passage of particles and macromolecules through cell membranes. *Symp. Soc. Gen. Microbiol.*, **15,** 89.

Hooke, R. (1665). *Micrographia*. London. Cited in Sharp (1934).

Hornby, W. E., M. D. Lilly, and E. M. Crook (1966). The preparation and properties of ficin chemically attached to carboxymethylcellulose. *Biochem. J.*, **98,** 420.

Horning, E. S. (1951). In *Cytology and Cell Physiology* (G. H. Bourne, ed.), 2nd edition, pp. 287–321. Oxford Univ. Press, Oxford.

Howard, A., and S. R. Pelc (1951). Nuclear incorporation of P^{32} as demonstrated by autoradiographs. *Exptl. Cell Res.*, **2,** 178.

Howard, A., and S. R. Pelc (1953). Synthesis of deoxyribonucleic acid in normal and irradiated cells and its relation to chromosome breakage. *Heredity*, **6,** 261.

Hughes, A. (1952). *The Mitotic Cycle*. Butterworths, London.

Huxley, H. E. (1960). In *The Cell* (J. Brachet and A. E. Mirsky, eds.), Vol. IV, pp. 365–481. Academic, New York.

Huxley, T. H. (1868). *The Physical Basis of Life*. Cited in Sharp (1934).

Inoué, S. (1953). Polarization optical studies of the mitotic spindle. I. The demonstration of spindle fibres in living cells. *Chromosoma*, **5,** 487.

Inoué, S. (1959). Motility of cilia and the mechanism of mitosis. *Rev. Mod. Phys.*, **31,** 402.

Inoué, S., and A. Bajer (1961). Birefringence in endosperm mitosis. *Chromosoma*, **12,** 48.

Inoué, S., and K. Dan (1951). Birefringence of the dividing cell. *J. Morphol.*, **89,** 423.

Jacobson, W. (1964). In *Cellular Injury* (A.V.S. de Reuck and J. Knight, eds.), pp. 136–161. Churchill, London.

Jacobson, W., and M. Webb (1952a). The two types of nucleoproteins during mitosis. *Exptl. Cell Res.*, **3,** 163.

Jacobson, W., and M. Webb (1952b). Nucleoproteins and cell division. *Endeavour*, **9**(44).

Jardetzky, C. D., and D. Glick (1956). Studies in histochemistry. 38. Determination of succinic dehydrogenase in microgram amounts of tissue and its distribution in rat adrenal. *J. Biol. Chem.*, **216,** 283.

Jones, A. L., and D. W. Fawcett (1966). Hypertrophy of the agranular endoplasmic reticulum in hamster liver induced by phenobarbital (with a review on the functions of this organelle in liver). *J. Histochem. Cytochem.*, **14,** 215.

Jones, G. R. N. (1963). Succinic dehydrogenase levels in livers of rats during early feeding of 4-dimethylaminoazobenzene: A reinterpretation of the biochemical data. *Brit. J. Cancer*, **17,** 153.

Jones, G. R. N. (1965). Quantitative histochemistry: Levels of certain oxidative enzymes in secondary neoplasia found in a rat after feeding 4-dimethylaminoazobenzene. *Brit. J. Cancer*, **19,** 360.

Jones, G. R. N., L. Bitensky, J. Chayen, and G. J. Cunningham (1961). Tissue dilution artefact: A re-interpretation of variations in levels of succinic dehydrogenase during chemical carcinogenesis. *Nature*, **191,** 1203.

Jones, G. R. N., A. J. Maple, E. K. Aves, J. Chayen, and G. J. Cunningham (1963). Quantitative histochemistry of succinate dehydrogenase in tissue sections. *Nature*, **197,** 568.

Judah, J. D. (1969). Biochemical disturbances in liver injury. *Brit. Med. Bull.*, **25,** 274.

Judah, J. D., K. Ahmed, and A. E. M. McLean (1964). In *Cellular Injury* (A. V. D. de Reuck and J. Knight, eds.), pp 187–205. Churchill, London.

Kakari, S. (1970). Observations on the use of nitro blue tetrazolium in the detection of early myocardial changes. *Histochem. J.*, **2,** 453.

Kamiya, N. (1959). *Protoplasmic Streaming*. Protoplasmatologia. Springer, Vienna.

Kaufmann, B. P., and N. de Deepesh (1956). Fine structure of chromosomes. *J. Biophys. Biochem. Cytol., Suppl.*, **2,** 419.

Kautz, J., and Q. B. De Marsh (1955). Fine structure of the nuclear membrane in cells from the chick embryo: On the nature of the so-called "pores" in the nuclear membrane. *Exptl. Cell Res.*, **8,** 394.

Kay, D. (ed) (1965). *Techniques for Electron Microscopy*. Blackwells, Oxford.

Kloster, F. E., J. D. Bristow, and H. E. Griswold (1965). Medical problems in mitral and multiple valve replacement. *Prog. Cardiovascular Diseases*, **7,** 504.

Krebs, H. A., and T. Gascoyne (1968). The redox state of the nicotinamide-adenine dinucleotides in rat liver homogenates. *Biochem. J.*, **108,** 513.

Kurnick, N. B. (1950). The quantitative estimation of desoxyribonucleic acid based on methyl green staining. *Exptl. Cell Res.*, **1,** 151.

La Cour, L. F. (1966). The internal structure of nucleoli. *Chromosomes Today*, **1,** 150.

La Cour, L. F., and J. Chayen (1958). A cyclic staining behaviour of the chromosomes during mitosis and meiosis. *Exptl. Cell Res.*, **14,** 462.

La Cour, L. F., E. M. Deeley, and J. Chayen (1956). Variations in the amount of Feulgen stain in nuclei of plants grown at different temperatures. *Nature*, **177,** 272.

La Cour, L. F., J. Chayen, and P. B. Gahan (1958). Evidence for lipid material in chromosomes. *Exptl. Cell Res.*, **14,** 469.

La Cour, L. F., and B. Wells (1967). The loops and ultrastructure of the nucleolus of *Ipheion uniflorum. Z. Zellforsch.*, **82,** 25.

Lacy, D., and C. E. Challice (1957). The structure of the Golgi apparatus in vertebrate cells examined by light and electron microscopy. *Symp. Soc. Exptl. Biol.*, **10,** 62.

Lamarck, J. B. P. (1809). *Philosophie Zoologique*. Paris. Cited in Sharp (1934).

Lane, N. J. (1968). In *Cell Structure and Its Interpretation* (S. M. McGee-Russell and K. F. A. Ross, eds.), pp. 169–182. Arnold, London.

Lannigan, R., A. Barabas, K. Peters, and J. Hardwicke (1969). Some experimental models of the nephrotic syndrome. *Methods Achievement Exptl. Pathol.*, **4,** 141.

Lazarow, A., and S. J. Cooperstein (1953a). Studies on the mechanism of Janus green B staining of mitochondria. I. Review of the literature. *Exptl. Cell Res.*, **5,** 56.

Lazarow, A., and S. J. Cooperstein (1953b). Studies on the enzymatic basis for the Janus green B staining reaction. *J. Histochem. Cytochem.*, **1,** 234.

Lehninger, A. L. (1964). *The Mitochondrion.* Benjamin, New York.

Lehninger, A. L. (1966). Supramolecular organization of enzyme and membrane systems. *Naturwissenschaften*, **53,** 57.

Lehninger, A. L. (1970). Mitochondria and calcium ion transport. *Biochem. J.*, **119,** 129.

Levine, R., and R. Luft (eds) (1964). *Advances in Metabolic Disorders*. Academic, New York.

Lewis, M. R. (1923). Reversible gelation in living cells. *Bull. Johns Hopkins. Hosp.*, **34,** 373.

Lewis, W. H. (1923). *Anat. Record*, **26,** 15, 97. Cited in Cameron (1952).

Lewis, W. H. (1931). Pinocytosis. *Bull. Johns Hopkins Hosp.*, **49,** 17.

Lewis, W. H., and M. R. Lewis (1924). In *General Cytology* (E. V. Cowdry, ed.), pp. 385–426. University Press, Chicago.

Leuchtenberger, C. (1950). A cytochemical study of pycnotic nuclear degeneration. *Chromosoma*, **3,** 449.

Leuchtenberger, C. (1960). The relation of the deoxyribose nucleic acid (DNA) of sperm cells to fertility. *J. Dairy Sci.*, **43,** 31.

Leuchtenberger, C., and R. Leuchtenberger (1960). Quantitative cytochemical studies on the relation of deoxyribonucleic acid of cells to various pathological conditions. *Biochem. Pharmacol.*, **4,** 128.

Leuchtenberger, C., D. R. Weir, F. Schrader, and L. Murmanis (1955). The desoxyribose nucleic acid (DNA) content in spermatozoa of repeated seminal fluids from fertile and infertile men. *J. Lab. Clin. Med.*, **45,** 851.

Leuchtenberger, C., R. Leuchtenberger, F. Schrader, and D. R. Weir (1956). Reduced amounts of desoxyribose nucleic acid in testicular germ cells of infertile men with active spermatogenesis. *Lab. Invest.*, **5,** 422.

Leuchtenberger, C., G. S. Boyer, and J. J. Strain (1959). Quantitative cytochemical investigations on the effect of virus on cells. *Ann. N.Y. Acad. Sci.*, **81,** 73.

Lima-de-Faria, A. (1958). Recent advances in the study of the kinetochore. *Intern. Rev. Cytol.*, **7,** 123.

Lima-de-Faria, A. (1959). Matrix and kinetochore in living material. *Hereditas*, **45,** 463.

Lima-de-Faria, A. (1962). Metabolic DNA in *Tipula oleracea. Chromosoma*, **13,** 47.

Littlefield, J. W., E. B. Keller, J. Gross, and P. C. Zamecnik (1955). Studies on cytoplasmic ribonucleoprotein particles from the liver of the rat. *J. Biol. Chem.*, **217,** 111.

Lowry, O. H. (1963). The chemical study of single neurons. *Harvey Lectures*, **58,** 1.

Lowry, O. H. (1964). *Microanalysis for Histochemical Purposes*. Proc. 2nd Intern. Congr. Histochem. Cytochem., pp. 62–69. Springer, Berlin.

Ludford, R. J. (1927). *Proc. Roy. Soc. (London)*, **B102,** 406.

Ludford, R. J. (1951). In *Cytology and Cell Physiology* (G. H. Bourne, ed.), 2nd edition, pp. 373–418. Oxford Univ. Press, Oxford.

Ludford, R. J., and J. Smiles (1954). Ultra-violet microscopy of living malignant cells. I. The interphase nucleus. *J. Roy. Microscop. Soc.*, **73,** 173.

Lynch, R., L. Bitensky, and J. Chayen (1965). On the possibility of supercooling in tissues. *J. Roy. Microscop. Soc.*, **85,** 213.

McClintock, B. (1934). The relation of a particular chromosomal element to the development of the nucleoli in *Zea mays*. *Z. Zellforsch.*, **21,** 294.

McLean, A. E. M., and E. K. McLean (1966). The effect of diet and 1,1,1-trichloro-2,2-*bis*-(*p*-chlorophenyl) ethane (DDT) on microsomal hydroxylating enzymes and on the sensitivity of rats to carbon tetrachloride poisoning. *Biochem. J.*, **100,** 564.

McLean, A. E. M., E. K. McLean, and J. D. Judah (1965). Cellular necrosis in the liver induced and modified by drugs. *Intern. Rev. Exptl. Pathol.*, **4,** 127.

McLeish, J. (1953). The action of maleic hydrazide in *Vicia*. *Hereditas, Suppl.*, **6,** 125.

McManus, J. F. A. (1944). The Golgi element in the cells of the first and second convoluted tubules of the cat kidney. *Quart. J. Microscop. Sci.*, **85,** 97.

Magee, P. N. (1964). In *Cellular Injury* (A. V. S. de Reuck and J. Knight, eds.), pp. 1–22. Churchill, London.

Magee, P. N. (1966). Naturally occurring alkylating agents. *Proc. Roy. Soc. Med.*, **59,** 751.

Magee, P. N., and E. Farber (1962). Toxic liver injury and carcinogenesis. Methylation of rat-liver nucleic acids by dimethylnitrosamine *in vivo*. *Biochem. J.*, **83,** 114.

Magee, P. N., and T. Hultin (1962). Toxic liver injury and carcinogenesis. Methylation of proteins of rat-liver slices by dimethylnitrosamine *in vitro*. *Biochem. J.*, **83,** 106.

Magee, P. N., and P. F. Swann (1969). Nitroso compounds. *Brit. Med. Bull.*, **25,** 240.

Mahler, H. R. (1953). Multienzyme sequences in soluble extracts. *Intern. Rev. Cytol.*, **2,** 201.

Makino, S. (1957). The chromosome cytology of the ascites tumours of rats, with special reference to the concept of the stem line cell. *Intern. Rev. Cytol.*, **6,** 25.

Marinozzi, V., and W. Bernhard (1963). Présence dans le nucléole de deux types de ribonucleoprotéines morphologiquement distinctes. *Exptl. Cell Res.*, **32,** 595.

Mather, K. (1949). *Biometrical Genetics*. Methuen, London.

Mazia, D. (1961). In *The Cell* (J. Brachet and A. E. Mirsky, eds.), Vol. III, pp. 77–412. Academic, New York.

Michaelis, P. (1959). The genetical interactions between nucleus and cytoplasm in *Epilobium*. *Exptl. Cell Res., Suppl.*, **6,** 236.

Mirsky, A. E., and S. Osawa (1961). In *The Cell* (J. Brachet and A. E. Mirsky, eds.), pp. 667–770. Academic, New York.

Mitchison, J. M. (1966). Some functions of the nucleus. *Intern. Rev. Cytol.*, **19,** 97.

Moldenhawer, J. J. P. (1812). *Beiträge zur Anatomie der Pflanzen*. Cited in Sharp (1934).

Müller, D., W. Sandritter, H. G. Schiemer, and K. Endres (1959a). Röntgenhistoradiographische und Interferenzmikroskopische Trockengewichtuntersuchungen an Zellaustrichen. *Histochemie*, **1,** 438.

Müller, D., W. Sandritter, and G. Schwaiger (1959b). Eine Methode zurRöntgenhistoradiographischen Trockengewichtsbestimmung ohne Verwendung eines Referenzsystems. *Histochemie*, **1,** 420.

Mueller, G. C. (1970). Oestrogen action: A study of the influence of steroid hormones on genetic expression. *Biochem. J.*, **118,** 1P. (Also see Biochem. Soc. Symp. No. 32; 1971).

Munro, T. R., M. R. Daniel, and J. T. Dingle (1964). Lysosomes in Chinese hamster fibroblasts in culture. *Exptl. Cell Res.*, **35,** 515.

Naora, H. (1957). Microspectrophotometry of cell nuclei stained with the Feulgen reaction. IV. Formation of tetraploid nuclei in rat liver cells during postnatal growth. *J. Biophys. Biochem. Cytol.*, **3,** 949.

Niles, N. R., and D. L. Barnhouse (1967). The acid haematein stain and myocardial damage. *Arch. Pathol.*, **83,** 407.

Niles, N. R., L. Bitensky, J. Chayen, G. J. Cunningham, and M. V. Braimbridge (1964). The value of histochemistry in the analysis of myocardial dysfunction. *Lancet*, **i,** 963.

Niles, N. R., L. Bitensky, M. V. Braimbridge, and J. Chayen (1966). Histochemical changes related to oxidation and phosphorylation in human heart muscle. *J. Roy. Microscop. Soc.*, **86,** 159.

Niles, N. R., J. D. Zavin, and R. N. Morikado (1968). Histochemical study of effects of hypoxia and isoproterenol on rat myocardium. *Am. J. Cardiol.*, **22,** 381.

Novikoff, A. B. (1961a). In *The Cell* (J. Brachet and A. E. Mirsky, eds.), Vol. II, pp. 299–421. Academic, New York.

Novikoff, A. B. (1961b). In *The Cell* (J. Brachet and A. E. Mirsky, eds.), Vol. II, pp. 424–488. Academic, New York.

Novikoff, A. B. (1963a). In *Lysosomes* (A. V. S. de Reuck and M. P. Cameron, eds.), pp. 36–73. Churchill, London.

Novikoff, A. B. (1963b). In *Histochemistry and Cytochemistry* (R. Wegmann, ed.), pp. 465–481. Pergamon, Oxford.

Orrenius, S., J. L. E. Ericsson, and L. Ernster (1965). Phenobarbital-induced synthesis of the microsomal drug-metabolizing enzyme system and its relationship to the proliferation of endoplasmic membranes. *J. Cell Biol.*, **25,** 627.

Östergren, G. (1950a). Considerations on some elementary features of mitosis. *Hereditas*, **37,** 1.

Östergren, G. (1950b). Cytological standards for the quantitative estimation of spindle disturbances. *Hereditas*, **37,** 371.

Östergren, G. (1951a). The mechanism of co-orientation in bivalents and multivalents. The theory of orientation by pulling. *Hereditas*, **37,** 85.

Östergren, G. (1951b). In *Mécanisme de la Narcose*. Coll. Int. Centr. Nat. Rech. Scient., Vol. 26, p. 77.

Östergren, G., J. Molé-Bajer, and A. Bajer (1960). An interpretation of transport phenomena at mitosis. *Ann. N.Y. Acad. Sci.*, **90,** 381.

Palade, G. E. (1956). The endoplasmic reticulum. *J. Biophys. Biochem. Cytol., Suppl.*, **2,** 85.

Palade, G. E., and P. Siekevitz (1956). Liver microsomes. An integrated morphological and biochemical study. *J. Biophys. Biochem. Cytol.*, **2,** 171.

Palmerio, C., and J. Fine (1969). The nature of resistance to shock. *Arch. Surg.*, **98,** 679.

Parke, D. V. (1968). *The Biochemistry of Foreign Compounds*. Pergamon, Oxford.

Parsons, D. F. (1965). Recent advances correlating structure and function in mitochondria. *Intern. Rev. Exptl. Pathol.*, **4,** 1.

Paul, J. (1967). In *Cell Differentiation* (A. V. S. de Reuck and J. Knight, eds.), pp. 196–207. Churchill, London.

Paul, J., R. S. Gilmour, H. Thomou, G. Threlfall, and D. Kohl (1971). Organ-specific gene masking in mammalian chromosomes. *Proc. Roy. Soc.* (*London*), **B176,** 277.

Pearse, A. G. E. (1960). *Histochemistry: Theoretical and Applied*, 2nd edition. Churchill, London.

Pease, D. (1964). *Histological Techniques for Electron Microscopy*, 2nd edition. Academic, New York.

Pelc, S. R. (1959). Metabolic activity of DNA as shown by autoradiography. *Lab. Invest.*, **8,** 225.

Pelc, S. R. (1970). Metabolic DNA and the problem of ageing. *Exptl. Gerontol.*, **5,** 217.

Pelc, S. R. (1972). Metabolic DNA in ciliated protozoa, salivary gland chromosomes and mammalian cells. *Intern Rev. Cytol.*, in press.

Pelc, S. R., and M. Viola-Magni (1969). Decrease of labelled DNA in cells of the adrenal medulla after intermittent exposure to cold. *J. Cell Biol.*, **42,** 460.

Peters, R. A. (1963). *Biochemical Lesions and Lethal Synthesis*. Macmillan, New York.

Peters, R. A. (1969). The biochemical lesion and its historical development. *Brit. Med. Bull.*, **25,** 223.

Pinto, R. E., and W. Bartley (1969). The nature of the sex linked differences in glutathione peroxidase activity and aerobic oxidation of glutathione in male and female rat liver. *Biochem. J.*, **115,** 449.

Pomerat, C. M. (1961). Cinematography, indispensable tool for cytology. *Intern. Rev. Cytol.*, **11,** 307.

Porter, K. R. (1961). In *The Cell* (J. Brachet and A. E. Mirsky, eds.), Vol. II, pp. 621–676. Academic, New York.

Poulter, L. W., L. Bitensky, B. Cashman, and J. Chayen (1970). The maintenance of human synovial tissue *in vitro*. *Virchows Arch.*, **B4,** 303.

Prescott, D. M. (1961). The growth-duplication cycle of the cell. *Intern. Rev. Cytol.*, **11,** 255.

Preston, R. D. (1951). *Molecular Architecture in Plants*. Chapman and Hall, London.

Quarles, R. H., and R. M. C. Dawson (1969). A shift in the optimum pH of phospholipase D produced by activating long-chain anions. *Biochem. J.*, **112,** 795.

Redman, C. M., P. Siekevitz, and G. E. Palade (1966). Synthesis and transfer of amylase in pigeon pancreatic microsomes. *J. Biol. Chem.*, **241,** 1150.

Rees, K. R. (1966). The mechanism of action of aflatoxin in producing liver necrosis. *Proc. Roy. Soc. Med.*, **59,** 755.

Rees, K. R. (1968a). In *The Biological Basis of Medicine* (E. E. Bittar and N. Bittar, eds.), Vol. II, pp. 447–467.

Rees, K. R. (1968b). In *The Interaction of Drugs and Subcellular Components in Animal Cells* (P. N. Campbell, ed.), Churchill, London.

Rees, K. R., and M. J. Tarlow (1967). The hepatotoxin action of allyl formate. *Biochem. J.*, **104,** 757.

Reich, E., and I. H. Goldberg (1964). Actinomycin and nucleic acid function. *Prog. Nucleic Acid Res. Mol. Biol.*, **13,** 184.

Remmer, H. (1965). The fate of drugs in the organism. *Ann. Rev. Pharmacol.*, **5,** 405.

Remmer, H., and H. J. Merker (1963). Drug-induced changes in the liver endoplasmic reticulum: Association with drug-metabolizing enzymes. *Science*, **142,** 1657.

Rhoades, M. M. (1946). Plastid mutations. *Cold Spring Harbor Symp. Quant. Biol.*, **11,** 202.

Richards, B. M. (1960). In *The Cell Nucleus* (J. S. Mitchell, ed.), pp. 138–140. Butterworths, London.

Richards, B. M., P. M. B. Walker, and E. M. Deeley (1956). Changes in nuclear DNA in normal and ascites tumour cells. *Ann. N.Y. Acad. Sci.*, **63,** 831.

Ris, H., and A. E. Mirsky (1949). The state of the chromosomes in the interphase nucleus. *J. Gen. Physiol.*, **32,** 489.

Robertis, E. D. P. de, W. W. Nowinski, and M. A. Saez (1948). *General Cytology*, 1st edition, pp. 48 and 85. Saunders, Philadelphia.

Roels, H. (1966). Metabolic DNA: A cytochemical study. *Intern. Rev. Cytol.*, **19,** 1.

Roodyn, D. B. (1968). In *The Biological Basis of Medicine* (E. E. Bittar and N. Bittar, eds.), Vol. 1, pp. 123-178. Academic, New York.

Ross, K. F. A. (1967). *Phase Contrast and Interference Microscopy for Cell Biologists*. Arnold, London.

Rutishauser, A., and L. F. La Cour (1956). Spontaneous chromosome breakage in hybrid endosperm. *Chromosoma*, **8,** 317.

Sampson, M., and D. D. Davies (1966a). Metabolically labile DNA in mitotic and non-mitotic cells of *Zea mays*. *Life Sci.*, **5,** 1239.

Sampson, M., and D. D. Davies (1966b). Synthesis of a metabolically labile DNA in the maturing root cells of *Vicia faba*. *Exptl. Cell Res.*, **43,** 669.

Sampson, M., A. Katch, Y. Hotta, and H. Stern (1963). Metabolically labile deoxyribonucleic acid *Proc. Natl. Acad. Sci., U.S.*, **50,** 459.

Sandritter, W., M. Carl, and W. Ritter (1966). Cytophotometric measurements of the DNA content of human malignant tumours by means of the Feulgen reaction. *Acta Cytol.*, **10,** 26.

Sandritter, W., G. Kiefer, and W. Rick (1963). Über die Stöchiometrie von Gallocyaninchromalaun mit Desoxyribonukleinsäure. *Histochemie*, **3,** 315.

Sandritter, W., and D. Kleinhans (1964). Über das Trockengewicht, den DNS- und Histonproteingehalt von menslichen Tumoren. *Z. Krebsforsch.*, **66,** 333.

Sandritter, W., and D. Müller (1959). Vergleichende röntgenhistoradiographische und interferenzmikroskopische Trockengewichtsbestimmungen. *Experientia*, **15,** 158.

Sandritter, W., D. Müller, H. Schäfer, and H. G. Schiemer (1958). Histochemie von Sputumzellen.

II. Quantitative ultraviolett mikrospektrophotometrische Nucleinsäurebestimmungen. *Frankfurt Z. Pathol.*, **68,** 710.

Sandritter, W., H. G. Schiemer, H. Kraus, and U. Dörrien (1960). Interferenzmikroskopische Untersuchungen über das Wachstum von Einzelzellen (HeLa Zellen) in der Gewebekultur. *Frankfurt Z. Pathol.*, **70,** 271.

Schleiden, M. J. (1838). Beitrage zur phytogenesis. *Arch. Anat. Physiol. Wiss. Med.*, 137. Cited in Sharp (1934).

Schrader, F. (1944). *Mitosis*. Columbia Univ. Press, New York.

Schultz, J. (1941). The evidence for the nucleoprotein nature of the gene. *Cold Spring Harbor Symp. Quant. Biol.*, **9,** 55.

Schwann, Th. (1839). *Mikroskopische Untersuchungen über die Uebereinstimmung in der Struktur und dem Wachsthum der Thieren und Pflanzen*. Berlin. Cited in Sharp (1934).

Seligman, A. M. (1963). In *Histochemistry and Cytochemistry* (R. Wegmann, ed.), pp. 203–212. Pergamon, Oxford.

Serra, J. A. (1949). A cyto-physiological theory of the gene, gene mutation and position effect. *Port. Acta Biol.*, **A,** R. B. Goldschmidt Volume, 401.

Shafiq, S. A., and M. Gorycki (1968). In *Cell Structure and Its Interpretation* (S. M. McGee-Russell and K. F. A. Ross, eds.), pp. 361–370. Arnold, London.

Sharp, L. W. (1934). *Introduction to Cytology*. McGraw Hill, New York.

Sibatani, A. (1966). Genetic transcription or DNA-dependent RNA synthesis. *Progr. Biophys.*, **16,** 15.

Siekevitz, P. (1962). In *The Molecular Control of Cellular Activity* (J. M. Allen, ed.), pp. 143–166. McGraw Hill, New York.

Simard, R. (1970). The nucleus: Action of chemical and physical agents. *Intern. Rev. Cytol.*, **28,** 169.

Simard, R., and W. Bernhard (1966). Le phénomène de la ségrégation nucléolaire: Spécificité d'action de certain antimétabolites. *Intern. J. Cancer.*, **1,** 463.

Sinsheimer, R. L. (1957). First steps towards a genetic chemistry. *Science*, **125,** 1123.

Sjöstrand, F. S. (1967). *Electron Microscopy of Cells and Tissues*, Academic, New York.

Slater, T. F. (1968). In *The Biological Basis of Medicine* (E. E. Bittar and N. Bittar, eds.), Vol. 1, pp. 369–414. Academic, New York.

Slater, T. F. (1969). Free radical mechanisms in tissue injury. *Methods Achievement Exptl. Pathol.*, **4,** 30.

Slater, T. F., and P. A. Riley (1966). Photosensitization and lysosomal damage. *Nature*, **209,** 151.

Slater, T. F., A. L. Greenbaum, and D. Y. Wang (1963). In *Lysosomes* (A. V. S. de Reuck and M. P. Cameron, eds.), pp. 311–334. Churchill, London.

Small, J. (1954). *Modern Aspects of pH*. Bailliere, Tindall and Cox, London.

Smith, J. T., A. J. Funckes, A. J. Barak, and L. E. Thomas (1957). Cellular lipoproteins. I. The insoluble lipoproteins of whole liver nuclei. *Exptl. Cell Res.*, **13,** 96.

Smuckler, E. A., and M. Arcasoy (1969). Structural and functional changes of the endoplasmic reticulum of hepatic parenchymal cells. *Intern. Rev. Exptl. Pathol.*, **7,** 305.

Spear, G. S., R. J. Sluffer, A. J. Tousinis, C. G. Taylor, and J. D. Shulman (1971). Cystinosis. *Arch. Pathol.*, **91,** 206.

Spriggs, A. I. (1970). In *Cytology Automation* (D. M. D. Evans, ed.), pp. 24–27. Livingstone, Edinburgh.

Stanbury, J. D., J. B. Wyngaarden, and D. S. Fredrickson (eds.) (1966). *The Metabolic Basis of Inherited Disease*. McGraw Hill, New York.

Sternberg, W. H., E. Farber, and C. E. Dunlap (1956). Histochemical localization of specific oxidative enzymes. II. Localization of diphosphopyridine nucleotide and triphosphopyridine nucleotide diaphorases and the succindehydrogenase system in the kidney. *J. Histochem. Cytochem.*, **4,** 266.

Strangeways, T. S. P. (1924). *Tissue Culture in Relation to Growth and Differentiation*. Cambridge Univ. Press, Cambridge.

Strangeways, T. S. P., and R. G. Canti (1927). The living cell *in vitro* as shown by dark-ground illumination and the changes induced in such cells by fixing reagents. *Quart. J. Microscop. Sci.*, **71,** 1.

Street, H. E. (1965). In *Cells and Tissues in Culture* (E. N. Willmer, ed.), Vol. 3, pp. 459–532. Academic, New York.

Strugger, S. (1938). Die Vitälfarbung des Protoplasmas mit Rhodamin B und G. *Protoplasma*, **30,** 85.

Stuart, J., L. Bitensky, and J. Chayen (1969). Quantitative enzyme cytochemistry of leukaemic cells. *J. Clin. Pathol.*, **22,** 563.

Stuart, J., and J. S. Simpson (1970). Dehydrogenase enzyme cytochemistry of unfixed leucocytes. *J. Clin. Pathol.*, **23,** 517.

Swann, M. M. (1951a). Protoplasmic structure and mitosis. I. The birefringence of the metaphase spindle and asters of the living sea-urchin egg. *J. Exptl. Biol.*, **28,** 417.

Swann, M. M. (1951b). Protoplasmic structure and mitosis. II. The nature and cause of birefringence changes in the sea-urchin egg at anaphase. *J. Exptl. Biol.*, **28,** 434.

Swann, M. M. (1957). The control of cell division: A review. I. General mechanisms. *Cancer Res.*, **17,** 727.

Swift, H. (1953). Quantitative aspects of nuclear nucleoproteins. *Intern. Rev. Cytol.*, **2,** 1.

Theron, J. J. (1965). Acute liver injury in ducklings as a result of aflatoxin poisoning. *Lab. Invest.*, **14,** 1586.

Thomas, A. (ed.) (1970). *Organ Culture*. Academic, New York.

Thomas, R. (1960). In *The Cell* (J. Brachet and A. E. Mirsky, eds.), Vol. 4, pp. 1–44. Academic, New York.

Thompson, R. P. H., G. M. Stathers, C. W. T. Pilcher, A. E. M. McLean, J. Robinson, and R. Williams (1969). Treatment of unconjugated jaundice by dicophane. *Lancet*, **ii,** 4.

Trowell, O. A. (1959). The culture of mature organs in a synthetic medium. *Exptl. Cell Res.*, **16,** 118.

Trump, B. F., and J. L. E. Ericsson (1965). In *The Inflammatory Process* (B. W. Zweifach, L. H. Grant, and R. T. McLuskey, eds.), pp. 35–120. Academic, New York.

Trump, B. F., and F. L. Ginn (1969). The pathogenesis of subcellular reaction to lethal injury. *Methods of Achievement Exptl. Pathol.*, **4,** 1.

Vendrely, R., M. Alfert, H. Matsudaira, and A. Knobloch (1958). The composition of nucleohistone from pycnotic nuclei. *Exptl. Cell Res.*, **14,** 295.

Vincent, W. S. (1955). Structure and chemistry of nucleoli. *Intern. Rev. Cytol.*, **4,** 269.

Viola-Magni, M. (1966). An analysis of DNA loss and synthesis in the rat adrenal medulla nuclei upon cold stimulation. *J. Cell Biol.*, **30,** 213.

Virchow, R. (1858). *Die Cellularpathologie*, Berlin. Cited in Sharp (1934).

Wada, B. (1950). The mechanism of mitosis based on studies of the submicroscopic structure and of the living state of the *Tradescantia* cell. *Cytologia* (*Tokyo*), **16,** 1.

Walker, P. M. B., and H. G. Davies (1950). Ultra-violet microspectrography of living tissue culture cells. II. Microspectrographic studies of living cells and ultraviolet-irradiated chick fibroblasts. *Discussions Faraday Soc.*, **9,** 461.

Walker, P. M. B., and H. B. Yates (1952a). Nuclear components of dividing cells. *Proc. Roy. Soc.* (*London*), **B140,** 274.

Walker, P. M. B., and H. B. Yates (1952b). Ultra-violet absorption of living cell nuclei during growth and division. *Symp. Soc. Exptl. Biol.*, **6,** 265.

Wang, T. Y., D. T. Mayer, and L. E. Thomas (1953). A lipoprotein of rat liver nuclei. *Exptl. Cell Res.*, **4,** 102.

Watts, R. W. E., J. T. Scott, R. A. Chalmers, L. Bitensky, and J. Chayen (1971). Microscopic studies on skeletal muscle in gout patients treated with allopurinol. *Quart. J. Med.*, **40,** 1.

Weissmann, G. (1966). Lysosomes and joint disease. *Arthritis Rheumat.*, **9,** 830.

Weissmann, G. (1967). In *Rheumatology* (J. Rotstein, ed.), Vol. 1, pp. 1–28. Karger, Basel.

Weissmann, G. (1969). In *Lysosomes in Biology and Pathology* (J. T. Dingle and H. B. Fell, eds.), Vol. 1, pp. 276–295. North Holland, Amsterdam.

Weissmann, G., and I. Spilberg (1968). Breakdown of cartilage protein polysaccharide by lysosomes. *Arthritis Rheumat.*, **9**, 162.

Weissmann, G., I. Spilberg, and K. Krakauer (1969). Arthritis induced in rabbits by lysates of granulocyte lysosomes. *Arthritis Rheumat.*, **12,** 103.

Wettstein, D. von (1959). In *Developmental Cytology* (D. Rudnick, ed.), pp. 123–160. Ronald, New York.

Whisson, M. E. (1967). In *Cell Differentiation* (A. V. S. de Reuck and J. Knight, eds.), p. 244. Churchill, London.

White, M. J. D. (1948). *Animal Cytology and Evolution*. Cambridge Univ. Press, Cambridge.

Williamson, A. R. (1969). The biosynthesis of multichain proteins. *Essays Biochem.*, **5,** 139.

Williamson, D. H., P. Lund, and H. A. Krebs (1967). The redox state of free nicotinamide-adenine dinucleotide in the cytoplasm and mitochondria of rat liver. *Biochem. J.*, **103,** 514.

Williamson, J. R., E. T. Browning, R. Scholz, R. A. Kreisberg, and I. B. Fritz (1968). Inhibition of fatty acid stimulation of gluconeogenesis by (+)-decanoyl-carnitine in perfused rat liver. *Diabetes*, **17,** 194.

Willmer, E. N. (ed.) (1965). *Cells and Tissue in Culture: Methods, Biology, and Physiology*, Vols. 1 and 2. Academic, New York.

Wilson, E. B. (1925). *The Cell in Development and Inheritance*, 3rd edition. Macmillan, New York.

Wolff, S. (1969). Strandedness of chromosomes. *Intern. Rev. Cytol.*, **25,** 279.

Wolman, M. (1955). Problems of fixation in cytology, histology and histochemistry. *Intern. Rev. Cytol.*, **4,** 79.

Zaimis, E., L. Papadaki, A. S. F. Ash, E. Larbi, S. Kakari, M. Matthew, and A. Paradelis (1969). Cardiovascular effects of thyroxine. *Cardiovascular Res.*, **3,** 118.

Zech, L. (1969). Investigation of metaphase chromosomes with DNA-binding fluorochromes. *Exptl. Cell Res.*, **58,** 463.

Zirkle, C. (1928). Nucleolus in root tip mitosis in *Zea mays*. *Botan. Gaz.*, **86,** 402.

PART 6

CELLS AND AGING

19

CELLS AND AGING

J. Maynard Smith

1. INTRODUCTION

Aging processes may be defined as those processes responsible for an increase in the force of mortality with age; that is, processes that make an individual more likely to die in unit time as it grows older. In contrast, a population whose members died exponentially, as the atoms in a radioactive substance decay, shows no evidence of aging processes. So defined, "aging" and "senescence" are synonymous. All vertebrates that have been studied, and many invertebrates, age in this sense; no metazoa are known not to age. Among plants, some fungi age, and this is discussed in Section 4.B in this Chapter. It is not clear whether or not higher plants should be regarded as showing aging processes; if they do, then the processes are organismic rather than cellular in nature, since cuttings from old trees are as capable of giving rise to a new individual as cuttings from young ones. For this reason the cells of higher plants are not discussed here, although some mention is made of aging processes in fungi.

I shall concentrate here on aging processes which are best thought of as originating at the cellular or subcellular level. It is as well to state at the outset that there is no general agreement as to the nature of these processes. All that can be done is to formulate some of the questions that must be asked and to give some of the evidence that may help to answer them. Before doing so, there is one general point that must be made. There is no reason to believe that there is only one fundamental type of cellular aging process. It is sometimes assumed that because there is one fundamental method of transmitting hereditary information, one method of synthesizing enzymes, and so on, there must be one fundamental aging process. This is like saying that because there is only one way of steering a motor car, one way of starting it, and one way of stopping it, there must be only one way in which it can wear out. It seems much more likely that there are many different ways in which changes in cells may be responsible for the aging of the individual.

2. AGING OF TISSUE CULTURE CELLS

Until about 10 yr ago, it was generally believed that cells in tissue culture are capable of indefinite multiplication. This belief was based originally on the observation by Alexis Carrel that fibroblasts from chick embryo heart were capable of indefinite division in tissue culture. This belief was strengthened by the establishment of a mouse cell line capable of indefinite multiplication (Gey and Gey, 1936; Earle, 1943). Since that time a large number of such lines have been established from many different species. It was therefore natural to presume that aging of individuals could not be attributable to a loss of capacity to multiply in tissues consisting of dividing cells, and therefore to look for the causes of aging elsewhere. For example: (1) changes in postmitotic cells; (2) changes in extracellular material (e.g., collagen); and (3) disproportionate multiplication of particular groups of cells at the expense of the whole organism (e.g., malignancy, autoimmunity).

These processes may still be important in aging, but it is no longer safe to rule out the possibility that stem line cells in dividing tissues may gradually lose the capacity to divide, and so contribute to aging. Hayflick and Moorhead (1961) demonstrated that human embryonic fibroblasts in tissue culture undergo approximately 50 doublings before division slows down and finally stops.

A further analysis of this phenomenon (Hayflick, 1965) made it unlikely that the limited *in vitro* lifetime could be explained either by the presence of latent virus or by inadequacy of the medium. When male cells close to the end of their limited *in vitro* life-span were mixed with rapidly dividing female cells, the female cells continued to divide rapidly but the male cells were not rejuvenated. Cells kept at subzero temperatures resume growth when returned to 37°C and complete the same number of doublings as they would have done if growth had not been interrupted. It is therefore possible to mix young and old cells from the same original strain. When this was done, it was found that the mixed population completed an additional number of doublings approximately equal to that expected of the younger component. These experiments suggest that this aging process is intrinsic, so that in mixtures of young and old cells each component behaves as it would have done in isolation.

The limited life-span of fibroblasts in tissue culture has subsequently been observed by many other workers. It contrasts sharply with the apparently indefinite growth of the cell lines mentioned earler. Hayflick (1965) accordingly suggested that there exist two contrasting types of tissue culture: (1) cell strains—"normal" cells with a finite life span; and (2) cell lines—abnormal cells with an indefinite life-span in tissue culture, often with an abnormal morphology, usually aneuploid with a variable chromosome number, and malignant.

According to this view cell lines may arise either as a culture of originally malignant cells, or as a result of the transformation in culture of a cell strain into a cell line.

That fibroblast tissue cultures can be classified into these two different categories now seems to be generally accepted. It is less clear whether or not a limited life-span is a universal characteristic of normal cells in tissue culture. The difficulty of course lies in the definition of normal. Cell morphology is an unsatisfactory guide to normality. Malignancy cannot always be tested, either because the cells are human in origin or because no antigenically suitable host exists into which the cells can be transplanted. In practice therefore the most widely used criterion of normality is diploidy. Judged by this criteria, fibroblast cell lines from many mammalian species are abnormal (R. Moore and Radley, 1968).

The most convincing evidence for the existence of normal cells with an indefinite life-span in tissue culture has been presented by Moore and his colleagues (G. E. Moore et al., 1967; G. E. Moore and McLimans, 1968). Suspension tissue cultures of lymphoblastoid cells of normal individuals appear to be made up of normal cells and to have an indefinite life-span.

The primary evidence for normality is that at least 85% of the cells have a normal diploid karyotype and that there are no "marker" chromosomes. Malignancy cannot be tested, but it is significant that there are differences between cell lines from normal individuals and lines derived from the peripheral blood of patients with various leukemias. The microscopic and submicroscopic appearances of normal and "malignant" cells were identical. However, normal cells consistently had less ability to develop clones from single cells. Further, some of the malignant cell lines had an abnormal karyotype, and some synthesized abnormal products. There was no difference in growth rate between normal and malignant lines. Some of the normal cell lines were maintained for over 20 months, with a doubling time of between 20 and 50 hr, without displaying any signs of senescence.

One other important difference between fibroblast and lymphoblast cultures must be mentioned. Fibroblast cultures usually grow consistently soon after the culture is started. In contrast, there was a lapse period of from 40 to 100 days before continuous growth of lymphoblast cultures was established. One possible explanation of this is that only one or a few stem cells in the original population were capable of continued multiplication.

This difference between the indefinite growth of apparently normal lymphoblast cell cultures and the limited life-span of fibroblasts makes it impossible as yet to generalize about the life-span of cells in tissue culture. The problem is made more difficult to solve by the fact that when a tissue culture is established it is not always easy to determine to which *in vivo* cell type the culture cells correspond. Thus in the early stages of a tissue culture there is considerable selection between the cell types originally present in the explant. Franks and Henzell (1970) ,who set up a large number of tissue cultures from many different organs from mice of different ages, found that the cells in all cultures were similar in structure, from whatever organ they were derived. They suggest that the cells were probably derived from the blood vascular system, some resembling pericytes and others endothelial cells.

Nevertheless, the limited life-span of fibroblasts has now been observed sufficiently often in different laboratories for it to be regarded as a phenomenon that calls for an explanation, and for it to be reasonable to suggest, as Hayflick (1965) has done, that the aging of individuals may be attributable to the limited life-span of stem line cells.

The relevance of these tissue culture studies to the aging of individuals is discussed in Section 3, and possible mechanisms of clonal aging in Section 4.

3. SERIAL TRANSPLANTATION AND THE STUDY OF AGING

If the limited life-span of cells in tissue culture proves to be characteristic of cell types other than fibroblasts and, more important, if the life-span in culture is representative of the life-span of stem line cells *in vivo*, then the phenomenon originally described by Hayflick and Moorhead (1961) may be an important cause of aging. It is convenient to use the term "clonal aging" to describe the processes responsible for the limited life-span of a population of dividing cells, and the term "clonal aging theory" for the view that clonal aging is responsible for the aging of an individual.

The first problem is whether, if the stem line cells in an embryo are capable only of a further 50 divisions, it is sufficient to account for the actual production of tissue cells during the lifetime of an individual. The answer depends on the pattern of divisions. At one extreme, if each division of a stem cell gave rise to a stem cell and a terminal postmitotic cell, 50 divisions could give rise only to 50 cells, a number quite inadequate to account for the actual production of cells during a lifetime. At the other extreme, if starting from a stem cell every lineage undergoes 50 divisions before reaching a terminal postmitotic cell, one stem cell can give rise to 2^{50} terminal cells, or approximately 20 million tons of cells. As noted by Kay (1965), if there is a variation in the rate of division of primitive stem cells, it is possible for a single stem cell to give rise to 2^{50} terminal cells which are released continuously throughout life and not all at one time. Thus 50 divisions are in principle sufficient to account for the actual production of cells during a lifetime.

If clonal aging is important in the aging of individuals, the expected *in vitro* life-span of cultures derived from old individuals should be less than that of cultures derived from young individuals. This expectation is only partly realized (Hayflick, 1965). Thus fibroblasts derived from fetal lung underwent an average of 48 doublings, with a range from 35 to 63, whereas fibroblasts derived from adult lung underwent an average of 20 doublings, with a range from 14 to 29. This is in accordance with expectation. However, there was no relationship between the age of the adult donor and the survival of the culture; in fact, the longest-surviving culture (29 doublings) was derived from the oldest donor, aged 87 years. Goldstein (1969) has reported that skin fibroblasts from a 9-year old boy with progeria underwent only 2 doublings in culture, in contrast to age-matched controls which underwent 20 to 30 doublings.

A more direct method of testing whether or not clonal aging is important to the aging of the individual is the technique of serial transplantation. In this technique an organ or tissue is repeatedly transplanted into a young individual when its current host begins to age. If aging is intrinsic to the tissue (as it must be according to the clonal aging theory), a serially transplantanted tissue will deteriorate at the same rate as it would have done at its original site; if there are no intrinsic aging processes in the tissue (and if the act of transplantation itself is not damaging), a serially transplanted tissue will survive indefinitely, or at least for very much longer than it would have done at its original site. The results of such experiments are now briefly reviewed. As will be seen, they do not tell an unambiguous and consistent story. In interpreting them one general point is worth remembering. An experiment demonstrating prolonged survival of a serially transplanted tissue is less ambiguous than one demonstrating a limited life-span because the latter may result from the insult of transplantation or the abnormal site to which the tissue is transplanted.

The earliest experiments of this kind, and the ones with the most consistent and least unexpected results, were performed by Krohn (1962), who transplanted ovaries from young mice into old ones and vice versa, the ovaries of the host female being first removed. In all cases the ovaries behaved as they would have done at their original site. Old mice who had received ovaries from young ones continued the estrus cycle beyond the normal age, but ovaries from old mice transplanted into young ones were not rejuvenated and did not maintain the estrus cycle for longer than they would have done in the donor. These results are expected because of the presence in the ovary of postmitotic oocyte cells which cannot be replaced and which are essential to ovarian function. Hence these experiments, although interesting, are not relevant to the clonal aging theory.

In a later series of experiments, Krohn serially transplanted mouse skin. Here difficulty arises in distinguishing between graft and host skin within an isogenic line. Krohn found that the most satisfactory method was to transplant ear skin to the flank of the host, at which site it maintained its characteristic morphology. Krohn first reported (1962) that serially transplanted skin had a limited life-span. Later, however (Krohn, 1965), he reported that the limited life-span was characteristic of skin from A-line mice and was less apparent in skin from CBA mice or from various F_1 hybrid mice. Several grafts were at that time on their third host and were between 6 and 7 yr of age. This result suggests either that there is no aging process intrinsic to skin, or at least that it proceeds more slowly than other aging processes in the mouse. However, Krohn observed some diminution in the size of the grafts with age. Probably, but not certainly, this diminution was attributable to the effects of transplantation rather than to aging; using more frequent transfers, Krohn found that even young skin could not survive four transfers at intervals of 42 days.

On balance, Krohn's experiments on skin argue against the clonal aging theory. A different conclusion emerged from the work of Daniel et al. (1968) on serial transplantation of mouse mammary glands. The growth rate of transplanted glands declined with time, and the oldest transplant line was lost after seven transplantations, at a total age of 2 yr. In contrast, the growth rate of neoplastic glands did not decline with age or transplantation, and two transplant lines were growing vigorously after more than 30 transplantations and 8 yr. These are precisely the results to be expected from the clonal aging theory.

Experimental results on the serial transplantation of hemopoietic tissue are less consistent and more difficult to interpret. Mouse bone marrow cells injected into a heavily irradiated host mouse form colonies in the spleen. It is reasonable to regard the cells responsible for forming these colonies as the stem cells of the hemopoietic tissue because cells from these "first-generation" spleen colonies can give rise to "second-generation" colonies injected into a second host. As a test of clonal aging, Siminovitch et al. (1964) compared the colony-forming ability of cells from first-generation and second-generation colonies. They found that second-generation colonies have a much lower proportion of colony-forming cells. Unfortunately, it is not clear whether or not a similar decline in capacity to multiply occurs in the absence of transplantation. It must be remembered that spleen colonies grow in mice that have received 800–1000 rads, and which are in some cases at the point of death. A similar conclusion, with similar reservations, emerged from the work of Cudkowicz et al. (1964), who serially transplanted mouse bone marrow cells and found a decline both in the ability of the transplanted cells to ensure the survival of heavily irradiated mice (apparent only after four transplantations) and in the ability of the cells to repopulate the spleen, as judged by DNA synthesis in the spleen. The latter ability declined successively after each of four successive transplantations.

These results on hemopoietic tissue are consistent with the clonal aging theory but are also open to the interpretation that the loss of capacity to multiply is attributable to transplantation into heavily irradiated hosts.

Finally, Franks (1970) has reported that mouse prostate grafted into the flank of host mice still survives after 6 yr. The grafts retain their usual histological structure and, as this investigator remarks, there seems to be no reason why these tissues should not be maintained for several years yet.

It is difficult to draw any very firm conclusion from these experiments. Krohn's experiments on ovaries show, as would be expected, that if the continued functioning of a tissue depends on postmitotic cells being used up and not replaced the life-span of the tissue cannot be extended by serial transplantation. In contrast, Krohn's experiments on skin and Franks' on the prostate show that at least some tissues when serially transplanted can survive for at least twice as long as the normal life span of the whole organism. However, experiments on mammary tissues and hemopoietic tissue suggest a limited life-span. It may be relevant that in both the latter cases (and particularly in hemopoietic cells) the frequency of transfer was much higher than in the case of skin or prostate. Had Krohn transplanted skin as frequently as Daniel et al. transplanted mammary tissue, it seems doubtful that he would have observed indefinite survival.

On balance, the results of serial transplantation seem to argue against the view that clonal aging of mitotic cells is an important cause of aging.

4. MECHANISMS OF CLONAL AGING

Whether or not it will one day be possible to keep normal fibroblasts indefinitely in tissue culture, the limited life-span observed at present is not easy to explain. The mere dilution of some molecule or organelle that cannot be synthesized in culture is an inadequate explanation because dilution by a factor of 2^{50}, as would be required to account for aging after 50 doublings, is implausible. Three theories (at least) have been suggested; many others may be possible. In this section I will explain these theories and review the evidence for and against them, not only as explanations of clonal aging in tissue culture, but as explanations for other aging processes also.

A. Multitarget Theories

A widespread theory is that aging is a result of damage to the genetic material of cells, that is, to DNA. The relevance of this theory as applied to postmitotic cells is discussed in Section 5. The difficulty in applying the theory to clonal aging is as follows. If mutations occur in tissue culture cells, reducing their capacity to divide, the cells in which they occur will be eliminated from the culture by natural selection. Hence mutation could cause clonal aging only if its rate were so high per division that mutants would accumulate more rapidly than they could be eliminated. It is easy to see that in a diploid cell with two (occasionally one) copies of each essential gene if mutation were frequent enough to cause clonal aging it would do so in many fewer than 50 fissions.

It has been noted (Maynard Smith, 1965) that this difficulty disappears if it is presumed that: (1) some essential genes are present in a large number of replicating copies, for each of which the mutation rate is high; and (2) the viability of the cell is not greatly altered until a threshold number of copies have been damaged, after which it declines rapidly.

According to these assumptions, damaged genes accumulate in the population without reducing fitness, and so without selection being able to eliminate them. If the threshold number is large, it will be reached nearly simultaneously in different members of the population, and the subsequent decline in fitness may be too rapid to be prevented by selection. A similar suggestion was made for different reasons by Hayflick (1965).

The difficulty in applying this theory to aging of tissue culture cells lies in identifying an essential gene with multiple copies. The theory could account for aging in *Paramecium aurelia.* In this ciliate the macronucleus periodically breaks down, either after sexual conjugation or after a form of self-fertilization known as autogamy, and a new macronucleus is derived from the micronucleus. When this process of macronuclear reconstruction is prevented, division slows down and the clone is ultimately lost (Sonneborne and Schneller, 1955). The mechanism of clonal aging is not understood, but presumably depends on the accumulation of irreversible damage in the macronucleus. Since the macronucleus is thought to consist of over 800 identical or similar subunits, clonal aging resulting from random macronuclear damage is almost inevitable; it does not follow that in fact it occurs rapidly enough to cause the observed clonal aging.

A multitarget theory is also probably appropriate as an explanation of aging in *Physarum*, which is discussed in Section 4.B.

B. Error Catastrophies

An ingenious theory of aging has been suggested by Orgel (1963). Errors in the incorporation of amino acids into proteins must occasionally occur, but do not normally matter because the erroneous protein in time is broken down and replaced. Suppose, however, that an error is made in one of the enzymes involved in protein synthesis, so as to alter the specificity of the enzyme without destroying its activity. Since each such enzyme must play a part in synthesizing many protein molecules, several hundreds of which will like itself be enzymes concerned with protein synthesis, it follows that one erroneous enzyme may synthesize several erroneous enzymes, each of which will in turn synthesize several erroneous enzymes, and so on. If so, there will be an exponential increase in the frequency of errors, leading ultimately to the death of the cell.

Hayflick (1970) has suggested that error catastrophies may be responsible for clonal aging of tissue cultures. It is not clear why we should expect cell death to be synchronous within a culture or tissue, unless erroneous enzymes can be passed from cell to cell. In the latter case it is difficult to see how the germ line escapes, and in any case Hayflick's own results on cell mixtures would be difficult to explain.

However, there is evidence for thinking that error catastropies may be responsible for the aging of the vegetative stage of fungi. In some fungi (*Aspergillus*, *Podospora*), vegative aging is a regular phenomenon. The growth of the mycelium slows down, an accumulation of yellow pigment occurs, and death ultimately follows. In *Neurospora* the mycelium normally shows indefinite growth, but the mutant *nd* (natural death) shows senescent phenomena similar to those found in *Aspergillus*. Holliday (1969) has suggested that vegetative aging is attributable to error catastrophies of the type suggested by Orgel, and that *nd* causes vegetative aging in *Neurospora* by increasing the rate of errors in protein synthesis. He has tested this hypothesis by combining the mutant *nd* with a suppressible adenineless mutant *ad*. The logic of this experiment is as follows. Mutants usually cannot grow in the absence of adenine but can do so if 2′-deoxy-5-fluorouridine (FUDR) is added, presumably because in the presence of FUDR sufficient errors are made in translating the gene to ensure that a sufficient number of good enzyme molecules are made from the erroneous *ad* gene. If this explanation is correct, and if *nd* also causes errors in protein synthesis, *nd* might be expected to suppress *ad*. Holliday found that some growth was restored to *ad* on an adenineles medium by the gene *nd*. This rather indirect argument does suggest that errors in protein synthesis may be involved in the mutant *nd*, and hence in normal vegetative aging in *Aspergillus* and *Podospora*.

More direct evidence is that a small quantity of cytoplasm from a senescent mycelium induces senescence in a young one. This is to be expected from Orgel's theory, because if errors increase exponentially a mixture of "old" and "young" cytoplasm will have a life expectation much closer to the old than to the young; the mixture is in effect already in the throes of an error catastrophy. Similar evidence exists for clonal aging in *Amoeba proteus*. Normally *Amoeba* does not show clonal aging, and indeed could not do so since there is no sexual process that could rejuvenate. However, Muggleton and Danielli (1958) showed that, if amebae are starved so as to prevent growth and then after several weeks restored to normal food, the cultures may subsequently show a limited life-span in one of two ways. In one case, of "stem line" growth, a stem cell divides to give a stem cell and a cell which does not divide again; in the other case, of "logarithmic senescent" growth, the culture first grows logarithmically but then dies out. In a more recent paper, Muggleton and Danielli (1968) have shown that when cytoplasm from either type of "spanned" ameba is injected into normal amebae it can sometimes produce clones of limited life-span.

If Orgel's theory is true, cytoplasm from an aging cell acts as an infective agent. By this criterion experiments both on amebas and on fungi are consistent with the theory. A different result has been obtained by Poulter (1971) on the acellular slime mold *Physarum*. This mold exists as free-living haploid amebae, or as a syncitium containing diploid nuclei formed by the fusion of two amebae of different mating type. Poulter found that the amebae had an indefinite life-span, but that the syncitium had a sharply defined life-span the length of which depended on the genetic strain. Depending on genetic similarity, two syncitia join to form a single one. Poulter showed that after fusion the two types of nuclei retained the same relative frequencies, regardless of the age of the syncitia before fusion; this is to be expected since mitoses are synchronized throughout a syncitium.

The crucial observations concern the life expectation of a syncitium formed by the fusion of a young and old syncitium. Poulter found the expectation to be the mean of the life expectations of the fusing syncitia weighted by the quantity of material contributed by the syncitia. This finding is inconsistent with the Orgel hypothesis. It is just what would be expected from the multiple-target theory considered above. A possible explanation is that the targets are the nuclear genes in different nuclei, in which mutations occur that destroy functions necessary for the survival of the syncitium without altering the ability of the nucleus to divide. In a syncitium with synchronized mitoses, natural selection cannot eliminate harmful mutations, as it can in a cellular population.

If error catastrophes occur, they would be expected to contribute to aging by killing postmitotic cells. Although not relevant to clonal aging, it is convenient to consider the evidence on this point now. Most of the evidence comes from a study of *Drosophila* in which there is no cell division, other than in the gonads, in the adult insect. Harrison and Holliday (1967) fed amino acid analogs to late larval *Drosophila melanogaster* and found that the adults that emerged sometimes had a lowered expectation of adult life. This they interpreted as favoring Orgel's theory; by inducing errors in protein synthesis, they suggested the development of error catastrophes might have been accelerated.

The results are, however, open to a different interpretation. Maynard Smith et al. (1970) have shown that approximately 80% of the proteins present in an adult *Drosophila* at emergence from the pupa are not replaced during adult life. Hence if analogs are fed to late larvae, an appreciable number of irreplaceable adult proteins may contain these analogs, which will impair their function and so contribute to aging. A more direct test of Orgel's hypothesis is to feed analogs to young adults when they will be incorporated only into those proteins that turn over in the adult, and which might therefore be involved in an error catastrophe. Accordingly, Dingley and Maynard Smith (1969) fed fluorophenylamine (FPA) in just sublethal doses to young adult *Drosophila suboscura*. This analog is incorporated into proteins in adult *Drosophila*, and there are good reasons to believe that it interferes with the functioning of these proteins. Dingley and Maynard Smith found no reduction in adult life-span after ingestion of FPA.

Further evidence against the occurrence of error catastrophes in *Drosophila* has been obtained by Shmookler (1972), who measured the ability of adult *D. subobscura* of different ages to discriminate between amino acids and their analogs in protein synthesis. He compared ethionine–methionine and fluorophenylalanine–phenylalanine by feeding a mixture of ^{3}H-labeled analog and ^{14}C-labeled amino acid; he also made a similar comparison of fluorouridine–uridine incorporation into RNA. In no case was there any significant alteration in the ability to discriminate with age. Hence in *Drosophila* the evidence is against the occurrence of error catastrophes.

C. Genic Imbalance

The two theories so far considered, the multiple-target and the error catastrophe theories, have received little direct support from studies on tissue cultures. Moore and Radley (1968) have suggested a theory which has the virtue of accounting for the association between aneuploidy and the transformation of a strain into a cell line and also of suggesting an explanation for differences in the ease with which such transformation can be produced in different mammalian species.

They note that in tissue culture fibroblasts may divide once every 20 hr, whereas *in vivo* the division time is probably about 30 days. It follows that the S phase lasts for approximately 40% of the time in culture instead of only 1% of the time *in vivo*. Now during the S phase, the DNA of early-replicating genes has doubled, whereas that of late-replicating genes has not. Since RNA synthesis continues during the S phase, this genic imbalance may lead to a metabolic imbalance. If there is genic imbalance for too large a part of the time, this may ultimately lead to cell death.

According to this theory, genic imbalance can be cured by appropriate types of aneuploidy, in which late-replicating regions of the genome are doubled.

In support of this theory, Moore and Radley refer to the observations of Ford et al. (1961) and of Yu and Sinclair (1964) to show that, in the hamster, transformation from a cell strain to a cell line occurs fairly readily and involves a relatively small increase in total DNA. This they suggest is because in the hamster the late-labeling regions of the chromosomes are usually either whole chromosomes or whole chromosome arms, whereas in the wallaby the late-labeling regions are widely dispersed throughout the genotype. Hence in the hamster a less extensive rearrangement of chromosomal material is required to bring about the required doubling of the late-labeling regions.

The suggestion is an attractive one, since it does suggest a reason for the association between transformation and aneuploidy. One simple prediction of the theory, which should be readily testable, is that the life-span of fibroblast tissue cultures could be extended if their rate of division could be slowed down.

To conclude, the mechanisms of clonal aging are as yet unknown. In future discussions two points should be remembered. First, we do not understand the process of differentiation in mammalian cells; until we do, it may be that we cannot hope to understand clonal aging. Second, we must always distinguish between situations in which the definition of clonal aging given above is appropriate, that is, in which there is a population of cells so that damaged cells can be eliminated by selection, and situations in which it is inappropriate, as for example in the case of *Physarum*, whose syncitial structure makes selective elimination impossible.

5. AGING IN POSTMITOTIC CELLS

In trying to find the causes of aging of multicellular organisms, it is natural to examine first the changes in tissues consisting of postmitotic cells which cannot be replaced from a stem line population. It is suggestive that higher plants, in which all tissues can be replaced from continuously dividing meristematic regions, appear not to age, at least as judged by the possibility of rejuvenation by taking cuttings of a part of the plant consisting of a meristem and/or of cells recently derived from a meristem.

Little is known about aging in lower animals in which there may be few or no unreplaceable postmitotic tissues. There is, however, one very suggestive experiment by Sonneborn (1930) on a free-living plathyhelminth *Stenostomum* which reproduces vegetatively by transverse fission. Sonneborn found that clones always continued from the posterior half after each fission could be continued indefinitely, whereas clones always continued from the anterior half died out. The obvious difference between the two types of clone is that in the former the brain must be regenerated after each fission, whereas in the latter the brain is never replaced. This suggests that the aging of "anterior" clones is attributable to changes in brain cells which are not being replaced.

The same logic that draws attention to change in postmitotic cells also draws attention to those molecules in postmitotic cells that do not turn over. Of these, the most important are the nuclear DNA. Hence it is natural to formulate theories of aging by "somatic mutation," that is, by changes in the DNA of somatic cells of a kind which, if they occurred in the germ line, would be recognized as genetic mutations. Although not always stated explicitly, it is usually a tacit assumption of such theories that the mutations occur in postmitotic cells.

The evidence for and against somatic mutational theories of aging is not discussed here at any length because I have outlined them on two previous occasions (Maynard Smith, 1962, 1965). The most thorough and clearly formulated theory is attributable to Szilard (1959a). In its original form this theory was certainly false, since it predicted that inbred animals should live longer than outbred ones (Maynard Smith, 1959). In the face of this criticism, Szilard (1959b) modified the theory so as to make what had been a constant into a variable; in this modified form the theory was proof against the original criticisms but is still open to the experimental objections mentioned in the next two paragraphs.

Experimental evidence on somatic mutation comes mainly from two groups, insects and mammals. In the case of insects the decisive experiment was performed by Clark and Rubin (1961), who showed that haploid male *Habrobracon* (a solitary wasp) have the same life-span as diploid males. It is true that the life-shortening effects of x rays are greater for haploid males but, as noted by Lamb and Maynard Smith (1964), the difference is much less than would be expected if the induction of recessive mutations is an important cause of radiation-induced life-shortening. In *Drosophila*, males (which are haploid for one-quarter of the genome) have much the same life-span as females and they are equally radiosensitive. These observations show that recessive somatic mutations are irrelevant to normal aging and comparatively unimportant in radiation-induced life-shortening. The only alternative to this conclusion is that the number of copies of metabolically active genes in the somatic cells of adult insects is unrelated to chromosomal ploidy in the germ line. This is unlikely but cannot be ruled out; some adult tissues are certainly polytene, but others (e.g., neurons) appear not to be. If recessive mutations are irrelevant to aging, this leaves the possibility that dominant mutations may be important; at present, there is no reason to favor this possibility.

In mammals, Curtis (1965 and earlier) has presented evidence in favor of somatic mutation as a cause of aging. He studied the incidence of chromosomal aberrations in regenerating livers of mice of various ages and in mice exposed to ionizing radiation. He showed that the incidence of aberrations increases with age and is higher in strains of mice with a shorter life-span. However, when considered quantitatively, his evidence appears not to support a somatic mutation theory (Maynard Smith, 1965).

The fact that male and female mice do not have greatly different life-spans cannot be used as evidence against the importance of recessive somatic mutation, since according to the Lyon hypothesis (1961) only one copy of each sex-linked locus is active in the somatic cells of female mammals. However, studies on chemical mutagens by Alexander and Connell (1960, 1963) provide strong evidence against the importance of somatic mutations in aging. They found that bifunctional alkylating agents, which are known to break chromosomes, shorten the life of mice when administered in sublethal doses, whereas a monofunctional agent, EMS, although highly mutagenic, does not break chromosomes and does not shorten the life of mice in sublethal doses. This suggests that point mutations are not important in aging, but that chromosome breaks may be, presumably because of their importance in dividing cells; this would also account for the life-shortening effects of ionizing radiation.

The evidence given above suggesting that somatic mutation is not an important cause of aging in insects has led us (Maynard Smith et al., 1970) to investigate the possibility that changes in nonturnover proteins may be important. If *Drosophila* larvae are kept on a medium containing labeled amino acids (leucine, phenylalanine), and the resulting adults are transferred to an unlabeled medium, it is possible to follow autoradiographically and by total extraction the fate of the proteins present in the young adult. Apart from the protein in the "fat body," which is broken down soon after emergence, approximately 80% of the protein present in a freshly emerged adult is not replaced during adult life. The 20% that is replaced has a turnover time of 5–10 days; it is essential for the survival of the fly and is known to include the ribosomal proteins. More important in the present context, the 80% of nonturnover protein includes the muscle proteins, including the muscle mitochondria.

There is therefore every possibility that in *Drosophila* aging may be attributable to damage to proteins not replaced in the adult. Accordingly, Tribe and Ashurst (1970) looked for functional changes in the flight muscle mitochondria of the blowfly *Calliphora*. They found a drop in the P/O ratio and a striking loss of respiratory control in the mitochondria of old flies. That this loss of control occurs *in vivo* is suggested by a rise in resting metabolic rate with age (Tribe, 1966). The loss of coupling between oxidation and phosphorylation may result either from changes in the tertiary structure of the enzymes, or in the loss of some high-order structure on which coupling depends.

In insects therefore, or at least in dipterans, damage to proteins not replaced during adult life is a sufficient explanation for aging; it does not follow that it is the only aging process in these animals. It is not clear whether or not any comparable process is possible in mammals. Some extracellular structural proteins (e.g., tendon collagen) are not normally replaced in adult life and are known to change in their properties; it is less clear that this change in properties contributes to aging. Little is known about the occurrence of intracellular proteins with long half-lives; some preliminary observations by Nordgren et al. (1969) suggest that such proteins may exist. If so, changes in these may play a role in mammalian aging similar to that played by changes in mitochondrial proteins in insect aging.

6. CONCLUSIONS

In the introduction to this chapter it was noted there is no reason to suppose that there is one fundamental and universal cause of aging. Any one organism may age for several different and partly independent reasons, and the important causes may well be different in fungi, slime molds, insects, and mammals. Some important causes of aging in mammals—in particular malignant and autoimmune processes—have not been discussed here at all.

I have paid considerable attention to clonal aging, that is, the processes responsible for the limited life-span of a population of dividing cells. There can be no doubt that this phenomenon, as exhibited, for example, by normal diploid fibroblasts in tissue culture, is an important one which merits much further study. However, it is still not clear whether or not comparable processes occur in other cell types in tissue culture. Further, there is at least a suspicion that clonal aging does not occur in fibroblasts *in vivo*, since no correlation has been demonstrated between the life expectation of a fibroblast tissue culture and the life expectation of the adult from which it was derived. Finally, the evidence from serial transplantation suggests that even if clonal aging does occur *in vivo* it does not proceed fast enough to be an important cause of aging, at least in mice.

In seeking the cause of aging of the whole organism, it seems at present more fruitful to concentrate on postmitotic cells. In dipteran insects there is no cell division in the adult, except in the gonads; more unexpected, a large part of the protein of the young adult must also last the 2–3 months of adult life. A sufficient (but not necessarily the only) explanation of aging in these insects lies in the uncoupling between oxidation and phosphorylation in the muscle mitochondria, in which no protein replacement occurs in the adult.

In mammals also, it may well be that a major cause of aging is the deterioration of postmitotic cells. For example, Franks (1970) has recently argued that changes in neurons and in cells in the microvascular system are the main cause of aging in mammals. If so, it is not clear why the neurons of mice last for 2–3 yr, whereas those of man last approximately 30 times as long. Nor is the nature of the changes in these cells clear. Somatic mutations or changes in nuclear DNA are a possibility, although present evidence suggests that they are not an important cause of aging. Except for structural proteins such as collagen, little is known about proteins, similar to those in dipterans, whose life-span is of the same order as that of the individual, so it is impossible to say whether or not changes in such proteins are involved in aging. A final possibility is that the changes in the control circuits of cells, which must be involved in their more-or-less stable and irreversible differentiation, may not be completely stable. If so, aging changes in cells may be attributable to a gradual breakdown of the state of differentiation, leading to a loss of function. This is not a particularly helpful suggestion at present, since we know too little about the nature of differentiation to investigate the suggestion further. However, it is as well to conclude a discussion of aging by noting that we are not likely to understand its causes fully until we know more about the nature of differentiation.

REFERENCES

Alexander, P., and D. I. Connell (1960). Shortening of the life span of mice by irradiation with X-rays and treatment with radiomimetic chemicals. *Radiation Res.*, **12,** 38.

Alexander, P., and D. I. Connell (1963). In *Cellular Basis and Aetiology of Late Somatic Effects of Ionising Radiation* (R. J. C. Harris, ed.), pp. 259–263. Academic, New York.

Clark, A. M., and M. A. Rubin (1961). The modification by X-irradiation of the life span of haploids and diploids of the wasp, *Habrobracon*. *Radiation Res.*, **15,** 244.

Cudkowicz, G., A. C. Upton, and G. M. Shearer (1964). Lymphocyte content and proliferative capacity of serially transplanted mouse bone marrow. *Nature*, **201,** 165.

Curtis, H. J. (1966). In *Topics in the Biology of Aging* (P. L. Krohn, ed.), pp. 63–82. Wiley, New York.

Daniel, C. W., K. B. DeOme, J. T. Young, P. B. Blair, and L. J. Faulkin (1968). The *in vivo* life span of normal and preneoplastic mouse mammary glands: A serial transplantation study. *Proc. Natl. Acad. Sci., U.S.*, **61,** 53.

Dingley, F., and J. Maynard Smith (1969). Absence of a life-shortening effect of amino-acid analogues on adult *Drosophila*. *Exptl. Gerontol.*, **4,** 145.

Earle, W. R. (1943). *J. Natl. Cancer Inst.*, **4,** 165.

Ford, D. K., C. Boguszewski, and N. Aversperg (1961). Chinese hamster cell strains *in vitro:* Spontaneous chromosome changes and latent polyoma-virus infection. *J. Natl. Cancer Inst.*, **26,** 691.

Franks, L. M. (1970). Cellular aspects of ageing. *Exptl. Gerontol.*, **5,** 281.

Franks, L. M., and S. Henzell (1970). The development of spontaneous neoplastic transformation *in vitro* of cells from young and old mice. *European J. Cancer*, **6,** 357.

Gey, G. O., and M. K. Gey (1936). Maintenance of human normal cells and tumor cells in continuous culture. *Am. J. Cancer*, **27,** 45.

Goldstein, S. (1969). Lifespan of cultured cells in progeria. *Lancet*, February 22, 424.

Harrison, B. J., and R. Holliday (1967). Senescence and the fidelity of protein synthesis in *Drosophila. Nature*, **213,** 990.

Hayflick, L. (1965). The limited *in vitro* lifetime of human diploid cell strains. *Exptl. Cell Res.*, **37,** 614.

Hayflick, L. (1970). Aging under glass. *Exptl. Gerontol.*, **5,** 291.

Hayflick, L., and P. S. Moorhead (1961). The serial cultivation of human diploid cell strains. *Exptl. Cell Res.*, **25,** 585.

Holliday, R. (1969). Errors in protein synthesis and clonal senescence in fungi. *Nature*, **221,** 1224.

Kay, H. E. M. (1965). How many cell-generations? *Lancet*, August 28, 418.

Krohn, P. L. (1962). Heterochronic transplantation in the study of ageing. *Proc. Roy. Soc.* (*London*), **B157,** 128.

Krohn, P. L. (1966). In *Topics in the Biology of Aging* (P. L. Krohn, ed.), pp. 125–148. Wiley, New York.

Lamb, M. J., and J. Maynard Smith (1964). Radiation and aging in insects. *Exptl. Gerontol.*, **1,** 11.

Lyon, M. F. (1961). Gene action in the X-chromosome of the mouse (*Mus musculus* L.) *Nature*, **190,** 372.

Maynard Smith, J. (1959). A theory of ageing. *Nature*, **184,** 956.

Maynard Smith, J. (1962). The causes of ageing. *Proc. Roy. Soc.* (*London*), **B157,** 115.

Maynard Smith, J. (1965). In *Topics in the Biology of Aging* (P. L. Krohn, ed.), pp. 1–35. Wiley, New York.

Maynard Smith, J., A. N. Bozcuk, and S. Tebbutt (1970). Protein turnover in adult *Drosophila. J. Insect Physiol.*, **16,** 601.

Moore, G. E., and W. F. McLimans (1968). The life span of the cultured normal cell: Concepts derived from studies of human lymphoblasts. *J. Theoret. Biol.*, **20,** 217.

Moore, R., and J. M. Radley (1968). The nature of aneuploid change in JU 56 cells, and its relationship to continued cell proliferation. *Exptl. Cell Res.*, **49,** 638.

Moore, G. E., R. E. Gerner, and H. A. Franklin (1967). Culture of normal human leucocytes. *J. Am. Med. Assoc.*, **199,** 519.

Muggleton, A., and J. F. Danielli (1958). Ageing of *Amoeba proteus* and *A. discoides* cells. *Nature*, **181,** 1378.

Muggleton, A., and J. F. Danielli (1968). Inheritance of the "life-spanning" phenomenon in *Amoeba proteus. Exptl. Cell Res.*, **49,** 116.

Nordgren, R. A., G. P. Hirsch, R. A. Menzies, D. D. Hendley, R. Kutsky, and B. Strehler (1969). Evidence for long lived components in developing mouse tissues labelled with leucine. *Exptl. Gerontol.*, **4,** 7.

Orgel, L. E. (1963). The maintenance of the accuracy of protein synthesis and its relevance to ageing. *Proc. Natl. Acad. Sci., U.S.*, **49,** 517.

Poulter, R. (1971). Senescence in *Physarum polycephalum. Nature*, in press.

Shmookler, R. (in preparation).

Siminovitch, L., J. E. Till, and E. A. McCulloch (1964). Decline in colony-forming ability of marrow cells subjected to serial transplantation into irradiated mice. *J. Cellular Comp. Physiol.*, **64,** 23.

Sonneborn, T. M. (1930). Genetic studies on *Stenostomum incaudatum* n. sp. I. The nature and origin of differences in individuals formed during vegetative reproduction. *J. Exptl. Zool.*, **57,** 57.

Sonneborne, T. M., and M. Schneller (1955). The basis of aging in variety 4 of *Paramecium aurelia. J. Protozool., Suppl.*, **2,** 6.

Szilard, L. (1959a). On the nature of the aging process. *Proc. Natl. Acad. Sci., U.S.*, **45,** 30.

Szilard, L. (1959b). A theory of ageing. *Nature*, **184,** 958.

Tribe, M. A. (1966). Some physiological studies in relation to age in the blowfly *Calliphora erythrocephala* Meig. *J. Insect Physiol.*, **12,** 1577.

Tribe, M. A., and D. Ashurst (1970). Structural and biochemical changes in flight muscle mitochondria, *J. Cell Sci.*, **10,** 443.

Yu, C. K., and W. K. Sinclair (1964). Homogeneity and stability of chromosomes of Chinese hamster cells *in vitro*. *Can. J. Genet. Cytol.*, **6,** 109.

INDEX